Norbert Wiener: Collected Works
Volume I

Mathematicians of Our Time

Gian-Carlo Rota, series editor

1. Charles Loewner: Theory of Continuous Groups
   notes by Harley Flanders and Murray H. Protter

2. Oscar Zariski: Collected Papers
   Volume I
   Foundations of Algebraic Geometry and Resolution of Singularities
   edited by H. Hironaka and D. Mumford

3. Collected Papers of Hans Rademacher
   Volume I
   edited by Emil Grosswald

4. Collected Papers of Hans Rademacher
   Volume II
   edited by Emil Grosswald

5. Paul Erdős: The Art of Counting
   edited by Joel Spencer

6. Oscar Zariski: Collected Papers
   Volume II
   Holomorphic Functions and Linear Systems
   edited by M. Artin and D. Mumford

7. George Pólya: Collected Papers
   Volume I
   Singularities of Analytic Functions
   edited by R. P. Boas

8. George Pólya: Collected Papers
   Volume II
   Location of Zeros
   edited by R. P. Boas

9. Stanislaw Ulam: Selected Works
   Volume I
   Sets, Numbers, and Universes
   edited by W. A. Beyer, J. Mycielski, and G.-C. Rota

10. Norbert Wiener: Collected Works
    Volume I
    Mathematical Philosophy and Foundations; Potential Theory; Brownian Movement, Wiener Integrals, Ergodic and Chaos Theories, Turbulence and Statistical Mechanics
    edited by P. Masani

Norbert Wiener

Collected Works
With commentaries

Volume I

Mathematical Philosophy and Foundations;
Potential Theory;
Brownian Movement, Wiener Integrals,
Ergodic and Chaos Theories,
Turbulence and Statistical Mechanics

edited by
P. Masani

The MIT Press
Cambridge, Massachusetts, and London, England

This book was printed on Finch Title 93 and bound in Columbia Millbank Vellum 4958 by The Colonial Press Inc. in the United States of America

Library of Congress Cataloging in Publication Data

Wiener, Norbert, 1894-1964.
Mathematical philosophy and foundations.

(His Collected works, v. 1) (Mathematicians of our time, 10)
Bibliography: p.
1. Mathematics—Collected works. I. Title.
II. Series: Mathematicians of our time, 10.
QA3. W53 vol. 1 510′.8s [510] 74-17362
ISBN 0-262-23070-4

Commentators for Volume I

E. J. Akutowicz, Professor of Mathematics, University of Montpellier

G. Deem, Computational Physics Research Department, Bell Laboratories, Whippany, New Jersey

J. Feldman, Professor of Mathematics, University of California at Berkeley

A. Fine, Professor of Philosophy, University of Illinois at Chicago Circle

L. Gross, Professor of Mathematics, Cornell University

F. E. Hohn, Professor of Mathematics, University of Illinois

K. Ito, Professor, Research Institute for Mathematical Sciences, Kyoto University

J. -P. Kahane, Professor of Mathematics, University of Paris, Centre d'Orsay

H. E. Kyburg, Chairman, Department of Philosophy, University of Rochester

R. J. Levit, Professor of Mathematics, Emeritus, San Francisco State College

L. Lumer, former Lecturer in Mathematics, University of Washington at Seattle

P. Masani, University Professor of Mathematics, University of Pittsburgh

B. McMillan, Vice President, Bell Laboratories, Whippany, New Jersey

W. V. Quine, Edgar Pierce Professor of Philosophy, Harvard University

E. M. Stein, Professor of Mathematics, Princeton University

W. A. Veech, Professor of Mathematics, Rice University

R. L. Wilder, Research Associate, Department of Mathematics, University of California at Santa Barbara

Norbert Wiener (1894-1964). Photograph taken in 1925.

Contents

(Bracketed numbers are from the Bibliography.)

IA
Mathematical Philosophy and Foundations

IB
Potential Theory

IC
Brownian Movement, Wiener Integrals Ergodic and Chaos Theories, Turbulence and Statistical Mechanics

Preface

Norbert Wiener's profound contributions encompassed not only several areas of mathematics, but mathematical philosophy, relativity and quantum mechanics, communication engineering, the physiology of the heart and the nervous system, brain wave encephalography, sensory prosthesis—the field now widely known as cybernetics. His scholarly work also included incisive social, educational, and literary essays. The object of these *Collected Works* is not only the reproduction of *all* his scholarly publications other than books, but the placement of the overwhelming bulk of these in the context of present-day research by means of commentaries, tracing both their genesis and aftermath, written by eminent contemporary scholars.

The papers are organized by coherence of subject matter rather than by chronology, so that Wiener's intellectual evolution may stand out. Volume I includes the three important categories of papers representative of his early work and its pursuance in later years. Its organization is described in the section Introduction and Acknowledgment (page 3). Volume II will deal with generalized and classical harmonic analysis and Tauberian theory. Volume III will cover the Hopf-Wiener equation, prediction and filtering, as well as relativity and quantum mechanics. Finally, Volume IV will include his philosophical, cybernetical, and sociological papers, commentaries, and reviews.

This volume and the ones to follow owe a great deal to a great many people. I have thanked the commentators for this volume for their very valuable contributions in appropriate places in the Introduction and Acknowledgment, and will continue this system of acknowledgment in the later volumes. But it remains to mention several other individuals and institutions without whose help this publication would have been severely handicapped.

The Indiana University Research Foundation provided grants-in-aid of publication during 1970-1972, and I wish to thank them very much, especially its then Director Harry C. Day for encouragement. The rather complicated secretarial work involved in the organization of these volumes was carried out by Theresa Klassen and Rosemarie Dawson at Indiana University, and by Susan Black and Mildred Kosanovich at the University of Pittsburgh. I am most grateful for their devoted services. Finally, I thank Margaret Wiener for providing me with photographs of Professor Wiener.

P. Masani
Pittsburgh, Pennsylvania
October 1975

Norbert Wiener: Collected Works
Volume I

The Classification of Wiener's Papers

I. Mathematical Papers
   A. Mathematical philosophy and foundations
   B. Potential theory
   C. Brownian movement, Wiener integrals, ergodic and chaos theories, turbulence and statistical mechanics
   D. Generalized harmonic analysis and Tauberian theory
   E. Classical harmonic and complex analysis (orthogonal developments, quasi-analyticity, gap theorems, and Fourier transforms in the complex domain)
   F. Hopf-Wiener integral equations
   G. Prediction and filtering
   H. Relativity and quantum theories
   I. Miscellaneous mathematical papers

II. Cybernetical and Philosophical Papers
   A. Philosophical papers
   B. Cybernetical papers

III. Social, Ethical, Educational, and Literary papers

IV. Book Reviews, Prefaces, and Obituaries
   A. Book reviews and prefaces
   B. Obituaries

V. Abstracts

VI. Books and Other Publications, Not Reprinted in These Collected Works

Introduction and Acknowledgment

P. Masani

In these Works we have classified Wiener's publications into six broad categories labeled I, . . ., VI, and have divided the larger categories into subcategories labeled A, B, C, ... This classification has been motivated by practical editorial considerations, but we hope that it reflects to some extent the natural stages in Wiener's intellectual evolution.

The publications are indexed by the official year of their appearance. The internal ordering of the publications appearing in a given year is not chronological but according to the categories mentioned in the last paragraph: a run down the list a, b, c . . . for a particular year entails a run down the list of categories IA, IB, . . ., usually with omissions and repetitions, as will be apparent from a glance at the right-hand columns in the bibliography (pp. 7-26). For instance, [33d] means "the dth paper, according to category, which appeared in a journal marked 1933."

The objective in these collected works is not only the reproduction of all papers in categories I through V, but the placement of the overwhelming bulk of these papers in the context of modern research. This is done by means of commentaries written by contemporary scholars on individual papers or on batches of related ones. The commentaries, long or short, review the papers and trace their genesis as well as their impact on future research. They form a very valuable part of these Works, and to all commentators go our very sincere thanks for the time and energy they have so generously dedicated.

In the ordering of the reprints of papers and the accompanying comments we have eschewed chronology in favor of coherence of subject matter and conceptual development. We proceed to explain the specific groupings and to record gratefully the services rendered by individual commentators.

Wiener's first scholarly production is his unpublished doctoral dissertation entitled "A comparison between the treatment of the logic of relations by Schroder and that by Whitehead and Russell," written under the direction of Professor Karl Schmidt (Harvard, 1913).* This climaxed a rather fluctuating university education: after graduating from Tufts College in 1909 with a B. A. in engineering oriented mathematics and spending an inconclusive year as a postgraduate in zoology at Harvard, he switched to the philosophy de-

*For an interesting account of this thesis, in content and organization and of Russell's views concerning it, see I. Grattan-Guinness: *Wiener on the logics of Russell and Schroeder. An account of his doctoral thesis, and of his discussion of it with Russell,* Ann. Sci. 32 (1975), 103-132.

partment at Cornell in 1910-1911 and then to the philosophy department at Harvard during 1911-1913. Among his teachers were the philosopher Josiah Royce and the mathematician E. V. Huntington, but his dissertation advisor was Professor Karl Schmidt of Tufts College. He earned his Ph.D. in June 1913 (age 18) and then went to Cambridge University to pursue postdoctoral studies with Bertrand Russell.

Wiener's initial publications (1913-1922 approx.) bear the impress of his philosophy and mathematics teachers and of his own scientific knowledge and predilections. The papers are in the area of mathematical philosophy, but already the abstractions he created show an awareness of the limitations to sharpness emanating from noise. These papers have been categorized IA in our classification.

Philosophers as well as mathematicians have helped with the papers in IA. We are grateful to Professor H. E. Kyburg for his commentary on the papers on relation theory and theory of measurement [14a, b], [15a], [21a],† and to Professor A. Fine for his commentary on the very long and difficult paper [22a] on the philosophy of geometry. We thank Professor W. V. Quine for his comments on the interesting paper [15b] on mathematical certainty and the rather more dated paper [16a] on implication. For comments on the essay [23a] on the psychology of mathematics we thank Professor R. L. Wilder. On the more mathematical and less philosophical side, we have in IA, apart from Wiener's first printed paper [13a] on ordinals, the papers [20a-d] and [21b] on field postulation based on a single binary operation, work no doubt influenced by Sheffer and Huntington, for comments on which we are grateful to Professor R. J. Levit. Stimulated by Fréchet's writings, Wiener wrote the papers [20e] and [22b, c], which attempt to introduce topological ideas in terms of "continuity" as the primitive concept, for comments on which we are again thankful to Professor R. L. Wilder. Finally in this category, we thank Professor F. E. Hohn for his comments on the somewhat isolated paper [17a] on Boolean algebras.

We now come to a very creative stage of Wiener's work. At Cambridge University, thanks to wise counsel from Bertrand Russell, Wiener relearned mathematical analysis by attending G. H. Hardy's courses, studied Einstein's work on relativity and Brownian motion, and took an interest in Bohr's quantum theory of the atom. He also read G. I. Taylor's writings on turbulence. He soon had a firm knowledge of the Lebesgue integral, good insights

†Bracketed numbers refer to the Bibliography of Norbert Wiener found on pp. 7-26 of this volume.

into physics, and the philosophical view of a harmonious but noise-infected and causally incomplete universe. These assets, together with hard work based on important hints from A. Barnett on integration in function-space, culminated in the fundamental papers [20f], [21c, d], [23d], and [24d] on Brownian motion, which have profoundly affected the development of analysis and probability theory. We are indebted to Professor K. Ito for his comprehensive commentary on these papers.

During the early twenties there occurred another important development. Wiener often consulted the Harvard authority on potential theory, O. D. Kellog. From this contact sprung the papers [23b, c], [24a-c], and [25a], for comments on which we are thankful to Dr. Mrs. L. Lumer. These very original papers have left a permanent mark on potential theory, but they seem to have had very little influence on Wiener's future work. On the other hand, his papers on Brownian motion were overwhelmingly influential in shaping his subsequent research. Because the Brownian motion papers have a long aftermath in Wiener's research, we have put them in category IC after the potential theory papers (IB) in the classification as well as in the ordering of reprints and commentaries.

After the fundamental papers on Brownian motion mentioned above and commented on by Professor K. Ito, come [23e], [24e], [33a], and [34a] on random series and functions, for comments on which we thank Professor J.—P. Kahane. Then come the somewhat difficult and controversial "chaos" papers [38a] and [43a] for reviews of which we are indebted to Professor L. Gross, Dr. B. McMillan and Professor J. Feldman. Behind some of this work lay Wiener's hopes of resolving the outstanding questions of statistical physics, especially in the theory of turbulence. Paper [39b] and the abstracts [39h] and [40d] are devoted to the exploration of these questions, and we are grateful to Drs. B. McMillan and G. S. Deem for their very comprehensive survey of these papers and of the physical bearing of [38a] and [43a]. Regarding the last two papers, we should add that their significance is also being felt increasingly in quantum field theory, but that the time is not yet ripe to appraise their full impact on this area.

All these papers of Wiener were offshoots in one way or another of the earlier fundamental papers on Brownian motion, and of subsequent advances in the area recorded in his book [34d] with Paley, especially the introduction of the (measure-preserving) flow of Brownian motion. This last advance was no doubt influenced by Birkhoff's discovery of the ergodic theorem in 1931. In the late thirties ergodic theory itself engaged Wiener's attention: he wrote the interesting papers [39a] and [41a, b], for commentaries on which we are indebted to Professors E. J. Stein and W. A. Veech. Finally we have paper [57a] on the stochastic adjoint, for reviews of which we thank both Professor K. Ito and the coauthor Dr. E. J. Akutowicz.

The next stage of Wiener's work has to do with his remarkable synthesis of stochastic ideas, generalized harmonic analysis and tauberian theory (Category ID). This material will be covered in Volume II of these collected works.

Bibliography of Norbert Wiener

(Entries preceded by an asterisk are reprinted in this volume)

	Title	Category
*[13a]	*On a method of rearranging the positive integers in a series of ordinal numbers greater than that of any given fundamental sequence of omegas,* Messenger of Math. 43(1913), 97-105.	IA
*[14a]	*A simplification of the logic of relations,* Proc. Cambridge Philos. Soc. 17(1914), 387-390.	IA
*[14b]	*A contribution to the theory of relative position,* Proc. Cambridge Philos. Soc. 17(1914), 441-449.	IA
[14c]	*The highest good,* J. Phil. Psych. and Sci. Method 11(1914), 512-520.	IIA
[14d]	*Relativism,* J. Phil. Psych. and Sci. Method 11(1914), 561-577.	IIA
*[15a]	*Studies in synthetic logic,* Proc. Cambridge Philos. Soc. 18(1915), 14-28.	IA
*[15b]	*Is mathematical certainty absolute?,* J. Phil. Psych. and Sci. Method 12(1915), 568-574.	IA
*[16a]	*Mr. Lewis and implication,* J. Phil. Psych. and Sci. Method 13(1916), 656-662.	IA
[16b]	*The shortest line dividing an area in a given ratio,* Proc. Cambridge Philos. Soc. 18(1916), 56-58.	I,I
[16c]	Review of Cassius J. Keyser, *Science and Religion: the Rational and the Superrational,* J. Phil. Psych. and Sci. Method 13(1916), 273-277.	IVA
[16d]	Review of A. A. Robb, *A Theory of Time and Space,* J. Phil. Psych. and Sci. Method 13(1916), 611-613.	IVA
*[17a]	*Certain formal invariances in Boolean algebras,* Trans. Amer. Math. Soc. 18(1917), 65-72.	IA
[17b]	Review of C. J. Keyser, *The Human Worth of Rigorous Thinking,* J. Phil. Psych. and Sci. Method 14(1917), 356-361.	IVA

[18a] Review of Edward V. Huntington, *The Continuum and Other Types of Serial Order,* J. Phil. Psych. and Sci. Method 15(1918), 78-80. IVA

*[20a] *Bilinear operations generating all operations rational in a domain Ω,* Ann. of Math. 21(1920), 157-165. IA

*[20b] *A set of postulates for fields,* Trans. Amer. Math. Soc. 21(1920), 237-246. IA

*[20c] *Certain iterative characteristics of bilinear operations,* Bull. Amer. Math. Soc. 27(1920), 6-10. IA

*[20d] *Certain iterative properties of bilinear operations,* G. R. Strasbourg Math. Congress, 1920, 176-178. IA

*[20e] *On the theory of sets of points in terms of continuous transformations,* G. R. Strasbourg Math. Congress, 1920, 312-315. IA

*[20f] *The mean of a functional of arbitrary elements,* Ann. of Math. (2) 22(1920), 66-72. IC

[20g] Review of C. I. Lewis, *A Survey of Symbolic Logic,* J. Phil. Psych. and Sci. Method 17(1920), 78-79. IVA

*[21a] *A new theory of measurement: A study in the logic of mathematics,* Proc. London Math. Soc. 19(1921), 181-205. IA

*[21b] *The isomorphisms of complex algebra,* Bull. Amer. Math. Soc. 27(1921), 443-445. IA

*[21c] *The average of an analytic functional,* Proc. Nat. Acad. Sci. U. S. A. 7(1921), 253-260. IC

*[21d] *The average of an analytic functional and the Brownian movement,* Proc. Nat. Acad. Sci. U. S. A. 7(1921), 294-298. IC

[21e] *A new vector method in integral equations* (with F. L. Hitchcock), J. Math. and Phys. 1(1921), 1-20. I,I

*[22a] *The relation of space and geometry to experience,* Monist 32(1922), 12-60, 200-247, 364-394. IA

*[22b] *The group of the linear continuum,* Proc. London Math. Soc. 20(1922), 329-346. IA

*[22c] *Limit in terms of continuous transformation,* Bull. Soc. IA

Math. France 50(1922), 119-134.

[22d] *The equivalence of expansions in terms of orthogonal functions* (with J. L. Walsh), J. Math. and Phys. 1(1922), 103-122. IE

[22e] *A new type of integral expansion,* J. Math. and Phys. 1(1922), 167-176. I,I

*[23a] *On the nature of mathematical thinking,* Austral. J. Psych. and Phil. 1(1923), 268-272. IA

*[23b] *Nets and the Dirichlet problem* (with H. B. Phillips), J. Math. and Phys. 2(1923), 105-124. (Reprinted in [64f].) IB

*[23c] *Discontinuous boundary conditions and the Dirichlet problem,* Trans. Amer. Math. Soc. 25(1923), 307-314. IB

*[23d] *Differential-space,* J. Math. and Phys. 2(1923), 131-174. (Reprinted in [64f].) IC

*[23e] *O szeregach* $\sum_{1}^{\infty}(\pm 1/n)$*—Note on the series* $\sum_{1}^{\infty}(\pm 1/n)$, Bull. Acad. Polon. Ser. A, 13(1923), 87-90. IC

[23f] *Note on a new type of summability,* Amer. J. Math. 45(1923), 83-86. ID

[23g] *Note on a paper of M. Banach,* Fund. Math. 4(1923), 136-143. I,I

*[24a] *Certain notions in potential theory,* J. Math. and Phys. 3(1924), 24-51. IB

*[24b] *The Dirichlet problem,* J. Math. and Phys. 3(1924), 127-146. (Reprinted in [64f].) IB

*[24c] *Une condition nécessaire et suffisante de possibilité pour le problème de Dirichlet,* C. R. Acad. Sci. Paris 178(1924), 1050-1054. IB

*[24d] *The average value of a functional,* Proc. London Math. Soc. 22(1924), 454-467. IC

*[24e] *Un problème de probabilités denombrables,* Bull. Soc. Math. France 11(1924), 569-578. IC

[24f] *The quadratic variation of a function and its Fourier coefficients,* J. Math. and Phys. 3(1924), 72-94. ID

[24g] Review of four books on space, Rudolf Carnap's *Der Raum: Ein Beitrag zur Wissenschaftslehre,* E. Study's *Mathematik und Physik: eine erkenntnistheoretische Untersuchung* and *Die realistische Weltansicht und die Lehre vom Raume: zweite Auglage; erster Teil,* and Hermann Weyl's *Mathematische Analyse des Raum-problems. Vorlesungen gehalten in Barcelona und Madrid,* Bull. Amer. Math. Soc. 30(1924), 258-262. — IVA

[24h] Review of E. Study, *Denken und Darstellung: Logik und Werte; Dingliches und Menschliches in Mathematik und Naturwissenschaften,* Bull. Amer. Math. Soc. 30(1924), 277. — IVA

[24i] *In memory of Joseph Lipka,* J. Math. and Phys. 3(1924), 63-65. — IVB

*[25a] *Note on a paper of O. Perron,* J. Math. and Phys. 4(1925), 21-32. — IB

[25b] *The solution of a difference equation by trigonometrical integrals,* J. Math. and Phys. 4(1925), 153-163. — ID

[25c] *On the representation of functions by trigonometrical integrals,* Math. Z. 24(1925), 575-616. — ID

[25d] *Verallgemeinerte trigonometrische Entwicklungen,* Göttingen Nachr. (1925), 151-158. — ID

[25e] *Note on quasi-analytic functions,* J. Math. and Phys. 4(1925), 193-199. — IE

[25f] *A contribution to the theory of interpolation,* Ann. of Math. (2) 26(1925), 212-216. — I,I

[26a] *The harmonic analysis of irregular motion,* J. Math and Phys. 5(1926), 99-121. — ID

[26b] *The harmonic analysis of irregular motion* (Second Paper), J. Math. and Phys. 5(1926), 158-189. — ID

[26c] *The operational calculus,* Math. Ann. 95(1926), 557-584. — ID

[26d] *A new formulation of the laws of quantization of periodic and aperiodic phenomena* (with M. Born), J. Math and Phys. 5(1926), 84-98. — IH

[26e] *Eine neue Formulierung der Quantengesetze für periodische und nicht periodische Vorgänge* (with M. Born), Z. Physik 36(1926), 174-187. IH

[26f] *Analytic approximations to topological transformations* (with P. Franklin), Trans. Amer. Math. Soc. 28(1926), 762-785. I,I

[27a] *The spectrum of an array and its application to the study of the translation properties of a simple class of arithmetical functions,* Part I, J. Math. and Phys. 6(1927), 145-157. (Part II: *On the translation of a simple class of arithmetical functions,* by K. Mahler, *ibid.,* pp. 158-163.) ID

[27b] *A new definition of almost periodic functions,* Ann. of Math. (2) 28(1927), 365-367. ID

[27c] *On a theorem of Bochner and Hardy,* J. London Math. Soc. 2(1927), 118-123. ID

[27d] *Une méthode nouvelle pour la démonstration des théorèmes de M. Tauber,* C. R. Acad. Sci. Paris 184(1927), 793-795. ID

[27e] *On the closure of certain assemblages of trigonometrical functions,* Proc. Nat. Acad. Sci., U.S.A. 13(1927), 27-29. IE

[27f] *Quantum theory and gravitational relativity* (with D. J. Struik), Nature 119(1927), 853-854. IH

[27g] *A relativistic theory of quanta* (with D. J. Struik), J. Math and Phys. 7(1927), 1-23. IH

[27h] *Sur la théorie relativiste des quanta* (with D. J. Struik), C. R. Acad. Sci. Paris 185(1927), 42-44. IH

[27i] *Sur la théorie relativiste des quanta* (Note), C. R. Acad. Sci. Paris 185(1927), 184-185. IH

[27j] *Laplacians and continuous linear functionals,* Acta Sci. Math. (Szeged) 3(1927), 7-16. II

[27k] *Une généralisation des fonctions à variation bornée,* C. R. Acad. Sci. Paris 185(1927), 65-67. I,I

[28a] *The spectrum of an arbitrary function,* Proc. London Math. Soc. (2) 27(1928), 483-496. ID

[28b] *A new method in Tauberian theorems,* J. Math. and Phys. 7(1928), 161-184. ID

[28c] *The fifth dimension in relativistic quantum theory* (with D. J. Struik), Proc. Nat. Acad. Sci. U. S. A. 14(1928), 262-268. IH

[28d] *Coherency matrices and quantum theory,* J. Math. and Phys. 7(1928), 109-125. IH

[29a] *Harmonic analysis and group theory,* J. Math. and Phys. 8(1929), 148-154. ID

[29b] *A type of Tauberian theorem applying to Fourier series,* Proc. London Math. Soc. (2) 30(1929), 1-8. ID

[29c] *Fourier analysis and asymptotic series.* Appendix to V. Bush, *Operational Circuit Analysis,* New York, John Wiley, 1929, 366-379. ID

[29d] *Hermitian polynomials and Fourier analysis,* J. Math. and Phys. 8(1929), 70-73. IE

[29e] *Harmonic analysis and the quantum theory,* J. Franklin Inst. 207(1929), 525-534. IH

[29f] *On the spherically symmetrical statical field in Einstein's unified theory of electricity and gravitation* (with M. S. Vallarta), Proc. Nat. Acad. Sci. U. S. A. 15(1929), 353-356. IH

[29g] *On the spherically symmetrical statical field in Einstein's unified theory: A correction* (with M. S. Vallarta), Proc. Nat. Acad. Sci. U. S. A. 15(1929), 802-804. IH

[29h] *Mathematics and art (Fundamental identities in the emotional aspects of each),* Tech. Rev. 32(1929), 129-132, 160, 162. IIA

[29i] *Einsteiniana (Facts and fancies about Dr. Einstein's famous theory),* Tech. Rev. 32(1929), 403-404. IIIE

[29j] *Murder and mathematics,* Tech. Rev. 32(1929), 271-272. IVA

[30a] *Generalized harmonic analysis,* Acta Math. 55(1930), 117-258. (Reprinted in [64f] and [66b].) ID

[30b] Review of A. Eddington's *Science and the Unseen World,* Tech. Rev. 33(1930), 150. IVA

[31a] *Über eine Klasse singulärer Integralgleichungen* (with E. Hopf), Sitzber. Preuss. Akad. Wiss. Berlin, Kl. Math. Phys. Tech., 1931, pp. 696-706. (Reprinted in [64f].) IF

[31b] *A new deduction of the Gaussian distribution,* J. Math. and Phys. 10(1931), 284-288. I,I

[31c] *Reports from Cambridge–1931,* Tech. Rev. 34(1931), 82-83, 131, 218, 220. IIIE

[32a] *Tauberian theorems,* Ann. of Math. 33(1932), 1-100. (Reprinted in [64f] and [66b].) ID

[32b] *A note on Tauberian theorems,* Ann. of Math. 33(1932), 787. ID

[32c] *Back to Leibniz! (Physics reoccupies an abandoned position),* Tech. Rev. 34(1932), 201-203, 222, 224. IIA

[32d] *Reports from Cambridge–1932,* Tech. Rev. 34(1932), 62, 74. IIIE

[32e] Review of A. S. Besicovitch, *Almost Periodic Functions,* Math. Gaz. 16(1932), 275-277. IVA

[32f] *Analytic properties of the characters of infinite Abelian groups* (with R. E. A. C. Paley), Abstract, Int'l. Math. Congr., Zürich, 1932, 95. V

*[33a] *Notes on random functions* (with R. E. A. C. Paley and A. Zygmund), Math. Z. 37(1933), 647-668. IC

[33b] *A one-sided Tauberian theorem,* Math. Z. 36(1933), 787-789. ID

[33c] *Characters of Abelian groups* (with R. E. A. C. Paley), Proc. Nat. Acad. Sci. U. S. A. 19(1933), 253-257. ID

[33d] *The total variation of* $g(x+h) - g(x)$ (with R. C. Young), Trans. Amer. Math. Soc. 35(1933), 327-340. ID

[33e] *Notes on the theory and application of Fourier transforms* (with R. E. A. C. Paley) I, II, Trans. Amer. Math. Soc. 35 (1933), 348-355; III, IV, V, VI, VII, Trans. Amer. Math. Soc. 35(1933), 761-791. IE

[33f] *Putting matter to work (The search for cheaper power),* Tech. Rev. 35(1933), 47-49, 70, 72. IIIA

[33g] Review of Harald Bohr, *Fastperiodische Funktionen,* Math. Gaz. 17(1933), 54. IVA

[33h] *R. E. A. C. Paley—In Memoriam, Jan. 7, 1907-Apr. 7, 1933,* Bull. Amer. Math. Soc. 39(1933), 476. IVB

[33i] *The Fourier Integral and Certain of Its Applications,* Cambridge University Press, New York, 1933; reprint, Dover, New York, 1959; review by E. C. Titchmarsh in Math. Gaz. 17(1933), 129. VI

*[34a] *Random functions,* J. Math. and Phys. 14(1934), 17-23. IC

[34b] *A class of gap theorems,* Ann. Scuola Norm. Sup. Pisa, E(1934-1936), 1-6. IE

[34c] *Quantum mechanics, Haldane, and Leibniz,* Philos. Sci. 1(1934), 479-482. IIA

[34d] *Fourier Transforms in the Complex Domain* (with R. E. A. C. Paley), Amer. Math. Soc. Colloq. Publ. 19, Amer. Math. Soc., Providence, R. I., 1934. VI

[35a] *Fabry's gap theorem,* Sci. Repts. of Nat'l. Tsing Hua Univ., Ser. A, 3(1935), 239-245. IE

[35b] *Limitations of science (The holiday fallacy and a response to the suggestion that scientists become sociologists),* Tech. Rev. 37(1935), 255-256, 268, 270, 272. IIIA

[35c] *The student agitator (Is he accepting radicalism as an opiate?)* (with Carl Bridenbaugh), Tech. Rev. 37(1935), 310-312, 344, 346. IIIC

[35d] *Mathematics in American secondary schools,* J. Math. Assoc. Japan for Secondary Education (Tokyo) 17(1935), 1-5. IIIC

[35e] *The closure of Bessel functions,* Abstract 66, Bull. Amer. Math. Soc. 41(1935), 35. V

[36a] *A theorem of Carleman,* Sci. Repts. of Nat'l. Tsing Hua Univ., Ser. A, 3(1936), 291-298. IE

[36b] *Sur les séries de Fourier lacunaires. Théorèmes directs* (with IE

S. Mandelbrojt), C. R. Acad. Sci. Paris 203(1936), 34-36.

[36c] *Séries de Fourier lacunaires. Théorèmes inverses* (with S. Mandelbrojt), C. R. Acad. Sci. Paris 203(1936), 233-234. IE

[36d] *Gap theorems,* C. R. de Congr. Int'l. des Math., 1936, 284-296. IE

[36e] *A Tauberian gap theorem of Hardy and Littlewood,* J. Chinese Math. Soc. 1(1936), 15-22. IE

[36f] *Notes on the Kron theory of tensors in electrical machinery,* J. Electr. Engrg., China 7(1936), 277-291. I,I

[36g] *The role of the observer,* Philos. Sci. 3(1936), 307-319. IIA

[37a] *Taylor's series of entire functions of smooth growth* (with W. T. Martin), Duke Math. J. 3(1937), 213-223. ID

[37b] *Random Waring's theorems,* Abstract (with N. Levinson), Science 85(1937), 439. V

*[38a] *The homogeneous chaos,* Amer. J. Math. 60(1938), 897-936. (Reprinted in [64f].) IC

[38b] *On absolutely convergent Fourier-Stieltjes transforms* (with H. R. Pitt), Duke Math. J. 4(1938), 420-440. ID

[38c] *Fourier-Stieltjes transforms and singular infinite convolutions* (with A. Wintner), Amer. J. Math. 60(1938), 513-522. ID

[38d] *Taylor's series of functions of smooth growth in the unit circle* (with W. T. Martin), Duke Math. J. 4(1938), 384-392. ID

[38e] *The historical background of harmonic analysis,* Amer. Math. Soc. Semicentennial Publications Vol. II, Semicentennial Addresses, Amer. Math. Soc., Providence, R. I., 1938, 513-522. ID

[38f] *Remarks on the classical inversion formula for the Laplace integral* (with D. V. Widder), Bull. Amer. Math. Soc. 44 (1938), 573-575. IE

[38g] *The decline of cookbook engineering,* Tech. Rev. 40(1938), 23. IIIE

[38h] Review of L. Hogben, *Science for the Citizen,* Tech. Rev. IVA

40(1938), 66-67.

*[39a] *The ergodic theorem,* Duke Math. J. 5(1939), 1-18. (Reprinted in [64f].) IC

*[39b] *The use of statistical theory in the study of turbulence,* Proc. 5th Int'l. Congr. of Applied Mechanics, Sept. 12-16, 1938, Wiley, New York, 1939, 356-358. IC

[39c] *On singular distributions* (with A. Wintner), J. Math. and Phys. 17(1939), 233-246. ID

[39d] *Convergence properties of analytic functions of Fourier-Stieltjes transforms* (with R. H. Cameron), Trans. Amer. Math. Soc. 46(1939), 97-109; Math. Rev. 1(1940), 13; rev. 400. ID

[39e] *A generalization of Ikehara's theorem* (with H. R. Pitt), J. Math. and Phys. 17(1939), 247-258. ID

[39f] Review of Roger Burlingame, *March of the Iron Men,* Tech. Rev. 41(1939), 115. IVA

[39g] Review of W. George, *The Scientist in Action,* Tech. Rev. 41(1939), 202. IVA

*[39h] *A new method in statistical mechanics,* Abstract 133 (with B. McMillan), Bull. Amer. Math. Soc. 45(1939), 234; Science 90(1939), 410-411. V

[40a] Review of M. Fukamiya, *On dominated ergodic theorems in L_p ($p \geqslant 1$),* Math. Rev. 1(1940), 148. IVA

[40b] Review of M. Fukamiya, *The Lipschitz condition of random function,* Math Rev. 1(1940), 149. IVA

[40c] Review of Th. De Donder, *L'énergétique déduite de la méchanique statistique générale,* Math. Rev. 1(1940), 192. IVA

*[40d] *A canonical series for symmetric functions in statistical mechanics,* Abstract 133, Bull. Amer. Math. Soc. 46(1940), 57. V

*[41a] *Harmonic analysis and ergodic theory* (with A. Wintner), Amer. J. Math. 63(1941), 415-426; Math. Rev. 2(1941), 319. IC

*[41b] *On the ergodic dynamics of almost periodic systems* (with A. Wintner), Amer. J. Math. 63(1941), 794-824; Math. Rev. 4(1943), 15. IC

[42a] *On the oscillation of the derivatives of a periodic function* (with G. Pólya), Trans. Amer. Math. Soc. 52(1942), 249-256. IE

*[43a] *The discrete chaos* (with A. Wintner), Amer. J. Math. 65(1943), 279-298; Math. Rev. 4(1943), 220. IC

[43b] *Behavior, purpose, and teleology* (with A. Rosenblueth and J. Bigelow), Philos. Sci. 10(1943), 18-24. IIA

[45a] *La teoria de la estrapolacion estadistica,* Bol. Soc. Mat. Mexicana 2(1945), 37-45; Math. Rev. 7(1946), 461. IG

[45b] *The role of models in science* (with A. Rosenblueth), Philos. Sci. 12(1945), 316-322. IIA

[46a] *A generalization of the Wiener-Hopf integral equation* (with A. E. Heins), Proc. Nat. Acad. Sci. U.S.A. 32(1946), 98-101; Math. Rev. 8(1947), 29. IF

[46b] *The mathematical formulation of the problem of conduction of impulses in a network of connected excitable elements, specifically in cardiac muscle* (with A. Rosenblueth), Arch. Inst. Cardiol. Méxicana 16(1946), 205-265; Bol. Soc. Mat. Mexicana 2(1945), 37-42; Math. Rev. 9(1948), 604. IIB

[47a] *Sur la fonctions indéfiniment dérivables sur une demi-droite* (with S. Mandelbrojt), C. R. Acad. Sci. Paris 225(1947), 978-980; Math. Rev. 9(1948), 230. IE

[47b] *A scientist rebels,* Atlantic Monthly 179(1946), 46; Bull. Atomic Scientists 3(1947), 31. IIIB

[48a] *Time, communication and the nervous system,* Ann. New York Acad. Sci. 50(1948), 197-220; Math. Rev. 10(1949), 133. IIB

[48b] *Cybernetics,* Scientific American 179(1948), 14-18. IIB

[48c] *An account of the spike potential of axons* (with A. Rosenblueth, W. Pitts, J. García Ramos, and the assistance IIB

of F. Webster), J. of Cellular and Comparative Physiol. 32(1948), 275-318.

[48d] *A rebellious scientist after two years,* Bull. Atomic Scientists 4(1948), 338-339. IIIB

[48e] Review of L. Infeld, *Whom the Gods Love. The Story of Evariste Galois,* Scripta Math. 14(1948), 273-274. IVA

[48f] *Cybernetics, or Control and Communication in the Animal and the Machine,* Actualités Sci. Ind., no. 1053; Hermann et Cie., Paris; The MIT Press, Cambridge, Mass., and Wiley, New York, 1948; Math. Rev. 9(1948), 598. Partly reprinted as Rigidity and learning: ants and men, in *Classics in Biology (A Course of Selected Reading by Authorities),* Philosophical Library, New York, 1960, pp. 205-213. VI

[49a] *Sur la théorie de la prévision statistique et du filtrage des ondes,* Analyse Harmonique, Colloques Internationaux du CNRS, No. 15, pp. 67-74. Centre National de la Recherche Scientifique, Paris, 1949; Math. Rev. 11(1950), 376. IG

[49b] *A statistical analysis of synaptic excitation* (with A. Rosenblueth, W. Pitts, and J. García Ramos), J. of Cellular and Comparative Physiol. 34(1949), 173-205. IIB

[49c] *A new concept of communication engineering,* Electronics 22(1949), 74-77. IIB

[49d] *Sound communication with the deaf,* Philos. Sci. 16(1949), 260-262. IIB

[49e] *Some problems in sensory prosynthesis* (with J. Wiesner and L. Levine), Science 110(1949), 512. IIB

[49f] *Obituary—Godfrey Harold Hardy, 1877-1947,* Bull. Amer. Math. Soc. 55(1949), 72-77. IVB

[49g] *Extrapolation, Interpolation, and Smoothing of Stationary Time Series with Engineering Applications,* The MIT Press, Cambridge, Mass.; Wiley, New York; Chapman & Hall, London, 1949; paperback edition with the title *Time Series,* The MIT Press, 1964; Math. Rev. 11(1950), 118. VI

[50a] *Some prime-number consequences of the Ikehara theorem* ID

(with L. Geller), Acta Sci. Math. (Szeged) 12(1950), 25-28, Leopoldo Fejer et Frederico Riesz LXX annos natis dedicatus, Pars B; Math. Rev. 11(1950), 644; Math. Rev. 12(1951), 1002.

[50b] *Comprehensive view of prediction theory,* Proceedings of the International Congress of Mathematicians, Cambridge, Mass., 1950, vol. 2, pp. 308-321; Amer. Math. Soc., Providence, R. I., 1952, Expository lecture; Math. Rev. 13(1952), 477. IG

[50c] *Some maxims for biologists and psychologists,* Dialectica 4(1950), 186-191. IIA

[50d] *Purposeful and non-purposeful behavior* (with A. Rosenblueth), Philos. Sci. 17(1950), 318-326. IIA

[50e] *Cybernetics,* Bull. Amer. Acad. Arts and Sci. 3(1950), 2-4. IIB

[50f] *Speech, language, and learning,* J. Acoust. Soc. Amer. 22(1950), 696-697. IIB

[50g] *Entropy and information,* Proc. Sympos. Appl. Math., vol. 2, Amer. Math. Soc., Providence, R. I., 1950, p. 89; Math. Rev. 11(1950), 305. IIB

[50h] *Too big for private enterprise,* Nation 170(1950), 496-497; IIIA

[50i] *Too damn close,* Atlantic 186(1950), 50-52. IIIA

[50j] *The Human Use of Human Beings,* Houghton Mifflin, Boston, 1950; paperback edition by Doubleday, Anchor, Garden City, N. Y., 1954. Chapter 3 reprinted in *Classics in Biology (A Course of Selected Reading by Authorities),* Philosophical Library, New York, 1960, pp. 205-213. VI

[50k] *The brain* (short story), Tech. Eng. News 31(1950), 14-15, 33-34, 44, 50. (reprinted in paperback anthology, *Cross-roads in Time,* ed. Groff Conklin, Doubleday, Garden City, N. Y., 1953.) VI

[51a] *Problems of sensory prosthesis,* Bull. Amer. Math. Soc. 57(1951), 27-35. (Reprinted in [64f].) IIB

[51b] *Homeostasis in the individual and society,* J. Franklin Inst. 251(1951), 65-68. (Reprinted in [64f].) IIB

[52a] *Cybernetics (Light and Maxwell's demon),* Scientia (Italy) 87(1952), 233-235. IIB

[52b] *The miracle of the broom closet* (short story), Tech. Eng. News 33(1952), 18-19, 50. (Reprinted in the Magazine of Fantasy and Science Fiction, ed. Anthony Boucher, February, 1954, pp. 59-63.) VI

[53a] *Optics and the theory of stochastic processes,* J. Opt. Soc. Amer. 43(1953), 225-228; Math. Rev. 17(1956), 33. IG

[53b] *A new form for the statistical postulate of quantum mechanics* (with A. Siegel), Phys. Rev. 91(1953), 1551-1560; Math. Rev. 15(1954), 273. IH

[53c] *Distributions quantiques dans l'espace différentiel pour les fonctions d'ondes dépendant du spin* (with A. Siegel), C. R. Acad. Sci. Paris 237(1953), 1640-1642; Math. Rev. 15(1954), 490. IH

[53d] *Les machines à calculer et la forme (Gestalt), Les machines à calculer et la pensee humaine,* Colloques Internationaux du Centre National de la Recherche Scientifique, Paris, 1953, pp. 461-463; Math. Rev. 16 (1955), 529. IIB

[53e] *The concept of homeostasis in medicine,* Transactions and Studies of the College of Physicians of Philadelphia (4) 20(1953), No. 3, 87-93. IIB

[53f] *Problems of organization,* Bull. Menninger Clinic 17(1953), 130-138. IIB

[53g] *The future of automatic machinery,* Mech. Engrg. 75 (1953), 130-132. IIB

[53h] *Ex-Prodigy: My Childhood and Youth,* Simon and Schuster, New York, 1953; The MIT Press, Cambridge, Mass., 1965 (paperback edition also by The MIT Press); Math. Rev. 15(1954), 277. VI

[54a] *Men, machines, and the world about,* in *Medicine and Science,* New York Academy of Medicine and Science, ed. I. Galderston, International Universities Press, New York, 1954, pp. 13-28. IIB

[54b] *Conspiracy of conformists,* Nation 178(1954), 375. IIIE

[55a] *Nonlinear prediction and dynamics,* Proc. Third Berkeley Symposium on Mathematical Statistics and Probability, University of California Press, Berkeley, Calif., 1954/5, pp. 247-252; Math. Rev. 18(1957), 949. IG

[55b] *On the factorization of matrices,* Comment. Math. Helv. 29(1955), 97-111; Math. Rev. 16(1955), 921. IG

[55c] *The differential-space theory of quantum systems* (with A. Siegel), Nuovo Cimento (10) 2(1955), 982-1003, No. 4, Suppl. IH

[55d] *Thermodynamics of the message,* in *Neurochemistry,* ed. K. E. C. Elliott, Thomas, Springfield, 1955, pp. 844-849. IIB

[55e] *Time and organization,* Second Fawley Foundation Lecture, University of Southampton, 1955, pp. 1-16. IIB

[56a] *On a local L^2-variant of Ikehara's theorem* (with A. Wintner), Rev. Math. Cuyana 2(1956), 53-59. ID

[56b] *The theory of prediction,* in *Modern Mathematics for the Engineer,* ed. E. F. Beckenbach, McGraw-Hill, New York, 1956, pp. 165-187. IG

[56c] *"Theory of Measurement" in differential-space quantum theory* (with A. Siegel), Phys. Rev. 101(1956), 429-432. IH

[56d] *Pure patterns in a natural world,* in *The New Landscape in Art and Science,* ed. G. Kepes, Paul Theobald and Co., Chicago, 1956, pp. 274-276. IIB

[56e] *Brain waves and the interferometer,* J. Phys. Soc. Japan 18(1956), 499-507. IIB

[56f] *Moral reflections of a mathematician,* Bull. Atomic Scientists 12(1956), 53-57. (Reprinted from [56g].) IIIB

[56g] *I Am a Mathematician. The Later Life of a Prodigy,* VI

Doubleday, Garden City, New York, 1956; paperback edition by The MIT Press, 1964; Math. Rev. 17(1956), 1037.

*[57a] *The definition and ergodic properties of the stochastic adjoint of a unitary transformation* (with E. J. Akutowicz), Rend. Circ. Mat. Palermo (2) 6(1957), 205-217, Addendum, 349; Math. Rev. 20(1959), rev.4328. IC

[57b] *Notes on Pólya's and Turán's hypotheses concerning Liouville's factor* (with A. Wintner), Rend. Circ. Mat. Palermo (2) 6(1957), 240-248; Math. Rev. 20(1959), rev. 5759. ID

[57c] *On the non-vanishing of Euler products* (with A. Wintner), Amer. J. Math. 79(1957), 801-808. ID

[57d] *The prediction theory of multivariate stochastic processes,* Part I (with P. Masani), Acta Math. 98(1957), 111-150; Math. Rev. 20(1959), rev. 4323. IG

[57e] *Rhythms in physiology with particular reference to encephalography,* Proceedings of the Rudolf Virchow Medical Society in the City of New York, vol. 16, 1957, pp. 109-124. IIB

[57f] *The role of the mathematician in a materialistic culture (A scientist's dilemma in a materialistic world),* Columbia Engineering Quarterly, Proceedings of the Second Combined Plan Conference, Arden House, October 6-9, 1957, pp. 22-24. IIIA

[57g] *The role of the small cultural college in education of the scientists,* a speech given at Wabash College, Indiana, October 10, 1957. IIIC

[58a] *Logique, probabilité et méthode des sciences physiques,* in *La Méthode dans les Sciences Modernes,* Editions Science et Industrie, ed. Francois Le Lionnais, Paris, 1958, pp. 111-112. IH

[58b] *The prediction theory of multivariate stochastic processes,* Part II (with P. Masani), Acta Math. 99(1958), 93-137; Math. Rev. 20(1959), rev. 4325. IG

[58c] *Random time* (with A. Wintner), Nature 181(1958), 561-562. IG

[58d] *Sur la prévision lineáire des processus stochastiques vectoriels à densité spectrale bornée.* I (with P. Masani), C. R. Acad. Sci. Paris 246(1958), 1492-1495; Math. Rev. 20(1959), rev. 4324a. IG

[58e] *Sur la prévision linéaire des processus stochastiques vectoriels à densité spectrale bornée.* II (with P. Masani), C. R. Acad. Sci. Paris 246(1958), 1655-1656; Math. Rev. 20(1959), rev. 4324b. IG

[58f] *My connection with cybernetics. Its origins and its future,* Cybernetica (Belgium) 1(1958), 1-14. IIB

[58g] *Time and the science of organization,* Part II, Scientia 93(1958), 225-230. IIB

[58h] *Science: The megabuck era,* New Republic 138(1958), 10-11. IIIA

[58i] *Nonlinear Problems in Random Theory,* The MIT Press, Cambridge, Mass., and Wiley, New York, 1958; paperback edition, The MIT Press, 1966. VI

[59a] *A factorization of positive Hermitian matrices* (with E. J. Akutowicz), J. Math. Mech. 8(1959), 111-120. IG

[59b] *Nonlinear prediction* (with P. Masani), in *Probability and Statistics,* The Harald Cramér Volume, ed. U. Grenander, Stockholm, 1959, 190-212. IG

[59c] *On bivariate stationary processes and the factorization of matrix-valued functions* (with P. Masani), Teor, Verojatnost. i Primenen. 4(1959), 322-331. (English transl. Theor. Probability Appl. 4(1959), 300-308.) IG

[59d] *Man and the machine* (Interview with N. Wiener), Challenge (The Magazine of Economic Affairs) 7(1959), 36-41. IIIA

[59e] *The Tempter* (novel), Random House, New York, 1959. VI

[60a] *The application of physics to medicine,* in *Medicine and Other Disciplines,* New York Academy of Medicine, ed. IIB

I. Galderston, International Universities Press, 1960, pp. 41-57.

[60b] *The brain and the machine* (Summary of an address), in *Dimensions of Mind,* ed. S. Hook, Collier Books, 1960, (Proceedings of Third Annual New York Univ. Institute of Philosophy held on May 15-16, 1959), pp. 113-117. IIB

[60c] *Kybernetik,* Contribution to *Wörterbuch der Soziologie,* F. Enke Verlag, Stuttgart, 1960, pp. 620-622. IIB

[60d] *Some moral and technical consequences of automation,* Science 131(1960), 1355-1358. IIIA

[60e] *The duty of the intellectual,* Tech. Rev. 62(1960), 26-27; reprinted almost in entirety in *The grand privilege,* Sat. Rev. 43(1960), 51-52; also in Technion 18(1961), 86-87—"A professor tells what a professor is." IIIB

[60f] Preface to *Cybernetics of Natural Systems,* by D. Stanley-Jones, Pergamon Press, London, 1960, pp. v-viii. IVA

[61a] *Über Informationstheorie,* Naturwissenschaften 48(1961), 174-176. IIB

[61b] *Science and society,* Voprosy Filosofii (1961), No. 7, 117-122; reprinted in Estratto Rivista Methodos 13(1961), 1-8, and in Tech. Rev. 63(1961), 49-52. Excerpts in Science 138(1962), 651. IIB

[61c] *Cybernetics,* Second edition of [48f] (revisions and two additional chapters), The MIT Press and Wiley, New York, 1961; paperback edition, The MIT Press, 1965. VI

[62a] A verbal Contribution to Proc. of the International Symposium on the Application of Automatic Control in Prosthetics Design, August 27-31, 1962, Opatija, Yugoslavia, pp. 132-133. IIB

[62b] *The mathematics of self-organizing systems,* in *Recent Developments in Information and Decision Processes,* MacMillan, New York, 1962, pp. 1-21. IIB

[63a] *Random theory in classical phase space and quantum mechanics* (with Giacomo Della Riccia), Proc. Internat. IH

Conference on Functional Analysis, Massachusetts Institute of Technology, Cambridge, Mass., June 9-13, 1963; *Analysis in Function Space,* The MIT Press, Cambridge, Mass., 1964, pp. 3-14.

[63b] *Introduction to neurocybernetics* (with J. P. Schadé) and *Epilogue,* in *Progress in Brain Research,* vol. 2 of Nerve, Brain and Memory Models, Elsevier Publishing Co., Amsterdam, 1963, pp. 1-7, 264-268. IIB

[63c] *The lonely nationalism of Rudyard Kipling* (with K. Deutsch), Yale Rev. 52(1963), 499-517. IIID

[64a] *On the oscillations of nonlinear systems,* Proc. Symposium on Stochastic Models in Medicine and Biology, Mathematics Research Center, U.S. Army, June 12-14, 1963, ed. John Gurland, University of Wisconsin Press, Madison, Wisconsin, 1964, pp. 167-177. IG

[64b] *Dynamical systems in physics and biology,* Contribution to series "Fundamental Science in 1984," The New Scientist (London) 21(1964), 211-212. IIB

[64c] *Machines smarter than men?* (Interview with N. Wiener), U.S. News and World Rept. 56(1964), 84-86; abbreviated in Reader's Digest 84(1964), 121-124. IIIA

[64d] *Intellectual honesty and the contemporary scientist* (Transcript of talk given to Hillel Group at Massachusetts Institute of Technology), Tech. Rev. 66(1964), 17-18, 44-45, 47. IIIA

[64e] *God, Golem, Inc.—A Comment on Certain Points Where Cybernetics Impinges on Religion,* The MIT Press, Cambridge, Mass., 1964; paperback edition, The MIT Press, 1966. VI

[64f] *Selected Papers of Norbert Wiener* with expository papers by Y. W. Lee, Norman Levinson, and W. T. Martin, The MIT Press, Cambridge, Mass., 1964. VI

[65a] L'homme et la machine, Proc. Colloques Philosophiques Internationaux de Royaumont, July, 1962; *Le concept d'information dans la science contemporaine,* Gauthier- IIB

Villars, Paris, 1965, pp. 99-132.

[66a] *Wave mechanics in classical phase space, Brownian motion, and quantum theory* (with G. Della Riccia), J. Math. Phys. 7(1966), 1372-1383. IH

[66b] *Differential Space, Quantum Systems and Prediction* (with A. Siegel, B. Rankin, W. T. Martin), The MIT Press, Cambridge, Mass., 1966. VI

[66c] *Generalized Harmonic Analysis and Tauberian Theorems* (paperback edition of [30a] and [32a]), The MIT Press, Cambridge, Mass., 1966. VI

IA
Mathematical Philosophy and Foundations

A Simplification of the Logic of Relations.

By N. WIENER, Ph.D. (Communicated by Mr G. H. Hardy.)

[*Read* 23 February 1914.]

Two axioms, known as the axioms of reducibility, are stated on page 174 of the first volume of the *Principia Mathematica* of Whitehead and Russell. One of these, ∗12·1, is essential to the treatment of identity, descriptions, classes, and relations: the other, ∗12·11, is involved only in the theory of relations. ∗12·11 is applied directly only in

∗20·701·702·703 and ∗21·12·13·151·3·701·702·703.

It states that, given any propositional function ϕ of two variable individuals, there is another propositional function of two variable individuals, involving no apparent variables, and having the same truth-value as ϕ for the same arguments, or in symbols:

$$\vdash : (\exists f) : \phi(x, y) . \equiv . f!(x, y).$$

In ∗20 and ∗21·701·702·703 all that is done with ∗12·11 is to extend it to cases where the arguments of ϕ and f are classes and relations: ∗12·11 is essential to the development of the calculus of relations only owing to its application in ∗21·12·13·151·3. Here it is needed to make the transition between the definition of a binary relation and its uses. This is due to the fact that a binary relation itself is not defined, but only propositions about it, and ∗12·11 is needed to assure us that these propositions about it behave as if there were a real object with which they concern themselves. The authors of the *Principia* wish to treat a binary relation as the extension of a propositional function of two variables: that is, when they speak about the relation between x and y when $\phi(x, y)$, they mean to speak of any propositional function which holds of those values of x and y, and only those values, of which ϕ holds. Now, as it leads one into vicious-circle paradoxes to speak directly of "any propositional function which holds of those values of x and y, and those only, of which ϕ holds," they first define a proposition concerning the relation between x and y when $\phi(x, y)$ as a proposition concerning *a propositional function involving no apparent variables* which holds of x and y when and only when $\phi(x, y)$. Then they need to use ∗12·11 to assure us that, whatever ϕ may be, there always is some such propositional function. Now, if we can discover a propositional function ψ of one variable so correlated with ϕ that its extension is determined uniquely by that of ϕ, and vice versa—if, to put it in symbols, when ψ' bears to ϕ' the same relation that ψ bears to ϕ, $\vdash :. \phi'(x, y) . \equiv_{x,y} . \phi(x, y) : \equiv : \psi'\alpha . \equiv_\alpha . \psi\alpha$—, we can entirely avoid the use of $*12{\cdot}11$, and interpret any proposition concerning the extension of ϕ as if it concerned the extension of ψ; for the existence of the extension of a propositional function of one variable is assured to us by $*12{\cdot}1$, quite as that of one of two variables is by $*12{\cdot}11$. Now, is such a ψ the propositional function

$$(\exists x, y) . \phi(x, y) . \alpha = \iota`(\iota`\iota`x \cup \iota`\Lambda) \cup \iota`\iota`\iota`y.$$

For it is clear that for each ordered pair of values of x and y there is one and only one value of α, and vice versa. On the one hand, as $\iota`(\iota`\iota`x \cup \iota`\Lambda)$ is determined uniquely by x, and $\iota`\iota`\iota`y$ is determined uniquely by $y$, $\iota`(\iota`\iota`x \cup \iota`\Lambda) \cup \iota`\iota`\iota`y$ is determined uniquely by x and y. On the other hand, if

$$\iota`(\iota`\iota`x \cup \iota`\Lambda) \cup \iota`\iota`\iota`y = \iota`(\iota`\iota`z \cup \iota`\Lambda) \cup \iota`\iota`\iota`w,$$

either $\iota`\iota`y = \iota`\iota`z \cup \iota`\Lambda$ or $\iota`\iota`y = \iota`\iota`w$. The former supposition is clearly impossible, for, as $\iota`z \neq \Lambda$, $\iota`\iota`z \cup \iota`\Lambda$ is not a unit class. From the latter alternative we conclude immediately that $y = w$. Similarly, $x = z$.

Therefore, when x and y are of the same type, we can make the following definition:

$$\hat{x}\hat{y}\phi(x, y) = \hat{\alpha}\{(\exists x, y) . \phi(x, y) . \alpha = \iota`(\iota`\iota`x \cup \iota`\Lambda) \cup \iota`\iota`\iota`y\} \quad \text{Df.*}$$

It will be seen that in this definition of $\hat{x}\hat{y}\phi(x, y)$ it is essential that the x and the y should be of the same type, for if they are not $\iota`(\iota`\iota`x \cup \iota`\Lambda)$ and $\iota`\iota`\iota`y$ will not be, and $\iota`(\iota`\iota`x \cup \iota`\Lambda) \cup \iota`\iota`\iota`y$ will be meaningless. To overcome this limitation, and secure typical ambiguity for domain and converse domain of $\hat{x}\hat{y}\phi(x, y)$ separately, we make the following definitions:

$$\hat{\alpha}\hat{y}\phi(\alpha, y) = \hat{\kappa}\{(\exists \alpha, y) . \phi(\alpha, y) . \kappa = \iota`(\iota`\iota`\alpha \cup \iota`\Lambda) \cup \iota`\iota`(\iota`\iota`y \cup \iota`\Lambda)\} \quad \text{Df.}$$

$$\hat{\kappa}\hat{y}\phi(\kappa, y) = \hat{\mu}\{(\exists \kappa, y) . \phi(\kappa, y) . \mu = \iota`(\iota`\iota`\kappa \cup \iota`\Lambda) \cup \iota`\iota`[\iota`(\iota`\iota`y \cup \iota`\Lambda) \cup \iota`\Lambda]\} \quad \text{Df.}$$

etc.

$$\hat{x}\hat{\beta}\phi(x, \beta) = \hat{\kappa}\{(\exists x, \beta) . \phi(x, \beta) . \kappa = \iota`[\iota`(\iota`\iota`x \cup \iota`\Lambda) \cup \iota`\Lambda] \cup \iota`\iota`\iota`\beta\} \quad \text{Df.}$$

$$\hat{x}\hat{\lambda}\phi(x, \lambda) = \hat{\mu}\{(\exists x, \lambda) . \mu = \iota`\{\iota`[\iota`(\iota`\iota`x \cup \iota`\Lambda) \cup \iota`\Lambda] \cup \iota`\Lambda\} \cup \iota`\iota`\iota`\lambda\} \quad \text{Df.}$$

etc.

* This may seem circular as ι is a relation, defined in the *Principia* as $\overrightarrow{I}$, but it really is not circular, for $\iota`x$ may be defined directly as the class, $\hat{y}(y = x)$.

Though these definitions may seem to conflict with one another, they really do not conflict, for where one of them is applicable, the others are meaningless, since they define relations between objects of different types. Moreover, it is easy to see that our definitions are so chosen that

$$\vdash : \hat{\mu}\hat{\nu}\phi(\mu, \nu) = \hat{\varpi}\hat{\rho}\psi(\varpi, \rho) \,.\supset .\, \iota^{\backprime}\mathrm{D}^{\backprime}\hat{\mu}\hat{\nu}\phi(\mu, \nu)$$
$$= \iota^{\backprime}\mathrm{D}^{\backprime}\hat{\varpi}\hat{\rho}\psi(\varpi, \rho) \,.\, \iota^{\backprime}\text{Ɑ}^{\backprime}\hat{\mu}\hat{\nu}\phi(\mu, \nu) = \iota^{\backprime}\text{Ɑ}^{\backprime}\hat{\varpi}\hat{\rho}\psi(\varpi, \rho).$$

This is important, as we might easily have defined relations so that they might have several domains or converse domains of different types. This is why we did not define $\hat{\alpha}\hat{y}\phi(\alpha, y)$ simply as

$$\hat{\kappa}\{(\exists\alpha, y) \,.\, \phi(\alpha, y) \,.\, \kappa = \iota^{\backprime}(\iota^{\backprime}\iota^{\backprime}\alpha \cup \iota^{\backprime}\Lambda) \cup \iota^{\backprime}\iota^{\backprime}\iota^{\backprime}\iota^{\backprime}y\},$$

for this would also represent

$$\hat{\alpha}\hat{\beta}\{(\exists y) \,.\, \phi(\alpha, y) \,.\, \beta = \iota^{\backprime}y\}.$$

It will be seen that what we have done is practically to revert to Schröder's treatment of a relation as a class of ordered couples. The complicated apparatus of ι's and Λ's of which we have made use is simply and solely devised for the purpose of constructing a class which shall depend only on an ordered pair of values of x and y, and which shall correspond to only one such pair. The particular method selected of doing this is largely a matter of choice: for example, I might have substituted V, or any other constant class not a unit class, and existing in every type of classes, in every place I have written Λ.

Our changed definition of $\hat{x}\hat{y}\phi(x, y)$ renders it necessary to give new definitions of several other symbols fundamental to the theory of relations. I give the following table of such definitions:

$$\mathrm{Rel} = \hat{\kappa}\{\kappa \subset \hat{x}\hat{y}\,(x = x \,.\, y = y)\} \quad \text{Df.}$$
$$xRy \,.=.\, \hat{z}\hat{w}\{z = x \,.\, w = y\} \subset R \,.\, R \in \mathrm{Rel} \quad \text{Df.}$$
$$\phi R \,.=.\, (\exists\alpha) \,.\, \alpha = R \,.\, \alpha \in \mathrm{Rel} \,.\, \phi\alpha \quad \text{Df.*}$$
$$(R) \,.\, \phi R :=: \alpha \in \mathrm{Rel} \,.\supset_\alpha .\, \phi\alpha \quad \text{Df.}$$
$$(\exists R) \,.\, \phi R :=. (\exists\alpha) \,.\, \alpha \in \mathrm{Rel} \,.\, \phi\alpha \quad \text{Df.}$$

The first two and the last two of these definitions replace *21·03·02 and *21·07·071 respectively. From these definitions and the laws

* We shall understand in this way any propositional functions containing capital letters in the positions proper to their arguments. Thus $\sim\phi R$ shall be understood as

$$(\exists\alpha) \,.\, \alpha = R \,.\, \alpha \in \mathrm{Rel} \,.\, \sim\phi\alpha,$$

and not as

$$\alpha = R \,.\, \alpha \in \mathrm{Rel} \,.\supset_\alpha .\, \sim\phi\alpha.$$

We make this definition as well as the two following ones because a propositional function of a class of the sort we have defined as a relation may significantly take as arguments classes of the same type which are not relations, and we wish to define propositional functions of relations in such a manner as to require that their arguments be relations.

of the calculus of classes it is an exceedingly simple matter to deduce any of the propositions of $*21$ which are not explicitly used for the purpose of deriving the properties of relations from the particular definition of relations given there, and from this it is easy to prove that the formal properties of the objects I call relations are essentially the same as those of the relations of the *Principia*.

But it is obvious that since they are also classes, our relations will possess some formal properties not possessed by those of the *Principia*. I give in conclusion a table of some of the more interesting of these:

$$\vdash . R \cup S = R \mathbin{\dot{\cup}} S$$
$$\vdash . R \cap S = R \mathbin{\dot{\cap}} S$$
$$\vdash : R \subset S . \equiv . R \mathbin{\dot{\subset}} S$$
$$\vdash . R - S . \equiv . R \mathbin{\dot{-}} S$$
$$\vdash . \dot{V} \subset V$$
$$\vdash . \Lambda = \dot{\Lambda}$$
$$\vdash . \mathrm{Rel} \subset \mathrm{Cls}$$
$$\vdash : Rp\kappa . \equiv . R\dot{p}\kappa$$
$$\vdash : Rs\kappa . \equiv . R\dot{s}\kappa$$
$$\vdash . \alpha + \beta \ \mathrm{sm}\ s\text{'}\alpha \uparrow \beta$$
$$\vdash . \alpha \times \beta \ \mathrm{sm}\ \alpha \uparrow \beta$$

Comments on [14a]

H. E. Kyburg

This fundamental and justly famous paper offered the first definition of the ordered pair within set theory and thus enormously enhanced the attractiveness and plausibility of the abstract and general approach to mathematics and logic embodied in Russell and Whitehead's *Principia Mathematica* and in set theory. Wiener's definition, having the form

$$\langle x,y \rangle = \iota\,(\iota\iota\, x \cup \iota\, \Lambda) \cup \iota\,\iota\,\iota\, y$$

for entities x and y of the same logical type, is as complicated as it is only because of the necessity for taking account of the rich array of types in *Principia Mathematica.* Kuratowski provided a simplification appropriate to set theory which in turn has been adopted by Quine in mathematical logic and by most set theoreticians:

$$\langle x,y \rangle = \{\{x\},\ \{x,y\}\}\,.$$

It is in this form now familiar to everyone.

Editor's note. For Wiener's own remarks on this paper see [53h], p. 191.

*A Contribution to the Theory of Relative Position**. By NORBERT WIENER, Ph.D. (Communicated by Mr G. H. Hardy.)

[*Received* 14 March 1914.]

The theory of relations is one of the most interesting departments of the new mathematical logic. The relations which have been most thoroughly studied are the *series*: that is, relations which are contained in diversity, transitive, and connected or, in Mr Russell's symbolism, those relations R of which the following proposition is true:

$$R \subset J \,.\, R^2 \subset R \,.\, R \cup \breve{R} \cup I \upharpoonright C`R = C`R \uparrow C`R.$$

Cantor, Dedekind, Frege, Schröder, Burali-Forti, Huntington, Whitehead, and Russell, are among those who have helped to give us an almost exhaustive account of the more fundamental properties of series. There is a class of relations closely allied to series, however, which has received very scant attention from the mathematical logicians. Examples of the sort of relation to which I am referring are the relation between two events in time when one completely precedes the other, or the relation between two intervals on a line when one lies to the left of the other, and does not overlap it, or, in general, the relation between two stretches α and β, of terms of a series R, when any term lying in α bears the relation R to any term lying in β. Relations of this sort, which I shall call relations of complete sequence, differ in general from series in not being *connected*: that is, for example, it is not necessary that of two distinct events, each of which wholly precedes or follows some other event, one should *wholly* precede the other, for the times of their occurrence may overlap. But in all the instances we have given, the relation of complete sequence is closely bound up with some serial relation: the relation of succession between the events of time is intimately related to the series of its instants, the relation between two intervals on a line one of which lies completely to the other's left is intimately related to the series of the points on the line, and so on. These considerations lead us to the general questions, (1) what are the formal properties which characterise relations of the sort we have

* The subject of this paper was suggested to me by Mr Bertrand Russell, and the paper itself is the result of an attempt to simplify and generalize certain notions used by him in his treatment of the relation between the series of events and the series of instants.

called relations of complete sequence? and (2) what is the nature of the connection between relations of complete sequence and series?

One very general property which belongs to relations of the sort we have called relations of complete sequence is that they never hold between a given term and itself. This property—that of being contained in diversity—they share with series proper. Writing cs for the class of relations of complete sequence, we can represent this fact in the symbolism of the *Principia Mathematica* of Whitehead and Russell by the formula

$$\mathrm{cs} \subset \mathrm{Rl}`J.$$

Another property they share with series is that of transitivity. If, for example, the event x wholly precedes the event y, while the event y wholly precedes the event z, the event x wholly precedes the event z. But they possess another property more powerful logically, which may be called a generalized form of transitivity. If the event x wholly precedes the event y, and the event y neither wholly precedes nor wholly follows the event z, while the event z wholly precedes the event w, then the event x will wholly precede the event w. All the other relations which we have mentioned as examples of relations of complete precedence will be found to possess the same property, which, moreover, will be satisfied by all those relations which we would naturally call relations of complete precedence. We may, then, so define "relations of complete precedence" as to regard this as a property common to all such relations. In symbols, we shall then have

$$\vdash . \mathrm{cs} \subset \hat{R}\{R \mid (\dot{-} R \dot{-} \breve{R}) \mid R \mathrel{\mathsf{G}} R\}.$$

The relation $(\dot{-} R \dot{-} \breve{R})$, with its field limited to that of R, is what we ordinarily know as simultaneity. In most theories of time and of relations of complete precedence, it has been thought necessary to treat precedence and simultaneity as coördinate primitive ideas. Nevertheless, those who hold such theories have to assume such propositions as the following, in order to make simultaneity and precedence possess the appropriate formal properties*:

$$\vdash . S \dot{\cap} P = \dot{\Lambda},$$

$$\vdash . S \mathrel{\dot{\cup}} P \mathrel{\dot{\cup}} \breve{P} = C`S \uparrow C`S,$$

$$\vdash . S \mathrel{\mathsf{G}} \breve{S},$$

$$\vdash . C`S = C`P.$$

* In the following list of propositions, S stands for 'is simultaneous with,' and P for 'precedes.'

From these it is an easy matter to deduce that

$$\vdash . S = (\dot{-} P \dot{-} \breve{P}) \upharpoonright C`P,$$

while on the hypothesis that $P \subset J$, the converse deduction can readily be made. Therefore, we may define simultaneity as that relation which holds between x and y when both either follow or precede something and neither precedes the other. The second property of relations of complete sequence may, then, be interpreted to state that if R is such a relation, then if xRy, y-is-simultaneous-with-respect-to-R to z*, and zRw, then xRw.

We shall find that most of the properties of relations of the sort of complete temporal succession between events follow from the two conditions which we have mentioned above—indeed, many of the most important ones follow from the second alone—so that we shall *define* a relation of complete succession as one which satisfies those two conditions: in other words, we shall make the following definition:

$*0{\cdot}01$†. $\mathrm{cs} = \mathrm{Rl}`J \cap \hat{R}\{R \mid (\dot{-} R \dot{-} \breve{R}) \mid R \subset R\}$ Df.

Moreover, as we shall have frequent cause to refer to the relation $(\dot{-} P \dot{-} \breve{P}) \upharpoonright C`P$, and as this expression is rather unwieldy, we shall abbreviate it as follows:

$*0{\cdot}02$. $P_{\mathrm{se}} = (\dot{-} P \dot{-} \breve{P}) \upharpoonright C`P$ Df.

Now the question arises, how are the members of cs related to series? How, for example, is the relation between an event and another that completely succeeds it related to the relation between an *instant* and another that follows it? Two methods of procedure are open to us; we may define an event as a class of instants, and derive succession between events from that between instants, or we may define an instant as the class of all the events that occur at it. Both these methods seem to have certain inherent disadvantages: if we choose the first method, then we cannot consider the possibility of several events occurring with the same times of beginning and ending, whereas if we choose the second alternative, we cannot consider the possibility of all the events of one moment happening also at another and vice versa. However, we shall choose the latter method of procedure, since cs is a more general notion than ser. This can be proved as follows:

$$\vdash . R \mid (\dot{-} R \dot{-} \breve{R}) \mid R = R \mid [(\dot{-} R \dot{-} \breve{R}) \upharpoonright C`R] \mid R$$
$$= R \mid \hat{x}\hat{y}\,(x \dot{-} Ry \,.\, y \dot{-} Rx \,.\, x, y \in C`R)\, R \qquad (1)$$

* In this paper, 'x-is-simultaneous-with-respect-to-R to y' will be interpreted as meaning $x\,[C\,(\dot{-} R \dot{-} \breve{R}) \upharpoonright C`R]\,y$.

† I follow the method of the *Principia Mathematica* of Russell and Whitehead.

$$\vdash : R \in \text{connex} \,.\, \supset \,.\, R \,|\, (\dot{-} R \dot{-} \breve{R}) \,|\, R = R \,|\, \hat{x}\hat{y}\,(x = y) \,|\, R$$
$$= R \,|\, I \,|\, R$$
$$= R \,|\, R \qquad (2)$$

$$\vdash : R \in \text{ser} \,.\, \supset \,.\, R \,|\, (\dot{-} R \dot{-} \breve{R}) \,|\, R \in R \,.\, R \in \text{Rl}`J \,.$$
$$\supset \,.\, R \in \text{cs} \qquad (3)$$

$\vdash . (3) . \supset \vdash . \text{ser} \subset \text{cs}.$

Moreover, it has been shown by Mr Russell that it is advantageous for purposes of methodological simplicity to regard the instants of time as constructions from its events. This is an additional reason for starting from the members of cs and forming certain members of ser as functions of them. Let us, then, agree that an instant, for example, is to be regarded as a class of events, and a point on a line as a class of the segments of the line, for the purposes of this paper. The question then arises, when is a class of events an instant, and when is a class of segments a point? It is obvious on inspection that not every class of events is an instant: all the events which make up a given instant must be simultaneous with one another, and all the events which are simultaneous with every member of the instant must belong to that instant. Moreover, Λ must not be an instant. It can also be seen readily that any class satisfying these conditions will be an instant. That is, if P is the relation of an event to an event which completely follows it, it is a simple matter to show that the class of all instants is

$$\hat{\alpha}\,\{\alpha = p`\overrightarrow{P_{\text{se}}}``\alpha\}^{*}.$$

One instant precedes another when and only when some event belonging to the one entirely precedes some event belonging to the other. That is, calling the relation of precedence between instants inst$`P$, we can easily show that we have

$$\vdash . \text{inst}`P = (\breve{\epsilon} \,;\, P) \upharpoonright \hat{\alpha}\,\{\alpha = p`\overrightarrow{P_{\text{se}}}``\alpha\}.$$

Let me now make the following definitions for any value of P:

$*0\cdot03.\quad \tau_P = \hat{\alpha}\,\{\alpha = p`\overrightarrow{P_{\text{se}}}``\alpha\} \qquad$ Df.

$*0\cdot04.\quad \text{inst} = \hat{Q}\hat{P}\,\{Q = (\breve{\epsilon} \,;\, P) \upharpoonright \tau_P\} \qquad$ Df.

I wish to show that

$$\vdash . \text{inst}``\hat{R}\,\{R \,|\, R_{\text{se}} \,|\, R \in R\} \subset \text{ser},$$

* This definition is due to Mr Russell.

and hence that

$$\vdash . \text{inst``cs} \subset \text{ser}.$$

This shows us how we can construct a serial relation from any relation of the same sort as complete succession; or, indeed, from any relation agreeing with it in only one respect.

$*0{\cdot}1.\quad \vdash . \text{inst``}\hat{R}\,\{R \mid R_{\text{se}} \mid R \mathrel{G} R\} \subset \text{ser}.$

Proof.

It is easy to show that

$$\vdash : \alpha\, \text{inst‘}P\beta . \equiv .$$
$$\alpha = p‘\overrightarrow{P_{\text{se}}}``\alpha . \beta = p‘\overrightarrow{P_{\text{se}}}``\beta . (\exists x, y) . x \in \alpha . y \in \beta . xPy \qquad (1)$$

from the definitions of inst and τ_P. From this we can deduce

$$\vdash : \alpha\, \text{inst‘}P\beta . \supset . \beta = p‘\overrightarrow{P_{\text{se}}}``\beta . (\exists x, y) . x \in \alpha . y \in \beta . \sim xP_{\text{se}}y,$$

since, by the definition of P_{se}, xPy and $xP_{\text{se}}y$ are incompatible. This reduces to

$$\vdash : \alpha\, \text{inst‘}P\beta . \supset . \beta = p‘\overrightarrow{P_{\text{se}}}``\beta . (\exists x) . x \in \alpha . \sim (x \in p‘\overrightarrow{P_{\text{se}}}``\beta),$$

from which we can deduce

$$\vdash : \alpha\, \text{inst‘}P\beta . \supset_{\alpha,\beta} . \alpha J \beta$$

or $\quad \vdash . \text{inst‘}P \in \text{Rl‘}J \qquad (2)$

Also, we find from (1) that

$$\vdash : \alpha\, \text{inst‘}P\beta . \beta\, \text{inst‘}P\gamma . \supset .$$
$$\alpha = p‘\overrightarrow{P_{\text{se}}}``\alpha . \beta = p‘\overrightarrow{P_{\text{se}}}``\beta . \gamma = p‘\overrightarrow{P_{\text{se}}}``\gamma .$$
$$(\exists x, y, u, v) . x \in \alpha . y, u \in \beta . v \in \gamma . xPy . uPv.$$

This implies

$$\vdash : \alpha\, \text{inst‘}P\beta . \beta\, \text{inst‘}P\gamma . \supset .$$
$$\alpha = p‘\overrightarrow{P_{\text{se}}}``\alpha . \gamma = p‘\overrightarrow{P_{\text{se}}}``\gamma . (\exists x, v) . x \in \alpha . v \in \gamma . xP \mid P_{\text{se}} \mid Pv.$$

This, together with (1), gives us

$$\vdash . \text{inst``}\hat{R}\,\{R \mid R_{\text{se}} \mid R \mathrel{G} R\} \subset \text{trans} \qquad (3)$$

By the definitions of inst and τ_P, we find that

$$\vdash : \alpha, \beta \in C‘\text{inst‘}P . \supset . \alpha = p‘\overrightarrow{P_{\text{se}}}``\alpha . \beta = p‘\overrightarrow{P_{\text{se}}}``\beta.$$

By an easy deduction, we can arrive, from this proposition and the definition of P_{se}, at the proposition

$$\vdash :: \alpha, \beta \in C‘\text{inst‘}P . \supset :. \alpha = p‘\overrightarrow{P_{\text{se}}}``\alpha . \beta = p‘\overrightarrow{P_{\text{se}}}``\beta :.$$
$$x \in \alpha . y \in \beta : \supset_{x,y} : xPy . \vee . yPx . \vee . xP_{\text{se}}y,$$

whence we get

$$\vdash :: \alpha, \beta \in C\text{‘inst‘}P \,.\supset :. \alpha = p\text{‘}\overrightarrow{P}_{\text{se}}\text{“}\alpha \,.\, \beta = p\text{‘}\overrightarrow{P}_{\text{se}}\text{“}\beta :.$$
$$x \in \alpha \,.\, y \in \beta \,.\supset_{x,y} .\, \sim xPy \,.\, \sim yPx : \supset : u \in \alpha \,.\supset_u .\, u \in p\text{‘}\overrightarrow{P}_{\text{se}}\text{“}\beta,$$

or $\vdash :: \alpha, \beta \in C\text{‘inst‘}P \,.\supset :. \alpha = p\text{‘}\overrightarrow{P}_{\text{se}}\text{“}\alpha \,.\, \beta = p\text{‘}\overrightarrow{P}_{\text{se}}\text{“}\beta :.$

$$x \in \alpha \,.\, y \in \beta \,.\supset_{x,y} .\, \sim xPy \,.\, \sim yPx : \supset .\, \alpha \subset \beta.$$

By an exactly similar argument,

$$\vdash :: \alpha, \beta \in C\text{‘inst‘}P \,.\supset :. \alpha = p\text{‘}\overrightarrow{P}_{\text{se}}\text{“}\alpha \,.\, \beta = p\text{‘}\overrightarrow{P}_{\text{se}}\text{“}\beta :.$$
$$x \in \alpha \,.\, y \in \beta \,.\supset_{x,y} .\, \sim xPy \,.\, \sim yPx : \supset .\, \beta \subset \alpha.$$

Combining these, we get

$$\vdash :: \alpha, \beta \in C\text{‘inst‘}P \,.\supset :. \alpha = p\text{‘}\overrightarrow{P}_{\text{se}}\text{“}\alpha \,.\, \beta = p\text{‘}\overrightarrow{P}_{\text{se}}\text{“}\beta :.$$
$$x \in \alpha \,.\, y \in \beta \,.\supset_{x,y} .\, \sim xPy \,.\, \sim yPx : \supset .\, \alpha = \beta.$$

This we may write as

$$\vdash :: \alpha, \beta \in C\text{‘inst‘}P \,.\supset :. \alpha = p\text{‘}\overrightarrow{P}_{\text{se}}\text{“}\alpha \,.\, \beta = p\text{‘}\overrightarrow{P}_{\text{se}}\text{“}\beta :.$$
$$(\exists x, y) \,.\, x \in \alpha \,.\, y \in \beta \,.\, xPy : \vee : (\exists x, y) \,.\, x \in \alpha \,.\, y \in \beta \,.\, yPx : \vee : \alpha = \beta.$$

By (1), this becomes

$\vdash :. \alpha, \beta \in C\text{‘inst‘}P \,.\supset : \alpha \,\text{inst‘}P\beta \,.\vee .\, \beta \,\text{inst‘}P\alpha \,.\vee .\, \alpha = \beta,$

or $\vdash .\, \text{inst‘}P \in \text{connex}$ (4)

Combining (2), (3), and (4), we get the desired conclusion: namely,

$\vdash .\, \text{inst“}\hat{R}\{R \mid R_{\text{se}} \mid R \in R\} \subset \text{ser}.$

From this we can easily conclude that

$\vdash .\, \text{inst“cs} \subset \text{ser}.$

It will be noticed that two of the three serial properties of inst‘P—its being contained in diversity and its connexity—are independent of the properties of P itself. It is especially noticeable that no use is made of $P \in J$ in proving inst‘$P \in J$, nor, indeed, in deducing any of the serial properties of inst‘P. inst is a valuable tool for what Mr Russell calls "fattening out" a relation: i.e. deriving from a non-serial relation a relation with many of the properties of series*.

It is interesting to consider under what conditions inst‘P will be compact. If we define csd as follows:

$$*0{\cdot}2.\quad \text{csd} = \text{cs} \cap \hat{R}\{R \in R \mid R_{\text{se}} \mid R \,.\, \breve{R} \mid R_{\text{se}} \in \breve{R} \mid \min_R \mid \overrightarrow{R}_{\text{se}}\}\quad \text{Df},$$

* Since writing this article, I have discovered an operation which will turn *any* relation into a series (though not necessarily an existent one) and will leave unchanged the relation-number of any series to which it is applied. It is the operation which transforms P into inst‘[(inst‘P)$_{pv}$].

we shall find that $R \epsilon \text{csd}$ is a sufficient condition for the density of inst'R. This condition says that (1) R is a relation of complete sequence, (2) if x precedes y by the relation R, there are two members of the field of R neither of which bears the relation R to the other, while x precedes the one by R, while the other precedes y by R, (3) if x follows by R some R-contemporary of y, it follows some *initial* R-contemporary of y. This latter condition, which was first formulated by Mr Russell, ensures that if $x \epsilon C'R$ and $R \epsilon \text{csd}$, $\overrightarrow{\min}_R{}'\overrightarrow{R}_{se}{}'x \epsilon \tau_R$. This I now wish to prove.

$*0{\cdot}21.\ \vdash : P \epsilon \text{csd} . x \epsilon C'P . \supset . \overrightarrow{\min}_P{}'\overrightarrow{P}_{se}{}'x \epsilon \tau_P.$

Proof.

It follows from the definition of $p'\kappa$ and $\overrightarrow{\min}_P{}'\alpha$ that

$$\vdash . p'\overrightarrow{P}_{se}{}''\overrightarrow{\min}_P{}'\overrightarrow{P}_{se}{}'x = \hat{y}\{\alpha \epsilon \overrightarrow{P}_{se}{}''[\overrightarrow{P}_{se}{}'x \cap C'P - \breve{P}''\overrightarrow{P}_{se}{}'x] . \supset_\alpha . y \epsilon \alpha\}.$$

Since it follows from the definition of P_{se} that $\vdash . C'P_{se} \subset C'P$, this reduces to

$$\vdash . p'\overrightarrow{P}_{se}{}''\overrightarrow{\min}_P{}'\overrightarrow{P}_{se}{}'x = \hat{y}\{\alpha \epsilon \overrightarrow{P}_{se}{}''[\overrightarrow{P}_{se}{}'x - \breve{P}''\overrightarrow{P}_{se}{}'x] . \supset_\alpha . y \epsilon \alpha\}.$$

This becomes by a little manipulation

$$\vdash . p'\overrightarrow{P}_{se}{}''\overrightarrow{\min}_P{}'\overrightarrow{P}_{se}{}'x = \hat{y}\{zP_{se}x . z \mathbin{\dot{-}} \breve{P} \mid P_{se}x . \supset_z . yP_{se}x\} \qquad (1)$$

On the other hand, it follows from the definition of $\overrightarrow{\min}_P{}'\alpha$ that

$$\vdash . \overrightarrow{\min}_P{}'\overrightarrow{P}_{se}{}'x = \hat{y}\{yP_{se}x . y \mathbin{\dot{-}} \breve{P} \mid P_{se}x\}.$$

Since by definition any R which belongs to csd satisfies the condition, $\breve{R} \mid R_{se} \subseteq \breve{R} \mid \min_R \mid \overrightarrow{R}_{se}$, we get

$$\vdash : P \epsilon \text{csd} . \supset . \overrightarrow{\min}_P{}'\overrightarrow{P}_{se}{}'x = \hat{y}\{yP_{se}x . y \mathbin{\dot{-}} \breve{P} \mid \min_P \mid \overrightarrow{P}_{se}x\}.$$

From this we may deduce

$$\vdash : P \epsilon \text{csd} . \supset . \overrightarrow{\min}_P{}'\overrightarrow{P}_{se}{}'x = \hat{y}\{yP_{se}x :. zP_{se}x . z \mathbin{\dot{-}} \breve{P} \mid P_{se}x : \supset_z : yPz . \vee . y \mathbin{\dot{-}} Pz . z \mathbin{\dot{-}} Py . y, z \epsilon C'P\}.$$

But when yPz is the correct alternative in the conclusion of the second proposition in the brackets, together with $yP_{se}x$, this gives us $z\breve{P} \mid P_{se}x$, which contradicts the hypothesis. Hence, by the definition of P_{se}, we have

$$\vdash : P \epsilon \text{csd} . \supset . \overrightarrow{\min}_P{}'\overrightarrow{P}_{se}{}'x = \hat{y}\{yP_{se}x : zP_{se}x . z \mathbin{\dot{-}} \breve{P} \mid P_{se}x . \supset_z . yP_{se}z\} \qquad (2)$$

Now, it is part of the hypothesis $P \in \mathrm{csd}$ that $P \mathrel{\dot\subset} J$. From this it is easy to deduce that $I \upharpoonright C`P \mathrel{\dot\subset} P_{\mathrm{se}}$, or that $x \in C`P \,.\supset_x .\, xP_{\mathrm{se}}x$. Moreover, it follows from the definition of P_{se} that yPx and $y\breve{P}_{\mathrm{se}}x$ are incompatible hypotheses, and hence that $x \mathrel{\dot{-}} \breve{P} \mid P_{\mathrm{se}}\, x$. This fact, combined with (1), gives us

$$\vdash : P \in \mathrm{csd} \,.\, x \in C`P \,.\supset .\, p`\overrightarrow{P_{\mathrm{se}}}``\mathrm{min}_P`\overrightarrow{P_{\mathrm{se}}}`x$$

$$= \hat{y}\,\{yP_{\mathrm{se}}\,x : zP_{\mathrm{se}}\,x \,.\, z \mathrel{\dot{-}} \breve{P} \mid P_{\mathrm{se}}\,x \,.\supset_z .\, yP_{\mathrm{se}}\,z\} \quad (3)$$

From (2), (3), and the definition of τ_P, we have

$$\vdash : P \in \mathrm{csd} \,.\, x \in C`P \,.\supset .$$

$$\mathrm{min}_P`\overrightarrow{P_{\mathrm{se}}}`x = p`\overrightarrow{P_{\mathrm{se}}}``\mathrm{min}_P`\overrightarrow{P_{\mathrm{se}}}`x \,.\supset .\, \mathrm{min}_P`\overrightarrow{P_{\mathrm{se}}}`x \in \tau_P.$$

This is the desired proposition.

It will be observed that the only portions of the hypothesis of $P \in \mathrm{csd}$ of which we actually make use in this theorem are $P \mathrel{\dot\subset} J$ and $\breve{P} \mid P_{\mathrm{se}} \mathrel{\dot\subset} \breve{P} \mid \mathrm{min}_P \mid \overrightarrow{P_{\mathrm{se}}}$. The theorem ensures us that $\vdash .\, C`P \subset s`\tau_P$: that is, in the case of time, that each event shall be at some instant—the instant at which it begins. For, since $P \mathrel{\dot\subset} J$, $I \upharpoonright C`P \mathrel{\dot\subset} P_{\mathrm{se}}$. This ensures that $x \in \overrightarrow{P_{\mathrm{se}}}`x$. Moreover, as we have just seen, $x \mathrel{\dot{-}} \breve{P} \mid P_{\mathrm{se}}\,x$, or $x \in -\breve{P}``\overrightarrow{P}`x$. Therefore, if $x \in C`P$, $x \in \overrightarrow{P_{\mathrm{se}}}`x \cap C`P - \breve{P}``\overrightarrow{P_{\mathrm{se}}}`x$, or $x \in \mathrm{min}_P`\overrightarrow{P_{\mathrm{se}}}`x$. As we have proved in *0·21 that $\mathrm{min}_P`\overrightarrow{P_{\mathrm{se}}}`x \in \tau_P$, we get the formula

$$\vdash .\, C`P \subset s`\tau_P.$$

I now wish to prove that $\mathrm{inst}``\mathrm{csd} \subset \mathrm{comp}$.

*0·22. $\vdash .\, \mathrm{inst}``\mathrm{csd} \subset \mathrm{comp}$.

Proof.

As we saw in *0·1, (1),

$$\vdash : \alpha\, \mathrm{inst}`P\beta \,.\equiv .\, \alpha = p`\overrightarrow{P_{\mathrm{se}}}``\alpha \,.\, \beta = p`\overrightarrow{P_{\mathrm{se}}}``\beta \,.$$

$$(\exists x, y) \,.\, x \in \alpha \,.\, y \in \beta \,.\, xPy.$$

Since $R \in \mathrm{csd}$, by definition, implies $R \mathrel{\dot\subset} R \mid R_{\mathrm{se}} \mid R$, this gives us

$$\vdash :.\, P \in \mathrm{csd} : \supset : \alpha\, \mathrm{inst}`P\beta \,.\supset .$$

$$\alpha = p`\overrightarrow{P_{\mathrm{se}}}``\alpha \,.\, \beta = p`\overrightarrow{P_{\mathrm{se}}}``\beta \,.\, (\exists x, y) \,.\, x \in \alpha \,.\, y \in \beta \,.\, xP \mid P_{\mathrm{se}} \mid Py.$$

Since $R \in \text{csd}$ also implies $\breve{R} \,|\, R \;\mathsf{G}\; \breve{R} \,|\, \min_R \,|\, \overrightarrow{R_{se}}$, this becomes

$$\vdash :. P \in \text{csd} : \supset : \alpha \,\text{inst}`P\beta \,.\supset.\, \alpha = p`\overrightarrow{P_{se}}``\alpha \,.\, \beta = p`\overrightarrow{P_{se}}``\beta \,.$$

$$(\exists x, y) \,.\, x \in \alpha \,.\, y \in \beta \,.\, x\widehat{P \,|\, [\min_P \,|\, \overrightarrow{P_{se}}] \,|\, P}y.$$

$x\widehat{P \,|\, [\min_P \,|\, \overrightarrow{P_{se}}] \,|\, P}y$ says that there are a u and a v such that xPu, vPy, and $v \min_P \,|\, \overrightarrow{P_{se}}\, u$. This latter proposition is equivalent to $v \in \overrightarrow{\min_P}`\overrightarrow{P_{se}}`u$. We have just seen, moreover, that $u \in \overrightarrow{\min_P}`\overrightarrow{P_{se}}\, u$, and that $\overrightarrow{\min_P}`\overrightarrow{P_{se}}\, u \in \tau_P$. This gives us

$$\vdash :: P \in \text{csd} :. \supset :. \alpha \,\text{inst}`P\beta : \supset :$$

$$(\exists u, v) : \alpha = p`\overrightarrow{P_{se}}``\alpha \,.\, \beta = p`\overrightarrow{P_{se}}``\beta \,.\, \overrightarrow{\min_P}`\overrightarrow{P_{se}}`u \in p`\overrightarrow{P_{se}}``\overrightarrow{\min_P}`\overrightarrow{P_{se}}`u :$$

$$(\exists x, y) \,.\, x \in \alpha \,.\, y \in \beta \,.\, u, v \in \overrightarrow{\min_P}`\overrightarrow{P_{se}}`u \,.\, xPu \,.\, yPv.$$

From this and $*0{\cdot}1$, (1) it is an easy matter to deduce that

$$\vdash :. P \in \text{csd} : \supset : \alpha \,\text{inst}`P\beta \,.\supset.\, (\exists \gamma) \,.\, \alpha \,\text{inst}`P\gamma \,.\, \gamma \,\text{inst}`P\beta.$$

This gives us immediately

$$\vdash .\, \text{inst}``\text{csd} \subset \text{comp}.$$

It will be noticed that it is not true that csd $\subset$ comp. For example, if P is the relation of complete succession between one-inch stretches on a line, P will be a member of csd, and an inch stretch beginning half an inch after the end of another will bear the relation P to it, yet there will be no inch stretch to which the first bears the relation P and which bears the relation P to the second. $P \;\mathsf{G}\; P \,|\, P_{se} \,|\, P$ is a weaker hypothesis than $P \;\mathsf{G}\; P^2$, which implies it if $P \;\mathsf{G}\; J$.

Studies in Synthetic Logic. By NORBERT WIENER, Ph.D. (Communicated by Mr G. H. Hardy.)

[*Received* 13 July 1914.]

§ 1. In a recent article of mine in the *Proceedings of the Cambridge Philosophical Society**, I showed how we can regard the series of the instants of time as a construction from the non-serial relation of complete temporal succession between events in time, and how only a few simple presuppositions concerning the formal character of this relation of complete temporal succession sufficed to establish the seriality of the relation of succession between instants; and, in a foot-note, I showed further how, *without making any assumptions* concerning the formal properties of a given relation, P, we can construct another relation from P in a perfectly determinate manner, so that this latter relation will always be a series.

In this article, I wish to extend this method of series-construction in two different directions. I first mean to bring the definitions of order through triadic and tetradic relations under a single very general heading, and to show that Frege's theory of hereditary relations and the theory of series-synthesis developed in my former article can be generalized so as to apply to these. Then I shall give an alternative method of constructing series from non-serial relations which bears much the same relation to the various series of sensation-*intensities* that the method of my previous article bears to the series of instants that constitutes one sort of *extension*, time.

In general, our symbolism will be that of the *Principia Mathematica* of Whitehead and Russell, and we shall take the theorems established in that book for granted. But as we shall have much to do with polyadic relations, and as the parts of the *Principia* which will treat of general polyadic relations are not yet in print, it will be necessary for us to develop a symbolism of our own here. Such properties of polyadic relations as have precise analogues in the theory of classes we shall take for granted. Moreover, as we shall want to speak of properties of relations among *any* number of terms, and as in Mr Russell's system†, relations among m terms belong to different types than relations among n terms, if $m \neq n$, so that no propositional functions whose arguments range over

* "A Contribution to the Theory of Relative Position," vol. XVII, Part 5, pp. 441—9.

† See, however, my article, "A Simplification in the Logic of Relations," *Proc. Camb. Phil. Soc.*, vol. XVII, Part 5, pp. 387—90. The method of this article can be extended to n-adic relations in general.

m-adic and n-adic relations exist, we shall have to permit a certain logical laxity in our symbolism. Though our theorems really demand a separate, though precisely parallel, proof when the relations dealt with are m-adic and when they are n-adic, we shall have to treat these proofs as one. Though every relation holds among a definite set of terms, we shall permit dots to fill the places of an indefinite number of these. Though the analogues of $\dot{\cap}$, $\dot{\cup}$, $\dot{s}$, etc. are different with each different sort of relation with which they have to do, we shall represent them all by the symbols we use in the case of binary relations. To the reader acquainted with symbolic logic, there will be no difficulty in reducing any particular case of the theorems I prove to a strictly rigorous form.

§ 2. Let us write the proposition, '$a_1, a_2, \ldots, a_n$ are in the n-adic relation R,' as $R\{a_1, a_2, \ldots, a_n\}$. I shall call a property of an n-adic relation, R, an *n-transitivity* of R when it can be written in the form

$$(1)\quad (\exists b_1, b_2, \ldots, b_k) \,.\, T_R\{a_1, a_2, a_3, \ldots, a_n, b_1, b_2, \ldots, b_k\} \,. \supset_{a_1, a_2, \ldots, a_n} . \, R\{a_1, a_2, \ldots, a_n\},$$

where T_R is the logical disjunction of a number of expressions in the form

$$R\{c_1, c_2, \ldots, c_n\} \,.\, R\{c_1', c_2', \ldots, c_n'\} \,.\, R\{c_1'', c_2'', \ldots, c_n''\} \ldots R\{c_1^{(l)}, c_2^{(l)}, \ldots, c_n^{(l)}\},$$

where l is not necessarily the same in each of these expressions, and $c_1, c_2, \ldots, c_n, c_1', c_2', \ldots, c_n', \ldots, c_1^{(l)}, c_2^{(l)}, \ldots, c_n^{(l)}$, which are not all distinct from one another, are to be found among $a_1, a_2, \ldots, a_n, b_1, b_2, \ldots, b_k$. Ordinary binary transitivity is an example of a 2-transitivity; the property of 'betweenness,' which may be written

$$(\exists d) : abd \,.\, bdc \,.\, \mathsf{v} \,.\, abd \,.\, bcd \,.\, \mathsf{v} \,.\, adc \,.\, dbc \,.\, \mathsf{v} \,.\, abd \,.\, acd \,.\, bac \,.\, \mathsf{v} \,.\, dab \,.\, dac \,.\, bac \,.\, \mathsf{v} \,.\, bca : \supset_{a,b,c} . \, abc,$$

is a 3-transitivity; the property of Vailati's separation-relation, which may be written

$$(\exists e) : ab \,\|\, dc \,.\, \mathsf{v} \,.\, cd \,\|\, ab \,.\, \mathsf{v} \,.\, ab \,\|\, ec \,.\, ae \,\|\, cd : \supset_{a,b,c,d} . \, ab \,\|\, cd,$$

is a 4-transitivity. From these examples it is obvious that the transitivity-properties of relations are of very great logical interest, and that a method which shall point out significant analogies between the various sorts of transitivity is not without importance.

One property which all sorts of n-transitivity have in common is this: if R is any n-adic relation whatever, then it is always possible, given any particular form of n-transitivity, to construct in a perfectly determinate manner a relation, R',

2—2

including R, forming a well-defined function of R, having the desired sort of transitivity.

This is proved as follows: let the n-transitivity in question be the one given in (1). Decompose $T_R\{a_1, \ldots, a_n, b_1, \ldots, b_k\}$, as indicated, into a sum of expressions of the form

$$R\{c_1, c_2, \ldots, c_n\} \,.\, R\{c_1', c_2', \ldots, c_n'\} \ldots R\{c_1^{(l)}, c_2^{(l)}, \ldots, c_n^{(l)}\}.$$

Let there be, say, f such expressions, the pth one always with l_p R's. Replace each of these R's by one and one only of the variable relations $X_1, X_2, \ldots, X_m$, with the same arguments as the R it replaces, and let $m = \sum_{p=1}^{p=f} l_p$. We shall thus transform T_R into a relation which is a function of the m variable relations $X_1, X_2, \ldots, X_m$. Let us call this relation $\dfrac{T}{X_1 X_2 \ldots X_m}$. Now, let us define the relation $\underset{T}{\underline{X_1 X_2 \ldots X_m}}$ as follows:

$$(2)\quad \underset{T}{\underline{X_1 X_2 \ldots X_m}}\{a_1, a_2, \ldots, a_n\} \,.=.\, (\exists b_1, b_2, \ldots, b_k) \,.\, \frac{T}{X_1 X_2 \ldots X_m}\{a_1, a_2, \ldots, a_n, b_1, b_2, \ldots, b_k\} \quad \text{Df.}$$

Like $\dfrac{T}{X_1 X_2 \ldots X_m}$, $\underset{T}{\underline{X_1 X_2 \ldots X_m}}$ is a function of $X_1, X_2, \ldots, X_m$, where the latter may assume any values which are n-adic relations. Now, I define the class of T-powers of R, or, as I write it, $\overrightarrow{T_{\text{pr}}}\text{‘}R$, as follows:

$$(3)\quad T_{\text{pr}} = \hat{S}\hat{R}\{X_1, X_2, \ldots, X_m \in \mu \,.\supset_{X_1, X_2, \ldots, X_m} . \underset{T}{\underline{X_1 X_2 \ldots X_m}} \in \mu : R \in \mu : \supset_\mu . S \in \mu\} \quad \text{Df.}$$

I make the further definition,

$$(4)\quad R_T = \dot{s}\text{‘}\overrightarrow{T_{\text{pr}}}\text{‘}R \quad \text{Df.}$$

Now, R_T includes R and is a function of it, and has the desired sort of n-transitivity.

First, R_T includes R. For, since, as may be seen on inspection, $R T_{\text{pr}} R$, $R \in \overrightarrow{T_{\text{pr}}}\text{‘}R$. Since every member of a class is included in the sum of the class, $R \mathrel{\dot{\subset}} \dot{s}\text{‘}\overrightarrow{T_{\text{pr}}}\text{‘}R \mathrel{\dot{\subset}} R_T$. Secondly, as R_T is derived from R by a process which is really perfectly definite (though I admit that some of the stages of the process by which I have derived R_T from R are not uniquely determined, a little reflection will convince one that all the possible determinations of $\dfrac{T}{X_1 X_2 \ldots X_m}$ yield the same value of R_T), it is a function of R, and of R alone, once T is determined. Thirdly, R_T has the desired sort of n-transitivity. For we can write

$$T_{R_T}\{a_1, a_2, \ldots, a_n, b_1, b_2, \ldots, b_k\}$$

as a sum of products of the form

$$R_T\{c_1, c_2, \ldots, c_n\} \,.\, R_T\{c_1', c_2', \ldots, c_n'\} \ldots R_T\{c_1^{(l)}, c_2^{(l)}, \ldots, c_n^{(l)}\}.$$

Now to say $R_T\{d_1, d_2, \ldots, d_n\}$ is, by the definition of R_T, the same as to say that there is some S such that $ST_{\text{pr}}R$, and $S\{d_1, d_2, \ldots, d_n\}$. Therefore

$$T_{R_T}\{a_1, a_2, \ldots, a_n, b_1, b_2, \ldots, b_k\}$$

is equivalent to

$$(\exists S_1, S_2, \ldots, S_m) \,.\, \frac{T}{S_1 S_2 \ldots S_m}\{a_1, a_2, \ldots, a_n, b_1, b_2, \ldots, b_k\} \,. \quad S_1 T_{\text{pr}} R \,.\, S_2 T_{\text{pr}} R \,.\, S_3 T_{\text{pr}} R \ldots S_m T_{\text{pr}} R.$$

Therefore

$$
\begin{aligned}
(5)\quad \vdash :::. & (\exists b_1, b_2, \ldots, b_k) \,.\, T_{R_T}\{a_1, a_2, \ldots, a_n, b_1, b_2, \ldots, b_k\} :: \\
\equiv :: & (\exists S_1, S_2, \ldots, S_m) \,.\, \underline{S_1 S_2 \ldots S_m}_{T}\{a_1, a_2, \ldots, a_n\} \,. \\
& \qquad S_1 T_{\text{pr}} R \,.\, S_2 T_{\text{pr}} R \,.\, S_3 T_{\text{pr}} R \ldots S_m T_{\text{pr}} R :: \\
\equiv :: & (\exists S_1, S_2, \ldots, S_m) :. \underline{S_1 S_2 \ldots S_m}_{T}\{a_1, a_2, \ldots, a_n\} :. \\
& \quad X_1, X_2, \ldots, X_m \in \mu \,.\, \supset_{X_1, X_2, \ldots, X_m} \,.\, \underline{X_1 X_2 \ldots X_m}_{T} \in \mu : \\
& \qquad R \in \mu : \supset_\mu \,.\, S_1, S_2, \ldots, S_m \in \mu :: \\
\supset :: & (\exists S_1, S_2, \ldots, S_m) :. \underline{S_1 S_2 \ldots S_m}_{T}\{a_1, a_2, \ldots, a_n\} :. \\
& \quad X_1, X_2, \ldots, X_m \in \mu \,.\, \supset_{X_1, X_2, \ldots, X_m} \,.\, \underline{X_1 X_2 \ldots X_m}_{T} \in \mu : \\
& \qquad R \in \mu : \supset_\mu \,.\, \underline{S_1 S_2 \ldots S_m}_{T} \in \mu :: \\
\supset :: & (\exists S_1, S_2, \ldots, S_m) \,.\, \underline{S_1 S_2 \ldots S_m}_{T}\{a_1, a_2, \ldots, a_n\} \,. \\
& \qquad \underline{S_1 S_2 \ldots S_m}_{T}\, T_{\text{pr}} R :: \\
\supset :: & (\exists U) \,.\, U\{a_1, a_2, \ldots, a_n\} \,.\, U \in \overrightarrow{T_{\text{pr}}}\text{‘}R :: \\
\supset :: & (\dot{s}\text{‘}\overrightarrow{T_{\text{pr}}}\text{‘}R)\{a_1, a_2, \ldots, a_n\} :: \\
\supset :: & R_T\{a_1, a_2, \ldots, a_n\}.
\end{aligned}
$$

This is what we wished to prove, for, if we compare this with (1), it shows that R_T has the desired sort of transitivity.

When the transitivity in question is ordinary binary transitivity R_T becomes R_{po}. In general, the appropriate form of R_T performs the function of R_{po} in systems whose order is given by a triadic or tetradic or other polyadic relation.

§ 3. There is another important sort of property which the ordinary serial relation, the 'between' relation on a given line, and the separation-relation have in common. For the binary serial relation, it is ordinary connexity; for the 'between' relation on a given line it may be expressed in symbols as

$$(\exists m, n) : amn \mathbin{.} \vee \mathbin{.} man \mathbin{.} \vee \mathbin{.} mna : bmn \mathbin{.} \vee \mathbin{.} mbn \mathbin{.} \vee \mathbin{.} mnb :$$
$$cmn \mathbin{.} \vee \mathbin{.} mcn \mathbin{.} \vee \mathbin{.} mnc : \supset_{a,b,c} :$$
$$a = b \mathbin{.} \vee \mathbin{.} b = c \mathbin{.} \vee \mathbin{.} c = a \mathbin{.} \vee \mathbin{.} abc \mathbin{.} \vee \mathbin{.} bca \mathbin{.} \vee \mathbin{.} cab;$$

for the separation-relation it is

$$(\exists m, n, o) : am \| no \mathbin{.} \vee \mathbin{.} ma \| no \mathbin{.} \vee \mathbin{.} mn \| ao \mathbin{.} \vee \mathbin{.} mn \| oa :$$
$$bm \| no \mathbin{.} \vee \mathbin{.} mb \| no \mathbin{.} \vee \mathbin{.} mn \| bo \mathbin{.} \vee \mathbin{.} mn \| ob :$$
$$cm \| no \mathbin{.} \vee \mathbin{.} mc \| no \mathbin{.} \vee \mathbin{.} mn \| co \mathbin{.} \vee \mathbin{.} mn \| oc :$$
$$dm \| no \mathbin{.} \vee \mathbin{.} md \| no \mathbin{.} \vee \mathbin{.} mn \| do \mathbin{.} \vee \mathbin{.} mn \| od :$$
$$\supset_{a,b,c,d} : a = b \mathbin{.} \vee \mathbin{.} b = c \mathbin{.} \vee \mathbin{.} c = d \mathbin{.} \vee \mathbin{.} d = a \mathbin{.} \vee \mathbin{.} a = c \mathbin{.} \vee \mathbin{.} b = d \mathbin{.} \vee \mathbin{.}$$
$$ab \| cd \mathbin{.} \vee \mathbin{.} ac \| bd \mathbin{.} \vee \mathbin{.} ad \| bc.$$

For the sake of brevity, let us generalize the notion of 'field' in the following manner:

$$(6) \quad C = \hat{\alpha}\hat{R}\,\{\alpha = \hat{x}\,\{(\exists a_1, a_2, \ldots, a_{n-1})\} :$$
$$R\,\{x, a_1, a_2, \ldots, a_{n-1}\} \mathbin{.} \vee \mathbin{.} R\,\{a_1, x, a_2, \ldots, a_{n-1}\} \mathbin{.} \vee \ldots$$
$$\vee \mathbin{.} R\,\{a_1, a_2, \ldots, a_{n-1}, x\}\} \quad \text{Df.}$$

Now, I shall define a property of an n-adic relation, R, as an *n-connexity* of that relation if it can be written in the form

$$(7) \quad a_1, a_2, \ldots, a_n \in C\text{‘}R : l \neq m \mathbin{.} \supset_{l,m} : \sim(a_l \neq a_m) : \supset_{a_1, a_2, \ldots, a_n} :$$
$$R\,\{a_1, a_2, \ldots, a_n\} \mathbin{.} \vee \mathbin{.} R\,\{a_1', a_2', \ldots, a_n'\} \mathbin{.} \vee \mathbin{.}$$
$$R\,\{a_1'', a_2'', \ldots, a_n''\} \mathbin{.} \vee \ldots \vee \mathbin{.} R\,\{a_1^{(p)}, a_2^{(p)}, \ldots, a_n^{(p)}\},$$

where $a_1' \ldots a_n', a_1'' \ldots a_n'', \ldots, a_1^{(p)} \ldots a_n^{(p)}$ are each definite permutations of $a_1 \ldots a_n$. It is obvious that ordinary binary connexity is, by this definition, a 2-connexity, and that the properties of 'between' and separation which we have just mentioned are, respectively, 3- and 4-connexities.

Now, I wish to raise with regard to n-connexities the precise analogue of the question which we raised with regard to n-transitivities in the last section: is it possible, given any n-adic relation and any n-connexity, to form by a perfectly definite method an n-adic relation genuinely dependent on this relation, having the desired sort of n-connexity?

As in the former case, I shall answer this question by actually constructing such a relation. I shall define the relation $R_{\sigma\lambda}$ as the relation such that $R\{a_1, a_2, \ldots, a_n\}$ when, and only when,

$$a_1, a_2, \ldots, a_n \,\epsilon\, C`R,$$

and the conclusion of (7) is false*.

I shall define the class, ϖ_R, as follows:

$$(8)\quad \varpi_R = \hat{\alpha}\{x, y \,\epsilon\, \alpha \,.\, a_1, a_2, \ldots, a_{n-2} \,\epsilon\, C`R \,.\, \supset_{x, y, a_1, a_2, \ldots, a_{n-2}} .$$
$$R_{\sigma\lambda}\{x, y, a_1, a_2, \ldots, a_{n-2}\} \,.\, R_{\sigma\lambda}\{x, a_1, y, a_2, \ldots, a_{n-2}\} \,.\, \ldots$$
$$R_{\sigma\lambda}\{x, a_1, a_2, \ldots, a_{n-2}, y\} \,.\, R_{\sigma\lambda}\{y, x, a_1, a_2, \ldots, a_{n-2}\} \,.$$
$$R_{\sigma\lambda}\{a_1, x, y, a_2, \ldots, a_{n-2}\} \,.\, \ldots\, R_{\sigma\lambda}\{a_1, x, a_2, \ldots, a_{n-2}, y\} \,.\, \ldots$$
$$R_{\sigma\lambda}\{y, a_1, a_2, \ldots, a_{n-2}, x\} \,.\, \ldots\, R_{\sigma\lambda}\{a_1, a_2, \ldots, a_{n-2}, x, y\} ::$$
$$c \,\epsilon\, \alpha \,.\, \supset_c : R_{\sigma\lambda}\{c, b_1, b_2, \ldots, b_{n-1}\} \,.\, \vee \,.\, R_{\sigma\lambda}\{b_1, c, b_2, \ldots, b_{n-1}\} \,.\, \vee \ldots$$
$$\vee \,.\, R_{\sigma\lambda}\{b_1, b_2, \ldots, b_{n-1}, c\} :. \supset_{b_1, b_2, \ldots, b_{n-1}} . \, b_1, b_2, \ldots, b_{n-1} \,\epsilon\, \alpha\} \quad \text{Df.}$$

Next, I define ins as follows:

$$(9)\quad \text{ins} = \hat{P}\hat{Q}\{P\{\alpha_1, \alpha_2, \ldots, \alpha_n\} \,.\, \equiv_{\alpha_1, \alpha_2, \ldots, \alpha_n} : \alpha_1, \alpha_2, \ldots, \alpha_n \,\epsilon\, \varpi_Q :$$
$$(\exists a_1, a_2, \ldots, a_n) \,.\, a_1 \,\epsilon\, \alpha_1 \,.\, a_2 \,\epsilon\, \alpha_2 \ldots a_n \,\epsilon\, \alpha_n \,.\, Q\{a_1, a_2, \ldots, a_n\}\} \quad \text{Df.}$$

Now, I claim, ins$`R$ possesses the desired sort of n-connexity, whatever R may be.

For did it not, by (7), it would be possible to find n distinct α's, say $\alpha_1, \alpha_2, \ldots, \alpha_n$, such that none of those relations hold between them which can be made from those in the conclusion of (7) by substituting ins$`R$ for R, and each α for the a with the same number; while, as we learn from (9), each α is a member of ϖ_R. That is to say, if we pick out one member from α_1, say x_1, one from α_2, say x_2, and so on till we come to α_n, from which we pick out x_n, then $x_1, x_2, \ldots, x_n$ will stand to one another in none of the relations mentioned in the conclusion of (7), and hence will stand to one another in the relation $R_{\sigma\lambda}$. This will be true whatever the values that x_1 takes in α_1, x_2 in α_2, etc. It is easy to see that from this and the second half of the proposition in the brackets in (8), we can conclude that $\alpha_1 = \alpha_2 = \ldots = \alpha_n$, which contradicts our hypothesis. Hence, ins$`R$ always possesses the n-connexity expressed in (7).

Another and equally important property possessed by ins$`R$ is that, if (ins$`R$) $\{\alpha_1, \alpha_2, \ldots, \alpha_n\}$, *$\alpha_1, \alpha_2, \ldots, \alpha_n$ are all distinct.* For suppose that (ins$`R$) $\{\alpha_1, \alpha_2, \ldots, \alpha_1, \ldots, \alpha_n\}$. Then we shall have to have, by the definition of ins, $R\{a_1, a_2, \ldots, b, \ldots, a_n\}$, where a_1 belongs to α_1, a_2 to α_2, etc., b to α_1, and so on till we get to a_n, which belongs to α_n; $\alpha_1, \alpha_2, \ldots, \alpha_n$ are all, by the definition of ins, members of ϖ_R. Therefore, by the definition of ϖ_R, we shall

* It will be seen, of course, that $R_{\sigma\lambda}$, ϖ_R, and ins are essentially functions of the particular sort of n-connexity asserted in (7).

have $R_{\sigma\lambda}\{a_1, a_2, \ldots, b, \ldots, a_n\}$. We are thus led into a contradiction. It will be noted that this property too is characteristic of ordinary binary serial relations, of ternary relations such as the 'between' relation, and although in this case not clearly stated, of Vailati's separation-relations.

§ 4. Now two interesting questions arise: first, what hypothesis is necessary concerning the n-adic relation R if ins'R is to have a given sort of n-transitivity? and secondly, is it possible to build a function of R which has any given sort of n-transitivity, any given sort of n-connexity, and is such that if this function holds between $\kappa_1, \kappa_2, \ldots, \kappa_n$, the κ's are all distinct? The first question is exceedingly easy to answer. Let the transitivity in question be that of (1), and the connexity that of (7). Modify (1) in the following manner: if in any of the products that, added, make up T_R, a term, say x, occurs as argument to several R's, replace it in all but one of its occurrences by some term, so that in no two occurrences is it replaced by the same term; multiply the product in which it occurs by all the expressions which can be formed by taking $R_{\sigma\lambda}$ [derived from the connexity expressed in (7)], and giving it as arguments any n (not all necessarily distinct) of the terms which replace x, including x itself; and introduce the terms, other than x itself, which replace x, as apparent variables, in such a manner that their range is the whole left side of (1), and that they are preceded by an $\exists$. If we transform (1) in this way, it is easy to see, though tedious to prove, that we obtain a sufficient condition for ins'R's possessing the sort of n-transitivity indicated in (1) and the sort of n-connectedness indicated in (7).

As to the second question, it is almost self-evident that ins'[(ins'$R)_T$] possesses the sort of n-transitivity indicated in (1), the sort of n-connexity indicated in (7), and that if

$$\{\text{ins}'[(\text{ins}'R)_T]\}\{\kappa_1, \kappa_2, \ldots, \kappa_n\},$$

and $\kappa_i, \kappa_j, i = j$. The two latter properties follow simply from the fact that this relation is an ins of something; the fact that it has the former quality follows obviously from the following considerations. If Q has any sort of n-connexity, and $Q \subset P$, then P, a fortiori, has the same sort of n-connexity, if its field is that of Q; for the hypothesis of (7) (with R changed throughout to Q), remains unchanged, while, if

$$Q\{a_1^{(8)}, a_2^{(8)}, \ldots, a_n^{(8)}\}, \text{ then } P\{a_1^{(8)}, a_2^{(8)}, \ldots, a_n^{(8)}\},$$

so that the conclusion of (7) is true of P if it is true of Q. Therefore, (ins'$R)_T$ has the desired sort of n-connexity and n-transitivity, though it may be possible for us to have $i \neq j$, $\kappa_i = \kappa_j$, and (ins'$R)_T\{\kappa_1, \kappa_2, \ldots, \kappa_n\}$. Since (ins'$R)_T$ is connected in the way determined by (7), $[(\text{ins}'R)_T]_{\sigma\lambda}$ can only hold between $\alpha_1, \alpha_2, \ldots, \alpha_\kappa$

when $\alpha_1 = \alpha_2 = \ldots = \alpha_\kappa$. Therefore, $\varpi_{(\text{ins}'R)_T}$ is made up exclusively of unit-classes. Now, we can write the condition for the n-transitivity of ins'$[(\text{ins}'R)_T]$ as follows:

$$(10)\quad (\exists\lambda_1, \lambda_2, \ldots, \lambda_k)\, T_{\text{ins}'[(\text{ins}'R)_T]}\{\kappa_1, \kappa_2, \ldots, \kappa_n, \lambda_1, \lambda_2, \ldots, \lambda_k\} \,.\, \supset_{\kappa_1, \kappa_2, \ldots, \kappa_n} .\, \text{ins}'[(\text{ins}'R)_T]\{\kappa_1, \kappa_2, \ldots, \kappa_n\}.$$

The expression in the form $T_{\text{ins}'[(\text{ins}'R)]}$ is here the sum of products of terms of the form ins'$[(\text{ins}'R)_T]\{\mu_1, \mu_2, \ldots, \mu_n\}$, where the μ's are to be found among the κ's and λ's, and all the λ's appear somewhere as arguments to ins'$[(\text{ins}'R)_T]$. Therefore, since

$$C'\text{ins}'[(\text{ins}'R)_T] \subset \varpi_{(\text{ins}'R)_T},$$

all the κ's and λ's are unit classes. Therefore, since

$$\{\text{ins}'[(\text{ins}'R)_T]\}\{\nu_1, \nu_2, \ldots, \nu_n\}$$

holds when and only when $\nu_1 \ldots \nu_n$ are members of $\varpi_{(\text{ins}'R)_T}$, and there is an α_1 belonging to ν_1, an α_2 belonging to ν_2, ..., an α_n belonging to ν_n, we may write (10) as follows:

$$(11)\quad (\exists\beta_1, \beta_2, \ldots, \beta_k) \,.\, \beta_1, \beta_2, \beta_k, \alpha_1, \alpha_2, \ldots, \alpha_n \in \breve{\iota}``\varpi_{(\text{ins}'R)_T} \,.\, T_{(\text{ins}'R)_T}\{\alpha_1, \alpha_2, \ldots, \alpha_n, \beta_1, \beta_2, \ldots, \beta_k\} \,.\, \supset_{\alpha_1, \alpha_2, \ldots, \alpha_n} .\, (\text{ins}'R)_T\{\alpha_1\, \alpha_2, \ldots, \alpha_n\}.$$

From (5) it follows that (11) is identically satisfied, and hence that ins'$[(\text{ins}'R)_T]$ has the desired sorts of n-connexity and n-transitivity, and never, to put it roughly, relates a member of its field to itself, *whatever R may be.* Hence, if we have a system whose postulates can be put in the form of three propositions, one asserting a certain n-transitivity, another a certain n-connexity of a given n-adic relation, P, and the third asserting that P never relates a member of its field to itself, then, given any n-adic relation, R, we can construct a function of R having the desired properties of P. Moreover, it is easy to see that if R itself has the desired properties, the constructed relation will be, so we may put it, of the same formal properties as R, but two types higher.

§ 5. Now, there are very important sorts of relations whose definitions may be put in the above form. The general 'between' relation between members of a series is, it is easy to see, *completely* determined as to its formal properties by the three propositions

$$(\exists d) : abd \,.\, bdc \,.\, \mathsf{v} \,.\, abd \,.\, bcd \,.\, \mathsf{v} \,.\, adc \,.\, dbc \,.\, \mathsf{v} \,.\, abd \,.\, acd \,.\, bac \,.\, \mathsf{v} \,.\, dab \,.\, dac \,.\, bac \,.\, \mathsf{v} \,.\, bca : \supset_{a,b,c} .\, abc,$$

$$(\exists m, n) : amn \,.\, \mathsf{v} \,.\, man \,.\, \mathsf{v} \,.\, mna : bmn \,.\, \mathsf{v} \,.\, mbn \,.\, \mathsf{v} \,.\, mnb : cmn \,.\, \mathsf{v} \,.\, mcn \,.\, \mathsf{v} \,.\, mnc : \supset_{a,b,c} : a = b \,.\, \mathsf{v} \,.\, b = c \,.\, \mathsf{v} \,.\, c = a \,.\, \mathsf{v} \,.\, abc \,.\, \mathsf{v} \,.\, bca \,.\, \mathsf{v} \,.\, cab,$$

$$abc \,.\, \supset_{a,b,c} .\, a \neq b \,.\, b \neq c \,.\, a \neq c.$$

Similarly, if we understand the separation-relation to hold only between four distinct terms, the general separation-relation is *completely* determined by the three following propositions:

$$(\exists e): ab \parallel dc \,.\vee.\, cd \parallel ab \,.\vee.\, ab \parallel ec \,.\, ae \parallel cd : \supset_{a,b,c,d} . \; ab \parallel cd,$$

$$\begin{aligned}(\exists m, n, o): &\; am \parallel no \,.\vee.\, ma \parallel no \,.\vee.\, mn \parallel ao \,.\vee.\, mn \parallel oa : \\ &\; bm \parallel no \,.\vee.\, mb \parallel no \,.\vee.\, mn \parallel bo \,.\vee.\, mn \parallel ob : \\ &\; cm \parallel no \,.\vee.\, mc \parallel no \,.\vee.\, mn \parallel co \,.\vee.\, mn \parallel oc : \\ &\; dm \parallel no \,.\vee.\, md \parallel no \,.\vee.\, mn \parallel do \,.\vee.\, mn \parallel od : \\ \supset_{a,b,c,d} : &\; a = b \,.\vee.\, b = c \,.\vee.\, c = d \,.\vee.\, d = a \,.\vee.\, a = c \,.\vee.\, b = d \,.\vee. \\ &\; ab \parallel cd \,.\vee.\, ac \parallel bd \,.\vee.\, ad \parallel bc,\end{aligned}$$

$$ab \parallel cd \,.\supset_{a,b,c,d} .\; a \neq b \,.\, a \neq c \,.\, a \neq d \,.\, b \neq c \,.\, c \neq d \,.\, b \neq d.$$

That is, *from any triadic or tetradic relation, we are able to construct a between-relation or a separation-relation, respectively.* This fact should play much the same part in explaining how the regular relations of space may be derived from the irregular relations to be found in our experience that the analogous fact concerning dyadic relations plays in showing how the serial relation of the instants of time may be derived from the non-serial relation of complete succession between events*. Logically too this fact has a considerable interest, for it gives a hint of another method of defining mathematical systems than by the use of postulates; given our fundamental logical postulates to start with, we may be able to select the fundamental 'indefinables' of a mathematical system in such a manner that whatever values they may assume within their range of significance, the fundamental formal properties of the system will remain invariant.

§ 6. Of course, *all* the formal properties of a triadic or tetradic relation are not determined when the relation is completely determined as a between or separation relation. Hence there remain interesting and important questions yet as to whether simple properties of R may be given which will give ins'R or R_T or ins'[(ins'R)$_T$] properties analogous to density or 'Dedekindianness,' etc. If density with respect to a given transitivity, say that of (1), be the property of a relation R which holds when the implication in (1) is converted, then it requires little proof to see that if the converse of (1), modified in the manner that (1) is modified in the first paragraph of § 4, is true of R, and if $C'R \subset s'\varpi_R$, then ins'R will have the required sort of density. I know of no simpler property of R, however, by which we can replace $C'R \subset s'\varpi_R$, and, at any rate, if R is a between or separation relation, this sort of density will not be the property which we would naturally call by that name. If $R\{a, b, c\}$ means 'b is between a and c,'

* See *Proc. Camb. Phil. Soc.*, vol. XVII, Part 5, pp. 441—9.

then what we would naturally call density would be the property of R which can be written

$$(a, c) :: a, c \in C{}^{\backprime}R \,.\, \supset :. (\exists b) \,.\, R\{a, b, c\} : \vee \,.\, a = c.$$

Provided that $C{}^{\backprime}P \subset s{}^{\backprime}\varpi_P$, then if P is any triadic relation having this property, then ins'P, and hence, as may be seen easily, (ins'$P)_T$ and ins'[(ins'$P)_T$], will have this property.

§ 7. Let us now turn to the second topic to be treated in this paper, the problem of the synthesis of the series of sensation-intensities from the relations between sensations given in experience. This problem, in itself, is not one of pure logic or of pure mathematics, but its solution depends upon the solution of a purely logical and mathematical problem. In my previous article*, as I said at the beginning of this paper, I showed how from the relation of complete succession between the events in time, we can construct the series of the instants in time. The method was the following: we make the definitions:

(12) $$P_{\text{se}} = (\dot{-} P \dot{-} \breve{P}) \upharpoonright C{}^{\backprime}P \qquad \text{Df.}$$

(13) $$\tau_P = \hat{\alpha}\{\alpha = p{}^{\backprime}\overrightarrow{P_{\text{se}}}{}^{\backprime\backprime}\alpha\} \qquad \text{Df.}$$

(14) $$\text{inst} = \hat{Q}\hat{P}\{Q = (\breve{\epsilon} \,;\, P) \upharpoonright \tau_P\} \qquad \text{Df.}$$

If P is the relation between two events, x and y, when x is over before y begins, then P_{se} is the relation between two events which occur together at some moment; τ_P is the class of all instants of time—that is, the class of all those classes, α, such that α is made up of events in such a manner that every two events in α occur together at some moment, and if an event occurs at the same moment with every member of α, then it belongs to α; and inst'P is the relation between two members of τ_P—that is, instants—when some event at the first instant is over before some event at the second instant begins: that is, it is the relation between an instant and a succeeding instant. If $P \mid P_{\text{se}} \mid P \in P$, whether P is a temporal relation or not, inst'P will be a series. Now, let P stand for the relation, say, between any coloured object and a noticeably brighter one. Then P_{se} will be the relation between two coloured objects when the first is apparently of the same brightness as the second, for it is the relation between two members of the field of P—that is, coloured objects—when neither is in the relation P to the other. Now, it is obvious that when $xP \mid P_{\text{se}}y$, x must be, noticeably or unnoticeably, more bright than y, for this proposition says that x is noticeably brighter than some object which, at the brightest, is indistinguishable from y. Therefore, it is obvious that if $xP \mid P_{\text{se}} \mid Py$, x is brighter than something noticeably brighter than y, and hence is noticeably brighter than

* See *Proc. Camb. Phil. Soc.*, vol. XVII, Part 5, pp. 441—9.

y, and $P \mid P_{se} \mid P \subset P$. inst'$P$ is therefore here also a series, and nothing would seem more natural than for us to call it the series of sensation intensities.

But there are serious objections against this method of procedure, and here a genuine logical problem arises. For, although it is natural to regard a sensation-intensity as a class of sense-objects—the class of sensations 'of a certain intensity'—we naturally consider the intensity of a given sensation as uniquely determined, and the relations between two sensations, x and y, when x is of the same intensity as y, as a transitive, symmetrical, reflexive relation. Now, in general, τ_P is not a class of mutually exclusive classes, and the relation between two terms which belong to the same member of τ_P is not transitive. The fact that a certain river was flowing during the Siege of Troy, and is flowing while I am writing this article, does not mean that I was writing this article during the Siege of Troy, yet if we take P as the relation between one event and another which completely follows it, my writing this article and the flowing of the river will both belong to some member of τ_P; the Siege of Troy and the flowing of the river will both belong to some other member of τ_P. So we have the definite mathematical problem before us: given a relation, P, fulfilling certain conditions, not sufficient to make it a series, we wish to construct from it a serial relation in such a manner that the terms of this series shall form a class of mutually exclusive classes.

I shall first give the method by which this series may be derived from the relation between x and y when x is of noticeably greater intensity than y; then I shall state a set of conditions sufficient to secure the serial character of the derived relation, and finally I shall interpret conditions and results. Perhaps the best method logically would be first to formulate all the conditions to which the original relation must be subject, and then to treat the problem as a purely formal one, but the logical gain would hardly compensate us for the loss in clarity. So I first make the following definitions:

(15) $P_s = (\breve{\overrightarrow{P}}_{se} \mid \overrightarrow{P}_{se}) \upharpoonright C'P$ Df.

(16) $\lambda_P = \mathrm{D}'\overrightarrow{P}_s$ Df.

(17) $\mathrm{int} = \hat{Q}\hat{P}\{Q = [\breve{\epsilon} ; (P_{se} \mid P)] \upharpoonright \lambda_P\}$ Df.

If P is the relation between x and y when x is, say, noticeably brighter than y, then P_{se} is the relation between two things which are not distinguishable as concerns their brightness, and P_s is the relation between two things possessing brightness when each of the things which is indistinguishable from the one in brightness is also indistinguishable from the other, and *vice versa.*

It follows at once from the definition of P_s that it is transitive, symmetrical, and reflexive, whatever P may be, and hence in this respect it satisfies the requirements we have set up for the relation between two members of a sensation-intensity.

λ_P is the class of brightness-intensities, where P is the relation 'noticeably brighter than.' Since λ is defined as $D' \overrightarrow{P_s}$, it follows that it must always be a class of mutually exclusive classes; for suppose that two members of λ_P, say $\overrightarrow{P_s}'x$ and $\overrightarrow{P_s}'y$, had the term z in common. Then we would have zP_sx and zP_sy. From the definition of P_s it is symmetrical, so we get xP_sz and zP_sy, which, on account of the transitivity of P_s, gives us xP_sy, and, hence, $\overrightarrow{P_s}'x \subset \overrightarrow{P_s}'y$. In just the same way, we get $\overrightarrow{P_s}'y \subset \overrightarrow{P_s}'x$, or, finally, $\overrightarrow{P_s}'y = \overrightarrow{P_s}'x$. int'$P$ is the relation between two members of λ_P when a member of one is in the relation $P_{se} | P$ with a member of the other. Whatever P is, int'$P \subseteq J$. For suppose that $\alpha\,(\text{int}'P)\,\alpha$. Then, since α must belong to λ_P, every term of α stands in the relation P_s to every term of α. However, from the definition of int'P, there must be two terms of α, x and y, such that $xP_{se} | Py$. This may be written as

$$(\exists z) \,.\, xP_{se}z \,.\, zPy.$$

From this and the definition of P_{se}, we get $(\exists z) \,.\, xP_{se}z \,.\, z \sim P_{se}y$, or $\overrightarrow{P_{se}}'x \neq \overrightarrow{P_{se}}'y$, which may be written $x \dot{-} P_sy$. Thus, the assumption that $\sim(\text{int}'P \subseteq J)$ is self-contradictory.

A condition which will ensure the transitivity of int'P is $P_{se} | P \,\epsilon\, \text{trans}$. For it follows from the definitions of P_s, λ_P, and int that if $\alpha\,(\text{int}'P)^2\,\beta$, $\alpha\,[\{\epsilon \,;(P_{se} \,|\, P \,|\, \overrightarrow{\breve{P}_{se}} \,|\, \overrightarrow{P_{se}} \,|\, P_{se} \,|\, P)\} \upharpoonright \lambda_P]\,\beta$.

Now,

$$(18) \quad \vdash . \overrightarrow{\breve{P}_{se}} \,|\, \overrightarrow{P_{se}} \,|\, P_{se} = \hat{x}\hat{z}\,\{(\exists \alpha, y) \,.\, \alpha = \overrightarrow{P_{se}}'y \,.\, \alpha = \overrightarrow{P_{se}}'x \,.\, yP_{se}z\}$$

$$= \hat{x}\hat{z}\,\{(\exists y) \,.\, \overrightarrow{P_{se}}'y = \overrightarrow{P_{se}}'x \,.\, z \,\epsilon\, \overrightarrow{P_{se}}'y\} = P_{se}$$

Therefore, $P_{se} | P \,|\, \overrightarrow{\breve{P}_{se}} \,|\, \overrightarrow{P_{se}} \,|\, P_{se} | P$ is simply $P_{se} | P \,|\, P_{se} | P$. If $P_{se}|P$ is transitive, then we find that $\alpha\,(\text{int}'P)^2\,\beta$ implies that $\alpha\,[\{\epsilon \,;(P_{se} \,|\, P)\} \upharpoonright \lambda_P]\,\beta$, which is simply $\alpha\,(\text{int}'P)\,\beta$. A hypothesis which will make $P_{se} | P$ transitive is $P \,|\, P_{se} \,|\, P \subseteq P$. This is the same condition which we found to suffice for the transitivity of inst'P.

When will int'P be connected? Under what conditions, that is, will it be true that

$$\alpha, \beta \,\epsilon\, C'\text{int}'P \,.\, \alpha \neq \beta \,.\, \supset_{\alpha,\beta} : \alpha\,(\text{int}'P)\,\beta \,.\, \vee \,.\, \beta\,(\text{int}'P)\,\alpha\,?$$

This condition is manifestly implied by

$$\alpha, \beta \in \lambda_P \,.\, \alpha \neq \beta \,.\, \supset_{\alpha,\beta} : \alpha (\text{int}`P) \beta \,.\, \mathsf{v} \,.\, \beta (\text{int}`P) \alpha.$$

Since $\alpha (\text{int}`P) \beta$ merely demands that $\alpha$ and $\beta$ should be members of $\lambda_P$, and that *some* member of $\alpha$ should bear the relation $P_{se} | P$ to *some* member of $\beta$, and since if $x$ and $y$ are both members of $\alpha$, and $\alpha \in \lambda_P$, $xP_s y$, int$`P$ will be connected if

$$x \dot{-} P_s y \,.\, \supset_{x,y} : x P_{se} | Py \,.\, \mathsf{v} \,.\, y P_{se} | Px.$$

Now,

$$(19) \quad \vdash :: x \dot{-} P_s y :. \supset_{x,y} :. \overrightarrow{P}_{se}`x \neq \overrightarrow{P}_{se}`y :.$$
$$\supset_{x,y} :. (\exists z) : zP_{se}\, x \,.\, z \dot{-} P_{se}\, y \,.\, \mathsf{v} \,.\, zP_{se} y \,.\, z \dot{-} P_{se}\, x :.$$
$$\supset_{x,y} :. (\exists z) : zP_{se} x \,.\, zPy \,.\, \mathsf{v} \,.\, zP_{se} x \,.\, yPz \,.\, \mathsf{v} \,.\, zP_{se} y \,.\, zPx \,.\, \mathsf{v} \,.\, zP_{se} y \,.\, xPz :.$$
$$\supset_{x,y} :. xP_{se} | Py \,.\, \mathsf{v} \,.\, yP | P_{se}\, x \,.\, \mathsf{v} \,.\, yP_{se} | Px \,.\, \mathsf{v} \,.\, xP | P_{se} y.$$

If $P | P_{se} \subset P_{se} | P$, this reduces at once to the condition that we have just shown to be sufficient for the connectedness of int$`P$.

§ 8. We have seen, then, that if

$$P_{se} | P \in \text{trans} \ \text{ and } \ P | P_{se} \subset P_{se} | P, \ \text{int}`P \in \text{ser}.$$

Now the questions arise, what do these conditions mean when P is, for example, the relation 'noticeably brighter than'? and, are they true of such relations? The meaning of $P_{se} | P \in$ trans in such a case is clear, as is also its truth; $P_{se} | P$ is the relation between two objects, x and y, when x is not merely apparently, but actually brighter than y, for $xP_{se} | Py$ says that x is only subliminally different, if at all different, in brightness from something that is supraliminally brighter than y. Now, the transitivity of the relation, 'brighter than,' is obvious: at least as obvious, at any rate, as the existence of a series of brightnesses.

The meaning of $P | P_{se} \subset P_{se} | P$, however, is not quite so obvious. This condition demands that if x be noticeably brighter than something indistinguishable from y, it shall be indistinguishable from something noticeably brighter than y. We may interpret this demand as saying: if x is noticeably brighter than everything noticeably less bright than y, then y is noticeably less bright than everything noticeably brighter than x. A little reflection will convince us that this proposition is probably true: moreover, it is easy to see that its truth, and the truth of analogous propositions concerning all sorts of sensory intensity, form necessary conditions for the truth of the Weber-Fechner law. For suppose that this proposition were false: we might then have, to put it crudely, x and y both just noticeably brighter than x, and u just noticeably brighter than x, but subliminally different from y. Let a be the objective strength of the stimulus produced by z; then, by Weber's law, the strength of the stimulus produced by x or y will be $a(1 + c)$, where c is a constant independent of the value of a. Since u is just noticeably brighter than x, the strength of stimulus produced by u will be

$$a(1+c)(1+c) = a(1+2c+c^2).$$

But since u is only subliminally different from y in brightness, the strength of the stimulus produced by u is less than $a(1+2c+c^2)$.

Hence, we are landed in the contradiction,

$$a(1+2c+c^2) < a(1+2c+c^2).$$

A little reflection will convince the reader that any other way of violating the condition, $P \mid P_{se} \subseteq P_{se} \mid P$, would likewise be incompatible with Weber's law.

This seems the proper place to call attention to the fact that if P be the relation of complete precedence between the events in time, $P \mid P_{se} \subseteq P_{se} \mid P$ is *false*. For suppose that at this present moment two events begin, one of which lasts five minutes and the other ten. It is clear that neither event can be simultaneous with an event which wholly precedes the other: that is, neither bears to the other the relation $P_{se} \mid P$. Now suppose that one minute after the shorter event is ended, some event begins. This bears the relation P_{se} to the longer event, and the shorter event bears to it the relation P. Therefore, the shorter event bears to the longer event the relation $P \mid P_{se} \dot{-} P_{se} \mid P$. So we have proved nothing in this article which entitles us to say that if P is the relation of complete precedence among the instants of time, int'P is a series. And, as a matter of fact, it is not a series. If, however, we limit the field of P to events, say, that last exactly five minutes, then $P \mid P_{se} \subseteq P_{se} \mid P$, and int'$P$ is a series.

In case P is the relation, 'noticeably brighter than,' one can readily see that int'P is not only a series, but the series we mean when we speak of the series of brightnesses. For, if Weber's law is true, or even if some quantitatively different law of the same general form is true, P_s is exactly the relation which holds between two things of the same brightness, for xP_sy says, practically, the limina of distinguishability from x are the limina of distinguishability from y, and it can be deduced from this and Weber's law that this is true when and only when x and y produce stimuli of the same intensity, and hence it follows further from Weber's law, x and y must be of the same sensation-intensity. λ_P is therefore the class of all classes containing all the things of the same brightness as a given thing, and hence can be fittingly called the class of all brightnesses; and what could be more natural than to say that a given brightness is greater than another when and only when a thing of the first brightness is brighter than a thing of the second?

If we want to secure the compactness of int‘P, it is sufficient to assume the compactness of $P_{se} | P$, though not, as far as I know, necessary. Similarly, $P_{se} | P \epsilon \text{Ded}$ is a condition sufficient to assure the Dedekindian character of int‘P.

The interest and importance of this work on sensation-intensities lies in the fact that it is often naively assumed by psychologists that the series of sensation-intensities is in some wise a datum of experience, and not a construction. As a result, they are led into the most grotesque interpretations of such numerical formulae as Weber's law. A series of sensation-intensities is often treated as if it were, in some sense or other, a series of sensation-*quantities*, without any analysis whatsoever of the basis on which this series is put into one-one correspondence with the series of 0 and the positive real numbers, in order of magnitude. It is at any rate a necessary preliminary to this exceedingly complex problem to know what the series of sensation-intensities really are, and what their relation to our experience is: without this analysis, no scientific psychophysics is possible.

A NEW THEORY OF MEASUREMENT: A STUDY IN THE LOGIC OF MATHEMATICS

By NORBERT WIENER.

[Read November 13th, 1919.]

Introduction.

It is a deeply rooted popular idea that mathematics is but another name for measurement. Notwithstanding the fact that the existence of such non-metrical branches of mathematics as projective and descriptive geometry, the theory of groups, the algebra of logic, &c., prove this notion false, it is nevertheless true that the applications of mathematics have, up to the present time, been, almost without exception, applications of measurement. The natural sciences, in so far as they have been regarded as at all amenable to a mathematical treatment, have reduced themselves to the correlation of different ranges of measurement—of space, time, and mass, in the case of physics, of intensity of stimulus and intensity of sensory experience, in the case of psychophysics, and so on. Now, things do not, in general, run around with their measures stamped on them like the capacity of a freight-car: it requires a certain amount of investigation to discover what their measures are. It is, then, a necessary preliminary to the most complete scientific work that we should possess an analysis of the process through which we go in measuring the magnitude of a thing.

Now, a very beautiful theory of measurement has been developed in the third volume of the *Principia Mathematica* of A. N. Whitehead and B. Russell. It depends on the consideration of "vector-families"; that is, of sets of vectors such that it is possible to go from any point of the field ordered by the vectors to one and only one point along any given vector. These vectors correspond to definite increments of the measures of the terms in the common field of the vectors: thus, in the "vector-family" representing distances to the right along a given line, from a given point, an increment of the distance of this given point from a point to the extent of one inch, or of two inches, &c., is a vector. Since it is always possible to leave a point by any vector one pleases, our system of measurement must contain magnitudes larger by any desired amount than any given magnitude. As a consequence, measurement by "vector-families" breaks down when we have to deal with ranges of quantities that are essentially limited. As the authors of the *Principia* say of their theory, "We exclude magnitudes which have a definite maximum, unless they are circular, like the angles at a point, or the distances on an elliptic straight line."* The authors of the *Principia* justify this omission by the further statement, "But, except when they are circular, such magnitudes are of little importance." This is, in general, true, but there is one exception which is well worth considering. Perhaps the least satisfactory and most discussed portion of the unsatisfactory and much discussed theory of measurement is that which deals with the measurement of the intensities and qualities of sense-data.† Now, the intensities and qualities of sense-data are not susceptible to increments of arbitrary magnitude. There is no degree of loudness as much greater than that of a foghorn at close range as the loudness of this is greater than that of the ticking of a watch at the distance of ten feet. There is no note as much higher than that of the cricket as the latter is higher than the lowing of an ox. There is no object as much more intensely red than a drop of blood in the sunlight as the latter is than a piece of grey flannel. Nevertheless, we often do speak of the measure of the loudness or pitch of a note or the intensity of a colour, and if Weber's law is to have any meaning in any other form than that in which it refers to "just noticeable" differences in intensity or quality,‡ we must be able to establish some intrinsic criterion whereby we can determine the ratio of one difference in sensation quality or intensity to another, which will not presuppose that such an interval can be increased by any numerical factor whatever.

The first steps which are essential to the measurement of sensory qualities and intensities are the determination of the fundamental experience by means of which this measurement is to be performed, and the derivation from this crude, uncouth experience of functions which will have certain comparatively neat properties, and which hence will form a mere convenient starting point for the process of measurement proper. Our measurement of sensation-intensities obviously has its origin in the consideration of intensity-intervals between sensations: that is sufficiently

---

* Vol. 3, p. 340, lines 1–4.

† Except in one respect, which we shall indicate later, the theories of qualities and intensities coincide.

‡ Cf. "Studies in Synthetic Logic," *Proc. Camb. Phil. Soc.*, Vol. 18, Part 1, pp. 24–28, §§ 7, 8, by Norbert Wiener.

indicated by the fact that our measurement of a sensation-intensity always reduces itself, sooner or later, to the determination of its ratio to some standard intensity, while such a proposition as "x is twice as intense as y," is simply a paraphrase for some such statement as "The interval of intensity between x and y equals that between y and some sensation of zero intensity." The fundamental experience for which we are searching is not, however, of the form "The interval between x and y equals that between u and v." Not every two intervals which seem equal are equal: two intervals may only be subliminally different. Nevertheless, our only direct method of determining with what intervals an interval is equal lies, as we shall show, in the determination, first, of with what intervals it *seems* equal. But even the seeming equality of two intervals is not quite what we want: two intervals seem equal, as far as we are concerned, when and only when neither seems greater than the other. We shall take, then, as the relation which forms the basis of the measurement of sensation-intensities, "The interval between x and y seems less than that between u and v," where intervals are regarded as possessing signs, ascending intervals being regarded as positive, and "less than" is to be interpreted as "*algebraically* less than." In taking this as our primitive experience, we do not mean to assert—in fact we should categorically deny—that this relation is given as such in our experience, and that no further analysis of it is possible: what we *do* assert is that it represents a much more minute analysis of the basis of our measurements of sensation-intensities than any yet given, and forms a convenient starting point for a theory of sensation-intensities.

Although our initial relation enables us to give a sort of order to all sensation-intervals which seem greater or less than other intervals in some definite way (*i.e.* as loudness-intervals or as brightness-intervals, &c.), we shall, in our subsequent work, limit its range of application to positive supraliminal intervals, or intervals which seem greater than some interval between a thing and itself. We do this, because zero-intervals have certain properties which interfere with our subsequent theory. Later on, we shall use as our criterion of the genuine equality of two intervals the fact that all the intervals which are indistinguishable from (*i.e.* neither noticeably greater than nor noticeably less than) either are indistinguishable from the other. It will be possible, therefore, for us to say that two intervals may be indistinguishable in magnitude, yet that one is greater than the other. Let the interval between x and y and the interval between x and z be a pair of intervals of this sort. Then y must be indistinguishable from z by direct comparison, yet we must be able to say that y and z are not really of the same intensity, and hence that the interval between y and z, though indistinguishable from the zero-interval, is not of measure zero, or else it will not seem natural to call the interval between x and y of really different size than that between x and z. Now, it appears that an interval seems either to be one of difference or of identity. If it seems to be one of difference, it is indistinguishable from intervals of difference, and from those only, while all intervals of identity seem of the same magnitude. Hence, if our criterion of the genuine equality of two intervals is that all the intervals which seem identical with either seem identical with the other, if an interval be subliminal, it is genuinely identical in magnitude with the zero-interval. We thus obtain the result that the interval between y and z, whose difference from zero we wish to secure, is genuinely identical in magnitude with a zero-interval. To avoid this, we limit our discussion at first to positive supraliminal intervals.

As we said above, we regard two intervals as genuinely equal* when and only when all the intervals which are indistinguishable from (*i.e.* seem neither greater nor less than) either are indistinguishable from the other. Genuine equality is, as may readily be seen, a reflexive, symmetrical and transitive relation. It is possible, therefore, to group all positive supraliminal intervals into naturally exclusive sets, such that no member of any set is genuinely equal to a member of another set, and every member of a set is genuinely equal to every member of the set. The relation between two terms which consists in their being separated by some interval belonging to one of these sets we shall call the *vector* associated with the set. We are able, then, to regard any positive supraliminal interval as an instance of one of these vectors. We can next define the vector corresponding to a subliminal interval as follows: since all intervals representing a vector are equal, we should naturally regard the difference of an interval belonging to one vector from an interval belonging to another vector as independent with respect to its magnitude of the particular intervals chosen from these vectors. Now, the difference of two intervals may be readily defined when their upper ends coincide as the interval from the lower end of the interval from which the subtraction is to be made to the lower end of the subtracted interval. Furthermore, every subliminal interval, if its upper end does not lie in the neighbourhood of the maximum possible magnitude of sensations of the appropriate sort, may be regarded in the above manner as the difference of two supraliminal intervals with coincident upper ends. In the neighbourhood of the maximum possible magnitude of sensations of the appropriate sort, every subliminal interval may be regarded as the difference in an analogous

* Cf. "Studies in Synthetic Logic.

sense of two intervals with coincident *lower* ends. It will be found, then, that the vector corresponding to a subliminal interval may be defined in terms of two supraliminal vectors whose difference it forms, as the relation between two terms when one either first ascends an interval belonging to one, and from that point descends an interval belonging to the other, or first descends an interval belonging to the second, and then ascends an interval belonging to the first. Now, we wish to confine our discussion to vectors made up of positive intervals. To this end, it is necessary that the subtracted vector should be smaller than that from which it is to be subtracted. We need next, therefore, a criterion of the relative magnitudes of two vectors.

In looking for such a criterion, we shall suppose it to be axiomatic that in any intrinsic comparison of differences between intervals, a subliminal or unnoticeable difference is always to be treated as less than a noticeable one. This being the case, if an interval be indistinguishable from some interval noticeably greater than another, it is necessarily greater than the other one—for, if the interval R be indistinguishable from the interval S, which is noticeably greater than T, either R is subliminally greater than S, or equal to it, or only subliminally less than it. In the first two cases, the naturalness of supposing R greater than T is obvious at once; in the latter case, the above principle will render it obvious. Another similar condition which determines that R is greater than T is that there should be an interval S, indistinguishable from T, but noticeably less than R. It is logically demonstrable that one of these two criteria determines that an interval R is greater than an interval T, or else that T is greater than R when, and only when, R and T are not genuinely equal, in accordance with our former definition. We may *define* R as greater than T, then, when one of the two conditions just stated is satisfied.

We have now a complete definition of the class of intensity-vectors belonging to any given range of sensations. In the next section of this paper, we shall cover the same ground we have just been covering in a stricter manner, with the aid of the symbolism of the *Principia Mathematica*. If $\phi(x, y, u, v)$ stand for, "The difference between x and y in a given respect seems less than that between u and v," then we shall use Id_ϕ to stand for the relation of genuine equality in the appropriate set between positive supraliminal intervals;* $s``\mathrm{D}`\overrightarrow{\mathrm{Id}_\phi}$ will be the class of

* We shall regard an interval as simply the ordered couple formed by its upper and lower extremity.

positive supraliminal vectors: Dc_ϕ stands for the relation, "less than," among positive supraliminal intervals, and Vc_ϕ is the class of all intensity-vectors. Vs_ϕ, which we shall later define in terms of Vc_ϕ is a class of relations which will form part of Vc_ϕ when ϕ has the particular kind of value just attributed to it, but which, unlike Vc_ϕ, will always consist of mutually exclusive vectors—*i.e.* of vectors such that no two distinct ones have a common beginning from which they reach to a common end—whatever properties ϕ may have.

We are now in a position to consider the problem of measurement itself. To this end, we define two vectors, R and S, as having the ratio μ/ν in a class of vectors κ (which in the case of the measurement of sensory-intensities, will be the Vs_ϕ just considered) if there is a vector T belonging to κ such that if we start from a member of the field of T and take successively μ steps belonging to T, we sometimes take one step belonging to R, while similarly, ν successive steps belonging to T sometimes cover the the same ground as one step belonging to S. This method of defining the ratio of two vectors in terms of a common submultiple, instead of in terms of a common multiple, as in the *Principia*, is chosen for the reason that we do not wish, for example, the existence of a loudness $\frac{9,999,999,999}{10,000,000,000}$ as great as that of the falling of a pin to depend on that of a loudness 9,999,999,999 times as great as that of the falling of a pin.

Each vector will bear various ratios to other vectors: it will be ten times this vector, twice that, half the other, and so on. We have seen, however, that in the case of sensation-intensities, no non-zero interval can be multiplied by an arbitrarily great numerical factor. The class of ratios, that is, which a given vector bears to other vectors, has, in general, a lower limit or minimum in the scale of real numbers which will be distinct from zero. In case there is a maximum vector to which the vector to be measured bears a ratio, this ratio will be the minimum of the ratios which the vector to be measured bears to other vectors, but, in general, a vector will not bear to any vector a ratio which is the minimum or lower limit of the ratios which it bears to other vectors, or, as we shall call it, its index. This index we shall represent by Ind_κ'R, where R is the vector to be measured, and κ the class of vectors in terms of which the measurement takes place. The index of a vector may roughly be taken to represent its measure in terms of the greatest possible vector of the set, but a vector can be measured by its index whether such a greatest possible vector exist or not. Since the series of real numbers is Dedekindian—*i.e.* since, whenever we divide it into two classes, so that every term of the one is greater than every term of the other, either the former class will have a minimum, or the second a maximum—every vector will have an index, and it is easy to show that no vector can have more than one index.

We have now found a way to measure such things as sensation-intervals; our next task is to discover a way to measure such things as sensations. Now, the natural measure of a sensation is the index of an interval stretching to it from a sensation of zero intensity. There may not be, however, a sensation of zero intensity. In such a case, the natural thing to do would seem to be to approximate to what the interval between this non-existent sense-datum of zero intensity and the given sensation would be, supposing we were wrong in judging data of zero intensity not to exist, by taking successively less and less intense sensations, measuring the intervals between these and the given sensation, and taking the upper limit or maximum of the values so obtained. Now, a sensation is less intense than a given sensation when the interval from it to the given sensation is positive, and the degree of their difference in intensity is measured by the index of their interval, so the natural measure of the intensity of a given sensation is the upper limit of the indices of intervals having it as their upper boundary. This we shall represent by $\text{Meas}_\kappa\text{'}x$, where x is the sensation to be measured, and κ is the class of vectors by which it is to be measured. It should be noted that in *any* system of measurement bounded at both ends, Meas_κ will enable us to measure the position of any given term—it is not confined in its application to sensation-intensities. It should also be noted that the measure of any term in an unbounded system of measurement is zero, as every vector can be repeated an infinite number of times, and hence the lower limit of the ratios it bears to other vectors is zero. Therefore, since the measure of a term is the limit of a set of indices of vectors, it also must be zero. In any system each term will have one, and only one measure, but the same measure may belong to different terms. In such a case, if κ is of the form Vc_ϕ or Vs_ϕ, where ϕ is the relation between x, y, u, and v, when the intensity-interval between x and y seems algebraically less than that between u and v, we shall say that two different terms having the same measure are of the same intensity, and that a sensation-intensity is the class of all terms having some given measure. The measure of an intensity is the measure of its terms. The relation between an intensity and its measure is one-one.

It will not be a necessary consequence of the definition of the measure of a vector that the measure of a vector containing an interval formed by taking steps belonging to two vectors R and S, successively will be the sum of the measures of R and S. We can easily obtain a new definition of the measure of a vector, however, which will always have this desired property. This requires first the formation of a new class of vectors, which we shall term Reg_κ, as a function of the class of vectors κ with which we started. The measure of one of these, say R, will be designated $\overset{\smile}{\text{Dist}_\kappa}$ 'R. We shall also define μ_κ as the relation between a member of Reg_κ and an μ-th submultiple of it (μ being any real number). We shall prove that it follows from our definitions that any vector which is the μ-th multiple of a ν-th multiple of another vector is the $(\mu.\nu)$-th multiple of the second vector. The details of the definitions of these various notions are, however, of no special interest to the general reader, so I shall reserve them for the technically logistical portion of this paper.

Finally, we shall show how it is possible to remove the limitation of our system of measurement to systems of measurement with definite maxima or upper limits, and at the same time construct a method of measurement of sensation-qualities and intensities which is in certain cases more natural than that just given. Since all ranges of sensation-qualities and intensities are bounded above, it is always possible to make the "maximum possible interval" our standard of measurement, and in the case of certain sensation-*qualities*, such as chroma, this seems the most natural standard to take—for example, we say that this patch of colour is of the highest possible degree of saturation, that one is only half saturated, and so on. In the case of most ranges of sensation-intensity, such as the scale of loudnesses, such a method of measurement seems highly artificial. It does not seem natural to measure the ticking of a watch in millionths of a boiler-factory-power. The interval which most psychologists have taken as a standard in such instances as these is the just-noticeable interval. We do not know, however, that all just-noticeable intervals are equal, nor yet that there are any just-noticeable intervals. We shall choose the first just-noticeable interval as our standard of measurement, and we shall avoid the assumption that there is some single definite just-noticeable interval, just as we avoided the assumption that there was any single, definite, greatest possible interval, by making the class of intervals which are less than just-noticeable—that is, are subliminal—our real standard of measurement. Now, all intervals which relate sense-data not noticeably more intense than any other sense data* are subliminal. We have now on our hands, therefore, the task of finding a way to measure terms of the fields of a class of vectors in terms of a cer-

* It may be necessary in some cases to exclude sensations of zero intensity from consideration here and elsewhere for reasons analogous to those which previously led us to defer our consideration of subliminal intervals till after supraliminal intervals had been discussed.

tain portion of these fields—namely, in the case of sense-intensities, in terms of the class of subliminal sense-data.

Now, we can use the system of measurement already developed to find the indices of all those portions of vectors which relate subliminal data to subliminal data, or, in general, data of a given portion of the fields of the class of vectors to data of the given portion of the fields of the class of vectors, in terms of the class of all such vectors. Let us regard the index of such a portion of the vector as a property of the whole vector, and not merely of the part for which it is primarily defined. Let us measure any vector R of the original set by finding some μ-th part of it to which an index has already been given by the method just indicated, and associating with R a quantity μ times the value of this index. This quantity will, in general, depend on μ. Let us call the relation of any such quantity to R, $\mathrm{Inx}_{\kappa, \alpha}$, where κ is the class of vectors originally taken, and α is the class of members of the fields of these vectors in terms of which all the members of the fields of these vectors are to be measured. Just as we previously defined the measure of a term to be the maximum or upper limit of the indices of intervals leading up to it, so we now define the measure of a term in terms of κ and α as the maximum or upper limit of the values of the $\mathrm{Inx}_{\kappa, \alpha}$'s of intervals leading up to it. This we shall call the $\mathrm{Meas}_{\kappa, \alpha}$ of the term in question. From this point on, the development of this theory of measurement runs precisely parallel to that of our previous theory.

At the end of this paper, we shall give a proof that under certain conditions which we there state, our method of measurement gives substantially the same results as that of the *Principia*. This portion of the paper is of a merely technical interest, in that it correlates this work with what has previously been done on the subject. It had better be omitted on a first reading of this paper.

1. We shall now cover the same ground we have already covered in the introduction in a strictly rigorous and logical manner, with the help of the symbolism of the *Principia Mathematica*. As we saw in the introduction, an experience which, as far as we are concerned, may be regarded as lying at the foundation of all our measurements of, for example, brightnesses, is, "The interval between x and y, considered with reference to their brightness, seems algebraically less than that between u and v," where all ascending supraliminal intervals—*i.e.* intervals between a sense-datum and another sense-datum of noticeably greater intensity, are regarded as positive. Let us call the above proposition $\phi(x, y, u, v)$. The thing that we should normally call the interval between x and y is the

ordinal couple $x \downarrow y$. Now, we shall have occasion to regard ϕ as a dyadic relation between intervals rather than as a tetradic relation among brightness-sensations. On this account we shall make the following definition :

$$(1) \quad \mathrm{Cp}_\phi = \hat{R}\hat{S}\{(\exists x, y, u, v).\mathrm{R} = x \downarrow y.\mathrm{S} = u \downarrow v.\phi(x, y, u, v)\} \quad \mathrm{Df}$$

If ϕ is the relation mentioned above, Cp_ϕ is the relation between a given brightness-interval and a brightness-interval which seems greater than it—*algebraically* greater, that is. We wish, however, to limit ourselves to the discussion of positive intervals. We obtain this result as follows : we first limit Cp_ϕ to positive *supraliminal* intervals—that is, to intervals which bear the relation Cnv ‘Cp_ϕ to some interval of the form $x \downarrow x$. Then we form from this relation the one defined by means of the following definitions as Id_ϕ

$$(2)^* \quad \mathrm{P}_{se} = (\dot{-}\,\mathrm{P}\,\dot{-}\,\breve{\mathrm{P}}) \upharpoonright \mathrm{C}‘\mathrm{P} \quad \mathrm{Df}$$

$$(3)^* \quad \mathrm{P}_s = (\overrightarrow{\breve{\mathrm{P}}_{se}} \,|\, \overrightarrow{\mathrm{P}_{se}}) \upharpoonright \mathrm{C}‘\mathrm{P} \quad \mathrm{Df}$$

$$(4) \quad \mathrm{Id}_\phi = \{\mathrm{Cp}_\phi \upharpoonright \hat{\mathrm{R}}[(\exists x).(x \downarrow x)\mathrm{Cp}_\phi \mathrm{R}]\}_s \quad \mathrm{Df}$$

Id_ϕ is, then, the relation between two supraliminal positive brightness-intervals when they agree in every respect when they are compared with other brightness-intervals in respect to their magnitude. $\mathrm{D}‘\overrightarrow{\mathrm{Id}_\phi}$ will, therefore, be the result of sorting out the class of all positive supraliminal brightness-intervals into classes each containing all the intervals of a given magnitude. $\dot{s}‘‘\mathrm{D}‘\overrightarrow{\mathrm{Id}_\phi}$ will be the class of all positive supraliminal vectors—of all relations, that is, which connect pairs of sensations such that the brightness-interval between the members of such a pair is of a given fixed size for each member of $\dot{s}‘‘\mathrm{D}‘\overrightarrow{\mathrm{Id}_\phi}$ chosen, and is positive and supraliminal.

Now, one brightness-interval is less than another if it is either noticeably less than some brightness-interval indistinguishable from (*i.e.* neither noticeably greater than nor noticeably less than) the other, or is

* Cf. the discussion of indistinguishability and genuine identity in the introduction and *Studies in Synthetic Logic.*

indistinguishable from something noticeably less than the other.* Let us define this relation, as it applies to supraliminal positive brightness-intervals, as follows:—

(5) $$P_{dc} = P \mid P_{se} \cup P_{se} \mid P \qquad \text{Df}$$

(6) $$Dc_\phi = \{Cp_\phi \upharpoonright \hat{R}[(\exists x).(x \downarrow x)\, Cp_\phi R]\}_{dc} \qquad \text{Df}$$

Dc_ϕ is the desired relation. Now, one criterion of the equality of two subliminal or supraliminal positive intervals is, as we have seen, that they both can be formed by going up from some sensation by a step of a certain determinate size, and then down by a smaller step of another determinate size, or else by first taking a downward step of the smaller size, and by then ascending from the point just reached by a step of the larger size: that is, the relation between two terms connected by such an interval will be of the form $\breve{S} \mid R \cup R \mid \breve{S}$ where R and S are both members of $\dot{s}``D`\overrightarrow{Id}_\phi$, and S bears to R the relation $\breve{\in};Dc_\phi$. In symbols, we have the following definition:

(7) $$Vc_\phi = \hat{T}\{(\exists R, S),\ S[(\breve{\in};Dc_\phi) \upharpoonright \dot{s}``D`\overrightarrow{Id}_\phi]\, R\,.\, T = \breve{S} \mid R \cup R \mid \breve{S}\} \qquad \text{Df}$$

Vc_ϕ is, then, the class of brightness-vectors. We need to write

$$T = \breve{S} \mid R \cup R \mid \breve{S},$$

and not $$T = \breve{S} \mid R \quad \text{nor} \quad T = R \mid \breve{S},$$

* These two criteria are mutually irreducible. A sensation-interval of maximal magnitude will not be indistinguishable from something noticeably greater than a subliminally smaller interval, while it will be noticeably greater than something indistinguishable from the latter interval. In the same way, a sensation-interval of minimal magnitude will not be indistinguishable from any interval noticeably less than a subliminally greater interval, but will be noticeably less than something indistinguishable from the latter interval. The same sort of a statement may be made with reference to sensations, considered as to their intensity. The statement to the contrary in the *Studies in Synthetic Logic* already referred to is simply false, and arose from the author's not considering the fact that all ranges of sensory qualities or intensities have, roughly speaking, a maximum. If the int ' R of that paper be defined as $\{\epsilon ; (R \mid R_{se} \cup R_{se} \mid R)\} \upharpoonright \lambda_R$ instead of as $\{\epsilon ; (R_{se} \mid R)\} \upharpoonright \lambda_R$, the hypothesis

$$R \mid R_{se} \cup R_{se} \mid R_\epsilon \text{ Trans}$$

be substituted for $$R_{se} \mid R_\epsilon \text{ Trans} \quad \text{or} \quad R \mid R_{se} \mid R \in R,$$

and everything written about the hypothesis $R \mid R_{se} \in R_{se} \mid R$ be struck out as unnecessary, since int ' R will, as now defined, always be connected, the paper will be correct. The hypothesis $R \mid R_{se} \in R_{se} \mid R$ will be satisfied everywhere except, so to speak, in the neighbourhood of the top of the scale of sensation-intensities, if R be the relation, "noticeably more intense than."

since, to put it crudely, we may be able to take a step of the size $a-b$ downwards or upwards when we are unable to take a step of the size a in the given direction, where a is less than half the length of the whole sensation-range, while we are always able to take a step of size a either upwards or downwards from any given sense-datum. But even so, Vc_ϕ is not yet quite the class of sensation-intensity vectors we are looking for in order to apply our later theory of measurement, although it could, it is true, be used for that purpose as it stands. It will not be made up of mutually exclusive vectors in the case where ϕ is the relation which we have assumed it to be in the case of sensation-intensities. That is, it will be possible to go from one sensation to another, in general, by several distinct vectors. We would naturally suppose, however, that as the vectors which form the basis of our further theory of measurement represent, roughly, distances, that it would be impossible for two vectors to overlap. We get around this difficulty by defining in terms of Vc_ϕ a class of vectors which would naturally be called intensity-vectors, or, in the case now under consideration, brightness-vectors, which will, by its very definition, be made up of mutually exclusive vectors. To this end, we first define the overlapping of two relations as the relation which holds between them when they both relate a given pair of terms together. In symbols, this becomes

(8) $$\mathrm{Ov} = \hat{P}\hat{Q}\{\exists ! P \dot{\frown} Q\} \quad \mathrm{Df}$$

Then we fuse together any members of Vc_ϕ which can be connected by a chain of overlappings, by the process indicated in the following definition:

(9) $$\mathrm{Vs}_\phi = \dot{s} \text{``} D \text{`} \overrightarrow{(\mathrm{Ov} \upharpoonright \mathrm{Vc}_\phi)_*} \quad \mathrm{Df}$$

Vs_ϕ is called the class of separated vectors (*i.e.* in the present case, of separated brightness-vectors). It is useful on account of such conditions as the following: it is possible under all circumstances to go from a given sensation to one whose intensity is greater by a units,* by going up $a+b$ units in the scale of sensations, arranged as to their intensity, and then descending b units, or else by going down b units, and then ascending $a+b$ units, if $a+b$ is less than $\frac{1}{2}c$, where c is the greatest number of units by which two sensations can differ in intensity in the appropriate respect, while, if $a+b > \frac{1}{2}c$, this is only possible under certain condi-

* I speak as if we were able to measure sensations at this stage, although in fact we are not, since to express this point in strictly exact language would involve an intolerable and confusing prolixity.

tions. Nevertheless, if it sometimes is possible, with a certain determination of b, to make a step a units in length by going b units down and $a+b$ units up, and it always is possible to make a step a units in length by the same process when b is given another determination, it would be most unnatural to regard these two processes as determining distinct vectors.

2. We now come to that part of our discussion which has most directly to do with measurement. We shall first define the ratio between two relations belonging to a given class (*i.e.* in the case of sensation-intensities, between two intensity-vectors) in the sense which will be relevant in this paper, and derive a few allied classes and relations from it. I shall say that if R and S both belong to κ, they have the κ-ratio $(\mu/\nu)_\kappa$, if there is a relation P which belongs to κ, such that

$$\exists ! P^\mu \dot{\cap} R \quad \text{and} \quad \dot{\exists} ! P^\nu \dot{\cap} S$$

In symbols, we have

$$(10) \qquad (\mu/\nu)_\kappa = \hat{R}\hat{S}\{(\exists P) . P, R, S \epsilon \kappa . \exists ! P^\mu \dot{\cap} R . \exists ! P^\nu \dot{\cap} S\} \quad \text{Df*}$$

The definition of μ/ν in the *Principia* will not do for us, because there, if R is to have to S the relation μ/ν, where μ/ν is expressed in its lowest terms, we must have $\dot{\exists} ! R^\nu \dot{\cap} S^\mu$, so that it would be impossible for us to say, for example, that a fog-horn sounds two-thirds as loud as a boiler-factory, without assuming that there is something sounding twice as loud as a boiler-factory. On the other hand, it will be seen that, on our definition, it may very well be that a fog-horn makes a noise two-thirds as loud as a boiler-factory, without there being any noise twice as loud as that of a boiler-factory.

An important notion connected with $(\mu/\nu)_\kappa$ is that of the class of κ-ratios which a given relation R bears to other relations. I shall define Rt_κ as follows:

$$(11) \qquad \mathrm{Rt}_\kappa = \hat{S}\hat{R}\{(\exists M, \mu, \nu) . S = (\mu/\nu)_\kappa . \mu, \nu \neq 0 . R (\mu/\nu)_\kappa M\} \quad \text{Df}$$

Then $\overrightarrow{\mathrm{Rt}_\kappa}$ ‘R is this class. In a vector family,† the analogue of $\overrightarrow{\mathrm{Rt}_\kappa}$ ‘R —*i.e.* in a submultipliable connected vector-family, $\overrightarrow{\mathrm{Rt}_\kappa}$ ‘R itself—contains ratios less than any given ratio. In a class of relations which serve as

* This is not the $(\mu/\nu)_\kappa$ used in the discussion of cyclical systems in the *Principia*.

† Cf. *Principia Mathematica*, Part VI, Section B.

the basis of a system of measurement with a maximum, the ratios, taken in the sense of the *Principia*, which are the analogues of $\overrightarrow{\mathrm{Rt}_\kappa}$ ' R, determine as their "lower limit" in the series of real numbers a number which, in general, will not be zero, which we shall call the κ-*index* of R, or Ind_κ ' R. To define this, we need first to determine the analogue of one of our ratios in the system of the *Principia*. If $(\mu/\nu)_\kappa$ be one of our ratios, its analogue will be Eq_κ ' $(\mu/\nu)_\kappa$, where Eq_κ is defined by

$$(12) \qquad \mathrm{Eq}_\kappa = \hat{\mathrm{R}}\hat{\mathrm{S}}\,\{(\exists \mu, \nu).\, \mathrm{R} = \mu/\nu .\, \mathrm{S} = (\mu/\nu)_\kappa\} \quad \mathrm{Df}$$

Then Ind_κ will be defined by

$$(13) \qquad \mathrm{Ind}_\kappa = p \mid (\overrightarrow{\mathrm{H}_\epsilon}) \mid (\mathrm{Eq}_\kappa)_\epsilon \mid \overrightarrow{\mathrm{Rt}_\kappa} \quad \mathrm{Df}$$

This will always make Ind_κ ' R a real number or Λ, by the definition of a real number given in ✱310* of the *Principia*. It follows from ✱72—12—15—16, ✱71—25 that Ind_κ is a one-many relation: that is, no relation has more than one κ-index. We are enabled to compare incommensurable members of κ by their κ-indices, for their κ-indices express, roughly speaking, their magnitude in terms of that of the maximum of all the κ-vectors. It should be noted, however, that we have nowhere assumed that κ contains any greatest κ-vector: indeed, we have assumed nothing at all about κ except that it is a class of relations.

Ind_κ enables us to assign a real number to every member of κ—that is, for example, to every vector between sensations considered with regard to their intensities or qualities. We do not merely wish, however, to measure vectors between sensations considered with regard to their intensities and qualities, but also to measure these intensities and qualities themselves. The measure of the intensity of a sensation is, roughly speaking, the index of its difference from a sensation of zero intensity—that is, it is the upper limit or maximum (taken in the series of real numbers, in their natural order) of the indices of the vectors leading up to it from sense-data of smaller intensity. If κ stands for the class of separated intensity-vectors, and Meas_κ is to stand for the relation between the measure of the intensity of a sense-datum and a sense-datum possessing this measure, we get, in symbols

$$(14) \qquad \mathrm{Meas}_\kappa = \{\mathrm{limax}_{\Theta'} \mid (\mathrm{Ind}_\kappa)_\epsilon \mid (\overleftarrow{\mathrm{Q}})_\epsilon \mid \varepsilon\} \upharpoonright s\text{'C''}\kappa \quad \mathrm{Df}$$

* References beginning with a ✱ are to paragraphs and theorems in the *Principia Mathematica*.

By $*310-1$, $*207-41$, $*204-1$, $*72-12-15$, $*71-25-26$ of the *Principia*, Meas_κ is a one-many relation. It obviously correlates with the whole of $s\text{`}C\text{``}\kappa$ a part of $C\text{`}\Theta'$, or the class of all positive real numbers and zero. Meas_κ will not, in general, be one-one.*

From Meas_κ we may derive a class of vectors or relations connecting members of $s\text{`}C\text{``}\kappa$ which will always have certain very important properties, quite independent of those of κ. I shall define this class, which I shall call Reg_κ, as follows

(15) $$\text{Dist}_\kappa = \hat{R}\hat{\mu}\ \{R = \breve{\text{Meas}}_\kappa \,;\, \mu +_a\} \quad \text{Df}$$

(16) $$\text{Reg}_\kappa = D\text{`}\text{Dist}_\kappa - \iota\text{`}\dot{\Lambda}$$

The measure of a member of Reg_κ is the value of μ for which this member is $\text{Dist}_\kappa\text{`}\mu$. This is $\breve{\text{Dist}}_\kappa\text{`}R$, since $\text{Reg}_\kappa \rceil \breve{\text{Dist}}_\kappa \in 1 \to 1$. That this is so may be proved as follows: it results from the form of the definition of Dist_κ that $\text{Dist}_\kappa \in 1 \to \text{Cls}$, and hence that $\text{Reg}_\kappa \rceil \text{Dist}_\kappa \in 1 \to \text{Cls}$. Now let us suppose that $\text{Dist}_\kappa\text{`}\mu = \text{Dist}_\kappa\text{`}\nu \neq \dot{\Lambda}$: *i.e.* that $\text{Dist}_\kappa\text{`}\mu$ and $\text{Dist}_\kappa\text{`}\nu$ are identical, and belong to Reg_κ. This gives, by definition,

$$\breve{\text{Meas}}_\kappa \,;\, \mu +_a = \breve{\text{Meas}}_\kappa \,;\, \nu +_a \neq \dot{\Lambda}$$

From this we deduce

$$(\exists R, S) : (\exists \varpi, \rho) \,.\, R\, \breve{\text{Meas}}_\kappa\, \varpi \,.\, \varpi (\mu +_a) \rho \,.\, \rho\, \text{Meas}_\kappa\, S :$$

$$(\exists \sigma, \tau) \,.\, R\, \breve{\text{Meas}}_\kappa\, \sigma \,.\, \sigma (\nu +_a) \tau \,.\, \tau\, \text{Meas}_\kappa\, S,$$

whence we obtain

$$(\exists R, S, \varpi, \rho, \sigma, \tau) \,.\, \varpi\, \text{Meas}_\kappa\, R \,.\, \sigma\, \text{Meas}_\kappa\, R \,.\, \rho\, \text{Meas}_\kappa\, S \,.\, \tau\, \text{Meas}_\kappa\, S \,.$$

$$\varpi = \mu +_a \rho \,.\, \sigma = \nu +_a \tau \,.$$

Since $\text{Meas}_\kappa \in 1 \to \text{Cls}$, and since $D\text{`}\text{Meas}_\kappa \subset C\text{`}\Theta'$, by definition, it follows from $*310-123$, $*312-55-41$ of the *Principia* that $\mu = \nu$. Hence

* A condition which will make Meas_κ one-one is

$$\text{Ind}_\kappa \upharpoonright \kappa \in 1 \to 1 :: (\exists x) :. \, y \in s\text{`}C\text{``}\kappa \,.\, \supset_y : (\exists R) : xRy : zSy \,.\, S \in \kappa \,.\, \supset_{z,S} \,.\, S\, (\breve{\text{Ind}}_\kappa \,;\, \Theta')\, R$$

$\text{Ind}_\kappa \upharpoonright \kappa \in 1 \to 1$ is true if, for example, the members of κ, arranged in the natural order of their indices, form a series, as we should naturally expect when they are *e.g.* intervals between sensation-intensities defined as in the "Studies in Synthetic Logic" (not between sensations). The second part of this condition amounts to the assumption that there is a member of $s\text{`}C\text{``}\kappa$ such that if $y$ is another member of $s\text{`}C\text{``}\kappa$, there is a member of κ which relates this term to y, whose κ-index is less than that of y.

O 2

$\mathrm{Reg}_\kappa \upharpoonright \overset{\smile}{\mathrm{Dist}_\kappa} \epsilon 1-1$. Since Reg_κ is the domain of this relation we can always speak of $\overset{\smile}{\mathrm{Dist}_\kappa}$ ' R if $R \epsilon \mathrm{Reg}_\kappa$.

It is easy to prove that if R, S and R|S are all members of Reg_κ, the measure of R|S is the sum of the measures of R and of S, or, in symbols,

$$(17) \quad \vdash : R, S, R|S \,\epsilon\, \mathrm{Reg}_\kappa . \supset_{R,S,\kappa} . \overset{\smile}{\mathrm{Dist}_\kappa} \text{'} (R|S) = \overset{\smile}{\mathrm{Dist}_\kappa} \text{'} R +_a \overset{\smile}{\mathrm{Dist}_\kappa} \text{'} S$$

As a consequence of this, since $\mathrm{Reg}_\kappa \upharpoonright \mathrm{Dist}_\kappa \epsilon 1 \rightarrow 1$ we get

$$(18) \quad \vdash : R, S, R|S, S|R \,\epsilon\, \mathrm{Reg}_\kappa . \supset . R|S = S|R$$

Relative multiplication, then, applied to members of Reg_κ, is not only an associative operation, but a commutative operation, and may be regarded, roughly, as a kind of addition. This shows that the old opinion that only extensive magnitudes are subject to a commutative, associative operation of addition is erroneous. We can not only get a commutative, associative operation of addition which will apply to vectors such as the members of Reg_κ, but also one which will apply to sensation-intensities, as we may now call the members of $D \text{'} \overleftarrow{\mathrm{Meas}_\kappa}$, where κ is a class of separated intensity-vectors, since they are classes of all the sense-data whose intensity has a given measure. This operation is simply the operation on α and β which gives the γ such that the measure of the members of γ is the sum of the measure of the members of α and the measure of the members of β.

Another consequence of (17) is that if we define μ_κ as the relation which holds between R and S when they both belong to Reg_κ and

$$\overset{\smile}{\mathrm{Dist}_\kappa} \text{'} R = \mu \times_a \overset{\smile}{\mathrm{Dist}_\kappa} \text{'} S$$

—that is, if we put

$$(19) \quad \mu_\kappa = \{\mathrm{Dist}_\kappa ; \mu \times_a\} \upharpoonright \mathrm{Reg}_\kappa \quad \mathrm{Df}$$

we shall have

$$(20) \quad \vdash : (\mu_\kappa \text{'} R) | (\nu_\kappa \text{'} R) = \varpi_\kappa \text{'} R . \supset . \mu +_a \nu = \varpi$$

or, more generally,

$$(21) \quad \vdash : \exists ! [(\mu_\kappa \text{'} R) | (\nu_\kappa \text{'} R)] \dot{\cap} \varpi_\kappa \text{'} R . \supset . \mu +_a \nu = \varpi$$

Moreover, since $\mathrm{Reg}_\kappa \upharpoonright \mathrm{Dist}_\kappa \, \varepsilon \, 1 \rightarrow 1$, and $(\mu \times_a) | (\nu \times_a) = (\mu \times_a \nu) \times_a$, we get

$$(22) \quad \vdash : \exists \, \mu_\kappa | \nu_\kappa \dot{\cap} \varpi_\kappa . \supset . \mu \times_a \nu = \varpi$$

These propositions are the rough analogues of $*$ 356—33—54 in the *Principia*.

Let us next bring the discussion of measurement that we have just finished into correlation with that part of our theory that derived Vs_ϕ from ϕ. If ϕ be such a relation as, "The interval of saturation in color between x and y seems less than that between u and v," the natural unit of measurement, as we said in the introduction, would seem to be a complete saturation, and the method of measurement which we have just elaborated would seem fairly natural and appropriate. The measure of the saturation of a given sensation x, will then be $Meas_{Vs_\phi}{}^{\backprime}x$. It will be natural to say that two sensations have the same degree of chroma if they bear to one another the relation $\breve{Meas}_{Vs_\phi} | Meas_{Vs_\phi}$. This leads to the following definitions

$$(23) \qquad Qual_\phi = D{}^{\backprime}\overleftarrow{Meas}_{Vs_\phi} \quad Df$$

$$(24) \qquad Mag_\phi = Meas_{Vs_\phi} \, \epsilon \upharpoonright Qual_\phi \quad Df$$

$Qual_\phi$ is the class of all degrees of saturation, in the particular case just discussed. It consists, however, of degrees of saturation considered as quantities, and though it may *de facto* coincide with the λ_R of the *Studies in Synthetic Logic*, if R be taken to be the relation, "noticeably more saturated than," or with $D{}^{\backprime}sg{}^{\backprime}(s{}^{\backprime}Vs_\phi)_{se}$, it cannot, without the aid of complicated logical hypotheses, be proved identical with either.

Mag_ϕ is the relation of the magnitude of a degree of saturation to the degree of saturation, and is one-one. If we so desire, we can define $Dist_\phi$ as $\hat{R}\hat{\mu}\{R = \breve{Mag}_\phi ; \mu +_a\}$ Reg_ϕ as $D{}^{\backprime}Dist_\phi - \iota{}^{\backprime}\dot{\Lambda}$, and μ_ϕ as

$$\{Dist_\phi ; \mu \times_a\} \upharpoonright Reg_\phi$$

and the propositions

$$\vdash : \dot{\exists} ! [(\mu_\phi{}^{\backprime}R) | (\nu_\phi{}^{\backprime}R)] \dot{\cap} \varpi_\phi{}^{\backprime}R . \supset . \mu +_a \nu = \varpi,$$

and

$$\vdash : \dot{\exists} ! \mu_\phi | \nu_\phi \dot{\cap} \varpi_\phi . \supset . \mu \times_a \nu = \varpi,$$

will be universally valid. Reg_ϕ will consist of all intervals, positive, zero, and negative, between degrees of saturation, and if R be any such interval, its measure will be $\breve{Dist}_\phi{}^{\backprime}R$. μ_ϕ will be the proportion between a member of Reg_ϕ and another, one μ-th of its size.

3. In our previous work we have, roughly speaking, made the maximum of the magnitudes of all members of κ our real unit of measurement: let us see how this measurement may be adapted to measurement by smaller units. The way to do this is to select a certain region of our

scale of measurement which has a definite maximum, to measure the intervals lying in this region by the method given above, and to measure all intervals in general by finding fractions of them lying in this interval. To this end, we make the following definitions

(25) $\mu \div \lambda = \mu \times_{\alpha} \mathrm{H}``(1/\lambda)$ Df

(26) $\mathrm{Inx}_{\kappa, \alpha} = \hat{\mu}\hat{\mathrm{R}}\{(\exists \lambda, \mathrm{S}) . (\mu \div \lambda) \, \mathrm{Ind}_{\upharpoonright \alpha `` \kappa - \iota ` \Lambda} \, \mathrm{S} \upharpoonright \alpha . \mathrm{S}^{\lambda} = \mathrm{R}\}$ Df

Here we use as our standard of measurement the greatest distance between members of α, measured by members of κ, and not the greatest distance between members of $s`\mathrm{C}``\kappa$. $\mathrm{Inx}_{\kappa, \alpha}$ is not, in general, a one-many relation: its converse domain is $s`\mathrm{Pot}``\{\kappa \cap \hat{\mathrm{R}}\{\exists ! \mathrm{R} \upharpoonright \alpha\}\}$ and its domain is included in the class of real numbers. We next define $\mathrm{Meas}_{\kappa, \alpha}$ in terms of $\mathrm{Inx}_{\kappa, \alpha}$, as we defined Meas_{κ} in terms of Ind_{κ}, in the following manner

$$\mathrm{Meas}_{\kappa, \alpha} = \{\mathrm{limax}_{\Theta} \mid (\mathrm{Inx}_{\kappa, \alpha})_{\epsilon} \mid (\breve{\mathrm{G}})_{\epsilon} \mid \overleftarrow{\varepsilon}\} \upharpoonright s`\mathrm{C}``\mathrm{Pot}``\{\kappa \cap \hat{\mathrm{R}}\{\dot{\exists} ! \mathrm{R} \upharpoonright \alpha\}\} \quad \mathrm{Df} \qquad (27)$$

Notwithstanding the fact that $\mathrm{Inx}_{\kappa, \alpha}$ is not, in general, one-many, $\mathrm{Meas}_{\kappa, \alpha}$ is always necessarily one-many.

We are now in a position to define $\mathrm{Dist}_{\kappa, \alpha}$, $\mathrm{Reg}_{\kappa, \alpha}$, and $\mu_{\kappa, \alpha}$ just as we defined Dist_{κ}, Reg_{κ}, and μ_{κ}: that is, as follows

(28) $\mathrm{Dist}_{\kappa, \alpha} = \hat{\mathrm{R}}\hat{\mu}\{\mathrm{R} = \breve{\mathrm{Meas}}_{\kappa, \alpha} ; \mu +_{\alpha}\}$ Df

(29) $\mathrm{Reg}_{\kappa, \alpha} = \mathrm{D}`\mathrm{Dist}_{\kappa, \alpha} - \iota`\Lambda$ Df

(30) $\mu_{\kappa, \alpha} = \{\mathrm{Dist}_{\kappa, \alpha} ; \mu x_{\alpha}\} \upharpoonright \mathrm{Reg}_{\kappa, \alpha}$ Df

All the theorems which we proved concerning Meas_{κ}, Dist_{κ}, Reg_{κ}, and μ_{κ} will remain valid if we alter κ every time that it occurs as a subscript to κ, α.

Now, if $\phi(x, y, u, v)$ stands for "the brightness-interval between x and y seems less than that between u and v," the theory of *sensation*-measurement developed in § 2, though it is still applicable, does not seem appropriate nor natural: we do not usually take the greatest possible brightness, say, that of the sun, as our standard of brightness. Our usual way of measuring brightness seems to be, perhaps, to count the number of "thresholds" which lie between it and utter darkness. In all ordinary psychological work, this method of measurement is adequate, but it will not enable us to subdivide the step from one limen to the next into equal parts. Nevertheless, the limen is a very natural standard to use in the measurement of brightnesses. Since we do not wish to assume that all steps from a given sense-datum to a just brighter one are equal, we shall make the first interliminal difference our standard of measurement. We do this by taking Vs_ϕ as the κ and the class of sense-data lying in the first interliminal space as the α of the theory of measurement which we have just developed. Now, a sense-datum lies in the first interliminal space with respect to brightness, or is of subliminal brightness, if no interval leading to it from below is supraliminal: that is, if it belongs to the field of $\dot{s}\,`\hat{\mathrm{R}}\{(\exists x).(x \downarrow x)\,\mathrm{Cp}_\phi\,\mathrm{R}\}$, but not to its converse domain. Let us define this class, and certain functions of it, as follows

(31) $$\mathrm{Sbl}_\phi = \overrightarrow{\mathrm{B}}\,`\dot{s}\,`\hat{\mathrm{R}}\{(\exists x).(x \downarrow x)\,\mathrm{Cp}_\phi\,\mathrm{R}\} \quad \mathrm{Df}$$

(32) $$\mathrm{Quant}_\phi = \mathrm{D}\,`\mathrm{Gs}\,`\mathrm{Meas}_{\mathrm{Vs}_\phi,\,\mathrm{Sbl}_\phi} \quad \mathrm{Df}$$

(33) $$\mathrm{Mgn}_\phi = \mathrm{Meas}_{\mathrm{Vs}_\phi,\,\mathrm{Sbl}_\phi} \mid \epsilon \upharpoonright \mathrm{Quant}_\phi \quad \mathrm{Df}$$

Sbl_ϕ is, in the case just considered, the class of all sensations of subliminal brightness. Qnant_ϕ is the class of all degrees of brightness. $\mathrm{Mgn}_\phi\,`\xi$ is the quantitative measure of ξ in terms of the brightness-limen, if ξ is a brightness. The further development of this theory of measurement runs precisely parallel to that of measurement in terms of the maximum interval.

4. The question which we have next to settle is that of the relation of our theory of measurement to that of the *Principia*. It is obvious that since the theory of the *Principia* applies only to kinds of measurement without definite maxima, while our theory of measurement, in the form it takes in § 2, only applies to kinds of measurement with definite maxima, the form of our theory with which we must compare that of the *Principia* is that developed in § 3, which need not only apply to kinds of measurement with definite maxima. To carry out this comparison, we need to set some standard problem, and compare the answers which the two systems give to it. Now, the theory in the *Principia* starts from a certain sort of class of relations or vectors κ, and culminates in the definition of the real proportion X among members of κ as X_κ. Our definition of the real proportion μ, among members of $\mathrm{Reg}_{\kappa,\alpha}$ is $\mu_{\kappa,\alpha}$, and therefore a natural definition of the real proportion μ, among members of κ will be $\mu_{\kappa,\alpha} \upharpoonright \kappa$. The fact that the real proportion in the *Principia* appears in

the form of a function of a relation, while ours appears as a function of a class, is due to the fact that the real numbers which are applied there are the $\dot{s}$'s of the real numbers we apply. The natural way to put the question, "When does the theory of measurement developed in the *Principia* give the same results as that developed in § 3 of the paper?" is "When is the formula $\mu_{\kappa, a} \upharpoonright \kappa = (\dot{s}\,`\,\mu)_\kappa$ valid?"

Now, if κ is a connected family, containing all the powers of its members, it follows from $*331-24-42$ that

$$(\mu/\nu)_\kappa = \hat{R}\hat{S}\{(\exists P) \cdot P\epsilon\kappa \cdot R = P^\mu \cdot S = P^\nu\},$$

or, *in extenso*,

$$(34) \qquad \vdash : \kappa\epsilon \mathrm{FM\ conx} \,.\, s\,`\,\mathrm{Pot}``\kappa \subset \kappa \,.\supset\cdot$$

$$(\mu/\nu)_\kappa = \hat{R}\hat{S}\{(\exists P) . P\epsilon\kappa \cdot R = P^\mu . S = P^\nu\}.$$

Since, if $R = P^\mu$ and $S = P^\nu$, $R^\nu = P^{\mu \times \nu} = S^\mu$, and since, under the above conditions, R, S, and R^ν, being powers of a member of κ, are members of κ, and hence not $\dot{\Lambda}$, we derive from $*302-02$ and $*303-01$ the conclusion,

$$(35) \qquad \vdash : \mathrm{Hp}\,(34) \,.\supset\cdot\, (\mu/\nu)_\kappa \,\mathsf{G}\, (\mu/\nu) \upharpoonright \kappa$$

From (34), (35), and $*333-42$, $*351-1$, we get

$$(36) \qquad \vdash : \mathrm{Hp}\,(34) \cdot \kappa\epsilon \mathrm{FM\ ap\ submult} \,.\supset\cdot\, (\mu/\nu)_\kappa = (\mu/\nu) \upharpoonright \kappa$$

Furthermore, it results from (34) and $*336-41$ that

$$(37) \qquad \vdash : \mathrm{Hp}\,(34) \cdot R\epsilon\kappa_\partial \,.\supset .\, (\mu/\nu)_{\overleftarrow{U}_\kappa `R} = (\mu/\nu)_\kappa \upharpoonright U_\kappa\,`R$$

Moreover, it is easy to prove that if we call $p\,`\,\overleftarrow{\dot{s}}\,`\kappa``D\,`R$ by the name $a$, and if $P \upharpoonright a$ and $Q \upharpoonright a$ exist, then to say that P bears the relation $(\mu/\nu)_\kappa \upharpoonright \overleftarrow{U}_\kappa\,`R$ to Q is equivalent to saying that $P \upharpoonright a$ bears the relation $(\mu/\nu)_{\upharpoonright a``\overleftarrow{U}_\kappa `R}$ to $Q \upharpoonright a$, under the hypotheses that κ is a connected family containing all the powers of its members, and that R belongs to κ_∂. In symbols we have

$$(38) \qquad \vdash :. \mathrm{Hp}\,(34) \,.\, R\epsilon\kappa_\partial \cdot a = p\,`\,\overleftarrow{\dot{s}}\,`\kappa_\partial``D\,`R \,.\, \dot{\exists}\,!\, P \upharpoonright a \cdot \dot{\exists}\,!\, Q \upharpoonright a :\supset:$$

$$P\{(\mu/\nu)_\kappa \upharpoonright \overleftarrow{U}_\kappa\,`R\}\,Q \,.\equiv\cdot\, (P \upharpoonright a)(\mu/\nu)_{\upharpoonright a``\overleftarrow{U}_\kappa `R}(Q \upharpoonright a).$$

Putting (38) and (36) together, we get

(39) $\vdash :. \mathrm{Hp}(38) . \mathrm{Hp}(36) :\supset: \mathrm{P} \{(\mu/\nu) \upharpoonright \overleftarrow{\mathrm{U}_\kappa} ‘\mathrm{R}\} \mathrm{Q} .$

$\equiv . (\mathrm{P} \upharpoonright \alpha)(\mu/\nu)_{\upharpoonright\alpha“\overleftarrow{\mathrm{U}_\kappa}\mathrm{R}} (\mathrm{Q} \upharpoonright \alpha).$

With the end of (39), (11), and (12), we obtain

(40) $\vdash :. \mathrm{Hp}(39) :\supset: \mathrm{P} \{[\mathrm{Eq}_{\upharpoonright\alpha“\overleftarrow{\mathrm{U}_\kappa}\mathrm{R}} ‘(\mu/\nu)_{\upharpoonright\alpha“\overleftarrow{\mathrm{U}_\kappa}\mathrm{R}}] \upharpoonright \overleftarrow{\mathrm{U}_\kappa}‘\mathrm{R}\} \mathrm{Q} .$

$\equiv . \mathrm{P} \{\upharpoonright\alpha^{\dagger}(\mu/\nu)_{\upharpoonright\alpha“\overleftarrow{\mathrm{U}_\kappa}‘\mathrm{R}}\} \mathrm{Q}$

whence we deduce

(41) $\vdash : \mathrm{Hp}(36) . \mathrm{R}\epsilon\kappa_\partial . \alpha = p‘\overleftarrow{\dot{s}}‘\kappa_\partial“\mathrm{D}‘\mathrm{R} . \supset .$

$(\mathrm{Eq}_{\upharpoonright\alpha“\overleftarrow{\mathrm{U}_\kappa}‘\mathrm{R}}‘\mathrm{S}) \upharpoonright \{\overleftarrow{\mathrm{U}_\kappa}‘\mathrm{R} \cap \hat{\mathrm{X}}\{\dot{\exists} ! \mathrm{X} \upharpoonright \alpha\}\} = \{(\upharpoonright\alpha)^{\dagger}\lceil\mathrm{D}‘\mathrm{R}t_{\upharpoonright\alpha“\overleftarrow{\mathrm{U}_\kappa}‘\mathrm{R}}\}‘\mathrm{S}$

The question now arises, when will it be true that if $\mathrm{RU}_\kappa\mathrm{T}$,

$$\dot{\exists} ! \mathrm{T} \upharpoonright p‘\overleftarrow{\dot{s}}‘\kappa_\partial“\mathrm{D}‘\mathrm{R}?$$

Now, if κ is an initial family, it is not difficult to deduce from $*335-15$ and the definition of a vector-family that $p‘\overleftarrow{\dot{s}}‘\kappa_\partial“\mathrm{D}‘\mathrm{R}$ is made up of all those things which bear such a relation as $\breve{\mathrm{P}}$ to everything bearing such a relation as $\mathrm{R}|\mathrm{Q}$ to $\mathrm{init}‘\kappa$, where $\mathrm{P}\epsilon\kappa_\partial$ and $\mathrm{Q}\epsilon\kappa$. Now, by $*335-17$, the relative product of two members of an initial family is a member of the same family (see also $*331-23$), and hence, since

$$x\breve{\mathrm{P}}|\mathrm{R}\ \mathrm{init}‘\kappa . \supset . x\breve{\mathrm{P}}|\breve{\mathrm{Q}}|\mathrm{R}|\mathrm{Q}\ \mathrm{init}‘\kappa,$$

if $x\breve{\mathrm{P}}|\mathrm{R}\,\mathrm{init}‘\kappa$ and if whenever $\mathrm{Q}\epsilon\kappa$, $\mathrm{P}|\mathrm{Q} \subset \mathrm{J}$, then $x \epsilon p‘\overleftarrow{\dot{s}}‘\kappa_\partial“\mathrm{D}‘\mathrm{R}$. Therefore, by $*335-21$, $\overrightarrow{\breve{\mathrm{P}}|\mathrm{R}}‘\mathrm{init}‘\kappa \subset p‘\overleftarrow{\dot{s}}‘\kappa_\partial“\mathrm{D}‘\mathrm{R}$. Therefore, since $\mathrm{R}\epsilon\kappa_\partial$, $\mathrm{init}‘\kappa \epsilon p‘\overleftarrow{\dot{s}}‘\kappa“\mathrm{D}‘\mathrm{R}$. Moreover, since all initial families are connected by $*336-41$, if $\mathrm{RU}_\kappa\mathrm{T}$ there is a member of κ_∂, say S, such that $\mathrm{R} = \mathrm{S}|\mathrm{T}$. Since $\kappa \subset 1 \to 1$ and all the members of κ have a common converse domain, $\mathrm{T} = \breve{\mathrm{S}}|\mathrm{R}$ and hence $\mathrm{T}‘\mathrm{init}‘\kappa$ is the same as $\breve{\mathrm{S}}‘\mathrm{R}‘\mathrm{init}\,\kappa$, and consequently belongs to $p‘\overleftarrow{\dot{s}}‘\kappa_\partial“\mathrm{D}‘\mathrm{R}$. Therefore $\dot{\exists} ! \mathrm{T} \upharpoonright p‘\overleftarrow{\dot{s}}‘\kappa_\partial“\mathrm{D}‘\mathrm{R}$. This gives us, by (41), $*335-17$, and the definition of FM init,

(42) $\vdash : \kappa\epsilon\mathrm{FM\ ap\ init\ submult} . \mathrm{R}\epsilon\kappa_\partial . \alpha = p‘\overleftarrow{\dot{s}}‘\kappa_\partial“\mathrm{D}‘\mathrm{R} . \supset .$

$\upharpoonright\overleftarrow{\mathrm{U}_\kappa}‘\mathrm{R}|\,\mathrm{Eq}_{\upharpoonright\alpha“\overleftarrow{\mathrm{U}_\kappa}‘\mathrm{R}} = (\upharpoonright\alpha)^{\dagger}\lceil\mathrm{D}‘\mathrm{R}t_{\upharpoonright\alpha“\overleftarrow{\mathrm{U}_\kappa}‘\mathrm{R}}$

If $\dot{s}$ 'κ_∂ is a series, then no term can at once belong to D'R and to p '$\overleftarrow{\dot{s}}$ 'κ_∂ "D'R, by *202—503. It may be deduced from *336—41, moreover, that if P$\epsilon\overleftarrow{\mathrm{U}}_\kappa$'R, D'P$\subset$D'R. From this and what we have said above, we may conclude that $\lceil\alpha$"$\overleftarrow{\mathrm{U}}_\kappa$'R $= \lceil\alpha$"$\kappa-\iota$'$\dot{\Lambda}$. From this, (42), and *334—32, we get

(43) $\vdash : \kappa\epsilon$FM sr init submult. R$\epsilon\kappa_\partial$. $\alpha = p$ '$\overleftarrow{\dot{s}}\kappa$"D'R . $\supset$.

$$\lceil\overleftarrow{\mathrm{U}}_\kappa\text{'R} \mid \mathrm{Eq}_{\lceil\alpha\text{"}\kappa-\iota\text{'}\dot{\Lambda}} = (\lceil\alpha)^\dagger \upharpoonright \text{D'R}t_{\lceil\alpha\text{'}\kappa-\iota\text{'}\dot{\Lambda}}$$

Now, it follows from this and (13) that Ind$_{\lceil\alpha\text{"}\kappa-\iota\text{'}\dot{\Lambda}}$'S is the class of all those ratios which are less than any ratio which the member of $\overleftarrow{\mathrm{U}}_\kappa$'R of which S forms a part bears to any member of $\overleftarrow{\mathrm{U}}_\kappa$'R: by *352—72, then, if κ be a serial family, as we have supposed in (43), the member of $\overleftarrow{\mathrm{U}}_\kappa$'R containing S bears to R (or to any member of $\overleftarrow{\mathrm{U}}_\kappa$'R) no ratio which is a member of Ind$_{\lceil\alpha\text{"}\kappa-\iota\text{'}\dot{\Lambda}}$'S, and no member of κ preceding the one containing S bears R such a ratio, while, by the definition of Ind$_{\lceil\alpha\text{"}\kappa-\iota\text{'}\dot{\Lambda}}$ and *352—72, any member of κ following the one containing S does bear R such a ratio, if any. Hence it is easy to prove that if the rational multiples of R form a median class (*271) of U$_\kappa$, ($\dot{s}$'Ind$_{\lceil\alpha\text{"}\kappa-\iota\text{'}\dot{\Lambda}}$)$_\kappa$ is a relation which holds between the member of κ containing S and R, if we understand the use of κ as a subscript in the sense defined in *356—01. With the aid of *351—11, *336—44, *270—4, *271—15, we arrive at the conclusion,

(44) $\vdash : \kappa\epsilon$FM sr init. (C'R$_\kappa$) med U$_\kappa$. Cnv'$\dot{s}$'$\kappa_\partial\,\epsilon$ semi Ded.

$$\alpha = p\text{'}\overleftarrow{\dot{s}}\text{'}\kappa_\partial\text{"D'R} . \mathrm{Ind}_{\lceil\alpha\text{"}\kappa-\iota\text{'}\dot{\Lambda}}\text{'S} \neq \Lambda . \mathrm{T}\epsilon\kappa_\partial . \mathrm{S}\,\mathrm{G}\,\mathrm{T} . \supset .$$

$$\mathrm{T}\,(\dot{s}\text{'}\mathrm{Ind}_{\lceil\alpha\text{"}\kappa-\iota\text{'}\dot{\Lambda}}\text{'S})\,\mathrm{R}$$

Now, if it is true that by repeating any vector belonging to κ_∂ a sufficient number of times, we can get beyond any given vector belonging to κ, we may replace Ind$_{\lceil\alpha\text{"}\kappa-\iota\text{'}\dot{\Lambda}}$'S $\neq \Lambda$. S G T in the above formula by

$$\mathrm{S} = \mathrm{T}\lceil\alpha \neq \Lambda,$$

as follows readily enough from *352—72, *336—64. That this repetition is possible is a consequence of *337—13 and *336—01—011. We thus obtain the following theorem:

(45) $\vdash : \kappa \epsilon \mathrm{FM} \text{ sr init} \cdot (\mathrm{C}`\mathrm{R}_\kappa) \text{ med } \mathrm{U}_\kappa \cdot \mathrm{Cnv}`s`\kappa_\partial \epsilon \text{ semi Ded} .$

$$\alpha = p`\overleftarrow{\dot{s}}`\kappa_\partial``\mathrm{D}`\mathrm{R} . \mathrm{T}\epsilon\kappa_\partial \cdot \mathrm{S} = \mathrm{T} \upharpoonright \alpha \neq \dot{\Lambda} . \supset .$$

$$\mathrm{T}(\dot{s}`\mathrm{Ind}_{\upharpoonright \alpha``\kappa - \iota`\dot{\Lambda}}`\mathrm{S})_\kappa \mathrm{R}$$

Now, we have seen that $\mathrm{T} \upharpoonright \alpha \neq \dot{\Lambda} . \equiv . \mathrm{RU}_\kappa \mathrm{T}$, under the above conditions. I wish to show that, given any vector belonging to κ, there is some member of $\overleftarrow{\mathrm{U}_\kappa}`\mathrm{R}$, say Q, such that $\mathrm{P}\epsilon\mathrm{Pot}`\mathrm{Q}$. We prove this as follows: by $*337-13$ and $*336-01-011$, there is some power of R, say R^ν, which bears the relation U_κ to P. By $*337-27$, there is a member of κ, say M, such that $\mathrm{M}^\nu = \mathrm{P}$. This M is the Q we want, since by $*336$, if κ is a serial family, and ν a positive integer, $\mathrm{RU}_\kappa \mathrm{S}$ is equivalent to $\mathrm{R}^\nu \mathrm{U}_\kappa \mathrm{S}^\nu$. Hence, if $\mathrm{P}\epsilon\kappa_\partial$, it follows from $*356-33$ and a little elementary arithmetic that the class $\overrightarrow{\mathrm{Inx}_{\kappa, \alpha}}`\mathrm{P}$ contains just two members—the real number which, when applied, relates P to R—and that if $\mathrm{P} \sim \epsilon \kappa_\partial$, $\mathrm{Inx}_{\kappa, \alpha}`\mathrm{P}$ contains Λ alone, $\mathrm{Meas}_{\kappa, \alpha}`x$ will then be the upper limit or maximum of the measures in terms of R of the vectors leading up to x. Since these measures, by $*356-63$, are themselves each the limit or maxima of the class applied rational numbers which the rational multiples of R less than some vector leading up to x bear to R, $\mathrm{Meas}_{\kappa, \alpha}`x$ may easily be shown to be the class of all the ratios which vectors which connect members of $\overleftarrow{\dot{s}}`\kappa_\partial`x$ bear to R, and consequently to be a real number (cf. $*336-41$, $*352-72$), such that, when we apply it to $\kappa$, the vector leading from $x$ to init$`\kappa$ bears it to R. In symbols, we have

(46) $\vdash : \mathrm{Hp}\,(45) \cdot x \epsilon s`\mathrm{C}``\kappa . \supset .$

$$[(\kappa \rceil \breve{\mathrm{C}})`(x \downarrow \mathrm{init}`\kappa)](\dot{s}`\mathrm{Meas}_{\kappa, \alpha}`x)_\kappa \mathrm{R}$$

Now, it follows from the assumption $(\mathrm{C}`\mathrm{R}_\kappa)$ med $\mathrm{U}_\kappa$ that every member of $\kappa$ has a real measure in terms of R. Since $\dot{s}`\kappa_\partial$ is serial, and since, by $*331-22$, $\mathrm{I} \upharpoonright s`\mathrm{G}``\kappa\epsilon\kappa$, it follows from the fact that $s`\mathrm{G}``\kappa = s`\mathrm{C}``\kappa$ in every vector family that any two members of $s`\mathrm{C}``\kappa$ are connected by a member of κ. Hence we may deduce from $*356-26-54$ that if μ is positive, $\mathrm{Dist}_{\kappa, \alpha}`\mu \epsilon \kappa_\partial$ if $\mu$ is zero, $\mathrm{Dist}_{\kappa, \alpha}`\mu = \mathrm{I} \upharpoonright s`\mathrm{G}``\kappa$, while if $\mu$ is negative, $\mathrm{Dist}_{\kappa, \alpha}`\mu \epsilon \mathrm{Cnv}``\kappa_\partial$. From this, by $*356-26$, and the fact that vectors standing in the relation x_κ to R always exist if κ is initial, serial, submultipliable, and semi-Dedekindian, and x is a non-negative relational real number, we may deduce the conclusion

$$(47) \qquad \vdash : \mathrm{Hp}\,(45) . \supset . \mathrm{Reg}_{\kappa, a} = \kappa \cup \mathrm{Cnv}\,``\,\kappa$$

Another easily proved theorem is

$$(48) \qquad \vdash : \mathrm{Hp}\,(45) . \supset . \mathrm{P}(s\,\text{`}\,\overbrace{\mathrm{Dist}_{\kappa, a}}\,\text{`}\,\mathrm{P})_\kappa\, \mathrm{R}$$

By (30) and $*356-33-26$, if μ is a positive or zero real number, and $\mathrm{P}\,(\mu_{\kappa, a} \upharpoonright \kappa)\,\mathrm{Q}$, then we have, in the sense of $*356$, $\mathrm{P}\,(\dot{s}\,\text{`}\,\mu)_\kappa\,\mathrm{Q}$. The converse of this is easily proved by the same theorems, since P and Q always bear some applied real number to R: that is, we have

$$(49) \qquad \vdash : \mathrm{Hp}\,(45) . \mu \epsilon \mathrm{C}\,\text{`}\,\Theta' . \supset . \mu_{\kappa, a} \upharpoonright \kappa = (\dot{s}\,\text{`}\,\mu)_\kappa$$

This is the theorem we set out to prove, and it establishes the fact that in an initial, serial, semi-Dedekindian family, if the rational multiples of a given vector form a median class of the series of vectors, the system of measurement defined in the *Principia* gives substantially the same results as the system defined in this paper, if, to put it crudely, we take this given vector whose rational multiples form a median class of the series of vectors as a unit, by making the class of all the things which follow every member of its domain in the series generated by the vectors the a of our previous work.*

5. In conclusion, let us consider what bearing all this work of ours can have on experimental psychology. One of the great defects under which the latter science at present labours is its propensity to try to answer questions without first trying to find out just what they ask. The experimental investigation of Weber's law is a case in point: what most experimenters do take for granted before they begin their experiments is infinitely more important and interesting than any results to which their experiments lead. One of these unconscious assumptions is that sensations or sensation-intervals can be measured, and that this process of measurement can be carried out in one way only. As a result, each new experimenter would seem to have devoted his whole energies to the invention of a method of procedure logically irrelevant to everything that had gone before: one man asks his subject to state when two intervals between sensations of a given kind appear different; another bases his

* It will be noticed that in such a vector family, the series of vectors, arranged in order of magnitude, is of the form $(\overset{\smile}{+\dot{1}})\,\vartheta$.

whole work on an experiment where the observer's only problem is to divide a given colour-interval into two equal parts, and so on indefinitely, while even where the experiments are exactly alike, no two people choose quite the same method for working up their results. Now, if we make a large number of comparisons of sensation-intervals of a given sort with reference merely to whether one seems larger than another, the methods of measurement given in this paper indicate perfectly unambiguous ways of working up the results so as to obtain some quantitative law such as that of Weber, without introducing such bits of mathematical stupidity as treating a "just noticeable difference" as an "infinitesimal," and have the further merit of always indicating *some* tangible mathematical conclusion, no matter what the outcome of the comparisons may be.

MASSACHUSETTS INSTITUTE OF TECHNOLOGY.

Comments on [14b], [15a], [21a]

H. E. Kyburg

These three papers are concerned primarily with the problem of deriving a series of magnitudes characterizing a kind of phenomenon from such primitive and epistemologically unproblematical relations as "seems louder than."

Consider a series (ordered set) such as the real numbers, the points on a line, or the instants of time. Wiener in the first of these papers [14b] explores the relationship between the properties of the usual ordering relation R (for example, greater than) on such a set, and the relation that obtains between two intervals in such a series when every member of one bears R to every member of the other. This relation of complete sequence is interesting in its own right and also has important consequences for the theory of measurement.

In the first paper [14b], the relation of complete sequence is defined as any relation R satisfying

(i) $x\,R\,y \rightarrow x \neq y,$

(ii) $x\,R\,y \wedge (\sim y\,R\,z \wedge \sim z\,R\,y) \wedge z\,R\,w \rightarrow x\,R\,w.$

The relation $\sim y\,R\,z \wedge \sim z\,R\,y$, corresponding to the intuitive relation of "overlapping," is called "simultaneity" and denoted by R_{se}.

From a philosophical point of view, it may be maintained that it is more fundamental to derive the series (of instants of time, of points on a line) from consideration of intervals and the relation of complete sequence, than the converse.

This Wiener proceeds to do. An "instant" x is a set of intervals which is the intersection of all the sets of sets of intervals each of which bears R_{se} to some member of x:

$$\tau_R = \hat{x}(x = p\,\text{`}R_{se}\,\text{``}x).$$

In more familiar notation, this would be

$$\tau_R = \{x : x = \cap\, \{y : \vee\, w(\omega \epsilon x \wedge y = \{z : z\,R_{se}\,w\}\,).$$

One instant precedes another when some interval belonging to the one wholly precedes (bears R to) some interval belonging to the other. Thus if inst‘R is the relation of precedence between instants,

$$\text{inst}‘R = (\breve{\epsilon}; R) \upharpoonright \hat{\alpha} \left\{\alpha = p ‘\overrightarrow{R}_{\text{se}} “\alpha\right\},$$

or, more familiarly,

$$\text{inst}\ ‘R = \breve{\epsilon}|R|\, \epsilon \cap \tau_R \times \tau_R .$$

The function *inst* is thus definable in general as

$$\text{inst} = \hat{Q}\hat{P} \left\{Q = (\breve{\epsilon}\,; P) \upharpoonright \tau_P\right\},$$

or

$$\text{inst} = \left\{\langle Q, P\rangle : Q = (\breve{\epsilon}\ |P|\, \epsilon) \cap \tau_P \times \tau_P\right\}.$$

The main theorem of this paper is that the image under *inst* of a relation of complete precedence (that is, a relation satisfying (i) and (ii) above, though only (ii) is required in the formal statement of the theorem) is a series in Russell's sense:

$$\text{inst}“\ \hat{R} \left\{R|R_{\text{se}}|\ R \Subset R\right\} \subset \text{ser}.$$

Two further theorems give conditions under which relations of the sort considered give compact series.

The first part of the second article [15a] is concerned to extend the notions and notations developed in *Principia Mathematica* for binary relations to the case of polyadic relations. When n is greater than 2, various forms of n-transitivity may be specified. Similarly, one may devise various forms of n-connexity for n-adic relations. The net result of Wiener's investigations concerning n-transitivity and n-connexity is this:

Given an n-adic relation P, such that
(a) it has a certain form of n-transitivity
(b) it has a certain form of n-connexity
(c) it never relates a member of its field to itself,
then given any n-adic relation R, we can construct a function of R having these same properties.

This is related to the theory of measurement in that it suggests that we may begin with rather poorly behaved perceptible relations (for example, noticeably louder than), and construct functions that will serve the same purpose in ordering experience, but which will also have such desirable properties as irreflexivity, transitivity, and connexity in some interesting sense. It is these constructed, well-behaved relations, that can serve as the basis for measurement.

The second part of the article [15a] continues to pursue the program of the first article. Construing P now as the relation between two things when one is noticeably brighter than the other, P_s is defined as the relation of being equally bright:

$$P_s = (\overset{\smile}{\overrightarrow{P}}_{se} | \overrightarrow{P}_{se}) \upharpoonright C\text{'}P,$$

or

$$P_s = \{\langle x,y\rangle : \wedge\hat{z}(xP_{se}z \leftrightarrow yP_{se}z)\}.$$

P_s is clearly transitive, reflexive, and symmetrical. λ_P is just the set of equivalence classes generated by P_s:

$$\lambda_P = D\text{'}\overrightarrow{P}_s,$$

or

$$\lambda_P = \{x : \vee y(x = \quad z : zP_s y\})\},$$

or, more intuitively, the set of brightness intensities. Finally, int'P is the relation that holds between two members of λ_P, x and y, when something indistinguishable from a member of x is noticeably brighter than a member of y, that is, when x $(P_{se}|P)$ y. Formally,

$$\text{int} = \hat{Q}\hat{P}\{Q = [\breve{\epsilon}; (P_{se}|P)] \upharpoonright \lambda_P\},$$

or

$$\text{int} = \{\langle Q,P\rangle : Q = \breve{\epsilon}|(P_{se}|P)|\epsilon \cap \lambda_P \times \lambda_P\}.$$

The relation int'P will be transitive when $P_{se}|P$ is transitive, and connected if $P|P_{se} \Subset P_{se}|P$; and thus if both these conditions are satisfied int'P will be a series. Wiener argues that both conditions are in fact met in the case of brightness; but the main interest of his construction is that it reveals certain very important assumptions made implicitly, if not unconsciously, by psychologists who propose to measure the intensities of stimuli.

The third paper [21a] of the set provides a detailed analysis of the way in which one could proceed from apparent relations of greater or less to quantitative measures. Although this work seems extremely suggestive, its suggestions have only relatively rarely been accepted. In many respects the current work in the theory of measurement proceeds in ignorance of the seminal importance of these papers.

Editor's note. For Wiener's own remarks on [14b] and [15a], see [53h], pp. 201 and 212.

From the research of Dr. Grattan-Guinness, referred to in my Introduction, we now learn that Dr. Kyburg's view of the "seminal importance" of these three papers was shared by Bertrand Russell. His referee's report on [21a] to the London Mathematical Society reads as follows:

"This is a paper of very considerable importance, since it establishes a completely valid method for the numerical measurement of various kinds of quantity which have hitherto not been amenable to measurement except by very faulty methods.

Although Dr. Wiener's principles can be applied (as he shows in the later portions of his paper) to quantities of any kind, their chief importance is in respect of such things as intensities, which cannot be increased indefinitely. Much experimental work in psychology, especially in connection with Weber's Law, has been done with regard to intensities and their differences, but owing to lack of the required mathematical conceptions its results have often been needlessly vague and doubtful. So far as I am aware, Dr. Wiener is the first to consider, with the necessary apparatus of mathematical logic, the possibility of obtaining numerical measures of such quantities. His solution of the problem is, so far as I can see, complete and entirely satisfactory. His work displays abilities of high order, both technically and in general grasp of the problem; and I consider it in the highest degree desirable that it should be printed."

THE RELATION OF SPACE AND GEOMETRY TO EXPERIENCE*

I. Geometry as a Schematization of Experience

GEOMETRY is considered by every one to rank among the most certain of sciences. One can have grave doubts, for example, as to the universal validity of any theory in biology, or even honest misgivings concerning the absolute precision of the law of the conservation of energy, but it is hard to imagine a man who is really sincere in questioning the theorem of Pythagoras, that the square on the hypotenuse of a right-angled triangle is equal in area to the sum of the two squares on the legs of the triangle. This conviction which we possess that the theorems of geometry are valid seems essentially independent of any confirmation or substantiation by experience. After we are really initiated into the processes of geometrical reasoning, our certainty of the truth of the theorem of Pythagoras cannot be augmented nor diminshed one jot nor tittle by any actual measurement of a figure illustrating the theorem, if the figure should not substantiate the theorem, so much the worse for the figure, we should say.

It is a highly significant fact, however, that this very science of geometry, which seems to keep itself so independent of experience, is one of the most useful of all sciences in our daily life of experience. The surveyor, the navigator, the carpenter, all make continual use of geom-

* This sequence of lectures was read at Harvard University in the Fall Semester of 1915.

etry in the course of their every-day pursuits, and not only do they do so, but they have an implicit confidence, which always proves to be justified, that the results of their geometrical reasonings—provided only that these are correct in a purely intrinsic, geometrical sense and are based on correctly gathered data—will lead them to perfectly correct conclusions with regard to the world of things experienced with which they deal in their daily lives. The surveyor knows that if his observations are correct, and if he has committed no error of geometry in his computations, the map which he has designed in accordance with a few elementary geometrical laws will be a good map of the region it represents. We have thus the interesting spectacle of a science which seems to scorn experience as its basis, yet furnishes results of the utmost empirical application and value. The question at once occurs to us: How does this happen?

Several theories of the nature of geometry have been devised to bridge this gap. Let us first consider Kant's discussion of geometry. I do not propose to consider here the whole of Kant's treatment of this topic, but only a certain aspect of it—that aspect, namely, which is expressed in the following passage:[1] "Geometry is a science which determines the properties of space synthetically, and yet *a priori*. What, then, must be our representation of space, in order that such a cognition of it may be possible? It must be originally intuition. . . . But this intuition must be found in the mind *a priori*, that is, before any perception of objects, consequently must be pure, not empirical, intuition. For geometrical principles are always apodeictic, that is, united with the consciousness of their necessity, as, 'Space has three dimensions.' But propositions of this kind cannot be empirical judgments, nor conclusions from them."

[1] *Critique of Pure Reason,* Transcendental Aesthetic, §3, Meiklejohn's translation.

That is, Kant says not only that geometry is known *a priori,* but also that our whole original knowledge of *spacc,* the subject-matter of geometry, is *a priori,* and he regards these two assertions as practically tantamount to one another. It seems to the casual observer, however, as though spatial properties could also be given to us *a posteriori,* in experience. It seems as if the straightness of a stick or its length were known quite as empirically as its color or its hardness. Whatever we may say about space, there is no question possible with regard to the statement that spatial qualities are capable of being experienced. Now, it is not with space in any ulterior sense, but with *spatial qualities* that geometry, as used by the surveyor or the navigator, deals. It is not lines in any purely abstract meaning of the term, but the hair-lines in his telescope, or the path of a light-ray, that concern him, and he knows that if his measurements of the lengths, straightness, angles, etc., of these are correct, his computations will also be correct, provided only that he has made proper use of geometrical reasoning. It is such lines as these that form part of his space—and yet he feels the need of no experiment to substantiate the result of his geometrical reasoning. The *a priori* certainty which Kant attributes to geometry is one which is utterly irrelevant to its applications in our life; the abyss between his space, to which geometry applies, and the concrete spatial properties of concrete things, remains unbridged in his system, notwithstanding the fact that he calls space the form of our external experience since the apriority of the geometry which we apply must be the apriority of an empirical intuition, not that of a pure intuition. The geometry which he discusses is one which applies to an entirely non-empirical realm, and which he nowhere brings into touch with those fields of experience in which our every-day geometry plays so great a role.

One of the chief motives which leads Kant to this somewhat incomplete if not positively unsatisfactory treatment of geometry, as one can readily see from this paragraph which we have quoted, is that he considers the apriority of geometry impossible unless our knowledge of its subject-matter is also *a priori.* It is clear, then, that *if we can consistently hold that it is possible for geometry to be a priori, and yet to have an empirical subject-matter,* one strong argument in favor of Kant's view of space has vanished, and we are able to formulate a theory of the relation of the non-empirical science of geometry to the objects of our experience as surveyors or navigators, etc., which is more consonant with the views of our every-day common sense than that of Kant. It is this view of the relation between experience and geometry— the view, namely, that geometry, though *a priori,* deals with an empirical subject-matter— which I intend to suggest as a possibility in what follows.

Before I go on, however, to my discussion of this theory, I wish to devote a little time and attention to a third theory, different both from that of Kant and from that which forms the thesis of this course of lectures. This theory is that of Ernst Mach, as expounded in his little book, *Space and Geometry.*[2] Professor Mach's views form the precise antithesis of those of Kant, both with respect to space and to geometry. As to space, he says:[3]

"If for Kant space is not a 'concept,' but a 'pure (mere?) intuition a priori,' modern inquiries on the other hand are inclined to regard space as a concept, and in addition as a concept which has been derived from experience. We cannot intuit our system of space-sensations *per se;* but we may neglect sensations of objects as something subsidiary; and if we overlook what we have done, the notion may easily arise that we are actually concerned

[2] Translated by T. J. McCormack, Open Court Publishing Company, Chicago.

[3] *Op. cit.,* p. 34.

with a pure intuition. If our sensations of space are independent of the quality of the stimuli which go to produce them, then we may make predications concerning the former independently of external or physical experience. It is the imperishable merit of Kant to have called attention to this point. But this basis is unquestionably inadequate to the complete development of a geometry, inasmuch as concepts, and in addition thereto concepts derived from experience, are also requisite to this purpose."

Mach claims, in other words, that space is essentially a system of *space-sensations* or *space--experiences*, which seems to take the form of a "pure intuition" merely because in our geometrical considerations we confine our attention to one particular phase of the objects with which we are concerned, and neglect all those aspects of our experiences which, though they are necessarily present, are not spatial in their nature. According to him, he says, we are enabled thereby to consider the interrelations of the spatial aspects of our experience with entire disregard of what the other sides of our experience may be. Nevertheless, he holds, space is given to us in a completely empirical manner. Or, as Mach puts it in another book of his,[4] "Space and time are well-ordered systems of sets of sensations."

It seems obvious to the common-sense of us all that Mach is at bottom correct in this statement, for space is somehow or other, we all should say, a system of experiences. Everything looks promising, therefore, for a satisfactory account of the sources and nature of our geometrical certainty. Let us see what the explanation of this is which Mach offers us. He expounds his view as follows:[5]

"The knowledge that the angle-sum of the plane triangle is equal to a *determinate quantity* has thus been reached

[4] *The Science of Mechanics*, translated by T. J. McCormack, Open Court Publishing Co., p. 506.

[5] *Space and Geometry*, p. 58.

by experience, not otherwise than the law of the lever or Boyle and Mariotte's law of gases. It is true that neither the unaided eye nor measurements with the most delicate instruments can demonstrate *absolutely* that the sum of the angles of a plane triangle is *exactly* equal to two right angles. But the case is precisely the same with the law of the lever and with Boyle's law. All these theorems are therefore idealized and schematized experiences: for real measurements will always show slight deviations from them. But whereas the law of gases has been proved by further experimentation to be approximate only and to stand in need of modification when the facts are to be represented with great exactness, the law of the lever and the theorem regarding the angle-sum of a triangle have remained in as exact accord with the facts as the inevitable errors of experimenting would lead us to expect; and the same statement may be made of all the consequences that have been based on these two laws as preliminary assumptions."

This result—namely, that Mach regards the certainty of geometry as of empirical origin, and simply due to the fact that our experiments with lines and angles, etc., by means of paper-folding and similar methods have always substantiated our geometrical predictions as well as could be expected when we take into consideration the inherent inaccuracies of the experiments—this result, I say, is by no means satisfactory. Nobody would ever think of testing the theorems of Pythagoras by means of a foot rule or a protractor; the only things which would be tested by such an attempt and which would have to be rejected in case of a non-verification of the theorem would be the foot rule or the protractor. However useful paper-folding and similar pursuits may be in leading our interest toward things geometrical and in giving us the first dawning ideas about what it is with which geometry concerns itself, geometry deals *directly* with points, lines, planes and angles, and not, except in some periphrastic sense, with such gross topics as folded bits of paper, rules, and micrometers. Whatever the edge of a piece of paper may do or be, a *line* is the shortest distance between two points, does not cut any other line in more than one point, and has all the other properties which are attributed to lines in a text-book of geometry. If a crease in a piece of paper fails to have these properties, why—it simply is not a line. However useful geometry may be in the theory of paper-folding or navigation or astronomy, *prima facie* geometry is *not* the study of paper-folding nor of navigation nor of astronomy, and the accuracy or correctness of any part of any of these studies may be impeached without involving as a corollary the impeachment of any portion of geometry or theorem belonging to it. The geometry of which Mach talks is simply not the geometry of the mathematician; Mach solves the problem of space and geometry to his own satisfaction by flatly ignoring the non-experimental nature of geometry, just as Kant solves it by not entering into a discussion of that empirical character which actually pertains to space. Both positions are unnatural; what is the natural alternative which avoids the objections besetting each of them?

I have already stated that the view which I maintain in this course of lectures is that geometry, though *a priori*, deals with an empirical subject-matter. How is this, however, possible? How can our study of a subject which is known in a manner open to all the uncertainties and inaccuracies which beset empirical knowledge in all its manifestations—namely, space—be possessed of an *a priori* and purely intrinsic certainty, not rooted at all in experience? The answer to this question is by no means as difficult as it might seem at first sight. It will be noted that Mach does not make geometry deal with raw, undigested experience, but, as he says, "All these theorems are . . . idealized and schematized experiences."

Now, the study of an idealized or schematized experience differs from that of a raw or crude experience in that it has to take account of two distinct factors—the experience, and the mode of schematization employed. To illustrate how this is the case, suppose that I am considering a set of statistical tables of the death-rate of Boston from year to year. I may regard these tables from several different standpoints. I may be interested, for example, in the seasonal fluctuations of the death-rate. In this case the table of statistics gives me information which could not have been predicted with more than approximate accuracy and certainty, and which is completely dependent upon concrete experience. On the other hand, I may be primarily interested in the method of tabulating statistical data which is used in these tables; in this case, when I have once grasped the principle underlying the method, I am quite as well able to predict anything you please in the next year's tables which concerns details that are dependent solely on the method of tabulation employed as I am to yield the same information concerning this year's tables or concerning last year's tables. The method of tabulation employed may and should be made as suggestive as possible of the actual empirical laws of the death-rate of Boston and as useful as possible in the handling of the data tabulated, but once it has been chosen, it is entirely independent of the particular empirical properties of these data, and remains essentially incapable of substantiation or of contradiction by them. Thus, though the study of his tables from the standpoint of the form of tabulation employed is of immense practical use to the statistician for the handling of his empirical material, once that form is definitively fixed, it is really an *a priori* science, notwithstanding the fact that the data expressed in the tables are themselves known *a posteriori.*

It is possible to regard geometry in a way quite parallel to a set of statistical tables—though I do not mean to suggest that statistics play any part whatsoever in geometrical reasoning. We may regard a point, for instance, not as a direct object of experience, but as a certain arrangement or collection of objects of experience, in a manner which I shall explain in detail in the subsequent lectures of the present course. A point of this sort will, in general, depend for its actual properties on the concrete natures of the experiences of which it is constructed, but it will also have certain properties which, unlike its other attributes, are independent of the concrete natures of these particular experiences, and are predictable on the basis of a knowledge merely of the principle in accordance with which the points of our space have been synthetized from our experience. These latter properties of points are studied in geometry, while those which are dependent on concrete experiences belong rather to physics or to the other natural sciences. Thus space, which is made up of points, lines, etc., constitutes a kind of tabulation of the experiences of our outer senses; yet geometry, which has space as its subject-matter, since it depends on the method of tabulation alone, as I claim in this course of lectures, is an *a priori,* not an experimental, science. This is the view for the possibility of which I am here pleading.

My view might be stated as follows: Geometry is the science of a *form* into which we cast our spatial experiences. I shall not express my view in this manner, for I wish to keep it clearly distinct from two other views which might with equal justice lay claim to this mode of expression. These views are that of Kant, upon which we have already touched, on the one hand, and the view of those mathematicians, on the other, who hold that the only species of geometry which can possess *a priori* certainty is that geometry which concerns itself, not with the actual points and lines of the world in which we live, but with the *laws* in accordance with which a great many of the properties of these points and lines can be deduced from a small number of properties which they seem to possess, or at any rate seem to possess approximately.

Let us first see wherein our view differs from that of Kant. Kant says that geometry is the synthetical science *a priori* of the form of the external sense, whereas we say that geometry deals with the intrinsic properties of a schematism into which we cast our external experiences; wherein lies the real difference between these two very similar views, and what is its significance? The difference is this: Kant regards geometry as the study of a schematism imposed on the world by our external senses themselves, before any act of experience, and utterly independently of any such act. On the other hand, we maintain that geometry deals with an experience schematized after it has come into existence, and with concrete practical ends in view, even though this schematism may be permanent once it has come into existence and been accepted by us. As a consequence, Kant is unable, as we have previously indicated, to explain how it is that we are able to apply geometry to experience *in a certain concrete and definite manner,* as it is applied by the sailor and the surveyor, or at least he fails to give any hint of how this application is to take place, for the schematism which constitutes the subject-matter of geometry is made, he tells us, before and without reference to the concrete experiences of the surveyor and the sailor, by the essential nature of the outer senses, themselves, and would be the same were there no such particular experiences as those of the sailor or the surveyor. We, on the other hand, are able to maintain that the schematism of geometry is useful for the surveyor and the sailor just because it is designed with the purposes of the surveyor and the sailor in view. This is still true even though that schematism remains just what it is forever, once it has been selected. For example, we choose the schematism "line" in such a manner that some particular line of geometry will be determined as unambiguously as possible in a certain easily recognizable manner by every ruler edge or plumb-line or line of vision in our actual experience. Then, while the lines we have chosen in our schematism may have a host of interesting and valuable properties which are determined by the schematism alone, we may make certain of our geometrical objects standing hostage, as it were, for the physical objects mentioned above, and make our reasonings and experiments refer to these lines rather than to the physical objects themselves, so that our reasonings and experiments may be facilitated by the manifold transformations and systematizations suggested by pure geometry. Since our geometrical lines, though constructions and schemata, are constructions and schemata made on the basis of concrete experiences, we are able to recognize empirically this correspondence between geometrical lines and certain physical entities to which we have just referred, and hence make the former take the place of the latter in the formulation of scientific laws. This cannot be done on the basis of Kant's theory—and this is its fatal defect—because space, according to him, though the form of our external experience, is completely prior to any concrete experience, and hence no correspondence between certain spatial entities and certain physical entities can be recognized empirically, if we accept his theory of the matter.

So much for Kant; let us now consider the pros and cons of the view of those who hold that the only sort of geometry which can possess *a priori* certainty is that geometry which concerns itself, not with the actual points and lines of the world in which we live, but with the *laws* in accordance with which a great many of the properties of these points and lines can be deduced from a few laws which we observe that these points and lines possess, or very nearly possess. This view, that is, says that the real subject-matter of geometry is the formal deduction of its theorems from its axioms, which are not self-evident statements concerning the space in which we live, but mere hypotheses which may perchance be satisfied by an infinity of systems, and it claims further that geometry is not at all concerned with the question whether these axioms and theorems apply to any particular objects or constructions in the world of sense. This latter application, it maintains, must be determined by experience alone, and depends on experience for its validity. Now, it is perfectly true that there is a legitimate non-empirical science, which has as good a claim to the name of geometry as the discipline which we are discussing here, which is concerned with the deduction of the theorems of Euclid from the axioms of Euclidean geometry. I doubt, however, whether this mere abstract logical deduction constitutes the whole of what we ordinarily call geometry, or even the whole of that part of geometry which can lay claims to *a priori* certainty. There certainly appear to be such things as lines, which are more than mere blank spaces in the scheme of symbolism or of logical deduction by means of which the appropriate theorems are obtained from any set of truths which can be put into the form expressed in Euclid's axioms. It seems as if these lines must, from the very necessity of their nature, satisfy the laws of Euclidean geometry, while certain particular lines bear an intimate association with such concrete empirically known things as straight edges and light-rays. This association seems to be presupposed in our every-day life when we say, "This is more nearly a true line than that," as if the true line were a sort of a criterion with which we could empirically compare certain empirical objects. This two-faced aspect of geometry, which is *a priori,* yet deals with an empirical subject-matter, is not explained by those who hold the view we criticize, and is explained on our view.

We hold, then, that geometry is an *a priori* science, which deals with a certain schematization of experience, which we may call space, in so far as its properties depend on the method of schematization alone. This schematization has a superficial appearance quite different from that of the experiences of which it is composed before they are schematized. Experience presents us only with objects that have extension, while a point has no extension. Experience never gives us a perfectly straight line, nor a precise circle, nor an absolutely accurate sphere. All these things, however, form topics dealt with in geometry. Now, we have claimed in this paper that geometry is a schematization of experience, not in the sense that it is a kind of approximate copy of experience with all the roughnesses left out, but in the sense that it is formed from experience by the application of some principle, just as a table of statistics represents the facts it concerns in accordance with a certain principle of tabulation. Just as, notwithstanding the fact that a table of statistics does not resemble the matters tabulated, a statement about the former is but a periphrasis for a statement about the latter, so a geometrical proposition is really concerned with experience, notwithstanding the fact that its direct subject-matter has an appearance differing in many respects from that of experience.

After all this talk of geometry as a method of tabulation, many of you will want to see a concrete example of this sort of tabulation, taken from the field of geometry. It is rather difficult, however, to exhibit such an example in the limited portion of this lecture which remains. I can, nevertheless, give you an example of a similar tabulation employed in a field very analogous to geometry—the study of the formal properties of time.

There are certain laws which we always unquestioningly accept as valid concerning time in quite the same spirit that we hold geometry to be *a priori*. We believe, for example, that time is composed of instants which are timeless, that no two instants are contemporaneous, that of two distinct instants, one must precede the other, and that if the instant *a* precedes the instant *b*, and the instant *b* precedes the instant *c*, that *a* precedes *c*. We consider these statements as quite as truly *a priori* as the theorem of Pythagoras, and regard the former and the latter as quite analogous with one another. The events which we experience, however, always occupy time, an event may neither precede nor follow another, and so on indefinitely. How are we able, the question is, to regard instants as tabulations of events of such a sort that we can be sure, from a knowledge of the method of tabulation alone, without any concrete empirical knowledge of events, that instants will have the formal properties we have attributed to them? I shall give such a method of tabulation in the following paragraphs, though I cannot spare the time to show, as is the case, that it is a consequence of the method itself that instants have the formal properties we have attributed to them.

Experienced events are said to happen at certain instants: what do we mean by such statements? When I say that I see this patch of red here at, say, noon, what do I mean? My first meaning is, perhaps, that I have taken out my watch, looked at it, and have seen both the hands pointing at the figure XII, and that I have experienced this as simultaneous with my experience of the patch of red. But if I look into the matter more thoroughly, I find that this is not all I mean to assert when I say that I see this patch of red at noon. The time during which I see the hands of my watch in a certain position is always of a duration not zero. If the watch had suddenly passed out of existence while I was looking at it, I should still have continued to have seen it for a fraction of a second, during which noon would have passed beyond recall. Now, I can approach more nearly to a precise formulation of what I meant by the proposition that I saw the patch of red at noon if I name still other events which were experienced as simultaneous with the position of the hands of the watch, but which did not endure in experience for the whole period that the hands of the watch were experienced to remain in their position. For example, I can say, perhaps, that this patch was not only experienced as simultaneous with the position of the hands indicating noon, but that the experience of their indicating 11.59′59½″ had not yet died out while I saw the patch. By noting more and more events, each experienced as simultaneous with the patch, and each experienced as simultaneous with each other (for they all, we should say, are experienced as being at noon), we can finally arrive at the specification of a given instant without duration at which the patch was seen, though all the events used in fixing this instant may have consumed time, and have been possessed of all the other gross properties characteristic of *experienced* events, and the relation of simultaneity among them may have been given in experience.

Perhaps I can best illustrate our method of determining noon by a diagram.

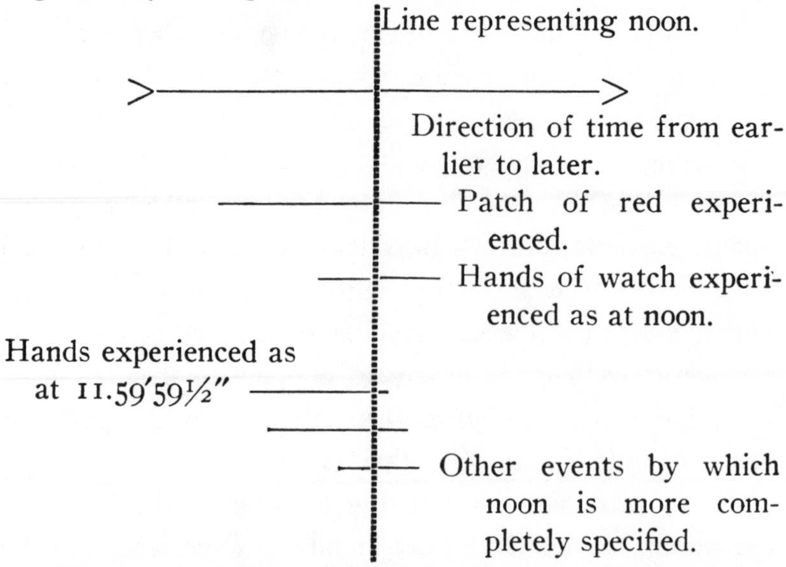

We are able to continue this process further and further by the adjunction of more and more events to the set by which we determine noon. Every such event will be experienced as simultaneous with every other. Finally we shall come to a stage where no more new terms can be found which we can adjoin to our set—that is, there will be no other events which will be experienced as simultaneous with *all* the members of our set. In such a case, we shall have given as complete a determination of noon as is possible on the basis of experiences. The patch of red, which we wish to say is seen at noon, is one of these.

But what *is* noon, which seems to form the subject-matter of the proposition, "This patch of red is seen at noon," which we have been considering? We wish to interpret noon as a sort of tabulation of experience: the answer we give shall therefore read, "Noon is the whole class of events, each of which is experienced as simultaneous with each other, which contains every event experienced as simultaneous with all its constituent events, by means of which we have dated our patch of red." This definition may seem to be circular, for it may seem that the class of events in question could not be specified except with reference to a pre-existing notion of noon. This criticism is, however, invalid, for it may be shown that if instants and events have the formal properties and interrelations that we universally attribute to them, every instant, such as noon, will determine uniquely and be uniquely determined by some set of events of which every two members are experienced as simultaneous, and which is such that it contains every event which is experienced as simultaneous with all its members. The definition of these latter entities involves no circularity, for it depends merely upon a previous acquaintance with events, the relation of simultaneity and a few elementary logical notions such as that of a collection, and not at all upon any specific acquaintance with noon, or with any other instant.

Now, the relation of simultaneity among events, whether as such it can be experienced or not, is certainly far closer to experience than an instant. In all this work, the point we are making is not that the terms and relations with which we start and which we take to represent experience are immediately given—we do not even assume that there are immediately given terms and relations—but that, if I may use such a phrase, they are closer to givenness, that they are less elaborated, that they are the results of a lesser degree of sophistication than the ordinary notions of science. Whether the "experience" with which we started in this lecture is itself already schematized or not does not concern us here; it is enough that space and time mark a degree of schematism greater in intricacy than what we here call experience.

To sum up: we have contrasted the aloofness of geometry from empirical verification with its tremendous value when applied to experience, and have noted the problem which this situation creates. We have discussed Kant's views on the relation between geometry and experience, and have seen that his statement that geometry deals with space, which is given prior to any concrete experience whatever, is hard to reconcile with our empirical recognition of geometrical forms. We have seen that a view which should hold that geometry, though *a priori*, deals with an empirically known subject-matter, would avoid this particular difficulty. Then, taking up Mach's standpoint, it became clear to us that his view, that geometrical certainty is of experimental and empirical origin, is in direct conflict with the practice of all mankind in matters geometrical, and that we all should hold that any geometrical experiment was a test rather of the instruments of measurement used than of the geometrical theorem involved. We noticed the suggestiveness of Mach's notion of geometry as the science of a schematized experience, but saw that in the study of statistical tables, for instance, certain aspects of the study of a schematized experience may be independent of the matter schematized, and depend only on the form of the schematism. We held geometry to be of a similar nature. We observed that our view lent itself to the formulation, "Geometry is the science of a *form* into which we cast our spatial experiences," but we observed that such a formulation would also cover Kant's view that geometry is the study of the form of the external sense, and the view that geometry is merely concerned with the deduction of geometrical theorems from geometrical axioms, so that the certainty which we usually attribute to geometry is entirely dependent on the fact that it is a science of abstract deduction. Kant's view, we found, differs from our view in the fact that it makes the form of the external sense prior to all experience, and consequently cannot explain the empirical identification of spatial entities, while we hold that the schematism with which geometry deals is imposed only after and on the basis of concrete experience. The view that geometry is only concerned with a certain deductive chain did not explain why we act as if geometry were the *a priori* study of a certain concrete system which we can apply to experience as a criterion of straightness or of circularity or of any similar geometrical property. We saw that geometrical propositions, though they seem to deal with such entities as points, lines, etc., are mere paraphrases for propositions about experience in some more direct sense. We finally gave an example of the sort of schematization or tabulation of which geometry makes use, taking this example from *Our Knowledge of the External World as a Field for Scientific Method in Philosophy*, by Mr. Bertrand Russell.

The remaining lectures will be devoted to a more or less tentative discussion of the details of the methods of tabulation and schematization used in geometry. They will very often involve the use of simple geometrical reasoning, but only to prove that the methods of tabulation we here employ yield results similar to those yielded by the methods of schematization which we must tactily use in building up the entities of our every-day geometry.

II. The Point as a Tabulation of Solids

In our last lecture we put forward the view that geometry is concerned with the study of a certain tabulation or arrangement of experience, in so far as this arrangement is determined, not by the nature of the material arranged, but by some already fixed principle of arrangement. As an example of what such a tabulation or arrangement would be like, we gave a brief discussion of the definition of instants as arrangements of simultaneous events. In this and the following lectures we shall attempt, in a similar manner, to define the subject-matter of every-day geometry as a system of tabulations of things which can be experienced and their relations—that is, to exhibit the methods of schematism employed in geometry. In our last lecture we made the further claim that the ordinary theorems of Euclidean geometry could be regarded as consequences solely of the methods of tabulation and arrangement employed in geometry. In the ensuing portion of this course we shall try to show, as far as we are able, how points and the relations between points may be so defined as complexes of objects which can be experienced and of their empirically knowable relations that, though space will be dependent on sense, the geometrical properties of space will be independent of sense, and will follow solely from the schematism by which space is obtained from sense. We shall aim to show that, just as a cube does not depend for its cubical properties on the material from which it is made, just as a wooden, a stone and an iron cube all have eight apices, twelve edges and six faces, so a geometry, although its propositions may have relevance to the actual world in which we live, has a validity independent of the particular nature of the world to which it applies. We maintain that geometry has this universal validity, not only in the sense in which it says that *if* any system satisfies a certain set of premises geometry is applicable to it, but also in the sense in which it asserts categorically that geometrical theorems must apply to the entities which we define as points, whatever the concrete nature of the world in terms of which they are defined may be.

The first task which we have before us is the determination of the fundamental spatial experiences in terms of which our subsequent schematizations and definitions are to be made. The first essential condition which these fundamental experiences must satisfy is that they should be genuine experiences. This excludes at once the possibility that they should deal with such essentially non-experienceable entities as points without magnitude or curves without thickness and so forth. It demands that the fundamental spatial experiences should concern such things as visible patches of color or tangible solids. This necessary condition which these experiences must satisfy leaves us still a great possibility of ambiguity as to their nature. As we are humanly unable to do what is perhaps the most natural thing in this situation and make our method of schematization apply to all experiences which we should ordinarily claim to have a spatial import, on account of the immense technical difficulties such a task would involve, we are obliged to introduce a certain degree of arbitrariness and artificiality into the selection of the fundamental experiences from which we shall build up our geometry. Whether the experience of the solid be primitive in experience or not, this much is certain, that it belongs to a much lower stratum of schematization and synthesis of experience than such unextended things as points, lines and other geometrical entities, and that things or solids are the last word in primitiveness and immediate givenness for the man unsophisticated by psychology. Since our discussion in these lectures is only tentative anyway, and since solids offer a very convenient starting-point for the development of a schematism leading to geometry, we shall regard our primitive spatial experience as one dealing with solids. We have not yet completely specified the nature of our primitive spatial experience, however, as it is possible that there are many different kinds of facts concerning solids which can be experienced. One of the simplest to handle—although possibly not one of the simplest in the order of experience—of these facts is the fact that a certain solid is observed to contain a part in common with another solid. We shall, therefore, select an experience of the intersection or overlapping of two solids as the fundamental spatial experience. Two solids, we shall say, are experienced as intersecting or overlapping or having a part in common with one another if they both seem to contain some solid or if one seems to contain the other or if they seem to come into contact.

The experience of the overlapping of solids is not, however, as it stands, a sufficient point of departure for a schematization which is to lead to geometry. We wish to be able to define a straight line as a sort of a tabulation of solids. Now, if all we know about solids is the relations of overlapping that hold among them, we will be unable to discriminate between a straight line and a tortuous one. The whole of space could be kneaded like a lump of putty without changing a solid into anything else or altering the relations of overlapping which solids bear to one another, but by such a transformation you could deform a straight line into a curve as tortuous as you please. It is obvious, then, that if we are to be able to define straight lines in terms of the experience of the intersection of solids, we must put some kind of a limitation on the kind of solids considered. We shall put upon them the limitation that they are to be *convex*. Now, a convex solid is one such that any two points which belong to it can be connected by a piece of a line which nowhere passes outside of it. Thus a solid sphere is convex, a cylinder is convex, a cone is convex, and a cube is convex, while a solid in the shape of an hour-glass is not convex, a doughnut is not convex, a bowl is not convex, and no figure which is hollow is convex. As a matter of fact, convexity is synonymous with the absence of hollowness in any sense, and since hollowness can roughly be judged by the eye and the finger without reference to straight lines, convexity may also be determined by a more or less direct reference to experience. We know what it means to say that a bowl is hollow and that a billiard-ball is convex long before we ever think of correlating these properties of solids with the definition of convexity just given. We can, therefore, make our fundamental experience that of the intersection of *convex* solids, and be sure that it is near to genuine experience. Further, it may be shown by a simple bit of geometry that if the world were, say, made of clay, and were so squeezed out of shape that all convex solids and their relations of overlapping should remain unchanged, every straight line would remain straight. Consequently, once the set of all convex solids in space has been identified, the set of all lines in space is determined, and it seems very probable, to say the least, that lines can be defined in terms of convex solids.

The sort of fact from the schematization of which we shall obtain space is, "This convex solid is *experienced* to intersect that one," and not simply, "This convex solid intersects that one." The formal properties of experienced intersection and of actual intersection are probably, however, closely analogous in most respects. Each solid may be regarded as having, outside of its physical extension, a sort of *aura,* of definite extent, such that two solids are experienced as overlapping when, and only when, the solids formed out of each by adjoining to it its aura actually intersect. For example, two spheres a hundredth of an inch apart may seem to be in contact, as far as our unaided senses can tell: then we shall say that the aura of each extends at least one two-hundredth of an inch beyond its physical extension. The difference between the relation of apparent or experienced intersection among convex solids and that of their actual physical intersection is to all intents and purposes, then, a difference in the solids chosen as intersecting rather than in the formal properties of the relation of intersection itself, for if we replace convex solids by convex solids plus their auræ, we can interpret the apparent intersection of the former as the actual intersection of the latter.

We are now in a position to define our points—that is, to exhibit them as tabulations of convex solids. We shall define our points as collections or aggregates of solids. This may seem curious to many of you. "What!" you may think, "Is not a point small and a solid large? Is not a class of solids even larger than a solid? Then how can a point be a class of solids? How can the part be greater than the whole? How can points be made of solids, as you say, and solids also be made of points, as the mathematician says?" Now, all these questions result from a confusion of the relation of a member of a collection to the collection of which it is a member with the relation of an object filling a given space to an object filling a space including that which the first object fills. One tends to think, for example, that because Harvard University is a class of men, Harvard University fills more space than a single man. But, when one comes to think of this example more thoroughly, one sees that *in the sense in which a man fills a certain space,* it is nonsense to talk about Harvard University as filling any space. It is only in a metonymous sense that Harvard University can be said to fill the space occupied by all its members. Harvard University has only such properties as belong to different logical dimensions from those of its members. In fact, it is a general proposition of logic that no collection can have any properties that can in precisely the same sense be significantly asserted—or denied, for that matter—of any of its members. Thus, Harvard University, although it has certain intimate connections with certain portions of space, cannot be said to occupy any space at all in the sense in which I now occupy, the space vertically above this platform, and in an analogous way, in the sense in which a solid can occupy space, a class of solids cannot occupy space, and in the sense in which a class of solids can occupy space, a solid cannot occupy space. It is, therefore, nonsense to speak of a class of solids as either smaller or larger than a solid. Hence we do not, in defining a point as a class of solids, make the part larger than the whole, for the point and the solid are rendered by such a definition incomparable as to magnitude.

The second paradoxical feature of our definition of a point—that we define a point as a collection of solids, whereas in ordinary geometry, a solid is regarded as a class of points—is eliminated still more easily. A solid, in the sense in which points are classes of solids, is an entirely different thing from a solid, in the sense in which a solid is a class of points. They are no more identical than the collection of clubs to which John Smith belongs is identical with John Smith himself. The only thing that entitles us to call both solids is that the world in which we live is probably so organized that corresponding to each solid in our first sense there is a class of points uniquely determined by it and representing no other solid than it, which we may call "the same solid as it," just as it might be that in some town one could identify every man by the list of clubs to which he belongs, and could say, whenever one should see a list of clubs to which some man belongs, "That's John Smith," or "That's William Jones," or whoever else it might be.

Our definition of points in terms of solids is to be justified, as are all definitions in this kind of work, by its fruits. We shall so define a point that if the things we call convex solids are really the convex solids of an ordinary Euclidean space, the things we call points will correspond in a certain determinate manner to the points of ordinary Euclidean space; the things we shall later call lines will have all the nice properties that lines should have; and finally, the whole space we shall obtain as the end of our discussion will have all the attributes that pertain to our every-day space. On the basis of this first definition of points and of lines we shall give a second and finally a third definition of points and of lines which will, on the one hand, make each point of the first sort determine a single point of the second or third sort and each line of the first sort determine a line of the second or third sort in such a manner that the geometrical properties of a figure made up of points and lines of the first sort will be substantially unchanged if each point and line of the figure be replaced by the analogous point or line of the second or third sort—which will, I repeat, do all this *if the points and lines of our original system form a set satisfying the axioms of ordinary geometry*, or, to put it in a more elementary manner, if two lines in our first sense have a point in common when and only when two decent and well-behaved lines ought to have a point in common. On the other hand, we shall so frame our definitions of points and of lines of our third kind that, however irregular the formal properties of the points and lines of our first sort may be, however often lines that should intersect, did our original system obey the laws of geometry, fail to intersect, or lines that should fail to intersect do intersect, our lines and points of the third sort must, so long as logic is logic, have all the properties appertaining to lines and to points in a Euclidean geometry. Furthermore, we shall develop a theory of measurement in this third space that we finally attain which will be consonant, on the one hand, with our usual ideas of the operations performed in actual physical measurement, and which, on the other, will be in perfect harmony with the laws of measurement laid down in ordinary Euclidean geometry. All this is done on the basis of our original definition of a point, and constitutes an ample justification for it.

After this rather long-winded apology for the definition of a point as a class of solids, let us state this definition in precise terms. *A point is a collection of convex solids such that* (1) *any two convex solids belonging to it are experienced as intersecting, and* (2) *if a convex solid is experienced as intersecting EVERY member of such a collection, it can only be itself a member of the collection.* We saw previously that the relation of experienced intersection among convex solids reduces itself to the relation of actual intersection among other solids—namely, those formed out of convex solids by adjoining their auræ to them, or as we shall hereafter call them, *a-solids.* Our definition is therefore practically equivalent to one which should read as follows: a point is a class of a-solids such that (1) any two members of the set intersect, and (2) any a-solid that intersects every member of the class must itself be a member of the class. Now, what does this mean?

Let us consider the class of all the a-solids which, as we should say in our every-day life, contain a given point x on their surface or in their interior. In the first place, every two members of this set intersect, for earlier in this lecture we have taken the term intersection to cover contact or tangency, and two figures with a point in common either intersect bodily if the point in question lies in the interior of one of them, or come into contact with one another if the point lies on the surface of each of them. In the second place, it may readily be shown that if an a-solid intersects every a-solid that contains x, it must itself contain x. This proof depends upon the fact that if an a-solid does not contain a given point, another a-solid can be found which contains the point, but does not intersect the first a-solid. Taking this principle for granted—its truth can very easily be established on the hypothesis that the aural layer of a convex solid is of a uniform thickness throughout space, or on many similar hypotheses which do not assume so much—the desired consequence follows in this way: if an a-solid intersects every a-solid that contains x, but does not itself contain x, we get a contradiction, for by the principle which we have just enunciated, there must be a second a-solid, not intersecting our first a-solid, but containing x, while, by hypothesis, this is impossible. Consequently, if an a-solid intersects every a-solid that contains x, it must itself contain x. We have thus shown that a collection of all the a-solids which, as we should ordinarily put it, contain some point, satisfies both the criteria which a class of a-solids must fulfil to be a point by our definition, since any two members of it intersect, and any a-solid which intersects all its members belongs to it.

To give a completely satisfactory justification of my definition of a point as a class of a-solids whereof any two intersect and which are such that any a-solid intersecting every member of the set belongs to the set, it is not enough to show, as I have just shown, that every collection of all the a-solids containing some point, which may be said to represent or even to be that point, is a point in accordance with our definition; we must also show that no other collections of a-solids are points in accordance with our definition. We must show that if, on the one hand, a collection of a-solids does not exhaust those which, as we should ordinarily state it, contain some point in common, or if, on the other, there is no point common to all its members, the collection of a-solids in question fails to satisfy one or both of the two criteria which determine whether a given collection of a-solids is or is not a point in accordance with our definition of a point. Now, it is easy enough to show that if all the members of a collection *a* of a-solids contain a given point, but do not exhaust the collection of the a-solids which contain the point, there are other a-solids—i. e., the other a-solids containing the point in question—which do not belong to the collection *a,* but intersect every member of *a,* so that *a* is not a point in accordance with our definition. It is not easy to show, however, that if a collection of a-solids is of such a nature that there is no point, to use ordinary geometrical language, which all its members contain in common, this collection of solids fails to satisfy at least one of the two criteria both of which a collection of a-solids must satisfy if we are to call it a point in accordance with the definition we have given. In fact, I have not yet succeeded in proving this theorem, and I have nowhere seen any proof given for it, yet I am convinced that it is true and that it can be proved. I am convinced of this because, notwithstanding a considerable amount of effort, I have been unable to discover a single collection of a-solids, except the collection of all the a-solids that contain some given point, which satisfies both of the two conditions which all the things that are points by our definition must satisfy. Therefore, notwithstanding the gap in my chain of reasoning, I shall go on from this point as if I had proved that our definition of a point is perfectly adequate, and that the only collections of a-solids which satisfy our definition of a point are such as are made up from all the a-solids which, as we should naturally put it, contain some point. If we suppose that this is proved, *provided that our experience of the relation of intersection among convex solids is to receive the geometrical interpretation in terms of a-solids which we have given it,* since our first definition of points in terms of the experience of the intersection of convex solids will then be practically equivalent to our second definition of a point in terms of a-solids, our points in our first sense, though defined in terms of an experience, will well deserve the name of points.

Our next task is to define what is perhaps the next most fundamental notion in geometry—the notion of a line—in terms of our experience of the intersection of convex solids. It will be remembered that convex solids stand in a very close relation to straight lines, for a convex solid is one that contains the whole of a bit of any straight line whose ends lie inside it. Now, this fact enables us to define a bit of a straight line in terms of our experience of the intersection of convex solids as follows. We have just seen how a point may be regarded as a class of a-solids which is what we should ordinarily call the class of all the a-solids containing that point. An assumption which we shall make at this point is that all a-solids are convex and that we can thus regard a point as a class of all of a certain kind of convex solids which contain a given point. This assumption is extremely natural. It is a consequence of the other assumption which we suggested previously, to the effect that the aural layer of a convex solid is of uniform thickness throughout space, but does not presuppose the latter assumption. From the hypothesis we have stated we can readily draw the conclusion that if a and b are any two points *qua* classes of a-solids, then every a-solid which forms a member both of a and of b contains, in ordinary geometrical phraseology, the whole piece of a straight line intercepted between a and b. That this is true follows from the fact that *a* and *b* are points inside any a-solid which belongs to them both, since a member of a point is an a-solid which contains it. Consequently, since an a-solid is convex, any a-solid which belongs both to *a* and to *b* contains the whole linear segment or bit of line between them, and consequently every point on this segment. Therefore, every point on this segment possesses as a member any a-solid within which *a* and *b* lie. This is another application of the principle that the members of a point are the a-solids which spatially contain it. It is thus a necessary condition if *c* is to lie on the linear segment between *a* and *b* that all those a-solids which belong both to *a* and to *b* should also belong to *c*. That this condition is also sufficient may be proved on the hypotheses that the thickness of the aural layer of all a-solids is constant and that an a-solid can be transported to any part of space, and yet remain an a-solid. Both these hypotheses are very probably true—at least within that part of space whereof we have any experience at all. The deduction of the sufficiency of our condition from these hypotheses, though easy, is a little too intricate for us to give here.

We have, then, given a necessary and sufficient condition that one point, *qua* class of a-solids, should lie on the bit of line between two other points of the sort. Let us reinterpret this statement in terms of points consisting, not of a-solids, but of general convex solids. If three points, *a, b* and *c,* consisting of a-solids, are so arranged that *c* lies on the linear segment between *a* and *b*, and if *a′*, *b′*, and *c′* are, respectively, the points consisting of general convex solids corresponding to *a, b* and *c*, then it will be natural for us to say that *c′* is between *a′* and *b′* and on the line determined by them. That is, *c′* will be between *a′* and *b′* when and only when *c* contains all the a-solids common to *a* and to *b*. Now, *a* contains a given a-solid as a member when and only when *a′* contains the convex solid from which this a-solid is formed by the adjunction of its aura, and a similar relation subsists between b and b', and between c and c'. Therefore, c contains all the a-solids common to a and b when and only when c' contains all the convex solids common to a' and to b'. Consequently c' lies on the linear segment between a' and b' when and only when c' contains all the members common to a' and to b'. Now, we have not yet defined linear segments or any such things, and this property of a', b' and c', when c' contains the common part of a' and b', is defined in purely logical terms introducing only such notions as those of part and class, involving no concrete geometrical notion, except such, of course, as are involved already in the notion of a point, which we have already defined in terms of our experience of the intersection of convex solids. We may therefore *define* a point c' to lie between two others, a' and b', when and only when c' contains the common portion of a' and b', and we shall be sure, on the one hand, that if our experience of the intersection of convex solids has the properties that are to be expected of it, this relation of betweenness will not have been misnamed, and, on the other, that this definition involves no notions other than that of our experience of the intersection of convex solids and certain general logical notions.

I wish now to define the notions of segment, end-point and line, in terms of the relation of betweenness just defined, and hence ultimately in terms of our experience of the intersection of convex solids. If a and b are distinct points, the class of all the points c which are such that c is between a and b constitutes the linear segment ab, and a and b are its end-points. The *line ab* is the class of all points belonging to linear segments which have at least two points in common with the linear segment ab. The agreement of all of these notions with the conventional notions of segments, end-points and lines, subject to a certain reservation which we shall make in the next two lectures, will be obvious on a brief reflection. The adequacy of our definition of a line will be apparent if we reflect that any two linear segments which have two points in common are segments on the same line, while if *x* is any point on a line *l*, and *s* is any segment on *l*, a segment *t* can be discovered which will contain *x* and have at least two points in common with *s*.

To sum up what we have said in this lecture, we first defined a point as a class of convex solids, whereof any two are experienced to intersect, and which is further such that it contains as members all those convex solids which are experienced as intersecting all its members. We justified this definition of a point and showed that the entities which are thus defined as points are such things as one could naturally call points, providing that our experience of the intersection of convex solids has such formal properties as one would naturally attribute to it, since under this hypothesis each of our points will be a collection of all the convex solids which are experienced as containing some point, and may, since the notion of a point is only now defined for the first time, be identified with the latter point, which they are experienced as containing. We have defined a point *a* as between a point *b* and a point *c* when *a* contains the common part of *b* and *c*. From this definition alone we have derived definitions of a linear segment, of the end-points of a linear segment, and of a line. All these definitions have been made solely in terms of the experience which we have chosen as fundamental—that of the intersection of convex solids.

The work in this lecture is based on that of Dr. A. N. Whitehead and Mr. Bertrand Russell on space and time, as given in Mr. Russell's *Scientific Method in Philosophy,* Chapter IV. The definitions of betweenness and of a line are borrowed from Prof. Huntington's article in the *Mathematische Annalen* for 1912, but go back to the work of Kempe and Prof. Royce.

III. The Extension of Space Beyond the Bounds of Experience

In our last lecture, you will remember, we arrived at the definition of a point as a class of convex solids, and of a line as a class of points, in terms of our experience of the intersection of convex solids. These definitions, however, and, indeed, any definitions that start directly from our experience of the intersection of convex solids, must suffer from certain rather obvious defects. We intend to use our definitions of lines and of points to set up a theory of spatial measurement. To do this, we shall make much use of the construction of parallelograms: for example, we shall define the distance AB on a given line as equal to the distance CD on the same line if it is possible to construct a linear segment or piece of a straight line EF parallel to AB in such a manner that AE is parallel to BF and EC is parallel to FD. The following diagram will represent such a situation.

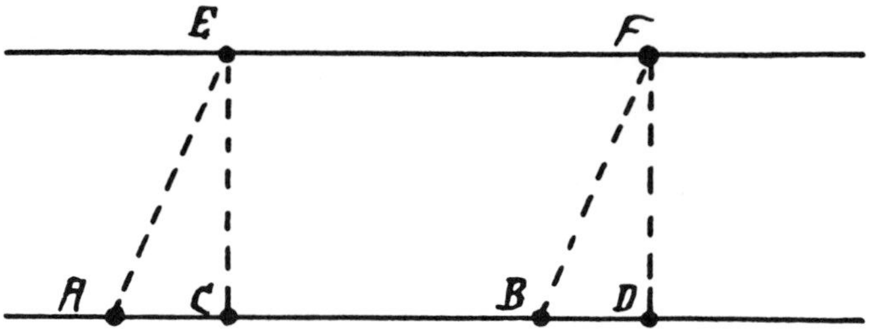

This demands that we are in possession of a definition of parallelism. In Euclidean geometry, to say that two lines are parallel is equivalent to saying that they lie in the same plane and do not intersect one another. We may define two lines as being in the same plane if and only if they both have a point in common with each of a pair of intersecting lines *l* and *m*, and do not pass through the point of intersection of *l* and *m*; thus in the following diagram, *p* and *q* are in the same plane, or, as mathematicians say, are coplanar.

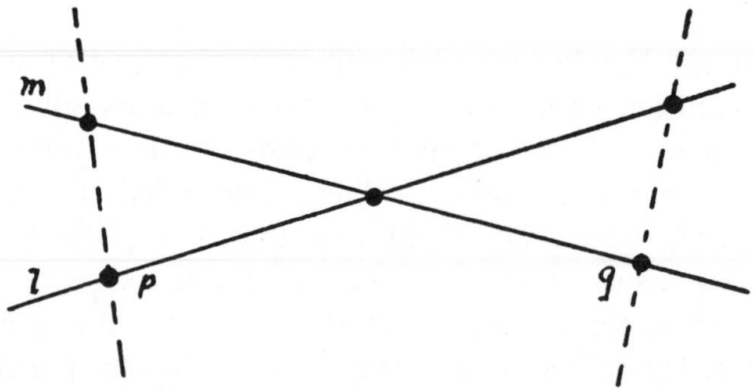

We might, therefore, define two lines as parallel if they are coplanar and do not intersect, without introducing any new fundamental notions into our system. But, as we have said, there are certain defects inherent in the definitions of lines and of planes that we have already given and these defects make such a definition of parallelism undesirable. These defects are due to the fact that our experience of the intersection of convex solids does not record the intersection of convex solids at the uttermost confines of space; beyond a certain extreme distance, whatever it may be, the intersection of convex solids is not experienced. As a matter of fact, I can hardly be said to have any experience of the intersection of convex solids except in the immediate neighborhood of my own body. As a consequence, any lines which would naturally be said to meet at a point lying outside the very limited region within which convex solids are experienced to intersect would, in accordance with the definition which we are considering, be parallel lines, for, by our definition of a point, a point is a collection of convex solids of which every two are experienced as intersecting, and which contains every convex solid experienced as intersecting all its members, and consequently there can be no points outside the region of the apparent intersection of convex solids, so that any two coplanar lines which fail to have a point in common within this region must fail to have any point whatsoever in common, and must, by the definition which we are considering, be parallel. Indeed, since our points are only such points as lie within a certain region of space, the lines which we defined in our last lecture are merely such parts of lines as lie within this region. Not only would this whole condition of affairs not be consonant with our natural notions of lines and of parallelism, but it would further fail to give parallelism even the most important formal properties which it possesses in ordinary geometry. For instance, it would be impossible to use the parallelogram construction as a criterion of the equality of two different linear segments on the same straight line, as a very simple geometrical construction, for it may easily be shown that our definition will make all distances along a given line equal. Hence, our definition of parallelism is at fault, and we must look around for a new one.

In what direction are we to look for this new definition of parallelism? The defect of the definition that we have just rejected—that two coplanar lines without a common point are parallel—is due to the fact that two lines which ought to have a point in common do not always have a point in common, if we define points and lines in the manner indicated in the last chapter. If we are able, then, to introduce new definitions of a point and of a line such that every two lines which ought, as ordinary, commonsense, decently-behaved lines, to have a point in common, will have a point in common, we shall have brought the problem of finding a definition of the parallelism of two lines a great deal nearer solution. This new definition of a point and this new definition of a line should, if possible, be made in terms of our previous definitions of a point and a line alone, without introducing any new notions. The question we now ask is, can new definitions of a point and of a line, which make two lines contain points in common just when they ought to, be made in terms of our previous definitions of a point and of a line?

Now, we have seen that our previous definition of points gives us only the points within a certain region of space. Let us assume, for simplicity's sake—for, though this assumption is almost certainly false, it is not, as such, essential to our further argument, and it enables one to picture in his mind's eye what I have to say much better than any other hypothesis—that our points, in the sense in which points have been already defined, are *all* the points in the interior of some closed convex solid, and that our straight lines are consequently all the segments of straight lines intercepted by the surface of this solid. How are we able to recognize in terms of the points and lines inside this solid the entities, whatever they may be, that we should naturally call points outside of the solid? The problem is closely analogous to that of the recognition that a certain set of astronomical observations from various points on the surface of the earth all are observations of a single star, even though that star is utterly inaccessible to us. The problem to which I allude is not that of recognizing the star as the same star from observations at different *times;* it is the far simpler one of discovering that many observations made at the same time pertain to a single object. Let us suppose, for example, that we have four observers, a, b, c, and d, all looking at a star or planet x from different points on the surface of the earth. How do the observers *a, b, c* and *d* know that it is a single star or planet at which they are looking?

The whole and complete answer to this question would involve considerations which are irrelevant here; it is obvious that one of the things that our four astronomers must know, however, is that they are all looking at the same *place*—that, in other words the axes of their telescopes converge on one point, or if the object at which they are all looking is sufficiently far away to be considered, for all practical optical purposes, as at an infinite distance, they must know that the axes of their telescopes point, to all intents, in one direction, or to put it otherwise, that they are all parallel. This knowledge, moreover, must be attained and is attained independently of any direct knowledge the astronomers have concerning the point to which all the axes of their telescopes converge, for this point is exceedingly remote from them, and is known by them in no other way than by these very observations concerning which we are now trying to find out why it is that the astronomers regard them as observations of a single point. When the astronomer says that at such-and-such an instant this point of space has this or that property—as for example that of being occupied by a planet—all that he means or has a right to assert must concern the observations in which the telescope is directed towards this point, for if the observations should remain the same, but the whole remainder of the universe should be changed in any manner whatsoever, the astronomer would still be entitled to make the same assertion concerning this point as formerly. The knowledge of the convergence to *x* of the optical axes of the telescopes at *a, b, c,* and *d* is attained by a measurement of the angles which the lines between *a, b, c, d,* and *x* make with one another and a measurement of the distances of *a, b, c,* and *d* from one another. These observations do not require any direct knowledge of *x*, but only of the positions of the telescopes at *a, b, c,* and *d,* for if we know the latitude and longitude of *a, b, c,* and *d,* if we know the compass-bearing of each telescope—that is, whether it is pointing east or west or southeast or north-northeast-by-north, etc.,—and if we know the slope of each of the telescopes, we know all the angles which any two lines connecting two of the points *a, b, c, d,* and *x* make with one another, and the mutual distances of the points *a, b, c,* and *d.* What we realy talk about, then, when we discuss the position of the planet is the aggregate of the positions of the telescopes by which it is observed.

Another thing to notice is that if we know that the telescopes at *a, b,* and *c* are all directed to one point, and that the telescopes at *b, c,* and *d,* are all directed to one point, we know that the telescopes at *a, b, c,* and *d* are all directed to one and the same point. This is rendered obvious by a simple diagram. The significance of this fact will appear if we consider that if the telescopes at *a* and at *b* are directed at one point, and the telescopes at *b* and at *c* are both directed at one point, all three telescopes need not be directed at any single point. This also is shown by a diagram.

We are now able to return to the discussion of our real problem—the problem, namely, how we are to recognize points outside that convex region of space within which all the points that we have already defined are located in terms of the points and portions of lines lying inside this region. The portions of lines lying inside our convex region—i. e., the class of all the lines that we defined in our last lecture—are the exact analogues of the telescopes or the astronomers whom we have just discussed. Just as the astronomer's statements about the position of a star really concern the positions of certain telescopes, so propositions which seem to deal with points lying beyond the bounds of our experience really concern certain collections of our lines: namely, with such as are made up of all those lines that "point at" some point lying outside our region. As far as we are concerned, such collections of lines, since they correspond uniquely to the points at which they are directed or from which they spread out, may be regarded as *constituting* these points. This situation can easily be rendered obvious by a diagram.

This definition gives rise to many problems. In the first place, how is it possible to get along in a system in which some points—those within the region of space directly accessible to experience—are the elements of which lines are classes, while other points in space ZZ those not in that region directly accessible to experience—are classes of these self-same lines? In the second place, is it possible to define the property which a class of lines has when every member of the class is directed towards some given point beyond the bounds of experience in terms of that experience which we have taken as primitive—in terms, namely, of our experience of the intersection of convex solids—without the introduction of any new experience or concept not derivable from that experience of intersection? If such a definition is possible, how are we to proceed to discover it? In the third place, how are we to tell when three or more of our new points are situated on a single line, and how are we to define such a line? These three problems will form the chief subject-matter of the remainder of this lecture and of the following lecture.

Let us take them up in the order just indicated. How, we asked, is it possible to get along in a system in which some points—those within the region directly accessible to our experience—are the elements of which lines are made up as classes, while the remainder of the points of space are classes of lines? The answer is—it is not possible, and we do not intend to try to do so in this paper. Not only is it highly inconvenient and unnatural for one point in a system of geometry to be an aggregate of aggregates of other points, but there are good philosophical reasons—indicated by Mr. Russell in that part of the *Principia Mathematica* which deals with the Theory of Types, but too complicated and foreign to the subject-matter of this course of lectures for us to discuss here—there are good philosophical reasons, I say, for holding that no assertion which can be made significantly concerning a given entity, say *x*, can also be made significantly concerning a collection of collections which has some member of which *x* is in turn a member. Therefore, since a line, in the sense defined in our last lecture, is a class of the points which we then defined, it is impossible for one to assert any proposition significantly concerning these points, on the one hand, and also concerning the classes of lines that we intend in the future to call points, on the other. Now it is, to say the least, extremely awkward to have to phrase every proposition that concerns itself with points in one manner when it concerns itself with the points inside a given region and in an entirely different manner when it deals with the points outside this region. We shall consequently define the points inside the region directly accessible to experience as well as those outside it as classes of the lines that we defined in our last lecture, and we shall term all points *qua* classes of straight lines *generalized points,* in order that we may not confuse them with the points defined in our last lecture. Just as we agreed to regard each of the generalized points lying beyond the bounds of our direct experience as the class of all the lines which, as we should say in every-day language, are directed towards the point, so we shall agree to regard those generalized points lying within the region directly accessible to our experience as a class of all the lines which we should usually consider to pass through some point inside this region. If such a class of lines happens to be the class of all the lines which contain in common a point, in the sense defined in our last lecture, then if x be the point which they all have in common and a be the class of lines, we shall say that a is the generalized point corresponding to x, but it must be clearly understood that *a is not x*.

We are now in a position to deal with our second question: is it possible to define the property which a class of lines has when every member of it, as we should usually put it, is directed towards a given point, in terms of our experience of the intersection of convex solids? It should be noticed that this is the crucial question of this entire lecture, and that our reduction of generalized points to classes of lines having some property which we should usually call "passing through a given point," but which, as a matter of fact, we wish to define without reference to any point through which the lines are supposed to pass, is in the unpleasant situation of Mahomet's coffin until we find a way of identifying this property. In the analogous instance of the astronomers and the star or planet, a collection of telescopes all pointing at a certain point in space is, as we saw, distinguished from a collection of telescopes not all pointing at any one point in space by the fact that when all the telescopes are directed towards a single point certain trigonometrical formulae connecting the latitudes, longitudes, geographical directions, and slopes of the several telescopes hold good which do not hold good when the telescopes are not all aimed at any one point. Such a method of determining whether or not all the lines of a given collection are directed to a single point is inapplicable to the case where we are to define the generalized points of space solely in terms of its points and lines in the sense of our last lecture, for we have as yet no definition of an angle or of a distance nor of a slope: all that we have defined up to this point is the set of all the points and linear segments that lie within a given region. Our problem hence reduces itself to that of the determination of such classes of lines as are made up from all the lines that pass through a given region—the region, namely, within which we experience the intersection of convex solids—and some chosen point inside or outside of this region, in terms of the intersection-relations of those portions of lines lying inside the region.

We can, however, narrow our problem still further, and indicate the method of its solution with still greater definiteness, if we remember a certain fact about straight lines which we pointed out when we were discussing the case where several observers are looking at a single star. It will be remembered that we showed that if the axes of the telescopes *a, b*, and *c* converge to a single point and the axes of the telescopes *b, c*, and *d* likewise converge to a single point then the axes of all the four telescopes *a, b, c*, and *d* all converge to the *same* point. As we may easily show by a diagram, we may generalize this statement and say that, given any collection of telescopes, if there are two among them, say *a* and *b*, such that if *x* be any member of the collection of telescopes the axes of *a, b*, and *x* all converge to one point, then all the axes of the telescopes of the collection converge to a single point. The converse of this statement is even more obviously true. We can thus define a collection of telescopes as one, the axes of all of whose members converge to some one point provided that it contains two members the axes of which intersect and that the collection is made up of all telescopes which form, taken together with these two, a triad of telescopes whose axes converge to a single point, and in addition of these two telescopes themselves. If, that is, we are in possession of a criterion of the convergence of a triad of telescope-axes, we are able to define the collections of all telescope-axes converging to some point or other. In exactly the same manner we are able to define certain classes of the lines we discussed in our last lecture as classes of all the lines which, we should ordinarily say, pass through some point or other, whether that point is or is not within that region which is directly accessible to our experience, or in other words, as generalized points. If we have a criterion which enables us to discover when any three lines converge to any point whatsoever in space, for the property of line-triads which reads, "If two line-triads each of which is made up of three lines converging to a point possess two lines in common, all four lines making up the two triads pass through some single point," is not confined in its application to the axes of telescopes but applies equally well to all kinds of lines. Therefore, if we are already in the possession of a definition of a convergent triad of lines, we may define a generalized point as the class of all lines, in the sense in which we defined lines in our last lecture, which either are one of two given lines, say l and m, or form together with l and m triads in which the three members of the triad stand to one another in the relation which is ordinarily denominated 'all passing through the same point,' provided only that l and m are distinct intersecting lines—that is, distinct lines which form two of the members of some triad of lines which all would naturally be said to pass through some point. If, then, we are able to give a definition of the relation among three portions of lines lying inside a given region of space which we should naturally call, that of all being directed towards some one point and which the mathematician terms the relation of concurrence, which involves only such notions as we can define in terms of the points and lines of our last lecture, we are in a position to define the class of all generalized points in space, wherever they may be situated. One of the notions which it is permitted for us to use in the definition of the concurrence of three lines is that of the relation which two of the lines of our last lecture bear to one another when they possess in common one of the points of our last lecture as a member, for this notion can be defined in terms of points, lines, and notions of pure logic alone, and consequently ultimately in terms of the experience of the intersection of convex solids and of in addition only such notions as belong to pure logic and not to concrete experience.

This demand will be satisfied if we give an adequate definition of the concurrence of any three lines which do not all lie in a single plane, for we can define in terms of this relation the concurrence of any three lines whatever, whether they are concurrent or not, in the following manner: the lines *a, b,* and *c* are said to be concurrent if *d* and *e* are two lines such that each of the three lines *a, b,* and *c* forms with *d* and *e* a triad whose members are concurrent *and do not lie in the same plane.* Now, it is an extremely easy task to give a definition of a relation between three lines not all in the same plane, which, though it is slightly more general than the relation of concurrence, includes the latter as a special case, and is only *slightly* more general than it. This new relation is that which holds among three coplanar lines when they are either all concurrent or all parallel. We shall introduce into the definition of this no notion which we have not already defined in terms of the experience that we have taken as fundamental—the experience, namely, of the intersection of convex solids. We can define a plane, readily enough, as the class of all those points, in the sense in which we have already defined points, which lie on any line which has two distinct points in common with some given pair of lines that themselves have a point in common, but do not coincide. As the lines that we have already defined really represent those segments of the lines of ordinary geometry intercepted by the surface of that region of space within which convex solids appear to intersect, our planes, as they are now defined, will actually represent the planar areas intercepted by the surface of this region, provided that it is possible to draw from any point of such an area a line cutting any two given linear segments intercepted by the surface of the area in two distinct points. That this is possible we may readily show to be the case under the hypothesis, which we have every reason to believe satisfied, that the region of space accessible to our experience is convex. Now, it is a familiar theorem of elementary solid geometry that if *p*, *q*, and *r* be any three distinct planes of which no two are parallel and which do not all possess any line in common, then the intersection of *p* and *q*, the intersection of *q* and *r*, and the intersection of *r* and *p* will form a triad of concurrent or parallel lines. The proof of this theorem is simple, and the situation it represents is illustrated by the corner of a room, where the walls and floor represent *p*, *q*, and *r*, and the three edges of the room that meet at the corner are the lines of intersection of pairs of the planes *p*, *q*, and *r*. The case where the three lines are parallel is represented by the three faces and the three edges of a triangular prism. From these examples, it is further easy to guess the truth of the converse theorem of that which we have just stated: three lines not all in one plane are concurrent or parallel *only* when they are the three lines of intersection of pairs of the planes belonging to a certain triad. If we apply these theorems to the lines of our last lecture, under the hypothesis that these represent the linear segments intercepted by the surface of a certain convex region, we shall obtain the result that three of the lines of our last lecture that do not all lie in one plane are concurrent, wherever the point of their concurrence may be situated, or parallel *when and only when they are coplanar by pairs*. Since the relation of coplanarity among the lines of our last lecture has been already defined and the concurrence or parallelism of three lines not all coplanar has not yet been defined in terms of our experience of the intersection of convex solids, we may regard the equivalence expressed in our last sentence as a *definition* of the concurrence or parallelism of three lines not all coplanar.

We may now go on and say that three of our lines, *l, m,* and *n* are concurrent or parallel, whether they are all coplanar or not, when and only when they are all distinct and there are two of our lines, *a* and *b,* let us say, which are such that *a, b,* and *l, a, b,* and *m,* and *a, b,* and *n* form three triads, respectively, each made up of three concurrent or parallel lines that are not all coplanar, in the manner that we just defined in the last paragraph. This definition *resembles* the definition of the concurrence of three lines, whether they are all in the same plane or not, which we suggested earlier in this lecture, but it differs from the latter in that it defines the *concurrence or parallelism* of any three lines, and not their simple concurrence. To prove the adequacy of this definition is a simple matter, and reduces itself to the proof that if *a* and *b* intersect, and *l, m,* and *n* each form with *a* and *b* a triad of concurrent lines, *l, m,* and *n* are concurrent, and that further if *a* and *b* are parallel, and *l, m,* and *n* each form with *a* and *b* a triad of parallel lines, then *l, m,* and *n* are all parallel to one another. These two propositions are obvious on inspection. We thus see that our definition of parallel or concurrent triads of lines covers those triads, and only those triads, of the lines of our last lecture that we should naturally call concurrent or parallel triads.

To sum up what we have said in this lecture: we saw that the definitions of our last lecture yield us only those points and linear segments within a certain limited region of space. We found it necessary, therefore, to search for a definition of all the points and lines of space in terms of those lying inside this region, and found our problem analogous to that of the astronomer in the location of a planet by its parallax. We learned from this example that if we were in possession of a definition of the concurrence of three lines, we could define a point anew as a class of concurrent lines of the sort defined in our last lecture, and thus obtain a system of points extending throughout space. We searched for such a definition of the concurrence of three lines, but found instead a definition of the concurrence or parallelism of three of the lines of our last lecture, involving no concrete notion other than that of our experience of the intersection of convex solids. The problems that remain before us in the next lecture are first, that of observing what effects the difference between the relation of concurrence, for which we sought, and that of concurrence or parallelism, which we obtained, will involve with respect to our new points and their definition, and secondly, that of the definition of the lines that connect our new points.

NORBERT WIENER.

MASSACHUSETTS INSTITUTE OF TECHNOLOGY.

THE RELATION OF SPACE AND GEOMETRY TO EXPERIENCE

IV

The Extension of Space Beyond the Bounds of Experience*

IN our last lecture, we arrived at a definition of a general triad of parallel or concurrent lines, instead of what we were looking for—a definition of a general triad of concurrent, not of parallel lines. We wish now to ascertain to what extent the definition we obtained may take the place of that for which we searched in the definition of a point as a class of concurrent lines. Now, it is one of the commonplaces of elementary geometry that parallel lines share many of the most important properties of intersecting lines. Two coplanar lines, for example, must either be parallel or intersect; three planes not all possessing a line in common may intersect either in three concurrent or in three parallel lines; and so on indefinitely. We have just given a definition of a generalized point in terms of the relation of concurrence among three lines: we said that a generalized point was a class a of lines, in the sense in which lines have been already defined, such that there are

* Professor G. A. Pfeiffer of Columbia has called to my attention the fact that the definition of a point as a set of convex solids each intersecting each and excluding no solid intersecting all, is too inclusive. It includes the set of all convex solids intersecting all the sides of a given triangle. The definition should be amended to read: A point is a set of convex solids each intersecting each, and such that if every solid of the set intersects some convex solid of a set σ, then some solid of σ belongs to the set.

two intersecting lines, *l* and *m*, which are so related to *a* that *a* is the class made up (1) of all the lines *x* which are of such a nature that *l*, *m*, and *x* are concurrent, and (2) of the lines *l* and *m* themselves. The analogy between intersection or concurrence, on the one hand, and parallelism, on the other, suggests to us an interesting logical experiment: let us substitute the word "coplanar" (i. e., parallel or intersecting) for "intersecting" in the above definition, and the phrase "concurrent or parallel" for the word "concurrent," and let us see what we shall obtain as the result. The amended definition will now read as follows: A generalized point is a class *a* of lines, in the sense in which lines have been already defined—that is, to all intents and purposes, a class of all the linear segments which span that region of space wherein convex solids are experienced to intersect—such that two such lines, *l* and *m*, which lie in the same plane can be assigned which determine *a* in such a way that it is made up (1) of all lines *x* such that *l*, *m*, and *x* are either concurrent or parallel and (2) of the two lines *l* and *m* themselves. Let us consider this definition thoroughly, and see what it means. Two alternatives and only two are open: either the *l* and *m* by which *a* was determined intersect one another, or they do not intersect one another, although they are by hypothesis coplanar—that is, they are parallel. First, suppose that *l* and *m* intersect. Then there can be no lines parallel to both, for no line can be parallel to both of two intersecting lines in a space in which the lines we have defined have the properties we have attributed to them—that is, in a space in which our initial experience of the intersection of convex solids has the properties that we should naturally attribute to it. Consequently, all the lines which are either parallel or concurrent with both *l* and *m* must be concurrent with both *l* and *m*. For this reason, the generalized point *a* is made up of all the lines which, when taken together with *l* and *m* form concurrent triads—or in other words, which pass through the intersection of l and m but are distinct from either—together with the lines l and m themselves. a, that is, is the class of all the lines, in the sense in which we have already defined lines, which pass through what we should ordinarily call a certain point—namely, the intersection of l and m. That is, all the things that we wished to call generalized points remain generalized points when we replace our original definition of a generalized point by the one just given. There are, however, things that are generalized points in accordance with our new definition which would not be called generalized points under our old definition. If the l and the m by which our generalized point a is defined do not intersect, then since they are coplanar, they must be parallel. It is manifestly impossible for three lines, two of which are parallel, to be concurrent, so that all the lines x which are parallel or concurrent with l and m both must be parallel with l and m, and consequently the generalized point a reduces to l and m and all the lines parallel to them: that is, to all of those lines that we have defined which point in a single direction. These generalized points which consist in all the lines in the region accessible to our experience pointing in some direction and are generalized points by our second definition and not by our first tentative one we shall call *irregular* generalized points, for example, a direction may be regarded as an irregular generalized point. If our experience of the intersection of convex solids behaves as it should, every generalized point either corresponds to a point of ordinary geometrical space or is a direction.

It will be noticed that the definition of generalized points on which we finally agreed satisfies that criterion which we have insisted that every definition in this paper should fill—that it introduces no notions other than that of our experience of the intersection of convex solids or notions which have already been defined in terms of this. This is the case because it involves only the notions of a plane and of concurrence or parallelism, which have already been defined in the requisite manner. Here is perhaps as good a place as any to repeat at some little length the reason why we are always so insistent that our definition shall involve no other notion than that of our experience of the intersection of convex solids and notions already defined in terms thereof, except, of course, such purely abstract and logical notions as those of class, member, disjunction, etc. It will be remembered that our object in this course of lectures is to show that the geometrical properties of space are due, not to any peculiar formal properties of our original spatial experience, but to the method in which the points, lines, planes, distances, etc., of geometry are defined as constructions from or tabulations of our original spatial experience. Therefore, while it is perfectly in order for us to make use of the conventional geometrical notions of lines, points, planes, etc., and of the theorems of geometry concerning them that we may show that the things within our system of definitions have such not purely formal properties as we should naturally expect entities of their respective names to have, we must not introduce into our definitions any notions which we do not either explicitly start from or reach by some chain of definitions starting from these explicit primitive notions alone. Otherwise, we have not given an adequate characterization of the method by which we attain to our final geometrical concepts, and are hence not able to give a satisfactory proof that the fact that these notions obey the laws of geometry results simply and solely from the method by which they are reached and defined in terms of the notions which we have explicitly taken as primitive. It is for this reason that we do not make use, for example, of the first definition of generalized points which we gave; we had in our possession no criterion of the intersection of two lines or of the concurrence of three lines which was independent of our every-day geometrical notions and depended solely on our experience of the intersection of convex solids, which we have taken as primitive, so that our first definition of a generalized point, which was essentially dependent on the notions of intersection and of concurrence, failed to satisfy our demand that the purely geometrical properties of the entities defined by it should result wholly from the method in accordance with which they were derived from our fundamental experience.

Our final definition of generalized points is, then, an adequate definition, and covers all the entities which one would naturally call points, but it still suffers from a defect which, it is true, is essentially different from the one that constituted an objection to our very first definition of points, which this new definition of generalized points was designed to replace. Our first definition of points was defective in that it did not give us enough points—instead of yielding us all points in space, it only yielded us all those points in a certain region of space—while our definition of generalized points has the disadvantage of covering, not only such entities as our good ordinary every-day points, but also such entities as directions, which would not be termed in points in ordinary Euclidean geometry. The problems then arise, how can we distinguish our good, ordinary, every-day points from the things which we have called directions or irregular generalized points? and, can this be done by finding some property, possessed by all directions, not possessed by any other generalized points, and definable in terrms of our experience of the inetrsection of convex solids alone? We shall discuss both these questions in the next lecture and give reason why we should answer the second of them in the affirmative. We shall thereby have given a definition of a point which, to say the least, apply to a set of entities correlated in a one-one manner with the points of ordinary space (providing that the experience of the intersection of convex solids has those properties which we should naturally expect it to have), which might possibly be said to *be* the set of all the points of ordinary space, seeing that we do not yet know what the points of ordinary space are. This set is the set of all generalized points, with the irregular generalized points removed from it. The definition of these points will not, however, be that which we shall accept as final in this course of lectures, for this definition of "point" will not secure to space its ordinary geometrical properties, whatever the formal attributes of our experience of the intersection of convex solids may be, whereas we desire to arrive ultimately at definitions of points, lines, distances, etc., which will of themselves secure to space its geometrical properties, and will entail as consequences the usual theorems of geometry.

Another thing to which I desire to call attention again in this place is the essential difference between the things which we called points in accordance with our first definition of that term and generalized points. The class of all our generalized points is not made up simply of all the points, in the first sense in which we defined points, together with certain additional entities; all our generalized points are classes of the lines that we have already defined and all such lines are classes of points, in the first sense we gave to that term. Consequently, none of these latter points can any more be generalized points than a collection of collections of tables can be a table, or a group of families of human beings can be a human being. Nevertheless, to a point in our first sense, there may happen to correspond a certain generalized point. If our experience of the intersection of convex solids has the properties that it would naturally be expected to have, the class of all the lines, in the sense in which we have already defined lines, which contain as a member a certain point, in our first sense, constitutes a generalized point which, we may say, corresponds to or represents the point, in our first sense. It is not by any means true, however, that, whatever formal properties our initial experience of the intersection of convex solids might have, it would follow *merely from the definitions of points, lines, and generalized points in terms of our initial experience* that the class of all the lines passing through some point, in our first sense, is a generalized point. Now, we desire to give a definition of the generalized points corresponding to given points, in our first sense, which, while it will yield the same result as the definition just suggested in the case where our fundamental experience behaves itself, and will make the generalized point consisting of all the lines that we have defined which contain a given point, in our first sense, correspond to this point, will not be dependent for its importance, to all intents and purposes, on the hypothesis that our initial experience of intersection has not been misbranded and possesses all those properties which we should naturally associate with its name. This result we secure by saying that if a generalized point contains two distinct lines as members, while these lines themselves both contain a certain point of our first sort as members, the generalized point shall correspond to the latter point. If points in our first sense are analogues of those of the points of an ordinary geometrical space which lie within a certain closed convex region, as they are when our initial experience behaves itself, and generalized points are analogues of pencils of all the linear segments intercepted by the surface of this region which are concurrent or parallel with a given coplanar pair of such segments, including the linear segments belonging to this pair, then, since if any such pencil contains two segments both containing a given point it contains all of our segments that pass through this point, any generalized point which contains two distinct lines passing through a given point, in our first sense, will consist of all of those of our lines that pass through this point. If our initial experience does not behave itself, however, and has other formal properties than those which we have attributed to it in this and previous lectures, it will be a much more common thing for a generalized point to contain two lines passing through a given point than for it to contain all the lines passing through that point, so that our new definition of the correspondence between a point, in our first sense, and a generalized point will be much more frequently applicable than the one which we have rejected. It should be noticed that our new definition of this relation of correspondence offers us no security that to each of the points, in our first sense, there shall correspond one point and one only, nor that the converse correspondence is unequivocal. To one generalized point there may correspond many points in our first sense, and to one point in our first sense there may correspond many generalized points. We can define unequivocally, however, the *class* of generalized points corresponding to a given class of points in our first sense. This class is made up of all the generalized points which correspond to any member of the class of points in our first sense, whether this correspondence is one-one or not. If we possess a definition of the class of all the points, in the first sense, in some region, we may define in this manner the class of all the generalized points in the same region.

We have completed, then, our discussion of generalized points. To be able to make any use of generalized points, however, we must be in a position to assert geometrical statements concerning them. The first and most important of all geometrical statements are those which concern the collinearity of points or the concurrence of lines. To be able to make such statements we must know what a line is and when three points are collinear. Since our generalized points are classes of lines, in a sense of the term which we have already defined, such that all the lines belonging to a given generalized point are either parallel or concurrent, and since, when these lines are concurrent, the ordinary geometrical point to which our generalized point corresponds is the apex of the pencil of lines constituting the generalized point, it might be thought that a necessary and sufficient condition for the collinearity of three generalized points would be that they should contain in common a member—i. e., a line in our previously defined sense. One should bear in mind, however, that the things which we called lines in accordance with our former definition did not correspond to the complete lines of ordinary geometry, but rather to the linear segments intercepted by the surface of a certain closed convex region of space to which we shall refer briefly as R, whereas there are generalized points representing all the points in space. Since R is closed, there are lines in space which do not intersect R. Let *l* be one of these, and let A, B, and C be three distinct points on *l*. Since, to put it crudely, a generalized point is made up of all the linear segments intercepted by the surface of R which are aimed at some point or in some given direction, if we should adopt the definition of collinearity which would make three distinct generalized points collinear when and only when they all contain some member (i. e., some line in the sense already defined) in common, A, B, and C cannot be collinear, for there is no segment of a line intercepted by the surface of R that points at all three of them. We must therefore search about for another definition of collinearity.

The definition of collinearity that we have just rejected is, however, a sufficient condition of collinearity and an adequate preliminary definition of it, provided that the three generalized points A, B, and C, whose collinearity we assert, are on a line which passes through the region R. It is fairly obvious, too, that if A, B, and C be any three generalized points, if *l* is a member of A, *m* is a member of B, and *n* is a member of C, and if *l*, *m*, and *n* are coplanar, in the sense in which we defined the coplanarity of our lines in our last lecture, we should naturally say that A, B, and C are all in the plane of *l*, *m*, and *n*. Furthermore, if R is any solid region in space and *l* is any set of points in space, it is obvious that if at least two planes, *p* and *q*, can be contructed in such a manner that each contains *l* and each possesses a planar area in common with R, *l* is made up solely of the points in some line. These two planes have only the points of the single line *l* in common. A, B, and C, then, are collinear, if they both belong to two distinct planes passing through R in planar regions. That is, by virtue of the fact which we stated just a second or so ago, if *l* and *l′* are members of A, if *m* and *m′* are members of B, and if *n* and *n′* are members of C, while *l*, *m*, and *n*, on the one hand, and *l′*, *m′*, and *n′*, on the other, form triads of coplanar lines, A, B, and C are collinear, provided that the plane of *l*, *m*, and *n* is different from that of *l′*, *m′*, and *n′*. This latter condition may be formulated as follows: there shall not be any two distinct intersecting lines of our first sort meeting at a point *a* of our first sort which are each cut in a point other than *a* by every one of the six lines, *l*, *m*, *n*, *l′*, *m′*, and *n′*. On account of this fact and on account of the fact that the coplanarity of the lines of our first sort can be defined solely in terms of our experience of the intersection of convex solids, the necessary condition which we have given for the collinearity of three generalized points is formulable purely in terms of notions that can themselves be defined ultimately in terms of our experience of the intersection of convex solids alone. To show that we can regard this condition as a definition of the collinearity of three generalized points, it simply remains for us to prove that this criterion is also a sufficient criterion of the collinearity of three of those generalized points that correspond to ordinary geometrical points, and to investigate what is the natural interpretation in the language of ordinary geometry of the relation of collinearity thus defined when it relates directions to one another or to ordinary generalized points. Then we may be sure that the relation which we have defined as collinearity will have the properties one would naturally associate with its name when it connects ordinary generalized points, and we shall know the translation into ordinary geometrical language of the relation we call collinearity, when it holds among directions or between them and other generalized points. We shall finally close this chapter with a definition of "generalized lines"—i. e., lines made up of generalized points—in terms of the relation of collinearity.

Our first remaining task is, as we have said, to show that the condition we just gave for the collinearity of three points is a sufficient one. We wish to prove, that is, that if A, B, and C are three generalized points which we should naturally call collinear, then six lines in our first sense of the term—namely, l, l', m, m', n, and n'—can be assigned in such a manner that l and l' shall belong to A, m and m' to B, and n and n' to C, while l, m, and n, on the one hand, and l', m', and n', on the other, form coplanar triads of lines in distinct planes. Returning to the geometrical interpretation of the lines in our first sense, we see that this condition will be fulfilled if it is always possible, given a convex solid region in space, and a set of three collinear points anywhere in space, to draw six lines, two passing through each of the points in question, in such a manner that all pass through the solid region, and that there are two distinct planes, each containing one line through each point. This is practically equivalent to the obvious proposition that two distinct planes can be drawn, passing through any line you please, and cutting a given solid region of space in planar areas. Consequently, if our experience of the intersection of convex solids lives up to its name, the relation which we have defined as the collinearity of generalized points, in so far as it applies to generalized points other than directions, is at least the precise analogue of the relation known by the same name in ordinary geometry. All that remains now is to consider the properties of the relation we have defined as collinearity when it relates irregular generalized points to one another or to ordinary generalized points.

There seem, at first sight, to be three possible ways in which directions may enter into relations of collinearity. It seems *a priori* possible that either (1) three directions should be collinear, or that (2) two directions and one ordinary generalized points should be collinear, or that (3) one irregular and two ordinary generalized points should be collinear. However, we shall see that, provided our initial relation of apparent intersection behaves itself, the second alternative just suggested cannot occur. For, if it does occur, let us suppose that the two directions in question are A and B, and that the ordinary generalized point is C. Then, by the definition of collinearity, there are two distinct planes, each containing a member of A, a member of B, and a member of C. Since A and B are each sets of parallel lines, and represent different directions in space, the two planes in question are parallel, for it is an elementary geometrical theorem that two distinct planes such that one of them contains two given lines in different directions and the other of them contains two lines parallel, respectively, to these lines are parallel. Since, however, each of the two planes contains a member of C, and since C, being an ordinary generalized point, is made up of segments of intersecting lines, the two planes in question intersect. We thus obtain a flat contradiction in the case where two of three collinear generalized points are directions and the third is not, and where our initial experience has the formal properties that are attributed to it in ordinary geometry.

Let us, then, consider the first of the two remaining alternatives. What are the conditions under which three directions will be collinear in accordance with our definition? Let these irregular generalized points be A, B, and C. By our definition, they will be collinear if there are two distinct planes such that each of our three points contains a line in each of the planes as a member. By what we said in the last paragraph, these two planes must be parallel. Three directions, that is, are collinear when and only when each of them possesses as a member a line, in the sense already defined, in each of two given parallel planes. Since, however, a direction is made up of all the lines, in our first sense, parallel to a given line of that sort, and since if two planes are parallel each of them contains some line parallel to any line you please of the other, the condition that we have just given for the collinearity of three directions is equivalent, provided that our points, lines, etc., have the ordinary geometrical properties, to the condition that some member of A, some member of B, and some member of C be coplanar, or in other words, since we should naturally say that a point lies in the same plane as a line through it—that is, as one of its members—it is equivalent to the condition that A, B, and C all belong to some ordinary plane. To sum up, in a space that obeys the laws of geometry, the collinearity of directions is equivalent to coplanarity, in some ordinary geometrical plane.

The only remaining alternative is that the three points A, B, and C should consist of two ordinary generalized points and one direction. Let A and B be the ordinary generalized points, and let C be the direction. If A, B, and C are to be collinear, there must be two distinct planes, each of which contains a member of A, a member of B, and a member of C. We should naturally say that A and B are on the intersection of these two planes, since each is situated on (i. e., contains as a member) a line in each plane. Since the two members of C lying, respectively, in each of these planes are parallel, as C is made up of all those of our lines that point in some single direction, both of these lines must, to put it crudely, be parallel with the intersection of the two planes, the line AB, for it is a theorem of elementary solid geometry that if *p* and *q* be any two intersecting planes, if *l* is a line in *p*, and if *m* is a line in *q*, the only condition under which *l* and *m* can be parallel is that both should be parallel with the line of intersection of *p* and *q*, unless one of them coincides with this line, and the other is, of course, parallel with it. For suppose the contrary to be the case. Then the line of intersection of *p* and *q*, since, by the definition of l, it is coplanar with *l*, and since, by hypothesis, it neither is parallel to it nor coincides with it, must cut *l* in a single point. *m* and *l* are coplanar, since they are parallel, and consequently the intersection of *p* and *q*, which cuts them both, lies in the same plane as *l* and *m*. Therefore the plane which *l* determines together with this intersection—namely, *p*—is identical with the plane that *m* determines together with this intersection—that is, *q*, and since we assumed that *p* and *q* were *two* planes, and not a single plane, we get a contradiction. This proves our theorem. As a consequence, if space is decently behaved, two ordinary generalized points A and B are collinear with a direction C when and only when C may be said to be made up of lines parallel to the line AB. In this case we may very naturally call C the direction of AB.

We are now in a position to define generalized lines. The generalized line AB is the class of all generalized points that are collinear with A and B, together with the two generalized points A and B themselves. In a properly behaved space generalized lines, like generalized points may be divided into two classes: ordinary generalized lines and irregular generalized lines. An ordinary generalized line is made up of all the ordinary generalized points which we should usually regard as lying in that line, together with the irregular generalized point which is the direction of the line. As a consequence, two parallel ordinary generalized lines always intersect in an irregular generalized point. An irregular generalized line is made up of all the irregular generalized points on these ordinary generalized lines which, we should naturally say, lie in a certain plane. This one may readily show. Every generalized line, therefore, contains irregular generalized points, and since even parallel generalized lines intersect—in irregular generalized points—it is clear that we must sooner or later find a way of distinguishing irregular generalized points from ordinary generalized points, in order that we may remove the former from our system, and obtain a set of generalized points which corresponds completely to the set of all the points in ordinary geometrical space; and hence, since we are still unaware just what the set of all the points in our ordinary geometrical space is, by accomplishing the task whose necessity has just been indicated, we shall obtain a set of entities which may be said to *be* the set of all the points in our ordinary geometrical space. To do this with the introduction of few or no new sorts of experience is, as we have already said, the task of our next lecture.

We have succeeded, then, in defining all the points and lines of space, together with certain additional entities which we have called *irregular* points and lines—they really include the points at infinity and lines at infinity, respectively, of projective geometry—in terms indirectly of our experience of the intersection of convex solids, but directly in terms of the points in a given convex region of space and of the linear segments intercepted by the surface of this region. The method which we have used here in the solution of this latter problem is not original, but is due to Bonola, who makes use of it in an article that appeared about 1903 in the *Giornale di Matematice.* We have omitted the theorems which he proves there, however, and have only made use of his definitions and method, because it is only these that are relevant to the main purpose of our paper. The general notion of defining points as classes of lines dates back at least as far as Pasch, *Vorlesungen über der neueren Geometrie*, and has since been used by many mathematical writers, notably Schur and Whitehead.

V

Parallel Lines and their Empirical Basis

IN OUR previous lectures we have arrived at certain definitions which enable us to express the lines and points of our ordinary, every-day space, or at least entities that correspond to them in a one-one fashion, under a few quite natural hypotheses, in terms of our experience of the intersection of convex solids, where only such solids as come nearer to us than a certain maximum distance are experienced as intersecting at all. These definitions, moreover, yield us entities corresponding to *all* the points and lines of our every-day space. As we showed in our last two lectures, the reason that a definition of *all* the points and lines of space is such a necessary step in the definition of geometrical notions in terms of our experiences is that unless we have definitions of points and lines which will give us all the points and lines of space, an adequate definition of parallelism is impossible; and without an adequate definition of parallelism one cannot introduce measurement and such notions as depend on measurement (e. g., distance, angle, area, and volume) into geometry, notwithstanding the fact that the study of these is one of the most important portions of ordinary Euclidean geometry. Now that we have given definitions that cover all the points and lines of geometry, it might be thought that it would be a very easy task to define the relation between two lines which consists in their being parallel. Two lines, one would naturally say, are parallel if and only if they lie in the same plane, but have no point in common. This definition, however, is objectionable for a reason different from any which has led us to discard certain of the definitions which we have formerly made: the space of points and lines which we reached in our last lecture is not, it is true, too poor and niggardly in points and in lines, but—and this is almost as bad—it is too *rich* in points and in lines. It will be remembered that we showed in our last lecture how, under the hypothesis that our experience of the intersection of convex solids has such formal properties as the name we have given it suggests, the entities that we defined as "generalized points" will not, to put it crudely, all turn out to be the points of ordinary geometry, for to some of our generalized points there will correspond in our everyday space, not points, but such *directions* as east or north or up. We further showed how, under the same hypothesis any two of the things we defined as "generalized lines"—i. e., lines made up of generalized points, always intersect if they are in the same plane, even if they are parallel, since a generalized line contains its direction as if it were a point upon it, and two parallel lines always have the same direction—for we do not regard east and west, or up and down, or any other two opposite directions as distinct. Since coplanar generalized lines intersect, even if they are parallel, it is impossible to define two generalized lines as parallel when and only when they are coplanar and do not intersect. How, then, are we to define parallelism in terms solely of our experience of the intersection of convex solids, which we have chosen to regard as primitive in this course of lectures? Manifestly, the simplest way to do this is to find some intrinsic peculiarity of those generalized points which represent directions, so that we may recognize these intruders and remove them from our system. But before we can execute the sentence of death on these anarchistic invaders, we must find out which of our generalized points are the culprits and which are not. Remember, what we are trying to do is to define all geometrical notions in terms of our experience of the intersection of convex solids *alone*. Either the diagrams we draw on the blackboard and the references we make to ordinary Euclidean geometry are only aids to intuition, to enable us to picture to ourselves what we are doing, and to keep us from getting tangled up in the prolixity and dullness that are inseparable from any purely abstract chain of definitions, or else these diagrams and geometrical phrases and chains of reasoning are just so many assurances that, *if* the experience of the intersection of convex solids with which we have started has such properties as one would naturally attribute to an experience of that name, the entities that we have called points and lines will have many properties belonging to the lines and points of ordinary geometry, and might be regarded as actually identical with the points and lines of ordinary geometry, without any very extravagant alteration of the usual meanings of these terms. Except in such merely illustrative and secondary ways, no use whatever of geometrical theorems, notions, or figures is to be made in this course of lectures. It is consequently not possible to us to adduce the usual geometrical notion of parallel lines for the purpose of distinguishing ordinary generalized points from directions, in any such way as to render this distinction an integral part of that geometry which we are trying to build up from the beginning, on the basis of experience alone. Until we reach a definition of parallel lines dependent upon our experience of the intersection of convex solids and upon no other notion, or else explicitly augment our list of those experiences which we take as primitive in such a manner as to be able to give a completely adequate definition of parallel lines—and we have done neither of these things so far—we are not entitled to define a generalized point as a direction when and only when it is made up of parallel lines of our first sort, even though we thus defined directions in our last lecture. We have thus the problem before us: how are we to distinguish directions from other generalized points, either by means of a definition of the parallelism of the lines which we defined in our second lecture, which definition must be made entirely in terms of experience, or through some method not dependent on a previous definition of parallelism?

We shall develop in this lecture two alternative definitions of the undesirable irregular generalized points or directions. One of these definitions will involve no notion not already defined by us in terms of our experience of the intersection of convex solids, and will be to that extent superior to our other definition, which will demand the introduction of new ideas, entirely foreign to anything that has already been considered in this course of lectures. Notwithstanding this fact, the second method which we shall give for the definition of parallel lines will have certain marked advantages over the other one in that, for example, the notion which we define as parallelism by means of the latter is likely to depart further from our usual notion of parallelism than that which we obtain through the method that introduces notions not definable in terms of those already presented in this course. We shall first take up the method of definition of parallel lines which involves no new primitive experiences.

You all remember how our experience of the intersection of convex solids practically reduced itself to the actual intersection of other convex solids. These other solids we called a-solids, and we regarded the a-solid corresponding to a given convex solid as made up of the convex solid itself, together with what we called its *aura*—a certain layer surrounding it in such a manner that one a-solid is experienced as intersecting another when and only when the aurae of the two actually intersect one another. On several occasions we made the more or less natural assumption, which, it is true, was nowhere *absolutely* essential to our reaoning, that the aurae of all convex solids, and all parts of the aura of any convex solid, are of the same uniform thickness throughout. Let us suppose in all that now follows that this assumption is correct, and that the uniform thickness of all the aurae of convex solids is *t*. This being the case, it is obvious that no a-solid can be smaller or equal to a sphere of radius *t*. It is further obvious that a-solids can be found which approximate as closely as you please in shape and size to spheres of radius *t*, supposing that *t* exists, since convex solids may be constructed as small, and consequently as near to a point in size, as you please, whereas these spheres of radius *t* may be regarded as if they were formed by the adjunction to a point of an aura of thickness *t*. Now, I wish to define the class of all the points, in the first sense in which we used that term, which lie inside or on the surface of any of these spheres of radius *t* in terms of our experience of the interection of convex solids and of notions already derived from this experience alone. How am I to go about it, on the basis of what I have already said concerning such spheres?

Let it be remembered that when we defined a point as a class of convex solids, we so defined it that every convex solid which is experienced as passing through a point is a member of that point. As a consequence, the class of points which lie in the a-solid of a given convex solid—i. e., the class of all those points which are experienced as if they belonged to the convex solid, for the aura of a convex solid, which transforms it into an a-solid is that part of space in which it is experienced as lying, but does not lie—this class, I say, is precisely the class of all the points of which the convex solid is a member. It is a consequence of what we said in the last paragraph that if we take a sufficiently small convex solid, the class of all the points contained in its a-solid will approach as closely as we wish to the class of points in some sphere of radius t—or, as we shall say, to the class of all the points in some *a-sphere*. On the other hand, an a-sphere is always smaller than an a-solid. Since we can construct a-solids approximating as closely as we wish to any given a-sphere, though they never become quite as small as the latter, it is possible to determine any given a-sphere by a sequence of a-solids approximating to it. By a judicious choice of the members of such a sequence, it is possible to make each approximation to a given a-sphere surround the next more accurate one. In such a case, those points which belong to the a-sphere will be precisely those which lie within *every one* of the a-solids belonging to the sequence. We are consequently in a position to give a definition of an a-sphere, *qua* set of points, in our first sense, without introducing any other notions than such as are definable in terms of our experience of the intersection of convex solids, provided we are able to give a definition of a sequence of a-solids leading to an a-sphere. It is obvious that a partial list of these properties which a sequence leading to an a-sphere must have includes (1) the property which consists in there being no a-solid contained in every member of the sequence (for then the members of the sequence would not approach as near as you please to an a-sphere, but rather to an a-solid), and (2) the property which consists in the fact that each member of the sequence is contained in the previous one. The phrase "is contained in" which is used in the above statements has the following interpretation: one a-solid is contained in another when and only when all those points which lie in the first of the a-solids in the manner defined above also lie in the second. Using the phrase, "is contained in," in this sense, what are the various sorts of sequences of a solids which are such that (1) there is no a-solid contained in every member of the sequence, and (2) each member of the sequence is contained in the previous one? A little reflection will convince us that such sequences of a-solids approach and contain as the set of all the points common to all their members either (1) the set of all the points in an a-sphere, or (2) a cylinder of radius *t*, with hemispherical caps at its ends, or (3) a solid consisting of all the points whose distance from a given bit of a plane is not greater than *t*. It is easy to see that solids of classes (2) and (3) contain a-spheres, whereas no a-sphere contains another. We may therefore *define* a set of points, in our first sense, as the set of all the points in some a-sphere, when and only when it is made up of all the points common to all the members of a sequence of a-solids such that each term of the sequence contains the next, though there is no a-solid contained in every member of the sequence, provided that our set of points contains as a part no other set satisfying the conditions just formulated. To avoid certain exceptional possibilities which may arise when the a-spheres which we have just defined lie on the boundary of the region within which we experience the intersection of convex solids, we shall recast our definition so as to exclude all such sets of points as contain either but a single point or no point at all from the sets of points which we call a-spheres. If you follow this definition of an a-sphere step by step, you will see that no notion is involved in it other than that of our experience of the intersection of convex solids and notions which have already been defined in terms of this by various logical artifices.

Now that we have defined our a-spheres, how are we to make use of them in attaining the true goal of this chapter—the definition of the irregular generalized points? We are still a long way off from the solution of this latter question. Let it be noted, however, that we have defined, practically speaking, the class of all those spheres with radius *t* which lie in a certain region of space. Furthermore, any two equal spheres determine uniquely a right circular cylinder into which they both fit as a marble fits into a paper cylinder wrapped tightly about it, while any two lines lying wholly within the surface of such a cylinder are both parallel to the axis of this cylinder, and consequently to one another. Consequently, a generalized point will be a direction whenever it contains two members —i. e., two lines, in our first sense of the term—which both belong to a single cylindrical surface of the sort just indicated. It seems further likely that we can find a cylinder determined by two a-spheres pointing in any desired direction. Under this hypothesis, we may regard a generalized point as a direction when and only when it contains as members two distinct lines both situated within the surface of some cylinder determined by two a-spheres. As a consequence, if we are able to give a general definition of what it means to say that two lines, in our first sense, both lie within the surface of some cylinder determined by two a-spheres, we possess a completely adequate criterion by which we may distinguish directions from ordinary generalized points. What remains for us now is the definition of the surface of a cylinder determined by two a-spheres in terms of notions which we have already carried back to experience.

In our second lecture, we gave a definition of a linear segment made up of points, in our first sense of that term, and of the end points of such a segment. We shall say that a point p is *properly between* two other points, q and r, if p is distinct from both q and r, but lies on a linear segment of which they are the end-points. If A and B are two classes of points, we shall say that the region between A and B is the class of all of those points between some member of A and some member of B. Since a-spheres are so defined as classes of points as to include the points on their surfaces, the region between two a-spheres is the class of all points inside (not on the surface of) the two spheres, together with the class of all points *inside or on the surface of* the right circular cylinder stretching from a great circle of one a-sphere to a great circle of the other. Now, we wish to be able to define the surface of this cylinder, and, in general, the surface of the region between any two classes of points. But what intrinsic difference is there between the surface of a region and that part of it which is not surface? One would naively say that the surface of a region is that part of it which one can reach without digging down into it. Let us analyse this notion and see what the essence of it really is. To say that one can only reach a certain point of a region by digging down into it really means, to put it crudely, that if we probe into the region with some straight thing, such as a wire, the wire must touch the region some time *before* it reaches the point in question. Stated in precise language, this reads: if a linear segment contains a point in a region, but not on its surface, it contains at least one other point in the region besides. Since the surface of the region consists of those points of it which we can reach *without* digging down into the region, a strict definition of the surface of a region reads: the surface of a region R is the class of all those points which belong to R and also belong to some linear segment, taken as including its end-points, which has no other point in common with R. It will be found that this definition is in complete accord with our usual notions of the properties which the surface of a region should have. Let it be noticed, too, that we have so framed our definitions of the region between two solids that the surface of the region between two a-spheres consists only of the portion of a right circular cylinder lying between a great circle on one a-sphere and a great circle on the other, and does not contain the surface of the two projecting hemispherical caps which must be adjoined to this cylinder to make the complete region between the two a-spheres—for the hemispherical caps do not include their outer surface, as no point of this outer surface is situated between some point of one a-sphere and some point of the other, so that any linear segment which contains a single point of one of these caps contains other points of the cap as well.

Now that we have arrived at the definition of the surface of a cylinder stretching from a great circle of one a-sphere to a great circle of another, we are in a position to define a line lying in this cylinder, or, as mathematicians say, an *element* of this cylinder. We shall say that a straight line, in our first sense, is an element of one of the cylinders which we have just defined if and only if it contains a linear segment in common with the cylinder, for then it is obviously a prolongation of a piece of line lying entirely within the cylinder. As we have already said, all that remains for us to do if we desire a definition of irregular generalized points in terms of our experience of the intersection of convex solids is to define a direction as a generalized point which contains two distinct elements of some one of the cylinders that we have defined in terms of a-spheres. We can then go on and define two *generalized* lines as parallel when and only when they both contain some particular direction, or irregular generalized point, in the sense in which we have just now defined such entities. Although this definition of the parallelism of two generalized lines is made in terms of directions, which are themselves defined in terms of the parallel elements of a cylinder determined by two a-spheres, we cannot eliminate the intermediate step of defining directions, first, on account of the fact that the elements of a cylinder such as we have described are lines of our first sort, while we desire to define the parallelism of generalized lines, and secondly, because only such lines as are not further from one another than twice the thickness of the aural layer of a convex solid can both be elements of one of our cylinders, even though they be generalized lines.

Another point to notice concerning the definition of parallelism that we have just given is that the a-spheres which play so large a part in it are really nothing but the *minima sensibilia* of the philosophers, for an a-sphere is simply the region which a point seems to fill, to put it crudely. Of course, we do not mean to say that a-spheres are actually perceived; they may or may not be, but whichever is the case, they mark, like the *minima sensibilia* of Hume, some sort of a lower boundary of the perceivable, with respect to its spatial extension. The assumption which we have made that a-spheres are equal—that is, that the aural layer of all a-spheres and all parts of the aural layer of each a-sphere are of the same thickness—is, then, practically the same assumption as that which the philosophers of the English school made when they degraded the magnitude of a spatial region as determined by the number of *minima sensibilia* or "points" in it, which amounted to the same thing as supposing that all *minima sensibilia* were equal. Our formulation of this view, however, avoids many of the crudities of the orthodox Humian position. We have not avoided the difficulty, however, that it does not seem quite likely that the acuteness of our discrimination between bodies which do and bodies which do not intersect is the same throughout space. As a consequence, it would appear, contrary to what we have assumed, that the *minima sensibilia* remote from us and towards the boundary of that part of space which is open to our experience at all are larger than those more centrally situated, for the cruder our experience is, the larger must a thing be if it is to be perceived. This fact would make our "cylinders" determined by two a-spheres more or less conical in form, so that it would be possible for one of them to contain two non-parallel lines, or even not to contain any two parallel lines at all; and consequently our "cylinders" would be of no avail to us in the discrimination of irregular generalized points from ordinary generalized points. Therefore, though the definition of directions and that of parallelism which we have given define certain things and relations in terms of our experience of the intersection of convex solids alone, and in so far satisfy the requirements that must be satisfied by all our definitions, we have not any sufficient reason, to say the least, for supposing that the directions and parallelism that we have just defined represent precisely the directions and parallelism which we meet in the space of ordinary geometry. Since this is the case, an alternative definition of an irregular generalized point is desirable, as is also a definition of parallelism dependent thereon, and we shall try so to frame these new definitions that they shall not be entirely dependent for their applicability to the space of ordinary geometry on the assumption that all aurae of convex solids are of the same uniform thickness.

The first thing to notice is that any method which gives us three distinct directions that are not all directions of lines in any one plane—such directions as east, north, and up, for intance, satisfy this condition—that any method which does this, I say, yields us all the directions in space. We have already seen that any three directions lie in a single generalized line, in our every-day space, when and only when they are all, like east, north-east, and north, directions of lines in some single plane. This being the case, given any two distinct lines made up entirely of irregular generalized points and any particular irregular generalized point whatever, it is possible to construct a generalized line passing through the generalized point in question and cutting our two given irregular generalized lines in two distinct points; for, given any two non-parallel planes in ordinary space, and any direction whatever, one can construct a plane containing that direction—that is, parallel to any line pointing in that direction—and either cutting our two given planes in two distinct lines which point in different directions or else parallel to one of our two given planes, so that it contains every direction belonging to that one of our given planes. The reason why we have insisted that our two given planes shall not be parallel is that two parallel planes represent the same generalized line, for they contain the same directions. The facts which we have just stated, together with the fact that, in our ordinary space, only a direction can be collinear with two directions, entitle us to define the set of all irregular generalized points or directions in terms of any already known non-collinear triad of directions in the following manner: if A, B, and C are three known directions which are not all collinear, then a direction will be a generalized point which is collinear with a generalized point on AB and a distinct one on AC.

How are we to determine our A, B, and C, however?

Here we shall for the first time introduce into our definitions a notion not defined in terms of our experience of the intersection of convex solids. Let us suppose that we have somehow or other, by means into which we shall not now inquire, obtained the knowledge, not of the general notion of a sphere, but of four given sets of points, in our first sense, which we shall call unit spheres, whose centers are not all coplanar. What is required here can easily be given by experience, as it is not a general criterion of sphericity which we postulate, but an empirical acquaintance with a concrete set of entities. Let these four spheres be called X, Y, U, and V, respectively. The definitions of cylinders, their surfaces, and directions determined by them, which we gave in connection with our first theory of parallelism had no essential dependence on the fact that the spheres to which we applied them were a-spheres

—they can be applied without any alteration to the determination of an irregular generalized point in terms of any two equal spherical collections of points, in our first sense. As a consequence, since X, Y, U, and V, being all unit-spheres, are all equal, the cylinders XY, XU, and XV all determine irregular generalized points. These three points are not all collinear, as each is the direction of the axis of the corresponding cylinder, and these three axes are, by hypothesis, not all coplanar. As a consequence, we are able to define the class of all irregular generalized points in terms of the three directions, XY, XU, and XV, and therefore ultimately in terms of the four points X, Y, U, and V, and our experience of the intersection of convex solids, and of no other notion. We can now go on, as before, and define two lines as parallel when and only when they both contain a direction in common. This definition of parallelism does not, like our first definition, presuppose some such probably false assumption as that of the uniform thickness throughout all space of the aural layer of a convex solid, and moreover, as we shall see later, lends itself more readily than our first definition of parallelism to the introduction of metrical considerations into geometry, but it labors under the grave disadvantage that it does not depend exclusively on our experience of the intersection of convex solids, as our former definition does, but demands in addition the selection of four distinct convex solids from all those in the universe, not merely as spheres, but as the particular set of equal, non-coplanar spheres on which our whole subsequent theory of measurement will depend; while we are given in experience no four such spheres, singled out from among all convex solids. Thus neither of our two definitions of parallelism is entirely satisfactory. That a much better definition of parallelism than either, combining their advantages and eliminating their shortcomings, can be obtained without any very great difficulty,

I do not doubt in the least, but I have not been able to formulate such a definition up to the present. You all see the philosophical interest of the solution of this problem: on it depends a satisfactory knowledge of what space really is, and of what the true meaning of geometrical propositions is, since many of the most important geometrical propositions deal either with parallelism, or with the measurement of distances, angles, etc., all of which depend in an ordinary Euclidean space upon parallelism.

To conclude: we have developed two distinct definitions of parallelism. Both of them start from the notion of the cylinder enwrapping two spheres. One of them involves ultimately no other concrete experience than that of the intersection of convex solids, but is not completely satisfactory, since, if we accept it, we must make an assumption concerning the nature of this experience which the experience probably fails to satisfy, in order to secure that the relation we call parallelism is really essentially the same as the relation usually known by that name. On the other hand, we have given a definition of parallelism which avoids this difficulty, but involves other experiences than that of the intersection of convex solids, namely, those of a certain set of four non-coplanar equal unit spheres. The present lecture covers the weakest portion of the theory which we are developing in this course, but there is every reason to believe that its weakness is only temporary, being due rather to the fact that the problems discussed have hitherto been little investigated than to the inherent nature of the subject.

In our next lecture we shall begin the investigation of how we are able to develop a theory of measurement on the basis of either of the definitions that we have given of parallel lines.

VI. The Logic of Distances

In our last lecture, we arrived at two alternative definitions of parallel generalized lines. On the basis of either of these definitions we are able to define what is ordinarily called by mathematicians a vector. I can best illustrate what a vector is by introducing a few familiar examples. In a region so small that the sphericity of the earth is not noticeable, the relations, "ten feet to the north of," "a mile south-west of," "two yards up," etc., are, to all intents and purposes, vectors. A vector, that is, is the relation which holds between one point and another when they are separated by a given distance in a given direction. Let it be noticed, by the way, that a vector leading from A to B does not, in general, lead from B to A—for example, if A is ten feet east of B, then B is not ten feet east of A, but ten feet *west* of A.

One property of vectors in ordinary geometry—the property, indeed, which we shall use in defining them—is that if A and B are separated by a given vector, if C and D are separated by the same vector, and if the points, A, B, C, and D, do not all lie on a single line, then the line AC is parallel to the line BD. It is conversely true that if AC is parallel to BD, and if, further, AB is parallel to BC, the vector which separates A from B also separates C from D. These facts follow from the very elementary geometrical theorems that if a quadrilateral has a given pair of opposite sides equal and parallel, the other two sides are parallel, and that if each side of a quadrilateral is parallel to the opposite side, ech side is equal to the opposite side. From these facts and the fact that a vector always points in a given direction, it is easy to deduce the conclusion that to say that the vector that stretches from A to B is the same as the vector that stretches from C to D is precisely equivalent to the statement that AB is parallel to CD and that AC is parallel to BD, provided that A, B, C, and D are not all collinear. Furthermore, if A, B, C, and D are all collinear, and the vector from A to B is the same as the vector from C to D, then there must be some pair of points, E and F, which do not lie on AB, such that the vector from E to F is identical both with that from A to B and with that from C to D, as there obviously exist on any line pointing in a given direction all possible vectors in that direction. We may, therefore, *define* the vector separating A and B (which we shall write (AB)) as the relation which a point C bears to another point D when and only when either (1) AC is parallel with BD and AB is parallel with CD, or (2) A, B, C, and D are all on some single straight line, and there are two points, E and F, which are not on this line, and which are such that A, B, E, and F, on the one hand, and C, D, E, and F, on the other satisfy the condition that we laid down in one for A, B, C, and D: that is, if AB is parallel to EF, if AE is parallel to BF, and if CE is parallel to DF. Furthermore, a relation is defined as a vector in general if there are two points, A and B, such that the relation is the vector (AB). It will be noticed that this definition, like all those which have already been accepted by us, involves no notions which have not either been taken explicitly as primitive experiences by us, or else been defined in terms of these primitive experiences alone. When we have given, according to which of our definitions of parallelism we choose, either our experience of the intersection of convex solids alone, or this experience together with four selected solids which we choose as unit spheres, it is unambiguously determined whether any given thing is or is not a vector. I want you all to notice that we have not presupposed in any manner any notion already involving measurement in the definition of a vector which we have just given.

The next notion which we have to define is that of all vectors lying within a given region. Suppose that α is the class of all the points in a certain region of space. Let us consider only such vectors as separate point lying in α from points lying in α. Furthermore, let us consider even these vectors as relations which hold between points in α only. The relations thus obtained will be called the vectors-in-α. To give a concrete illustration of what I mean by the vectors-in-α, suppose that we give to our vectors the geographical interpretation which we gave to them in a previous illustration of the notion of vector, but that we limit our discussion to vectors on the surface of the earth—to vectors in such directions, namely, as are represented on a compass-card—excluding those, for instance, that poinit vertically upward, or 45° downward and north-east, etc. Let α be the class of all the points on the surface of some island in the ocean. We wish to consider only such vectors as lead from points on the island to points on the island. These vectors will be the class of vectors on the island, provided that we only consider them in so far as they join points on the island to one another. Suppose, for example, that our island is circular and one mile in radius. "Three miles to the north of" will not, then, determine a vector-in-the-island, for one will be unable to find any two points in the island such that one is three miles to the north of the other. The relation, "one mile to the north of," however, will determine a vector lying in the island, for it will be possible to select two points on the island such that one is one mile to the north of the other. The *vector-in-the-island* determined by the relation "one mile to the north of" will differ from the ordinary vector "one mile to the north of" in that, if A and B are not both points in the island, while A may be separated from B by the ordinary vector, "one mile to the north of"—i. e., A may be one mile to the north of B—it is impossible for A to be separated from B by the vector-in-the-island determined by the relation " one mile to the north of." This is the sole difference between the ordinary vector "one mile to the north of" and the vector-in-the-island known by the same name. The ordinary vector "one mile to the north of" will be called in our subsequent discussion the *extension* of the vector-in-the-island of the same name, and in general we shall call the ordinary vector R, from which the vector-in-α S is obtained by considering R only in so far as it relates members of α to members of α, the *extension* of S.

The next point which we shall take up is what it means to say that one vector R is *n* times as great as another vector S. Suppose that I go from a point A to a point B separated from A by the vector S. Suppose that from B, I go still further on to a point C, which is separated from B by the same vector S. Further, suppose that I might have gone directly from A to C by the vector R, or to put it a little differently, that A is separated from C by the vector R. Then we should naturally say that R is twice as great as S: that is, if by going *n* miles to the north from A to B, and by going *n* miles to the north from B to C, we find that we have gone *p* miles to the north in going from A to C, we should naturally say that *p* miles is twice as great a distance as *n* miles. In a precisely similar manner, we can define a vector R as 3, 4, 5, 6, ... , or *k* times as great as a vector S, if by taking 3, 4, 5, 6, ... , or *k* steps of the vector S we can sometimes take one step of the vector R. Let it be noticed that I only say that *sometimes* we can take one step of R by taking *k* steps of S, and not that this is always possible. In ordinary geometry, it is true that if one vector can sometimes be obtained by repeating another vector *k* times end on end, it can always be so obtained.

We do not wish, however, to have either the intrinsic adequacy or the generality of application of our definitions depend on the axioms or theorems of ordinary geometry, and it would be easy, by substituting other relations and entities than such as we should naturally call the relation of experienced intersection among convex solids or spheres, respectively, for our experience of the intersection of convex solids and for the four selected spheres in terms of which, by our second definition, we determined the class of all the irregular generalized points—it would be easy, I say, to obtain systems in which, for some values of A and B, one could go from A to B either by one R-step or by *k* S-steps, while for other values of A and B, it might be impossible to go from A to B by a single R-step, but still possible to go from A to B in *k* S-steps. Here, of course, we assume R and S to be vectors, in the sense in which we have already defined vectors, and since our definition of one vector as a multiple of another still applies, we shall say even in this case that S is one *k*th of R. We shall further define a multiple of a vector-in-a-region in a manner precisely parallel to that in which we have just defined a multiple of a general vector: that is, we shall say that if R and S are both vectors-in-α, and it is sometimes possible to accomplish a journey consisting of one step of R by *k* steps of S, we shall say that R is *k* times as great as S. We have thus given a definition of the relation which one vector-in-α bears to another *k* times as large, in terms only of our original experience of the intersection of convex solids, and perhaps of certain selected convex solids.

We shall now define what it means to say that one vector-in-a-given-region bears a ratio to another. A vector-in-α R bears the ratio m/n to another vector-in-α S if there is a vector-in-α T such that R is m times as large as T and such that S is n times as large as T. Thus, for example, the relation, "two miles to the north of," on an island is, by our definition, two-thirds as great as the relation, "three miles to the north of," on the same island, since there is a certain relation—namely, "one mile to the north of" on the same island—which is half as large as the former relation or vector and a third as large as the second. This definition now enables us to compare different vectors-in-a-region with respect to their magnitude. Be it noted, however, that these vectors which we compare must be vectors in the same direction, for it is impossible, for example, for us to go ten miles to the east by taking any number of successive one-mile steps to the north, nor is there any step which, when repeated a certain integral number of times, will give you a one-mile step to the north, and which, when repeated another integral number of times, will yield a ten-mile step to the east. Furthermore, we are not yet in a position to compare *any* two vectors in a given direction with one another with respect to magnitude, for two distances along one and the same line may be incommensurable. Suppose that I draw on this blackboard a line one foot long vertically upward from some selected point O, and another line of the same length running horizontally to the left from the same point. Connect the two free extremities of the linear segments so obtained by a linear segment *l*. Lay off a linear segment as long as *l* stretching upward from O. It is a simple matter to give a strict mathematical proof that there will be no distance which, when multiplied by some integer, will be equal to this latter distance, while it will equal one foot, when multiplied by some other integer. As a consequence, our definition of the ratio of two vectors will not enable us to ascertain what ratio our one-foot vector upwards from O bears to our vector in the same direction of length equal to *l*, notwithstanding the fact that they are vectors in the same direction. What we have just pointed out concerning vectors in a blackboard holds true, of course, for vectors in any region whatsoever: our criterion for determining the ratio of two vectors in the same direction does not enable us to compare the magnitudes of incommensurable ratios directly.

We wish, however, to be able to define the quotient of any vector—at least in a given region—by any other vector whatsoever. We wish to be able to regard a vector as a *distance* alone, regardless of its direction. We wish to find some way of comparing the magnitudes of incommensurable vectors. These things must be carried out on the basis of such experiences and notions alone as we have explicitly taken as primitive, if we are to obtain in our treatment of geometry any satisfactory theory of measurement. How are we to accomplish all this?

I shall first give a rough sketch of the method by which we shall obtain this desired result, and then I shall take this method up step by step and give precise definitions of everything I shall do. Roughly speaking, my method is to take some set of generalized points which may be regarded as a sphere, and to find a way of measuring every vector in space in terms of its diameter. As there is a diameter of our sphere in the same direction as any vector we please in space, one can see that the methods which we have already introduced for the comparison of the magnitudes of vectors in the same direction will be applicable in this instance. We shall consequently compare vectors in different directions by referring them to the diameter of our sphere as a standard, for that diameter is of a constant and determinate length, which is independent of the direction in which this diameter is taken. However, in our precise formulation of this method of comparing distances in different directions, there will appear no explicit reference to any diameters of our standard sphere. Our definitions will not depend for their intrinsic logical value as definitions upon any particular assumptions to be made concerning the set of generalized points which we take as our sphere—

we shall not even assume that it possesses anything such as one would naturally call a diameter. Such notions expressing peculiar properties of spheres as that of the center of a sphere or that of the radius of a sphere will likewise fail to make their appearance in the ensuing discussion.

The only spheres which we have already considered *qua* sets of points are sets of points in the first sense which we have given to the word "point" in this course of lectures. The spheres which we wish to consider now are sets of *generalized* points, for we wish to be able to consider our points as termini of vectors, and vectors, since they depend upon parallel lines, have been defined in the space of generalized points, and it was only in this space that we were able to arrive at the notion of parallel lines. We wish to avoid in our present lecture the necessity of supposing arbitrarily that any more sets of points are spheres than it is abolutely necessary for us to so consider. As a consequence of this, a definition of the sphere of generalized points corresponding to a sphere of points in our first sense is a desideratum. It will be remembered that we have already given a definition of the correspondence between a generalized point and a point, in our first sense, in terms of the experience of the intersection of convex solids alone. It will be further remembered that we had no reason to regard this correspondence as one-one except under the hypothesis that our experience of the intersection of convex solids has such properties as we should naturally attribute to it. Be this as it may, I wish to call attention to the fact, which is obvious upon a very slight reflection, that, given a set of points in our first sense, the set of all the generalized points which correspond to any of its members, whether univocally or not, is uniquely determined by the set of points in our first sense. By applying this fact to spheres, it is easy to see that, given any sphere of points in our first sense, there is always a certain single set of generalized points which we may define as "the same sphere" in the space of generalized points. This identification of these two spheres presupposes no other primitive notion than that of our experience of the intersection of convex solids.

Let us suppose that we have already singled out some sphere of points, in our first sense, and that we have obtained from it the corresponding sphere of generalized points. Let us discuss the class of all vectors-in-this-sphere—that is, by our definition, the class of all vectors in space, in so far as they connect points in this sphere. Since the sphere has a finite radius, if R is a vector-in-the-sphere, we cannot multiply R by a coefficient as large as we please, and yet have the resulting vector remain within the sphere. If we are on an island ten miles in diameter, it is clear that we can somehow manage to take one step of four miles to the north, or even two successive steps of four miles to the north, but it is impossible to make a journey in the island consisting of three consecutive steps of four miles to the north. For the same sort of reason as that which makes this latter task impossible—it involves a straight journey of twelve miles on an island only ten miles across—it is impossible, in general, to multiply a vector in a given spherical region by an arbitrarily large coefficient, and to obtain a vector in the same region as a result.

The question now arises, granted that it is true that a vector in a finite spherical region cannot be multiplied by an arbitrarily large numerical coefficient, just how is one to determine by how large a coefficient it may be multiplied, and still remain in the region under consideration? In particular, what relation does this coefficient bear to what one would ordinarily call the magnitude of the vector in question? The answer is fairly obvious if it is ordinary Euclidean geometry with which we are dealing, and if the "sphere" inside of which our vectors are confined is actually a sphere. It may be seen on inspection that, by our very definition of a vector, a vector not limited to a given region can be laid off on any line which points in the direction of the vector. Furthermore, not only can our vector always be laid off on such a line, but if *l* is such a line, and if A is any point on *l*, there is some point B on *l* which is separated from A by the vector in question. For example, if *l* is any line going north and south, and A is any point on this line, then if we choose any such vector as "five miles south of," there is a point on *l* which is separated by this vector from A—that is, which is five miles south of A. Furthermore, it is not only obvious that, as a consequence of this, every vector can be laid out on some line passing through the center of our sphere—for a line can be drawn in any desired direction through any point you please—but it is also true that every vector-in-our-sphere can be laid off on some diameter of the sphere. This is the same as saying, for instance, that if we are on a circular island, and can somewhere make a journey of ten miles to the north without leaving the island, we certainly can take such a journey along a diameter of the island. All this is the consequence of the familiar fact that a diameter is the longest linear segment which can be drawn within a circle or a sphere. From this principle it results that if R is a vector inside a given sphere, all vectors inside the sphere can be laid off on the diameter of the sphere pointing in the same direction as R. If the fraction which we have just defined as the ratio of two vectors be what we should naturally consider their ratio to be—and we have already shown that this is actually the case if our primitive notions live up to their names—it is obvious that what one would ordinarily call the ratio of the diameter of our sphere to any vector which we wish to measure is at least as great as the ratio of *any* vector on the diameter, in the appropriate direction to the vector which we wish to measure, and consequently, from what we have just seen, it is as great as the ratio borne to the vector which we desire to measure by *any vector whatsoever in the sphere.*

It is an easy matter to show that the notion of ratio which we have defined among the vectors in our sphere agrees precisely, as we have already said, with our usual notions of ratio, under the hypothesis that our fundamental experiences have not been misnamed. We are therefore justified in making use of the ordinary geometrical properties of ratio to find out what the magnitude of a vector-in-a-sphere has to do with the extent to which it can be multiplied by a ratio. Now, it is a commonplace of geometry that if we have given to us two linear segments, say AB and CD, either they are commensurable—that is to say, they bear a ratio to one another which is the quotient of one integer by another—or, given any distance *d,* no matter how small, it is possible to find on AB a point E, such that AE is commensurable with CD, while EB is less than *d.* The proof of this is at once simple and instructive, so perhaps I may be permitted to give it here. All of you know that when two quantities, such as distances, bear no ratio to one another of the form m/n, where m and n are integers, but are yet comparable with respect to their magnitudes, we say that the ratio of one quantity to the other is irrational, and we represent it by a non-terminating decimal fraction, such as 3.141592. ., which is π, or the ratio of the circumference of a circle to its diameter. Now, what does it mean, for example, to say that the circumference of a circle is 3.141592. . . . . times as large as its diameter? It means that if we take a distance 31/10 as large as the diameter of a circle, we shall fall less than 1/10 of the diameter short of the circumference, that if we take a distance 314/100 as large as the diameter, we shall fall less than 1/100 of the diameter short of the circumference, that if we take a distance 3141/1000 of the diameter, we shall fall less than 1/1000 of the diameter short of the circumference, and so on indefinitely We can thus define a distance which is commensurable with the diameter of a circle, and yet differs from its circumference by less than $1/10^k$ of its diameter, for any assigned positive integral value of k. It is obvious that we can make $1/10^k$ as small as we please by making k sufficiently large. By using precisely this line of reasoning, we may prove our thesis: that, given any two incommensurable linear segments, AB and CD, and any distance d as small as you please, there is a point E on the segment AB such that AE is commensurable with CD—that is, the vector AE bears a ratio, in the sense already defined in terms of our fundamental notions, to the vector CD—while the distance BE is less than d. As a consequence, it is possible to lay off on some diameter of our standard sphere vectors which are commensurable with any given vector in the sphere, say R, and which, if they do not completely fill up the diameter, leave a remainder which can be made smaller than any assignable vector. Therefore, it is natural for us to say that, although the ratios which other vectors bear to a given vector R can never exceed what we should properly call the ratio of the diameter of the sphere to the given vector R, there is no smaller ratio which is not exceeded by some of the ratios which other vectors in the sphere bear to R. That is, if we call the class of all the ratios which other vectors in the sphere bear to R by the name α, the ratio which the diameter of our sphere bears to our vector R will be larger than or as large as any member of α, yet there will be ratios which belong to α yet differ from this latter ratio which the diameter bears to R by less than ε, when ε is any assigned quantity, however small.

In the preceding paragraph, we came across a notion which is of the highest importance for the determination of the magnitude of a vector. This notion is that of a number n which is larger than or as large as all the members of a given class K of numbers, yet is such that there are members of K which differ from n in magnitude by as little as you please. It is obvious that there is only one value of n which can have this property, for a given class K, since if m also had the property in question, it would, of necessity, be either larger or smaller than n. If larger, since some members of K would differ from it by less than ε, however small ε might be, they would differ from it by less than m-n, and hence, contrary to our hypothesis, would be greater than n. If m is less than n, a precisely similar contradiction arises. As a result, the ratio which a vector R bears to the diameter of our sphere of measurement is uniquely determined by the class of all the ratios which other vectors in our sphere bear to it. Since this is the case, there is no objection to our *defining* the notion—previously undefined—of the ratio which the diameter of our sphere bears to the vector-in-the-sphere R as the number which is greater than all the ratios which vectors-in-the-sphere bear to R, but to which these latter ratios approach as near as you please. This definition will involve no notions not already incorporated into our system, and will be sufficient to secure the unicity of the ratio which the diameter of our sphere bears to our given vector R. If there are ratios which other vectors bear to R and these are all less than some fixed finite number, it is possible to prove by means of our definition alone that there is some number which represents the ratio which the diameter of our sphere bears to our given vector, although this proof involves certain elementary mathematical considerations, drawn from the modern theory of aggregates, which would be a little too intricate for us to consider here.

In ordinary Euclidean space, unless R is a vector that connects a point to itself, all the ratios which other vectors-in-our-sphere bear to R will be less than some fixed finite number, so that the ratio which the diameter of our sphere bears to R is actually determined as some particular number, if there are any vectors in our sphere that bear ratios to R. Now, it follows from our definitions alone that R bears the ratio 1 to itself, so that there are always vectors which bear the ratio 1, at least, to R. From this it follows, first, that every vector which cannot be repeated an indefinite number of times end on end and still remain within our sphere of measurement determines some single number that represents the ratio of the diameter of the sphere to R, and second, that this latter number is always greater than or equal to one, since it is greater than or equal to any ratio which R bears to another vector-in-the-sphere, and one of these ratios is equal to one. Where a vector R can be repeated end on end an indefinite number of times, and there are vectors smaller than the diameter of our sphere—i. e., vectors-in-our-sphere—that bear as large a ratio as you please to R, the natural thing to say would seem to be that the ratio of the diameter of our sphere to R is infinite. In a space living up to the axioms and theorems of ordinary geometry, if our sphere is a genuine sphere, this latter situation can only arise when R is the vector that connects a point with itself. It is obvious that since a step of this vector leaves you just where you were, any number of consecutive steps of this vector will likewise leave you just where you were, and will not take you outside of our sphere of measurement.

Let us consider for a while the ratio which a given vector-in-our-sphere bears to the diameter of our sphere. Every vector-in-our-sphere bears to its diameter a ratio which may naturally be regarded as the reciprocal of the ratio which the diameter of our sphere bears to the vector in question, and which we may *define* as this reciprocal as the notion of this ratio has not been already defined. In this context, we shall regard zero as the reciprocal of ∞. We shall, for short, call the ratio which the vector-in-our-sphere R bears to the diameter of this sphere the *index* of R with respect to the sphere of measurement in question. We may naturally regard the index of R in a certain sphere as the expression of the magnitude of R in terms of the diameter of our standard sphere as a unit. It follows from the fact that the ratio of the diameter of our standard sphere to any vector-in-the-sphere cannot be less than one, that the index of any vector-in-our-sphere, being the reciprocal of the ratio which the diameter bears to it, cannot be greater than one. This is as we should expect—no vector-in-a-sphere can exceed in magnitude the diameter of the sphere. Further, the index of a vector need not be a rational number, as may be seen without much difficulty by a mathematical analysis of the stages through which we have gone in defining it, for its reciprocal, the ratio which the diameter of our standard sphere bears to the vector-in-the sphere in question is defined, not as a ratio, but as a limit of ratios, and a limit of ratios may be an irrational number. The importance of this fact arises from the fact that it makes it possible for us to measure by their indices not only such vectors as are commensurable with the diameter of our sphere of measurement, but also vectors which are incommensurable with this diameter. This measurement of incommensurable distances by a common unit would have been impossible if we had tried to found a theory of measurement directly upon the notion of ratios among vectors, without the introduction of our standard sphere. Another fact which makes us introduce the standard sphere into our theory of measurement is that it enables us to compare vectors in different directions, as we have already seen, and makes it possible for us to consider them as mere distances, and not as vectors—i. e., *directed* distances. The index of a vector represents it in terms of the diameter of the sphere in its direction as a unit. However, the length of one diameter of a sphere is the same as the length of another diameter of the same sphere, no matter in what directions the two diameters of the sphere point. As a consequence of this, the index of a vector expresses its length in terms of a unit which is independent of the direction of the vector. We are thus enabled to say that a vector-in-our-sphere in one direction is as long as, or half as long as, or twice as long as, etc., a vector in another direction, according to whether the index of the first vector is equal to, or half as great as, or twice as great as the index of the second, respectively.

To conclude, we have defined in this lecture the notions of a vector, and of a vector-in-a-region. We saw that these could be regarded as magnitudes, and we defined ratios among them in terms solely of notions that we have already taken as fundamental—that is, in terms only of the experience of the intersection of convex solids, or in terms of this, together with our experience of four selected spheres, according to the one of our two alternative definitions of parallelism that we choose. Then, given any region whatever, we saw how we could define a system of measurement in terms of it which would always enable us to measure all the vectors-inside-the-region in terms of a common unit, and which, in case our space should satisfy the axioms of ordinary Euclidean geometry, and in case our selected standard region should be an ordinary sphere, would give us an ordinary, every-day Euclidean system of measurement. In our next lecture we shall consider how this theory of measurement may be extended so as to cover, not only all distances within a certain sphere, but all distances in space, and we shall consider on what principle our standard sphere of measurement shall be selected.

NORBERT WEINER.

MASSACHUSETTS INSTITUTE OF TECHNOLOGY.

THE RELATION OF SPACE AND GEOMETRY TO EXPERIENCE

VII. Conflicting Measurements

In our last lecture we defined vectors—i. e., directed distances—in terms of such notions as we had already taken as primitive in the definition of parallelism, and of no other notions. We then defined in terms of these notions and that of some particular spatial region alone, which we suppose to be spherical, in all future applications of this definition, the set of all vectors-in-the-sphere: that is, the class of all directed distances inside our sphere. We defined, again introducing no new notions whatever into our definition, the magnitude of each vector inside a given sphere in terms of the diameter of the sphere as a unit, or as we put it, the *index* of a vector-in-the-sphere. This definition of the index of a vector was so framed as to give perfectly unambiguous indices of the vectors in any region whatever, no matter how it might be shaped, in terms of that region, without involving, for example, that this region should have anything at all analogous with a diameter, but it was also so framed that, in an ordinary Euclidean space, if the "sphere of measurement" is really spherical in shape, the system of measurement of vectors-in-itself defined by it will agree completely with the system of measurement that is characteristic of ordinary Euclidean space.

Our purpose in this lecture is to obtain from the theory of measurement developed in the last lecture a general system of measurement by which we can determine the distance between any two points in space. In our last lecture, we left this problem still unsolved in two distinct ways. In the first place, the theory of measurement which we developed in the last lecture only told us how to measure the distance between two points both of which lie within our sphere of measurement. This leaves open the problem how we are to measure distances between two points of which one or both lie outside our standard sphere. In the second place, we have not discussed in any manner whatever the problem, how we are to determine what our standard sphere of measurement is to be, and how it is to be discriminated from other sets of generalized points. These two questions and the further problems to which they give rise form the subject-matter of this and a large part of the following lecture.

As we have just said, our first task is to find a method of extending the system of measurement which we have developed for all distances inside a given sphere to all distances in space, so that we may be able to compare any distance in space with the diameter of our standard sphere. You will remember that we defined a certain vector throughout space as the extension of a certain vector-in-a-sphere R, when and only when in all the cases where two points that are separated by R they are separated by the vector throughout space in question. It is clear that it is only natural to assign to a given vector throughout space that is the extension of R the index of R as an expression of the magnitude of the vector throughout space. Of course, we have said nothing which makes it a consequence of the definition of the extension of a vector-in-a-region that a vector in space cannot be the extension of vectors-in-a-given-sphere having distinct indices, though this cannot happen if our fundamental notions live up to their names. When this happens, we shall say that the vector in space in question possesses *both* of these indices, at least for the present. Since by this attribution of an index to the extension of a vector in our standard sphere we have found a way of measuring distances between points in any part of space, it might seem that we have obtained a satisfactory definition of a distance, which will apply to any distances whatever. A little reflection, however, will convince one that this is not the case. There are certain vectors which are not the extensions of any vectors in our standard sphere, in general. If our standard sphere is one inch in radius, it is obvious that no distance of two inches can be the extension of any vector inside our standard sphere, for there can be no two points both inside our standard sphere separated by a vector two inches long. The question is therefore before us, how are we to measure in terms of the diameter of our standard sphere distances larger than this diameter? This problem is a particular case of the more general one as to how we measure any magnitude with a scale smaller than the magnitude itself. We have that problem on hand, for example, when we wish to measure a yard and a half of cloth with a footrule. We solve this problem, as a matter of practice, as follows: we first apply the footrule to one end of the piece of cloth, and make a step of one foot. We then start from the end of the piece already measured and make another step of one foot. We find that after making in this manner four distinct successive steps, each a foot in length, provided that all these steps have been taken in a straight line pointing directly towards the further end of the piece of cloth, if we now take a step of only half a foot in length, we shall precisely reach the end of the strip of cloth. That is, if we take four successive steps of the whole length of the footrule, and so dispose them that we are led by them as near as possible to the end of the strip of cloth, we shall have to take a further step of the length of half a foot, measured by the rule, to reach the end of the strip of cloth. It is not essential, however, that our first four steps should each be a foot in length: we might first have taken a step seven inches in length, measured by the footrule, then one eight inches in length, then one eleven inches in length, then one ten inches in length, then one six inches in length, then one five inches in length, and finally, another step seven inches in length. That is, if we cover the space from one end of the strip of cloth to the other in a finite number of steps which are measurable by our footrule and which are so disposed that the sum of the lengths of these measurable steps is as small as possible, we call this sum of all the lengths of these steps the total length of the strip of cloth.

Let us now return to the problem of the measurement of distances which are too large to fit into our sphere of measurement. Suppose that the distance between the point A and the point B is of this kind. Then it will be of course impossible to go from A to B by a single step which belongs to the extension of some vector situated inside our standard sphere, but it may be possible to make the transition from A to B by the intervention of a finite sequence of successive steps each of which is an instance of the extension of some vector in our standard sphere. Let these steps be instances of the vectors which form the extensions of the vectors-in-our-sphere S, S', S'', ..., $S^{(n)}$, and let the index of $S^{(k)}$ be $i^{(k)}$, where $S^{(k)}$ stands for the *k*th step in the chain connecting A and B. Let the sum $i+i'+i''+\ldots+i^{(n)}$ be called I. I represents, then the total length of the chain of linear segments representing the steps by which the transition from A to B is made. The length I evidently depends on the particular chain which we select to connect A and B, and on what particular vector-in-our-sphere we regard as furnishing each member of our chain as its extension, when a member of our chain represents the extension of two distinct vectors-in-our-sphere. In ordinary geometry, there is a certain minimum length which chains connecting A and B actually possess—that is, I has a certain minimum value, and there are actually chains of a finite number of vectors which are extensions of vectors in our standard sphere which have this value of I as the sum of the indices of their members. Such chains stretch in a straight line from A to B: they exist, for every distance between two points which is in magnitude less than the diameter of our standard sphere may easily be shown to be an instance of the extension of some vector-in-our-sphere, and Archimedes' axiom holds in ordinary geometry—that is, since when any two distances *l* and *m* be given, there is some integer *k* such that the distance *kl* is greater than *m,* so that we may get anywhere by a finite number of steps as small as we please. If, however, the collection of generalized points which we select as spherical should turn out, after all, not to be spherical, or if our initial relation of apparent intersection among convex solids should belie its name, we have no proof at hand that there exists a chain connecting A and B whose length is actually the shortest that such a chain can have: there may be chains, for example, in which I may be made to assume any value you please greater than two, while there may be no chain for which I assumes precisely the value "two." It would seem highly unnatural to say that in this situation, which, as we have seen, can never occur in ordinary geometry, A is at no distance from B. We wish, therefore, to obtain a definition of the distance between A and B which will be the least possible value of I when such a value exists, and which will be, to put it crudely, sufficiently like that number to be called naturally the distance from A to B when there is no single value of

I which is smaller than all its other values. The quantity which we thus define as the distance between A and B must further be such as always to exist when A and B are ordinary generalized points. Furthermore, it must not be such as to make our system of measurement trivial too often, by causing too many distances equal to zero, or in some similar manner.

In our last lecture, we took notice of the fact that if we are given any set S of positive real numbers, and if all the members of S are less than some given positive number *n*, there is some single positive real number which we may call *x*, which is at least as great as any member of S, but which is also such that if *e* be any positive real number, however small it may be, there is some member of S greater than *n*—*e*. x is called the upper limit or maximum of S. Whether S be made up of numbers all of which are smaller than some fixed real positive number or not, it may be shown in a similar manner that S also determines a single positive real number *v*--which may be zero—which is smaller than any member of S, but which is such that if *e* be a positive real number as small as you please, $y+e$ is greater than some member of S. This number *y*, which is uniquely determined by the class S is called the *lower limit or minimum of* S. The existence of the lower limit or minimum of any class of positive or zero real numbers may be proved as follows: let S be the class in question, and let S′ be the class of all positive zero real numbers smaller than any member of S. S or S′ may or may not contain any terms. Let us suppose that S is not the null-class, the class with no members, and that it actually contains some terms. In this case, S′ may or may not contain some members. If S′ contains no member, it is obvious that *e*, however small it may be, is greater than some member of S: that is, $0+e$ is greater than some member of S. This is true because, by hypothesis, there is no positive or zero number smaller than every member of S, as such a number would belong to S′, which we suppose without members. Since S is made up entirely of positive or zero real numbers, it can contain no member less than zero. Consequently 0 is the lower limit or minimum of S. If S′ contains members, let y be the upper limit or maximum of S′. Then y is at least as great as any member of S′, and it follows, since we can easily show that S′ contains members which approach as closely as we choose to the members of S, that there are members of S smaller than $y+e$, however small a positive real number e may be. On the other hand, there can be no member of S smaller than y. For suppose that z is such a member of S. Then, in accordance with the way in which we have determined S′, since it is made up of all positive or zero real numbers less than every member of S, there can be no member of S′ greater than or equal to z: that is, there can be no member of S′ greater than or equal to $y—e$, where e is the positive number $y—z$. Consequently y fails to satisfy the definition of the upper limit or maximum of S′. But y is by definition the upper limit or maximum of S′, so that our supposition that there is a member of S smaller than y engenders a contradiction. We see as a result of this that y satisfies both of the conditions which go to make up the complete definition of the minimum or lower limit of S, so that there exists a minimum or lower limit of S in this case, as well as in that where S′ has no members. If we now consider the remaining alternative concerning the natures of S and S′—that is, if S contains no members whatever, as is the case when it is made up of all odd multiples of ten or of all integers that are commensurable with π —we shall, to simplify matters, make an *ad hoc* definition of the lower limit or minimum of S, and shall say that this lower limit or minimum is zero. This latter definition is, it is true, somewhat artificial. If we make this convention, it follows from what we have said that any conceivable collection of numbers has at least one minimum or lower limit: the uniqueness of this limit, to which we have already referred, may be demonstrated by a very simple proof, quite analogous to that which we gave last time for the uniqueness of the *maximum or upper limit* of a class of positive numbers.

We have just seen how it is always possible to assign to a given set of positive or zero real numbers one and only one positive or zero real number which is smaller than or at least as small as any member of the set, but to which the members of the set approach as near as we please, and we have called this number the minimum or lower limit of the set. Let us see whether we can define the distance between two points as the minimum or lower limit of the set of values of I for different values of I for the various paths connecting the points in question. In the first place, we must show that, provided that there is a smallest value of I, this must be the minimum or lower limit of all the possible values of I for the two points, since this condition is necessary if we are to regard the distance between two points as the length of the shortest path between them, as we do in ordinary space. This is true because the smallest of a set of numbers is at least as small as any member of the set, while there is no degree of approximation with which you cannot make it represent some member of the set, since it is itself a member of the set; consequently the smallest of a set of numbers, provided there is such a number, is the minimum or lower limit of the set. It results from this that the length of the shortest path from A to B —i. e., the least value of I for the two points in question— which we should naturally call the true distance between A and B, if it exists, is precisely that lower limit or minimum of the possible values of I which we have just agreed to call the distance between A and B. However, if there is no shortest distance between A and B along any path made up of the vectors that we have already measured—that is, if there is no least value of I—there must always, by what we have just seen, be a minimum or lower limit of the values of I, which will have many of the properties that are characteristic of a minimum value of I, and which will satisfy the definition which we have just formulated of the distance between A and B.

Given our sphere of comparison, then, we have thus been able to define in terms only of those fundamental notions which we have already explicitly formulated the distance between any two of those generalized points which correspond to ordinary geometrical points. Upon a slight investigation, we should find that we can prove, independently of any assumptions concerning the formal properties of the objects exemplifying our fundamental notions, a few simple geometrical theorems concerning distances. We can prove, for instance, that if A, B, and C are any three points, the sum of the distance from A to B and that from B to C is not less than that from A to C. We can also prove that the distance from A to B equals that from B to A. We cannot, however, prove from our definitions alone, that the distances that we have so far defined have *all* the formal properties of distances in ordinary geometry. In ordinary geometry, for example, if the mutual distances of four distinct points are given, and the distances of a fifth point from three of these are known, the distance of this fifth point from the remaining vertex of the tetrahedron formed by the four original points is determined to have one of only two possible values, while our definitions do not secure to any distance any precise quantitative relation to other distances, in general.

We have given the definition of distance just developed with the original intention of using it only for those distances that do not fit into our sphere of reference—the diameter of which, by the way, is still the unit of our sphere of measurement. If, however, we examine into the definition, we see that it may also naturally be applied to distances which fit into our sphere of measurement—for any two generalized points which are separated by a vector lying in our sphere of measurement, and consequently by the extension of this vector, are thereby connected by a chain consisting of that single step, and consequently have a distance in the sense in which a distance is defined in terms of a chain of vectors. It is further obvious that in a system of generalized points which behave in a decent geometrical manner, a chain connecting A and B and consisting of a single vector is at least as short as any other chain connecting A and B, so that the magnitude of the vector from A to B is the same as the distance from A to B, in the sense of the minimum length of a chain from A to B. We can consequently throw away our first definition of the distance from A to B entirely, and be sure that if we define all distances in space by chains, our definition will be natural—at least as natural, that is, as our direct definition of distances by means of vectors. We have thus attained a definition of all the distances in space which involves besides the notions that we have taken as primitive only that of a certain standard sphere of measurement.

The question now arises, what sphere is the actual unit sphere of our measurements? If it turns out that we can select no set of generalized points as the unique unit sphere, how are we to determine the various spheres which shall serve as the bases of our measurements, and if these spheres give us different values as expressions of the magnitude of a given distance, how can we reduce the various results which we obtain to a single internally coherent system of measurement? It is easy to indicate two directions in which we may approach this problem, corresponding respectively to the two definitions which we gave of parallel lines in a preceding lecture. It will be remembered that in the process of obtaining one of these definitions of parallel lines we defined a certain collection of sets of generalized points called a-spheres entirely in terms of our experience of the intersection of convex solids. We saw that if our experience of the intersection of convex solids records two convex solids as intersecting when and only when they both are situated in a certain region of space and approach one another closer than a certain distance which is the same everywhere in space, our a-spheres, *qua* sets of generalized points, will be spheres of a certain uniform size. This result will seem more familiar to you if I put it in another form. Many of you have seen the various processes of the differential and integral treated in the textbooks of the more old-fashioned sort as if they dealt with operations of division or summation among certain entities called infinitesimals. These infinitesimals were regarded as magnitudes, all very small, but equal, in general. By the English mathematicians of the eighteenth century, who, like their philosophical colleagues, were of a more empiricist attitude to their subject than those of the Continent, these infinitesimals with which the calculus seems to deal were regarded as if they were the smallest objects accessible to our direct sensory experience. Since it was an essential property of the infinitesimals, in their mathematical use, that they should be equal, in general, those mathematicians who held this view were driven to the supposition that all just-noticeable sensible objects—all *minima sensibilia*—are equal. Now, it will be remembered that our a-spheres bear a very close analogy to the *minima sensibilia* of the philosophers, in that they mark the lower boundary of the sensibilia we are considering—namely, convex solids—with respect to their magnitude. Our assumption that all a-spheres are equal is consequently one of those things which was involved as an unquestioned presupposition in the empiricist treatment of the calculus, although it involves the admission of far less than is necessary for the support of that view. The fact that it is such a natural hypothesis, and that it is an element in a view of the greatest historical importance, makes the consideration of a theory of measurement based on this hypothesis a thing of the utmost interest, whether it be strictly true or not that all *minima sensibilia* are what we should ordinarily regard as of equal magnitude. Let us first notice, however, that it is extremely improbable that the delicacy of our sensory discrimination, and as a corollary the size of our *minima sensibilia,* is in any ordinary sense the same throughout all those parts of space which are more directly accessible to our sense-experience.

On the assumption that we could naturally call all a-spheres equal, and that our space has all the normal geometrical properties, it is a matter of indifference which a-sphere we use as a unit of measurement, and all the diameters of all a-spheres are equal. It will not, however, be in general a matter of indifference which a-sphere we take as our standard sphere, if this assumption is not fulfilled. In such a case, it may be possible to get several numbers, all of which express with equal right the distance between two points, A and B, according to the a-sphere which we choose as our unit of measurement. If we are to be able to regard some single number as the only true distance AB, as measured in terms of the diameter common to all a-spheres as a unit, we must find some way of obtaining from all these different measures of this distance some single quantity, which is uniquely determined by all these various measures, and which coincides with their common value if they agree. There is, of course, a certain degree of arbitrariness in our selection of this distance, but one may easily show that if we regard the true distance AB as the minimum or lower limit of all the values which we can get for the distance AB by using various a-spheres as standard spheres, we shall obtain an entirely unequivocal definition of the distance AB. In the first place, this distance will always exist, be finite, and uniquely determined, since, as we have seen, all these things are true of the minimum or lower limit of any set of positive or zero numbers. In the second place, if all the distances of A from B, measured by a-spheres, agree, they will all coincide with their minimum or lower limit, as one may see on inspection. We have thus obtained a definition of the distance between any two points in space which involves no notion other than that of our experience of the intersection of convex solids, which secures that any two points in space shall be at one and only one distance from one another, and which will completely agree with our usual notions of the distance between two points, provided that our experience of the intersection of convex solids has such properties as we should naturally expect it to have, and provided, moreover, that we can naturally call all *minima sensibilia* equal.

In the system of measurement thus obtained, some of the theorems of ordinary geometry will still hold good, irrespective of any assumptions about the nature of our fundamental experience; for example: the distance AB will equal the distance BA, the sum of the distances AB and BC will be at least as large as the distance AC, and so on indefinitely. For the most part, however, the geometrical properties of our distances will be dependent on the nature of our experience of the intersection of convex solids, and will not, in general, agree with those characteristics of ordinary Euclidean geometry, if our fundamental experience has not such properties as we should naturally expect it to have. For example, the theorem that, if any tetrahedron is given, the distance of a point from one of its corners must assume one of a certain pair of values, once the distance of the point from each of the other three corners is known, which is true in ordinary geometry, cannot be deduced from our definition of distances alone.

However, the object of this course of lectures is to show how we can eliminate from geometry all presuppositions concerning the particular formal properties of our original notions, and yet define our points, lines, distances, etc., in terms of these in such a manner that all the ordinary formal geometrical properties of these latter entities should follow from their definitions alone. To do this, we must be able to obtain a definition of distance which will depend on our experience of the intersection of convex solids alone, and which will of itself be a sufficient guarantee that our distances satisfy the usual geometrical laws, and it may further be shown that, once we have a perfectly satisfactory definition of all the distances in space, we have it in our power to give perfectly satisfactory definitions of all geometrical entities. As we have just seen, we have not yet succeeded in completely performing this task; the problem, however, is an extremely interesting one, and one of the utmost philosophical importance. Let us turn our attention to it for a little while.

Our last definition of distance has practically indicated to us how we should make a survey of the universe, for it correlates with every pair of ordinary generalized points the distance between them. However, it will be a bad survey, in general. By this I mean distances which it causes to separate points will not check up, will not "gee" with one another. You all know that a surveyor always makes more observations than are merely sufficient to indicate the position of each point he is mapping a single time only: he observes the same point five or six times over from various observation-stations, and records all his observations. In an ideally perfect survey, all these observations would come out in complete accord with one another, but as a matter of fact, they seldom or never come out in complete accord with one another in any actual survey. The several determinations of each triangulation-station indicate on the map, not one point, but a number of points, situated in more or less close proximity to one another. It is a part of the task of the surveyor to fix a single point, which may be said to be the best representative of these several points, in terms of the collection made up of all of them. This he does by means of a certain mathematical theory known as the theory of least squares. This theory does not concern us here, except in so far as it fulfills the function of reducing a set of unharmonious measurements to a harmonious system; we shall have nothing to say concerning its technical details. Its function is to make every single point which forms a station in a certain survey correspond to one point on the map representing the results of this survey, and to one only. Then the surveyor can go and take the distances between these uniquely determined points on his map, and say that they represent—when the appropriate alterations of scale are made—the distances between the points on the earth's surface corresponding to the points on the map, and we may be *a priori* sure that the proper geometrical relations will hold among the distances so determined.

In order to make the discussion of surveying which we have just given more simple to grasp, we have been guilty of a slight inaccuracy of statement. We said that the several determinations of each triangulation station indicate on the map, not one point, but a number of points. As a matter of fact, however, the several determinations of each triangulation-station indicate no specific point whatever on the map until we have already determined a certain correspondence between points on the map and points on the surface of the earth. This presupposes that we are able to locate certain points on the map before others and to discriminate between such as we shall locate first and those which we locate later. We do not possess, however, any criterion for such a location. It is consequently necessary for us to possess some means of deriving at one blow from our confused measurements of the distances between our stations quantities which we can regard as the expression of the magnitudes of these distances, among which the proper geometrical relations will of necessity hold, if we are to be able to map our triangulation-stations in such a manner that to each triangulation station there will correspond one and only one point on the map. This task can also be accomplished with the aid of the method of least squares, in the case of any survey involving only *a finite number of points.* From any survey, then, however bad it may be, provided that it involves only a finite number of points, a set of finite quantities may be obtained, one of which is correlated with each pair of the points surveyed, and among these quantities the formulae which correlate distances in an ordinary Euclidean space will hold.

However, the bad survey of the universe which was made by our definition of distances in terms of a-spheres involves the measurement, not of a finite number of distances, but of an infinite number of distances, provided that the space that we have defined in terms of our experience of the intersection of convex solids is comparable in richness with the space of ordinary geometry. Now, the method of least squares is unable, as far as I know, to bring order into surveys that are bad at an infinite number of points. We do not at present possess a method of deriving a well-behaved Euclidean set of distances from the hodge-podge set of distances which we obtain, in general, from our definiton of distances in terms of a-spheres. I see no reason, however, for supposing that the problem of obtaining a method which should perform the same function for infinitely irregular systems that the theory of least squares fills for finitely irregular systems should be essentially insoluble. The solution of this problem is of absolutely vital importance for the philosophy of space, for however we may define distance, nothing is more certain than that the distances of points in space, as we first learn and observe them, only approximately satisfy the laws of geometry, yet we are absolutely sure that there are actual distances which our observed distances represent and which satisfy these laws precisely. If we were enabled by a theory analogous to that of least squares, to derive from our chaotic observed distances, distances whose geometrical properties and relations should be secured *a priori,* we could immediately explain this phenomenon, on which so much importance has been laid by philosophers.

In this lecture, we have extended the system of measurement determined by a sphere from its inferior to the whole of space. On the basis of that definition of parallelism which starts with a-spheres, we saw that a system of measurement could be developed, but that this would have certain defects. We saw that this system of measurement does not secure *a priori* the geometrical properties of space, but that it would only require the formation of a branch of mathematics which should deal with problems essentially similar in character to those dealt with in an already existing branch to give us a space whose geometrical properties would be *a priori.* In our next lecture we shall take up the problem of defining distances on the basis of our alternative definition of parallel lines, and we shall finally discuss the philosophic problems to which both these systems of measurement give rise.

VIII. Geometry an a Priori Science

IN OUR last lecture, we developed a theory of measurement upon the basis of that definition of parallel lines which involved the notion of a-spheres. In this lecture, we shall first develop an alternative theory of measurement, in which we follow out that definiton of parallelism which involved, not only the relation of apparent intersection among convex solids, but also four selected convex solids, which we agreed to call "unit spheres." If the distances of any point in space from the centers of these four spheres are known, and our four spheres, to put it in ordinary geometrical terms, are four spheres whose centers are not all coplanar, the position in space of the point in question is completely determined, and if the positions of two points are determined in this manner, their distance from one another is completely determined. As we intend in what follows to secure the proper geometrical interrelations of our distances, and to cause the theorems of geometry to be satisfied, by means of defining the position of any point in space in terms of its distances from the four centers of our standard spheres, the first notion which we must define is that of the distance of a point from the center of a sphere. Now, it is obvious that, whether a point x is inside a sphere S or not, the furthest point in S from x lies on the line through x passing through the center of S, and is that one of the points on the surface of S and also on the line in question which is further from x. If y be the point in S furthest from x, it therefore follows that the distance from x to y is the distance from x to the center of S, plus the radius of S. One might consequently think that we might define the distance of x from the center of S as the distance of x from the most remote point of S, minus the radius of the sphere. We wish, however, to have our definition applicable even where S is not precisely a sphere, and we consequently do not wish to make the assumption that there is any single point of S most remote from x an essential condition of the usefulness of our definition. Now, if there is any single point in S remote from x, the distance of this point from x will be the maximum or upper limit of the distances of points in S from x. However, it will be much more usual, in general, for there to exist a maximum or upper limit of the distances of points in S from x than for there to exist some single point at precisely that distance from x and belonging to S. We shall consequently regard the distance of a point from the center of S as the maximum or upper limit of the distances of points in S from the point in question, minus the radius of S. Now, if we make S itself our standard sphere of measurement, in the sense explained in our last two lectures, since the diameter of S will be our unit, the radius of S will be one-half. Consequently, it will be natural to regard the distance of a point from the center of S as the maximum or upper limit of the distances of points in S from the point in question, minus one-half. This definition, as may be readily seen, involves no other notions than such as we have already taken explicitly as primitive, together with such notions as "one-half," or "limit," which, as the modern logicians have shown, are purely logical notions.

Given the four equal, non-coplanar spheres that served as the foundation of our second theory of parallelism, we are now able to define the distances of any point we please from the center of each of them, in terms only of such notions as we have already taken explicitly as primitive. It is interesting to note that our definition involves absolutely no reference to such entities as the centers of these spheres. Be that as it may, we can treat our definitions as if they yielded us the distances of a point in space from four given points in space—the centers of our four spheres. We intend to use these four distances as the coördinates of our points, and to determine the position of any point in space by means of these four distances. We finally desire to determine the distances of any two points in space from one another in terms of the distances of each of the points from each of the four centers of our equal spheres. Our actual coördinates will not, however, be the four distances of a point from our four centers, since we wish our coördinates to be absolutely independent of one another, as this will facilitate such a definition of the distances in space as will secure automatically their satisfaction of the laws that bind distances to one another in ordinary geometry. Now, in ordinary Euclidean space, if we know the distance of a point from each of the vertices of a triangle, we know that its distance from any fourth point in space can assume one of only two possible values. We shall consequently determine a point, not by its distance from each of our four centers, but by its distance from three of our centers, and by whether its distance from the fourth center has its greatest possible value, the other three distances being given, or not. We shall, for example, write the coördinates of a point whose distance from the center of the standard sphere A is a, whose distance from the center of the standard sphere B is b, whose distance from the center of the standard sphere C is c, and whose distance from the center of the standard sphere D has its maximum value, a, b, and c being fixed, in the form $(a, b, c, +)$, while if the distance of our point from the center of D has not its maximum value, and everything else remained unchanged, we should represent our point by $(a, b, c, -)$. In these definitions, we have introduced no notion that we have not already taken explicitly as primitive.

The three coördinates and a sign by which we determine a point are completely independent of one another if our space already obeys the laws of ordinary geometry. If, however, we know the coördinates of two points, we do not yet know, by these data alone, the distance by which the two points are separated. This knowledge depends further upon a knowledge of the shape and size of the tetrahedron formed by the four centers of our spheres of reference—that is, on a knowledge of the remoteness of the four centers from one another. To discover this, we must already possess a definition of the distance between the centers of two unit spheres. A definition which is in every way analogous to the definition which we have already given of the distance of a point from the center of a sphere reads as follows: the distance between the centers of two spheres, S and T is the maximum or upper limit of the distances between points in S and points in T, whether measured with reference to S or to T as the standard sphere, minus the radius of S plus the radius of T, which is one, since the diameters of S and of T are regarded as possessing a common value, and this common value is taken as our unit of distance.

We now possess sufficient data to transform the system of coördinates which we have already obtained into an ordinary Cartesian system of coördinates. If we form certain functions of the coördinates already defined for a point p in a perfectly determinate manner—which, it is true, are a little too intricate for us to exhibit here—and call these X, Y, and Z, we may be sure that, if our points, lines, distances, etc., have such properties as we should naturally associate with the names we give them; and if our four "spheres" are actually four equal spheres whose centers do not all lie in the same plane, our coördinates X, Y, and Z, will be an ordinary set of rectangular Cartesian coördinates whose axes are determined in a certain definite manner by our four spheres of reference, and whose unit of distance is the common diameter of all four standard spheres. We may also be sure that even if our space is not so completely subject to the ordinary laws of geometry, many of our points will, in general, still determine coördinates, and any point that determines the coördinates X, Y, and Z will determine them uniquely. We shall call the values of X, Y, and Z that are determined by a certain point *p* the *fundamental coördinates* of *p*.

It is by no means a necessary consequence of the definition of the fundamental coördinates of a generalized point that to a given point there must always correspond some set of coördinates. Let us call those generalized points that have fundamental coördinates *proper* generalized points. By the definition of a proper generalized point, it must have one set of fundamental coördinates and one only. It does not follow from this, however, or from anything else we have yet said, that no two distinct proper generalized points have the same set of fundamental coördinates, whatever the formal properties of our fundamental notions may be. We desire, however, to obtain ultimately a definition of a spatial point which will secure by itself that no two points hold the same position in space—that is, that no two points have the same fundamental coördinates. How shall we do this? We shall do it by introducing a new definition of a point in place of our definition of a generalized point, just as we formerly introduced the definition of a generalized point in place of our very first definition of a point. Our final definition reads as follows: a *revised point,* we shall say, is a class other than the null class, or class without any members—of all the proper generalized points that have a given set of fundamental coördinates. Obviously, we assign this set of coördinates to the revised point in question. It is evident, then, that no two distinct revised points can have the same set of coördinates, for otherwise they would coincide. A revised point, then, must determine a set of coördinates different from that of any other revised point; it further follows from the mode of formation of a revised point that every revised point has a set of coördinates, and that no revised point can have more than one set of coördinates. The appropriateness of calling revised points points will become obvious if you reflect that in an ordinary, well-behaved space a revised point will contain only one member —a certain generalized point—and that the revised point will consequently represent the same point, in our everyday sense of the term (whatever that sense may be) as the single generalized point which is its member. We have thus obtained a set of entities which can naturally be called points, between which and certain sets of coördinates or triads of numbers there subsists a certain correlation which is one-one, and extends over all of our revised points. In ordinary geometry, the correlation between points and their Cartesian coördinates is not only one-one, but also connects all the points in space with all possible sets of coördinates. Consequently, if our revised points are to have all the formal properties of the points in ordinary geometry, we must show, over and above what has been already shown, that every set of Cartesian coördinates—every triad of real numbers—determines some point. Now, it does not follow from the definition of a revised point that we have just given that every set of Cartesian coördinates determines a point. The manifest and obvious way to remove this imperfection in our system of definition is to find some entities that will fill the gaps in our system of points which are left by the absence of a revised point corresponding to a given set of coördinates. We can do this in the following manner: if to a given set of coördinates there corresponds a revised point, we shall say that the revised point is the point, in our final sense of the word, that corresponds to it, but if there is no revised point corresponding to a given set of coördinates, we shall call the set of coördinates itself the point that fills the position indicated by our set of coördinates. If we thus interpolate sets of coördinates as points into our system, we shall find that our complete set of points will be in one-one correspondence with our complete set of coördinate triads. We shall thus have obtained a space which agrees perfectly in this respect with the space of ordinary geometry.

Now, it is familiar to all those of you that have had an elementary mathematical training that all the theorems of geometry my be reduced to purely algebraic theorems, entirely independent of space and dependent only on the properties of number and quantity, when once a system of Cartesian coördinates has been set up. It has been shown by the modern mathematical logicians that all theorems that deal with number and quantity alone are theorems of pure logic, and are independent of any concrete experience whatever. Since these things are the case, then once a system of Cartesian coördinates has been defined in terms of our fundamental notions, we have done all that is necessary to prove the *a priori* character of geometry, for we can now so define lines, planes, circles, angles, etc., in terms of our set of coördinates—and ultimately in terms of those notions that we have explicitly taken as primitive —that all the theorems of geometry shall result from these definitions and the laws of algebra alone. Consequently, if the space of our every-day life is constituted of the entities that I have just called revised points and of the number-triads of the space that I have just defined, the theorems of geometry are *a priori* true, even though space is actually a function of experience.

This brings me to the end of the technically logical and mathematical portion of the present course. Let us consider what we have accomplished, and let us see in how far we have fulfilled the purpose which this course was designed to fulfill, and have proved or rendered more probable the thesis which we set out to prove. The thesis of this course of lectures is that, whereas space is a function of experience, the geometrical properties of space are *a priori* certain, and may be proved without involving any reference to the concrete nature of the experience of which space is a function. We claimed that the geometrical properties of space were due to the method of schematization by which space is obtained as a function of experience. To show that such a situation is possible, one thing that must be done is to exhibit a system in which geometrical properties are the results of a method of schematization applied to an arbitrary subject-matter, and of this method of schematization alone. We can fairly claim to have accomplished this task by exhibiting the spatial system resulting from our second definition of parallelism, for in this we have introduced no notion pertaining to a certain spatial system—no notion, indeed, other than that of a certain relation, which we *call* the relation of experienced intersection among convex solids, and four arbitrary terms or sets of terms that enter into this relation, which we call four equal spheres whose centers do not all lie in a single plane—and notwithstanding the arbitrariness of the formal properties of our fundamental notions, we have so framed our definitions that certain of our entities must of necessity possess the formal properties characteristic of geometrical objects. We can further claim that we have pointed out that it would not require the solution of any problems generally different in character from those solved in already existing branches of mathematics (such as the theory of least squares), to derive a system of entities obeying all the formal properties of geometrical entities from certain relations or facts essentially similar to those more or less directly accessible to our experience, in such a manner that no presuppositions concerning the formal properties of these relations and facts are involved. That is, we have made it seem very probable that, even if we did already possess a geometry, and had no knowledge whatever of points, lines, planes, etc., we could build up from those facts ascertainable by means of a more or less direct experience, by a method involving only such notions as belong to logic and are not dependent on any concrete experience, a system of entities which one could naturally call lines, points, planes, etc., in such a manner that one could be *a priori* sure that the formal properties of these lines, points, planes, etc., would be those laid down by the laws of ordinary geometry. All this, though suggestive of the actual relations that subsist between geometry and experience, is not in any way conclusive evidence as to the nature of these relations. We have given no reason to justify us in supposing that the entities that we have called points, lines, planes, etc., are the same entities that we call by those names in our every-day life. Indeed, the feeling of artificiality that haunts us at every step of the ground we have covered seems to forbid this view, and I think that one will be perfectly justified if he categorically denies that the particular entities that we have called by various geometrical names are those that are called by those names in our every-day life. But the fact that there is nothing inherently impossible in the formatioin of a space whose geometrical properties shall be *a priori* certain, yet which will be a function of experience alone, and the absence of any other existing view of the relation between space and experience that will explain the association of certain geometrical entities with certain empirically known physical entities, and in addition thereto the non-experimental nature of geometry, entitles us to say that we have rendered the view that geometry deals with some schematization, some systematization, some arrangement of experience highly probable. We have given good reasons for supposing that the apriority of geometry is genuine, but that it is an apriority of method, not of subject-matter, and we have illustrated how an apriority of method which is not apriority of subject-matter is possible in the field of geometry.

We have the question still before us: if space, as we have said, is a schematization of experience, what sort of a schematization is it actually, and how is this schematization related to that which we have exhibited in this course of lectures? The first thing to notice is that it is very highly probable that the schematization by which the space of our every-day life is formed is probably not a fixed, immutable schematism, for which there is one analysis that is always right, while all the other accounts of its structure are always wrong. For instance, the schematization by which the space of a carpenter is reached is necessarily different from that by which a physicist attains his space: one might almost say that with the physicist, his straight lines are his light-rays, for a light-ray in a vacuum is the criterion by which he tests the straightness of anything else, and consequently it is at least reasonable to suppose that the complicated synthesis and organization of experience through which he must go in order to obtain a light-ray—since light-rays, as such, are not given in experience—must play a part in the construction of the lines of his geometry; while this obviously cannot be the case with the carpenter, whose criteria of straightness are simply chalk-lines and T-squares. It seems likely, moreover, that even with one and the same person, the schematism he uses may vary with the problem which he is attacking: on one occasion, the physicist may find it more convenient to regard a point as if it were built up by some process of organization similar to that by which he organizes his experiences under the form of a light-ray; on another occasion, his method of schematization may be more analogous to that by which he obtains a gravitational line of force. It may be incorrect to speak of "space" as something unique: all that we ought to mention may be "spaces."

On the other hand, I feel fairly sure that certain tasks will have to be performed by all the divers methods of schematization that may lead us from experience to space. For example, the problem to which we devoted our third and fourth lectures—the problem, namely, of proceeding beyond that portion of space which is more or less directly accessible to our experience, and of obtaining definitions of points and of lines which will yield us all the points and lines in space—is one that is inseparable from any method of schematization by which we obtain space from experience, and I believe that the method of making this extension of space that we there developed is essentially similar to that which we use in our actual processes of obtaining space. The problem of an infinite theory of least squares, of turning a bad survey of the universe into a good one, is one that we must meet in almost any process of deriving space from experience. The problem of distinguishing parallel lines from intersecting lines is another problem that is not confined to the system developed in this course. It will thus be true that though there are many different systems of schematization which, on certain occasions, yield us space as a function of experience, and while the method which we have described in the preceding chapters may be different from all of these, there will be certain problems that run through all these methods and the one which we have given, so that an exposition of the method which we have given cannot but throw light on the methods of schematization by which space is actually obtained.

It may seem to you that these tasks which I have just mentioned as essential steps in the attainment of space as a function of experience are, as our introspection shows us, not performed consciously, while they are too intricate to be performed unconsciously. I think, however, that both of these statements are open to question. In the first place, whenever a man thinks clearly enough to give a definite criterion for the straightness of a line, or the intersection of two lines, or whatever other geometrical property you please of physical, sensible objects, he is really simply rendering explicit some stage in the synthesis of experience which constitutes his space, and in so far as he has succeeded in forming his criteria of straightness, intersection, etc., into a system which must, by an internal necessity, be coherent, he has succeeded in making the complete method of schematization of his space determinate. We have already referred in this course of lectures to the fact that the surveyor, when he obtains a coherent map by a definite method from a mass of disharmonious data, is doing something of essentially the same nature as what we have been trying to do in this course of lectures, for he possesses a method by which he can deduce a map in which the laws of geometry hold from the most hopelessly incoherent and lawless set of observations or experiences. In the second place, it is a well-known fact that the complexity of a mental process is in itself no absolute bar to its unconsciousness. A man may have the most complicated set of criteria by which he determines the straightness of a given line or the flatness of a surface, but he may never have introspectively considered the nature of this process. It may turn out that many or all of the geometrical properties of his space result from these criteria alone, and that these criteria or definitions of straightness, flatness, etc., are so formulated that they constitute a perfectly coherent system, without his ever explicitly knowing that he uses these criteria or definitions at all. It does not seem to me at all unlikely that the mathematician or physicist, who unconsciously performs such intricate processes of reasoning as differentiation or integration, should perform unconsciously some comparatively simple synthesis whereby space, with all its geometrical properties, is obtained as a function of experience. As to the non-mathematician, who is unable to follow a complicated train of reasoning even consciously, and who can formulate only vague and unclear definitions, there is no reason to suppose that in the space which he obtains as a result of his own process of synthesis from experience the laws of geometry hold in any but a rough and vague and rough way. In short, it is perfectly possible for the methods of schematization by which space may be obtained from experience to be unconscious, at least in the sense that we never see it as a systematic whole.

In closing this course, I wish to make a few remarks about the manner in which the theory of space we have developed has answered that problem which we found the Kantian view of space unable to meet. I refer to the problem of the correspondence of certain physical objects with certain spatial entities. How can we use one of the geometrical lines that we might define in terms of our revised points, for instance, as a criterion of the straightness of such a convex solid as a mark on the blackboard with a piece of chalk may be? The answer to this question is very simple. We have already seen how we can regard a convex solid as a set of points, in our first sense, and we have also seen how we may make a set of points in our first sense determine uniquely a set of generalized points. We are thus able to regard a convex solid as if it were a certain set of generalized points. It is easy to carry this process a little further, and to regard a convex solid as a certain set of revised points determined uniquely by it, which may be regarded as the set of all the revised points inside the convex solid in question. We can then give a simple mathematical definition for the accuracy with which the convex solid in question represents a straight line: we can say, for example, that the linearity of a set of points is the ratio of the longest linear segment connecting two points of the region or set to the longest perpendicular segment that contains two points of the set. On Kant's view, it is essentially impossible that anything of the sort should be done, since his lines, forming portions of space, can be given to us only *a priori*, so that we cannot recognize them in an empirical situation, and associate them in a definite manner with empirical objects. We have thus completely established our case against Kant; our case against Mach and the whole Empiricist school of philosophers of mathematics has already been made out.

NORBERT WIENER.

MASSACHUSETTS INSTITUTE OF TECHNOLOGY.

Commentary on [22a]

A. Fine

The seven papers collected here from *The Monist* constitute a set of lectures on the philosophy of geometry given at Harvard University in 1915. The major pattern of philosophical discussion of geometry has been set by the seventeenth century debate between Leibniz and Newton over a relational versus an absolute conception of space. Both sides in that debate represent an empirical approach to geometry. That is, both Leibniz and Newton approach questions of space with a framework that is susceptible to empirical tests. The major difference between them is that for the relational view this framework is basically topological whereas for the absolute view the framework is a metric one. By contrast there is a third historically important competitor in the philosophy of geometry, Immanuel Kant, who represents an a priori approach. In Kant's view the concepts and relations of geometry are prior to experience and indeed constitute the means by virtue of which spatial experience is possible. This position, however, generates a puzzle that runs throughout the corpus of Kant's philosophy: how can concepts which do not arise from experience nevertheless apply to it?

It is to this puzzle, as it relates to the concepts of geometry, that Wiener's lectures are addressed. Wiener holds that the relations among geometrical concepts must be fixed a priori, for they are certain. (He makes this slide from the certain to the a priori without displaying any concern about the warrant for such a move.) In order to bridge what Wiener sees as a gap between these a priori relations and ordinary experience, Wiener adopts a method expounded by Bertrand Russell in his Harvard lectures of 1914, published as *Our Knowledge of the External World.* Russell's method, which is derived from A. N. Whitehead, consists in moving from what he takes to be given directly in experience, namely, sense-data, by means of fairly elaborate logical constructions to the objects and concepts of the world of physics. Thus Russell constructs the "points" of physical geometry as certain sets of sense-data, corresponding to our observation of enclosure relations among solids, subject to fairly elaborate logical constraints. It is this method of logical constructions from appearances (Wiener calls them "schematizations") that Wiener uses to bridge the gap. For if all the relevant geometric concepts can be defined by means of such logical constructions, then the connection with ordinary experience will involve nothing more than the move between a defined concept and its defining base in experience.

I should point out that Russell's program of logical constructions from a sense-data base did not fare well from 1920 on. This century's two best-known philosophers (next to Russell himself), Rudolph Carnap and Ludwig Wittgenstein, were both proponents of the program for a time and both subsequently abandoned and criticized it. Internally, with respect to the very introduction of sense-data and the feasibility of the technical constructions under the given logical constraints, and externally, with respect to the intelligibility and and cogency of such a reconstruction of ordinary concepts, the program has seemed seriously defective. Its companion within mathematics, the philosophical school known as logicism, has had a similar fate. Thus contemporary philosophical discussions of geometry do not follow the trail marked out in these lectures by Wiener, but rather they have returned to the empirical tradition of Leibniz and Newton.

From a philosophical point of view these articles by Wiener have mainly a historical interest. From the point of view of foundations of geometry it should be noted that Wiener himself claims no novelty in the constructive techniques that he employs. With regard to the moves from betweenness to linear segments to lines, the introduction of parallels, vectors and the procedure of coordinization, I think this is correct. But Wiener's construction of points as classes of convex solids (as amended by the footnote to Part IV, p. 200) does seem new. This is interesting for there were a fair number of such constructive definitions about at that time, notably those of Whitehead, Huntington, Nicod, and, somewhat later, R. L. Moore. (K. Menger, "Topology without Points," *Rice Institute Pamphlets,* 1940, contains a brief survey of attempts to construe points as classes of more massive entities.)

It was originally Wiener's intention in these papers to frame definitions of the basic geometrical concepts from which the ordinary Euclidean properties would follow logically, thus rendering Euclidean geometry secure a priori and yet applicable to the physical universe. He includes in the program the metric as well as the topological concepts. At just this point, however, he is forced to modify this ambitious program, for he finds himself unable to derive the properties of the Euclidean metric (in particular a version of the law of cosines) from his definitions. The difficulty here was only pinpointed much later by Tarski[1] who showed the impossibility of defining Euclidean "congruence" in terms of "betweenness" and "point." Moreover in framing his coordinization procedure Wiener assumes, without mention, the continuity of his lines (that is, that each line is connected in the topology induced on the line by the betweenness relation). But as Riemann had argued in his inaugural address,[2] this continuity, together with the homogeneity of the spatial points, is sufficient to preclude the existence of an internally definable Riemannian metric.

As for the a priori certainty along with the applicability of Euclidean geometry, it is surely ironic that these lectures were delivered just one year before the publication of Einstein's famous 1916 paper on general relativity, a paper that undercut the long standing hegemony of Euclid.

Editor's Note. Wiener was in his mid-twenties when he wrote the papers [22a]. His ideas on geometry changed thereafter. Although we have no written record of these changes, it is clear to those who knew him in his later years that he had abandoned the a priori view of geometry expressed in [22a] in favor of the logical empiricist position, preeminently and succinctly expressed by Einstein's dictums:

> "Geometry (G) predicates nothing about the relations of real things, but only geometry together with the purport (P) of physical laws can do so. Using symbols, we may say that only the sum of (G) + (P) is subject to the control of experience."
>
> And again:
>
> "As far as the laws of mathematics refer to reality, they are not certain; and as far as they are certain, they do not refer to reality."

See A. Einstein, *Sidelights on Relativity,* London, 1922, pp. 35, 28.

References

1. A. Tarski, Erkenntnis, 35 (1935), 80.

2. G. F. B. Riemann, *On the hypotheses which lie at the foundations of geometry,* reprinted in *A Source Book in Mathematics,* edited by D. E. Smith, Dover, New York, 1929.

IS MATHEMATICAL CERTAINTY ABSOLUTE?

THE place where most people would look for absolute certainty is in pure mathematics or logic. Indeed, "mathematical certainty" has become a byword. Now, just as Aristides was ostracized because people were tired of hearing him called "The Just" so much, so we become somewhat suspicious of the absolute certainty of mathematics through hearing it continually dwelt upon. Is, then, mathematics absolutely certain? To answer this question we must first consider a few points concerning the nature of pure mathematics.

Pure mathematics (or logic, which is merely the same discipline under another name) is defined by Mr. Russell as "the science whose propositions contain no constants." That is, all the "things" about which logic and mathematics seem to assert specific propositions—the truth-values, universes of discourse, classes, syllogisms, etc., with which logic deals, and the numbers, integral, fractional, real, and complex which form the subject-matter of arithmetic and algebra, the points, lines, and planes of geometry, and the functions, definite integrals, etc., of analysis—are mere constructions, made to help us express and explain what certain sorts of propositions have in common, and not at all things of the real world. According to Professor Frege and Mr. Russell, a proposition such as "two plus two equals four" does not really involve such objects as two or four might be supposed to be, but merely asserts that if one considers a property that belongs to a thing *a*, and another thing *b*, distinct from *a*, and to nothing else, and another property that belongs to a thing *c*, and another thing *d*, distinct from *c*, and to nothing else, then the property consisting of the disjunction of these two properties is possessed by objects which we may term *m*, *n*, *o*, and *p*, which are all distinct from one another, and if it be possessed by an object *x*, then *x* is either *m*, *n*, *o*, or *p*.

But even if the things with which mathematics deals are fictions, it must be admitted that we can handle these fictions without knowing how they are put together. The average mathematician neither knows, nor, I grieve to say, cares, what a number is. You may say if you like that his analysis is blunted and his work rendered unrigorous by this deficiency, but the fact remains that not only can he attain to a very great degree of comprehension of his subject, but he can make advances in it, and discover mathematical laws previously unknown. The whole logical analysis of the concept of number scarcely dates back forty years, yet the first mathematical use of numbers is lost in prehistoric antiquity.

Now, if mathematics is essentially the science of propositions involving no constants, how is it that there were mathematicians before Frege? How is it that mankind was able to handle the notion of number for myriads of years with hardly the ghost of an idea of what a number was? It is almost infinitely improbable, as one sees at once from the illustration given above, that we have Frege's notion of number before we study Frege's work, for it is so unfamiliar to us when we first learn it, and it can not be argued that since Frege's numbers have the formal properties of our every-day numbers they are identical with them, for a little reflection will convince us that on the basis of Frege's and Mr. Russell's own work, we can produce other constructions different from those to which they give the name of number, yet having formal properties which, as far as we are interested in them from the standpoint of a definition of number, are identical with those of Frege's numbers, and that it will be, in general, impossible to say that one of these constructions more truly represents the proper analysis of our naïve notion of number than another, for all of them will seem almost equally unfamiliar to us when we first become acquainted with them. With regard to the ordinary integers with which elementary arithmetic deals, for example, it is even impossible to say whether, in the strict mathematical sense of the word, they are ordinal or cardinal numbers—that is, whether or not they imply an arrangement of the collections of objects to which they refer.

We can not, then, regard naïve mathematics, whether it be the naïve mathematics of a schoolboy or of a Leibniz, as merely a less explicit statement of what the modern analyst expresses with the aid of his involved technique and symbolism: whence, then, does it draw what certainty it possesses? Perhaps I can explain this best by reminding the reader of an experience which very many people must have had while they were learning mathematics. Every one, or almost every one, at any rate, must remember what agony his first lessons in geometry gave him when he was a schoolboy. The theorems seemed obvious enough to him, but how on earth, he probably wondered, can one get the theorems out of the axioms? No doubt, he thought to himself, two straight angles are always equal, but how is it that one is justified in proving it by superposing one on the other? The axioms did not tell him just when he could superpose one figure on another and when he could not. On the other hand, if he were to go by common sense, and not by his axioms, in proving his theorems, how did it happen, he must have puzzled, that he was not allowed to make use of such eminently sensible methods of proof as measuring the lengths of the lines occurring in his figures, determining the perimeter of a circle by rolling it along a straight line, etc.? After several months' practise in geometry, however, although he was still unable to give a formulation of the principles by which he worked which would satisfy the demand for rigor of the modern student of the axioms of geometry, he céased to ask these questions, yet seldom went wrong in his geometrical reasonings. He used proofs involving superposition where, and only where, they led to valid results, and never tried to solve a problem by measuring his lines and angles, or by rolling a circle along a line. In short, although he was by no means able to analyze his geometrical proofs in detail, he had formed *habits* of handling the ideas of geometry which, as the time went on, became less and less likely to lead him astray. It was in the uniformity of these habits that all the certainty of his geometrical demonstrations lay—at any rate, until he had begun to correlate his geometry with arithmetic or logic—and the postulates and axioms of geometry served merely to help him fix these habits and render them uniform. As it was by no means impossible that these habits should have broken down in some particular instance—though, after he had studied geometry for years, it was extremely unlikely—the certainty of his geometrical demonstrations was not absolute.

Now, it is not merely in the schoolboy's study of geometry that habit plays a large part: the life of every branch of mathematics lies in a habit. Let us suppose the schoolboy of the previous example replaced by a practised mathematician, and the garbled collection of so-called "axioms" which form the introduction to most school geometries replaced by a genuine set of postulates, made as rigorous as any yet devised. How is the mathematician ever to apply his postulates to one another? His postulates themselves can not tell him how they should be applied, for then he would have to make a proposition form a part of its own subject-matter, and he would be involved in vicious circle paradoxes. He can not solve the problem by merely adjoining new postulates to his set, telling us how to use the old ones, for either he has still the problem before him, how is he to use these new postulates, or he has an infinite regress of postulates, each depending for the rule by which it is to be applied on the preceding one. The only alternative which seems to me really open is that he should apply his postulates to one another in some way, the uniformity of which is secured by the fact that he has got the habit of handling certain sorts of combinations of symbols and of ideas in a certain manner. He feels instinctively, as it were, that here one can substitute this term for that, there one can leave off that parenthesis, etc. And this habit of using his symbols and compounding his ideas in such a way as to produce the results which other mathematicians have produced, and of obtaining new propositions in a certain determinate manner, is so ingrained in him and so uniform that the chances of his being led to deduce different and conflicting theorems from the same premises are very nearly *nil.*

Yet that these chances are not necessarily entirely absent is best shown by the fact that in many cases, where mathematicians had uniformly deduced certain conclusions from certain premises for, perhaps, centuries, great mathematicians have been able to change deliberately the habits with which they drew conclusions from these premises, and to deduce an absolutely different set of consequences from the original postulates, conflicting with the former conclusions, by bringing to expression as an additional postulate part of what was latent in the original habit, and contradicting it. This is the way the non-Euclidean geometries were first discovered, and the way that, after them, a whole family of systems such as finite spaces, non-Archimedean geometries, etc., have been constructed. This is the way negative numbers, fractions, irrational numbers, and complex numbers were first introduced. Now, although it perhaps never happened before the recognition of the axiom of parallels that a mathematician ever introduced a proposition only true in non-Euclidean geometry in a chain of reasonings about Euclidean geometry, it is by no means certain on *a priori* grounds that such a slip could not have occurred. Therefore, the demonstrations in geometry before the days of non-Euclidean geometry were only relatively certain.

Some of the mathematicians among my readers will object, in all probability, that our habits of geometrical reasoning are now absolutely determined, because the sets of postulates recently set up for geometry are what is called *perfect* or *categorical*: that is, that any new postulate, involving no non-geometrical notions, adjoined to the set, would either be a consequence of the other propositions of the set, or would contradict them. This is perfectly true, as far as it goes, but to understand its implications we must ask, how does one prove it? and what does it mean? Now, the simplest of the modern ways of defining a system as geometrical is by expressing all the notions involved by it in terms of some fundamental relation, and stating certain limiting propositions about this. These defining propositions are of such a sort as to hold of all relations which are what is called *similar* to any given relation about which they hold. A set of defining propositions, or postulates is then perfect, if no proposition which will apply to any relation similar to R if it applies to R, and which will still further limit the class of relations to which the set applies, can be asserted.

It will be seen, then, that to prove the perfectness of a set of postulates of, say, geometry, we already need a theory of relations, which will, among other things, explain the notion of similarity, and that the certainty of the perfectness of the set, on which depends our knowledge that our way of compounding the postulates of the set needs no habit to make it unambiguous, is itself dependent on the certainty of the formal calculus of relations. Moreover, if one deduces the theorems of the relational calculus directly or indirectly from certain premises, one can not claim, without aruging in a vicious circle, that he can prove that these premises form a perfect set, and that therefore our habits of using them can not be ambiguous.

Yet the theory of relations, like every mathematical theory, must be grounded either in postulates or in some other mathematical theory. The best foundation which has yet been given for it is that expounded in the "Principia Mathematica" of Dr. Whitehead and Mr. Russell. In this work, the theory of relations is deduced indirectly from certain postulates about propositions and "propositional functions" or concepts. The first postulate stated by Mr. Russell is very interesting in this connection: it says, "Any proposition implied by a true proposition is itself true." Unlike most of its successors, this is stated in words, and not in symbols. This fact is not without importance. Mr. Russell intends to use this proposition to justify himself in leaving off a true hypothesis from an implication. Now, if the proposition justifying this appeared in a tangible form as a premise in such a case, we should need to assume it a second time to justify its elimination in its first occurrence, and so on *in infinitum.* We should never, that is, be able to make a single deduction, for we could never separate a conclusion from its premises. We must be able to drop true premises in a definite manner, and this first postulate of Mr. Russell's is expressed in words, and not in symbols in recognition of the fact that, while this is the case, our power of doing so resides, not in the formulæ of logic themselves, but in our habit of using them. Now it is not only possible, but highly probable, that there are habits in accordance with which we might deduce different results from Mr. Russell's postulates, and possible, but almost infinitely improbable, that we might at any time mistake one of these habits for the proper one. It seems also possible to me that this chance of uncertainty might be reduced to any desired degree by the insertion of new postulates in Mr. Russell's system defining the mode of application of the previous ones. The negation of these would lead to non-Russellian logics much as the negation of the postulate of parallels leads one to non-Euclidean geometries. It appears to me unlikely that such an amplification of Mr. Russell's set of postulates would ever render it possible for us to prove that no further ambiguities in the habits according to which we use these postulates would be possible.

Apparently, then, it is in any case highly probable that we can get no certainty that is absolute in the propositions of logic and mathematics, at any rate in those that derive their vadility from the postulates of logic. But are not the postulates themselves absolutely certain? Is there any conceivable room for uncertainty in the law of contradiction, or in the other axioms of logic? It appears to me that even here dogmatism is not the proper position to maintain. It seems a just maxim that we can not be absolutely sure that a proposition is true until we have a perfectly adequate knowledge of what it says—such a statement as, "Abracadabra, and I am sure of it" remains pure nonsense until one knows definitely what is meant by "Abracadabra," while even when we come to the relatively definite propositions of physics, such as the law of the conservation of energy, one of the chief sources of doubt as to their absolute validity is, in many cases, our lack of certainty as to what they really assert. Now, such "laws of thought" as the law of contradiction, or the law of identity, have already undergone a considerable change in their meaning on account of the analysis to which the new mathematical logic has subjected them—the law of contradiction, "Everything is either *A* or *not-A*," has been rendered a rather late inference in the "Principia Mathematica," limited in its meaning by the theory of types, and not derivable from any single one of the set of postulates there given. The law of identity has been shown to be a consequence of the definition of identity, which requires an elaborate logic for its very formulation. Even if one accepts "*p* is true or false" as the same proposition as the law of contradiction and "*p* is equivalent to *p*" as the law of identity, these may come in at a stage when the theory of propositions has already reached a high level of development, if we accept Sheffer's analysis of the calculus of propositions, and it is by no means inconceivable that this should make a certain difference in their complete meanings. Moreover, it is not impossible that the notion of a "proposition," in the sense in which this word is used in the "Principia," may itself be capable of analysis in terms of some more simple notion—it is part of mathematical and logical progress not only that our sets of postulates should be rendered more precise by the adjunction of new postulates, but that the "habit" by which we use a set of postulates pertaining to a certain mathematical or logical system we use should be made more unambiguous by the reference of the system as a whole to a finer system, which gives us a smaller opportunity for ambiguity in the habit by which we use its postulates, as a center of orientation, as it were. There is no need, then, of supposing that even the axioms of the "Principia" or any similar set we shall ever come to are not subject to further analysis, and that we have an absolutely adequate knowledge of the meaning of any logical proposition whatever. Hence, although our degree of uncertainty in logic is so infinitesimal as not to enter at all in the allowance we make for error in our scientific reasonings, we have no reason to suppose it is altogether absent.

NORBERT WIENER.

HARVARD UNIVERSITY.

Comments on [15b]

W. V. Quine

Since our idea of mathematical certainty antedates the latter-day sophisticated analyses of the meaning of mathematical propositions, Wiener infers that our sense of mathematical certainty does not rest on meanings. It rests rather, he thinks, on habits of deduction and symbolic manipulation that have gained our confidence by not engendering conflict. This force of unformalized habit never submits completely to explicit axiomatization, even when the axioms are categorical. After all, Wiener remarks, categoricity itself is just another mathematical concept, subject to axiomatization in turn. There is a vicious regress, he argues, in the idea of axiomatic formalization without remainder. It is the same regress that was entertainingly expounded by Lewis Carroll in "What the tortoise said to Achilles" (*Mind,* 1895)—a paper that the youthful Wiener had missed.

Some such sense of the limitations of formalism dates back to Wiener's boyhood, as recalled in *I Am a Mathematician* [56g], p. 324. In an earlier place (*Ex-Prodigy* [53h], p. 193) he cites Gödel's incompleteness theorem as confirming this view. I know of no anticipation by Wiener of the considerations that underlie Gödel's proof. But the paper that is now before us foreshadows, in Wiener's remark about categoricity, Skolem's set-theoretic relativity: the awareness that isomorphism is not absolute, but relative always to some preassigned universe of sets.

Three times, in his last paragraph, Wiener refers to the law of excluded middle as the law of contradiction; an obvious inadvertency.

MR. LEWIS AND IMPLICATION

THE theory of implication developed by the symbolic logicians seems to have aroused a considerable degree of antagonism among certain students of Logic. There are many philosophers to whom you can not mention the name "Russell," without evoking such comments as, "His logic is purely artificial, for it is nonsense to suppose that a false proposition implies any proposition, or that any proposition implies any true proposition," or, "Who could ever reasonably maintain that, 'The moon is made of green cheese,' implies, 'Caesar died in his bed?' " Most of these critics have not expressed their objections to Mr. Russell's position in black and white, so that it is impossible for us to see just in what the strength and the weakness of their arguments consist; Mr. C. I. Lewis, however, has had the courage of his convictions and has developed his criticisms of the views of Mr. Russell together with certain very interesting logical theories of his own in a series of articles which have appeared partly in *Mind* and partly in this JOURNAL.

The sum and substance of Mr. Lewis's objections to Mr. Russell is this: Mr. Russell, following the older symbolic logicians, holds that a false proposition implies any proposition and that any proposition implies any true proposition. That is, p implies q if either p is false or q is true. Mr. Lewis claims, and not without a certain degree of justice, that this is not what we ordinarily mean by implication. We do not, for example, usually say that, "Socrates was a solar myth," implies, "All triangles have two or more sides." Therefore, as Mr. Lewis tells us, "Not only does the calculus of implication contain false theorems, but all its theorems are not proved. For the theorems are implied by the postulates in the sense of 'implies' which the system uses. Hence *it has not been demonstrated* that the theorems can be inferred from the postulates, even if all the postulates are granted. The assumptions, *e. g.,* of the 'Principia Mathematica' imply the theorems in the same sense that a false proposition implies anything."[1]

Mr. Lewis's reasoning here is fallacious, and the fallacy he commits is that of denying the antecedent. From the fact that if a set of postulates deals correctly with our ordinary relation of inference, it will yield us a correct logic, he infers that if a set of postulates fails to deal with this relation, and, like the Russellian logic, seizes upon some other relation as its fundamental notion, the logic to which it leads must be faulty and incorrect. This is a manifest and a grave error; it is conceivable that we may develop a valid theory

[1] This JOURNAL, Vol. X., p. 242.

of demonstration, the fundamental notion of which is other than what we ordinarily call inference, which is correctly derived from its own postulates. It is not necessary that a theory whose purpose it is to yield us a norm of valid inference should itself in the first instance be a theory of inference. We say that one proposition can be inferred from another if there is a certain relation between them such that we are compelled to accept the former proposition as true if we accept the latter one. The purpose of Logic, in so far as Logic is a norm of inference, is to provide us with certain methods which, when applied to any true proposition of a suitable sort, will yield us other true propositions. These methods need not of themselves involve any reference to the concept of inference, and may not lead us to realize that they are methods of inference. Indeed, they can not lead us to realize that they are methods of inference, even if they actually concern themselves with such methods, for then they would form a portion of their own subject-matter, and we should be involved in that philosphical lifting of oneself up by one's boot-straps, the pernicious consequences of which have been pointed out so well by Mr. Bertrand Russell in that part of the "Principia Mathematica" which deals with the theory of types. If the natural history of the process of inference is a branch of Logic, it is a Logic of a very different type from that which it is the purpose of the logisticians to develop, and there is no reason under the sun why this latter Logic should be doomed under penalty of death to make use of our every-day notion of implication.

The only questions, then, which can reasonably be asked concerning the correctness of the Russellian Logic are, Is it actually a correct norm of valid inference? and, Is the coherence and self-consistency which it claims for itself, and tries to justify by an orderly derivation of its theorems from a small and simple set of postulates, genuine or factitious? As we have already seen, Mr. Lewis answers both of these questions in a manner adverse to the claims of Mr. Russell, but we have found his arguments to be fallacious. To arrive at the true answer to these questions, we must discuss what the function of postulates is in a deductive system such as the Russellian logic. Now, the main function of the postulates of any system is to stand as hostages for the system: they must be statements, the acceptance of which commits one to the acceptance of the entire system as true. That is, we must be able to affirm that the system of propositions is true, unless, by chance, the postulates should fail to be satisfied. This is not to be taken, as Mr. Lewis seems to be in danger of taking it, as the assertion of any occult connection or motive force acting between the postulates and the system: there is no need for us to suppose that the truth of the propositions is conditioned in any causal way by the truth of the postulates. That either the postulates are false, or the system of propositions is true[2]—this is all we need know concerning the relation which the postulates of Logic bear to the system of logical truths, since this is enough to secure for us that the truth of the propositions of Logic must be maintained by any one who believes in its postulates. It matters not whether the disjunction expressed in this last proposition be what Mr. Lewis calls "extensional" or what he calls "intensional":—if it is a mere extensional disjunction, such as can hold between two unrelated and mutually irrelevant propositions, then we may have attained our knowledge of the truth of the propositions of logic independently of our postulates, but we shall have attained it, nevertheless. So long as the conditions of the truth of our theorems are not incorrectly extended and expanded, the question of the genesis of such propositions as we accept as true has no interest for us, except in so far as it is bound up with the question of their validity. The knowledge of the postulates must be a sufficient ground for a knowledge of the theorems:—if we find that it is more than sufficient, and that we do not need a previous knowledge of the postulates to attain to that of the theorems, so much the better. If, that is, we interpret Mr. Lewis as maintaining that we are justified in inferring one proposition from another whenever we are able to proceed to the first from the second by a valid process of reasoning, then, since we are clearly bound to accept the Russellian theorems in the Algebra of Logic if we accept the Russellian postulates, we must maintain that the Russellian postulates

[2] In an article entitled, "A Too Brief Set of Postulates for the Algebra of Logic" (This JOURNAL, Volume XII., p. 523), Mr. Lewis makes what practically amounts to the claim that the postulate, "Any true proposition implies all true propositions," is a sufficient basis for the whole of logic, or rather, that the methods of the Russellians should lead them to this conclusion, which Mr. Lewis regards as very objectionable. The grounds on which Mr. Lewis bases this claim are that since this postulate is true, and since it tells us that any true proposition implies any true proposition, it implies (in its own sense, which is also that of Mr. Russell) any true proposition whatever. There is no question that Mr. Lewis's postulate does actually imply any true proposition, but this is not the entire function which a postulate must fill. The fact that any proposition in a mathematical system, so it seems, can be made a member of some set of postulates or other, simply goes to show that the primacy of the postulates of a mathematical system is a primacy in the order of knowledge, not in the order of existence. I need not say, however, that even this primacy of the postulates of a system refers only to their status within a given investigation. Now, it is obvious that this priority in the order of knowledge can not be claimed for Mr. Lewis's postulate. It is for this reason that I have said that the fact that either a theorem must be true or a set of postulates false is all that we must *know* for us to be able to say that they imply the theorem, and not that it is all that need be true to render the latter statement true in its usual sense.

imply the theorems, not only according to their own peculiar definition of the relation of implication (which may, indeed, have but little in common with the ordinary definition of that relation), but precisely according to our usual understanding of the relation of implication. Now, this is taken by Mr. Lewis himself as the ultimate criterion of the true nature of implication.

It may be objected by some that the relation of implication by which we obtain the theorems of logic from its postulates—or, indeed, any theorems from any postulates—is not merely the Russellian relation of material implication, which a false proposition bears to any proposition and any proposition to a true one. As Mr. Lewis says, "Euclid's parallel postulate or Lobachevski's postulate about coplanars is—one of them—false. Nevertheless, he errs who would take either postulate to imply anything and everything. Logical consequences follow regardless of truth or falsity of premises."[3] (Mr. Lewis's statement here is unquestionably true. Mr. Russell's material implication is not adequate to the deduction of the theorems of geometry from their premises, and it might be thought that it is still less adequate to the deduction of the theorems and propositions of logic from Mr. Russell's set of primitive propositions. This, however, is not the case. There is, as Mr. Lewis himself recognizes, a vast difference between the postulates of logic and the postulates of geometry. "Pure mathematics is not concerned with the truth either of postulates or of theorems: so much is an old story .... Indeed, the attempt to separate formal consistency and material truth is, in the case of the logic, peculiarly difficult."[4]) That is, pure mathematics is concerned only with the question whether the theorems follow from the postulates, while logic must take its postulates and theorems as both true. A mathematical set of postulates can be investigated without any reference to an actual system which in reality embodies them, while the postulates of logic, in so far as they remain postulates of logic, must be regarded as actually embodied in the constitution of every system. The postulates of geometry are hypotheses, or *types of possible truths;* the postulates of logic, if they are correctly stated, are *truths.* Now, a hypothesis or type of truths is not a proposition. The postulates of geometry, *qua* postulates of geometry, may apply indifferently to points in space or to number-triads or to any other sort of entity you please, and are on this very account neither true nor false in themselves, but only in their particular manifestations. Since they are neither true nor false, and hence not propositions, it is nonsense to speak of their implying anything or being implied by anything in the same way in which we can speak of propositions,

[3] This JOURNAL, Vol. X., p. 432.

[4] *Ibid.*, p. 429.

such as the postulates of logic, as implying or being implied by something. It is only natural, then, that it should be permissible to use methods in deducing the theorems of logic from their postulates which are prohibited in the case of the postulates and theorems of geometry. It is by virtue of the very distinction between logic and other branches of mathematics, which Mr. Lewis stresses so strongly, that the objections which he raises against Mr. Russell's employment of the relation of material implication in the development of his theorems may be shown to be irrelevant.

The relation of implication which the postulates of geometry bear to its theorems deserves a further consideration, for a misapprehension of the nature of this relation is at the bottom of most of Mr. Lewis's errors. The postulates of geometry, as we have seen, are not propositions, but blank forms of propositions, which may be filled in in a countless number of different ways. Thus the postulate, "Any two points are connected by one and only one line," may be filled in with a content consisting of actual spatial entities, or of number-triads (in place of points) and pairs of linear equations in three unknown quantities (in place of lines), or with any one of an infinity of other possible specific determinations. Now, Mr. Russell denotes a blank form for propositions by the names, "universal" and "propositional function." Moreover, he has given a definite and complete discussion of the relation of implication between propositional functions, which he calls the relation of *formal implication.* A propositional function ϕ is said to imply another ψ formally if every entity which fills out the blank form ϕ into a true proposition also fills out the blank form ψ into a true proposition.[5] Thus to say that the postulates of geometry imply the theorems formally is to say that every system which satisfies the postulates of geometry also satisfies the theorems of geometry. This is manifestly true of the postulates and theorems of our geometry, and constitutes a necessary condition for the validity of the latter. A little reflection will convince us that it is also a *sufficient* condition, for otherwise we should have theorems not implied by the postulates of geometry, but true of all geometrical systems. This means of all *possible* geometrical systems, for a geometrical system is a universal, and the possibility of a universal is identical with its actuality. Now, we should ordinarily say that any proposition follows from a given set of postulates, if it is true of every possible system which satisfies these postulates, on the ground of the very nature of universals themselves. As a consequence, we see that Mr. Russell's notion of formal implication among universals is in every respect in harmony with our every-day use of the term.

[5] Whitehead and Russell, "Principia Mathematica," p. 21.

It is this relation of formal implication which Mr. Russell always uses when he derives the theorems of any mathematical system, such as that of linear order,[6] from their postulates, and the correctness of his view of the nature of implication is substantiated by the fact that in every case his results have agreed with those of other mathematicians who have made use of the same postulates.

It may be objected to this analysis of the relation of implication between postulate and theorem, that the relation of implication still holds, and differs from that of material implication, when our postulates are propositions concerning specific objects, and not mere propositional functions which may apply to anything at all. I think, however, that it is only in so far as they are taken as the representatives of propositional functions that propositions imply anything in this sense. When I deduce the properties of actual space from Euclid's postulates, I am really deducing conclusions from certain laws which an infinity of systems may satisfy, and which our space does satisfy, and I am applying them to our space. The possibility of a reference to other systems plays an essential part in the deduction. So, too, when I say, "If it rains, I shall get wet," the real implication which I desire to assert is, "*In any situation such as the present,* when it is raining on me, I get wet." The tacit, "other things being equal," which may always be prefixed to such implications, points out the universality of reference which such implications are intended to have, as will be seen if we write it, "whenever other things are equal." This explains the unnaturalness of such material implications as, "If two and two are four, Caesar is still alive," for there is no obvious general law or formal implication to which they may be reduced as instances.

The reason why Mr. Lewis seems unable to understand the significance or even the nature of Mr. Russell's formal implication appears to be that he ignores the distinction between propositions and propositional functions. Hence he regards $(x) : \phi x \cdot \mathsf{U} \cdot \psi x$, the relation of formal implication between two laws, as but a particular instance of what he calls "strict implication," which relates propositions.[7] Further on in the same article, he speaks of "cases" of the truth or falsity of a proposition, while a proposition, unlike a propositional function, is either simply true or simply false, and can have no instances of truth or falsity. These slips are particularly regrettable in the case of a man who has published so extensively on logistical matters as Mr. Lewis has. If Mr. Lewis disagrees with Mr. Russell's distinction between propositions and propositional functions, he should have made the fact of this disagreement clear before setting forth on his

[6] "Principia Mathematica," Vol. II., Part V.

[7] This JOURNAL, Vol. X., p. 430, note.

criticism of Mr. Russell's logistical views, as this distinction is the heart and soul of the Russellian logic.

Mr. Lewis's arguments against Mr. Russell have little logical cogency, and one feels that they were developed to give an excuse for his constructive work in the definition of his system of "strict implication," which really requires no such apology for its existence. It is not the place here for me to comment upon this exceedingly valuable and interesting piece of work, except to remark that its logical worth is utterly independent of an acceptance of Mr. Lewis's contentions against Mr. Russell. One may grant that the Russellian system is both a logic and a self-subsistent logic, and yet realize the obvious fact, which it is to Mr. Lewis's credit to have noticed, that it is unable to distinguish between the notion of truth, pure and simple, and the notion of that truth which results as a consequence of the laws of logic alone.[8] The logic developed by Mr. Lewis is able to give an account of this notion, and is, in so far, more complete in its apparatus—though not necessarily more *correct* in any way—than that of Mr. Russell. In the form which it finally assumes, it starts with a distinction between *de facto* and necessary falsity, from which a natural transition leads us to the notions of necessary and *de facto* truth, disjunction, and implication. The whole theory is carried out with great patience and ingenuity, and taking it all in all, with logical correctness.[9] It is, however, a supplement to the Russellian logic, and not a refutation of it.

NORBERT WIENER.

HARVARD UNIVERSITY.

Comments on [16a]

W. V. Quine

Lewis's paper, however wrong-headed, was historic. It opened the campaign against material implication—a campaign that gave rise in the ensuing half century, for better or worse, to a lot of literature on modal logic.

Wiener quotes Lewis as questioning not only the postulates of Russell's logic, but also the steps that carry Russell from his postulates to his theorems. He quotes Lewis as arguing that if the postulates imply the theorems only in Russell's spurious sense of implication, they are too weak really to assure the truth of those theorems. Wiener defends Russell on this point, observing that even if 'implication' is a misnomer for Russell's material relation, still this relation is not one that a true postulate will ever bear to a false theorem. Insofar as we are assured that this relation holds between the postulates and the theorems, and are assured of the truth of the postulates, we are assured of the truth of the theorems.

Wiener remarks further that when we assure ourselves that this material relation holds between the postulates and the theorems of an uninterpreted system, we thereby assure ourselves of a relation stronger than material implication after all, namely, universally quantified material implication, or what Russell calls formal implication. This, Wiener urges, is indeed implication in a reasonable sense of the term.

When we properly use the verb 'implies', we talk about the implying sentence and the implied sentence by name or description. When we use the conditional particle 'if-then', on the other hand, we join the two sentences themselves into a compound sentence. If Russell had heeded this distinction between mention and use, and had talked not of material implication but of a material conditional, he would have roused far less antagonism and would have had a much more defensible position. If Lewis had noted this point, the literature on modal logic might have been leaner. Wiener did not note it either.

ON THE NATURE OF MATHEMATICAL THINKING.

By

NORBERT WIENER, Ph.D., Massachusetts Institute of Technology, U.S.A.

IF you divide the various sciences and learned disciplines in accordance with their subject matter, you will find the first and deepest line of cleavage between mathematics on the one side and the whole remaining body of human knowledge on the other. In what concerns its proved results, mathematics stands alone in its qualities of rigour, logical concatenation, precision, and conclusiveness. When to these marks of the subject matter of mathematics is conjoined the marvellous perfection of form possessed by that most familiar of all mathematical sciences, the geometry of the Greeks, it is entirely natural that the technique involved in obtaining such results and the mental processes of the mathematician should seem to the layman awful and mysterious. The latter attributes to the order of invention the characteristics of the order of presentation, and assigns to the nascent thoughts of the investigator something of the logical accuracy and sequence which appear in his published memoirs.

Now, there is perhaps no place where the order of being and the order of thinking need to be differentiated with such care as in mathematics. It needs but little reflection to see that any account of mathematics which makes logic not only the norm of the validity of its processes but also its chief heuristic tool is absurd on the face of it. The theory of the syllogism will tell you that when you have the propositions, "All A is B," and, "All B is C," you can derive the further proposition, "All A is C," but you will not find among all the works of Aristotle and Bertrand Russell combined, with the Schoolmen thrown in for good measure, one iota of informathe worse off. So long as he keeps his published writings tion which will, without any further act of thought on your part, tell you when to use the syllogistic method, or what particular propositions to employ as major and minor premises. Logic will never answer a question for you until you have put it a definite question. Even then it will never volunteer any information. It has but two words in its language, and those are "yes" and "no." Logic is a critic, not a creator, even as regards its own laws of criticism. While a man endowed with logic alone would assuredly never do any bad mathematics, he would just as assuredly never do any good mathematics.

Mathematics is every bit as much an imaginative art as a logical science. As has just been said, if you wish to know the answer to a question, you must first ask it, and the art of mathematics is the art of asking the right questions. From any set of postulates or premises or assumptions there may be derived an infinite set of lemmata and theorems and conclusions, every one as sound in its logical deduction as any other. Some of these will be recognised by any mathematician as of transcendent importance, more will constitute the ordinary stock in trade of the mathematical journals, but by far the greatest part will be by common acceptance nugatory and trivial. This charge is entirely beyond the jurisdiction of logic, but the ability to discriminate between such trivial theorems and the really vital conclusions of a mathematical science is precisely that quality which the competent mathematician has and the incompetent mathematician lacks.

What is an important theorem? Some theorems are important because of their direct physical and technical applications, others because of their position in the development of a further theory which is of interest, and yet others because of the beauty, symmetry, and richness of the theory of which they form a part. These latter qualities are of a nature essentially aesthetic, and are of course bound up with individual and personal judgment after the fashion of all aesthetic qualities. In the general recognition of varying degrees of beauty and importance, together with the lack of any permanent and universally accepted norm of these characters, in the existence of fashions, of local and national standards and of individual eccentricity, mathematical taste shows its essential kinship with taste in the arts.

In order to do good mathematical work, then, and in fact to do *any* mathematical work, it is not enough to grind out mechanically the conclusions to be derived from a given set of axioms, as by some super-Babbage computing machine. We must select. The postulates with which we start contain our conclusions only in the sense in which the keyboard of the pianoforte contains a sonata, in the sense in which a yard of canvas and tubes of paint contain a painting, or a block of marble a statue.

The imagination is the mainspring of mathematical work, while logic is its balance-wheel. As in a watch, it is not until the mainspring has been wound up to a certain extent that the balance-wheel starts to move. It is not until after we have put ourselves a mathematical question, and have propounded at least a tentative answer to this question that there is any possibility of logical reasoning. Our tentative answer may be vague to begin with—very vague, and of a nature totally repugnant to logical thinking, for it may not even be in a form determinate enough to put down in black and white on paper. There is nothing more surprising than the power of the mind to formulate these vague yet useful hypotheses concerning a subject matter abstract and logical in character. What is it, I wonder, that forms the real content of our consciousness at one of the moments of reflective reverie which constitute so large a part of our periods of research? What we have can scarcely be a dim and confused image of the theorem at the end of our investigation, for the dim light of intuition may be a will-of-the-wisp, and our investigation may end frustrate. Those psychologies of meaning which see the psychological counterpart of the binomial theorem in an obscure strain at the back of the eyeballs are surely not very helpful in their analysis of the mental state of the mathematician.

This mathematical day-dreaming (in the midst of a difficult research, not a little of it is ordinary dreaming by night) is perhaps easiest to understand in the case of geometry, where it is largely dependent on a carefully cultivated power of spatial imagination. Even here it is remarkable how a crude two or three-dimensional image can do service as the vehicle of a notion in four or five dimensions, or even in space of infinite dimensionality. In the highly rarefied regions of modern analysis, however, such aid as the spatial imagination can furnish, though it is of undeniable value, is fitful and occasional. No picture of an everywhere dense denumerable set of points, or of a continuous curve lacking a tangent at every point, is in the least adequate to the complexity of the situation which it represents. Throughout function theory, postulate theory, and the theory of assemblages, the whole mass of habits of thought which makes possible any imagination whatever is as much a new acquisition of the human mind as the body and organs of the butterfly are new acquisitions of the caterpillar.

Habits of thought—it is these rather than the sensory and imaginational content of the mind which constitute what is vital in mathematical imagining. Inasmuch as the mathematical imagination must sooner or later submit to the criticism of logic, it is essential that these habits should accord with logic. First and foremost among these habits is the habit that the mathematician should continually subject his ideas to trial by logic. He must incessantly try to draw the consequences inherent in his notions, and must instantly recognise when he is proving too much, and is drawing a conclusion which is manifestly false. He must arrange the steps of his proposed theory in a tentative logical order, narrowing the unproved gaps until his results cohere from beginning to end. He must revolve his system in his mind, trying it by all the examples his ingenuity can muster. When he finds a flaw, he must consider whether it is inherent in the very nature of his ideas, or merely adventitious and to be circumvented by a more ingenious approach. Whatever he builds up he must try to tear down, and whatever he tears down he must strive to build up again.

Not only must the mathematician employ his imagination in the invention of new problems and the discovery of *experimenta crucis* to test his answers to these problems, but he must ever be on the alert to see the widest consequences of the methods which lead to his conclusions. Many a theory is encumbered by restrictions which are either altogether unnecessary or are easily replaced by others of a more fundamental character. Many a research answers half a problem when it might just as readily answer the whole. In every branch of mathematics there is one plane of generality on which the theorems are easiest to prove, and needless complication arises as quickly by falling short of this as by exceeding it. It is a mark of the great mathematician to have taken a number of separate theories, fragmentary, intricate and tortuous, and by a profound perception of the true bearing and weight of their methods to have welded them into a single whole, clear, luminous, and simple.

Mathematics is an experimental science. The formulation and testing of hypotheses play in mathematics a part not other than in chemistry, physics, astronomy, or botany. Just as in the science of nature, old ways of regarding things are compared, tried against the facts, worn down by mutual attrition, until they take on a new and unfamiliar aspect. It matters little in what concerns scientific method and the mental processes of the investigator that the mathematician experiments with pencil and paper while the chemist uses test-tube and retort, or the biologist stains and the microscope. An experiment is the confronting of preconceived notions with hard facts, and the notions of the scientist are just as much the result of preconception, the facts just as hard, in mathematics as anywhere else. The only great point of divergence between mathematics and the other sciences lies in the far greater permanence of mathematical knowledge, in the circumstance that experience only whispers "yes" or "no" in reply to our questions, while logic shouts.

Since, however, pencil and paper are cheaper than retorts and microscopes, and since there are no long periods of waiting in mathematical research such as are incurred in the other sciences by the construction of apparatus, or the time-consuming propensities of chemical reactions, or any of the thousand and one petty worries which make the hair of the laboratory worker turn gray before its time, there is one great advantage with the mathematician: he may blunder to his heart's content, waste time in asking questions which he cannot answer, fumble and bungle and muddle, and if he can salvage one or two good ideas from this wreckage, neither he nor anyone else is a penny clear of error, if he welcomes every ghost of a shadow of an idea that comes his way, and tries it before casting it aside, he suffers no harm but great good; for it is just these waifs of notions that may furnish the new point of view which will found a new discipline or reanimate an old. He who lets his sense of the mathematically decorous inhibit the free flow of his imagination cuts off his own right hand.

Comments on [23a]

R. L. Wilder

This popular exposition of the nature of mathematics and mathematical creation shows an unusual perception of the place of logic in mathematics, standing in stark contrast to the view (popular at the time) that logic dominates mathematics and the attendant "Mathematics is the science of p implies q" definition. Published when mathematics was still conceived of as "truth arrived at by deduction," its insistence that "Mathematics is an experimental science" should have acted as a timely antidote. It was perhaps unfortunate that it was published in a foreign psychological-philosophical journal rather than in a periodical more within the purview of the mathematical community. It is certainly as accurate a description today of how many creative mathematicians conceive of their field as it was when published.

Correction. Line -10 on p. 268 should be transferred to lie between lines 14 and 15 on p. 272.

ON A METHOD OF REARRANGING THE POSITIVE INTEGERS IN A SERIES OF ORDINAL NUMBER GREATER THAN THAT OF ANY GIVEN FUNDAMENTAL SEQUENCE OF Ω.

By *N. Wiener*.

1. LET I represent the series of positive integers greater than 1 in their order of magnitude.

2. Let p_n stand for the nth prime in order of magnitude. Let A represent a series of positive integers, not necessarily in order of magnitude. Let a and b, respectively, be the ath and bth integers in order of magnitude. Let $a \underset{\to}{A} b$ mean, "a precedes b in the order determined by A." Construct, now, the series P of the primes of the form p_n, where $p_a \underset{\to}{P} p_b$, when, and only when, $a \underset{\to}{A} b$. Let us call this series $P(A)$. It will be seen immediately that $P(A)$ is by definition ordinally similar to A, and hence must have the same ordinal number.

For example, if A be the series

$$1, 3, 5, 7, 9, \ldots, 2, 4, 6, 8, 10, \ldots,$$

$P(A)$ will be the series

$$1, 3, 7, 13, 19, \ldots, 2, 5, 11, 17, 23, \ldots.$$

If A be the series

$$1, 3, 5, 7, 9, \ldots, 2, 6, 10, 14, 18, \ldots, 4, 12, 20, 28, 36, \ldots, 8, 24, \ldots,$$

$P(A)$ will be the series

$$1, 3, 7, 13, 19, \ldots, 2, 11, 23, 41, 59, \ldots, 5, 31, 67, 103, \ldots, 17, 83, \ldots.$$

3. Given a well-ordered series P of primes, it will have a first term, a second term, ..., an nth term. Let the nth term be represented by the symbol P_n. Take, now, those products of k terms of P satisfying the following conditions:

(a) Every such product contains at least one P_n, where n is finite.

(b) If P_a is a factor of such a product, and if when P_b is another factor of that product $a < b$, the product contains a distinct factors, none of which are equal to 1.

(*c*) No factor of the product occurs more than once in the product.

Since all the factors by which the products in question are determined are primes, and since no factor occurs twice in any product, it follows that each group of n factors satisfying (*a*), (*b*), and (*c*) determines one product, and one only, and *vice versa*.

Let us represent a product of the form in question by $P_n . p' . p'' . p''' \ldots p^{(n-1)}$, where $p', p'', \ldots, p^{(n-1)}$ are distinct members of P, and, if $p^{(k)} = P_l$, $n < l$. It will be seen on inspection that any product which will satisfy (*a*), (*b*), and (*c*) may be expressed in this form, and *vice versa*. It is also clear that the order of the terms in the product is a matter of indifference. We may then, without any loss of generality, assume that $p' < p'' < p''' < \ldots < p^{(n-1)}$.

I shall now arrange these products in a series $p(P)$ in accordance with the following rules:

(1)
$$P_n . p' . p'' \ldots p^{(n-1)} \underset{\rightarrow}{p(P)} P_m . q' . q'' \ldots q^{(m-1)}, \text{ if } n < m;$$

(2)
$$P_n . p' . p'' \ldots p^{(n-1)} \underset{\rightarrow}{p(P)} P_n . q' . q'' \ldots q^{(n-1)}, \text{ if } p' \underset{\rightarrow}{P} q';$$

(3)
$$P_n . p' . p'' \ldots p^{(k-1)} . p^{(k)} . p^{(k+1)} \ldots p^{(n-1)} \underset{\rightarrow}{p(P)} P_n . p' . p'' \ldots$$
$$\ldots p^{(k-1)} . q^{(k)} . q^{(k+1)} \ldots q^{(n-1)}, \text{ if } p^{(k)} \underset{\rightarrow}{P} q^{(k)}.$$

I now wish to prove that if the ordinal number of P is α, that of $p(P)$ is α^ω, provided α is a number with no immediate predecessor.

By rule 3, if $p \underset{\rightarrow}{P} q$,

$$P_n . p' . p'' \ldots p^{(n-2)} p \underset{\rightarrow}{p(P)} P_n . p' . p'' \ldots p^{(n-2)} q.$$

For this to be true, however, it is necessary that (1) neither p nor q should be a P_m, where $m < n$, and (2) that

$$p > p^{(n-2)} > \ldots > p'' > p', \quad q > p^{(n-2)} > \ldots > p'' > p'$$

by the conventions we decided on in representing a product in the form $P_n . p' . p'' \ldots p^{(n-1)}$. That is, p and q are excluded from (1) the $(n-1)$ terms preceding P_n in p, and (2) the finite number of members of P not greater in numerical value than the largest $p^{(k)}$. Except for this finite group of values, p and q may assume any other value in P, and the order of the products $P_n.p'.p''...p^{(n-2)}.p$ and $P_n.p'.p''...p^{(n-2)}.q$, which will actually exist, will be the same as the order in P of p and q. That is, the series of the $P_n.p'.p''...p^{(n-2)}.p$'s will be similar to that of the p's, with a finite number of terms of the latter removed. Since, however, the ordinal number of the p's has no immediate predecessor, it can be shown readily that the removal of a finite number of terms from P will not alter its number, and therefore that the series of the $P_n.p'.p''...p^{(n-2)}.p$'s, arranged as they occur in $p(P)$, where $P_n, p', p'', ..., p^{(n-2)}$ are assigned, and p is allowed to take all possible values, has the ordinal number α.

In a precisely parallel manner, it may be shown that the series of the $P_n.p'.p''...p^{(n-3)}.q.r$'s, arranged as they occur in $p(P)$, where $P_n, p', p'', ..., p^{(n-3)}$ are assigned, q is allowed to take all possible values, and r is given some particular appropriate value for each value of q, has the ordinal number q.

Therefore, the series of the $P_n.p'.p''...p^{(n-3)}.q.r$'s, arranged as they occur in $p(P)$, where $P_n, p', p'', ..., p^{(n-3)}$ are assigned and *both q and r are allowed to take all possible values*, forms a series of the number α of series of the number α, or a series of the number α^2 by the definition of α^2.

Similarly, the series of the $P_n.p'.p''...p^{(n-4)}.q.r.s$'s, arranged as they occur in $p(P)$, where $P_n, p', p'', ..., p^{(n-4)}$ are assigned, and q, r, and s are allowed to take all possible values, forms a series of the number α of the series of the number α^2, or a series of the number α^3.

In a similar manner it can be shown that it follows in general from rules (2) and (3) that if the series of the $P_n.p'.p''...p^{(k)}.p^{(k+1)}...p^{(n-1)}$'s, where $P_n, p', p'', ..., p^{(k)}$ are assigned, and $p^{(k+1)}, ..., p^{(n-1)}$ are allowed to take all possible values, when arranged as they occur in $p(P)$, has the umber α^{n-k-1}, the series of the $P_n.p'.p''...p^{(k-1)}.p^{(k)}...p^{(n-1)}$'s where $P_n, p', p'', ..., p^{(k-1)}$ are assigned, and $p^{(k)}, ..., p^{(n-1)}$ are allowed to take all possible values when arranged as they occur in $p(P)$, has the number α^{n-k}.

Therefore, by mathematical induction, the number of the series of terms $P_n.p'.p''...p^{(n-1)}$, arranged as they occur in $p(P)$, where P_n is given and $p', p'', ..., p^{(n-1)}$ are allowed to assume all possible values, is α^{n-1}.

Now, by (1), $p(P)$ consists in the various series of terms $P_n.p'.p''...p^{(n-1)}$, where $p', p'', ..., p^{(n-1)}$ are allowed to assume all possible values, arranged in the order of magnitude of n.

Therefore the ordinal number of $p(P)$ is

$$\alpha^1+\alpha^2+\alpha^3+\alpha^4+\ldots+\alpha^n+\ldots=\alpha^\omega.$$

As an example of $p(P)$, let P be the series

1, 3, 7, 13, 19, 29, ..., 2, 5, 11, 17, 23, 31, ...,

where every prime whose position in the series of primes is odd belongs in the first part of the series, and every prime whose position in the series of primes is even belongs in the second part. Then $p(P)$ will be the series

3.7, 3.13, 3.19, 3.29, ..., 3.2, 3.5, 3.11, 3.17, 3.23, 3.31, ...

7.13.19, 7.13.29, ..., 7.13.17, 7.13.23, 7.13.31, ...

7.19.29, 7.19.37, ..., 7.19.23, 7.19.31, 7.19.41, ...

7.29.37, ..., 7.29.31, ...

....

....

7.2.13, 7.2.19, 7.2.29, ..., 7.2.5, 7.2.11, 7.2.17, ...

7.5.13, 7.5.19, 7.5.29, ..., 7.5.11, 7.5.17, 7.5.23, ...

....

....

13.19.29.37, 13.19.29.43, ..., 13.19.29.41, 13.19.29.47, ...

13.19.37.43, 13.19.37.53, ..., 13.19.37.47, 13.19.37.59, ...

....

....

13.19.23.29, 13.19.23.37, ..., 13.19.23.31, 13.19.23.41, ...

13.19.31.37, 13.19.31.43, ..., 13.19.31.41, 13.19.31.47, ...

....

....

13.29.37.43, ..., 13.29.37.41, ...

13.29.43.53, ..., 13.29.43.47, ...

....

....

13.29.31.41, ..., 13.29.31.37, ...

and so on indefinitely.

4. Given a set of series $A_1, A_2, A_3, \ldots, A_n, \ldots$, whose numbers are positive integers, construct the series of numbers, S, such that, if $a \underset{\rightarrow}{A_n} b$, $2^n(2a-1) \underset{\rightarrow}{S} 2^n(2b-1)$, and, if a belongs to A_m and b to A_n (if $m<n$), $2^n(2a-1) \underset{\rightarrow}{S} 2^m(2b-1)$.

It is clear that no term in S is repeated, for, if $a \neq b$, $2^n(2a-1) \neq 2^n(2b-1)$, and, if $m > n$, there are no pairs of terms a and b such that $2^n(2a-1) = 2^m(2b-1)$, for, if this could happen, an odd number $2a-1$ would equal an even number $2^{m-n}(2b-1)$. Let us call the series S, obtained from $A_1, A_2, \ldots, A_n, \ldots$, $S_n(A_n)$. Representing the number of each A_k by α_k, it is easy to see that the ordinal number of $S_n(A_n)$ is $\alpha_1 + \alpha_2 + \alpha_3 + \ldots + \alpha_n + \ldots$. As $\alpha \geqq \alpha_1$, $\alpha_1 + \alpha_2 \geqq \alpha_2$, $\alpha_1 + \alpha_2 + \alpha_3 \geqq \alpha_3$, ..., $\alpha_1 + \alpha_2 + \ldots + \alpha_n \geqq \alpha_n$, ..., it is obvious that that the ordinal number of $S_n(A_n) \geqq$ the upper limit of the ordinal number of A_n. As an example of $S_n(A_n)$, let A_n be the series

$$2^{n!}.1,\ 2^{n!}.3,\ 2^{n!}.5,\ 2^{n!}.7,\ \ldots,\ 2^{n!-1}.1,\ 2^{n!-1}.3,\ 2^{n!-1}.5,\ \ldots$$

$$\ldots,\ 2^{n!-2}.1,\ 2^{n!-2}.3,\ 2^{n!-2}.5,\ \ldots,\ \ldots,$$

$$\ldots,\ \ldots,\ 2^{(n-1)!+1}.1,\ 2^{(n-1)!+1}.3,\ \ldots,$$

whose ordinal number is obviously

$$\omega\,[n! - (n-1)!] = \omega\,[(n-1)\,(n-1)!],$$

whose upper limit is $\omega . \omega$. Then $S_n(A_n)$ will be the series

$$2\,(2.2-1),\ 2\,(2.2.3-1),\ 2\,(2.2.5-1),\ \ldots,$$

$$2^2(2.2^2-1),\ 2^2(2.2^2.3-1),\ \ldots,$$

$$2^2(2.2-1),\ 2^2(2.2.3-1),\ \ldots,$$

$$2^3(2.2^6-1),\ 2^3(2.2^6.3-1),\ \ldots,$$

$$2^3(2.2^5-1),\ 2^3(2.2^5.3-1),\ \ldots,$$

$$2^3(2.2^4-1),\ 2^3(2.2^4.3-1),\ \ldots,$$

$$2^3(2.2^3-1),\ 2^3(2.2^3.3-1),\ \ldots,\ 2^4(2.2^{24}-1),\ \ldots.$$

Its number will be ω^2, which is $\geqq \omega^2$.

5. The number of I, by the definition of ω, is ω.

Let us write $\Phi(A)$ for $p\,\{P(A)\}$. Then, by (2), (3), the number of $\Phi(I)$ is

$$\omega^{\omega}.$$

Then, by (2), (3), the number of $\Phi^2(I)$ is

$$(\omega^{\omega})^{\omega} = \omega^{\omega^2}.$$

H 2

Then, by (2), (3), the number of $\Phi^3(I)$ is

$$(\omega^{\omega^2})^\omega = \omega^{\omega^3}.$$

Then, by (2), (3), the number of $\Phi^n(I)$ is

$$\omega^{\omega^n}.$$

Then, by (4), the number of $S_n\{\Phi^n(I)\}$ is

$$\omega^\omega + \omega^{\omega^2} + \omega^{\omega^3} + \ldots + \omega^{\omega^n} + \ldots = \omega^{\omega^\omega}.$$

Then, by (2), (3), the number of $\Phi[S_n\{\Phi^n(I)\}]$ is*

$$\left(\omega^{\omega^\omega}\right)^\omega = \omega^{\omega^{\omega+1}}.$$

Then, by (2), (3), the number of $\Phi^2[S_n\{\Phi^n(I)\}]$ is

$$\left(\omega^{\omega^{\omega+1}}\right)^\omega = \omega^{\omega^{\omega+2}}.$$

Then, by (2), (3), the number of $\Phi^m[S_n\{\Phi^n(I)\}]$ is

$$\omega^{\omega^{\omega+m}}.$$

Then, by (4), the number of $S_m(\Phi^m[S_n\{\Phi^n(I)\}])$ is

$$\omega^{\omega^{\omega+1}} + \omega^{\omega^{\omega+2}} + \omega^{\omega^{\omega+3}} + \ldots + \omega^{\omega^{\omega+n}} + \ldots = \omega^{\omega^{\omega.2}}.$$

Let us write $\Psi(A)$ for $S_m\}\Phi^m(A)\}$.

We have shown that the number of $\Psi(I)$ is ω^{ω^ω}, and that that of $\Psi^2(I)$ is $\omega^{\omega^{\omega.2}}$. Similarly, it can be shown that the number of $\Psi^n(I)$ is $\omega^{\omega^{\omega.n}}$. Therefore, by (4), the number of $S_m\{\Psi^m(I)\}$ is equal to $\omega^{\omega^\omega} + \omega^{\omega^{\omega.2}} + \omega^{\omega^{\omega.3}} + \ldots + \omega^{\omega^{\omega.n}} + \ldots$, and is at least $\omega^{\omega^{\omega^2}}$.

∴ the number of		$\Phi[S_m\{\Psi^m(I)\}]$	is at least		$\left(\omega^{\omega^{\omega^2}}\right)^\omega = \omega^{\omega^{\omega^2+1}}$.
∴	,, ,,	$\Phi^n[S_m\{\Psi^m(I)\}]$	,,	,,	$\omega^{\omega^{\omega^2+n}}$.
∴	,, ,,	$\Psi[S_m\{\Psi^m(I)\}]$	,,	,,	$\omega^{\omega^{\omega^2+\omega}}$.
Similarly,	,,	$\Psi^2[S_m\{\Psi^m(I)\}]$	,,	,,	$\omega^{\omega^{\omega^2+\omega.2}}$.
,,	,,	$\Psi^n[S_m\{\Psi^m(I)\}]$	,,	,,	$\omega^{\omega^{\omega^2+\omega.n}}$.
,,	,,	$S_n(\Psi^n[S_m\{\Psi^m(I)\}])$	,,	,,	$\omega^{\omega^{\omega^2.2}}$.

* We can always apply Φ to any S_n [for S_n, and in consequence $P(S)$, has no last term] and be sure of raising the number of the S to the ωth power.

Let us write $F(A)$ for $S_n\{\Psi^n(A)\}$. We have shown that the number of $F(I)$ is at least $\omega^{\omega^{\omega^2}}$, and that that of $F^2(I)$ is at least $\omega^{\omega^{\omega^2.2}}$. Similarly, we may prove that the number of $F^n(I)$ is at least $\omega^{\omega^{\omega^2.n}}$.

∴ the number of $S_n\{F^n(I)\}$ is at least $\omega^{\omega^{\omega^2.\omega}} = \omega^{\omega^{\omega^3}}$.

" " $\Phi[S_n\{F^n(I)\}]$ " " $\omega^{\omega^{\omega^3+1}}$.

" " $\Phi^m[S_n\{F^n(I)\}]$ " " $\omega^{\omega^{\omega^3+m}}$.

" " $\Psi[S_n\{F^n(I)\}]$ " " $\omega^{\omega^{\omega^3+\omega}}$.

" " $\Psi^m[S_n\{F^n(I)\}]$ " " $\omega^{\omega^{\omega^3+\omega.m}}$.

" " $F[S_n\{F^n(I)\}]$ " " $\omega^{\omega^{\omega^3+\omega^2}}$.

" " $F^m[S_n\{F^n(I)\}]$ " " $\omega^{\omega^{\omega^3+\omega^2.m}}$.

" " $S_m(F^m[S_n\{F^n(I)\}])$ " " $\omega^{\omega^{\omega^3.2}}$.

Let us write $G(A)$ for $S_n\{F^n(A)\}$. We have shown that the number of $G(I)$ is at least $\omega^{\omega^{\omega^3}}$ and that that of $G^2(I)$ is at least $\omega^{\omega^{\omega^3.2}}$. It can be shown in the same manner that the number of $G^n(I)$ is at least $\omega^{\omega^{\omega^3.n}}$. Let us write $H(A)$ for $S_n\{G^n(I)\}$. Then it is obvious that the ordinal number of $H(I)$ is at least $\omega^{\omega^{\omega^4}}$.

In a precisely analogous manner we can construct a series of number at least $\omega^{\omega^{\omega^n}}$, whatever n may be. Let us call this series, in general, $K_n(I)$, where $K_1(I)=\Psi(I)$, $K_2(I)=F(I)$, $K_3(I)=G(I)$. $K_n(I)$ is always constructed according to a perfectly definite method, leaving no possible doubt what step to take after any given step, for after any series you have obtained you form the Φ of that series, and after any series of series, you form its S. Therefore no implicit postulation of Zermelo's axiom is to be found in any of my constructions, so that I can be sure that they always exist. Therefore I can form $S_n\{K_n(I)\}$, and its number will be $\omega^{\omega^{\omega^\omega}}$ at least.

In a precisely parallel manner we can construct a rearrangement not less than $\omega^{\omega^{\omega^{\omega^\omega}}}$, etc. Given rearrangements of I, which will be at least as large as ω, ω^ω, ω^{ω^ω}, $\omega^{\omega^{\omega^\omega}}$, etc., we can, by means of S, construct a rearrangement of I at least as large as an

$$\omega \text{ times } \left\{\omega^{\omega^{\omega^{\omega^{\omega^{\cdots}}}}}\right.,$$

or an ϵ.

In general, given a rearrangement L of I, such that the ordinal number of $L(I)$ is at least as large as ω^{ω^α}, $\Phi\{L(I)\}$ will have an ordinal number at least as large as $\omega^{\omega^{\alpha+1}}$. Also, given a sequence of rearrangements $L_1, L_2, L_3, \ldots, L_n, \ldots$, of I of ordinal numbers at least as large as $\omega^{\omega^{\alpha_1}}, \omega^{\omega^{\alpha_2}}, \ldots, \omega^{\omega^{\alpha_n}}, \ldots$, respectively, if α_ω be the limit of the sequence $\alpha_1, \alpha_2, \alpha_3, \ldots, \alpha_n, \ldots$, then the ordinal number of $S_n(\omega^{\omega^{\alpha_n}}) \geqq \alpha_\omega$. Therefore we can construct a number at least as large as ω^{ω^α}, where α denotes any ordinal number which can be formed from 1 by the repetition of the operations (1) of adding 1 to a previously given ordinal number, and (2) of taking the number of any given infinite sequence of numbers previously obtained. But the class of such numbers is the class of numbers of fundamental sequences of Ω. Therefore, if α is the number of a fundamental sequence of Ω, we can get by our method a not smaller rearrangement of the number-series than ω^{ω^α}. If, then, $\omega^{\omega^\alpha} \geqq \alpha$, the proposition I set out to prove is obviously proved. This is clearly true if $\omega^\alpha \geqq \alpha$. This can be proved in the following manner:

(1) $\omega^1 \geqq 1$.

(2) Let $\omega^\alpha \geqq \alpha$. Then

$$\omega^{\alpha+1} = \omega^\alpha . \omega = \omega^\alpha + \omega^\alpha . \omega \geqq \alpha + \omega^\alpha . \omega \geqq \alpha + 1.$$

(3) Let $\omega^{\alpha_n} \geqq \alpha_n$, when α_n takes any one of the infinite series of values $\alpha_1, \alpha_2, \alpha_3, \ldots, \alpha_m, \ldots$, whose upper limit is α_ω. Since $\alpha_\omega > \alpha_n$, it can be shown that

$$\omega^{\alpha_\omega} = \omega^{\alpha_n+\beta} = \omega^{\alpha_n}\omega^\beta = \omega^{\alpha_n} + \omega^{\alpha_n}\omega^\beta \geqq \omega^{\alpha_n}.$$

Therefore, $\omega^{\alpha_\omega} \geqq \alpha_n$, whatever n is. Therefore, $\omega^{\alpha_\omega} \geqq \alpha_\omega$.

Therefore, when α is the number of a fundamental sequence of Ω, we have a method of rearranging the positive integers in a series of number $\geqq \alpha + 1$, and hence $> \alpha$.

It will be noted that the method I have developed enables me directly to reorder, not the whole, but a part of the series of the integers in a series of number greater than the number ω^ω, but this is of no importance, for let the part of I so arranged be A. Let a_1 be the numerically smallest member of A, a_2 the next, and so on. Then replace each a_n by n. This will give a rearrangement of I similar to the already obtained rearrangement of A.

The interest of the construction of rearrangements of I lies in the fact that all the proofs hitherto given of the existence of numbers greater than those of any given fundamental sequence of Ω have involved the multiplicative axiom.* By actually rearranging I in a series of such a number, we avoid this.

It should be noted that the particular nature of the process Φ we have chosen of increasing the number of a series by rearranging it is a matter of more or less indifference; any other process which, when applied to a series, always gives a larger or equal series would have done quite as well, logically. For example, if $\Phi'(B) = B$, the number of $S_n\{\Phi'^n(B)\} = \omega$, multiplied by the number of B; and, as it can be shown that $\omega . \alpha > \alpha$, it is clear that, by the same sort of proof which we used to show that Φ and S together enable us to construct a rearrangement of I larger in number than any given fundamental sequence of Ω, S alone will enable us to do it. However, at least at first, the use of Φ enables us to increase the ordinal number of the rearrangement of I more rapidly than that of S alone would.

Comments on [13a]

P. Masani

This is the first printed paper of Wiener,written at the age of 18 while attending G. H. Hardy's course. The result proved does not seem to have appeared in the literature on the subject, at least in its general form; for example, it is not mentioned in W. Sierpinski's *Cardinal and Ordinal Numbers,* Warsaw, 1958. However, according to Professor W. Gustin it is known to workers in the field.

Wiener's own comment on it reads: "Looking back on this paper, I do not think it was particularly good. . . . Still, it gave me my first taste of printer's ink, and this is a powerful stimulant for a rising young scholar" [53h] , p. 190.

BILINEAR OPERATIONS GENERATING ALL OPERATIONS RATIONAL IN A DOMAIN Ω.

By Norbert Wiener.

The notion of a corpus, or domain of rationality, is familiar to mathematicians. It can easily be extended to cover sets, not of numbers, but of functions of one or more variables (which may, indeed, degenerate into numbers), exhibiting the group property with reference to the four fundamental operations—addition, subtraction, multiplication, and division—barring only division by 0. Now, it should be noted that these four operations are not only methods of combining functions, but are themselves functions of two variables. Thus $x + y$ is a function of x and y. To combine two functions $f(x)$ and $g(x)$ by addition is to eliminate u and v from the three equations

$$z = u + v,$$
$$u = f(x),$$
$$v = g(x),$$

and thus to define z in terms of x.

It is thus possible to regard a function-corpus containing addition, subtraction, multiplication, and division, as a set Σ of functions or operations which contains a sub-set Σ' such that every combination of operations of Σ by operations of Σ' belongs to Σ. In certain cases, Σ' and Σ will coincide. Thus let Σ be the set of all rational functions with coefficients rational in a domain Ω: if $F(x_1, x_2, \cdots, x_n)$, $f_1(y_{11}, y_{12}, \cdots, y_{1k})$, $f_2(y_{21}, y_{22}, \cdots, y_{2l})$, $\cdots$, $f_n(y_{n1}, y_{n2}, \cdots, y_{nm})$ belong to Σ, $F(f_1, f_2, \cdots, f_n)$ will also belong to Σ, and this whether the y's are all distinct or not. A set of operations, whether a functional corpus or not, which possesses this invariancy under iteration, may be called an *iterative field*.

Iterative fields constitute one of the generalizations of substitution-groups—in fact, substitution-groups are iterative fields of functions of one variable. Now, in a group it is often possible to pick out a very restricted set of elements—a so-called basis—which by their combination generate all the elements of the group. Iterative fields likewise have bases—addition, multiplication, subtraction, and division constitute a basis for the iterative field consisting of all rational operations with rational coefficients. It thus becomes a matter of some interest to determine the smallest possible basis of an iterative field—in particular to determine if and how the field can be generated by the iteration of a single operation. If this is possible, it corresponds to a cyclical group.

In this paper the question at issue is how an iterative field consisting of all rational operations with their coefficients rational in a domain Ω may be generated by the iteration of a single operation of the form

$$x\,|\,y = \frac{A + Bx + Cy + Dxy}{E + Fx + Gy + Hxy}.$$

The chief result obtained is that if Ω is the domain of rationals, the necessary and sufficient condition that $x\,|\,y$ constitute a basis is that there is a linear transformation T such that $T^{-1}\{T(x)\,|\,T(y)\}$ or $T^{-1}\{T(y)\,|\,T(x)\}$ is of the form

$$\frac{x - y}{x + Ay},$$

where A is any rational number, or else of the form

$$\frac{n(x - y + xy)}{ny + xy},$$

when n is any integer other than 0. Thus $(x - y)/(3x + 6y)$ and $(xy + 3x - 3y - 2)/(xy + x + 3y + 3)$ both serve as bases for all rational operations with rational coefficients, since the first is of the form

$$\frac{1}{3}\left(\frac{3x - 3y}{3x + 2(3y)}\right),$$

and is a transform of

$$\frac{x - y}{x + 2y};$$

while the second may be written

$$\frac{2((x + 1) - (y + 1) + (x + 1)(y + 1))}{2(y + 1) + (x + 1)(y + 1)} - 1,$$

and is a transform of

$$\frac{2(x - y + xy)}{2y + xy}.$$

1. **Definitions.** A *bilinear* operation is a binary operation of the form

$$\frac{A + Bx + Cy + Dxy}{E + Fx + Gy + Hxy}.$$

An operation *rational in a domain* Ω is a rational operation whose coefficients belong to Ω.

An operation $f(x_1, x_2, \cdots, x_k)$ is said to *generate* a class K of operations and numbers which may be considered as operations on no variables, when it has the following four properties.

(1) K contains f.

(2) If K contains $g(x_1, x_2, \cdots, x_{i-1}, x_i, x_{i+1}, \cdots, x_{j-1}, x_j, x_{j+1}, \cdots, x_k)$, K contains $g(x_1, x_2, \cdots, x_{i-1}, x_i, x_{i+1}, \cdots, x_{j-1}, x_i, x_{j+1}, \cdots, x_k)$, and also $g(x_1, x_2, \cdots, x_{i-1}, x_j, x_{i+1}, \cdots, x_{j-1}, x_i, x_{j+1}, \cdots, x_k)$.

(3) If K contains $g(x_1, x_2, \cdots, x_k)$ and $h(y_1, y_2, \cdots, y_l)$, K contains the operation on $x_1, x_2, \cdots, x_{i-1}, x_{i+1}, \cdots, x_k, y_1, y_2, \cdots, y_l$ obtained by substituting h for x_i.

(4) K contains no proper sub-set K' with properties (1), (2), and (3).

Infinity will be considered a proper argument for an operation, and $f(\infty)$ will be defined as $\lim_{x \doteq \infty} f(x)$, if this latter expression has a unique significance. Infinity will be considered as a member of every domain.

If T is the non-singular linear transformation

$$x' = \frac{a + bx}{c + dx},$$

the *transform* of $f(x_1, x_2, \cdots, x_k)$ by T is defined as the operation which expresses y in terms of $x_1, x_2, \cdots, x_k$, and is expressed in implicit form in the following set of simultaneous equations.

$$y' = f(x_1', x_2', \cdots, x_k'),$$

$$y' = \frac{a + by}{c + dy},$$

$$x_i' = \frac{a + bx_i}{c + dx_i} \qquad (1 \leq i \leq k).$$

2. **Theorem.*** *Every transform of* $(x - y)/(x + Gy)$ *by a* T *rational in* $\Omega_{(G)}$ *generates all operations rational in* $\Omega_{(G)}$ *if only* $G \neq -1$.

Proof. For the sake of brevity, we shall represent the operation $(x - y)/(x + Gy)$ by the symbol $x|y$. If we can show that whenever $G \neq -1$, we can build up a chain of definitions wherein addition, subtraction, multiplication, division, and G are derived from iterations of $x|y$, it is clear that every operation rational in $\Omega_{(G)}$ can be derived from $x|y$ by iteration. The formation of *ad hoc* definitions to cover all cases where the iterations given become indeterminate offers no difficulty.

There are two cases. First, let $G \neq \pm 1$. Our chain of definitions will read as follows.

* In this and the following theorems, the operations generated by the bilinear operations in question may be undefined for a set of isolated values of their arguments.

$$
\begin{aligned}
0 &= x \mid x, \\
1 &= x \mid 0, \\
x/y &= 1 \mid (y \mid x), \\
xy &= x/(1/y), \\
A(x) &= 1/[(x \mid 1) \mid 1] \text{ [in ordinary symbolism, } A(x) = (G + 1 - x], \\
x - y &= \{xA[(y/x)A(0)]\}/A(0), \\
x + y &= x - [(a - y) - a], \\
G &= [a - 1/(0 \mid x)] - a.
\end{aligned}
$$

If $G = 1$, this chain is replaced by the following sequence of definitions:

$$
\begin{aligned}
0 &= x \mid x, \\
1 &= x \mid 0, \\
-1 &= 0 \mid x, \\
x/y &= 1 \mid (y \mid x), \\
xy &= x/(1/y), \\
A(x) &= x^2\{[x \mid (1 \mid x)] \mid [(x \mid 1) \cdot (x^2 \mid 1)]\} \text{ [i.e., } 1/(2x - 1)], \\
B(x) &= x^2\{\{(-1)A\{(1 \mid x^2)[(x \mid 1) \mid (-1 \mid x)]\}\} \mid A\{(x \mid 1) \mid (-1 \mid x)\}\} \\
&\qquad \left[\text{i.e., } 1 - \frac{x}{y}\right], \\
x - y &= xB(y/x), \\
x + y &= x - (-1)y.
\end{aligned}
$$

Since $x \mid y$ is rational in $\Omega_{(G)}$, it only generates operations rational in $\Omega_{(G)}$, while we have just seen that it generates all operations rational in $\Omega_{(G)}$. A transformation T rational in $\Omega_{(G)}$ is simply a permutation of the numbers and operations of $\Omega_{(G)}$ among themselves. Hence the T-transform of $x \mid y$ generates all operations rational in $\Omega_{(G)}$, and no others.

3. **Theorem.** *Every transform of*

$$\frac{n(x - y + xy)}{ny + xy}$$

by a T rational in $\Omega_{(1)}$ generates all operations rational in $\Omega_{(1)}$, if n is a non-0 integer in $\Omega_{(1)}$.

Proof. The discussion of the last theorem applies *in toto*, except that the chair of definitions must be altered . We shall denote

$$\frac{n(x - y + xy)}{ny + xy}$$

by $x \mid y$. Three cases arise.

(1) $n = 1$. Consider the following definitions:

$$
\begin{aligned}
0 &= (x \mid x) \mid x, \\
-1 &= 0 \mid x, \\
\infty &= x \mid 0, \\
-x &= \infty \mid (x \mid -1),
\end{aligned}
$$

$$A(x, y) = -\{-(-x|-x)|-[(-y|-y)|(-y|-y)]\}\left[\text{i.e., } 1 - \frac{x}{y}\right],$$
$$1 = \infty | \infty,$$
$$x/y = A\{A(x, y), 1\},$$
$$xy = x/(1/y),$$
$$x - y = xA(y, x),$$
$$x + y = x - [(a - y) - a].$$

(2) $n > 1$. Consider the following definitions:

$$a^1(x) = x|x,$$
$$a^{k+1}(x) = a^1[a^k(x)] \ [\text{i.e., } a^k(x) = nx/(kx + n)],$$
$$0 = a^n(x)|x,$$
$$-1 = 0|x,$$
$$\infty = x|0,$$
$$P(x, y) = a^{n-1}(x)|a^{n-1}(y) \ [\text{i.e., } (x - y + xy)/(y + xy)],$$
$$-x = \infty|(x|-1),$$
$$A(x, y) = -P\{-P(-x, -x), -P[P(-y, -y), P(-y, -y)]\}$$

[as before],

$$1 = P(\infty, \infty),$$
$$x/y = A\{A(x, y), 1\},$$
$$xy = x/(1|y),$$
$$x - y = xA(y, x),$$
$$x + y = x - [(a - y) - a].$$

(3) n is negative. $x|y$ is a transform by $y' = y/(y + 1)$ of $-n(y - x + xy)/(-nx + xy)$, which differs from the cases considered in (1) and (2) merely by the interchange of x and y.

4. **Theorem.** *If $x|y$ is a bilinear operation generating every operation rational in any domain, it is a transform by a linear transformation rational in some $\Omega_{(G)}$ either of*

$$\frac{x - y}{x + \alpha y} \text{ or of } \frac{y - x}{y + \alpha x},$$

where α is a number in $\Omega_{(G)}$ other than ∞ or -1, or of

$$\frac{n(x - y + xy)}{ny + xy},$$

where n is a number in $\Omega_{(G)}$ other than ∞ or 0.

Proof. Let

$$x|y = \frac{A + Bx + Cy + Dxy}{E + Fx + Gy + Hxy}.$$

Consider the roots of the cubic equation $x|x = x$. If for any of these roots, r_1, r_2, or r_3, $r_i|r_i$ is determinate, then no sequence of iterations of $|$ on r_i can generate anything but r_i. As the condition we have supposed is equivalent to the assumption that $x|y$ be continuous when x and y are in the neighborhood of r_i, it follows that when x and y are in the neighborhood of r_i, $x|y$ is also in that neighborhood. It results that if $f(x_1, x_2, \cdots, x_n)$ is an operation derived by the iteration of $x|y$, and if $x_1, x_2, \cdots, x_n$ are all in the neighborhood of r_i, $f(x_1, x_2, \cdots, x_n)$ will also be in that neighborhood. This restricts f to a relatively narrow range of variation. In particular, it cannot be any constant other than r_i itself. Since the four fundamental operations $+$, $-$, $\times$, and $\div$ generate all rational constants, it follows that $x|y$ cannot generate these, and hence cannot generate all rational operations with rational coefficients.

Consequently every root of $x|x = x$, be it finite or infinite, is also a finite or infinite root of both the quadratics,

$$Hx^2 + (F + G)x + E = 0, \tag{1}$$

and

$$Dx^2 + (B + C)x + A = 0. \tag{2}$$

If these two equations are not equivalent, the algorithm of the greatest common factor leads to the common root

$$r = -\frac{\begin{vmatrix} E & H \\ A & D \end{vmatrix}}{\begin{vmatrix} F + G & H \\ B + C & D \end{vmatrix}}.$$

This is rational in the domain of the coefficients, and is furthermore, from what has been said, the only finite or infinite root of $x|x = x$. This we shall call case A.

If (1) and (2) are equivalent, it may be seen on inspection that each of their roots is also a root of $x|x = x$, which is *in extenso*

$$Hx^3 + (F + G - D)x^2 + (E - B - C)x - A = 0. \tag{3}$$

This will then have the root $r_1 = D/H$, which is rational in the domain of the coefficients. It follows from what has been said that r_1 is a root of (1). The other root of (1) is $r_2 = (E/H)/(D/H)$, or E/D, which is also rational in the domain of the coefficients. The roots of (3) are then the double root r_1 and the single root r_2. This we shall call case B.

We have seen that a linear transform of an operation has essentially the same iterative properties as its original. Accordingly we may subject the $x|y$ of case A to the transformation $u' = u + r$, thus obtaining the operation

$$x \| y = \frac{A' + B'x + C'y + D'xy}{E' + F'x + G'y + H'xy}.$$

We shall now have 0 as a root of the quadratic

$$(1') \qquad H'x^2 + (F' + G')x + E',$$

and a triple root of

$$(3') \qquad H'x^3 + (F' + G' - D')x^2 + CE' - B' - c')x - A' = 0.$$

It follows that $A' = E' = B' + C' = F' + G' - D' = 0$, whence

$$x \,\|\, y = \frac{B'x - B'y + (F' + G')xy}{F'x + G'y + H'xy}$$

The transformation $u' = B'u/[F'u + (F' + G')]$ reduces this to the form*

$$x \,\|\|\, y = \frac{n(x - y + xy)}{ny + xy}.$$

Exceptional cases might seem to arise where $B' = 0$ or $F' + G' = 0$. If $B' = 0$, on transforming $x \,\|\, y$ by $u' = 1/u$, we get

$$x \,\|\|\, y = G'x + F'y + H',$$

and it is manifest that the iteration of this can only lead to linear operations. If $F' + G' = 0$, $x \,\|\, y$ reduces to the form

$$\frac{x - y}{P(x - y) + Qxy}.$$

If we transform this by $u' = u/(Pu + Q)$ we get

$$x \,\|\|\, y = \frac{x - y}{xy}.$$

Be it noted that the degree of every term in the numerator is odd, while the power of every term in the denominator is even. It is easy to show that if f and g be two rational operations each in the form either of the quotient of a polynomial whose terms are all of odd degree by a polynomial whose terms are all of even degree, or of the quotient of a polynomial whose terms are all of even degree by a polynomial whose terms are all of odd degree, the operation obtained by substituting f as one of the arguments of g will share in this property. Thus if $f(x, y) = (x - y)/xy$ and $g(x, y, z) = (x^2 + y^2 + z^2)/(x + yz^2)$, $g(x, y, f(u, v))$ will be $(x^4y^2 + x^2y^4 + x^2 - 2xy + y^2)/(x^3y^2 + x^2y - 2xy^2 + y^3)$, where all the terms of the numerator are of even degree and all the terms in the denominator are of odd degree. Hence the iteration of an operation in this class can never lead beyond the confines of the class, and must consequently fail to generate all operations rational in any domain.

* $n \neq 0$, as the expression would otherwise reduce to a constant. $n \neq \infty$; otherwise, whatever x, $x \,\|\|\, x = x$.

In case B, submit $x|y$ to the transformation $u' = (r_2u - r_1)/(u - 1)$. $x|y$ will go into an operation

$$x \| y = \frac{A' + B'x + C'y + D'xy}{E' + F'x + G'y + H'xy}.$$

The double root of $x \| x = x$, or of

$$H'x^3 + (F' + G' - D')x^2 + (E' - B' - C')x - A' = 0,$$

will be 0, while the single root will be ∞. Hence

$$A' = H' = E' - B' - C' = 0.$$

Since 0 and ∞ are also roots of

$$H'x^2 + (F' + G')x + E' = 0,$$

and of

$$D'x^2 + (B' + C')x + A' = 0,$$

D' and E' also vanish. From these facts concerning the coefficients, it may be seen that $x \| y$ reduces to

$$\frac{x - y}{Mx + Ny}.$$

Transforming this operation by $u' = u/M$, the result is of the form

$$\frac{x - y}{x + \alpha y}.$$

If $M = 0$, the above transformation is impossible, but if we transform by $u' = -u/N$, we get $(y - x)/y$, of the form $(y - x)/(y + \alpha x)$.

We have thus actually carried through the transformations which reduce $x|y$ to one of the standard forms, and have found them to be rational in the domain of the coefficients of $x|y$.

5. **Theorem.** *If a transform of $n(x - y + xy)/(ny + xy)$ generates all operations rational in any domain, then n is of the form p/q, where p is an integer in $\Omega_{(n)}$ and q is a natural number sharing with p no numerical factor, while there is some natural number k such that q is a factor of p^k.* (An example of the situation contemplated in this theorem is $p = 1 + \sqrt{-5}$; $q = 2$; $k = 2$; $p^k = -4 + 2\sqrt{-5} = q(-2 + \sqrt{-5})$.)

Proof. Consider the class of all numbers in $\Omega_{(n)}$ of the form

$$\frac{p^k + aq}{p^k + bq},$$

where q and k are natural numbers, a and b are integers in $\Omega_{(n)}$, and p is an integer in $\Omega_{(n)}$ sharing no numerical factor with q. Suppose that every number in $\Omega_{(n)}$ can be represented in this form. Let r/s be any number in

$\Omega_{(n)}$ and let r and s be integers in $\Omega_{(n)}$. Then

$$(r - s)p^k = qt,$$

where t is an integer in $\Omega_{(n)}$, namely $sa - rb$. By properly choosing r and s, $r - s$ may be made to contain no factor, actual or ideal, in common with q. Hence, if all numbers in $\Omega_{(n)}$ are reducible to the above form, for some value of k, q is a factor of p^k.

Now, if $n = p/q$, and

$$x \mid y = \frac{n(x - y + xy)}{ny + xy}$$

generates all operations rational in $\Omega_{(n)}$, $(p^k + aq)/(p^k + bq)$ must assume all values in $\Omega_{(n)}$. For if we let x and y be $(p^k + aq)/(p^k + bq)$ and $(p^l + \alpha q)/(p^l + \beta q)$, it follows that

$$\begin{aligned} x \mid y &= \frac{(p/q)(x - y + xy)}{(p/q)y + xy} \\ &= \frac{\dfrac{p}{q}\left\{\dfrac{p^k + aq}{p^k + bq} - \dfrac{p^l + \alpha q}{p^l + \beta q} + \dfrac{p^k + aq}{p^k + bq} \cdot \dfrac{p^l + \alpha q}{p^l + \beta q}\right\}}{\dfrac{p}{q} \cdot \dfrac{p^l + \alpha q}{p^l + \beta q} + \dfrac{p^k + aq}{p^k + bq} \cdot \dfrac{p^l + \alpha q}{p^l + \beta q}} \\ &= \frac{p^{k+l+1} + Aq}{p^{k+l+1} + Bq}, \end{aligned}$$

where A and B are integers in $\Omega_{(n)}$. Hence if K is the class of all members of $\Omega_{(n)}$ of the form $(p^k + aq)/(p^k + bq)$, when x and y belong to this class, $x \mid y$ will also belong to this class. If $x \mid y$ is to generate all operations rational in $\Omega_{(n)}$, and *a fortiori* all numbers in $\Omega_{(n)}$, K must coincide with $\Omega_{(n)}$. In the previous paragraph we have seen this condition to be equivalent to that expressed in the formulation of the theorem.

6. **Theorem.** *The necessary and sufficient condition that the set of all operations generated by the rational bilinear operation $x \mid y$ be the set of all rational operations with rational coefficients is that $x \mid y$ be a rational transform of*

$$\frac{x - y}{x + Ay} \quad \textit{or} \quad \frac{y - x}{y + Ax},$$

where A is any rational number other than -1, or of

$$\frac{n(x - y + xy)}{ny + xy},$$

where n is any integer other than 0.

Proof. This results immediately from 2–5, if the domain in question be made that of the rationals.

MASSACHUSETTS INSTITUTE OF TECHNOLOGY,
CAMBRIDGE, MASS.

A SET OF POSTULATES FOR FIELDS*

BY

NORBERT WIENER

Introduction

Huntington† has developed a set of thirteen independent postulates for fields. These postulates make use of the undefined notions of multiplication and addition. They thus represent a state of the theory of postulates for fields analogous to that prevailing in boolean algebras before the work of Sheffer.‡ Sheffer showed that the three operations of disjunction, conjunction, and negation could be derived by iterating the operation $\bar{a} \odot \bar{b}$, and that the logical constants z and u could be obtained in like manner. It is the purpose of this paper to perform an analogous task for ordinary algebra. Our undefined operation, which we shall symbolize by $x @ y$, will turn out to have the formal properties characteristic of $1 - x/y$.

Postulate set for fields

We assume:

I. A class K,

II. A binary K-rule of combination @.

III. The following properties of K and @:

1. Whatever x may be, there is a K-element y such that $x @ (y @ y)$ is not a K-element.

2. If x and y are K-elements, but $x @ y$ is not a K-element, there is a K-element z such that $y = z @ z$.

3. Whenever x and y are distinct K-elements, either $x @ y$ or $y @ x$ is a K-element.

4. Whenever x, y, u, v, and their indicated combinations are K-elements, and $x @ y = u @ v$, then $x @ u = y @ v$.

Definition. $Z = x @ x$, if x and $x @ x$ belong to K, and $x @ x$ is unique in the system.

---

* Presented to the Society, December 30, 1919.

† E. V. Huntington, *Note on the definitions of abstract groups and fields by sets of independent postulates*, these Transactions, vol. 6 (1905), pp. 181–193.

‡ H. M. Sheffer, *A set of five independent postulates for boolean algebras, with application to logical constants*, these Transactions, vol. 14 (1913), pp. 481–488. Sheffer's terminology and method of exposition are followed closely here.

5. If x, y, $x @ y$, and Z are K-elements, and $x @ y = Z$, then $x = y$.
6. If x, y, z, Z, and their indicated combinations belong to K,

$$[(x @ y) @ (Z @ y)] @ z = [(x @ z) @ (Z @ z)] @ y.$$

Definition. $U = Z @ x$, if x, Z, and $Z @ x$ belong to K, and $Z @ x$ is unique in the system.

Definition. $x \odot y = \{U @ \{[(\overline{U @ y} @ U) @ x] @ U\}\} @ U$, if all the indicated expressions belong to K. Otherwise, if Z belongs to K, $x \odot y = Z$.

Definition. $x \sim y = x \odot (y @ x)$.

7. If x, y, u, v, and all their indicated combinations belong to K, we have $(x \sim y) \sim (u \sim v) = (x \sim u) \sim (y \sim v)$.

Definition. $x \oplus y$ is defined as u, the unique K-element for which $x = u \sim y$, if such a unique K-element exists, and as Z, if Z is a K-element, and if there is no K-element u for which $x = u \sim y$.

Classification of postulates 1–7

Postulate 1 is a simple existence-postulate. Postulates 2 and 3, while in form hypothetical existence-postulates, function as K-closing postulates in a limited sense, for they narrow the class of cases where x and y are K-elements, while $x @ y$ fails to be a K-element. Postulates 6 and 7 are simple equivalence-postulates, for they assert that two expressions shall always have the same meaning when they are significant. Postulates 4 and 5 are hypothetical equivalence-postulates, with equivalences for hypotheses. Thus our set consists of one simple existence-postulate, two hypothetical existence-K-closing postulates, two simple equivalence-postulates, and two hypothetical equivalence-postulates.

Consistency of postulates 1–7

With the following interpretation of K and @ postulates 1–7 are satisfied: K consists of all rational numbers, and $x @ y = 1 - x/y$.

Independence of postulates 1–7

With each of the interpretations of K and @ given in (1)–(7) below, all postulates except the one correspondingly numbered are satisfied; that postulate is therefore independent of the remaining six.

(1) K contains only one element m; $m @ m = m$.

(2) K contains only two elements, m and n; $m @ m \neq m$; $n @ m = n$; $m @ n$ and $n @ n$ do not belong to K.

(3) K contains only two elements, m and n; $m @ m = m$; $n @ n = n$; $m @ n$ and $n @ m$ do not belong to K.

(4). K contains only three elements, $l_4 = l_1$, l_2, and l_3; $l_k @ l_{k+1}$ does not belong to K; otherwise $l_j @ l_k = l_{k+1}$.

(5) K contains only two elements, m and n; $m @ m = m @ n = n @ m = n$; $n @ n$ does not belong to K.

(6) K contains only three elements, l, m, and n; @ is defined by the following table (for example: $l @ m = n$; $l @ l$ does not belong to K).

@	l	m	n
l		n	m
m		l	n
n		m	l

(7) K consists of all complex numbers;

$$(x + iy) @ (u + iv) = 1 - \frac{x - iy}{u - iv}.$$

Deductions from the postulates

Theorem I. *K contains at least two elements.*

Theorem II. *K contains an element answering to the definition of Z.*

Theorem III. *If x and y belong to K, and $y \neq Z$, $x @ y$ belongs to K.*

Theorem IV. *K contains no elements of the form $x @ Z$.*

Theorem V. *K contains an element distinct from Z, answering to the definition of U.*

Theorem VI. *If x, y, and z belong to K and $y \neq Z$, $z \neq Z$, then*

$$[(x @ y) @ U] @ z = [(x @ z) @ U] @ y.$$

Theorem VII. *If x is a K-element, $(x @ U) @ U = x$.*

Theorem VIII. *If $y \neq Z$ and $x = y \odot z$, then $z = (x @ y) @ U$, and conversely.*

Theorem IX. *If x and y belong to K, $x \odot y$ belongs to K.*

Theorem X. $x \odot y = y \odot x$.

Theorem XI. *If $x \odot y = Z$, $x = Z$ or $y = Z$, and vice versa.*

Theorem XII. *If x and y belong to K, $x \oplus y$ belongs to K.*

Theorem XIII. *If $x \oplus x = x$, $x = Z$ and vice versa.*

Theorem XIV. *If $u \odot x = u \odot y$ and $u \neq Z$, then $x = y$.*

Theorem XV. $(x \odot y) \odot z = x \odot (y \odot z)$.

Theorem XVI. $x \odot (y \oplus z) = (x \odot y) \oplus (x \odot z)$.

Theorem XVII. $x \oplus y = y \oplus x$.

Theorem XVIII. $(x \oplus y) \oplus z = x \oplus (y \oplus z)$.

Theorem XIX. *If x is a K-element, there is a K-element y such that $y \oplus x = Z$.*

Theorem XX. *If x and y are K-elements, there is a K-element u such that $u \oplus x = y$.*

Proofs of the preceding theorems

Proof of I. If K contains only one element x, it follows by 2 that $x @ x = x$. Hence $x @ (x @ x) = x$ and is a K-element. As this contradicts 1, K contains at least two elements.

Proof of II. (1) K contains an element of the form $x @ x$ [by 1 and 2].

If x and y are K-elements, $x @ y$ or $y @ x$ is a K-element [by 3].

If $x @ y$ is a K-element, $x @ y = x @ y$. Then, if $x @ x$ and $y @ y$ are K-elements,

$$x @ x = y @ y. \tag{2}$$

Similarly, if $y @ x$, $x @ x$, and $y @ y$ are K-elements,

$$x @ x = y @ y. \tag{3}$$

Hence K contains a unique term Z [by (1), (2), (3) and the definition of Z].

Proof of III. This theorem follows directly from II and 2.

Proof of IV. This theorem follows directly from II and 1.

Proof of V. $[(Z @ y) @ (Z @ y)] @ z$ belongs to K unless $y = Z$ or $z = Z$ [by III]. Therefore, if $y \neq Z$, $z \neq Z$,

$$[(Z @ y) @ (Z @ y)] @ z = [(Z @ z) @ (Z @ z)] @ y \text{ [by 6 and II].}$$

(1) $\therefore\ Z @ z = Z @ y$ [by II].

(2) $Z @ Z$ is not a member of K [by IV].

(3) U is a member of K [by (1), (2) and the definition of U].

If $Z = U$, then if x is a K-element other than Z, $Z @ x = Z$ [by III and the definition of U]. Then $Z = x$, which is impossible [by 5]. But K contains an element other than Z [by I].

(4) $\therefore\ Z \neq U$.

Statements (3) and (4) constitute the theorem.

Theorems I–V inclusive will be used in the following proofs without explicit citation.

Proof of VI. This theorem follows directly from V and 6.

Proof of VII. $[(x @ U) @ U] @ x = [(x @ x) @ U] @ U$ [by VI], $= (Z @ U) @ U = U @ U = Z$.

$$\therefore\ (x @ U) @ U = x.$$

Proof of VIII. Let $x = y \odot z$ and $x \neq Z$, $y \neq Z$, $z \neq Z$. Then

$$Z @ U \neq z @ z \text{ [by V, 5].}$$

$$\therefore\ U @ z \neq Z @ z \text{ [by 4].}$$

$$\therefore\ (U @ z) @ U \neq Z \text{ [by 5].} \tag{1}$$

$$\therefore\ [(\overline{U @ z} @ U) @ y] @ U \neq Z.$$

(2) $\therefore\ \{U @ \{[(\overline{U @ z} @ U) @ y] @ U\}\} @ U$ belongs to K.

$$x = \{U @ \{[(\overline{U @ z} @ U) @ y] @ U\}\} @ U \text{ [Definition of } \odot\text{]}.$$

$$x @ U = \{\{U @ \{[(\overline{U @ z} @ U) @ y] @ U\}\} @ U\} @ U$$

$$= U @ \{[(\overline{U @ z} @ U) @ y] @ U\}.$$

$$\therefore\ U @ x = [(U @ z) @ U] @ y \text{ [by 4, VII]}.$$

$$\therefore\ U @ z = y @ x \text{ [by 4, VII]}.$$

$$\therefore\ x @ y = z @ U \text{ [by 4]}.$$

(3) $\therefore\ (x @ y) @ U = (z @ U) @ U = z$ [by VII].

(4) If $y = Z$ or $z = Z$, then $x = Z$ [by the definition of $\odot$].

If $x = Z$, then $\{U @ \{[(\overline{U @ z} @ U) @ y] @ U\}\} @ U$ is not a K-element [by (1)].

(5) Then $y = Z$ or $z = Z$ [by (2)]. $\therefore$ If $y \neq Z$, $x = z = Z$ [by (4) and (5)].

(6) Then $(x @ y) @ U = Z = z$.

Statements (3) and (6) constitute the theorem, and all the steps taken are reversible.

Proof of IX. This follows immediately from II and the definition of $\odot$.

Proof of X. Let $x = y \odot z$ and $y \neq Z$, $z \neq Z$, then

$$Z = [(x @ y) @ U] @ z \text{ [by VIII]},$$

$$= [(x @ z) @ U] @ y \text{ [by VI]}.$$

(1) Consequently $x = z \odot y$ [by VIII, 5].

If $y = Z$ or $z = Z$, $\{U @ \{[(\overline{U @ z} @ U) @ y] @ U\}\} @ U$ is not a K-element, and by the definition of $\odot$, $y \odot z = z \odot y = Z$.

This fact and (1) constitute the proof of X.

Proof of XI. Suppose $x \odot y = Z$. Then $x = Z$ or

$$y = (Z @ x) @ U = U @ U = Z \text{ [by VIII]}.$$

The converse of this proposition follows from the definition of $\odot$.

Proof of XII. If $x \oplus y$ does not belong to K, there are two distinct K-elements, u and v, such that

(1) $x = u \sim y = v \sim y$ [by definition of $\oplus$].

Then either $u \sim y = Z$ or $(u \sim y) \sim (v \sim y) = Z$ [by XI, definition of $\sim$].

Then either $u \sim y = Z$ or $u \sim v = Z$ or $y = Z$ or

$$(u \sim v) \sim (y \sim y) = Z \text{ [by 7]}.$$

If $m \sim n = Z$, $m = n$, and if $m = n \neq Z$, $m \sim n = Z$ [by definition of $\sim$].

$$\therefore \quad u = y \quad \text{or} \quad u = v \quad \text{or} \quad y = Z.$$

$$v = y \quad \text{or} \quad v = u \quad \text{or} \quad y = Z.$$

(2) $$\therefore \quad u = v \quad \text{or} \quad y = Z.$$

If $y = Z$, $x = u \odot U = v \odot U$ [by (1)].

(3) Then $u = (u @ U) @ U = (v @ U) @ U = v$ [by VII, VIII, X].

Statements (2) and (3) constitute the theorem.

Proof of XIII. If $x \oplus x = x$ and $x \neq Z$, then $x \sim x = x$ [by definitions of $\odot$, $\sim$, and $\oplus$].

(1) Then $x = x \odot (x @ x) = x \odot Z = Z$ [by XI].

(2) $Z \oplus Z = Z$ [by XI and the definitions of $\sim$ and of $\oplus$].

Statements (1) and (2) constitute the theorem.

Proof of XIV. Suppose $z = u \odot x = u \odot y$ and u is a K-element other than Z. Then $x = (z @ u) @ U = y$ [by VIII].

Proof of XV. Suppose $u = (x \odot y) \odot z$, $v = x \odot y$, and $x \neq Z$, $y \neq Z$, $z \neq Z$. Then $y = (v @ x) @ U$ and $v = (u @ z) @ U$ [by VIII, X].

$$\therefore \quad y = \{[(u @ z) @ U] @ x\} @ U$$

$$= \{[(u @ x) @ U] @ z\} @ U \text{ [by VI].}$$

The steps already taken are reversible, and by retracing them after the interchange of x and z, it appears that if $x \neq Z$, $y \neq Z$, $z \neq Z$,

(1) $$(x \odot y) \odot z = (z \odot y) \odot x = x \odot (y \odot z) \text{ [by X].}$$

If x, y, or $z = Z$, by XI,

(2) $$(x \odot y) \odot z = Z = x \odot (y \odot z).$$

Statements (1) and (2) constitute the theorem.

Proof of XVI.

LEMMA. $x \odot (u \sim v) = (x \odot u) \sim (x \odot v)$, if both sides belong to K, unless $x = Z$.

Proof. Suppose $u \neq Z$, $v \neq Z$. Let $x \odot u = m$ and $x \odot v = n$. Then $m \neq Z$, $n \neq Z$ [by XI]. Then $u = (m @ x) @ U$ and $v = (n @ x) @ U$ [by VIII]. Hence $u @ m = [(m @ x) @ U] @ m = [(m @ m) @ U] @ x$ [by VI] $= U @ x = v @ n$ [by VI, as before].

(1) Therefore $v @ u = n @ m$ [by 4].

$$x \odot (u \sim v) = x \odot [u \odot (v @ u)] \text{ [by definition of } \sim],$$

$$= (x \odot u) \odot (v @ u) \text{ [by XV]},$$

$$= m \odot (n @ m) \text{ [by (1)]},$$

(2) $$= m \sim n \text{ [by definition of } \sim].$$

(3) If $u = Z$, $m = Z$, and neither $x \odot (u \sim v)$ nor $m \sim n$ is a K-element, by the definition of $\sim$.

If $u \neq Z$, and $v = Z$, then $n = Z$, so that $x \odot (u \sim v) = m \odot (v @ u)$ [as before], $= m \odot U = m \odot (n @ m)$

$$(4) \qquad = m \sim n \text{ [as before].}$$

Statements (2), (3), and (4) constitute the lemma.

Let $y \oplus z = p$. If $p \neq Z$, then $y = p \sim z$ [def. of $\oplus$, (1) of VIII]. Then $x \odot y = (x \odot p) \sim (x \odot z)$ [by (2) and (4)].

$$(5) \qquad \therefore \quad x \odot p = (x \odot y) \oplus (x \odot z) \text{ [by XII, def. of } \oplus \text{]}.$$

If $p = Z$, there is no K-element q such that $y = q \sim z$ [def. of $\oplus$]. Then if $x \neq Z$ there is no q such that $x \odot y = (x \odot q) \sim (x \odot z)$ [by the Lemma, XI, XIV].

(6) Then $x \odot p = (x \odot y) \oplus (x \odot z)$ [by the def. of $\oplus$].

(7) If $x = Z$, the theorem becomes, by XII, $Z = Z \oplus Z$, which is an immediate consequence of the definition of $\oplus$.

Statements (5), (6), and (7) constitute the theorem.

Proof of XVII. Suppose $x \oplus y = u$, $u \neq Z$. Then $x = u \sim y$ [def. of $\oplus$], $= u \odot (y @ u)$ [def. of $\sim$].

This proposition is equivalent to $(x @ u) @ U = y @ u$, or to [by VII, VIII]; $(y @ u) @ U = x @ u$ [by VI].

Reversing our former reasoning, $y = u \sim x$, and

$$(1) \qquad u = y \oplus x \text{ [by def. of } \oplus \text{, XII].}$$

(2) If $x \oplus y = Z$, $y \oplus x = Z$ [by def. of $\oplus$, XII, (1)].

Statements (1) and (2) constitute the theorem.

Proof of XVIII. Let $x \oplus y = m$, $y \oplus z = n$, $m \oplus z = u$, and $x \oplus n = v$.

Case I. Suppose x, y, z, m, n, u, and v are all distinct from Z. Then $x = m \sim y$, $z = n \sim y$, $x = v \sim n$, $z = u \sim m$ [by def. of $\oplus$, XVII].

$$\therefore \quad x \sim z = (m \sim y) \sim (n \sim y) = (v \sim n) \sim (u \sim m) \text{ [def. of } \sim \text{]}.$$

$$\therefore \quad m \sim n = (v \sim n) \sim (u \sim m) \text{ [by 7, def. of } \sim \text{]}.$$

$$\therefore \quad u \sim m = (v \sim n) \sim (m \sim n) \text{ [by XVII, defs. of } \oplus \text{ and } \sim \text{]}.$$

$$= v \sim m \text{ [by 7, def. of } \sim \text{]}.$$

$$\therefore \quad Z = (u \sim m) \sim (v \sim m) \text{ [by def. of } \sim \text{]}$$

$$= (u \sim v) \sim (m \sim m) \text{ [by 7, def. of } \sim \text{]}$$

$$= u \sim v \text{ [by def. of } \sim \text{]}$$

$$= u @ v \text{ [by def. of } \sim \text{, XI]}.$$

$$(1) \qquad \therefore \quad v = u \text{ [by 5]}.$$

Case II. Let x, y, or $z = Z$.

(2) $a \oplus Z = Z \oplus a = a$ [by XVII and the defs. of $\oplus$ and $\sim$].

(3) Combining this proposition with XVII, the proof is obvious.

Case III. Let x, y, z, n, u, and v be all distinct from Z, and let $m = Z$. Then $y = n \sim z$, $x = v \sim n$, $z = u \sim Z = u$ [by defs. of $\oplus$ and $\sim$; XVII].

(4) Whatever p, $x \neq p \sim y$. Suppose $u \neq v$. Then

$$x = v \sim n$$
$$= (v \sim n) \sim (u \sim u) \text{ [by def. of } \sim],$$
$$= (v \sim u) \sim (n \sim u) \text{ [by 7]},$$

(5) $$= (v \sim u) \sim y.$$

(6) Statements (5) and (4) contradict one another. Hence $u = v$.

(7) *Case IV.* Similarly, if n alone is Z, $u = v$.

(8) *Case V.* It follows from the cases already considered by XII and the definitions of $\sim$ and $\oplus$ that if $u = Z$, $v = Z$, and vice versa.

Statements (1), (3), (6), (7), and (8) constitute the theorem.

Proof of XIX. If a, b, and x are distinct from Z and from one another,

$$(a \sim x) \sim (b \sim x) = (a \sim b) \sim (x \sim x) \text{ [by 7]},$$
$$= (a \sim b) \odot U \text{ [def. of } \sim, U],$$
$$= \{U @ \{[(\overline{U @ U} @ U) @ (a \sim b)] @ U\}\} @ U$$
$$\text{[def. of } \odot],$$
$$= \{U @ \{[U @ (a \sim b)] @ U\}\} @ U \text{ [def. of } U],$$
$$= \{(a \sim b) @ \{[U @ U] @ U\}\} @ U \text{ [by 4, VI]},$$

(1) $$= a \sim b \text{ [by VII]}.$$

(2) Then $(a \sim x) \sim a = (b \sim x) \sim b$ [by 7]. Now, $(a \sim x) \sim a$ can never be expressed in the form $u \sim x$. For, suppose $(a \sim x) \sim a = u \sim x$. Then $(a \sim x) \sim u = a \sim x$ [by 7].

(3) $$\therefore \quad a \sim x = (a \sim x) \odot [u @ (a \sim x)] \text{ [def. of } \sim].$$

But we have just seen in (1) how it can be shown that

(4) $$(a \sim x) = (a \sim x) \odot U.$$

Hence, $u @ (a \sim x) = U$ [(3), (4), X, XIV], $= Z @ (a \sim x)$ [def. of U].

$$\therefore \quad U = Z @ u = (a \sim x) @ (a \sim x) = Z.$$

Since this contradicts V,

(5) $$(a \sim x) \sim a \neq u \sim x.$$

From (2) and (5) we conclude that, if $x \neq Z$, there is a unique term

$$\bar{x} = (a \sim x) \sim a,$$

such that whatever u, $\bar{x} \neq u \sim x$.

(6) $\therefore\ Z = \bar{x} \oplus x$ [def. of $\oplus$, XII]

(7) $Z = Z \oplus Z$ [def. of $\oplus$, U, $\sim$, argument of (1)].

Statements (6) and (7) constitute the theorem, unless it is impossible for an a to be found distinct from Z and x. In this case K contains only Z and U. Here the same result may be obtained by a simple computation.

Proof of XX. Let x and y be K-elements. Then either $y = Z$ or $y \sim x$ is a K-element v [by IX, def. of $\sim$].

(1) If $y = Z$, there is a K-element u such that $u \oplus x = y$ [by XIX].

(2) If $y \neq Z$, since $v = y \sim x$, $v \oplus x = y$ [by def. of $\oplus$, XII].

POSTULATES 1–7 AND THE DEFINITION OF A FIELD

In the article already cited, Huntington gives the following reduced set of postulates for a field, the sufficiency of which he proves, but not their independence.

1. If a and b belong to K, $a \oplus b$ belongs to K.
2. $(a \oplus b) \oplus c = a \oplus (b \oplus c)$.
3. $a \oplus b = b \oplus a$.
4. If a and b belong to K, $a \odot b$ belongs to K.
5. $(a \odot b) \odot c = a \odot (b \odot c)$.
6. $a \odot b = b \odot a$.
7. Either $a \odot (b \oplus c) = (a \odot b) \oplus (a \odot c)$, or
$$(b \oplus c) \odot a = (b \odot a) \oplus (c \odot a).$$
8. Given a and b, there is an x such that $a \oplus x = b$.
9. Given a and b, and $a \oplus a \neq a$, there is ay such that $a \odot y = b$.
10. There are at least two elements in the class.

That the postulates of Huntington are deducible from those of this paper will be seen at once from an inspection of the following table of correspondences. The figures beneath the line refer to Huntington's postulates; those above, to the present article.

Corresponds to	XII	XVIII	XVII	IX	XV	X	XVI	XX	VII, X	I
	1	2	3	4	5	6	7	8	9	10

If for any elements a and b of Huntington's class, we write a @ b for $1 - a/b$, Huntington's set implies set 1–7.

A SINGLE UNDEFINED NOTION FOR COMPLEX ALGEBRA

The laws of complex algebra cannot be expressed entirely in terms of $+$ and $\times$. Huntington* has given a set of postulates for complex algebra in which the undefined notions are addition, multiplication, the class of all complex numbers, the class of all real numbers, and the order of magnitude among real numbers. He has suggested† how all these notions may be derived from that of the absolute value of the difference between two complex numbers, though he has given no set of postulates for this operation. Apart from this operation, which is rather unlike those on which we are accustomed to base sets of postulates in that it always results in a number of a very special sort—a real number, no single operation has been developed which may serve as the sole foundation of complex algebra.

There are other such operations, however, and one of them is closely allied to the one taken for the fundamental notion of this set of postulates. This operation has already been employed to demonstrate the independence of postulate 7. Denoting $a - ib$, the conjugate‡ imaginary of $a + ib$, by $C(a + ib)$, our operation, which we shall term $x * y$, is

$$1 - C\left(\frac{x}{y}\right).$$

We define Z, U, and multiplication in terms of $*$ in the same manner in which they have already been defined in terms of @. $C(x)$ is defined as

$$U * [U * (U * x)].$$

We say that x is real if $x = C(x)$. We define $x \wedge y$ as $C(x * y)$, and addition is defined in terms of $\wedge$ as previously in terms of @. If x and y are real, x is said to be larger than y when there is a K-element z, the product of a real by itself, which when added to y gives x.

The formulation of a set of postulates for $*$ is a very difficult matter. Postulates 1–6 of this paper may stand unchanged, but the properties of addition and of the series of reals are so intricate as not to lend themselves readily to independence proofs. In general, independence proofs that are satisfactory when addition and multiplication are assumed separately break down on more than one postulate when translated in terms of $*$. The system of $*$, though fundamentally the same as that of the familiar complex algebra, has a technique of its own of considerable difficulty.

* E. V. Huntington, *A set of postulates for ordinary complex algebra*, these Transactions, vol. 6 (1905), pp. 209–229.

† Ibid., note on p. 210.

‡ This use of the conjugate was suggested to me by an unpublished paper of Sheffer.

MASSACHUSETTS INSTITUTE OF TECHNOLOGY,
June, 1919.

CERTAIN ITERATIVE CHARACTERISTICS OF BILINEAR OPERATIONS.

BY DR. NORBERT WIENER.

Introduction.

IN a recent paper‡ the author has developed the necessary and sufficient condition that a bilinear operation in two variables should generate by iteration all rational operations with

* Vallée-Poussin, loc. cit., p. 118.

† Hausdorff, loc. cit., p. 211. See also Fréchet, "Les notions de limite et de distance," *Transactions Amer. Math. Society*, vol. 19 (1918), p. 54.

‡ *Annals of Mathematics*, vol. 21, No. 3 (March, 1920), pp. 157–165.

rational coefficients. This is a particular case of the general problem of determining just what operations any given bilinear operation will generate by iteration. While the complete solution of this problem is still to be accomplished, it is the purpose of this paper to develop methods of attack which will yield, in particular, an important necessary condition that each of two operations generate the other by iteration.

Definitions.

A bilinear operation is an operation $\varphi(x, y)$ of the form

$$\frac{A_\phi + B_\phi x + C_\phi y + D_\phi xy}{E_\phi + F_\phi x + G_\phi y + H_\phi xy}.$$

In this paper the coefficients will be supposed to be rational, ∞ will be considered a proper value and argument for a bilinear operation, and if $\varphi(x_1, y_1)$ is indeterminate as it stands, it shall be given the value $\lim_{x \doteq x_1,\, y \doteq y_1} \varphi(x, y)$ if this is determinate.

A bilinear operation will be said to be reduced if A_ϕ, $B_\phi + C_\phi$, D_ϕ, E_ϕ, $F_\phi + G_\phi$, and H_ϕ form a relatively prime set of integers.

Δ_ϕ is defined as

$$\begin{vmatrix} \begin{vmatrix} A_\phi & B_\phi + C_\phi \\ E_\phi & F_\phi + G_\phi \end{vmatrix} & \begin{vmatrix} A_\phi & D_\phi \\ E_\phi & H_\phi \end{vmatrix} \\ \begin{vmatrix} A_\phi & D_\phi \\ E_\phi & H_\phi \end{vmatrix} & \begin{vmatrix} B_\phi + C_\phi & D_\phi \\ F_\phi + G_\phi & H_\phi \end{vmatrix} \end{vmatrix}.$$

The following definitions are made:

$$J_\phi(u, v) = A_\phi v^2 + (B_\phi + C_\phi)uv + D_\phi u^2,$$

$$K_\phi(u, v) = E_\phi v^2 + (F_\phi + G_\phi)uv + H_\phi u^2,$$

$$L_\phi = vJ_\phi - uK_\phi.$$

We shall term the roots of $L_\phi(u, 1) = 0$, $r_1(\varphi)$, $r_2(\varphi)$, $r_3(\varphi)$. We shall call a root of $L_\phi(u, 1) = 0$ free if it is not a simultaneous solution of $J_\phi(u, 1) = 0$ and of $K_\phi(u, 1) = 0$. If $\Delta_\phi \neq 0$, and $r_1(\varphi) \neq r_2(\varphi) \neq r_3(\varphi)$, $r_1(\varphi) \neq r_3(\varphi)$, φ will be said to be general, otherwise special. Clearly if φ is general r_1, r_2, and r_3 are all free.

An iterative field is a set of functions of n variables which is closed with respect to the following operations:

(1) permutation of any two arguments;

(2) substitution of two identical arguments for two different arguments;

(3) substitution of a function of the set for any argument.

Two operations are said to be iteratively equivalent if any iterative field containing either contains the other.

An iterative characteristic of an operation is a property which belongs to all operations iteratively equivalent to that operation.

Theorems.

I. *The free roots of $L_\phi(u, 1) = 0$ constitute iterative characteristics of φ.*

For $L_\phi(u, 1) = 0$ is equivalent to $\varphi(u, u) = u$, unless u is a simultaneous root of $J_\phi(u, 1) = 0$ and of $K_\phi(u, 1) = 0$. Now, $\varphi(u, u) = u$ is a property of φ invariant under iteration. If r is a root of $L_\phi(u, 1) = 0$ that is not free, then in general $\varphi(r, r) = r$ is not significant and determinate. If this latter equation is significant, let us subject our number system to a linear transformation that changes r to ∞. It may readily be seen that two operations will retain their iterative relationships unchanged under such transformations, that roots of $L_\phi(u, 1) = 0$ will remain roots, and that free roots will remain free. φ will then become a ψ of the form $A + Bx + Cy$. This operation can only be iteratively equivalent to operations of its own sort, for all of which ∞ is a root that is not free. This completes the proof of our theorem. As a corollary we obtain

II. *No general bilinear operation is equivalent to any special bilinear operation.*

III. *If φ and ψ are reduced equivalent general bilinear operations, then*

$$A_\phi = A_\psi,$$

$$B_\phi + C_\phi - E_\phi = B_\psi + C_\psi - E_\psi,$$

$$F_\phi + G_\phi - D_\phi = F_\psi + G_\psi - D_\psi,$$

$$H_\phi = H_\psi.$$

Lemma 1. *If φ is reduced, u and v can be chosen relatively prime in such a manner that $J_\phi(u, v)$ is prime to $K_\phi(u, v)$.*

Proof. Let N be the H.C.F. of A_ϕ, $B_\phi + C_\phi$, and D_ϕ. Let M be the product of all prime factors in Δ_ϕ but not in N. Let P be the product of all the primes in M, but not in A_ϕ nor in D_ϕ. Let Q be the product of all primes in M and A_ϕ, but not in D_ϕ. Let R be the product of all primes in M and D_ϕ, but not in A_ϕ. Let S be the product of all primes in N, but neither in E_ϕ nor in H_ϕ. Let T be the product of all primes in N and E_ϕ, but not in H_ϕ. Let U be the product of all primes in N and H_ϕ, but not in E_ϕ. Let $u = PRSU$, $v = QT$. We get as a result that J_ϕ is prime to M, while K_ϕ is prime to N. Now, it may readily be shown that any factor common to J_ϕ and K_ϕ belongs to Δ_ϕ. Since, however, any prime factor of Δ_ϕ belongs either to M or to N, it is either prime to J_ϕ or to K_ϕ.*

Lemma 2. *If a and b are any two numbers not both containing a prime p and if c and d, e and f, are couples of the same sort, while*

$$ad - bc \equiv 0 \pmod{p^k}, \qquad af - be \equiv 0 \pmod{p^k},$$

then

$$cf - de \equiv 0 \pmod{p^k}.$$

This lemma, whose proof is immediate, enables us to classify fractions into sets modulo p^k. We shall say that

$$a/b \equiv c/d \pmod{p^k}$$

if

$$ad - bc \equiv 0 \pmod{p^k}$$

while a and b do not contain p in common, nor do c and d.

Lemma 3. *If u/v and w/z are in their lowest terms, and*

$$u/v \equiv w/z \pmod{p^k},$$

then if φ is any bilinear operation such that

$$\varphi(u/v,\, u/v) \equiv u/v \pmod{p^k},$$

it follows that

$$\varphi(w/z,\, w/z) \equiv w/z \pmod{p^k}.$$

The proof of lemma 3 offers no difficulty.

Combining lemmas 2 and 3, it is at once clear that the equation

$$\varphi(u/v,\, u/v) \equiv u/v \pmod{p^k}$$

for a given u/v and p^k is an iterative characteristic of φ.

* Cf. Mathews, Theory of Numbers, vol. 1, p. 132.

We are now in a position to take up the proof of Theorem III. Clearly $\varphi(u/v, u/v)$ is equal to $J_\phi(u, v)/K_\phi(u, v)$. Let u and v be two numbers such that $J_\phi(u, v)$ and $K_\phi(u, v)$ are prime to one another. By lemma 1, such numbers exist. Then to say that

$$J_\phi(u, v)/K_\phi(u, v) \equiv u/v \pmod{p^k}$$

is equivalent to saying

$$L_\phi(u, v) \equiv 0 \pmod{p^k},$$

or that p^k is a factor of $L_\phi(u, v)$. If φ is iteratively equivalent to ψ, clearly we must have

$$L_\psi(u, v) \equiv 0 \pmod{p^k}.$$

That is, every prime power that divides L_ϕ must also divide L_ψ, or in other words, L_ϕ is a factor of L_ψ for a particular set of values of u and v. However, since by Theorem I the equations $L_\phi(u, 1) = 0$ and $L_\psi(u, 1) = 0$ have the same three distinct roots, L_ψ is a constant multiple of L_ϕ. Hence L_ϕ is always a factor of L_ψ. Likewise, L_ψ is always a factor of L_ϕ. Hence, apart from a possible difference of sign, L_ϕ and L_ψ are identical. This proves Theorem III.

MASSACHUSETTS INSTITUTE OF TECHNOLOGY,
April 16, 1920.

CERTAIN ITERATIVE PROPERTIES OF BILINEAR OPERATIONS

By Norbert WIENER

The study of the logical foundations of a given mathematical system is not complete when a single set of postulates is discovered. It demands, as a matter of fact, the study of all the sets of postulates in terms of which the system may be determined, and of the manner in which the system may be determined in terms of them. In particular, the question arises, what sets of undefined notions and postulates of a given sort enable the determination of a given system in a given manner.

We shall discuss a question of this sort with respect to algebra. Algebra, for the purposes of the present paper, will be regarded as the theory of all algebraic operations obtainable as implicit functions from rational operations with integral coefficients. Such operations may be obtained from the variables they contain by a finite sequence of operations of addition, subtraction, multiplication and division. The autor has pointed out that there is a single operation in terms of which addition, subtraction, multiplication, and division can be defined by a finite sequence of iterations(1), and has given a set of postulates for fields in terms of this. This is the operation $x \mid y = 1 - x/y$. With the appropriate set of conventions concerning exceptional cases, the four species may be defined in the manner indicated in the following table :

$$\begin{aligned} 0 &= x \mid x, \\ 1 &= 0 \mid x, \\ x/y &= (x \mid y) \mid 1, \\ xy &= x/(1/y), \end{aligned}$$

if this is significant, and 0 otherwise,

$$\begin{aligned} x \backsim y &= x(y \mid x), \\ x - y &= (a \backsim y) \backsim (a \backsim x), \\ x + y &= x - (0 - y). \end{aligned}$$

(1) *Trans. Am. Math. Soc.*, April, 1920.

The question which I now wish to raise is, what other operations will serve in the same manner to generate all rational operations with integral coefficients and will thus serve as undefined notions for a set of postulates for fields or for real algebra in a way analogous to $1 - x/y$.

I have solved this problem completely[1] for the case where the operation in question is a rational operation with rational coefficients, and is restricted to the form

$$\frac{A + Bx + Cy + Dxy}{E + Fx + Gy + Hxy}.$$

Such an operation will be called a rational bilinear operation. By a linear transformation with rational coefficients, every rational bilinear operation that generates by iteration all rational operations with rational coefficients may be transformed either into

$$\frac{x - y}{x + Ay} \quad \text{or} \quad \frac{y - x}{y + Ax}$$

where A is a rational number other than -1, or into

$$\frac{n(x - y + xy)}{ny + xy}$$

where n is any integer other than 0. Conversely, every operation of these forms, as well as every operation obtainable from them by a linear transformation with rational coefficients, generates by its iteration all rational operations with rational coefficients.

There is a certain equivalence between any two operations of the above forms which we may call iterative equivalence. We shall define two operations as iteratively equivalent if each can be generated by the iteration of the other. The general question thus arises, closely connected with the theory of sets of postulates for algebra, given two operations, how can it be determined from their form whether they are equivalent or not. This question is as yet unsolved, even in the simple case where both operation are bilinear. The author, however, has been able to find in one very particular case a necessary condition for the iterative equivalence of two bilinear operations.

We shall start with a few definitions. We shall write

$$\varphi(x, y) = \frac{A_\varphi + B_\varphi x + C_\varphi y + D_\varphi xy}{E_\varphi + F_\varphi x + G_\varphi y + H_\varphi xy}$$

(1) *Annals of Mathematics*, march, 1920.

Δ_φ will be defined as

$$\begin{vmatrix} \begin{vmatrix} A_\varphi & B_\varphi + C_\varphi \\ E_\varphi & F_\varphi + G_\varphi \end{vmatrix} & \begin{vmatrix} A_\varphi & D_\varphi \\ E_\varphi & H_\varphi \end{vmatrix} \\ \begin{vmatrix} A_\varphi & D_\varphi \\ E_\varphi & H_\varphi \end{vmatrix} & \begin{vmatrix} B_\varphi + C_\varphi & D_\varphi \\ F_\varphi + G_\varphi & H_\varphi \end{vmatrix} \end{vmatrix}$$

φ will be said to be general if $\Delta_\varphi \neq 0$ and if the equation

$$A_\varphi + (B_\varphi + C_\varphi - E_\varphi)x + (D_\varphi - F_\varphi - G_\varphi)x^2 - H_\varphi x^3 = 0$$

has three distincts roots. The author has demonstrated in a paper in the course of publication in the *Bulletin of the American Mathematical Society* that if two bilinear operations φ and ψ are iteratively equivalent, and if

$$A_\varphi,\ B_\varphi + C_\varphi,\ \ D_\varphi,\ E_\varphi,\ F_\varphi + G_\varphi, \text{ and } H_\varphi$$

on the one hand, and

$$A_\psi,\ B_\psi + C_\psi,\ \ D_\psi,\ E_\psi,\ F_\psi + G_\psi, \text{ and } H_\psi,$$

on the other, form sets of integers relatively prime among themselves, then

$$\begin{aligned} A_\varphi &= A_\psi, \\ B_\varphi + C_\varphi - E_\varphi &= B_\psi + C_\psi - E_\psi, \\ F_\varphi + G_\varphi - D_\varphi &= F_\psi + G_\psi - D_\psi, \\ H_\varphi &= H_\psi. \end{aligned}$$

As has been said, this only goes a small way toward the solution of the problem of iterative equivalence. The investigation of this problem has been rendered very difficult by the fact that it has so far proved impossible to find methods of more than a very limited range of application.

Toulouse. — Imp. et Lib. Edouard Privat. — 3677

THE ISOMORPHISMS OF COMPLEX ALGEBRA.

BY DR. NORBERT WIENER.

(Read before the American Mathematical Society December 28, 1917.)

Definitions and Conventions. By the *transform* by a function $\phi(x)$ of a function $F(x_1, x_2, \cdots, x_n)$ we mean the expression

$$\varphi\{F(\varphi^{-1}(x_1), \varphi^{-1}(x_2), \cdots, \varphi^{-1}(x_n))\}.$$

We shall consider ∞ as a possible argument for a function, and we shall define $F(\infty)$ to be $\lim_{x=\infty} F(x)$. Similarly if $\lim_{x=\infty} F(x) = \infty$, as $|x|$ grows without limit in every possible way, we shall say that $F(\infty)$ is ∞, while if $\lim_{x=a} F(x) = \infty$ as x approaches a in every possible way, we shall say that $F(a) = \infty$. Similar conventions will be established for functions of more than one variable.

A function F of the complex variable z is said to be *continuous* at z_1 if $\lim_{\epsilon=0} F(z_1 + \epsilon) = F(z_1)$.

A function $F(x_1, x_2, \cdots, x_n)$ will be said to be *continuous in general* if there are only a finite number of sets $(x_1, x_2, \cdots, x_n)$ for which $F(x_1, x_2, \cdots, x_n)$ fails to be continuous.

A variable μ is said to *depend uniquely* on $x_1, x_2, \cdots, x_n$ if there are only a finite number of sets $x_1, x_2, \cdots, x_n$ for which μ is not uniquely determined, and if for these μ is undefined.

THEOREM. *If a function Φ is single-valued, as well as its inverse, over the set of arguments consisting of all complex numbers and ∞, and if it is continuous in general, and transforms every algebraic function into an algebraic function, it is a linear function or its conjugate.* The proof of this will involve almost no considerations except those of elementary algebra.

To begin with, let us consider any function of the form

$$F(x, y) = \frac{\alpha + \beta x + \gamma y + \delta xy}{\epsilon + \xi x + \eta y + \vartheta xy}$$

which does not degenerate into a constant nor into a function of a single variable. Such a function has the following properties.

(1) $F(x, y)$ depends uniquely on x and y.

(2) x depends uniquely on y and $F(x, y)$.

(3) y depends uniquely on x and $F(x, y)$.

It is easy to show that these properties will also belong to any transform of F by a biunivocal function

Now, no algebraic function not of the form

$$\frac{\alpha + \beta x + \gamma y + \delta xy}{\epsilon + \xi x + \eta y + \vartheta xy}$$

has properties (1), (2), and (3). Any algebraic function with these properties must be obtained by solving for z an equation $P(x, y, z) = 0$, where P is some polynomial. Since z is uniquely determined by x and y, we may assume, without any real loss of generality, that P is of the form*

$$[g(x, y) + zh(x, y)]^m,$$

which we may write

$$g^m + mg^{m-1}hz + \text{terms in higher powers of } z.$$

Since P is a polynomial, g^m and $g^{m-1}h$ are polynomials. Let us call these Q and R, respectively. P is then of the form

$$\frac{1}{g^{m^2-m}}\{g^m + zhg^{m-1}\}^m = \psi(x, y)\{Q(x, y) + zR(x, y)\}^m,$$

where Q and R may be taken so as to be mutually prime, by removing any common factor and transferring it to the factor ψ. By considerations of symmetry, we may write

$$\begin{aligned} P(x, y, z) &= \psi'(y, z)\{Q'(y, z) + xR'(y, z)\}^{m'} \\ &= \psi''(x, z)\{Q''(x, z) + yR''(x, z)\}^{m''}. \end{aligned}$$

It follows from a consideration of the irreducible factors of P that we may write

$$P(x, y, z) = (\alpha + \beta x + \gamma y + \delta xy - \epsilon z - \xi xz - \eta yz - \vartheta xyz)^m,$$

where $\alpha, \cdots, \vartheta$ are constants. Hence the only algebraic functions satisfying conditions (1), (2), and (3) are of the form

$$F(x, y) = \frac{\alpha + \beta x + \gamma y + \delta xy}{\epsilon + \xi x + \eta y + \vartheta xy}.$$

We shall now state a lemma which may readily be established by the usual method of undetermined coefficients. The function $1 - x/y$ is the only function of the form

$$F(x, y) = \frac{\alpha + \beta x + \gamma y + \delta xy}{\epsilon + \xi x + \eta y + \vartheta xy}$$

which satisfies the four conditions

* Notice that this is only true in complex algebra.

(1) $F(x, x) = 0$ for all x other than 0 or ∞;
(2) $F(0, x) = 1$ for all x other than 0 or ∞;
(3) $F(x, 0) = \infty$ for all x other than 0 or ∞;
(4) $F\{1, F\{1, F(1, x)\}\} = x$ for all x other than 0 or ∞.

Now, consider a function $G(x, y)$ which is derived from $1 - x/y$ by a transformation Φ which is continuous in general, which is one-to-one, and which leaves all algebraic functions algebraic. We have already shown that any such function $G(x, y)$ must be of the form

$$\frac{\alpha + \beta x + \gamma y + \delta xy}{\epsilon + \xi x + \eta y + \vartheta xy}.$$

Let us now subject this function to a linear transformation φ which turns the transforms by Φ of 0, 1, and ∞ back into these respective numbers. The resulting function, which we shall call $H(x, y)$, satisfies conditions (1), (2), (3), and (4). Hence we have $H(x, y) = 1 - x/y$. Hence the transformation χ formed by performing first Φ and then φ leaves invariant the function $1 - x/y$. I have shown in an earlier paper* that addition and multiplication can be obtained by the iteration of the function $1 - x/y$. Hence the transformation χ leaves these functions invariant.

Now, any continuous transformation of the complex plane which leaves multiplication invariant must leave invariant the circle of the roots of unity. In a like manner, any continuous transformation of the number-plane that keeps addition invariant must keep invariant first the set of all sets each consisting of all the rational multiples of some number, then the set of all lines each consisting of the real multiples of some number, and finally must turn into a line every line in the complex plane, since every such line can be formed by adding the same number to the product of a given complex number by a variable real number. Hence our transformation χ is an affine transformation which keeps invariant the points 0, 1, and the unit circle. There is no difficulty in showing that any such transformation is either the identity or the conjugate transformation.

It follows that Φ is obtainable by applying after the identity or the conjugate operation the linear operation φ^{-1}. This is precisely the theorem which we set out to prove.

MASSACHUSETTS INSTITUTE OF TECHNOLOGY.

* *Bilinear operations generating all operations rational in a domain* Ω, ANNALS OF MATHEMATICS, March, 1920.

Comments on [20a, b, c, d], [21b]

R. J. Levit

Wiener mentions two of these papers in *Ex-prodigy* [53h] with the remark, "They were by far the best pieces of mathematical work I had yet written." Sheffer's reduction of the primitives of Boolean algebra to a single binary operation must have challenged many mathematicians to attempt the corresponding feat for ordinary algebra. Yet, so thoroughly did Wiener dispose of the problem in these papers that no one felt inclined to pursue the subject for some twenty years.

A field cannot be closed under an operation such as Wiener's which is designed to generate all rational operations by iteration alone. By admitting inversion along with iteration Levit gave a definition of field by means of a single operation which is class-closing.[1] This paper also contains a simplified postulate set in terms of Wiener's operation, and the doctoral dissertation (University of California, Berkeley, 1941) of which it is a condensation includes two other field definitions based on different single operations using iteration alone. Another characterization of fields utilizing a single class-closing operation by L. Borofsky allows a simpler expression for addition.[2] Both cited papers include discussions from slightly different viewpoints of equivalent field operations, that is, operations which are formally indistinguishable though they have different expressions in terms of the fundamental operation (or operations) of the field.

References

1. R. J. Levit. Trans. Amer. Math. Soc. 57 (1945), 426-440.

2. S. Borofsky, Amer. J. Math. 71(1949), 92-104.

ON THE THEORY

OF SETS OF POINTS IN TERMS OF CONTINUOUS TRANSFORMATIONS

By Norbert WIENER

BOSTON (MASSACHUSETTS)

In an ordinary n-space there is a set of concepts of a non-mětrical nature, and such that each may be defined in terms of the other. Among these are sequential limit, nonsequential limit, neighborhood, and biunivocal, bicontinuous transformation? In a more general space, these notions need be equivalent no longer. The question thus arises of determining the relations over a general range between biuniform, bicontinuous transformation and the other notions. This the author has done in a paper which is to appear in the *Bulletin de la Société Mathématique de France*.

The undefinable notions taken to start with, are a class of elements K and a set Σ of biunivocal transformations of the whole of K into itself. The derived set or set of limit-elements of a set E is defined as the set of all terms A belonging to K such that A remains invariant under all the transformations of Σ leaving E-A invariant. A closed set is one which contains its derived set. The author has proved that the necessary and sufficent condition that Σ consists precisely of all biunivocal transformations of K leaving derivative properties invariant is that it be a group, that if it contains an operation R, it contains the converse of R, and that it contains every biunivocal operation turning all closed sets into closed sets, and only turning closed sets into closed sets. In the terminology of my paper, a system of this sort is a (J_1).

It becomes a matter of some interest to determine what systems are systems (J_1). A vector system, or system (Ve), is defined as a system K of elements correlated with a system σ of entities and the operations $\oplus$, $\odot$, and $\| \quad \|$ in a manner indicated by the following propositions :

(1) If ξ and η belong to σ, $\xi \oplus \eta$ belongs to σ,

(2) If ξ belongs to σ, and n is a real number $\geqslant 0$, $n \odot \xi$ belongs to σ,

(3) If ξ belongs to σ, $\|\xi\|$ is a real number $\geqslant 0$,

(4) $n \odot (\xi \oplus \eta) = (n \odot \xi) \oplus (n \odot \eta)$,

(5) $(m \odot \xi) \oplus (n \odot \xi) \acute{=} (m+n) \odot \xi$,

(6) $\|m \odot \xi\| = m \|\xi\|$,

(7) $\|\xi \oplus \eta\| \leqslant \|\xi\| + \|\eta\|$.

(8) $m \odot (n \odot \xi) = mn \odot \xi$,

(9) If A and B belong to K, there is associated with them a single member AB of σ,

(10) $\|AB\| = \|BA\|$,

(11) Given an element A of K and an element ξ of σ, there is an element B of K such that $AB = \xi$,

(12) $AC = AB \oplus BC$,

(13) $\|AB\| = 0$ when and only when $A = B$,

(14) If $AB = CD$, $BA = DC$.

$\|AB\|$ is an écart between A and B in the sense developed by Fréchet in his thesis, so that we may define a limit point of a set of points E as a point A such that for any positive ε there is a member B of E such that $\|AB\| < \varepsilon$. I have proved that in a system (Vc) the set of all biunivocal transformations leaving limit properties invariant is the set Σ of a (J_1). Examples of such systems are Euclidean n-spaces, Hilbert space, the space whose elements are bounded sequences of real numbers, and in which limit is taken uniformly, and the space of functions continuous over a closed interval, with limit again taken uniformly. I have also shown that the space of all infinite sequences of real numbers, with limit taken in the non-uniform sense is a (J_1) though the question wether it is a (Vc) remains still undetermined.

The next question which arises is, when does sequential limit as defined in a space (J_1) satisfy the four conditions of Riesz, summarized by Fréchet into three. These are, as numbered by Fréchet :

a) The derivative of E + F is the sun of the derivatives of E and of F.

3° A set consisting of a single element has no limit-element, and

4° If A is a limit-element of a set E, and if B is distinct from A, then there is at least one set F which has A but not B for a limit-element.

40

In a system (J_1), these become :

(I) If there is a transformation from Σ changing the element A but leaving invariant all the elements of the class E, and there is also a transformation from Σ changing A but leaving invariant all the elements of a class F, then there is a transformation from Σ changing A but leaving invariant every element of E + F.

(II) Given any two elements A and B, there is a transformation from Σ changing A but leaving B invariant.

(III) If there is a set E not containing the element A, but such that every transformation from Σ that leaves all the elements of E invariant leaves A also invariant, then given any element B distinct from A there is a set F such that there is a transformation from Σ changing B but leaving each element of F invariant, while there is no transformation from Σ changing A but leaving each member of F invariant.

It is interesting to note that in every system satisfying (I) and (II), the derivative of any set of elements is closed.

It is a matter of a certain amount of interest to obtain a set of propositions completely defining the formal properties of K and Σ in some specific system. The author has done this in the case of the set of all bicontinuous biunivocal transformations of a line into itself. The undefinables are a set K of elements and a set Σ of biunivocal transformations. The set of postulates for the analysis situs group of a line is the following.

I K contains at least three distinct elements.

II, III, and IV are the three propositions which together are equivalent to the statement that Σ is a (J_1).

V Is (I) of the conditions that a (J_1) satisfy Riesz's set of postulates.

VI If A, B, C, and D belong to K and $A \neq B$, $C \neq D$, then there is a transformation from Σ changing A to B and C to D.

In the remaining postulates, a *connected* set is a set which cannot be divided into two non-null parts, neither of which has a limit-element in the other. *A boundary element* of a set is a limit-element of the set which is also a limit-element of the set of all elements not in the given set. *A segment,* is a closed connected set with at least two boundary elements.

VII If E is any sub-set of K, and A is an element of K not a limit-element of E, then there is a segment of which A is an interior element and which contains no element of E.

VIII There is an at most denumerable sub-set K′ of K such that no member of Σ except possibly the identity transformation leaves every member of K′ invariant.

IX If E and F are two connected sets, and two boundary elements of E are boundary elements of F, then every other element of E is an element of F.

Postulates I-IX inclusive form a categorical set of postulates for the analysis situs group of a line. Postulates I-VIII, on the other hand, apply to all Euclidean spaces and to a much wider group of systems. Indeed, they apply to every system (Ve) in which the sum of two members of σ is independent of their order and in which, if A, B, and C are any three distinct elements such that $||AB|| = ||AC||$, then there is a finite set $B_1, B_2, \ldots, B_n$ of elements such that :

(1) $B_1 = B$, $B_n = C$,

(2) For all k, $||AB_k|| = ||AB||$,

(3) For all k, $||B_kB_{k+1}|| < ||AB||$.

Examples of such systems are all Euclidean n-spaces, the space of all bounded sequences (x_n) with limit taken in the uniform sense, Hilbert space, and the space of all functions continuous over a closed interval, with limit taken in the uniform sense.

THE GROUP OF THE LINEAR CONTINUUM

By NORBERT WIENER.

[Received September 28th, 1920.—Read November 11th, 1920.]

1. The linear continuum has already received a complete characterization in terms of order* and of limit.† Now, the author has shown that over a wide range of cases the notion of limit may be defined in terms of that of bicontinuous biunivocal transformation.‡ It is the purpose of this paper to develop a categorical theory of the structure of the line in terms of bicontinuous, biunivocal transformations, or, in other words, to give a complete postulational characterization of the analysis situs group of the line.

The set of postulates will be so framed that only one will have any direct effect on the dimensionality. All the other postulates together determine an analysis situs property of space which is shared by a large number of systems of a finite or infinite dimension number. A number of necessary conditions and a sufficient condition for a system to possess this property will be formulated.

INDEFINABLES.

2. Our indefinables are two in number—a set K of elements and a set Σ of one-one transformations of the whole of K into itself.

DEFINITIONS.

3. A sub-set E of K is said to have a *limit-element* A if A is invariant under every transformation belonging to Σ that leaves invariant every member of E except possibly A.

---

* Cf. E. V. Huntington, "A Set of Postulates for Real Algebra," *Trans. Am. Math. Soc.* (1905); O. Veblen, "Definition in Terms of Order alone in the Linear Continuum," *ibid.*

† R. L. Moore, "The Linear Continuum in Terms of Point and Limit," *Annals of Mathematics* (1914–15).

‡ N. Wiener, "Limit in Terms of Continuous Transformation," *Bull. Soc. Math. de France* (1921–22).

A set E is *closed* if it contains all its limit-elements.

A set E is *connected* if, whenever it is divided into the two non-null sets, F and G, either F has a limit-element in G or G has a limit-element in F.

$\bar{E}$ is the set of all elements in K but not in E.

An *interior* element of E is one that is not a limit-element of $\bar{E}$.

An element A is *exterior* to E if it is interior to $\bar{E}$.

An element A is a *boundary-element** of E if it is at once a limit-element of E and of $\bar{E}$.

A *segment* is a closed, connected set with at least two boundary elements.

A *component*† of a set E is a greatest connected sub-set of E.

The transformation $\breve{R}$ is the inverse of R. $R|S$ is the transformation which consists in performing first S and then R.

POSTULATES.

4. I. K contains at least three distinct elements.

II. If R is a biunivocal transformation of the whole of K into itself that turns all closed sets into all closed sets and only into closed sets, then R belongs to Σ.

III. If R and S belong to Σ, so does $R|S$.

IV. If R belongs to Σ, so does $\breve{R}$.

V. If there is a transformation from Σ changing A and leaving every member of E invariant, while there is a transformation from Σ changing A and leaving every member of F invariant, then there is a transformation from Σ changing A and leaving every member of $E+F$ invariant.

VI. If A, B, C, and D belong to K, and $A \neq C$, $B \neq D$, then there is a transformation from Σ changing A to B and C to D.

VII. If E is any sub-set of K and A is an element of K not a limit-element of E, then there is a segment of which A is an interior element and which contains no element of E.

VIII. There is an at most denumerable sub-set K' of K such that no member of Σ except possibly the identity transformation leaves every member of K' invariant.

---

* Cf. F. Hausdorff, *Grundzüge der Mengenlehre*, p. 214. The notions of boundary element and *Randpunkt* are not identical.

† *Ibid.*, p. 245.

IX. If E and F are two connected sets, and two boundary elements of E are boundary elements of F, then every other element of E is an element of F.

DEFINITIONS OF SYSTEMS.

5. A system satisfying postulates I–IX inclusive will be called a system (Li). A system satisfying postulates I–VII inclusive will be called a system (Sp).

A system (T_1) will be defined as in the author's previous paper in the *Bull. Soc. Math. de France*, as a system satisfying Postulates II–IV. A system (R) will be defined as by Fréchet,* as a system satisfying the conditions of F. Riesz.

1. Every limit-element of a set E is a limit-element of every set containing E.

2. Every limit-element of the sum of two sets E and F is a limit-element of at least one of the two sets.

3. A set containing a single element has no limit-element.

4. If A is a limit-element of a set E and B is distinct from A, there is always at least one set which has A for a limit-element without having B for a limit-element.

It has been proved by the author† that in the case of a (T_1), the necessary and sufficient condition that the system should also be an (R) is that it should satisfy the following three conditions:—

2′. This is verbally identical with V.

3′. Given any two elements, A and B, there is a transformation from Σ changing A but leaving B invariant.

4′. If there is a set E not containing the element A, but such that every transformation from Σ that leaves all the elements of E invariant leaves A also invariant, then, given any element B distinct from A, there is a set F not containing A such that there is no transformation from Σ changing A but leaving each member of F invariant, while there is a transformation from Σ changing B but leaving F invariant.

---

* "Sur la notion de voisinage dans les ensembles abstraits," *Bulletin des Sciences Mathématiques*, May 1918.

† *Loc. cit.*

A system (H) is one in which neighbourhoods are so defined as to satisfy Hausdorff's "Umgebungsaxiome":*

(A) Given any point x, there is at least one neighbourhood U_x, of which x is a member.

(B) If U_x and V_x are two neighbourhoods of x, then there is a neighbourhood W_x contained in both.

(C) If y belongs to U_x, there is a neighbourhood of y, U_y, contained in U_x.

(D) If x and y are two points, then neighbourhoods U_x and Y_z can be so chosen as not to overlap.

In a system (H) a set E is said to have a limit-point A if every neighbourhood U_A of A contains an infinity of points of E.†

A *vector-system*, or system (Ve), is defined as in my previous paper‡ as a system K of elements (represented by capitals), associated with entities called vectors (represented by Greek letters), real numbers (represented by lower case letters), and the operations $\odot$, $\oplus$, and $\| \ \|$ by the following laws:—

(1) If ξ and η are vectors, $\xi \oplus \eta$ is a vector.

(2) If ξ is a vector and $n \geqslant 0$, $n \odot \xi$ is a vector.

(3) If ξ is a vector, $\| \xi \|$ is a non-negative real number.

(4) $n \odot (\xi \oplus \eta) = (n \odot \xi) \oplus (n \odot \eta)$.

(5) $m \odot (n \odot \xi) = mn \odot \xi$.

(6) $(m \odot \xi) \oplus (n \odot \xi) = (m+n) \odot \xi$.

(7) $\| m \odot \xi \| = m \| \xi \|$.

(8) $\| \xi \oplus \eta \| \leqslant \| \xi \| + \| \eta \|$.

(9) If A and B belong to K, there is associated with them a unique vector AB.

(10) $\| AB \| = \| BA \|$.

(11) Given an element A of K and a vector ξ, there is an element B of K such that $AB = \xi$.

---

* *Loc. cit.*, p. 213.

† *Ibid.*, p. 219, definition of *β-Punkt*.

‡ *Loc. cit.*

(12) $AC = AB \oplus BC$.

(13) $\| AB \| = 0$ when and only when $A = B$.

(14) If $AB = CD$, $DC = BA$.

A system (Vr), or a *restricted vector system* is a vector system of at least two elements in which the sum of two vectors is independent of their order, and in which, if A, B, and C are any three distinct elements such that $\| AB \| = \| AC \|$, then there is a finite set $B_1, B_2, \ldots, B_n$ of elements such that

(1) $B_1 = B$, $B_n = C$.

(2) For all k, $\| AB_k \| = \| AB \|$.

(3) For all k, $\| B_k B_{k+1} \| < \| AB \|$.

We shall say that a set E has A for a limit-element if, for all the B's that belong to E, the lower bound of $\| AB \|$ is zero.

RELATIONS OF SYSTEMS.

6. We shall say that a system of one of our classes belongs to another of our classes if a translation into the language of the second class is always possible in such a manner as to keep limit properties invariant. We have already seen that every (Sp) or (Li) is a (T_1), and every (Li) is clearly an (Sp); we shall prove the further relations:

(1) Every (Sp) is an (R).

(2) Every (Sp) is an (H).

(3) Every (Vr) is an (Sp).

Proof of (1).

All that we need to prove is contained in propositions 3′ and 4′ of § 5. If there are at least three elements, 3′ is a consequence of VI. Now, there are at least three elements, by I.

As to 4′ it is enough to show that, given any two elements A and B, there is a set E having A but not B as a limit-element. It follows from VII, I, and 3′, that there is at least one set F_1 which has limit-elements without having the whole of K for the class of its limit-elements.

Let A_1 be a limit-element of this set, and B_1 an element not a limit-element of the set. By VI, there is a transformation from Σ changing A_1 to A and B_1 to B. Let this transformation change F_1 to F. Then, as a result of III and IV, F will have A for a limit-element, but not B.

Proof of (2).

Let a neighbourhood U_A consist of all the interior elements of some set E of which A is an interior element. By I, 3′, and VII at least one element has a neighbourhood, and by the use of VI, III, and IV, as above, every element will have at least one neighbourhood. Indeed, it may be shown by I, 3′, and VII that there is at least one set with both interior and exterior elements, so that this same argument may be used to show that any two elements will have two mutually exclusive neighbourhoods, thus proving that Hausdorff's conditions (A) and (D) are satisfied. (C) is an obvious result of the definition of neighbourhood, for a neighbourhood is a neighbourhood of any of its elements. As to (B), the interior elements of a set E that are also interior to a set F are interior to the common part of E and F; this follows from condition 2 that our set be a set (R).

It remains to show that limit in a system (H) corresponds to limit in a system (Sp). It is a result of our definition of neighbourhood that if E is a set having A as a limit-element, every neighbourhood of A contains some element of E other than A. It results from Riesz's condition 2 that every neighbourhood of A contains a set of elements of E having A as a limit-element. From 2 and 3 together it follows that every such set is infinite. Hence every (Sp)-limit is an (H)-limit. The converse relation follows from VII.

Proof of (3).

Let Σ consist of all biunivocal, bicontinuous transformations in our system (Vr). That this will give the same notion of limit as that defined in a system (Sp) I have proved in my previous paper. Postulates I, II, III, IV, and V demand no discussion. VII will be obvious if we consider that a "sphere" with its boundary-elements will answer to our definition of a segment, for it is closed, has at least two boundary-elements, and is connected, for any point is connected with the centre by a radius. Moreover, the centre is an interior point. VII will then follow from our definition of limit.

There remains only condition VI. It is clear that any element A of K can be changed to any other member B of K by a transformation from Σ, for it will follow from II and the various properties of vectors that the transformation which turns C into the element D such that $CD = AB$ belongs to Σ. In a similar way, it may be shown that the transformation which consists in holding an element A fast and "multiplying" all the vectors AB by the same numerical factor also belongs to Σ. We shall establish our theorem, then, if we show that if AB and AC are two vectors such that $\| AB \| = \| AC \|$, there is a transformation belonging to Σ holding A fixed and changing B into C, for every transformation of a point-pair into another may be reduced, as in ordinary geometry, into a "translation," an "expansion," and a "rotation." Our special hypothesis for a (Vr) enables us, moreover, without essentially limiting our problem, to suppose $\| BC \| < \| AB \|$.

Let us consider the vector transformation which turns ξ into

$$\xi \oplus \left\{ \frac{\| \xi \|}{\| AB \|} \odot BC \right\}.$$

This transformation is clearly univocal; it is, moreover, biunivocal. To prove this, let us make use of the fact that it results from our assumptions that if $\xi \oplus \eta = \vartheta$, η is uniquely determined by ϑ and ξ, and may be written $\vartheta \ominus \xi$. Now suppose that

$$\xi \oplus \left\{ \frac{\| \xi \|}{\| AB \|} \odot BC \right\} = \eta \oplus \left\{ \frac{\| \eta \|}{\| AB \|} \odot BC \right\}.$$

It results that
$$\xi \ominus \eta = \frac{\| \xi \| - \| \eta \|}{\| AB \|} \odot BC,$$

or
$$\| \xi \ominus \eta \| = \{ \| \xi \| - \| \eta \| \} \frac{\| BC \|}{\| AC \|}.$$

Now, by our hypothesis, $\| BC \| / \| AC \| < 1$. Hence either

$$\| \xi \ominus \eta \| = 0, \quad \text{or} \quad \| \xi \ominus \eta \| < \| \xi \| - \| \eta \|.$$

If we write this latter proposition in the form

$$\| \xi \ominus \eta \| \dot{+} \| \eta \| < \| (\xi \ominus \eta) \oplus \eta \|,$$

it will be seen to contradict (8) in the definition of a (Ve). Hence

$$\| \xi \ominus \eta \| = 0,$$

or what results from (13), $\xi = \eta$.

Let us consider the point-transformation which retains A invariant and changes every other element P into the element P' such that

$$AP' = AP \oplus \left\{ \frac{\| AP \|}{\| AB \|} \odot BC \right\}.$$

It results from what has been said and the properties of vectors that this is biunivocal; let us consider how it affects the magnitude of vectors. If P is transformed into P' and Q into Q' by our transformation, we wish to determine a relation between PQ and $P'Q'$.

Now, as an immediate consequence of the commutative law and the definition of our transformation,

$$P'Q' = PQ \oplus \left\{ \frac{\| AQ \| - \| AP \|}{\| AB \|} \odot BC \right\}.*$$

As a consequence,

$$\| P'Q' \| \leqslant \| PQ \| + \frac{\| BC \|}{\| AB \|} \mid \| AQ \| - \| AP \| \mid$$

$$\leqslant 2 \| PQ \|.$$

On the other hand, it may readily be proved that

$$\| P'Q' \| \geqslant \left| \| PQ \| - \frac{\| BC \|}{\| AB \|} \left| \| AQ \| - \| AP \| \right| \right|$$

$$\geqslant \| PQ \| \left\{ 1 - \frac{\| BC \|}{\| AB \|} \right\}.$$

It follows from these inequalities that, to put it roughly, $P'Q'$ is small when and only when PQ is small, and that a set of elements approaching indefinitely close to a given element is transformed into a set approaching indefinitely close to the transform of the given element, and *vice versa*. In other words, our transformation leaves limit-properties invariant in both directions, and so belongs to Σ. Moreover, our transformation leaves A invariant and changes B into the element D such that

$$AD = AB \oplus \left\{ \frac{AB}{\| AB \|} \odot BC \right\} = AB \oplus BC = AC,$$

or, in other words, into C. We thus have completed our proof of the equivalence of point-pairs by the consideration of "rotations."

EXAMPLES OF SETS (Vr).

7. (1) The system consists of all n-partite numbers $(x_1, x_2, \ldots, x_n)$. If $A = (x_1, x_2, \ldots, x_n)$ and $B = (y_1, y_2, \ldots, y_n)$, AB shall be the n-partite number $(x_1 - y_1, y_2 - y_2, \ldots, x_n - y_n)$, and every n-partite number shall be a vector. If

$$\xi = (u_1, u_2, \ldots, u_n) \quad \text{and} \quad \eta = (v_1, v_2, \ldots, v_n),$$

$$\| \xi \| = \sqrt{(u_1^2 + u_2^2 + \ldots + u_n^2)}, \quad k \odot \xi = (ku_1, ku_2, \ldots, ku_n),$$

and

$$\xi \oplus \eta = (u_1 + v_1, u_2 + v_2, \ldots, u_n + v_n).$$

---

* $(-n) \odot UV$ is to be understood as $n \odot VU$.

The independence of addition on order is immediately obvious. The other specifically (Vr) property results from the fact that any arc of a circle can be traversed with a finite number of chords each less in length than ϵ, for any given ϵ.

(2) The system of elements and that of vectors alike consist in all ∞-partite numbers $(x_1, x_2, \ldots, x_k, \ldots)$ such that there is a finite X such that for all k, $|x_k| \leqslant X$. If

$$A = (x_1, x_2, \ldots, x_k, \ldots) \quad \text{and} \quad B = (y_1, y_2, \ldots, y_k, \ldots),$$

$$AB = (x_1 - y_1, x_2 - y_2, \ldots, x_k - y_k, \ldots).$$

If $$\xi = (u_1, u_2, \ldots, u_k, \ldots) \quad \text{and} \quad \eta = (v_1, v_2, \ldots, v_k, \ldots),$$

$$\|\xi\| = \text{least upper bound } |u_k|,$$

$$m \odot \xi = (mu_1, mu_2, \ldots, mu_k, \ldots),$$

and $$\xi \oplus \eta = (u_1 + v_1, u_2 + v_2, \ldots, u_k + v_k, \ldots).$$

The commutative law is obvious; the other condition for a (Vr) can be demonstrated if we show that given ξ and η such that $\|\xi\| = \|\eta\| \neq 0$, there is a chain of vectors, $\xi_1 = \xi, \xi_2, \ldots, \xi_n = \eta$, such that for all j,

$$\|\xi_j\| = \|\xi\| \quad \text{and} \quad \|\xi_{j+1} \ominus \xi_j\| < \|\xi\|.$$

Such a chain may be constructed as follows; let ζ be the vector $(z_1, z_2, \ldots, z_k, \ldots)$, such that for all k, z_k is the larger of the two quantities u_k and v_k if they differ, and their common value, if they agree. Then

$$\|\zeta\| = \|\xi\|.$$

Let $$\frac{\|\zeta \ominus \xi\|}{\|\xi\|} = p, \quad \text{and} \quad \frac{\|\zeta \ominus \eta\|}{\|\xi\|} = q.$$

Let r be any integer larger than both p and q. Then the sequence of vectors

$$\xi,\ \xi \oplus \left\{\frac{1}{r} \odot (\zeta \ominus \xi)\right\},\ \ldots,\ \xi \oplus \left\{\frac{h}{r} \odot (\zeta \oplus \xi)\right\},\ \ldots,$$

$$\zeta,\ \zeta \oplus \left\{\frac{1}{r} \odot (\eta \ominus \zeta)\right\},\ \ldots,\ \zeta \oplus \left\{\frac{h}{r} \odot (\eta \ominus \zeta)\right\},\ \ldots,\ \eta,$$

may readily be shown to satisfy the conditions for a chain $\{\xi_j\}$.

(3) The system of all points and the system of all vectors consist alike

in all ∞-partite numbers $(x_1, x_2, \ldots, x_k, \ldots)$ such that the series

$$x_1^2+x_2^2+\ldots+x_n^2+\ldots$$

converges. AB, $m \odot \xi$, and $\xi \oplus \eta$ are defined as in (2). If

$$\xi = (u_1, u_2, \ldots, u_k, \ldots),$$
$$\| \xi \| = \sqrt{(u_1^2+u_2^2+\ldots+u_k^2+\ldots)}.$$

To show that our system is a (Vr), let us introduce a few considerations from the trigonometry of infinitely many dimensions. If

$$\xi = (u_1, u_2, \ldots, u_k, \ldots) \quad \text{and} \quad \eta = (v_1, v_2, \ldots, v_k, \ldots),$$

let us define $< \xi\eta$ as

$$\cos^{-1} \frac{u_1 v_1 + u_2 v_2 + \ldots + u_n v_n + \ldots}{\| \xi \| . \| \eta \|}.$$

The first question to arise is under what circumstances $< \xi\eta$ will exist. It may easily be shown that if Σu_n^2 and Σv_n^2 converge, $\Sigma(u_n+v_n)^2$ and $\Sigma(u_n-v_n)^2$ will converge.* It results that $\Sigma \frac{1}{4} \{(u_n+v_n)^2-(u_n-v_n)^2\}$ will converge, or that $\Sigma u_n v_n$ will converge. Furthermore, it is obvious that to multiply ξ or η by a positive constant will not affect the magnitude or existence of $< \xi\eta$. We may thus assume $\| \xi \| = \| \eta \|$, which gives us

$$< \xi\eta = \cos^{-1} \frac{u_1 v_1 + u_2 v_2 + \ldots + u_n v_n + \ldots}{u_1^2+u_2^2+\ldots+u_n^2}.$$

Now, consider the inequality $\Sigma(u_n-v_n)^2 \geqslant 0$. We may write this

$$\Sigma u_n^2 - 2\Sigma u_n v_n + \Sigma v_n^2 \geqslant 0.$$

Making use of the fact that $\Sigma u_n^2 = \Sigma v_n^2$, this becomes

$$2\Sigma u_n v_n \leqslant 2\Sigma u_n^2.$$

It may be proved in precisely the same manner that

$$-2\Sigma u_n v_n \leqslant 2\Sigma u_n^2.$$

Hence $< \xi\eta$ is the anticosine of a number not greater in absolute value than 1, and consequently exists.

As in ordinary geometry,

$$\| \xi \ominus \eta \|^2 = \| \xi \|^2 + \| \eta \|^2 - 2 \| \xi \| . \| \eta \| \cos < \xi\eta.$$

* Cf. Hausdorff, *loc. cit*, p. 287.

This may be proved by writing the formula out at length, when it will reduce to an identity. All the series involved will be absolutely convergent, so there is no difficulty about changing the order of terms.

Let us suppose, as above, that $\| \xi \| = \| \eta \|$, and let us consider $\cos < \xi \{\xi \oplus \eta\}$. This will be

$$\frac{u_1(u_1+v_1)+u_2(u_2+v_2)+\ldots+u_n(u_n+v_n)+\ldots}{\sqrt{(u_1^2+u_2^2+\ldots+u_n^2+\ldots)}\sqrt{\{(u_1+v_1)^2+(u_2+v_2)^2+\ldots+(u_n+v_n)^2+\ldots\}}}.$$

By our previous remarks this is an essentially positive quantity. We shall moreover get the identity

$$\begin{aligned}
\cos 2 &< \xi \{\xi \oplus \eta\} \\
&= 2\cos^2 < \xi\{\xi \oplus \eta\} - 1 \\
&= \frac{2\{\Sigma(u_n^2+u_n v_n)\}^2 - [\Sigma u_n^2][\Sigma(u_n+v_n)^2]}{[\Sigma u_n^2][\Sigma(u_m+v_m)^2]} \\
&= \frac{2\Sigma u_n^2 \Sigma u_m^2 + 4\Sigma u_n^2 \Sigma u_m v_m + 2\Sigma u_n v_n \Sigma u_m v_m - \Sigma u_n^2 \Sigma u_m^2 - 2\Sigma u_n^2 \Sigma u_m v_m - \Sigma u_n^2 \Sigma v_m^2}{[\Sigma u_n^2][\Sigma(u_n+v_m)^2]} \\
&= \frac{2\Sigma u_n^2 \Sigma u_m v_m + 2\Sigma u_n v_n \Sigma u_m v_m}{[\Sigma u_n^2][\Sigma(u_m+v_m)^2]} \\
&= \frac{[\Sigma u_m v_m]\{\Sigma u_n^2 + 2\Sigma u_n v_n + \Sigma v_n^2\}}{[\Sigma u_n^2][\Sigma(u_m+v_m)^2]} \\
&= \frac{\Sigma u_m v_m}{\Sigma u_m^2} = \cos \xi\eta.
\end{aligned}$$

It results from this that $< \xi (\xi \oplus \eta)$ is the half of $< \xi\eta$ in the first or fourth quadrant.

Now, let ξ and η be any two vectors of equal magnitude, provided only that neither is made up entirely of 0's. Form the vector ξ_3, which shall be a positive multiple of $\xi \oplus \eta$ with the same magnitude as ξ. In a similar manner, interpolate ξ_2 between ξ and ξ_3, and ξ_4 between ξ_3 and η, and let us know ξ and η as ξ_1 and ξ_5, respectively. We have

$$\cos < \xi\xi_3 = \cos < \xi_3\eta = \sqrt{\{\tfrac{1}{2}(1+\cos < \xi\eta)\}} \geqslant 0.$$

Hence

$$\begin{aligned}
\cos < \xi_1\xi_2 = \cos < \xi_2\xi_3 = \cos < \xi_3\xi_4 = \cos < \xi_4\xi_5 \\
= \sqrt{\{\tfrac{1}{2}(1+\cos < \xi\xi_3)\}} \geqslant \tfrac{1}{2}\sqrt{2}.
\end{aligned}$$

z 2

It follows from the law of cosines that

$$\|\xi_h-\xi_{h+1}\| = \sqrt{(\|\xi_h\|^2+\|\xi_{h+1}\|^2-2\|\xi_h\|.\|\xi_{h+1}\|\cos\angle\xi_h\xi_{h+1})}$$
$$\leqslant \|\xi\|\sqrt{(2-\sqrt{2})}$$
$$< \|\xi\|.$$

(4) The system of all elements and the system of all vectors both consist of all continuous functions of a real variable defined over a given closed interval. The vector fg is the function $f(x)-g(x)$. If $\xi=f(x)$ and $\eta=g(x)$, $\|\xi\| = \max|f(x)|$, $k\odot\xi=kf(x)$, and $\xi\oplus\eta=f(x)+g(x)$. The proof that this system is a (Vr) proceeds as in (2).

It may be noted that systems (1), (3), and (4) satisfy VIII.*

CONSISTENCY OF POSTULATES I–IX.

8. The following system satisfies Postulates I–IX: K consists of all the points on a line, and Σ consists of all bicontinuous, biunivocal transformations of the whole line into itself.

DEDUCTIONS FROM POSTULATES I–IX.

9. THEOREM I.—*If A and B are any two distinct members of K, there is a unique closed set (A, B), completely characterized by the facts that it is connected and that A and B are boundary elements of it.*

Proof.

It follows from Postulates I, VI, and VII that there is at least one set with at least two boundary elements. By VI, these can be transformed by a transformation from Σ into A and B, and by III and IV, this transformation will leave every connected set connected. By IX, this set is uniquely determined except as to whether it contains A and B. Adjoin to it its limit-elements, and it will clearly remain connected, while it will contain A and B.

THEOREM II.—*A and B are the only boundary-elements of (A, B).*

Proof.

Let D be any element not in (A, B). Consider the component† E of

* Hausdorff, *loc. cit.*, pp. 288, 289.

† Since we have proved that our system satisfies Hausdorff's axioms, we may take advantage of his proof of the existence of components.

(*A*, *B*) to which *D* belongs. This must have a limit-element *P* in (*A*, *B*), for otherwise the segment (*D*, *A*), which exists, by Theorem I, would not be connected. *P* is then a boundary-element of (*A*, *B*) which is the limit of the connected set *E* in (*A*, *B*).

Now, let *C* be any boundary-element of (*A*, *B*) other than *A* and *B*. If *C* were the limit of a connected set *F* in (*A*, *B*), then *F* would either have *A* for a limit-element, or *B* for a limit-element, or neither *A* nor *B*. In the first two cases it results from IX that *F* must coincide with (*A*, *B*), which is impossible. In the third case, it follows from V that *A* and *B* are boundary-elements of the connected set $(A, B)+F$, which hence must coincide with (*A*, *B*), by IX. This is again impossible. It follows that there is no such set as *F*.

Let *Q* be any boundary-element of (*A*, *B*) other than *C* and *P*. By IX, we may write (*A*, *B*) as (*Q*, *C*) or as (*Q*, *P*). Now, by VI, there is a transformation from Σ leaving *Q* invariant and changing *P* into *C*. By III and IV, this changes (*Q*, *P*) into (*Q*, *C*), and changes every connected set in (*Q*, *P*) having *P* as a limit-element into a connected set in (*Q*, *C*) having *C* as a limit-element. As the existence of sets of the latter sort has been disproved, while the existence of sets of the former sort has been proved, it follows that either our assumption of the existence of *C* or our assumption of the existence of *P* is inadmissible. If either assumption is incorrect, (*A*, *B*) has only two boundary-elements, which must be *A* and *B*.

THEOREM III.—*If* (*A*, *B*) *and* (*A*, *C*) *have an element in common other than A*, *either* (*A*, *B*) *contains* (*A*, *C*) *or vice versa.*

Proof.

Let *E* consist of all elements in (*A*, *B*) but not in (*A*, *C*), and let *F* be the component of *E* containing *B*. As (*A*, *C*) is connected, *F* has some limit-element *D* in (*A*, *C*). If *A* is the only limit-element of *F* in (*A*, *C*), $A+F$ is a connected set containing the boundary-elements *A* and *B*, and hence coincides with (*A*, *B*), which hence, contrary to assumptions, contains no other term than *A* in common with (*A*, *C*). By Theorem II, the only other value which *D* can have is *C*. Now, consider the set $F+(A, C)$. It is connected, and, by V, has *A* as a boundary element. By V, either *B* is a boundary-element or *B* belongs to (*A*, *C*). If *B* belongs to (*A*, *C*), then every element of (*A*, *B*) does likewise, for otherwise, as (*A*, *C*) has only two boundary-elements, *E* can have only *A* and *C* as limit-elements in (*A*, *C*). If *B* differs from *C*, this is clearly impossible, while if *B* coincides with *C*, $(A, B) = (A, C)$.

The only other possibility is that E contains no elements. In this case, (A, C) is contained in (A, B).

THEOREM IV.—*If C is interior to (A, B), $(A, B) = (A, C) + (C, B)$, and (A, C) shares with (C, B) no other element than C.*

Proof.

By Theorem III, (A, B) contains (A, C) and (C, B). If B belonged to (A, C), by Theorem III, (A, C) would contain, and hence coincide with (A, B). This contradicts our assumption. Hence, by Postulate V, B is a boundary-element of $(A, C) + (C, B)$. The same argument applies to A. Moreover, being the sum of two overlapping, closed, connected sets, by V, $(A, C) + (C, B)$ is closed and connected. Hence, by Theorem II,

$$(A, C) + (C, B) = (A, B).$$

If (A, C) and (C, B) had in common any other element than C, then, by Theorem III, either (A, C) would contain (C, B), or *vice versa*. In this case, either (A, C) or (C, B) would contain (A, B). Hence, by Theorem II, C would coincide with either A or B, and would not be an interior element of (A, B).

Definition.—If C is interior to (A, B), we shall write ACB. It is obvious that if ABC, A, B, and C are all different, and it is also obvious that ABC and CBA are equivalent. Furthermore, by Theorem III, ABC and ACB are incompatible.

THEOREM V.—*If ABC and ACD, then BCD.*

Proof.—By Theorem IV, ABD. Hence, by Theorem IV, either ACB or BCD. ACB, however is incompatible with ABC, by Theorem III.

THEOREM VI.—*ABC, ABD, and CBD are incompatible.*

Proof.—By Theorem IV, either ACB, $B = C$, or BCD. As Theorem III excludes the first two suppositions, which are incompatible with ABC, there remains only the last possibility, which, by III, is incompatible with BCD.

THEOREM VI.—*Either ABC, BAC, or ACB, if A, B and C are distinct.*

Proof.—Suppose the first two alternatives are not fulfilled. Then, by Theorem III, (A, C) and (B, C) have only C in common, $(A, C) + (B, C)$ is connected, and by Postulate V, has A and B as boundary-elements. Hence $(A, C) + (C, B) = (A, B)$, or, in other words, ACB.

THEOREM VII.—*If* ABC *and* BCD, *then* ACD.

Proof.—By Theorem VI, we have DAC, ADC, or ACD. If DAC and ABC, then by Theorem IV, DBC, which, by Theorem III, contradicts BCD. If ADC, then, by Theorem IV, ABD or DBC. DBC, by Theorem III, contradicts BCD. If ABD and BCD, then, by Theorem IV, ACD.

Definition.—$AB \mid CD$ shall mean any one of the following sets of relations:

(1) ACD, ABD.

(2) ACD, $B = D$.

(3) ACD, ADB.

(4) $A = C$, ABD.

(5) $A = C$, $B = D$, $A \neq B$.

(6) $A = C$, ADB.

(7) CAD, CAB.

(8) $A = D$, CAB.

(9) CDA, CAB.

THEOREM VIII.—*If* $AB \mid CD$ *and* $BP \mid CD$, *then* $AP \mid CD$.

Proof.—This involves merely the tabulation of the 81 possible cases and the application of Theorems III–VII in the instances in which they are appropriate.

THEOREM IX.—*If* $AB \mid CD$, $A \neq B$ *and* $C \neq D$.

Proof.—This follows from the fact that if ABC, $A \neq B \neq C$, and the definition of $AB \mid CD$.

THEOREM X.—*If* $A \neq B$, $C \neq D$, *then either* $AB \mid CD$ *or* $BA \mid CD$.

Proof.—This follows from Theorems VI, IV, V, and VII, as may be shown by tabulating the relations between A, B, C, and D, which are possible on the basis of Theorem VI.

THEOREM XI.—*If* $AB \mid CD$ *and* APB, *then* $AP\ CD$ *and* $PB \mid CD$.

Proof.—As above, by tabulating the possible cases, and making use of Theorems IV–VII.

THEOREM XII.—*If* $AP \mid CD$ *and* $PB \mid CD$, *then* APB.

Proof.—As above, by tabulation.

THEOREM XIII.—*If M and N are two classes of elements exhausting K, and such that there are two fixed elements C and D such that if A belongs to M and B belongs to N, $AB|CD$, then there is an element P such that if Q belongs to M and R belongs to N and $Q \neq P \neq R$, QPR.*

Proof.

Suppose that X and Y belong to M, and that XZY. Either $XY|CD$ or $YX|CD$, by Theorem X. Similarly, either $XZ|CD$ or $ZX|CD$, and either $YZ|CD$ or $ZY|CD$. Making use of Theorems XII and VI, it turns out that the only admissible combinations of hypotheses are $XZ|CD$, $ZY|CD$, $XY|CD$ and $YZ|CD$, $ZX|CD$, $YX|CD$. Since we have $XB|CD$ and $YB|CD$ for all B in N, we have, by Theorem VIII, $ZB|CD$ in both cases. It follows then from Theorems VIII and IX that Z does not belong to N, so that it must belong to M. In other words, if M contains X and Y, it contains every element in (X, Y), so that M is connected. Likewise, N is connected.

It follows from Postulate IX and Theorem I that there is just one element P which is a limit-element of M belonging to N or a limit-element of N belonging to M. Let Q belong to M and R to N. As (Q, R) is connected, it must contain P.

THEOREM XIV.—*There is a denumerable set K' of elements such that if A and B are any two elements, there is an element C from K' such that ACB.*

Proof.

Let K' be the set to which reference is made in Postulate VIII. Then every element is a limit-element of K'. It follows from the fact that a single element has no limit-element and Postulate V that a segment has interior elements. Hence every segment contains at least one element of K'.

THEOREM XV.—*There is no element A such that for all $B \neq A$, $AB|CD$, and there is no element A such that for all $B \neq A$, $BA|CD$.*

Proof.—This follows directly from Postulate VI.

THEOREM XVI.—*K can be put into (1, 1)-correspondence with the set of all real numbers, in such a way that two elements C and D can be selected such that $AB|CD$ when and only when the correspondent of A is larger than the correspondent of B.*

Proof.—By Theorems VIII, IX, and X, order as defined by $AB \mid CD$ is serial. By Theorems XI, XII, and XIII, it is what Russell calls "Dedekindian." By XI, XII, and XIV, it contains a denumerable "median class." Hence, by a well known theorem,* it is ordinally similar to the series of reals.

THEOREM XVII.—*In the correspondence of Theorem XVI, Σ goes over into the set of all bicontinuous biunivocal transformations of the series of reals.*

Proof.—In the transformation of Theorem XVI, a segment goes over into a segment (Theorems XI, XII). Now, by Postulate VII, and Theorem I, the limit of a set E consists of all those elements A such that every segment (C, D) of which A is an element other than C and D contains a member of E. Hence limit goes over into limit, and in virtue of Postulates II, III, and IV, a transformation from Σ, which is precisely a transformation keeping limit-properties invariant, goes over into a bicontinuous, biunivocal transformation of the number-line, and every bicontinuous, biunivocal transformation of the number-line may be thus obtained.

Theorem XVII is equivalent to the statement that our set of postulates is a categorical set of postulates for the analysis-situs group of the line.

CONSIDERATIONS OF INDEPENDENCE.

10. Up to the present, the author has been unable to solve the question of the independence of Postulates IV, V, VII, and VIII. Each of the other postulates is independent of all the rest. The examples given below satisfy all the postulates except the one whose number they are given.

I. K consists of one element; Σ contains only the identity transformation.

II. K consists of all points on a line; Σ consists of all biunivocal, bicontinuous transformations that preserve direction.

III. K consists of all points on a line; Σ consists of all biunivocal, bicontinuous transformations, together with the transformations that displace all points with rational coordinates a rational distance in one direction, and all points with irrational coordinates a rational distance in the other.

* Whitehead and Russell, *Principia Mathematica*, Vol. 3, ✱ 275.

VI. K consists of all points on two mutually exclusive lines; Σ consists of all biunivocal, bicontinuous transformations of K.

IX. K consists of all points on a circle; Σ consists of all biunivocal, bicontinuous transformations of K.

It may be said that the independence of VIII would be proved if we could produce a closed homogeneous* series with a number of terms greater than $2^{\aleph_0}$. Homogeneous series with more than $2^{\aleph_0}$ terms are known, but they are not closed.

---

* Hausdorff, *loc. cit.*, p. 173.

MASSACHUSETTS INSTITUTE OF TECHNOLOGY.

LIMIT IN TERMS OF CONTINUOUS TRANSFORMATION;

By Norbert Wiener.

1. The *calcul fonctionnel*, or the study of the limit properties of an abstract assemblage, has been investigated in the course of the last fifteen years from a number of distinct standpoints. In addition to the notion of sequential limit (¹) which furnished the starting-point of Fréchet's well known thesis, and the more restrictive concept of *écart* (or *distance*, as Fréchet, now calls it), there is Riesz' non-sequential limit (²), which Fréchet, in turn, has discussed as a special case of an extremely general notion of neighborhood (³). If, however, we consider the *calcul fonctionnel* with reference to the obviously intimate bearing which it has on analysis situs — which has been defined essentially as the study of the invariants of the group of all bicontinuous biunivocal transformations (⁴) — there is another avenue of approach to which our attention is immediately directed. This paper will be devoted to the discussion of the derivation of limit-properties from those of continuous transformations.

This work has been carried out in France with the aid of much advice and many important suggestions from Professor Fréchet, to whom I wish to express my sincerest thanks.

2. Let us start with a class Σ of biunivocal transformations of all the elements of a class C. On the hypothesis that these are to be considered as bicontinuous, how should we naturally proceed

(¹) *Sur quelques points du Calcul fonctionnel* (*Rend. Cir. Mat. Palermo*, vol. XXII, p. 4). This article will in the future be referred to as *Thesis*.

(²) In a Communication read before the International Congress of Mathematicians of Rome.

(³) *Sur la notion de voisinage dans les ensembles abstraits* (*Bulletin des Sciences mathématiques*, 2ᵈ séries, vol. XLII). Hereinafter to be referred to as *V*.

(⁴) A definition to this effect is to be found in the article on *Analysis situs* by Dehn and Heegard, in the *Encyklopädie der Mathematischen Wissenschaften*.

to define the limit-elements of E, a sub-class of C? It may be shown that in an *n*-space if the operations of Σ are bicontinuous in any ordinary sense, if A is a limit-element of E, every transformation of Σ which leaves invariant every element of E except A will leave A invariant also. It may furthermore be shown that if A is not a limit-element of E, there is some little region containing A but no point of E which we may permute by some transformation of Σ in such a way as to change A but leave each point of E invariant. We shall follow out the obvious analogy and make the following definition :

An element A *of* C *will be said to be a limit-element of a sub-set* E *of* C, *when and only when every transformation of* Σ *that leaves invariant all the elements of* E, *except possibly* A, *also leaves* A *invariant.* That this definition is natural over a wide set of cases will appear in what follows.

3. A system in which limit-element is défined in this manner will be called a system (*s*). It becomes a matter of interest to discover when limit-properties, as defined in a system (*s*), are really invariant under all the transformations of Σ, for only then will the transformations of Σ be in any true sense continuous. Now, let us transform C by the transformation T of Σ, and let us represent the transform of A by T(A), the set of transforms of elements of E (which does not contain A) by T((E)). If limit-properties are to be left invariant, if S is any transformation of Σ, then if, whenever B belongs to E, S(T(B)) = T(B), we shall have S(T(A)) = T(A). In other words, if we write the transformation which changes A into R(S(A)), R | S, and the transformation which changes R(A) into A, $\breve{R}$, it will follow that *every transformation of the form* $\breve{T}|S|T$ *which leaves each term of* E *invariant will also have to leave* A *invariant*, if S and T belong to Σ. If the operations of Σ are to be bicontinuous, the same statement applies to transformations of the form $T|S|\breve{T}$ These two conditions together are necessary and *sufficient* that Σ consist of bicontinuous operations. We shall speak of any system (Σ) satisfying these conditions as a system (J).

4. It will be observed that if Σ is a group, an (*s*) is a (J). It will

also be observed that in a (J) there is always a group of bicontinuous, biunivocal operations generated by the members of Σ. It does not follow, however, that these cases are identical as to their limit-properties, for though a bicontinous, biunivocal transformation which keeps invariant every element of E will also transform every limit-element of E into a limit-element of E, it does not necessarily result that it will keep every limit-element of E invariant.

The case when Σ consists precisely of the group of *all* bicontinuous, biunivocal transformations is especially interesting; we shall denominate it (J_1). If we call a set E *closed* if it contains all its limit-elements — if, that is, it contains all those elements that are left invariant by every transformation which belongs to Σ and leaves every element of E invariant — then it is easy to see that a bicontinuous transformation is precisely one which leaves all closed sets closed. We thus get as the necessary and sufficient condition that Σ should *contain* all bicontinuous biunivocal transformations.

A. *If* R *is a biunivocal transformation of* C *such that* R *and* $\breve{R}$ *both leave all closed sets closed, it belongs to* Σ.

If Σ *consists* of all bicontinuous, biunivocal transformations, it must also satisfy the group-conditions.

B. *If* R *belongs to* Σ, *so does* $\breve{R}$, and.

C. *If* R *and* S *belong to* Σ, *so does* R | S.

A, B, and C are moreover *sufficient* to show that Σ consists in all bicontinuous, biunivocal transformations, for it results from B and C that if S and T belong to Σ, so do $S|T|\breve{S}$ and $\breve{S}|T|S$, so that S and $\breve{S}$ simply permute the operations of Σ and change no limit — properties. Needless to say, every (J_1) is a (J).

4. Not every system in which limit-properties are defined, nor even every system in which sequential limit is defined, nor finally every system in which distance is defined is a (J). Consider the set of points on a number-line consisting of $x = 0$ and $x = \frac{1}{n}$ for all integral values of n. Manifestly, every transformation leaving limit-properties invariant will leave $x = 0$ invariant. Hence any biunivocal, bicontinuous transformation leaving all members of any set E invariant leaves $x = 0$ invariant, and $x = 0$ is a limit-element of any set. However, this does not agree with our original notion of limit.

The next problem is therefore to determine under what circumstances a system in which limit-element is defined belongs to one of the classes (J) or (J_1). To begin with we shall search for the necessary condition that a (V) [1] or system in which neighborhood is defined, be a (J). We may make use of the fact that our notion of limit necessarily satisfies the two conditions :

1° Every limit-element of a set E is also a limit-element of every set containing E;

2° The fact that an element A is or *is* not a limit-element of E is not affected by adjoining A to E.

Fréchet has shown that under these conditions, if we define a neighborhood of A as a set V_A of elements such that the set of all elements not in V_A does not have A as a limit-element, then the necessary and sufficient condition for a set E to have A as a limit-element is for E to have elements in every V_A. In terms of Σ, a neighborhood V_A of A will be a set of elements such that there is at least one transformation of Σ leaving invariant every element of C not in V_A but changing A. Since in a (J) every member of Σ is bicontinuous, a *necessary* condition that our system be a (J) is that given any element A and any neighborhood of A, there is at least one biunivocal transformation R of the whole of C such that :

(1) If B does not belong to V_A, $R(B) = B$;

(2) $R(A) \neq A$;

(3) If D is a limit-element of E, R(D) is a limit-element of R((E));

(4) If R(D) is a limit-element of R((E)), D is a limit-element of E.

If to these conditions be adjoined the condition that, if R be a biunivocal transformation of the whole of C satisfying the conditions (3) and (4) just mentioned, then if E is any set of elements each of which remains unchanged under R [2] and A is a limit-

(1) *V.*, p. 4.
(2) *Thesis*, Also. *V*, p. 1.

element of E, $R(A) = A$, we get a *necessary* condition that our system be a (J_1). This is, moreover, *sufficient*, for clearly the new system Σ' consisting of all biunivocal transformations of C satisfying (3) and (4) will be a (J_1). It only remains to prove that it leads to our original notion of limit. Obviously any set having A as a limit in our new sense will have terms in each V_A, and will consequently have A as a limit in our original sense. Moreover, every set E having A as a limit in our original sense will have at least one term in each V_A, for otherwise there would be at least one biunivocal transformation satisfying (3) and (4) leaving each term of E invariant but changing A, which would be contrary to our new assumption.

5. The assumption that every biunivocal bicontinuous transformation of the whole of C, if it leaves every element of a set invariant, leaves every element of the derivative invariant, will clearly be satisfied if whenever A is a limit-element of E and B is any element, A is a limit-element of a sub-set of E which does not have B for a limit-element. Let it be observed that this is a sufficient condition for a (J), to be a (J_1). This property of limit will always occur when our system is what Fréchet calls an (L), in which the derivative of a set may be defined in terms of sequential limit. It is an interesting matter to find a whole wide set of cases which all fulfill both this and the other part of the necessary and sufficient condition for a (J_1).

We shall say that a set σ of entities is a *vector family* if there are associated with it operations $\oplus$, $\odot$, and $\| \ \|$ satisfying the following conditions :

(1) If ξ and η belong to σ, $\xi \oplus \eta$ belongs to σ,
(2) If ξ belongs to σ and n is a real number $\geqq 0$, $n \odot \xi$ belongs to σ,
(3) If ξ belongs to σ, $\|\xi\|$ is a real number $\geqq 0$,
(4) $n \odot (\xi \oplus \eta) = (n \odot \xi) \oplus (n \odot \eta)$,
(5) $(m \odot \xi) \oplus (n \odot \xi) = (m + n) \odot \xi$,
(6) $\|m \odot \xi\| = m . \|\xi\|$,
(7) $\|\xi \oplus \eta\| \leqq \|\xi\| + \|\eta\|$,
(8) $m \odot (n \odot \xi) = mn \odot \xi$.

We shall say that a set E of elements is a system (Ve) if there is a vector-family σ such that.

I. If A and B belong to, E, there is associated with them a single member AB of σ;

II. $\|AB\| = \|BA\|$;

III. Given an element A of E and an element ξ of σ, there is an element B of E such hat $AB = \xi$;

IV. $AC = AB \oplus BC$;

V. $\|AB\| = 0$ when and only when $A = B$;

VI. If $AB = CD$, $BA = DC$.

It will be seen that $\|AB\|$ is an écart in Fréchet's original sense (1), and that we can say that a sub-class F of E has the limit-element A when, given any positive number ε, there is always a member B of F other than A, such that $\|AB\| < \varepsilon$.

Let us suppose given some set F consisting of all the elements B such that $\|AB\| < \varepsilon$, where ε is some positive quantity. Clearly every neighborhood of A will contain the whole of some set of the sort. Let us consider the following transformation : if $\|AB\| \geq \varepsilon$, let B be unchanged, but if $\|AB\| < \varepsilon$, let B be changed into the element C such that $AC = \frac{\|AB\|}{\varepsilon} \odot AB$. The existence and biunivocality of this transformation will be guaranteed by our assumptions. It will also be bicontinuous, as the following argument will prove.

Let B and C be any two elements in F, and let their transforms by our transformation be B′ and C′. Let D be the element such that $AD = \frac{\|AB'\|}{\|AB\|} \odot AC$. I wish to find an upper bound for $\|B'C'\|$ in terms of $\|BC\|$. We have

$$\begin{aligned}
\|B'C'\| &\leq \|B'D\| + \|DC'\| \\
&\leq \|B'A \oplus AD\| + \|DC'\| \\
&\leq \left\| \frac{\|AB'\|}{\|AB\|} \odot (BA \oplus AC) \right\| + \left\| \left\{ \frac{\|AB'\|}{\|AB\|} \odot CA \right\} \oplus \left\{ \frac{\|AC'\|}{\|AC\|} \odot AC \right\} \right\| \\
&\leq \frac{\|AB\|}{\varepsilon} \|BC\| + \left| \frac{\|AB\| - \|AC\|}{\varepsilon} \right| \|AC\| \\
&\leq \|BC\| \left\{ \frac{\|AB\|}{\varepsilon} + \frac{\|AC\|}{\varepsilon} \right\} \\
&\leq 2\|BC\|
\end{aligned}$$

(1) *Thesis*, p. 30.

Evidently, for a given B, if we have a set G such that, for every positive η, there is a member C of G such that $\|BC\| < \eta$, then the same statement will hold true of B′ and the set G′ made up of transforms of the elements of G.

Let us now proceed to find an upper bound for $\|BC\|$ in terms of $\|B'C'\|$. We have, proceeding as before, supposing that

$$AD' = \frac{\|AB\|}{\|AB'\|} \odot AC'$$

$$\begin{aligned}
\|BC\| &\leqq \|BA \oplus AD'\| + \|D'C\| \\
&\leqq \left\| \frac{\|AB\|}{\|AB'\|} \odot (B'A \oplus AC') \right\| + \left\| \left\{ \frac{\|AB\|}{\|AB'\|} \odot C'A \right\} \oplus \left\{ \frac{\|AC\|}{\|AC'\|} \odot AC' \right\} \right\| \\
&\leqq \sqrt{\frac{\varepsilon}{\|AB'\|}} \|B'C'\| + \left| \sqrt{\frac{\varepsilon}{\|AB'\|}} - \sqrt{\frac{\varepsilon}{\|AC'\|}} \right| \|AC'\| \\
&\leqq \|B'C'\| \left\{ \sqrt{\frac{\varepsilon}{\|AB'\|}} + \frac{\varepsilon}{\|AB'\| \left(\sqrt{\frac{\varepsilon}{\|AB'\|}} + \sqrt{\frac{\varepsilon}{\|AC'\|}} \right)} \right\} \\
&\leqq \|B'C'\| \left\{ \sqrt{\frac{\varepsilon}{\|AB'\|}} + \frac{\varepsilon}{2\|AB'\|} \right\}.
\end{aligned}$$

Unless B′ is A, this shows that a set of transforms of members of G can have B′ for a limit-element only when G has B for a limit-element. In the special case when B and A coincide, it is easily seen that $\|AC\|$ is small when and only when $\|AC'\|$ is small. Our transformation is thus bicontinuous and biunivocal when we consider elements B such that $\|AB\| < \varepsilon$. There is no difficulty in showing that the bicontinuity is valid over the whole system, for if $\|AB\| = \varepsilon$, then $AB = \frac{\|AB\|}{\varepsilon} \odot AB$, so that our transformations within and without the (sphere) $\|AB\| = \varepsilon$ make a precise join on the sphere. No element in the sphere except A if left invariant.

Now, let B be any element of a system (Ve). I say that given any neighborhood V_B of B, it will be possible to find a transformation of the sort just discussed which will change B but leave invariant every point not in V_B. Clearly, there will be some positive number η such that V_B will contain all the elements C such that $\|BC\| < \eta$. Let A be some element other than B — there always will be some such element — such that $\|AB\| < \frac{\eta}{2}$. Let ε be some number such that $\|AB\| < \varepsilon < \frac{\eta}{2}$. Then all the elements whose distance from A is less than ε will lie in V_B, while B will be one of these elements. Establish a transformation such as was discussed in the last paragraph. This will change B and leave invariant every element not in V_B, and will be biunivocal and bicontinuous. Hence the first part of the condition that our system be a (J_1) is satisfied. Since every (Ve) is an (L), the second part is also satisfied, and every (Ve) is a (J_1).

6. Examples of systems (Ve) are the following.

(1) The system consists of all n-partite numbers, $(x_1, x_2, \ldots, x_n)$, σ likewise consists of all n-partite numbers. If $A = (x_1, x_2, \ldots, x_n)$, and $B = (y_1, y_2, \ldots, y_n)$, $AB = (x_1 - y_1, x_2 - y_2, \ldots, x_n - y_n)$. If $\xi = (u_1, u_2, \ldots, u_n)$, $\|\xi\| = \sqrt{u_1^2 + u_2^2 + \ldots + u_n^2}$, and

$$k \odot \xi = (ku_1, ku_2, \ldots, ku_n).$$

If, moreover,

$$\eta = (v_1, v_2, \ldots, v_n),$$

$$\xi \oplus \eta = (u_1 + v_1, u_2 + v_2, \ldots, u_n + v_n).$$

(2) The system consists of all ∞-partite numbers,

$$(x_1, x_2, \ldots, x_n, \ldots),$$

such that there is a finite X such that whatever n, $|x_n| \leqq X$. σ likewise consists of all such numbers. If $A = (x_1, x_2, \ldots, x_n, \ldots)$ and $B = (y_1, y_2, \ldots, y_n, \ldots)$, $AB = (x_1 - y_1, x_2 - y_2, \ldots, x_n - y_n, \ldots)$. If $\xi = (u_1, u_2, \ldots, u_n, \ldots)$ and $\eta = (v_1, v_2, \ldots, v_n, \ldots)$,

$$\|\xi\| = \text{upper bound}_n |u_n|,$$

$k \odot \xi = (ku_1, ku_2, \ldots, ku_n, \ldots)$, and

$$\xi \oplus \eta = (u_1 + v_1, u_2 + v_2, \ldots, u_n + v_n, \ldots).$$

(3) The system consists of all ∞-partite numbers

$$(x_1, x_2, \ldots, x_n, \ldots),$$

such that the series $x_1^2 + x_2^2 + \ldots + x_n^2 + \ldots$ converges. σ likewise consists of all such numbers AB, $k \odot \xi$, and $\xi \oplus \eta$ are defined as in (2). If

$$\xi = (u_1, u_2, \ldots, u_n, \ldots), \qquad \|\xi\| = \sqrt{u_1^2 + u_2^2 + \ldots + u_n^2 + \ldots}.$$

(4) The system consists of all continuous functions of a real variable defined over a given closed interval. σ likewise consists of all such functions. The vector fg is the function $f(x)-g(x)$. If $\xi=f(x)$ and $\eta=g(x)$, $\|\xi\|=\max|f(x)|$, $k\odot\xi=kf(x)$, and $\xi\oplus\eta=f(x)+g(x)$.

7. In addition to these systems which are systems (Ve) « im grossen », there are systems which may be said to be systems (Ve) « im kleinen » or as we shall say, systems (Ve'). We shall characterize them as follows : every point A has at least one neighborhood V_A which can be put into biunivocal bicontinuous correspondence with a set $V_{A'}$ consisting of all the elements B' in a (Ve) such that $\|A'B'\|\leqq\varepsilon$ (1), in such a manner that A will correspond to A' and the set of all points in V_A that are limit-points of the set of all elements not in V_A shall correspond to the set of elements consisting of all elements B' such that $\|A'B'\|=\varepsilon$. It is clear that our argument by which we proved that every (Ve) was a (J_1) will also prove that every (Ve') is a (J_1). The points on a sphere or on a torus are examples of sets (Ve').

8. Another example of a set (J_1) is Fréchet's E_ω (2). This consists of all ∞-partite numbers $(x_1, x_2, \ldots, x_n, \ldots)$, with limit defined non-uniformly. As Fréchet has shown, in this space limit may be defined in terms of distance, the distance between two elements $(x_1, x_2, \ldots, x_n, \ldots)$ and $(x'_1, x'_2, \ldots, x'_n, \ldots)$ being defined to be $\sum_{n=1}^{\infty}\frac{1}{n!}\frac{|x_n-x'_n|}{1+|x_n-x'_n|}$. Clearly, then, any neighborhood V_A of $(x_1, x_2, \ldots, x_n, \ldots)$ will contain for some $\varepsilon>0$ all of the points $(x'_1, x'_2, \ldots, x'_n, \ldots)$ such that $\sum_{n=1}^{\infty}\frac{1}{n!}\frac{1+|x_n-x'_n|}{|x_n-x'_n|}<\varepsilon$. Let us choose k in such a manner that $\sum_{n=k+1}^{\infty}\frac{1}{n!}>\frac{\varepsilon}{2}$. If, then, I choose a certain set of positive quantities $\eta_1, \eta_2, \ldots, \eta_k$ in such a manner

(1) Here $\varepsilon>0$.
(2) *Thesis*, p. 39.

that $\sum_{n=1}^{k} \frac{1}{n!} \frac{\eta_n}{1+\eta_n} < \frac{\varepsilon}{2}$, a thing which is manifestly always possible, it will follow that V_A will contain every element $(x'_1, x'_2, \ldots, x'_n, \ldots)$, satisfying the finite set of conditions

$$x_1 - \eta_1 \leqq x'_1 \leqq x_1 + \eta_1, \qquad x_2 - \eta_2 \leqq x'_2 \leqq x_2 + \eta_2, \qquad x_k - \eta_k \leqq x'_k \leqq x_k + \eta_k.$$

Hence, if whenever V'_A is a neighborhood determined by a finite set of intervals in the above manner, there is a biunivocal, bicontinuous transformation changing $(x_1, x_2, \ldots, x_n, \ldots)$ but leaving invariant any element not in V'_A, our E_ω is a (J_1).

Now, we know there is a bicontinuous biunivocal transformation in the k-space made up of all elements $(x'_1, x'_2, \ldots, x'_k)$ which changes $(x_1, x_2, \ldots, x_k)$ but leaves invariant every element $(x'_1, x'_2, \ldots, x'_k)$ not satisfying simultaneously the conditions $x_1 - \eta_1 \leqq x'_1 \leqq x_1 + \eta_1, \ldots, x_k - \eta_k \leqq x'_k \leqq x_k + \eta_k$. Let us consider the transformation which affects the firts k coordinates of a point in E_ω in the above manner, but leaves all the other coordinates invariant. It is easy to see that this is bicontinuous and biunivocal, that it changes $(x_1, x_2, \ldots, x_n, \ldots)$, and that it leaves invariant every point not in V'_A. Hence E_ω is a (J_1).

9. Up to this point we have been considering the conditions that a (V) be a (J) or (J_1). Let us now reverse our point of view, and ask, given a (J) or (J_1), what are the conditions that it belong in one or another of the categories of Fréchet and Riesz. We have already seen (§ 1) that every (J) is a (V); the next most restricted classes so far discussed are those that also satisfy certain of the following conditions, numbered by Fréchet.

2. Given any two sets, E and F, $(E+F)'$ is contained in $E'+F'$.

3. A set consisting of a single element has no limit-element.

4. If A is a limit element of a set E, and if B is distinct from A, there is always at least one set F which has A for a limit-element without having B for a limit-element.

5. Given any set E, $(E)'$ is contained in E'.

In these condition, E' means the set of all limit-elements of E. Conditions 2, 3, and 4 make a set, what Fréchet calls after F. Riesz, who first investigated such sets, a set (R). Conditions 2, 3, and 5 have been found by Fréchet to form a combination even more important, for in every class (V) satisfying them the necessary, and ufficient condition for a set E to possess Borel's property is that every infinite sub-set of E should have at least one limit-element belonging to E (1).

Condition 2 may be written in the form : if A is a term which is neither a limit-element of E nor a limit-element of F, then it is not a limit-element of E + F. Stating this in terms of transformations, we get.

2′ *If there is a transformation from* Σ *changing the element* A, *but leaving all the elements of the class* E *invariant, and there is also a transformation from* Σ *changing* A *but leaving all the elements of the class* F *invariant, then there is a transformation from* Σ *changing* A, *but leaving all the elements of* E + F *invariant.*

In terms of transformation, 3 simply becomes.

3′ *Given any two elements,* A *and* B, *there is a transformation from* Σ *changing* A *but leaving* B *invariant.*

Condition 4 is rather awkward to translate; translating it literally, we get

4′ *If there is a set* E *not containing the element* A, *but such that every transformation from* Σ *that leaves all the elements of* E *invariant leaves* A *also invariant, then given any element* B, *distinct from* A, *there is a set* F *such that there is a transformation from* Σ *changing* B *but leaving each element of* F *invariant, while there is no transformation from* Σ *changing* A *but leaving each member of* F *invariant.*

10. It is interesting to remark that in sets (T) condition 5 is a consequence of 2′ and 3′. It results from 2′ and 3′ that the set of limit-elements of a class E is not affected by removing an element from E. Let A be any term in (E′)′. As an imn diate consequence of the definition of A, it is invariant under every transformation that leaver every element of E″ invariant. If A belongs to E′, our theorem needs no proof. If A does not belong to E′, on the other hand, E′, wich consits of all those elements B wich remain

(1) *V.*, p. 2, 3, 7.

invariant under all those transformations of Σ which leave each term of E — B invariant, whill consist of all those elements B which remain invariant under those transformations wich leave each term of E-A-B invariant. Hence every transformation that leaves each term of E — A invariant will leave each term of E′ invariant, and will hence leave A invariant. Hence again, A belongs to E′.

11. The next thing to investigate is the relation between systems (J) or (J_1) and systems in which sequence can be defined in terms of sequential limit-the systems (L) of Fréchet. We have already seen (§ 5) that a (J) which is an (L) may be considered as a (J_1) without change of limit-properties. There are, however, systems (J_1) that are not systems (L). Frechet [1] has given an example of a class (R) in wich no limit-point is the unique limit of any set. In this, the universe of discourse consists of all points on a line, while the set of limit-points of a classs is the set of its points of condensation. We shall show that is a (J_1).

We shall show (1°) that every biuniform transformation in this system, that retains limit-properties invariant, when it leaves every member of a class invariant, leaves every limit-element invariant, and (2°) that if an element A is not a limit-element of a class E, there is some transformation that is biuniform, that leaves all limit-properties invariant, that changes A, and that leaves invariant every element of E.

(1) Suppose E is a class of elements, and let A be a limit-element of E. Let A′ be any term distinct from A. Now, let F be some interval not containing A′ either in its interior or as an end-point but containing A in its interior and let G be the common part of E and F. Clearly A will be a limit-point of G, while A′ will not Therefore, by § 5, any transformation, which is bicontinuous and keeps every member of G invariant, cannot interchange A and A′. We are thus led into a contradiction unless we suppose that every bicontinuous biunivocal transformation wich leaves invariant every member of a class E also leaves invariant every limit-point.

---

(1) *V.* p. 9.

(2) If an element A is not a limit-element of a class E, there is some interval F containing A in its interior and containing only a denumerable set of elements of E. Let us adjoin to this set all the rational points and end-points of F, always excluding A, however we thus get a dense denumerable series, forming a median class of F. It can hence be put into one-one correspondence with the set of rational numbers between, 0 and 1, inclusive, by a biunivocal transformation T which will determine a biunivocal, bicontinuous transformation of F into the whole interval from 0 to 1. This is a well-known theorem of Cantor. Now, let S be the following transformation of the interval $0 \geqq x \geqq 1$:

If $0 < x < \frac{1}{2}$ and x is irrational

$$x' = \frac{x}{2}.$$

If $\frac{1}{2} < x < 1$ and x is irrational

$$x' = \frac{(3x - 1)}{2}.$$

If x is rational

$$x' = x.$$

Let R be the transformation of our line which leaves invariant all the elements not in F, and in F is equivalent to $\breve{T}|S|T$. There is no difficulty in seeing that R is biunivocal and bicontinuous, that it leaves invariant every member of E except possibly A, and that it changes A. Hence, A is a limit of a set E when and only when every biunivocal, bicontinuous transformation which leaves invariant every element of E leaves A invariant. Our system is thus a (J_1).

12. An (R) that is a (J) may fail to be an (L) even though every limit-point of a set E is always a limit-point of a denumerable sub-set of E. For example, let C be made up of all points on a line L_1 and all but one of the points on a line L_2. Let this one point be Q. Let Σ consist of all bicontinuous biunivocal transformations of L_1, combined in all possible ways with all bicontinous, biunivocal transformations of L_2 leaving Q invariant, with the proviso that every transformation of L_2 wich leaves invariant points on every interval containing Q should be associated only with the identity transformation on L_1. This example may be shown to satisfy our conditions 2′, 3′, and 4′, and to be an (R). « Limit-point » will have the ordinary meaning, with the exception that when Q would ordinarily be a limit-point of a set on L_2, every point of L_1 will now be a limit-point of that set. It will be impossible to single out any point of L_1 as the unique limit of a sub-set of any such set. On the other hand, it will always be possible to single out a denumerable sub-set of any given set which approaches any given limit-point of the set.

A neighborhood of any point A on L_2 will be a set containing an interval containing A. A neighborhood of any point A on L_1 will consist of a set containing an interval on L_1 containing A and an interval on L_2 containing Q. It is interesting to observe that this set invalidates the condition given by Fréchet ([1]) as sufficient but not necessary to make an (R) set an (L) set — that every point A should determine a sequence $\{V_A^{(n)}\}$ of neighborhoods such that every neighborhood of A should contain at least one neighborhood belonging to this sequence — for we have here an (R) satisfying this sub-condition without being an (L).

This system is not a (J_1), for any transformation that is bicontinuous and biunivocal on L_1 but leaves L_2 untouched retains all limit-properties invariant, but does not always leave A invariant when it leaves every element of E invariant and A is a limit-element of E.

13. Though Fréchet's sufficient condition for an (R) to be an (L) breaks down, it becomes satisfactory if supplemented with a further condition. This condition is that given any two elements A and B, it is possible to find a neighborhood V_A of A and a neighborhood V_B of B that are mutually exclusive. For a class (R) subjected to no further condition, the two properties just mentioned will not be necessary to make it an (L). An example which is an (L) but does not satisfy the first condition has been given by Fréchet ([2]) — it consists of all continuous or

([1]) *V.* p. 11.

([2]) *Relations entre les Notions de Limite et de Distance* (*Trans. Am. Math. Soc.*, vol. XIX, n° 1, p. 59).

discontinuous functions on a closed segment, with limit taken as not necessarily uniform. However, as this system does not have all derivative-sets closed, it is not a (J). It still remains, then, a possibility that these two conditions may be necessary to guarantee that an (R) which is also a (J) should be an (L).

That this possibility is not fulfilled, however the following example will show : let our space consist of all points on an ordinary three-space, and let Σ consist of all transformations that are biunivocal and bicontinuous on every plane containing the line $x=y=0$, but which do not interchange points in any two such planes and which are not necessarily bicontinuous as between the several points in question. Here every point not on $x=y=0$ will have as its neighborhood an ordinary planar neighborhood together with an arbitrary set of points from other planes while a point on the z-axis will have as a neighborhood an arbitrary selection of neighborhoods from the various planes, subject only to the condition that they all contain a segment of the z-axis containing in its interior the point in question. As there is otherwise no correlation between the neighborhoods of an axial point in the different planes, it is clear that no denumerable set of neighborhoods of such a point can be found such that every neighborhood of the point shall contain one of the set. For let $V_1, V_2, \ldots, V_n, \ldots$, be such a sequence of neighborhoods. Then we can select a set $P_1, P_2, \ldots, P_n, \ldots$ of axial planes, and determine [1] a neighborhood of our point in P, not containing all the points of V_1 in P_1, a neighborhood in P_2 not containing all the points of V_2 in P_2, and so on. We can, moreover, choose all these neighborhoods so that they will contain a given segment on the axis. Grouping them together and associating them with the whole of all planes not of the form P_n, we get a neighborhood V not containing any V_n in its entirety.

It is worthy of mention that this system is a (J_1).

14. What the precise necessary and sufficient conditions are that a set which is at once a (J) and an (R) be an (L) has not yet been determined. It appears difficult to obtain conditions

(1) The dependence of this on Zermelo's axiom is only specious.

which are more than reiterations of the fact that the set is an (L). It is reasonable to expect, however, that when the conditions are obtained, they will be two in number, and that one will guarantee that every set E having a limit-point A will have a denumerable sub-set having A as a limit, while the other will provide that it shall have a sub-set with A as its only limit.

Massachusetts Institute of Technology, August 31, 1920.

Comments on [20e], [22b, c]

R. L. Wilder

These papers, written under the influence of Maurice Fréchet, seem to form somewhat of an anomaly among Wiener's research interests; almost as though they were a temporary "trying his hand" in the foundations of topology during a period when these foundations were in the forefront of mathematical interest. For a vivid description of the reasons for his not pursuing further the lines of thought here displayed, reference may be made to Wiener's autobiographical work, *I Am a Mathematician* [56g], pp. 60-64. The technical and conceptual competence exhibited in these papers leaves no doubt, however, that he would have been highly successful had he chosen to continue.

Fréchet, in his classical work of 1928, *Les Espaces Abstraits,*[1] devoted considerable attention to the abstract vector spaces which figure so prominently in these papers, and which were developed independently and virtually simultaneously by Banach in *Sur les opérations dans les ensembles abstraits et leur application aux équations intégrales.*[2] The subsequent history of these spaces is too well known to warrant details here.

Of greater interest for the topologist is the problem posed by Wiener regarding determination of the conditions under which a family Σ of (1-1)-transformations of a set C onto itself will constitute precisely the group of homeomorphisms of that space, when limit element has been defined in terms of Σ in the manner prescribed. To this problem Fréchet devoted special attention in his book, remarking that the ideas involved were far from exhausted and called for further investigation.[3] In this connection, it seems curious that so little attention has been paid to Wiener's characterization of the linear continuum and its group of homeomorphisms in terms of the family Σ of biunivocal transformations and the limit elements defined by them. Aside from the novelty of the approach, this characterization is interesting in that the burden of identifying the line among the large class of spaces "(Sp)" is essentially carried by the single Axiom IX (Axiom VIII imposing only the separability).

The prominence of the group character of the transformations in this characterization is reminiscent of its role in the work of Lie on the foundations of geometry and especially of Hilbert's characterization of the number plane in "Über die Grundlagen der Geometrie," (later extended to 3-space by Montgomery and Zippin[5]). However, in Wiener's characterization, the complete group of homeomorphisms of the line into itself is involved. Studies based on the group of homeomorphisms of a space have been numerous in recent years; special citation might be made of the characterization of n-manifolds, $n > 1$, by Wechsler[6] and Whittaker.[7]

Correction. In [22b], the "(T_1)" occurring on p. 331, lines 8 and 20, and thereafter, should be "(J_1)".

References

1. Fréchet, *Les espaces abstraits et leur théorie considérée comme introduction a l'analyse générale,* Gauthier, Paris, 1928.

2. S. Banach, *Sur les opérations dans les ensembles abstraits et leur application aux équations intégrales,* Fund. Math. 3(1922), 133-181.

3. Reference 1, pp. 196ff.

4. D. Hilbert, *Über die Grundlagen der Geometrie,* Math. Ann. 56(1902-1903), 381-422.

5. D. Montgomery and L. Zippin, *Topological group foundations of rigid space geometry,* Trans. Amer. Math. Soc. 48(1940), 21-49.

6. M. T. Wechsler, *Homeomorphism groups of certain topological spaces,* Ann. Math. 62(1955), 360-373.

7. J. V. Whittaker, *On isomorphism groups and homeomorphic spaces,* Ann. Math. 78(1963), 74-91.

CERTAIN FORMAL INVARIANCES IN BOOLEAN ALGEBRAS*

BY

NORBERT WIENER

Many mathematical systems are defined through postulates concerning (1) a *class* (K) of *elements* ($a, b, c, \cdots$), and (2) a certain definite group of *relations* or *rules of combination* joining these elements to one another.

These relations or rules of combination form only a part of those which may be said to belong to the system, for the latter may be considered to contain all those relations and rules which may be defined as logical functions of those with which the postulates deal and of certain selected elements of the system, and of nothing else. Thus, ordinary real algebra may be defined by postulates involving only the rules of combination $+$ and $\times$, but it also contains as operations the rules of combination which, when applied to x and y, yield us $x^2 + 2xy$, or $x + 1/y$, or x^y. It consequently becomes an interesting question whether the postulates of a given system deal with relations or rules of combination whose position in the system is absolutely different from that of any of the other relations or rules belonging to the system; and if this is not so, it is natural to ask what other relations or rules of the system have formal properties similar to those of the entities which form the subject-matter of the postulates; for it would seem that a system whose postulates deal with entities which occupy a unique position in it has in some sense received a more thoroughgoing analysis than one where this is not the case. As the formal properties of the relations and operations in terms of which a system is defined, in so far as they are determinate, are given in the postulates, this question reduces itself to the investigation of what relations or operations of the system satisfy the postulates. We shall discuss this question in the case of the boolean algebras, although in a somewhat limited form.

This limitation consists in a somewhat narrower definition of the statement that an operation belongs to a given boolean algebra. We shall say that an operation belongs to a boolean algebra if it is the result of the performance of any finite sequence of the operations of the algebra—logical addition, logical multiplication, and negation—on the operands and certain specified "constant" elements of the algebra. That is, if it can be represented in the

* Presented to the Society, December 27, 1916.

established symbolism of the algebra. It was pointed out by Boole* that if we represent the "logical sum" of x and y by $x + y$, their "logical product" by xy, and the negation of x by $\bar{x}$, any operation on x and y may be written in the form $Axy + Bx\bar{y} + C\bar{x}y + D\bar{x}\bar{y}$, where A, B, C, and D are any elements of the algebra you choose. E. V. Huntington has shown† that a set of postulates may be developed for a boolean algebra in terms of the relation of logical addition alone. Our investigation will hence consist in seeing what conditions the coefficients A, B, C, and D must fulfil in order that the operation $Axy + Bx\bar{y} + C\bar{x}y + D\bar{x}\bar{y}$ may satisfy the conditions expressed in these postulates.

In the paper referred to above, Huntington says: ". . . We take as the fundamental concepts a class, K, with a rule of combination, $\oplus$; and as the fundamental propositions, the following nine postulates:

"*A.* $a \oplus a = a$ *whenever* a *and* $a \oplus a$ *belong to the class.*

"*B.* $a \oplus b = b \oplus a$ *whenever* $a, b, a \oplus b$, *and* $b \oplus a$ *belong to the class.*

"*C.* $(a \oplus b) \oplus c = a \oplus (b \oplus c)$ *whenever* $a, b, c, a \oplus b, b \oplus c, (a \oplus b) \oplus c$, *and* $a \oplus (b \oplus c)$ *belong to the class.*

"*D. There is an element* $\wedge$ *such that* $a \oplus \wedge = a$ *for every element* a.

"*E. There is an element* $\vee$ *such that* $\vee \oplus a = \vee$ *for every element* a.

"*F. If* a *and* b *belong to the class, then* $a \oplus b$ *belongs to the class.*

"*G. If the elements* $\wedge$ *and* $\vee$ *in Postulates D and E exist and are unique, then for every element* a *there is an element* $\bar{a}$ *such that* (1°) *if* $x \oplus a = a$ *and* $x \oplus \bar{a} = \bar{a}$, *then* $x = \wedge$; *and* (2°) $a \oplus \bar{a} = \vee$.

"*H. If Postulates A, D, E, and G hold, and if* $a \oplus \bar{b} \neq \bar{b}$, *then there is an element* $x \neq \wedge$ *such that* $a \oplus x = a$ *and* $b \oplus x = b$.

"*J. There are at least two elements,* x *and* y, *such that* $x \neq y$."

Huntington's $\oplus$ is meant to represent the operation of logical addition. We shall use $\oplus$ as a symbol for any operation in the algebra of logic which possesses such formal properties as to satisfy Huntington's postulates, and shall express logical addition and multiplication proper by the same symbols and according to the same conventions as those used in ordinary algebra. We shall indicate negation by a superposed bar, and the operation which takes the place of negation when $\oplus$ is substituted for $+$ by an accent.

If we take $Axy + Bx\bar{y} + C\bar{x}y + D\bar{x}\bar{y}$ as our $x \oplus y$, Huntington's postulate A becomes:

$$Aa + D\bar{a} = a.$$

As this is true independently of the value of a, we may substitute for it the universe, 1, or the null-class, 0. We thus see that $A = 1$ and that $D = 0$. The relation $x \oplus y$ can consequently be written in the form $xy + Bx\bar{y} + C\bar{x}y$.

* *Laws of Thought*, pp. 73–78.

† These Transactions, vol. 5 (1904), pp. 306–308.

Postulate B tells us that our operation is symmetrical with regard to x and y. Consequently $B = C$, and $x \oplus y$ reduces itself to the form

$$xy + B(x+y).$$

Postulate C is the law of associativity. $a \oplus (b \oplus c)$ becomes

$$a[bc + B(b+c)] + B[a + bc + B(b+c)].$$

This reduces to $abc + B(a+b+c)$. As this is a symmetrical function of a, b, and c, it is equivalent to $c \oplus (a \oplus b)$, and this, by postulate B, is but another form of $(a \oplus b) \oplus c$. Postulate C consequently imposes no restriction on an operation in the algebra of logic that is not already imposed by postulates A and B.

Postulate D is satisfied by all functions of the form $ab + B(a+b)$, for $a\bar{B} + B(a+\bar{B}) = a\bar{B} + aB = a$. Thus $\bar{B}$ is a possible value of our $\wedge$. Furthermore, there is no other entity with this property. For let x be such an entity. Then we may substitute 1 for a, and we get

$$1 = 1x + B(1+x) = x + B.$$

Substituting 0 for a, then

$$0 = 0x + B(0+x) = Bx.$$

This proves that $\bar{B}$ is the only value which x can assume.

Postulate E is always satisfied by functions of the form $xy + B(x+y)$, for $aB + B(a+B) = B$. It is also true that, if $ax + B(a+x) = x$, whatever a may be, then $x = B$. For let $a = 1$. Then

$$x = 1x + B(1+x) = x + B.$$

If $a = 0$,

$$x = 0x + B(0+x) = Bx.$$

From the first equation,

$$Bx = B(x+B) = B.$$

Putting these results together, we see that x must equal B, which is therefore the only possible value of $\vee$.

Postulate F is satisfied by all functions in the algebra of logic, as one may see on inspection. This fact enables us to neglect the hypothesis which Huntington attributes to postulates A, B, and C, and thus proves that for $x \oplus y$ to satisfy Huntington's postulates not only is it sufficient that it be of the form $xy + B(x+y)$, but also necessary, as is shown by a brief consideration of the theorems we have already proved.

As the hypothesis of postulate G has been shown to be satisfied by what we have already proved, we need only consider its conclusion. We then find that this postulate too is satisfied by all expressions of the form

$$xy + B(x + y).$$

If we make our $\bar{a}$ of the $(+, \times)$ system the element which Huntington calls $\bar{a}$ and which we have agreed to call a' to avoid confusion, we discover that if $x \oplus a = a$ and $x \oplus \bar{a} = \bar{a}$, then

$$ax + B(x + a) = a, \qquad \text{and} \qquad \bar{a}x + B(x + \bar{a}) = \bar{a}.$$

When we transform these into equations to 0, they become, respectively,

$$Bx\bar{a} + \bar{B}\bar{x}a = 0 \qquad \text{and} \qquad Bxa + \bar{B}\bar{x}\bar{a} = 0.$$

Putting these together, we see that $Bx + \bar{B}\bar{x} = 0$, so that $x = \bar{B}$, which we have found to be our $\wedge$. Furthermore,

$$a \oplus \bar{a} = a\bar{a} + B(a + \bar{a}) = 0 + B1 = B = \vee.$$

Thus the negation of our original algebra may still remain the negation of any of our new boolean algebras, and all operations of the form $xy + B(x + y)$ satisfy postulates *A–G*, while no other operations in the algebra have this property. It is further true that the negation of our original algebra can be replaced by no other operation definable as an operation belonging to the algebra of logic. As is well known, all operations belonging to the algebra of logic, in the sense in which this phrase is usually taken, may be written in the form $Mx + N\bar{x}$, if they are operations on the single variable x. Let our a' be expressed thus as $Ma + N\bar{a}$. Postulate G tells us (1°) that if

$$xa + B(x + a) = a,$$

and

$$x(Ma + N\bar{a}) + B[x + (Ma + N\bar{a})] = Ma + N\bar{a},$$

then x must equal $\bar{B}$. The hypothesis may be rewritten in the combined form

$$Bx\bar{a} + \bar{B}\bar{x}a + Bx(\bar{M}a + \bar{N}\bar{a}) + \bar{B}\bar{x}(Ma + N\bar{a}) = 0,$$

or

$$x(B\bar{a} + B\bar{M}) + \bar{x}(\bar{B}a + \bar{B}N) = 0.$$

Since $(B\bar{a} + B\bar{M})(\bar{B}a + \bar{B}N) = 0$, this equation is always solvable for x. Solving it, we get

$$x = \bar{B}a + \bar{B}N + u(\bar{B} + aM) = \bar{B}(a + N + u) + uaM,$$

where u is indeterminate. As x must equal $\bar{B}$, we obtain the equation

$$\bar{B} = \bar{B}(a + N + u) + uaM, \qquad \text{or} \qquad BuaM + \bar{B}\bar{u}\bar{a}\bar{N} = 0.$$

As a and u are both indeterminate, we may assign them both the value 0, whence we get $\bar{B}\bar{N} = 0$, or the value 1, whence $BM = 0$. The second part of G yields us the formula

$$a(Ma + N\bar{a}) + B(a + Ma + N\bar{a}) = B, \qquad \text{or} \qquad Ma\bar{B} + \bar{N}\bar{a}B = 0.$$

As a is indeterminate, we can make it either 1 or 0, thus getting the results $M\bar{B} = 0$ and $\bar{N}B = 0$. Putting these results together with those which we have just reached from the first part of G, we see that $M = 0$, $N = 1$, and $a' = 0a + 1\bar{a} = \bar{a}$. We thus see that *the only operation in the algebra of logic which can have the characteristic properties of negation in a system whose operations are such that they can be expressed in the symbolism of the algebra of logic is the operation of negation itself.*

Postulate H imposes no further restriction on $\oplus$. It tells us that if

$$a\bar{b} + B(a + \bar{b}) \neq \bar{b},$$

there is some element x other than $\bar{B}$, such that

$$ax + B(a + x) = a, \qquad \text{and} \qquad bx + B(b + x) = b.$$

Now $ab + \bar{B}(a + b)$ is such an x. In the first place, it cannot equal $\bar{B}$, for if it did, the equation $ab + \bar{B}(a + b) = \bar{B}$, which we should obtain, would reduce to the form $Bab + \bar{B}\bar{a}\bar{b} = 0$, whence we should get the solution $a = \bar{B}\bar{b} + u(\bar{B} + \bar{b})$, where u is indeterminate. Substituting this in the inequality $a\bar{b} + B(a + \bar{b}) \neq \bar{b}$, we get

$$\bar{B}\bar{b} + B(\bar{B}\bar{b} + u(\bar{B} + \bar{b}) + \bar{b}) \neq \bar{b}.$$

This is reducible to the form $\bar{b} \neq \bar{B}\bar{b} + B\bar{b}$, which is manifestly false. Consequently $ab + \bar{B}(a + b)$ is a different element from $\bar{B}$. Furthermore, we have

$$ax + B(a + x) = a[ab + \bar{B}(a + b)] + B[a + ab + \bar{B}(a + b)]$$
$$= ab + \bar{B}a + Ba = a.$$

Similarly, $bx + B(b + x) = b$. This proves our point.

Postulate J is obviously satisfied by our operation $\oplus$, just because it is satisfied by our original boolean algebra. We have consequently proved that *the necessary and sufficient condition that an operation expressible in the symbolism of a boolean algebra should possess the formal properties of the + operation is that it should be of the form $xy + B(x + y)$, and that a transformation of the ordinary addition into this new operation leaves the operation of negation unaltered.* As a consequence the operation made to correspond to logical multiplication by this transformation will be that which, when applied to x and y, yields us

$$\overline{(\bar{x} \oplus \bar{y})} = \overline{[\bar{x}\bar{y} + B(\bar{x} + \bar{y})]} = (x + y)(\bar{B} + xy) = xy + \bar{B}(x + y).$$

It is a well-known fact that all operations of the form $\bar{A}x + A\bar{x}$ are one-one, and that these are the only one-one operations in a boolean algebra.* It is obvious that if the whole of a boolean algebra is transformed into itself by one of these transformations, all its operations will be changed into relations of similar formal properties. The question thus arises, what relation will hold between $\bar{A}x + A\bar{x}$, $\bar{A}y + A\bar{y}$, and $\bar{A}z + A\bar{z}$ if $x + y = z$? Let $\bar{A}x + A\bar{x} = u$, $\bar{A}y + A\bar{y} = v$, and $\bar{A}z + A\bar{z} = w$. Then $x = \bar{A}u + A\bar{u}$, $y = \bar{A}v + A\bar{v}$, and $z = \bar{A}w + A\bar{w}$. But, by hypothesis, $x + y = z$. Therefore $\bar{A}u + A\bar{u} + \bar{A}v + A\bar{v} = \bar{A}w + A\bar{w}$. Hence

$$\bar{A}u + \bar{A}v = \bar{A}w, \qquad \text{and} \qquad A\bar{u} + A\bar{v} = A\bar{w}.$$

But $Aw = A\overline{(A\bar{w})} = A\overline{(A\bar{u} + A\bar{v})} = A(\bar{A} + u)(\bar{A} + v) = Auv$. Hence

$$w = Aw + \bar{A}w = Auv + \bar{A}u + \bar{A}v = uv + \bar{A}(u + v).$$

That is, the one-one operation which changes x into $\bar{A}x + A\bar{x}$ transforms the operation of logical addition into the operation indicated by $uv + \bar{A}(u + v)$. *Since the latter is in the general form for an operation which satisfies the postulates of a boolean algebra and is expressible in the symbolism of a given boolean algebra, as we have already seen, we have shown that all such operations are "ordinally similar"† to the operation of logical addition.* This we might also have deduced from the fact that an operation definable in the symbolism of a boolean algebra and satisfying Huntington's postulates is completely determinate when the element corresponding to 1 is given, so that there is only one operation which establishes this correspondence, namely, that which results from + when the whole algebra is subjected to a one-one transformation which makes the indicated change in 1.

It is natural to consider those properties of the boolean algebras which are independent of the special $\oplus$-operation we take for our logical addition as more deep-rooted and fundamental than the rest, much as we assign some sort of a priority to those properties of a geometrical space which are unaltered by projection. It is to be noted that the invariant properties in the boolean algebras differ from those in geometry which are unaltered by projection in that the former are invariant with reference to all transformations expressed by formulas belonging to the algebra in question, so long as they leave its formal properties unchanged, while such an analytic-geometry transformation as $(x' = x, y' = y, z' = z^3)$ does not alter the intrinsic formal properties of real space, though it alters the projective properties of certain configurations. It is a natural question to ask, whether a given relation between the entities of a boolean algebra retains its formal properties with respect to the algebra

* Schröder, *Algebra der Logik*, Vol. I, p. 463.

† Cf. Whitehead and Russell, *Principia Mathematica*, Vol. II, * 151.

unaltered by all of the transformations definable in terms of the algebra which leave the formal character of the algebra unchanged, just as it is an important question in geometry whether a given relation remains unaltered by a projective transformation; but from what we have seen, the latter question is the more deep-rooted of the two. Now it is a well-known fact that every relation between entities belonging to a boolean algebra, which can be expressed at all in the symbolism of the algebra, can be expressed as a logical function of equations and inequations to 0. We may therefore give a precise formulation to our question thus: when will an expression of the form

$$\sum A_{a_1, a_2, \ldots, a_k} \prod_{l=1}^{l=k} (\bar{a}_l x_l + a_l \bar{x}_l) = 0$$

(where k and l are numerical subscripts and the summation sign extends over the a's which are either 1 or 0) retain its truth-value unaltered by any transformation of the x's which turns each x_k into $\bar{B}x_k + B\bar{x}_k$, where B is independent of the value of k? From what we have already proved, it may be seen that such expressions, and only logical functions of such expressions or their negations, represent relations among the x's which remain invariant with reference to all transformations definable in terms of the algebra, which leave it still a boolean algebra. If our expression is to remain invariant under all such transformations, it must retain its truth-value unaltered by the transformation of all the x's into their negations. This transforms our relation into

$$\sum A_{a_1, a_2, \ldots, a_k} \prod_{l=1}^{l=k} (a_l x_l + \bar{a}_l \bar{x}_l) = 0.$$

From this and our original equation we can derive an equivalent equation by adding the expressions equated to 0. This equation will read

$$\sum A_{a_1, a_2, \ldots, a_k} \left[\prod_{l=1}^{l=k} (\bar{a}_l x_l + a_l \bar{x}_l) + \prod_{l=1}^{l=k} (a_l x_l + \bar{a}_l \bar{x}_l) \right] = 0.$$

This expression remains unchanged by *all* transformations of the x's which may be expressed in the symbolism of the algebra and still leave it a boolean algebra. As we have seen, any such transformation turns every x into $\bar{B}x + B\bar{x}$, where B is the same throughout the algebra. Also $(\bar{a}_l x_l + a_l \bar{x}_l)$ would be changed by this transformation into

$$\bar{B}(\bar{a}_l x_l + a_l \bar{x}_l) + B(a_l x_l + \bar{a}_l \bar{x}_l),$$

and $(a_l x_l + \bar{a}_l \bar{x}_l)$ would become

$$B(\bar{a}_l x_l + a_l \bar{x}_l) + \bar{B}(a_l x_l + \bar{a}_l \bar{x}_l).$$

As a consequence,

$$\prod_{l=1}^{l=k}(\bar{a}_l x_l + a_l \bar{x}_l) + \prod_{l=1}^{l=k}(a_l x_l + \bar{a}_l \bar{x}_l)$$

becomes

$$\bar{B}\prod_{l=1}^{l=k}(\bar{a}_l x_l + a_l \bar{x}_l) + B\prod_{l=1}^{l=k}(a_l x_l + \bar{a}_l \bar{x}_l) + B\prod_{l=1}^{l=k}(\bar{a}_l x_l + a_l \bar{x}_l) + \bar{B}\prod_{l=1}^{l=k}(a_l x_l + \bar{a}_l \bar{x}_l),$$

which is precisely its original value. This proves our point, and enables us to formulate the theorem that *the necessary and sufficient condition that a relation between any number of the elements of a boolean algebra should remain invariant with reference to all transformations of the algebra into itself which may be expressed in its symbolism and leave it a boolean algebra is that the relation should remain invariant with regard to negation. Such a relation is fully characterized by the fact that it can be expressed as a logical function of equations or inequations to* 0 *of completely expanded functions of the related elements, in which the coefficient of each product of all the elements or their negations is identical with that of the product whose factors are the negations of those of the former one.*

It will be noticed that Kempe's between-relation and obverse-relation* both satisfy this criterion, for Kempe's $ab \cdot c$, which is his way of saying that c is between b and a, may be written $ab\bar{c} + \bar{a}\bar{b}c = 0$, while if a, b, c, etc., form an obverse collection, the sum of their product and that of their negations is 0, and vice versa. It was, of course, recognized by Kempe that these relations, and all that can be defined in terms of them alone, represent a level of greater generality and universality in the boolean algebras than the operations of logical addition and logical multiplication, for the former do not enable us to discriminate between 0, 1, and the other elements of the system. What this article proves in addition is that Kempe's relations represent the very highest level of generality attainable in a boolean algebra, for they remain invariant under all the transformations of the algebra into another boolean algebra which may be expressed in the symbolism of the first algebra.

HARVARD UNIVERSITY,
April 24, 1916

* A. B. Kempe, *On a relation between the logical theory of classes and the geometrical theory of points*, Proceedings of the London Mathematical Society, vol. 21 (1890), pp. 147–182.

Comments on [17a]

F. E. Hohn

Wiener's first result derives its initial interest from the nature of Huntington's postulates for a Boolean algebra.[1] These postulates are stated in terms of the single operation of logical addition and the constants zero and one.

Wiener shows that the necessary and sufficient conditions that a binary operation (also called "addition") expressible in the symbolism of the algebra have the same formal properties as those which Huntington postulates for logical addition is that it have the form $xy + B(x+y)$, where B is a fixed element of the algebra. The operation of negation is unaltered, the new unit element is B, the new zero element is $\bar{B}$, and the new multiplication must be defined by $xy + \bar{B}(x+y)$.

Wiener's second result relates to the fact that the only one-to-one mappings of a Boolean algebra expressible in the symbolism of the algebra have the form $x' = A\bar{x} + \bar{A}x$, where A is a fixed element of the algebra. Every such mapping is onto. The mapping is then an isomorphism if and only if the new operation of addition has the form $xy + \bar{A}(x+y)$, for which A is the zero element and $\bar{A}$ the unit element. The last observation reveals the true significance of Wiener's first result.

Wiener now shows that the only relations invariant with respect to *all* of the mappings just referred to are expressible as logical functions of equations and inequations to 0 of functions whose disjunctive normal forms have the property that the coefficient of each minimal polynomial is the same as that of the minimal polynomial all of whose factors are the negations of the factors of the former.

The importance of this paper is that the last-mentioned result encompasses completely the classical invariant theory of Boolean algebras. Thus these results have no far-reaching consequences, and the paper does not appear to have influenced the subsequent literature of the subject to any significant extent.

Reference

1. E. V. Huntington, *Sets of independent postulates for the algebra of logic,* Trans. Amer. Math. Soc. 5(1904), 306-308.

IB
Potential Theory

NETS AND THE DIRICHLET PROBLEM

By H. B. PHILLIPS AND N. WIENER

1. **Introduction.** In this paper the Dirichlet problem on the plane and in three-space — in general, in n-space, as a matter of fact — is attacked through the analogous problem relating to the potential on a net of equally spaced wires of constant resistance. This method leads at once to the solution of the Dirichlet problem for any continuous boundary conditions on any polygonal or polyhedral boundary of a certain type. Then by means of an important lemma to the effect that if $\{R_n\}$ is any set of polygonal or polyhedral regions of this type, and potential functions U_n are assigned so as to be equicontinuous and uniformly bounded over the boundaries of the R_n's, they shall be equicontinuous over the interiors of the R_n's, we extend our solution of the Dirichlet problem to regions of a much more general character.

Let us start with some definitions. To begin with, a set of functions $\{f_n(P)\}$, each ranging over some corresponding region R_n, is said to be *equicontinuous* if there is a function $\phi(x)$ such that

$$\lim_{x \to 0} \phi(x) = 0,$$

and such that if P and Q lie on R_n,

$$|f_n(P) - f_n(Q)| < \phi(\overline{PQ}). \tag{1}$$

Be it noted that the regions R_n need not be regions having area (or volume) — they may, for example, be arcs of curves. Even in this case, the quantity $\overline{PQ}$ occurring in (1) is to be taken as distance measured *in a straight line*. Curvilinear distances play no part in this paper.

Given a region R in three-space, the *net of order n*[1] over R consists of those portions of the planes $x = a/2^n$, $y = b/2^n$, $z = c/2^n$ which lie within R; a, b, and c assuming all integral values. The *lines* of the net are the intersections of planes of two distinct systems. The *nodes* of the net are the points where planes of all three systems intersect. A function $f(P)$ is defined over the net

[1] Cf. C. Runge, *Göttingische Nachrichten*, Math.-Phys. Klasse 1911. Cf. also R. G. D. Richardson, *Trans. Am. Math. Soc.*, 18 (1917), pp. 489–521.

when it is defined at the nodes. A *boundary node* of the net is a node such that some adjacent node does not lie within R—that is, such that the net over R contains less than six nodes adjacent to the one in question. A function $f(P)$ defined over the net is defined within any mesh which contains in its interior no point not on R, as follows:

$$\begin{aligned}
f(x, y, z) &= f(x_1, y_1, z_1)+2^n[f(x_2, y_1, z_1)-f(x_1, y_1, z_1)](x-x_1) \\
&+2^n [f(x_1, y_2, z_1)-f(x_1, y_1, z_1)](y-y_1)+2^n[f(x_1,y_1,z_2)-f(x_1,y_1,z_1)](z-z_1) \\
&+2^{2n}[f(x_2, y_2, z_1)-f(x_1, y_2, z_1)-f(x_2, y_1, z_1)+f(x_1, y_1, z_1)](x-x_1)(y-y_1) \\
&+2^{2n}[f(x_1, y_2, z_2)-f(x_1, y_1, z_2)-f(x_1, y_2, z_1)+f(x_1, y_1, z_1)] (y-y_1)(z-z_1) \\
&+2^{2n}[f(x_2, y_1, z_2)-f(x_1, y_1, z_2)-f(x_2, y_1, z_1)+f(x_1, y_1, z_1)] (x-x_1)(z-z_1) \\
&+2^{3n}[f(x_2, y_2, z_2)-f(x_1, y_2, z_2)-f(x_2, y_1, z_2)-f(x_2, y_2, z_1) + f(x_1, y_1, z_2) \\
&\quad + f(x_1, y_2, z_1)+f(x_2, y_1, z_1)-f(x_1, y_1, z_1)](x-x_1)(y-y_1)(z-z_1)
\end{aligned} \tag{2}$$

A similar definition holds in the case of two or n dimensions.

2. **The Difference-Equation of Potential on a Net.** Consider a net of order ν in space of three dimensions with nodes at points having coördinates of the form

$$x=\frac{m}{2^\nu}, \quad y=\frac{n}{2^\nu}, \quad z=\frac{p}{2^\nu}, \tag{3}$$

m, n, p being integers. We shall now show that there exists a function $f(x, y, z)$ which has given values at the boundary nodes and has at each interior node a value which is the average of its values at the six adjacent nodes with which that one is connected by lines of the net. If the coördinates (3) represent such an interior node, the condition to be satisfied is

$$\begin{aligned}
6f\left(\frac{m}{2^\nu}, \frac{n}{2^\nu}, \frac{p}{2^\nu}\right) &= f\left(\frac{m-1}{2^\nu}, \frac{n}{2^\nu}, \frac{p}{2^\nu}\right)+f\left(\frac{m+1}{2^\nu}, \frac{n}{2^\nu}, \frac{p}{2^\nu}\right) \\
&+f\left(\frac{m}{2^\nu}, \frac{n-1}{2^\nu}, \frac{p}{2^\nu}\right)+f\left(\frac{m}{2^\nu}, \frac{n+1}{2^\nu}, \frac{p}{2^\nu}\right) \\
&+f\left(\frac{m}{2^\nu}, \frac{n}{2^\nu}, \frac{p-1}{2^\nu}\right)+f\left(\frac{m}{2^\nu}, \frac{n}{2^\nu}, \frac{p+1}{2^\nu}\right)
\end{aligned} \tag{4}$$

A function satisfying (4) will be called a potential function on the net. If adjacent nodes are connected by wires of equal resistance and given potentials are maintained at the boundary nodes, Kirchoff's laws show that the electric potential determined at the nodes of the net is such a potential function.

To show the existence of a potential function with given boundary values, consider the function

$$W=\sum\sum\sum\left[g\left(\frac{m+1}{2^\nu},\frac{n}{2^\nu},\frac{p}{2^\nu}\right)-g\left(\frac{m}{2^\nu},\frac{n}{2^\nu},\frac{p}{2^\nu}\right)\right]^2$$
$$+\sum\sum\sum\left[g\left(\frac{m}{2^\nu},\frac{n+1}{2^\nu},\frac{p}{2^\nu}\right)-g\left(\frac{m}{2^\nu},\frac{n}{2^\nu},\frac{p}{2^\nu}\right)\right]^2$$
$$+\sum\sum\sum\left[g\left(\frac{m}{2^\nu},\frac{n}{2^\nu},\frac{p+1}{2^\nu}\right)-g\left(\frac{m}{2^\nu},\frac{n}{2^\nu},\frac{p}{2^\nu}\right)\right]^2,$$

each summation being extended to all values of m, n, p for which the arguments used represent nodes of the net. Let $g(x, y, z)$ have the assigned values at the boundary nodes and values at the interior nodes which make W a minimum. Since W is a definite quadratic form in a finite number of variables (the values of g at the interior nodes) such a minimum exists. Differentiating W with respect to

$$g\left(\frac{m}{2^\nu},\frac{n}{2^\nu},\frac{p}{2^\nu}\right)$$

we find that the function $g(x, y, z)$ satisfies the equation

$$g\left(\frac{m}{2^\nu},\frac{n}{2^\nu},\frac{p}{2^\nu}\right)-g\left(\frac{m+1}{2^\nu},\frac{n}{2^\nu},\frac{p}{2^\nu}\right)$$
$$+g\left(\frac{m}{2^\nu},\frac{n}{2^\nu},\frac{p}{2^\nu}\right)-g\left(\frac{m-1}{2^\nu},\frac{n}{2^\nu},\frac{p}{2^\nu}\right)$$
$$+\ \ldots\ldots\ldots\ldots\ldots\ldots\ =0$$

which is equivalent to (4).

There cannot be two distinct solutions of (4) having the same values at the boundary points of the net. For, if there were, their difference would be zero at the boundary points and would satisfy (4) at all interior nodes. Now a function satisfying (4) cannot have a maximum numerical value at an interior node of the net. Hence the difference of the two assumed solutions has its maximum numerical value zero at the boundary, which proves the two solutions identical.

The equations (4) could be solved in the following way. Take the equation expressing

$$f\left(\frac{m}{2^\nu}, \frac{n}{2^\nu}, \frac{p}{2^\nu}\right)$$

in terms of values at adjacent nodes. For each term on the right of this equation corresponding to an interior node substitute its expression in terms of adjacent values by the proper equation of the type (4). In the result substitute again for all terms corresponding to interior nodes, and continue this process indefinitely. At each stage of this process

$$f\left(\frac{m}{2^\nu}, \frac{n}{2^\nu}, \frac{p}{2^\nu}\right)$$

is expressed as a linear function of values of $f(x, y, z)$ at interior and boundary nodes. From the form of equations (4) it is seen that the coefficients in this linear function are all positive and have a sum equal to 1. If r is the largest number of steps from a node to an adjacent node necessary to pass from any node to the boundary, after r substitutions the coefficients multiplying boundary values in the linear function have a sum not less than

$$\frac{1}{6^r}.$$

The sum of coefficients of interior values is then not greater than

$$1 - \frac{1}{6^r}.$$

After rs substitutions, the sum of coefficients of interior values is not greater than

$$\left(1 - \frac{1}{6^r}\right)^s$$

which approaches zero as s increases indefinitely. The coefficients of boundary values are less than 1 and never decrease. They therefore approach definite limits. In the limit

$$f\left(\frac{m}{2^\nu}, \frac{n}{2^\nu}, \frac{p}{2^\nu}\right)$$

is thus expressed as a linear function of the boundary values with coefficients all positive and having a sum equal to 1.

3. **The Potential on a Cubical Net.** Let us now consider a net of order ν extending over the interior of a cube which we may without essential restriction of generality consider to be the cube bounded by the coördinate planes and the planes $x=1$, $y=1$, $z=1$. Let us suppose, moreover, that the potential is everywhere zero over the surface of the cube except for the nodes in the xy-plane. On these nodes it is a given continuous function $f(x, y)$. We wish to determine the potential at all the nodes of the cube.

Following the method of Bernoulli let us seek solutions of (4) of the form $X(x)\,Y(y)\,Z(z)$. For such a function (4) becomes

$$\begin{aligned} 6X(x)\,Y(y)\,Z(z) = {} & X\left(x-\frac{1}{2^\nu}\right)Y(y)Z(z) + X\left(x+\frac{1}{2^\nu}\right)Y(y)Z(z) \\ & + X(x)\,Y\left(y-\frac{1}{2^\nu}\right)Z(z) + X(x)\,Y\left(y+\frac{1}{2^\nu}\right)Z(z) \\ & + X(x)\,Y(y)Z\left(z-\frac{1}{2^\nu}\right) + X(x)\,Y(y)Z\left(z+\frac{1}{2^\nu}\right). \end{aligned} \tag{5}$$

By division we get

$$\begin{aligned} 3 = {} & \frac{X\left(x-\frac{1}{2^\nu}\right)+X\left(x+\frac{1}{2^\nu}\right)}{2X(x)} + \frac{Y\left(y-\frac{1}{2^\nu}\right)+Y\left(y+\frac{1}{2^\nu}\right)}{2Y(y)} \\ & + \frac{Z\left(z-\frac{1}{2^\nu}\right)+Z\left(z+\frac{1}{2^\nu}\right)}{2Z(z)} \end{aligned} \tag{6}$$

Let us note that

$$\left.\begin{aligned}\frac{\sin\lambda\left(x-\dfrac{1}{2^\nu}\right)+\sin\lambda\left(x+\dfrac{1}{2^\nu}\right)}{2\sin\lambda x}&=\cos\frac{\lambda}{2^\nu},\\ \frac{\sinh\rho\left(z-\dfrac{1}{2^\nu}+a\right)+\sinh\rho\left(z+\dfrac{1}{2^\nu}+a\right)}{2\sinh\rho(z+a)}&=\cosh\frac{\rho}{2^\nu}.\end{aligned}\right\}\tag{7}$$

Hence

$$\sin(\lambda x)\sin(\mu y)\sinh\rho(z+a)\tag{8}$$

is a solution of (6) provided

$$\cos\frac{\lambda}{2^\nu}+\cos\frac{\mu}{2^\nu}+\cosh\frac{\rho}{2^\nu}=3.\tag{9}$$

If, moreover, $\lambda=r\pi$, $\mu=s\pi$, $a=-1$, where r and s are integers, (8) vanishes when $x=0$, $x=1$, $y=0$, $y=1$, or $z=1$. It thus vanishes over all the surface of the cube except the part in the xy-plane.

Equation (9) may be written in the form

$$1-\cos\frac{\lambda}{2^\nu}+1-\cos\frac{\mu}{2^\nu}=\cosh\frac{\rho}{2^\nu}-1,$$

whence

$$2\sin^2\frac{\lambda}{2^{\nu+1}}+2\sin^2\frac{\mu}{2^{\nu+1}}=2\sinh^2\frac{\rho}{2^{\nu+1}}.\tag{10}$$

From this it is clear that for any given λ and μ

$$\lim_{\nu\to\infty}\rho^2=\lambda^2+\mu^2.\tag{11}$$

Let k, l be positive integers and let m be the positive number such that

$$\lambda=k\pi,\quad \mu=l\pi,\quad \rho=m\pi\tag{12}$$

satisfy (9). Then

$$\sum_{k=1}^{2^\nu-1}\sum_{l=1}^{2^\nu-1}A_{kl}\sin(k\pi x)\sin(l\pi y)\sinh m\pi(z-1)\tag{13}$$

is a solution of (4) which vanishes over all the surface of the cube except the part in the xy-plane where it reduces to

$$-\sum_{k,l=1}^{2^\nu-1}\sum A_{kl}\sinh m\pi \sin (k\pi x) \sin (l\pi y). \tag{14}$$

It is to be noted that m is a function of k and l.

Let us suppose that (14) assumes the same values as the function $f(x, y)$ at the points

$$x=\frac{i}{2^\nu}, \quad y=\frac{j}{2^\nu},$$

where

$$0<i<2^\nu, \quad 0<j<2^\nu.$$

From theorems concerning finite Fourier's series, we get

$$-A_{k,l}\sinh m\pi=\frac{1}{2^{2\nu-2}}\sum_{i,j=1}^{2^\nu-1}\sum f\left(\frac{i}{2^\nu}, \frac{j}{2^\nu}\right)\sin\frac{ik\pi}{2^\nu}\sin\frac{jl\pi}{2^\nu}. \tag{15}$$

As ν increases indefinitely, this approaches

$$4\int_0^1\int_0^1 f(s, t) \sin (k\pi s) \sin (l\pi t)\, ds\, dt \tag{16}$$

since $f(s, t)$ is continuous. It is never greater in absolute value than 4 max. $|f(x, y)|$.

Using (15), expression (13) can be reduced to the form

$$\frac{1}{2^{2\nu-2}}\sum_{k,l=1}^{2^\nu-1}\sum\left\{\sum_{i,j=1}^{2^\nu-1}\sum f\left(\frac{i}{2^\nu}, \frac{j}{2^\nu}\right)\sin\frac{ik\pi}{2^\nu}\sin\frac{jl\pi}{2^\nu}\sin k\pi x\sin l\pi y \right.$$
$$\left.\frac{\sinh m\pi(1-z)}{\sinh (m\pi)}\right\}. \tag{17}$$

If $z\leq 1$,

$$\frac{\sinh m\pi(1-z)}{\sinh m\pi}=e^{-m\pi z}\left(\frac{e^{m\pi}-e^{-m\pi(1-2z)}}{e^{m\pi}-e^{-m\pi}}\right)\leq e^{-m\pi z}. \tag{18}$$

Consider (17) as a double summation with respect to the subscripts k, l. From (18) any term of this double sum is in absolute value not greater than the corresponding term in the series

$$\frac{1}{2^{2\nu-2}} \sum_{k,l=1}^{\infty}\sum \left\{ \sum_{i,j=1}^{2^\nu-1}\sum f\left(\frac{i}{2^\nu}, \frac{j}{2^\nu}\right) \sin\frac{ik\pi}{2^\nu} \sin\frac{jl\pi}{2^\nu} \right.$$

$$\left. \sin(k\pi x)\sin(l\pi y)e^{-m\pi z} \right\} . \qquad (19)$$

Each term in this double series is in absolute value not greater than the corresponding term in

$$\sum_{k,l=1}^{\infty}\sum 4 \max. |f(x, y)| e^{-m\pi z} \qquad (20)$$

which in the region $z > \epsilon > 0$ is absolutely and uniformly convergent.

As ν increases indefinitely, from (11) and (12), m approaches the limit

$$m = \sqrt{k^2 + l^2}. \qquad (21)$$

The k, l term of (17) approaches the limit

$$4 \int_0^1 \int_0^1 f(r, s) \sin(k\pi r) \sin(l\pi s) dr\, ds \sin(k\pi x) \sin(l\pi y) \frac{\sinh m\pi(1-z)}{\sinh m\pi} \qquad (22)$$

uniformly throughout the cube, where m has the value (21). The series

$$\sum_{k,l=1}^{\infty}\sum 4 \int_0^1 \int_0^1 f(r, s) \sin(k\pi r) \sin(l\pi s)\, dr\, ds \sin(k\pi x) \sin(l\pi y) \frac{\sinh m\pi(1-z)}{\sinh m\pi} \qquad (23)$$

is term by term in absolute value not greater than (20). It therefore converges absolutely and uniformly in the region

$$1 \geq z > \epsilon > 0. \qquad (24)$$

We have then shown that (13) approaches the limit (23) uniformly in the region (24); for the terms in which

$$k,\ l \leq n$$

approach uniformly the corresponding terms in (23) and as n increases indefinitely the remainder in each case approaches zero independently of ν.

Since k, l, m satisfy (21) the function of x, y, z defined by (23) is harmonic within the cube. Hence the net potential function approaches a bounded continuous harmonic function uniformly in the portion of the cube for which $z \geq \epsilon > 0$. If the boundary values are given by a function continuous over the whole surface of the cube, the net potential is the sum of six potentials of the kind just discussed. In that case it therefore has as limit a function harmonic within but not necessarily on the surface of the cube.

Suppose now

$$f(x, y) = 1. \tag{25}$$

Let $f(x, y, z)$ be the harmonic function defined by (23) and let $f_\nu(x, y, z)$ be the net potential on the net of order ν. In a similar way we could start with a function which is 1 on the plane $z=1$ and 0 on the other faces of the cube. The continuous function thus defined is

$$f(x, y, 1-z)$$

and the net potential

$$f_\nu(x, y, 1-z).$$

Similarly we define a pair of functions for each face of the cube. The sum of the six net potentials thus obtained is the net potential with value 1 at all the nodes on the surface of the cube. Therefore

$$\begin{aligned} f_\nu(x, y, z) + f_\nu(x, y, 1-z) + f_\nu(z, y, x) + f_\nu(z, y, 1-x) + f_\nu(x, z, y) \\ + f_\nu(x, z, 1-y) = 1. \end{aligned} \tag{26}$$

At any point inside the cube each of the net functions approaches the corresponding continuous function as limit. Hence, at any interior point,

$$\begin{aligned} f(x, y, z) + f(x, y, 1-z) + f(z, y, x) + f(z, y, 1-x) + f(x, z, y) \\ + f(x, z, 1-y) = 1. \end{aligned} \tag{27}$$

In the region

$$\left.\begin{array}{r} 0<\epsilon<x<1-\epsilon \\ \epsilon<y<1-\epsilon \\ 0\leq z\leq 1 \end{array}\right\} \tag{28}$$

all the functions of (27) except $f(x, y, z)$ approach 0 uniformly as z approaches 0. Hence $f(x, y, z)$ approaches 1 uniformly. In the same region each term of (26) except $f_\nu(x, y, z)$ approaches uniformly the corresponding term of (27). Hence $f_\nu(x, y, z)$ approaches the continuous harmonic function $f(x, y, z)$ uniformly in the region (28).

If the value 1 is assigned at the nodes of two or more faces of the cube and the value 0 at the others, we show in a similar way that in any closed region belonging to the cube and not containing points where the boundary values are discontinuous the net potential approaches uniformly a continuous potential which has the assigned boundary values.

4. **A Boundary Problem.** Let R be a region whose boundary C is made up entirely of a finite number of finite rectangles lying in the planes.

$$x=a_n, \quad y=b_n, \quad z=c_n, \tag{29}$$

where a_n, b_n, c_n are terminating binary numbers. With a point P of the surface as center construct a cube Γ_0 with sides parallel to the coördinate planes and edge δ so small that if a cube of this size or smaller is constructed with any point of C as center a part of its surface will lie outside and a part inside R. Let C_0 be the part of C within Γ_0. Let a function F be equal to 1 at the boundary nodes on C_0 and equal to 0 at all other boundary nodes. Let F_ν be the continuous function obtained by constructing the potential determined by F on the net of order ν and continuing it through the cells by trilinear interpolation. We wish to show that a cube can be constructed with P as center within which F_ν differs from 1 by less than an arbitrarily assigned quantity ϵ.

For this purpose construct a series of cubes $\Gamma_1, \Gamma_2, \ldots \Gamma_m$, with centers at P, sides parallel to the coördinate planes, and edges of length $\frac{1}{2}\delta, \frac{1}{4}\delta, \ldots, \frac{1}{2^m}\delta$. Let the part of C within

Γ_m be C_m and let the surface of Γ_m within R be I_m. On the surface I_0 the functions F_ν are zero or positive. We first show that these functions have on I_1 a value everywhere greater than a definite positive number $p>0$. For all finite values of ν these functions are positive on I_1, since they, like the net potentials, have their greatest and least values on the boundary. For sufficiently large values of ν we can construct a cube Γ' within Γ_0 with sides of the form (29) and lying as close to those of Γ_0 as we please. If we construct functions F_ν' which are zero on the surface I' of this cube within R and 1 on the part of C within Γ', we shall have

$$F_\nu \geq F_\nu'$$

at all the nodes on the surface of Γ' and so at all points of R within Γ'. As ν increases indefinitely, F_ν' approaches uniformly a function F' harmonic within Γ'. Since this function is not zero on I_1, the functions F_ν have a lower limit $p>0$ on I_1. This value p can be taken independently of the position of P on the surface, for the limit of such positions is again a position at which p is not zero. The same value can be used for all cubes of edge $\leq\delta$; for, when the edge is very small, the part of R within the cube is similar to one obtained with a larger cube.

Consider now the functions $F_\nu - p$. These functions are all zero or positive on I_1 and equal to $1-p$ on the part of C within Γ_1. These functions satisfy the same conditions on Γ_1 that F_ν did on Γ_0 except that the value on the surface is $1-p$ instead of 1. Hence on I_2

$$F_\nu - p \geq p(1-p),$$

or

$$F_\nu \geq 1-(1-p)^2.$$

By a continuation of this process we show that on I_m

$$F_\nu \geq 1-(1-p)^m. \tag{30}$$

Since these functions have the value 1 on the remainder of the surface of the region common to Γ_m and R, it follows that (30) is valid for all points of R within Γ_m.

5. **A Special Case of the Dirichlet Problem.** Let R be a region, as in § 4, whose boundary c is finite and made up entirely of rectangles in the planes $x=a_n$, $y=b_n$, $z=c_n$, where a_n, b_n, c_n are terminating binary numbers. Let the function $U(x, y, z)$ be determined on C and let it be continuous in the sense that if (x, y, z) and $(x+\Delta x, y+\Delta y, z+\Delta z)$ are points on C and

$$\text{Lim}\sqrt{(\Delta x)^2+(\Delta y)^2+(\Delta z)^2}=0,$$
$$\text{Lim}\ U(x+\Delta x, y+\Delta y, z+\Delta z)=\ U(x, y, z). \tag{31}$$

We shall construct a function $u(x, y, z)$ which has the given values on C, is harmonic over the interior of R, and continuous over $R+C$.

Let us first construct a net of order ν containing every edge and corner of C as line or node. This is clearly possible from the definition of C for all ν's from a certain μ on. Construct the net potential which reduces to $U(x, y, z)$ at the boundary nodes and extend this function through the interior of each cell by trilinear interpolation. Let the resulting function be $U_\nu(x, y, z)$, or briefly, $U_\nu(P)$, where P is the point (x, y, z).

We first show that the functions $U_\nu(x, y, z)$ are equicontinuous over the boundary C. Let

$$f(PQ)=\text{max.}|\,U(P_1)-U(Q_1)|, \tag{32}$$

where P_1, Q_1 is any pair of points on C such that

$$P_1\,Q_1\leq PQ.$$

Then $f(x)$ is an increasing function of x which by the continuity of $U(P)$ approaches zero with x. If P and Q are two points in the same mesh on C, we can pass from P to Q by two steps P to P_1, and P_1 to Q, each step being parallel to a side of the mesh and not greater than PQ. By the linearity of the interpolation

$$|\,U_\nu(P)-U_\nu(P_1)|$$

has its maximum value for a given distance PP_1 when P and P_1 are on one of the bounding lines of the mesh. Since U_ν and U have the same values at the ends of this line

$$|\,U_\nu(P)-U_\nu(P_1)|\leq\text{max.}|\,U(P)-U(Q)|\leq f(PQ).$$

Similarly

$$|U_\nu(P_1) - U_\nu(Q)| \leq f(PQ).$$

Hence

$$|U_\nu(P) - U_\nu(Q)| \leq 2f(PQ).$$

If P and Q lie in adjacent meshes, we can pass from P to a common edge and from that to Q by two steps of the kind just discussed each of length not greater than PQ. Hence in this case

$$|U_\nu(P) - U_\nu(Q)| \leq 4f(PQ).$$

Finally, if P and Q lie in meshes that are not adjacent, we can pass from P to a corner of its mesh, from that to a corner of the mesh in which Q lies, and from that to Q by three steps, no one of which is greater than PQ. Hence in this case

$$|U_\nu(P) - U_\nu(Q)| \leq 5f(PQ). \tag{33}$$

For any two points P and Q on C this last inequality is valid, which proves the equicontinuity of the functions U_ν on C.

We shall now extend the condition of equicontinuity to the case where P lies on the boundary and Q inside. In § 2 it was shown that the net potential in a linear function of values on the boundary with coefficients whose sum is 1. From the nature of the interpolation, this is evidently true also for the functions U_ν. A cube of side $2PQ$ with center at P contains Q. If $U_\nu(Q)$ is expressed as a linear function of boundary values, the argument of § 4 shows that the sum of coefficients associated with boundary points in a cube of side

$$2^m.2PQ$$

is not less than

$$1-(1-p)^m.$$

The sum of coefficients associated with points outside this cube is then not greater than $(1-p)^m$. By (33) the maximum variation of U_ν at boundary points within this cube is not greater than

$$5\, f(2^{m+1}.2PQ).$$

Since $U_\nu(P)$ is one of these boundary points

$$|U_\nu(P) - U_\nu(Q)| \leq 5f(2^{m+2}.PQ) + 2(1-p)^m \max.|U|. \tag{34}$$

Let

$$\frac{1}{2^{2m+1}} \leq PQ \leq \frac{1}{2^{2m}}.$$

As PQ approaches zero, m increases indefinitely, $2^{m+2}.PQ$ approaches zero and so the right side of (34) is an increasing function $f_1(PQ)$ which approaches zero with PQ.

Finally we wish to show the equicontinuity of the functions U_ν over the whole region $R+C$. If P and Q are nodes, the difference

$$U_\nu(P) - U_\nu(Q)$$

satisfies the net potential equation (4) and its maximum absolute value for a given distance PQ occurs when one of the points P or Q lies on the boundary. This maximum is then not greater than $f_1(PQ)$. If P and Q lie on the same edge of a cell, by the linearity of the interpolation the maximum difference is still not greater than $f_1(PQ)$. If P and Q lie in the same cell the step from P to Q can be broken into three steps parallel to the coördinate axes. In each case the maximum difference occurs when the points are on an edge. Hence in this case the maximum difference is not greater than $3f_1(PQ)$. The case where P and Q lie in adjacent cells can be resolved into two steps of the kind just discussed. Hence the maximum difference is not greater than $6f_1(PQ)$ Finally, if P and Q lie in nonadjacent cells, we can pass from P to a node of its cell, from that to a node of the cell in which Q lies and from that to Q, by three steps not greater than PQ. In any case the difference therefore satisfies the condition

$$|U_\nu(P) - U_\nu(Q)| \leq 13f_1(PQ).$$

Let

$$13\, f_1(PQ) = \phi(PQ).$$

Then

$$|U_\nu(P) - U_\nu(Q)| \leq \phi(PQ) \tag{35}$$

where P and Q are any points of $R+C$. This function $\phi(x)$ is an increasing function of x and

$$\lim_{x=0} \phi(x) = 0.$$

The functions U_ν are therefore equally continuous on $R+C$. They are also equally bounded, since they lie between the maximum and minimum values of U on the boundary. Hence it is possible[2] to select from an infinite set of the functions U_ν a sequence $\{V_n\}$ converging uniformly over $R+C$ to a continuous function $u(x, y, z)$, which clearly is equal to $U(x, y, z)$ on C.

Let Γ be any cube entirely in R with edges that belong to the net from some ν on. Over the boundary of this cube $\{V_n\}$ converges uniformly to u. The net potential function V_ν' corresponding to values of u on the surface of Γ is the sum of six net potentials of the kind discussed in § 3. In a region within Γ this net potential therefore converges uniformly to a harmonic function. Further when V_n differs from u by less than ϵ on the boundary of the cube, V_n' differs from V_n on the interior by less than ϵ. Hence V_n converges uniformly to a function harmonic in every cube Γ within R. Hence $u(x, y, z)$ is harmonic throughout R.

Any infinite set of the functions U_ν has a sub-set which converges to a function $u(x, y, z)$ harmonic in R and having the given boundary values on C. Since there is only one such harmonic function, U_ν converges uniformly to u. Otherwise there would be an $\epsilon > 0$ such that for an infinity of values of ν

$$\max. |U_\nu - u| > \epsilon,$$

implying that the infinite set of the functions U_ν does not have a sub-set converging to u.

6. **The Solution of the Dirichlet Problem.** Let R be an open region in n-space, bounded by a closed set of points C, and within some sphere of n dimensions about the origin. We shall say that R has the property (D) if, whenever a point O belongs to C, there are positive numbers a and b such that if $r < a$ and an n-sphere Γ of radius r is drawn with O as a center, then if Σ is the set of points in Γ but not in R, the projection[3] of Σ on at

[2] *Cf.* M. Fréchet, *Sur quelques points du calcul fonctionnel*, Rend. del Cir. Math. di Palermo, 1906.

[3] The use of the notion of projection in this paper is due to a suggestion of Prof. O. D. Kellogg.

least one of the $n-1$-spaces determined by a set of n perpendicular axes exceeds $b\,r^{n-1}$ in content. Our theorem is that if R has property (D), and $U\,(x, y, \ldots)$ is any continuous function defined for the points of C, then there is a function $u(x, y, \ldots)$, harmonic in R, continuous on $R+C$, and reducing to $U(x, y, \ldots)$ on C. We shall prove this theorem by parallel stages for $n=2$ and $n=3$.

To this end, suppose C overlaid with a net of order ν. The squares [or cubes] of this net which contain points of C will form a set of points K_ν. Then K_ν will, in general, have a part of its boundary within R and a part outside R. Let the part within R be C_ν. For a sufficiently large ν, C_ν will entirely bound a region (not necessarily connected) entirely within R. Let this region be R_ν.

Every node of the net of order ν on C_ν is not further than $d(\nu)=\sqrt{2}/2^\nu$ [or $\sqrt{3}/2^\nu$] from some point of C. Let $U_\nu\,(x, y)$ [$U_\nu, (x, y, z)$] be defined at a node P of C_ν as the value of $U(x, y)$ [$U(x, y, z)$] at some point Q of C such that $\overline{PQ}\leq d(\nu)$. On the edges [faces] of C_ν, let U_ν be defined by a process of linear [bilinear] interpolation between the adjacent nodes. Then $U_\nu(x, y)$ [$U_\nu(x, y, z)$] is continuously defined over C_ν. Now, we have proved that the Dirichlet problem is soluble for regions bounded by contours such as C_ν and for sets of boundary values such as U_ν Hence there is a function $u_\nu(x, y)$ harmonic over the interior of R_ν, continuous over $C_\nu+R_\nu$, and reducing to $U_\nu(x, y)$ [$U_\nu(x, y, z)$] on C_ν.

In defining the property (D), we have introduced certain positive numbers a and b related to a point O in C. Let $r<a$, and let Γ' be a circle [sphere] about O of radius $2r$. Choose r so small that if P is any point of C within Γ',

$$|U(O)-U(P)|<\epsilon.$$

If ν is so large that $2d(\nu)<r$, then if P lies on C_ν and in a circle Γ [sphere] of radius r about O, we shall have

$$|U(O)-U_\nu(P)|<\epsilon.$$

Let Γ_1 be a circle [sphere] of radius θr about O as center. Let Σ be the part of Γ_1 not in R, and Σ_ν the part of Γ_1 not in

R_ν. Clearly Σ is entirely enclosed in Σ_ν. It then follows from the character of the boundary of Σ_ν and from the fact that R has property (D), that in Σ_ν there can be drawn parallel to one of the coördinate axes [planes] a finite number of linear segments [of regions bounded by a finite number of arcs of circles and straight lines] of length [area] totalling to $b\theta r$ [to $b\theta^2 r^2$], and with projections which do not overlap. Let this set of segments [regions] be σ_ν.

Consider a distribution of matter of density 1 along σ_ν, generating a logarithmic [Newtonian] potential. On the periphery [surface] of Γ, this potential cannot exceed $-b\theta r \log[r(1-\theta)]$ $\left[\text{cannot exceed } \dfrac{b\theta^2 r^2}{r(1-\theta)}\right]$, since $r(1-\theta)$ is less than or equal to the minimum distance of σ_ν from the periphery [surface] of Γ. By a similar argument, the potential in Γ_1 can never be less than $-b\theta r \log 2\theta r$ $\left[\text{or } \dfrac{b\theta r}{2}\right]$. On σ_ν itself, the potential would be increased if the different parts of σ_ν were brought nearer by being brought into the same line [plane], and the maximum potential would be still further increased if they were compressed into a single segment [circle], and the potential taken at the center. This will give us as an upper bound to the potential on σ_ν in the two-dimensional case

$$-2\int_0^{\frac{b\theta r}{2}} \log x \, dx = b\theta r\left(1 - \log\frac{b\theta r}{2}\right),$$

and in the three-dimensional case

$$2\pi \int_0^{\theta r\sqrt{\frac{b}{\pi}}} \frac{1}{\rho}(\rho d\rho) = 2\theta r\sqrt{\pi b}.$$

Let the potential-function just defined be $V(P)$. Let a new harmonic function $W(P)$ be defined as

$$\frac{V(P) + b\theta r \log[r(1-\theta)]}{b\theta r\left\{1 + \log\dfrac{2(1-\theta)}{b\theta}\right\}}$$

in the two-dimensional case, and

$$\frac{V(P)-\dfrac{b\theta^2 r}{1-\theta}}{2\theta r\sqrt{\pi b}-\dfrac{b\theta^2 r}{1-\theta}}$$

in the three-dimensional case. Clearly $W(P)\leq 1$ on σ_ν, and $W(P)\leq 0$ on the periphery of Γ. Over Γ_1 we have in the two-dimensional case

$$W(P)\geq \frac{\log\dfrac{(1-\theta)}{2\theta}}{1+\log\dfrac{2(1-\theta)}{b\theta}},$$

and in the three-dimensional case

$$W(P)\geq \frac{\dfrac{\sqrt{b}}{2}-\dfrac{\theta}{1-\theta}\sqrt{b}}{2\sqrt{\pi}-\dfrac{\theta\sqrt{b}}{1-\theta}}.$$

In both cases, the expression on the right-hand side is positive if $\theta<\frac{1}{3}$. It is, moreover, constant at a given point O if θ is given a definite value, say 1/4. We shall use the symbol α for the expression on the right-hand side of these inequalities.

Now let Γ_2 be a circle [sphere] about O with radius $r/16$, let Γ_3 be a circle [sphere] about O with radius $r/64$, and in general let Γ_n be a circle [sphere] about O with radius $r/4^n$. Let $\xi(P)$ be any function harmonic over R_ν, positive over Γ, and at least 1 over the part of C_ν within Γ. Since $\xi(P)\geq W(P)$, we have $\xi(P)\geq \alpha$ over Γ_1. Form the new harmonic function $\dfrac{\xi(P)-\alpha}{1-\alpha}$.

This function is positive over Γ_1, and at least 1 over the part of C_ν within Γ_1. Hence over Γ_2,

$$\frac{\xi(P)-\alpha}{1-\alpha}\geq\alpha,$$

or

$$\xi(P) \geq 1-(1-\alpha)^2.$$

In the same way, over Γ_n,

$$\xi(P) \geq 1-(1-\alpha)^n.$$

Now, over the part of C_ν within Γ, we have

$$|U(O)-U_\nu(P)|<\epsilon.$$

Hence by a change in U_ν of less than ϵ, we can give to U_ν the constant value $U(O)$ over the part of C_ν within Γ. Let the modified function thus obtained from U_ν be V_ν, and let it give rise to a harmonic function v_ν over R_ν. Let the maximum possible absolute value of $U(P)$ be M. Then

$$\frac{v_\nu(P)+M+\epsilon}{U(O)+M+\epsilon} \text{ and } \frac{-v_\nu(P)+M+\epsilon}{-U(O)+M+\epsilon}$$

are functions satisfying the conditions we have laid down for $\xi(P)$. Hence over Γ_n, if we choose n so that $(1-\alpha)^n<\epsilon$

$$\frac{v_\nu(P)+M+\epsilon}{U(O)+M+\epsilon} \geq 1-\epsilon \text{ and } \frac{-v_\nu(P)+M+\epsilon}{-U(O)+M+\epsilon} > 1-\epsilon.$$

Hence over Γ_1

$$|v_\nu(P)-U(O)| \leq 2M\epsilon+\epsilon^2.$$

or

$$|u_\nu(P)-U(O)|<\epsilon^2+(2M+1)\epsilon.$$

By a proper choice of ϵ, we can make $\epsilon^2+(2M+1)\epsilon$ less than any assigned number, say η. Hence about every point O of C we can draw a circle [sphere] Γ_0 such that for a sufficiently large ν if P lies within this circle, and $u_\nu(P)$ exists,

$$|u_\nu(P)-U(O)|<\eta.$$

By the Heine-Borel theorem, every point in C can be made an interior point of one of a finite number of these circles [spheres] Γ_0, for a given value of η. If ν is made sufficiently large, C_ν will lie entirely within this finite set of the Γ_0. It will follow that from some value of μ on, two different u_μ's cannot differ by more than 2η on C_ν, and hence on R_ν. In other words, on any R_ν, the func tions $u_\nu(P)$ converge uniformly to a limit, which must thus be a harmonic function $u(P)$ defined over R.

Clearly over Γ_0,

$$|u(P) - U(O)| \leq \eta.$$

Now in Γ_0, η can be made as small as we please. Hence if $\{P_n\}$ is any sequence of points in R approaching a point O on C, we have

$$U(O) = \lim u(P_n).$$

In other words, if $u(P)$ be assigned the same values as $U(P)$ on C, it will constitute a continuous function. Hence we have solved the Dirichlet problem.[4]

[4] This paper was presented at the 1922 Christmas meeting of the American Mathematical Society at Cambridge. At the same meeting a paper involving a somewhat similar method of treatment was presented by Mr. Raynor of Princeton. That paper is to be offered for a doctoral thesis. It was further learned on that occasion that many of the results of this paper were obtained by different methods by Mr. Gleason of Princeton, in a paper likewise to be offered for a doctoral thesis.

Comments on [23b]

L. Lumer

This, the first of Wiener's papers on potential theory, shows clearly the electrical origin of Wiener's interest in the subject, at the same time as it displays practically all of the main ideas Wiener subsequently developed in all generality, after he found or introduced the appropriate needed tools.

The idea of using nets and linear difference-equations to approximate an ordinary or partial differential equation had already been seen much earlier (Runge, 1911[1] , and Richardson, 1917[2] , for ordinary, partial elliptic and tentatively hyperbolic, differential equations; Le Roux, 1914[3], for the Dirichlet problem). It had been applied by Phillips in a purely electrical context, but certainly was not very common, nor had it been pushed as precisely and successfully as here, up to the complete resolution of the Dirichlet problem for regular domains of a fairly general type. The idea is very much in use today for computers, but has left little trace in the abstract study of the Dirichlet problem; nor did any of the results in this paper have definite impact on the theory, due to their lack of generality.

But in the light of all further developments of potential theory and looked upon from the standpoint of the research it was suggesting and the results this leads to, this pioneering work should be considered as one of the most significant Wiener wrote on the subject.

Here appears the idea of dissociating the Dirichlet problem into first the unique determination of a harmonic function associated with the given continuous boundary conditions; next the study of that function at the boundary, which opened the path to the functional point of view with the use of integration theory and to the consideration of discontinuous boundary conditions ([23c] ; Perron, 1923[4] ; Lebesgue, 1924[5] ; Bouligand, 1924[6] ; later Brelot 1939[7] ; and many others), which also forced a deeper study of the notion of regularity of boundary points ([24c] ; Lebesgue, 1924[8] ; and many others). A sufficient condition given in that direction already senses the geometric and local character of regularity, and its equivalence with a certain "non-smallness" of the complement around the point.

Here also appears the definite technique of constructing the solution by the limiting process of approximation of the domain by regular ones and passage to the limit on the solutions corresponding to these domains and a continuous extension of the given continuous boundary function. Polycubic domains have ever since been used in approximating Euclidian domains. All Wiener needed at that point to be able to apply this technique to an arbitrary domain was the tool provided by the notion of capacity that he was to introduce himself shortly thereafter [24a].

References

1. C. Runge, Göttingen Nachr., 1911.
2. R. G. D. Richardson, Trans. Amer. Math. Soc. 18(1917), 489-518.
3. J. Le Roux, J. Math. Pures Appl. ser. 6, 10(1914), 189-230.
4. O. Perron, Math. Z. 18(1923), 42-54.
5. H. Lebesgue, C. R. Acad. Sci. Paris 178(1924), 349-354.
6. G. Bouligand, C. R. Acad. Sci. Paris 178(1924), 1054-1057.
7. M. Brelot, Acta Sci Math. (Szeged) 9(1939), 133-153.
8. H. Lebesgue, C. R. Acad. Sci. Paris 178(1924), 1053-1054.

DISCONTINUOUS BOUNDARY CONDITIONS AND THE DIRICHLET PROBLEM*

BY

NORBERT WIENER

It has been suggested that in the Dirichlet problem there is something essentially antagonistic between the utmost degree of generality attainable as regards the geometrical character of the boundary and the utmost of generality attainable as regards the boundary values assumed over this boundary. The purpose of this paper is to show that the solution of the Dirichlet problem merely for continuous boundary conditions at once entrains the unique determination of a harmonic function correlated with discontinuous boundary conditions of a very general sort. The logical tool employed to this end is the Daniell integral.† In the stress laid on generalized types of integration this paper is closely akin to one of G. C. Evans,‡ but it appears to the author that though the theory of Evans gives more detailed information concerning the character of the solutions of the Dirichlet problem in the neighborhood of the boundary, it is less direct than the present theory, and less extensible to regions of infinite connectivity or higher dimensionality.

I. The potential at a point as a linear functional of the boundary conditions

Let R be any open set of points in n-space connected in the sense that any two of its points form extremities of a polygonal line lying entirely within it, and not extending to infinity. Let the Dirichlet problem be solvable over R for any continuous boundary conditions on C. That is, if $U(P)$ is defined for every point P on C, and if

$$\lim_{\overline{PQ}\to 0} U(P) = U(Q)$$

* Presented to the Society, April 28, 1923.

† *A general form of integral*, Annals of Mathematics, ser. 2, vol. 19 (1918), p. 279.

‡ *Problems of potential theory*, Proceedings of the National Academy of Sciences, vol. 7 (1921), pp. 89–98.

for every Q on C, then there is a function $u(P)$, defined and continuous over $R+C$, harmonic over R, and reducing to U on C. With the aid of a form of generalized integral due to Daniell, we shall associate with any discontinuous function $V(P)$ of a very wide class of functions defined over C a unique function $v(P)$ defined and harmonic over R, which we might naturally call the solution of the boundary value problem corresponding to $V(P)$.

Let Q be a given point in the interior of R. Then $u(Q)$ may be regarded as a functional of $U(P)$. To symbolize this point of view, let us write

$$I_Q(U) = u(Q).$$

Since the functions harmonic over R form a linear set, we have

$$(C) \qquad I_Q(CU) = CI_Q(U),$$

and

$$(A) \qquad I_Q(U_1+U_2) = I_Q(U_1)+I_Q(U_2).$$

Moreover, because of the fact that no function harmonic over R has extrema over R,

$$(P) \qquad I_Q(U) \geqq 0 \quad \text{if} \quad U \geqq 0.$$

Another property of I_Q is that

(L) If $U_1, U_2, \ldots. U_n, \ldots$ is a sequence of continuous functions defined over C, and

$$U_1 \geqq U_2 \geqq \cdots \geqq 0 = \lim U_n$$

for every point on C, then

$$\lim I_Q(U_n) = 0.$$

To prove this, it is sufficient to show that

$$(1) \qquad \lim \max U_n = 0.$$

since

$$I_Q(U_n) \leqq \max U_n.$$

If (1) were false, it would be possible to find a positive number ε and a sequence $\{P_n\}$ of points on C such that for every n

$$U_n(P_n) > \varepsilon.$$

This sequence will have at least one limit element N, which will belong to C, as C is a closed, bounded set. By hypothesis

$$U_1(N) \geqq U_2(N) \geqq \cdots \geqq 0 = \lim U_n(N).$$

Hence there is a value of n, say ν, such that

(2) $$U_\nu(N) < \varepsilon/2.$$

Let $\{N_n\}$ be a sequence selected from the P_n's and approaching N as a limit. It follows from the hypothesis of (L) that for every N_n from some stage $n = \mu$ on

$$U_\nu(N_n) > \varepsilon.$$

Hence, since U_ν is continuous.

(3) $$U_\nu(N) \geqq \varepsilon.$$

Since formulas (2) and (3) contradict one another, (1) and hence (L) is proved.

The continuous functions defined over C form a linear set: that is, the sum of any two or a constant multiple of any one is also continuous. The absolute value of a continuous function is continuous. A continuous function is bounded over C. These properties of continuous functions are those attributed to a class T_0 by Daniell in the paper to which we have already referred. Moreover (C), (A), (P), and (L) show that the operator I_Q fulfils the conditions which he has laid down for an operator I on T_0. We are hence in a position to employ those definitions and theorems by which he enlarges the scope of the operator I.

II. THE DANIELL EXTENSION OF I_Q

In accordance with a definition of Daniell, a function U is of the class T_1 if $U_1 \leqq U_2 \leqq \cdots$ is a non-decreasing sequence of functions from T_0 and $U = \lim U_n$. Under these circumstances, by a theorem of Daniell,

$$I_Q(U_1) \leqq I_Q(U_2) \leqq \cdots.$$

and either $\lim I_Q(U_n)$ exists or else $I_Q(U_n)$ becomes infinite. We shall write

(4) $$u(Q) = I_Q(U) = \lim I_Q(U_n) = \lim u_n(Q).$$

In general, we shall preserve the usage here indicated for the correlation of corresponding small and capital letters, applying the latter to functions on C, the former to the functions thereby determined as generalized integrals on R.

Now $u_n(Q)$ is harmonic over R. Hence the sequence $\{u_n\}$ is a monotone sequence of harmonic functions. It follows from a theorem of Harnack* that if $u(Q)$ is finite for any Q in R, it is finite and harmonic for every Q in R, while $u_n(Q)$ converges uniformly to $u(Q)$ over any closed region S interior to R.

In accordance with Daniell's definition, if V is any function defined on C, we shall write $\dot{I}_Q(V)$ for the lower bound of $I_Q(U)$ for all the functions U of class T_1 such that $U \geqq V$. Similarly, we make the definition

$$\underset{\bullet}{I}_Q(V) = -\dot{I}_Q(-V).$$

When we have $\underset{\bullet}{I}_Q(V) = \dot{I}_Q(V)$, we say that V is summable, and with Daniell write what is in our notation

$$v(Q) = I_Q(V) = \underset{\bullet}{I}_Q(V) = \dot{I}_Q(V).$$

When V is summable, it is clearly possible to find, whatever positive number ϵ may be, a pair of functions U_1 and U_2, such that U_1 and $-U_2$ belong to T_1, while

$$U_1 \geqq V \geqq U_2,$$

and

$$I_Q(U_1) + I_Q(-U_2) = u_1(Q) - u_2(Q) < \epsilon.$$

The function $u_1(P) - u_2(P)$ is clearly harmonic. We now proceed as in the proof of Harnack's theorem.† Let us describe about Q any circle lying entirely within R. Let $u_1 - u_2$ assume the values W on the periphery of this circle, the radius of which we take to be a. Then over the interior of the circle

* Osgood, *Lehrbuch der Funktionentheorie*, Leipzig, 1907, p. 615. This theorem is there proved for the two-dimensional case, but the proof may be extended at once to n dimensions. The restriction of the theorem to "Bereiche S" is not essential, as it applies at once to all connected open regions.

† The proof is here given in the form appropriate to a space of two dimensions, but mutatis mutandis is valid in n dimensions. For the sake of explicitness the language of two dimensions is used in several subsequent passages.

$$u_1 - u_2 = \frac{1}{2\pi}\int_0^{2\pi} W \cdot \frac{a^2 - r^2}{a^2 - 2ar\cos(\theta - \psi) + r^2}\, d\psi.$$

Since

$$0 \leqq \frac{a^2 - r^2}{a^2 - 2ar\cos(\theta - \psi) + r^2} \leqq \frac{a + r}{a - r} \qquad (r < a)$$

we have

$$u_1 - u_2 \leqq \frac{1}{2\pi}\int_0^{2\pi} W \frac{a+r}{a-r}\, d\psi$$

$$= \{u_1(Q) - u_2(Q)\} \cdot \frac{a+r}{a-r} \tag{5}$$

$$< \varepsilon \frac{a+r}{a-r}.$$

It follows that within any circle about Q of radius less than a, by making ε approach 0, we shall make $u_1 - u_2$ approach 0 uniformly. Since $u_1(P) \geqq \dot{I}_P(V)$ and $u_2(P) \leqq \underset{\cdot}{I}_P(V)$, it follows that within this circle $v(P)$ exists, and is approached uniformly by $u_1(P)$ and $u_2(P)$ as ε approaches 0. Hence within this circle v is harmonic. By the formation of chains of circles such as those used in the theory of analytic continuation, it may be shown that if V is summable with regard to one point of R, it is summable with regard to all points of R, and that the function $v(P)$ thus defined is harmonic everywhere in R.

Daniell shows that the set of all summable functions is closed with regard to the operations of addition, multiplication by a constant, and taking the absolute value. He also shows that the extended operation I_Q satisfies (C), (A), and (P). Condition (L) is replaced by the important theorem which reads in our terminology as follows: if $\{U_n\}$ is a sequence of summable boundary functions such that $\lim U_n = U$, and if there is a summable function V such that over all C we have $|U_n| \leqq V$ for all n, then U is summable, and $u(P) = \lim u_n(P)$. As one consequence of these theorems, functions belonging to T_0 and T_1 are summable, and all the definitions given of $I_Q(U)$ agree when more than one of them is applicable.

III. The behavior of $I_Q(U)$ in the neighborhood of the boundary

If Q is a point of C, and K is a circle with Q as center, then any function $U(P)$ which is defined and summable on C, bounded, and 0 over that part of C

within K, determines a harmonic function $u(P)$ which approaches 0 as P approaches Q. To prove this, let us take a as the radius of K, and let K' be a concentric circle with radius $a/2$. Let the upper bound of $|U|$ be M. Then we shall define the function $V(P)$ as M outside of K, 0 within K', and $M((2r/a)-1)$ at those points in the annulus between K and K' at a distance r from Q. Clearly V is continuous. The boundary values V on C determine a function $v(P)$ continuous on $R+C$, harmonic on R, and reducing to V on C. By definition

$$|U(P)| \leqq V(P),$$

over C. Hence by a theorem of Daniell

$$|u(P)| \leqq I_P(|U|) \leqq v(P).$$

This holds over all R. Since

$$\lim_{P\to Q} v(P) = 0 = V(Q),$$

we have

$$\lim_{P\to Q} u(P) = 0 = U(Q).$$

This result is capable of immediate generalization. In the first place, by the addition of a constant to U, we get the result that if $W(P)$ is a bounded function summable on C and constant over K,

$$\lim_{P\to Q} w(P) = W(Q). \tag{6}$$

Next let W_1 be any bounded summable function defined over C, and let A and B be respectively the upper and lower bounds of the oscillation of W_1 in the neighborhood of the point Q of C. Then given any positive quantity ϵ, it is possible to describe a circle K about Q as center within which $A+\epsilon > W_1(P) > B-\epsilon$. It then follows from the extension of (P) to all summable functions and from (6) that if the functions formed from W_1 by substituting $A+\epsilon$ or $B-\epsilon$ within K are summable, the oscillation of $w_1(P)$ in the neighborhood of Q is not greater than from $A+\epsilon$ to $B-\epsilon$, and hence, since ϵ is arbitrary, is not greater than from A to B. If in particular W_1 is continuous at Q, then the function which is W_1 on C and w_1 on R is likewise continuous there.

Daniell has a theorem to the effect that a function equal for every argument to the greater or to the less of two summable functions is itself summable. The function which is the constant H over the part of C within K and 0 elsewhere may be shown without difficulty to belong to T_1, and hence to be summable. Hence the function obtained from W_1 by replacing its values within K by $A+\epsilon$ or $B-\epsilon$ is in fact summable.

I have so far been unable to eliminate without any further restriction the condition that W_1 be bounded, although I suspect that this condition is superfluous. The results obtained in this section are valid, of course, if for the word "circle" is substituted "n-sphere", in a space of n dimensions.

IV. Certain summable functions

A surface M in n-space is said to have capacity c if there is a positive function $f(P)$ defined over M such that

$$1 = \left(\int_M\right)^{n-1} f(P)\,(PQ)^{n-2}\,dS$$

(in the case $n=2$,

$$1 = -\int_M f(P)\log PQ\,dS)$$

independently of Q so long as Q remains on M, while

$$\left(\int_M\right)^{n-1} f(P)\,dS = c.$$

A set of points on C is said to have zero measure if the function which is 1 over the set and 0 over the rest of C is summable, and determines a harmonic function 0 over R. I say that if we have given a closed set of points S on the boundary C of an n-dimensional region R for which the Dirichlet problem is solvable, and if S can be included in the interior of a surface M of arbitrarily small capacity, then S is of zero measure.

To prove this, let us remark that we can distribute over the part of C within M a boundary potential that is continuous, 1 on S, and 0 on the boundary of M. This boundary potential, together with the boundary potential 0 over the rest of C, will determine a harmonic function $u(Q)$ over R, which will be uniformly less, within R and outside M, than

$$(7) \qquad \left(\int_M\right)^{n-1} f(P)\,(PQ)^{n-2}\,dS,$$

or its two-dimensional analogue.* Now let Q be a point of R at a distance a from the nearest point of C. Then (7) and hence $u(Q)$ will not exceed ca^{n-2} or $-c\log a$, as the case may be. Since C may be made arbitrarily small, the $\dot{I}_Q$ of our boundary function that is 0 except on S, where it is 1, cannot exceed 0. It follows at once that S is of zero measure. Thus a finite set of points, or, more generally, an $(n-2)$-spread on the $(n-1)$-spread bounding a region in the space of n dimensions is of zero measure.

A function which is bounded and has discontinuities only at a closed set of points of zero measure is summable. Let S be this set. Let M_a be the set of all points on C within a distance of less than a from some point of S. Let $U(P)$ be any function bounded on C and with all its discontinuities on S. It is then possible to form a function $U_a(P)$ continuous over the whole of C and differing from $U(P)$ only over M_a.† It is even possible to keep the set of functions $\{U_a\}$ uniformly bounded. These functions U_a remain summable even if any finite change is made in them over S, since any function finite over S and zero over the rest of C is less in absolute value than a function with Daniell integral everywhere zero, and hence is itself of zero Daniell integral. As a approaches 0, these functions U_a thus modified can be made to approach U boundedly over all C. Hence by a theorem of Daniell which we have already quoted, U_a is summable.

* In the two-dimensional case, to avoid difficulties due to the change of sign of the logarithm, the unit of measurement should be chosen greater than the greatest linear dimension of C.

† Given any set of points, it is possible to surround it within any assigned distance by a polyhedral boundary. It is hence possible to surround the set by a sequence of such boundaries, each including the next, approaching to the set within any assigned distance. Given any function uniformly continuous on the original set, it is possible to approximate to it by continuous functions on the polyhedral boundaries which all lie between the bounds of the original function, and which all constitute a function continuous over the set of points consisting of the original set, together with the polyhedra. The continuous functions thus obtained may be taken as the boundary values of an infinite set of harmonic functions defined between the successive polyhedra and outside the outermost one. Together these constitute a continuous function extending the original continuous function through the whole of space, and a fortiori over any surface in space.

MASSACHUSETTS INSTITUTE OF TECHNOLOGY.
CAMBRIDGE, MASS.

Comments on [23c]

L. Lumer

In this paper, the functional point of view prevails in the study of the Dirichlet problem, and the Daniell integral becomes an essential tool, at all possible degrees of generality.

The results are still partial: it took 16 more years (Brelot, 1939),[1] and far more advanced developments of potential theory to completely characterize the summable boundary functions, and thus provide a converse to Wiener's first result here (the determination of a harmonic function uniquely associated with each summable function on the boundary of a regular domain); it took much less to extend it to an arbitrary Euclidian domain [24a].

But the impact and generality are considerable: from the Euclidian or locally Euclidian domains, to the locally compact spaces of today's axiomatic theories of harmonic functions and partial differential equations, all with their various, compact metrizable or not even topological, ideal boundaries, the same approach has been used, the same result and the same proof have remained valid without modification.[2] So it is for the second result, on the behavior of the "solution" at a regular boundary point. This was remarkable at a time when no harmonic measure or Riesz representation theorem for a linear functional were yet available. Incidentally, it should be mentioned also that the question Wiener raises (p. 313) about possibly eliminating the boundedness hypothesis for the boundary function received much later a negative answer.[3]

References

1. M. Brelot, Acta Sci. Math. (Szeged) 9(1939), 133-153.

2. M. Brelot, Ann. Univ. Grenoble, Math. Phys. 22(1946), 167-219; M. Brelot and G. Choquet, Ann. Inst. Fourier (Grenoble) 2(1951) 199-263; M. Brelot, J. Math. Pures Appl., ser. 9, 35(1956), 297-335; C. Constantinescu and A. Cornea, Ergebnisse, Springer-Verlag, Berlin, vol. 32(1963); M. Brelot, Lectures on Mathematics, Tata Inst. of Fund. Research, Bombay, vol. 19(1960); H. Bauer, Lectures Notes in Mathematics, Springer-Verlag, Berlin, vol. 22(1966); and many others.

3. M. Brelot, Bull. Sci. Math., ser. 2, 69(1945), part 1, 153-156.

CERTAIN NOTIONS IN POTENTIAL THEORY[1]

By Norbert Wiener

Introduction. The Dirichlet problem — of determining a harmonic function assuming a given continuous set of values continuously on the boundary of a region R — does not always admit of a solution for every region R. For example, if we use polar cöordinates and take as our R the region $0<r<1$, we shall find that there is no harmonic function assuming the boundary values 0 for $r=1$ and 1 for $r=0$. In order to see why this is so, let us replace R by the slightly modified region $\epsilon<r<1$, and our boundary conditions by the stipulation that the harmonic function shall vanish for $r=1$, and shall equal 1 for $r=\epsilon$. The Dirichlet problem then becomes soluble, and has as its answer $u(r, \theta)=\dfrac{\log r}{\log \epsilon}$. Now, we have for every point in R

$$\lim_{\epsilon \to 0} u(r,\theta)=0.$$

It thus appears that by modifying the boundary slightly and then letting the harmonic function determined on the interior of the new boundary approach its limit as the modification of the boundary becomes progressively less and less, we get a definite

[1] The author wishes to express his thanks to Dr. J. L. Walsh and to Prof. O. D. Kellogg, both of Harvard University, with whom he has discussed matters in this paper, and to whom he owes much in the way of suggestion and stimulus. In spirit, in particular in the importance given to notions such as charge and capacity and in the use of Stieltjes integrals, this paper has much in common with one of Professor Kellogg. (*An Example in Potential Theory*, Proc. Amer. Acad. Arts and Sciences, Vol. 58, No. 14, June, 1923).

harmonic function as a limit — namely 0 — which fails, however, to assume continuously the assigned boundary values at $r=0$. The origin is, as it were, a part of the boundary of zero weight, and a boundary condition at the origin is null and void.

It will be seen, then, that the Dirichlet problem fails to be soluble in this particular case, not by the non-existence of a harmonic function corresponding to the boundary conditions, but by the fact that this harmonic function fails to assume continuously the desired value at the origin. We shall see that whenever the Dirichlet problem fails to be soluble, it is possible to determine a perfectly general function corresponding to our boundary conditions. In more precise language, our theorem reads:

THEOREM I. *Let R be a connected, bounded set of points in n-space, containing no point of its boundary. Let C be its boundary. Let $U(P)$ be any function continuous on the set of points C. Then*

1. *There is a function $v(P)$ continuous over $R+C$ and reducing to $U(P)$ on C;*
2. *There is a sequence $\{\Gamma_n\}$ of boundaries over the interior of which the Dirichlet problem is soluble for any continuous boundary condition, each contained in the interior of its successor, and together containing in their interiors every point of R;*
3. *The function $u_n(P)$ corresponding to the boundary conditions $v(P)$ over Γ_n and harmonic within Γ_n tends uniformly to a harmonic limit over any closed set of points interior to R as n increases without limit;*
4. *This limit, which we shall term $u(P)$, is independent of the choice of $v(P)$ and $\{\Gamma_n\}$.*

A precisely similar theorem will hold for the exterior Dirichlet problem. A particularly interesting case of the exterior Dirichlet problem arises when the assigned values on the boundary are constant over the entire boundary, while the assigned behavior of the desired harmonic function at infinity is the same as that of the potential due to a single charge. Physically, this problem is that of determining the potential due to a distributed charge on a conductor. However a charge on a conductor will not always distribute itself in such a way as to possess a surface-density.

Hence it will not always be possible to find a surface distribution of charge in the ordinary sense which will lead to a solution of this exterior Dirichlet problem. If we replace the ordinary notion of surface distribution, leading to a potential represented by a surface integral in the three-dimensional case, by a surface distribution corresponding to a three-dimensional Stieltjes integral, this difficulty will be removed, and we shall obtain a particular case of:

THEOREM II. *Let R be any bounded set of points in n-space ($n>2$) inside the rectangular parallelepiped with opposite vertices $(a_1, \ldots, a_n)$ and $(b_1, \ldots, b_n)$. Let $u(P)$ be a harmonic function with no singularities except on R, vanishing at infinity, and corresponding to boundary values 1 over R in the sense of Theorem I. Then over every point Q exterior to R we shall have*

$$u(Q)=\int_{a_1}^{b_1}\int_{a_2}^{b_2}\cdots\int_{a_n}^{b_n}(\overline{PQ})^{2-n}\,dM(P)$$

where $M(P)$ is an increasing function of the cartesian coördinates of P and the integral is in the sense of Daniell[3] *an n-dimensional Stieltjes integral. Moreover, if the parallelepiped with opposite vertices $(x_1, x_2, \ldots, x_n)$ and $(y_1, y_2, \ldots, y_n)$ is entirely exterior to R, then*

$$\int_{x_1}^{y_1}\cdots\int_{x_n}^{y_n}dM(P)=0.$$

An analogous theorem will be proved in the two-dimensional case.

Theorem II makes it possible to define the electrostatic capacity of a set of points. This is the total charge corresponding to a unit potential over the set. In other words, it is

$$\int_{-\infty}^{\infty}\cdots\int_{-\infty}^{\infty}dM(P).$$

The capacity of a portion of the boundary of a region gives a more precise definition than any yet given of the amount of boundary in that portion in the sense which is relevant with regard to the solubility of the Dirichlet problem. We shall establish:

[3] P. J. Daniell, Annals of Mathematics, Vol. 21 (1921), p. 30.

THEOREM III. *Let R, C, U, and u be defined as in Theorem I. Let Q be a point of C. Draw an n-sphere S with variable radius r and Q for center. Let the capacity of that part of C lying in or on the surface of S be C_1. Let the capacity of S be C_2. Let the outer Lebesgue measure of that part of S within R, in terms of S itself, be L. Then if there are a denumerable number of values of r tending to zero for which either*

$$L<N<1,$$

or

$$C_1>MC_2>0,$$

M and N being independent of R, then

$$\lim_{P\to Q} u(P)=U(Q).$$

In particular, if this condition is fulfilled for every point Q of C then the Dirichlet problem is soluble for any continuous boundary condition on C.[3]

As far as the knowledge of the author goes, this condition is satisfied by every example of an explicit solution of the Dirichlet problem yet described in the literature. It is not the opinion of the author, however, that it is a necessary condition for the solubility of the Dirichlet problem.

A further matter which comes within the scope of this paper is the effect of boundary conditions at a remote point in modifying the continuity of the $u(P)$ of Theorem I at a given point of the boundary. We shall prove:

THEOREM IV.[4] *Let R be a connected, bounded set of points in n-space, containing no point of its boundary. Let Q be a point of its boundary. Let S be a sphere with Q as center. Let R' be another connected, bounded set of points and let the portions of R and of R' within S concide. Let C' be the boundary of R'.*

[3] A theorem of content implying a part of the results of Theorem III is contained in Phillips and Wiener, *Nets and the Dirichlet Problem*, Jour. Math. Phys. M. I. T., Vol 2, pp. 105–124. Cf. also G. E. Raynor, *Dirichlet's Problem*, Annals of Mathematics, Second Series, Vol. 23, pp. 183–198.

[4] A part of the content of this theorem has been enunciated by Prof. O. D. Kellogg, in a forthcoming paper.

Let $u(P)$, defined as in Theorem I, satisfy the condition

$$\lim_{P \to Q} u(P) = U(Q)$$

for any function U continuous over C. Let $V(P)$ be any function continuous over C'. Let $v(P)$ be the function harmonic over R' corresponding to $V(P)$ after the fashion indicated in Theorem I. Then

$$\lim_{P \to Q} v(P) = V(Q).$$

§1. **The relative weights of two point-sets.** Let H and K be two closed bounded sets of points in the plane. Let the lower bound of the distance between a point of H and a point of K be the positive number d. Let $\{H_n\}$ be a set of boundaries, each of which divides the plane into an inside and an outside region, each containing all its successors, with only the points of H common to all. Let $\{K_n\}$ be a similar sequence about K. Let the exterior Dirichlet problem be soluble for any H_n or K_n, in the sense that it is possible to find a harmonic function over the exterior of the portion of boundary in question, assuming arbitrary given continuous values continuously on the boundary, and bounded at infinity. It will always be possible to construct sequences $\{H_n\}$ and $\{K_n\}$ for any H and K. For example, we may first divide the plane into little squares with corners of the form $(i/2^n, j/2^n)$, and take H_n as the boundary of the set of points lying within such of the squares as contain points of H.

It may be shown by a ready application of the alternating process that if m and n are chosen so large that H_m and K_n do not intersect, the exterior Dirichlet problem is soluble for the region excluded by H_m and K_n. Let the harmonic function assuming the values 1 on H_m and $f(P)$ on K_n be $u_{mn}(f; P)$. Then as m increases, the sequence of functions $u_{mn}(f; P)$ is non-increasing. Hence by a theorem of Harnack[5] the sequence $u_{mn}(f; P)$ converges uniformly to a harmonic limit, which we shall call $u_n(f;P)$, over any closed set of points outside H and K. Similarly, let the harmonic function assuming the values $g(P)$ on H_m and 0 on K_n be $\bar{u}_{mn}(g;P)$. Then as n increases, the sequence of func-

[5] Cf. Osgood, *Funktionentheorie*, p. 615.

tions $\bar{u}_{mn}(g;P)$ is non-decreasing. By the same theorem of Harnack, the sequence $\bar{u}_{mn}(g;P)$ converges uniformly to a harmonic limit. which we shall call $\bar{u}_m(g;P)$, over any closed set of points outside H_m and K.

Without any essential restriction, we may suppose that H_1 and K_1 are each exterior to the other. It follows from the considerations which we have just presented that from some value of m on, the functions $u_{mn}(0;P)$ will differ from $u_n(0;P)$ on H by less than ϵ, while the functions $\bar{u}_{mn}(1;P)$ will from some value of n on differ from $\bar{u}_m(1;P)$ on K by less than ϵ. Moreover, there will be a positive number b less than 1, such that for all m and n.

$$\left.\begin{array}{ll} u_{m1}(0;P)=\bar{u}_{m1}(1;P)>1-b & \text{on } H_1; \\ u_{1n}(0;P)=\bar{u}_{1n}(1;P)<b & \text{on } K_1. \end{array}\right\} \tag{1}$$

Now let us obtain a function harmonic outside H_m and K_n, assuming the value 1 on H_m and 0 on K_n, by the alternating process. Let us write $v_0(P)$ for $u_{m1}(0;P)$ and $w_0(P)$ for $u_{1n}(0;P)$. Let $w_k(P)$ be the harmonic function which assumes the value 0 on K_n and the same value as $v_{k-1}(P)$ on H_1. Let $v_k(P)$ be the harmonic function which assumes the value 1 on H_m and the same value as $w_{k-1}(P)$ on K_1. We shall have:

$$\left.\begin{array}{l} |v_1(P)-v_0(P)|<b, \\ |w_1(P)-w_0(P)|<b, \\ |v_2(P)-v_1(P)|<b^2, \\ |w_2(P)-w_1(P)|<b^2, \\ \quad\cdot\quad\cdot\quad\cdot\quad\cdot\quad\cdot \\ |v_k(P)-v_{k-1}(P)|<b^k, \\ |w_k(P)-w_{k-1}(P)|<b^k, \\ \quad\cdot\quad\cdot\quad\cdot\quad\cdot\quad\cdot \end{array}\right\} \tag{2}$$

Hence the series $\sum_1^\infty[v_k(P)-v_{k-1}(P)]$ and $\sum_1^\infty[w_k(P)-w_{k-1}(P)]$ are uniformly and absolutely convergent to a harmonic limit for every P for which they are defined and for every sufficiently late m and n. Over the region excluded by both H_1 and K_1 the functions $v_k(P)$ and $w_k(P)$ approach the same boundary values, and hence the same values everywhere. The limit approached by $v_k(P)$ and $w_k(P)$ as k becomes infinite we shall designate over the entire region exterior to H_1 and K_1 as $F_{mn}(P)$. That is,

$$\left.\begin{aligned} F_{mn}(P) &= v_o(P) + \sum_1^\infty [v_k(P) - v_{k-1}(P)], \text{ or} \\ F_{mn}(P) &= w_o(P) + \sum_1^\infty [w_k(P) - w_{k-1}(P)]; \end{aligned}\right\} \quad (3)$$

and by letting k be sufficiently great, we shall have uniformly

$$\left.\begin{aligned} |F_{mn}(P) - v_k(P)| &\leq \sum_{k+1}^\infty |v_j(P) - v_{j-1}(P)| < b^k/(1-b) < \eta/3, \text{ or} \\ |F_{mn}(P) - w_k(P)| &\leq \sum_{k+1}^\infty |w_j(P) - w_{j-1}(P)| < b^k/(1-b) < \eta/3. \end{aligned}\right\} \quad (4)$$

Here we shall introduce a lemma to the effect that

$$\left.\begin{aligned} |\bar{u}_{mn}(f;P) - u_{mn'}(f;P)| &\leq \max|f| \cdot |\bar{u}_{mn}(1;P) - \bar{u}_{mn}(1,P)|, \\ |u_{mn}(g;P) - u_{m'n}(g;P)| &\leq \max|1-g|\,|u_{mn}(0;P) - u_{m'n}(0;P). \end{aligned}\right\} \quad (5)$$

To prove the first part of this lemma, let $n' < n$, and let $f(P) \geq 0$. Then all we have to prove is

$$\bar{u}_{mn}(\max|f| - f;P) - \bar{u}_{mn'}(\max|f| - f;P) > 0.$$

This is easily demonstrated, however, since in general $\bar{u}_{mn}(\max|f| - f;P)$ is an increasing function of n. The extension of the theorem to non-positive f's and the demonstration of the second part of the theorem offer no difficulties.

Now let m, n, m', and n' all be so large that over the region R exterior to H and K and containing H and K

$$\left.\begin{aligned} &|u_{m1}(0;P) - u_1(0;P)| < \epsilon, \quad |u_{m'1}(0;P) - u_1(0;P)| < \epsilon, \\ &|\bar{u}_{1n}(1;P) - \bar{u}_1(1;P)| < \epsilon, \quad |\bar{u}_{1n'}(1;P) - u_1(1;P)| < \epsilon. \end{aligned}\right\} \quad (6)$$

Let $v_k'(P)$ and $w_k'(P)$ be the functions obtained, respectively, by substituting m' and n' for m and n in the definition of $v_k(P)$ and $w_k(P)$. Then by the definition of v_k and w_k, we obtain from (5)

$$\left.\begin{aligned} &|v_0(P) - v_0'(P)| < 2\epsilon, \\ &|w_0(P) - w_0'(P)| < 2\epsilon, \\ &|v_1(P) - v_1'(P)| < 4\epsilon, \\ &|w_1(P) - w_1'(P)| < 4\epsilon, \\ &\quad \cdot \quad \cdot \quad \cdot \quad \cdot \quad \cdot \quad \cdot \quad \cdot \\ &|v_k(P) - v_k'(P)| < 2k\epsilon, \\ &|w_k(P) - w_k'(P)| < 2k\epsilon, \\ &\quad \cdot \quad \cdot \quad \cdot \quad \cdot \quad \cdot \quad \cdot \end{aligned}\right\} \quad (7)$$

By choosing ϵ so that $2k\epsilon < \eta/3$, we have from (4) and (7)

$$|F_{mn}(P) - F_{m'n'}(P)| < \eta. \tag{8}$$

In other words, $F_{mn}(P)$ converges uniformly over R to a limit which of course is harmonic. We may term this limit, which we shall write $W_{HK}(H; P)$, the weight at P of H in the system H, K. It is easy to show that $W_{HK}(H; P) + W_{KH}(K; P) = 1$.

With the exception of this last sentence, all the results of this section extend at once to n dimensions. It is of course necessary to give a suitable interpretation to the exterior Dirichlet problem. As is familiar, the Dirichlet problem in three or more dimensions involves the additional condition that the harmonic function sought for shall vanish at infinity.

Another important remark is that $W_{HK}(H; P)$ is independent of the particular sequences $\{H_m\}$ and $\{K_n\}$ chosen in obtaining it. Let $\{H_m'\}$ be another sequence fulfilling the conditions laid down for $\{H_m\}$. H_1 will contain in its interior a term of $\{H_m'\}$, say H_2''. This will in turn contain in its interior a term of $\{H_m\}$, say H_3''. By continuing this process, we can construct a sequence $\{H_m''\}$, consisting alternately of terms from $\{H_m\}$ and $\{H_m'\}$. A similar remark holds of $\{K_n\}$. Clearly the value of $W_{HK}(H; P)$ formed by substituting $\{H_m''\}$ and the similarly derived sequence $\{K_n''\}$ for $\{H_m\}$ and $\{K_n\}$ respectively is at once identical with that corresponding to the sequences $\{H_m\}$ and $\{K_n\}$ and to that corresponding to the sequences $\{H_m'\}$ and $\{K_n'\}$.

§2. **The generalized Dirichlet problem.** Let R be a connected, bounded set of points in n-space, containing no point of its boundary, and let C be its boundary. Let C_1 and C_2 be two sub-sets of C, such that the lower bound of the distance from a point of C_1 to a point of C_2 is the positive number d. Then the number $W_{C_1C_2}(C_1; P)$ will exist for every point in R, and will lie between 0 and 1. If C_1 be kept fixed and C_2 be allowed to vary, $W_{C_1C_2}(C_1; P)$ will have a non-negative lower bound. This we shall term the *outer weight* of C_1 at P, and shall represent by the symbol $W(C_1; P)$. Clearly if C_1 and C_2 are defined as above,

$$W(C_1; P) + W(C_2; P) = W(C_1 + C_2; P), \tag{9}$$

while if C_1 and C_2 are any two sub-sets of C,

$$W(C_1; P)+W(C_2; P) \geq W(C_1+C_2; P). \tag{10}$$

For reasons of simplicity we shall now use two-dimensional language. Let C_x be the set of all points of C for which the X-coördinate does not exceed x. Then $W(C_x; P)$ is an increasing function of x. It therefore has no discontinuities other than a denumerable set of finite jumps. It follows at once that if $\{\xi_n\}$ is a denumerable set of values of x, there is only a denumerable set of values of a for which $\{\xi_n+a\}$ contains a discontinuity of $W(C_x; P)$. Hence it is possible to determine the origin of coördinates in such a manner that no value of x that is a terminating binary number is a discontinuity of $W(C_x; P)$. Similarly, if we let C_y' be the set of all points of C for which the Y-coördinate does not exceed y, and suitably choose the vertical level of the origin of coördinates, no value of y that is a terminating binary number will be a discontinuity of $W(C_y'; P)$.

Let us choose our origin in the manner indicated above, and let x be a terminating binary number. Let D be the set of points of C with an X-coördinate not less than $x-\epsilon$ and not greater than $x+\epsilon$. Then:

$$W(C_{x-2\epsilon}; P)+W(D; P) \leq W(C_{x+\epsilon}; P). \tag{11}$$

This is an immediate result of the fact that the weight of a set of points is at least as great as that of an included set of points, and of (9). We deduce at once

$$W(D; P) \leq W(C_{x+\epsilon}; P)-W(C_{x-2\epsilon}; P), \tag{12}$$

whence we obtain

$$\lim_{\epsilon \to 0} W(D; P)=0. \tag{13}$$

This last formula holds uniformly over any closed set of points belonging to R. This follows from the fact that $W(D; P)$ is a non-negative harmonic function. Hence if C is a circle with P as center lying entirely within R, if r is the radius of C, and if Q is a point with polar coördinates (ρ, θ) with respect to P as origin, then

$$W(D;Q)=\frac{1}{2\pi}\int_0^{2\pi} W(D;Q')\frac{1-\rho^2/r^2}{1-2\rho\cos(\theta-\theta')/r+\rho^2/r^2}d\theta' \tag{14}$$

$$\leq\frac{1}{2\pi}\int_0^{2\pi} W(D;Q')d\theta\frac{r+\rho}{r-\rho}=W(D;P)\frac{r+\rho}{r-\rho},$$

Q' being understood to be (r, θ'). It follows that (13) holds uniformly over any circle interior and concentric to C, and hence, by a process such as that used in the analytic continuation of a function, it may be proved that (13) holds uniformly over any closed set of points interior to R.

Now let $U(P)$ be any function continuous on C. There is no essential restriction on U if we assume $|U(P)|<1$. It is then possible to divide up the plane into a network of squares of vertical and horizontal lines, with opposite corners of the forms $(j/2^q, k/2^q)$ and $((j+1)/2^q, (k+1)/2^q)$, such that the oscillation of U in none of these squares exceeds θ. We may then remove from C a set D, consisting of all points in a set of vertical and horizontal strips containing all the boundary points of these squares, and in total outer weight less than ϵ. There will be left a set of sets of points of C, each consisting in all the points of C in one of the squares and not in D. We shall call this set $\{C_n\}$.

We shall now prove a lemma to the following effect: *if from each C_n we select a point P_n, and if R' is a closed set of points interior to R, then there is a function $u(P)$ harmonic over R, and such that if θ and ϵ tend to 0 in any manner whatever, then*

$$u(P)=\lim_{\vartheta,\epsilon\to 0}\Sigma U(P_n)W(C_n;P) \tag{15}$$

uniformly over R'.

To prove this, let us form sets $\{C_n'\}$ and points $\{P_n'\}$ corresponding to a value ϵ' of ϵ and a value θ' of θ each less than the values corresponding to $\{C_n\}$ and $\{P_n\}$. Then the removal from the sets $\{C_n'\}$ of the points of D will leave a finite number of disconnected sets $\{C_n''\}$. Let P_n'' be a point from C_n''. We shall have

$$|\Sigma U(P_n'')W(C_n'';P)-\Sigma U(Q_n)W(C_n'';P)|\leq\theta,$$

where Q_n is the value of the P_m from the C_m containing C_n''. Moreover,

$$\left| \Sigma U(Q_n)W(C_n''; P) - \Sigma U(P_n)W(C_n; P) \right|$$

$$\leq \Sigma W(C_n; P) - \Sigma W(C_n''; P) < \epsilon.$$

This holds uniformly over R'. Hence over R' we have

$$\left| \Sigma U(P_n'')W(C_n''; P) - \Sigma U(P_n)W(C_n; P) \right| < \epsilon + \theta,$$

and likewise

$$\left| \Sigma U(P_n'')W(C_n''; P) - \Sigma U(P_n')W(C_n'; P) \right| < \epsilon + \theta;$$

whence

$$\left| \Sigma U(P_n')W(C_n'; P) - \Sigma U(P_n)W(C_n; P) \right| < 2\epsilon + 2\theta.$$

From this (15) may be deduced at once.

It is manifest from this last demonstration that $u(P)$ is quite independent of the particular sequences of values of C_n and P_n employed in attaining to it. We now wish to prove that $u(P)$ is the function represented by the same symbol in Theorem I. In the following discussion, Γ_ν, $v(P)$, and $u_\nu(P)$ receive the same interpretations as in the formulation of Theorem I.

Since D is of outer weight less than ϵ, it is possible to find a boundary D' for the exterior of which the Dirichlet problem is soluble, enclosing D in its interior, and such that over R' its outer weight in the system D', C is less than ϵ. Now, the outer weight of D' over R' in the system D', Γ_ν is no greater than the outer weight of D' over R' in the system D', C, provided only ν is so large that R' lies entirely within Γ_ν. To prove this, we must first notice that it is possible to find a D'' containing D' in its interior and a contour containing a closed sub-set of C, say C_1, exterior to D'', such that the Dirichlet problem is soluble over C_1, D'', and

$$\left| W_{D''C_1}(D''; P) - W'(D'; P) \right| < \eta, \tag{16}$$

over R', where W' is used as the symbol for outer weight in the system C, D'. It is also possible to find a D''' interior to D'' and a Γ_ν' exterior to C_1 and containing all points of Γ_ν exterior to or on D'', such that the Dirichlet problem is soluble within the boundary D''', Γ_ν', and such that over R'

$$\left| W_{D'''\Gamma'_\nu}(D'''; P) - W''(D'; P) \right| < \eta, \tag{17}$$

where W'' is used as the symbol for outer weight in the system Γ, D'. Now, on Γ_ν'

$$W_{D'''\Gamma'_\nu}(D'''; P)=0,$$

$$W_{D''C_1}(D''; P)\geq 0;$$

whereas on D''

$$W_{D'''\Gamma'_\nu}(D'''; P)\leq 1,$$

$$W_{D''C_1}(D''; P)=1.$$

Since D'' and Γ_ν' entirely include R', we have

$$W_{D''C_1}(D''; P)\geq W_{D'''\Gamma'_\nu}(D'''; P) \tag{18}$$

over R'. Combining (16), (17), and (18), we get

$$W'(D'; P)\geq W''(D'; P), \tag{19}$$

which is our desired result.

If we remove from Γ_ν all the points interior to D', and if ν is sufficiently large, Γ_ν will fall apart into a set of separated sets of points $\overline{C}_n$, each consisting of all points of Γ_ν not within D' and within the same square used in defining C_n. We now wish to establish an upper bound for

$$|W'(C_n; P)-W''(\overline{C}_n; P)|$$

for all values of P in R'.

To this end, let us remark that

$$|W'(C_n; P)-W_{C_nF}(C_n; P)|<\epsilon$$

where F is the set of all points neither in C_n nor in D. If then G and H are contours surrounding C_n and F, respectively, sufficiently closely, and the Dirichlet problem is soluble for the region exterior to H and G, while $\Phi(P)$ is the harmonic function corresponding to the boundary conditions 1 on G and 0 on H, then over R' we have

$$|\Phi(P)-W'(C_n; P)|<\epsilon. \tag{20}$$

Let G be so chosen, for some H, and let ν be so large that G contains on its interior $\overline{C}_n$. Now let H be so chosen that (20) holds, while H is entirely exterior to Γ_ν. Let us compare the functions $\Phi(P)$ and $W_{\overline{C}_n\overline{F}}(\overline{C}_n; P)$, where $\overline{F}$ consists of all points in Γ_ν but not in $\overline{C}_n$ nor within or on D'. We can approximate to $W_{\overline{C}_n\overline{F}}(\overline{C}_n; P)$ as closely as we wish by the harmonic function $\Phi_1(P)$, which is determined by surrounding $\overline{C}_n$ by a contour G_1 and $\overline{F}$ by a contour H_1 outside of which the Dirichlet problem is soluble, and assigning boundary conditions 1 on G_1 and 0 on H_1. We may then take G_1 entirely within G and H_1 entirely without H. R' will be included by a contour formed of G, H_1, and D'. Over G and H_1 we have $\Phi \geq \Phi_1$, while neither Φ nor Φ_1 exceeds 1 on D'. It may then readily be deduced with the aid of a result similar to (19) that over R'

$$\Phi_1(P) - \Phi(P) \leq 2\epsilon. \tag{21}$$

Combining this result with (20), we can show without difficulty that

$$W_{\overline{C}_n\overline{F}}(\overline{C}_n; P) - W'(C_n; P) < 3\epsilon. \tag{22}$$

A precisely similar argument in which the rôles of C_n and F are interchanged throughout, with the correlated changes made in all the other sets of points, will lead to the result

$$W'(C_n; P) - W_{\overline{C}_n\overline{F}}(\overline{C}_n; P) < 3\epsilon. \tag{23}$$

Combining (22) and (23), we get

$$\left| W'(C_n; P) - W_{\overline{C}_n\overline{F}}(\overline{C}_n; P) \right| < 3\epsilon. \tag{24}$$

This holds uniformly over R'. Combining this with (19), we get the result

$$\left| W'(C_n; P) - W''(\overline{C}_n; P \right| < 4\epsilon \tag{25}$$

over R', since

$$W''(\overline{C}_n; P) + W''(D'; P) \geq W_{\overline{C}_n\overline{F}}(\overline{C}_n; P) \geq W''(\overline{C}_n; P)$$

There is no difficulty in proving that

$$\left| W'(C_n; P) - W(C_n; P) \right| < \epsilon, \tag{26}$$

and that, if W''' is used as the symbol for weight in the system Γ_ν, then

$$\left| W''(\overline{C}_n; P) - W'''(\overline{C}_n; P) \right| < \epsilon. \tag{27}$$

Combining (25), (26), and (27), we get over R'

$$\left| W(C_n; P) - W'''(\overline{C}_n; P) \right| < 6\epsilon. \tag{28}$$

If we now form a continuous function on Γ_ν (as we can do without difficulty in a way that we shall indicate later), which shall assume the value 1 on $\overline{C}_n$, 0 on $\overline{F}$, and intermediate values elsewhere, then it may be shown without difficulty that the harmonic function corresponding to these boundary conditions will differ from $W'''(\bar{C}_n; P)$ by less than ϵ over R'. Let this harmonic function be $\Phi_n(P)$. We then have

$$|\Sigma\, U(P_n)W(C_n; P) - \Sigma\, U(P_n)\Phi_n(P)| < 7N\epsilon \tag{29}$$

over R', where N is the number of distinct sets C_n.

Now let us compare $u_\nu(P)$ and $\Sigma\, U(P_n)\Phi_n(P)$. By making q sufficiently large, if there is any function $v(P)$, which we may without essential restriction suppose less than or equal to 1 in absolute value, continuous over $R+C$ and reducing to $U(P)$ on C, we can make the oscillation of $v(P)$ over the interior of each square into which $R+C$ is divided by the lines $x=j/2^q$, $y=k/2^q$ less than θ. We shall then have over $\bar{F}$, for ν sufficiently large,

$$|\Sigma U(P_n)\Phi_n(P) - u_\nu(P)| < \theta, \tag{30}$$

and over the part of Γ_ν not exterior to D'

$$|\Sigma U(P_n)\Phi_n(P) - u_\nu(P)| \leq N+1. \tag{31}$$

Now let $G(P)$ be a function defined over Γ_ν, equal to $\Sigma U(P_n)\Phi_n(P) - u_\nu(P)$ when this latter quantity is less than θ in absolute value, equal to θ when $\Sigma U(P_n)\Phi_n(P) \geq \theta$, and equal to $-\theta$ when $\Sigma U(P_n)\Phi_n(P) - u_\nu(P) \leq -\theta$. Let this give rise to a function $g(P)$ harmonic over R. Clearly the function $\Sigma U(P_n)\Phi_n(P) - u_\nu(P) - G(P)$ is zero over $\bar{F}$, and nowhere exceeds $N+1+\theta$ in absolute value. Hence by comparison of this latter function with $W''(D'; P)$, we get the result

$$|\Sigma U(P_n)\Phi_n(P) - u_\nu(P)| \leq \theta + (N+1+\theta)\epsilon \tag{32}$$

uniformly over R' Combining this result with (15), we get

$$|u(P) - u_\nu(P)| < \theta + (N+1+\theta)\epsilon + \sigma(\epsilon, \theta), \tag{33}$$

where σ vanishes as θ and ϵ tend to zero independently. Relation (33) will hold for all sufficiently great values of ν.

Now let ϵ and θ be chosen so small that $\theta < \eta/3$ and $\sigma(\epsilon, \theta) < \eta/3$. N will be determined by θ. We may then impose the further condition on ϵ that $(N+1+\theta)\epsilon < \eta/3$. We shall then have

$$|u(P) - u_\nu(P)| < \eta \tag{34}$$

over R' for all sufficiently large ν. That is,

$$u(P) = \lim_{\nu \to \infty} u_\nu(P) \tag{35}$$

uniformly over R', and Theorem I is proved in the two-dimensional case, provided only that we show the existence of boundaries Γ_ν and of an interpolation function $v(P)$.

The boundaries Γ_ν can be constructed with the greatest ease. All we need to do is to form the meshwork of lines $x = j/2^\nu$, $y = k/2^\nu$ in the plane, and to take as our Γ_ν that part of the boundary of the set of all points on squares of the net containing C which lies in R. We may then take for the value of $v(P)$ at each of the nodes of this meshwork lying on Γ_ν the value of $U(P)$ at some point on a mesh on which the node lies. Between the nodes, we may define $v(P)$ on Γ_ν by linear interpolation. The Dirichlet problem is soluble over the region bounded exteriorly by $\Gamma_{\nu+1}$ and interiorly by Γ_ν. Let $v(P)$ be defined over this region as the harmonic function corresponding to the assigned values of $v(P)$ on the periphery. There will be no difficulty in showing[6] that the $v(P)$ thus defined is continuous over $R+C$, if we give it the values $U(P)$ on C. It is to be noted that the $v(P)$ thus defined assumes its extreme values on C.

We have already obtained $u(P)$ in a manner independent of $v(P)$ and the boundaries Γ_ν. We have therefore given a complete demonstration of the validity of Theorem I in the interior two-dimensional case. However, neither the extension to n dimensions nor the obvious verbal alterations which are necessary

[6] Let P be a point on C and let Q be a point of R at a distance not exceeding ϵ from P. Draw a circle S of radius 2ϵ with P as center. Let Q lie on the zone between Γ_ν and $\Gamma_{\nu+1}$. Let the angular measure of the points in this zone and on the periphery of S be M_ν. Since $\Sigma M_\nu \leq 2\pi$, it follows that $\lim M_\nu = O$. Let B_1 be the upper bound of $|U(P) - U(P_1)|$ if $\overline{PP_1} \leq 2+1/2^{\nu-1}$ and let B_2 be the upper bound of $|U(P) - U(P_2)|$ for any P_2 on C. Then $|U(P) - v(Q)|$ is not greater than the potential at Q due to boundary values on S not exceeding B_2 anywhere and not exceeding B_1 except on a set of measure M_ν. That is,

$$|U(P) - v(Q)| \leq 3M_\nu B_2 + B_1.$$

It follows at once that $\lim_{Q \to P} v(Q) = U(P)$.

With regard to the entire theorem, cf. Lebesgue, *Sur le problème de Dirichlet*, Rendiconti di Palermo, v. 24 (1907), pp. 371–402.

to make Theorem I apply to the exterior Dirichlet problem changes the demonstration of Theorem I in a single essential respect.

The generalized solution which we have obtained for the Dirichlet problem has the following properties, which are merely generalizations of the familiar properties of the solution of the ordinary Dirichlet problem and which may be proved without any difficulty:

(1) The generalized solution of the Dirichlet problem lies between the upper and lower bounds of the given boundary values;

(2) The generalized solution of the Dirichlet problem is at every point an additive functional of the boundary values;

(3) The generalized solution of the Dirichlet problem coincides with the solution of the ordinary Dirichlet problem, if the latter exists;

(4) If the boundary C of R breaks up into two separated pieces, C_1 and C_2, and if $v(P)$ and Γ_n are defined as in the formulation of Theorem I, then if Γ_n' represents that part of Γ_n approximating to C_1, and the Dirichlet problem is soluble over the region within $\Gamma_n'+C_2$, then the function $u(P)$ of Theorem I is the uniform limit over any set of points R' closed and interior to R of the harmonic function corresponding to the boundary conditions $v(P)$ over Γ_n' and $U(P)$ over C_2. It is not indeed necessary that the Dirichlet problem be soluble over the region within $\Gamma_n'+C_2$ for boundary conditions other than these special ones.

It results immediately from (4) that if the generalized Dirichlet problem has the solution $u(P)$ for the boundary condition $U(P)$ on C, and C' is a boundary entirely interior to C, then $u(P)$ will be the solution of the generalized Dirichlet problem corresponding to boundary conditions $U(P)$ on C and $u(P)$ on C'.

These results will be assumed without explicit reference in the sequel.[7]

[7] The generalization of the Dirichlet problem in the present paper consists in a generalization of the boundary permissible, the boundary conditions remaining continuous. In a paper forthcoming in the *Trans. Am. Math. Soc.*, the author of this paper shows how for a region for which the Dirichlet problem is soluble for continuous boundary conditions, a generalized Dirichlet problem is soluble for a wide class of discontinuous boundary conditions. The two types of generalization of the Dirichlet problem thus are in two

§3. **The distribution of Charge on an isolated conductor.** It is a familiar physical fact that if an electric charge is released on an isolated conductor, it will spread over the surface of the conductor without gain or loss until the latter reaches a state of equilibrium. The potential caused by this charge will then be constant over the surface of the conductor. It accordingly becomes a matter of interest to ascertain whether the general exterior boundary value problem of Laplace's equation for boundary values 1 can be solved by the potential due to a surface distribution of charge. Of course this problem in its unmodified form is only of interest in three or more dimensions. For purposes of definiteness we shall deal with the three-dimensional case.

As is well known, if S is any closed surface of finite area possessing at every point a normal, and if $U(P)$ is any function harmonic within and on S, we have

$$\iint_S \partial U/\partial n \; dS=0, \tag{36}$$

where $\partial U/\partial n$ is the derivative of U along the internal normal. It is accordingly easy to prove that if $U(P)$ is a harmonic function over the whole of space, except for sets of points $A_1, A_2, \ldots,$ A_n, and if S is a closed boundary with a normal everywhere and over which U is harmonic, then if A_1 is within S while all other A's are exterior to S, $\iint_S \partial U/\partial n \; dS$ is independent of S, and is determined by U and A_1 alone. We shall call $1/4\pi \iint_S \partial U/\partial n \; dS$ the *charge* on A_1.

Now let H and K be two boundaries entirely exterior to one another, and let $v(P)$ be the harmonic function which corresponds in the sense of the last section to the boundary values $V(P)$ on H and K and 0 at ∞. Let $v_1(P)$ correspond to the boundary values $V(P)$ on H and 0 at ∞, K not being considered as part of the boundary. Let $v_2(P)$ correspond to the boundary values

essentially distinct directions. The methods of the *Transactions* paper, however, are as directly applicable to regions for which the Dirichlet problem is only soluble for continuous boundary conditions in the extended sense of this paper as to the cases there treated. Indeed all of that paper, except for so much as relates to the continuity of the solution of the generalized Dirichlet problem near the boundary, remains unchanged if the generalized Dirichlet problem in the sense of this paper is substituted for the ordinary Dirichlet problem.

$V(P)-v_1(P)$ on K and 0 at ∞, H not being regarded as part of the boundary. In general, let v_{2n+1} correspond to the boundary values $V(P)-\sum_1^{2n} v_k(P)$ on K and 0 at ∞, while V_{2n} corresponds to boundary values $V(P)-\sum_1^{2n-1} v_k(P)$ on H. It may be shown with the aid of considerations in no wise different from those familiar in the ordinary instances of the alternating process that

$$v(P)=\sum_1^{\infty} v_n(P) \tag{37}$$

uniformly over the entire exterior of H and K, as will result from the fact that there is a convergent geometric progression dominant to (37). It hence follows that the series

$$\left.\begin{aligned} v'(P) &= \sum v_{2n}(P) \\ v''(P) &= \sum v_{2n+1}(P) \end{aligned}\right\} \tag{38}$$

converge uniformly over the region exterior to K and that exterior to H, respectively, and are harmonic over these regions. They vanish at ∞.

It is not possible to break up $v(P)$ in any other way into two functions vanishing at infinity, one harmonic outside H, and the other harmonic outside K. For suppose that

$$v(P)=u'(P)+u''(P),$$

where u' is harmonic outside H and u'' is harmonic outside K. Then outside $H+K$

$$u'(P)-v'(P)+u''(P)-v''(P)=0.$$

It follows that $u'(P)-v'(P)$ and $v''(P)-u''(P)$ are part of the same harmonic function, which thus has no singularities in the finite plane and vanishes at infinity. It is therefore identically zero.

We may conclude at once that if $U(P)$ is the harmonic function corresponding to boundary values zero at infinity and 1 on the boundary made up of the non-intersecting closed portions A_1, A_2, . . . , A_n, then there are uniquely determined functions U_1, U_2, . . . , U_n, such that $U(P)=\sum U_k(P)$, while $U_k(P)$ is harmonic outside A_k and vanishes at infinity. We may verify the fact that the charge on A_k corresponding to the potential

U_k is the same as the charge on A_k corresponding to the potential U. We may indeed regard U_k as that part of U generated by charges on A_k.

U_k is positive or 0 on A_k. To show this, let us call A_k H and the sum total of the other A's K, and let us take 1 for $V(P)$ in the process leading to (37). Then $v''(P) = U_k(P)$. We may readily verify that

$$\left.\begin{array}{l} v_1 \text{ corresponds to a boundary condition } 1 \text{ on } H. \\ v_2 \text{ corresponds to a boundary condition } 1-v_1 \text{ on } K, \\ v_3 \text{ corresponds to a boundary condition } -v_2 \text{ on } H, \\ v_4 \text{ corresponds to a boundary condition } -v_3 \text{ on } K, \\ \ldots\ldots\ldots\ldots\ldots\ldots \\ v_{2n} \text{ corresponds to a boundary condition } -v_{2n-1} \text{ on } K, \\ v_{2n+1} \text{ corresponds to a boundary condition } -v_{2n} \text{ on } H. \end{array}\right\} \quad (39)$$

Hence

$$\left.\begin{array}{l} v_1+v_3+\ldots+v_{2n+1} \text{ corresponds to } 1-v_2-\ldots-v_{2-n} \text{ on } H, \\ v_2+v_4+\ldots+v_{2n} \text{ corresponds to } 1-v_1-\ldots-v_{2n-1} \text{ on } K. \end{array}\right\} \quad (40)$$

It is hence easy to prove by mathematical induction that the partial sums of the odd v's and also the partial sums of the even v's always lie between 0 and 1. Hence both v' and v'' are non-negative. Since U_k thus corresponds to boundary values non-negative on A_k, we may conclude at once that the charge on A_k corresponding to U is likewise non-negative, as a consideration of the gradient of U_k along any large sphere containing A_k will show.

Now let C be the boundary of an open region R extending to infinity, let C be bounded, and let U be the harmonic function corresponding to the boundary values 1 on C and 0 at infinity. Let A be a sub-set of C, and let B be a sub-set sharing no point or limit-point with A. Let U' be the harmonic function corresponding to the boundary conditions 1 on $A+B$ and 0 at infinity. Let σ' be the charge on A corresponding to U'. Then as B increases in such a manner as to make $A+B$ approach C, σ' will approach a definite positive limit σ, which we shall term the outer charge on A corresponding to the potential distribution U.

To show this, let B' be a set of points containing B as a sub-set, but forming part of C and sharing no point or limit-point with A. Let U'' be the harmonic function corresponding to the boundary values 1 on $A+B'$ and 0 at infinity. Then $U''-U'$ will correspond to boundary values non-negative on B', 0 on A, and 0 at infinity. In the alternating process leading to (37) let H be A, let K be B', and let $V(P)$ be $U''-U'$ on B' and 0 on A. Then

$$\left.\begin{array}{l} v_1 \text{ corresponds to a boundary condition } 0 \text{ on } H, \\ \qquad \text{and vanishes,} \\ v_2 \text{ corresponds to a boundary condition } V \text{ on } K, \\ v_3 \text{ corresponds to a boundary condition } -v_2 \text{ on } H, \\ v_4 \text{ corresponds to a boundary condition } -v_3 \text{ on } K, \\ \cdots\cdots\cdots\cdots\cdots \\ v_{2n} \text{ corresponds to a boundary condition } -v_{2n-1} \text{ on } K, \\ v_{2n+1} \text{ corresponds to a boundary condition } -v_{2n} \text{ on } H. \end{array}\right\} \quad (41)$$

Hence

$$\left.\begin{array}{l} v_1+v_3+\ldots+v_{2n+1} \text{ corresponds to } -v_2-\ldots-v_{2n} \text{ on } H, \\ v_2+v_4+\ldots+v_{2n} \text{ corresponds to } V-v_3-\ldots-v_{2n-1} \text{ on } K. \end{array}\right\} \quad (42)$$

We may then conclude by mathematical induction that the sums of the odd v's are non-positive on H, and also that the sums of the even v's exceed or equal V on K. Hence $v''(P)$ is negative, and σ'', the charge on A corresponding to U'', is less than σ'.

It follows that if $B_1, B_2, \ldots$ is a sequence of B's, each including its predecessor, while $\sigma_1, \sigma_2, \ldots$ are the corresponding σ's, then the sequence of σ's is a decreasing sequence of positive numbers and hence has a limit. If finally every point of C not a point or limit point of A is in some B_n, it is easy to show that the value obtained for $\sigma = \lim \sigma_n$ is unique and independent of the sequence of B_n's.

A theorem which is important, and which can be verified without difficulty is the following: if A and B are two sub-sets of C, the outer charge on $A+B$ does not exceed the outer charge on A, added to the outer charge on B. This is an immediate corollary of the fact that if D is a set of points sharing no point or limit-point with $A+B$, the charge on A corresponding to a potential of 1 on $A+B+D$ added to that on B is at least that on $A+B$.

Let C be defined as before. Let $C(x)$ be that part of C made up of those points with an X-coördinate not exceeding x. If $\sigma(x)$ is the outer charge on $C(x)$ corresponding to a potential 1 on C, it is an increasing function of x. As such it has only a denumerable set of discontinuities. As in Section 2, it is then possible to choose the origin of coördinates in such a manner that $\sigma(x)$ will have no discontinuities for any terminating binary x. Making a similar choice of the origin with regard to y and z, we find it possible to enclose the planes $x=i/2^n$, $y=j/2^n$, $z=k/2^n$ in so far as they consist of points of C in the interior of a closed set of points of arbitrarily small outer charge — say of outer charge less than ϵ. We may simultaneously make the outer weight of this set $<\epsilon$ over a closed set R' exterior to C.

The portion of C that remains will be divided up into a number of separated sets of points C_m, each of maximum linear dimensions not exceeding $1/2^{n-1}$. Let us investigate the distribution of charge on these sets corresponding to a boundary potential of 1 on ΣC_m. We shall find on each C_m a charge σ_m, and the potential due to the charge on C_m alone will be positive over all C_m. It will hence be positive over a sphere of radius $1/2^n$ containing C_m. It will of course not exceed 1. Let us expand the potential on the exterior of this sphere S in the Poisson integral.

$$U_m(P)=\frac{1}{4\pi a}\iint_S \frac{U_m(Q)(r^2-a^2)dS}{[r^2+a^2-2ar\cos(a,r)]^{3/2}}, \tag{43}$$

where P is a point on the exterior of S, Q is a point on the element dS of S, P has the polar coördinates (r, θ, ϕ), Q the coördinates (a, θ', ϕ'), U_m represents the potential on the surface of the sphere, and (a, r) is the angle between OQ and OP, O being the origin and the center of S. Now,

$$\frac{r-a)}{(r+a)^2}\leq\frac{r^2-a^2}{[r^2+a^2-2ar\cos(a,r)]^{3/2}}\leq\frac{r+a}{(r-a)^2}, \tag{44}$$

from which it follows that

$$\left|\frac{1}{r}-\frac{r^2-a^2}{[r^2+a^2-2ar\cos(a,r)]^{3/2}}\right| \leq \frac{4a}{(r-a)^2}. \tag{45}$$

Therefore

$$\left|u_m(P)-\frac{1}{4\pi a}\iint\limits_S \frac{U_m(Q)dS}{r}\right| \leq \frac{4a}{(r-a)^2}\frac{1}{4\pi a}\iint\limits_S U_m(Q)dS. \tag{46}$$

However, $1/4\pi a \iint\limits_S U_m(Q)dS$ is the value of the charge on the sphere, so that we see that the change in potential at P which would result from concentrating the entire charge of C_m at O cannot exceed the fraction $\dfrac{4ar}{(r-a)^2}$ of that potential.

Now let R' be a set of points at a minimum distance of d from C. If $w(P)$ is the harmonic function corresponding to a boundary potential of 1 on ΣC_m, and if n is so large that

$$\frac{4ad}{(d-a)^2}<\eta, \tag{47}$$

then

$$|w(P)-\Sigma\sigma_m/r_m|<2\eta, \tag{48}$$

where r_m is the distance from P to some point Q_m in C_m.

We know that by making ϵ sufficiently small we can make $w(P)$ uniformly approach over R' the potential corresponding to boundary conditions 1 over C. This we have substantially shown in Section 2. Let us now investigate the change in $\Sigma\sigma_m/r_m$ due to the decrease of the region D within which we have enclosed our planes $x=i/2^n$, $y=j/2^n$, $z=k/2^n$. We know that the outer charge on C_m+D corresponding to a potential 1 on C is at least as great as σ_m. On the other hand, the argument of (41) and (42) has shown that the outer charge on C_m corresponding to a potential 1 on C does not exceed σ_m. From this and the theorem which we have given concerning the outer charge on the logical sum of two sets, it results at once that if C_m' is the set of all points of C in the cube bounded by planes $x=i/2^n$, $y=j/2^n$, $z=k/2^n$ containing C_m, then the outer charge on C_m' corresponding to a potential 1 on C does not differ from σ_m by more than ϵ. If then this outer charge be σ_m', and if there be N cubes containing C_m's, we shall have

$$|U(P)-\Sigma\sigma_m'/r_m|<2\eta+N\epsilon+\phi(\epsilon), \tag{49}$$

where $U(P)$ corresponds to boundary values 1 on C and O at infinity, and $\phi(x)$ vanishes with x. Since ϵ is not involved in the determination of the C_m's', (49) reduces to

$$|U(P)-\Sigma\sigma_m'/r_m|\leq 2\eta. \tag{50}$$

Let $M(P)$ stand for the outer charge on the set of all points Q of C such that

$$x<x',\ y<y',\quad z<z'; \tag{51}$$

(x, y, z) and (x', y', z') being the coördinates of Q and P, respectively. The $M(P)$ is an increasing function of all the coördinates of P, and is continuous whenever the coördinates of P are terminating binary numbers. In the language of Daniell[8] it is of positive type. While Daniell is not very explicit as to his definition of a multiple Stieltjes integral, it is clear that since

$$\sigma_m'=\overset{3}{\Delta}_{ijk}M \tag{52}$$

in his notation, (x_{1i}, x_{2j}, x_{3k}) and $(x_{1,i+1}, x_{2,j+1}, x_{3,k+1})$ being opposite vertices of the cube containing C_m, we shall have as a consequence of (50) and (47)

$$U(P)=\int_{a_1}^{b_1}\int_{a_2}^{b_2}\int_{a_3}^{b_3}(PQ)^{-1}dM(Q). \tag{53}$$

The rest of Theorem II is obvious in the three-dimensional case. In spaces of dimensionality higher than 3, the proof requires no substantial modification. In space of two dimensions, we meet the complication that the potential due to an isolated unit point charge, instead of vanishing at infinity, has a negative logarithmic infinity. The specification of the exterior Dirichlet problem is accordingly a more difficult matter. We cannot require that the potential shall remain finite at infinity without rendering our entire problem nugatory, as then the exterior Dirichlet problem corresponding to boundary conditions 1 will be a constant. On

[8] P. J. Daniell, *Functions of Limited Variation in an Infinite Number of Dimensions*, Annals of Mathematics, vol. 21 (1919–1920), p. 31.

the other hand, there are an infinity of harmonic functions corresponding to the value 1 on the boundary and with a logarithmic infinity at infinity.

The direct two-dimensional analogue of Theorem II reads: *let R be any bounded set of points in the plane, entirely within a rectangle with sides parallel to the axes and with opposite corners (a_1, a_2) and (b_1, b_2). Let $u(P)$ be a harmonic function with no singularities in the entire plane except on R, except that at infinity there is a logarithmic singularity such that*

$$\lim_{p\to\infty} \frac{-\log r}{u(P)} = 1,$$

r being the distance of P from the origin. Let $u(P)$ correspond to boundary values 1 over R. Then there will be a function $M(P)$ representing the charge on that portion of the boundary of R consisting of points with both coördinates less than those of P. $M(P)$ will be an increasing function of the coördinates of P. Over every point Q exterior to R we shall have

$$u(Q) = \left\{ \int_{a_1}^{b_1} \int_{a_2}^{b_2} (-\log \overline{PQ}) dM(P) \right\} \Big/ \int_{a_1}^{b_1} \int_{a_2}^{b_2} dM(P). \tag{53}$$

Moreover, if the rectangle with opposite vertices (x_1, x_2) and (y_1, y_2) is entirely exterior to R, then

$$\int_{x_1}^{y_1} \int_{x_2}^{y_2} dM(P) = 0. \tag{54}$$

In order to make the formulation of this theorem definite, we must determine what we mean by the harmonic function behaving like $-\log r$ at infinity and corresponding to boundary values 1 over R. We shall mean by this the harmonic function $w(P) - \log r$, where $w(P)$ is finite at infinity, and corresponds in the sense of Theorem I to the boundary conditions $1+\log r$ on the exterior of R.

Let $v(P)$ be the harmonic function corresponding to the boundary conditions 1 on R and $-\log X$ on Σ, a circle with radius X about the origin. Let R' be a bounded set of points exterior to R. It may then be shown without difficulty that

$$u(P) = \lim_{x\to\infty} v(P) \tag{55}$$

uniformly over R'. Moreover, $t(P)=v(P)+\log X$ is a harmonic function corresponding to boundary conditions $1+\log X$ on R and O on Σ.

If R breaks up into the separated portions $A_1, A_2, \ldots, A_n$, it is possible to prove just as in the earlier part of this section that $t(P)$ may be decomposed in a unique fashion into harmonic functions $t_k(P)$, where $t_k(P)$ corresponds to a zero boundary value on Σ and has no other singularity except on A_k. The charge σ_k' on A_k corresponding to t, represented by the integral around A_k of the normal derivative of t, will be the same as the charge on A_k corresponding to t_k, and will approach a limit σ_k as X becomes infinite. As before, it may be shown that t_k corresponds to positive boundary conditions on the exterior of A_k. As before, moreover, by substituting for the exterior Poisson expansion the expansion for the solution of the Dirichlet problem over the region included between two circles, it may be shown that the difference between t_k and the potential corresponding to a charge σ_k' on a point of A_k and vanishing on Σ approaches O in comparison with σ_k uniformly over R' for any system of A_k's as the maximum dimension of any A_k approaches zero. Let the potential function corresponding to a unit charge on the point Q_m of A_m and O on Σ be M_m. Then

$$\left| t(P)-\Sigma\sigma_m' M_m \right|$$

admits a uniformly dominant expression vanishing with the greatest dimension of any A_m, over all R'. The same is therefore true of

$$\left| v(P)-\Sigma\sigma_m'\left(M_m-\frac{\log X}{\Sigma\sigma_m'}\right) \right| .$$

Now, $M_m-\dfrac{\log X}{\Sigma\sigma_m'}$ is the harmonic function corresponding to a charge of σ_m' at Q_m and assuming the boundary value $-\dfrac{\log X}{\Sigma\sigma_m'}$ on Σ. It may be shown with ease by an actual evaluation that

$$-\sigma_m \log \overline{PQ_m} = \lim\left(M_m-\frac{\log X}{\Sigma\sigma_m'}\right) . \tag{56}$$

Hence over R' $\quad |u(P)+\Sigma\sigma_m \log r_m|$

admits a uniformly dominant expression vanishing with the greatest dimension of any A_m.

From this point on the proof of our two-dimensional theorem follows lines in no essential wise different from those already used in proving Theorem II.

§4. **Capacity, and a sufficient condition for the solubility of the Dirichlet problem.** The capacity of a region R in space of more than two dimensions is the charge corresponding to a potential 1 on R and 0 at infinity. The capacity of a region R in two dimensions is the charge corresponding to a potential 1 on R, which is the boundary-value of a harmonic function behaving like $-\log r$ at infinity. In both cases, the total charge will be represented by

$$\int_{-\infty}^{\infty}\cdots\int_{-\infty}^{\infty} dM(P)$$

in the terminology of Theorem II or its two-dimensional analogue. It results from these theorems that every region that is bounded will have a capacity, and that the potential corresponding to boundary conditions c on a region R of capacity C at any point P is that due to a charge cC at a distance from P between that of the nearest and the most remote point of R.

Let us now proceed to the demonstration of Theorem III. While the proof is in no essential manner dependent on the dimensionality of our space, we shall for purposes of definiteness treat the three-dimensional case. Using the terminology of the enunciation of our theorem, let us consider the potential within S corresponding to a potential 1 on the part of C within S. This cannot be less than $C_1/2r = C_1/2C_2$. On S', a sphere about Q with radius ar, the same harmonic function cannot exceed $C_1/(a-1)r = C_1/(a-1)C_2$. It may be concluded without difficulty that there is a harmonic function not exceeding 1 on the part of C within S' for its boundary conditions, nor 0 on the part of S' within or on C, which is at least

$$\frac{C_1(a-3)}{C_2 2(a-1)}$$

within S. If $a=4$, this fraction will be $C_1/6C_2$.

On the other hand, the potential in some neighborhood of Q corresponding to a potential at least 1 on the part of C within S and at least 0 on the part of C without or on S will be at least the potential corresponding to a potential at least 1 on that part of S without C and 0 on that part of S within C, as may be shown by a consideration of the functions $u_n(P)$ of Theorem I. Hence a consideration of the Poisson integral shows that there is a sphere S'' about Q over which the potential function in question will be at least $1-L-\epsilon$, for any positive given ϵ.

Let us now investigate the potential function corresponding to a set of continuous boundary values on C always positive, nowhere exceeding 1, and 1 on the part of C within a certain sphere Σ about the point Q of C. It follows from the hypothesis of Theorem II and the remarks we have just made that there is a sphere Σ_1 about Q within which this harmonic function, which we shall call $W(P)$, is at least either $1-N-\epsilon$ or $M/6$, whichever is less. Let us call the lesser of these quantities d. $W(P)-d$ is non-negative in and on the boundary of Σ_1, and corresponds to boundary values at least $1-d$ on the part of C within Σ_1. An exactly similar argument to that which we have already carried through will show that there is a sphere Σ_2 about Q over which

$$W(P) \geq d+d(1-d) = 1-(1-d)^2. \tag{57}$$

It is in the same way possible to show that there is a sphere Σ_n about Q over which

$$W(P) \geq 1-(1-d)^n. \tag{58}$$

Hence $W(P)$ approaches the value 1 as p approaches Q by any route in R.

It may be observed that it is not strictly necessary for this that d be absolutely independent of r. Provided the denumerable set of values of r mentioned in the formulation of Theorem III diminish sufficiently rapidly, if d_k is the d corresponding to the k-th r, and $\prod (1-d_k)$ diverges to 0, $W(P)$ will still tend to 1 as P tends to Q.

It is at once possible to proceed to Theorem III in all its generality. In the first place, it is possible to describe about Q a sphere so small that the oscillation of U within it is less than ϵ. Hence a modification of U and consequently of u less than ϵ will render U constant over the part of C within this sphere. This modified $u(P)$ will lie between two bounds which our demonstration has already proved to be continuous at Q, with $U(Q)$ for their limit. Hence the oscillation of u in the neighborhood of Q cannot exceed ϵ, and vanishes.

§5. **The im Kleinen character of the Dirichlet problem.** We shall now proceed to the proof of Theorem IV. Let U be a non-negative function vanishing at Q and assuming the values 1 on the part of C exterior to a sphere Σ within Q as center and interior to S. The removal of the part of C outside Σ will not increase u, and will leave it continuous at Q. It follows that the function corresponding to *any* continuous boundary condition on the remaining part of C is continuous at Q.

By first removing a small portion of C' near Q, and then employing considerations similar to those in the paragraph preceding (15), it may be shown that we can choose a sphere Σ' interior to S so that the outer weight with respect to C' of that part of C' in the neighborhood of Σ' can be made arbitrarily small. If we discard the portions of C' in progressively smaller neighborhoods of Σ', we shall therefore replace $v(P)$ by functions of a sequence $v_n(P)$ converging uniformly to v over the part of R near Q. Now, $v_n(P)$ may be obtained by the alternating process as a uniformly convergent series of functions which are harmonic and either have continuous boundary conditions on the part of C' within Σ', and are hence by the last paragraph continuous at Q if they are given their nominal boundary values on C', or are functions which are continuous at Q because Q is interior to their boundary. Hence every v_n, and consequently v, is continuously convergent to $V(Q)$ as we approach Q by a path interior to R'.

This theorem may be summed up in the statement that the property of a boundary that the solution of the generalized Dirichlet problem be continuous in the neighborhood of a given point, is an *im Kleinen* property of that portion of the boundary in the neighborhood of the point in question.

Comments on [24a]

L. Lumer

This is generally considered as Wiener's most fundamental paper on potential theory. The main object is the determination of a generalized solution of the Dirichlet problem, for any continuous function on the boundary of an arbitrary Euclidian domain (Theorem I), which constitutes a major and definite breakthrough in the theory.

The limiting process used had already been suggested by a previous paper [23b]; the appropriate needed tool—the "outer weight," which is in fact the harmonic measure on the boundary—was introduced there. With a very deep insight in the subject (and no measure theory!), Wiener in reality produces his solution as the integral of the boundary function with respect to this "outer weight" measure (limit of the "Riemann" sums $\Sigma U(P_n)W(C_n,P)$), a type of representation he needs in order to apply his limiting process; and all his proof actually amounts to is the characterization of the harmonic measure on the boundary as the weak*-limit of the harmonic measures on the boundaries of the approximating domains.

This is the one aspect of Theorem I which prevails in further generalizations of the Dirichlet problem, where Wiener's method has its weaknesses and has to be replaced by the more flexible Perron's method of envelopes (nontopological ideal boundaries, axiomatic theories which have no exhaustion of the domain by regular ones,[1] Dirichlet problem for plurisubharmonic functions[2]). Wiener's proof is very complicated and difficult to follow; a simple and general proof, using the approximation of continuous functions by differences of subharmonic functions and the Stone-Weierstrass theorem, is given as early as 1929 in Kellog's book;[3] still it offers beautiful examples of the use of Schwarz's alternating process.

Theorem II introduces the notion of capacity through the resolution of an exterior Dirichlet problem, an approach which will permit numerous and diverse generalizations: replacement of the constant function 1 by any other potential or positive superharmonic function,[4] of the Euclidian space and Newtonian kernel by other spaces and kernels,[5] but also extension to potential theories without kernels, where this leads to the notion of reduced function, so important in today's axiomatic theories of harmonic functions and partial differential equations. Further researches done on capacity (namely, geometric interpretations with the transfinite diameter of Fekete-Polya-Szego,[6] Cartan's work using the energy, 1945,[7] the sets of capacity zero and Choquet's general theory of capacities 1954,[8] and many more) all suggest, almost 50 years later, that the notion of capacity is perhaps Wiener's most important and long-lasting contribution to potential theory.

References

1. H. Bauer, Lecture Notes in Mathematics, Springer-Verlag, Berlin, vol. 22(1966).
2. H. J. Bremermann, Trans. Amer. Math. Soc. 91(1959), 246-276.
3. O. D. Kellog, Grundlehren, Springer-Verlag, Berlin, vol. 31(1929).
4. M. Brelot, J. Math. Pures Appl., ser. 9, 24(1945), 1-32.
5. M. Brelot and G. Choquet, Ann. Inst. Fourier (Grenoble), 3(1951), 199-263.
6. M. Fekete, J. Reine Angew. Math. 165(1931), 217-224; G. Pólya and G. Szegö, ibid., 4-49.
7. H. Cartan, Bull. Soc. Math. France 73(1945), 74-101.
8. G. Choquet, Ann. Inst. Fourier (Grenoble) 5(1954), 131-296.

THE DIRICHLET PROBLEM

By Norbert Wiener

In a recent paper in the *Comptes rendus*,[1] Lebesgue points out that the Dirichlet problem divides itself into two parts, the first of which is the determination of a harmonic function corresponding to certain boundary conditions, while the second is the investigation of the behaviour of this function in the neighborhood of the boundary. In a paper appearing the same week[2] the author of the present paper made the same remark independently, and went on to give a precise definition of the sense in which the harmonic function depends on the boundary conditions. He proved, moreover, that his method assigns a definite harmonic function to any continuous boundary condition on any bounded set of points in any number of dimensions. In particular, he proved the theorem that the potential corresponding to boundary values 1 on a given bounded set of points and (in three or more dimensions) 0 at infinity determines in his sense a harmonic function generated by a distribution of charge (represented by a Stieltjes integral) over the bounded set. The charge on no region will be negative. The total amount of the charge will be known as the capacity of the boundary set of points. Thus every bounded set of points, whether a surface or not, and indeed whether measurable or not, will have a definite, finite capacity. Moreover, if C_2 is a set of points containing as a part C_1, the capacity of C_2 will at least equal that of C_1.

To return to Lebesgue: like the author, he points out that the solubility of the Dirichlet problem in the classical sense for a given region depends only on the *im Kleinen* properties of the boundary, and may be reduced to the investigation of what he terms the regularity of the individual points of the boundary. A point O of the boundary B of a set of points D is termed regular

[1] January 21, 1924.

[2] *Certain Notions in Potential Theory, Jour. Math. Phys. M. I. T.*, January, 1924.

if whenever $F(P)$ is a function continuous on B, and $f(P)$ is the corresponding function harmonic over D,

$$\lim_{P \to O} f(P) = F(O).$$

This must hold for *every* function f continuous on B.

The oldest fairly general condition for the regularity of a point on a boundary in three-space is that of Poincaré and Zaremba.[3] This is stated by Lebesgue as follows: *A point O of the boundary of a domain D is regular if it is the vertex of a closed conical surface the interior of which is exterior to D.* A result obtained by G. E. Raynor[4] may be stated as follows: *A point O of the boundary of a domain D is regular if there is a number λ greater than zero such that there exist an infinity of spheres S with O as center and such that the measure of the set of points on S and exterior to D exceeds λ times the measure of the surface of S.* This anticipates and contains condition A of Lebesgue, which reads: *The point O is regular if on each sphere with center O and radius r there is a point exterior to D and distant from D by a quantity at least equal to Kr, K being a given positive quantity.* Lebesgue also quotes a condition due to Bouligand,[5] which reads: *The point O is regular if it is the vertex of a conical surface, open or closed, exterior to D.*

All the conditions so far stated, with the exception of that of Raynor, are special cases of one published by H. B. Phillips and the author of the present paper[6] prior to all these conditions except that of Poincaré and Zaremba In its three-dimensional form, this reads as follows: *The point O is regular whenever there are positive numbers a and b such that if $r<a$ and a sphere S of radius r is constructed with O as center, the content of the projection of the points in S and not in D on some plane exceeds br^2.* In the proof of this theorem, only a denumerable set of spheres S are actually used. If the theorem be stated in the somewhat generalized

[3] Poincaré, *Sur les équations aux derivées partielles de la physique mathématique*, *Am. J. of M.*, V. 12, 1890; Zaremba, *Sur le principe du minimum*, *Bull. de l'Ac. des Sc. de Cracovie*, 1909.

[4] *Dirichlet's Problem*, *Annals of Mathematics*, dated March, 1922, but not actually published until the end of March, 1923.

[5] *Sur le problème de Dirichlet harmonique*, *Comptes rendus*, January 2, 1924

[6] *Nets and the Dirichlet Problem*, *Jour. Math. Phys. M. I. T.*, March, 1923.

form involving only a denumerable set of spheres, it will include that of Raynor.

In his paper on *Certain Notions in Potential Theory*, the author gives the following sufficient condition for the regularity of a point: *The point O of the boundary of a domain D is regular if there is a sequence of values of r tending to 0 such that either:*

(*a*) *The inner Lebesgue measure of the points on a sphere of radius r exterior to D in terms of the surface of the sphere itself; exceeds a positive quantity independent of r; or*

(*b*) *The capacity of the part of the boundary of D interior to a sphere of radius r exceeds M times the capacity of the sphere, M being independent of r.*

This condition includes as special cases all those previously mentioned. It is to be noted that it introduces the concept of capacity. This is rendered possible by an antecedent proof that every bounded set of points has a capacity.

Up to this point the theory has dealt with conical points and their immediate generalizations. We next turn to the discussion of cuspidal points and their generalizations. The published work on this subject, so far as is known by the author, is entirely due to Lebesgue,[7] who gave the first example[8] of a simply connected boundary in three-space with an irregular point. Lebesgue's results in this direction have been collected in his recent paper, and read as follows: *If there is a segment of a straight line (or analytic curve) terminating at the point O of the boundary of the domain D, if the segment is entirely exterior to D, and if there exist positive numbers A and B such that if P is a point of the segment, its distance from D exceeds $A\,\overline{OP}^B$, O is regular. If there exist positive numbers A and B such that when P lies on the segment, $\overline{OP}=\overline{OQ}$, and*

$$\overline{PQ}>Ae^{\frac{-B}{\overline{OP}}},$$

Q lies in D, O is irregular. Lebesgue states that he has been unable to generalize these results as he has generalized that of Zaremba, by constructing an infinite set of spheres about the

[7] Mr. Gleason of Princeton, however, verbally communicated to me last year several very interesting results in this direction which he obtained several years ago, before Lebesgue had published similar results.

[8] *Comptes rendus des séances de la société mathématique de France*, **1913**.

point O, marking on each the region excluded by D, and rotating each independently about O. He also states no results on cuspidal points reducing to flat tongues.

We now come to conditions of a much more general character. It was stated by Poincaré and repeated by Lebesgue that if a point is regular for a given boundary, it is regular for a boundary entirely interior to the first in the neighborhood of this point. Lebesgue gives the following necessary and sufficient condition for the regularity of the point O. *For a point O of the boundary of a domain D to be regular, it is necessary and sufficient that there exist a function $F(x, y, z)$ continuous at O, attaining its lower bound at O and O' only, and such that everywhere in D*

$$\frac{\partial^2 F}{\partial x^2}+\frac{\partial^2 F}{\partial y^2}+\frac{\partial^2 F}{\partial z^2}\leq 0.$$

Other forms of this condition are stated by Lebesgue and Kellogg.[9] They all suffer from the defect of involving the geometrical character of the boundary only in a very indirect and devious manner. From the geometrical point of view, indeed, they are scarcely more than restatements of the regularity of O.

It is the purpose of this paper to develop a complete necessary and sufficient characterization of regular points which shall be at least quasi-geometrical. Its statement reads as follows: *Let O be a point of C, the boundary of an open set of points D. Let λ be any positive quantity less than 1. Let γ_n be the capacity of the set of all points Q not belonging to D such that*

$$\lambda^n \leq \overline{OQ} \leq \lambda^{n-1}.$$

Then O is regular or irregular according as

$$\frac{\gamma_1}{\lambda}+\frac{\gamma_2}{\lambda^2}+\ .\ .\ .\ +\frac{\gamma_n}{\lambda^n}+\ .\ .\ .$$

diverges or converges.

In order to prove this theorem, it will be necessary to repeat from the author's previous paper a few theorems relating to the determination of a harmonic function by continuous boundary conditions on an arbitrary boundary. He there proved that if

[9] *An Example in Potential Theory, Proc. Am. Ac. Arts and Sciences*, **1923.**

C is any closed bounded set of points, and $F(P)$ any function continuous on this set of points, there is a function $f(P)$ harmonic except on C, and determined by F in the following manner. We first construct[10] a continuous function $g(P)$ defined throughout space and reducing to $F(P)$ on C. We then construct a sequence of boundaries for which the Dirichlet problem is soluble, each within its predecessor, and together sharing no point not on C. This is always possible. We finally form the sequence of harmonic functions assuming the boundary values $g(P)$ on these boundaries. Outside C this sequence will converge to a harmonic function independent of $g(P)$ and the sequence of boundaries, and entirely determined by C and $F(P)$. In the particular case where $F(P)$ is identically 1, $g(P)$ may also be taken as identically 1.

It may be remarked at once that if $f(P)$ and $g(P)$ are any two harmonic functions corresponding to boundary values $F(P)$ and $G(P)$, respectively, on the same boundary C, and if throughout C we have $F(P) \geq G(P)$, then we shall always have $f(P) \geq g(P)$. This may be considerably generalized: let Q be a point interior to the boundary C, and let C' be a boundary containing Q in its interior. Let $F(P) \geq G(P)$ at every point of C in or on C', and let $f(P) \geq g(P)$ at every point of C' interior to C. Then $f(Q) \geq g(Q)$.

If $f(P)$ is the harmonic function corresponding to 1 on the boundary C, if C' is another boundary such that every point of C is on or exterior to C', and if $g(P)$ is the harmonic function corresponding to 1 on C', while both f and g correspond to a potential 0 at infinity, in case the region over which they are defined is unbounded, then $g(P) \geq f(P)$ at any point where both are defined. From this it may be concluded at once that if a given set of points completely surrounds another set, the former set has a capacity at least as large as the latter.

The lemmas of the last two paragraphs are readily proved from the definition given by the author of the sense in which the generalized Dirichlet problem is soluble. They will be presupposed without explicit reference in all that follows.

[10] For the possibility of this construction, see Lebesgue, *Sur le problème de Dirichlet, Rendiconti di Palermo*, v. 24 (1907), pp. 371–402.

We now proceed to the proof of our main theorem. We first wish to show that if

$$\frac{\gamma_1}{\lambda}+\frac{\gamma_2}{\lambda^2}+\ .\ .\ .\ +\frac{\gamma_n}{\lambda^n}+\ .\ .\ .$$

diverges, O is regular. It is to be noted that under this condition either

$$\frac{\gamma_1}{\lambda}+\frac{\gamma_3}{\lambda^3}+\ .\ .\ .\ +\frac{\gamma_{2n+1}}{\lambda^{2n+1}}+\ .\ .\ .$$

or

$$\frac{\gamma_2}{\lambda^2}+\frac{\gamma_4}{\lambda^4}+\ .\ .\ .\ +\frac{\gamma_{2n}}{\lambda^{2n}}+\ .\ .\ .$$

diverges.

Let C_n be the set of all points exterior to D in the zone between a sphere of radius λ^n about O as center and a sphere of radius λ^{n-1} Its capacity, as we have seen, will be γ_n. Now consider the set of points consisting of all points contained in a sphere of radius $\lambda^{n-3/2}$ about O as center and exterior to C_n. We wish to consider the harmonic function corresponding to boundary values O over the sphere and 1 over C_n. This function we shall term $F_n(P)$.

To begin with let C_n consist entirely of regular points. Let the harmonic function $V_n(P)=1$ on C_n and 0 at infinity be represented by the Stieltjes integral

$$\int\int\int \overline{PQ}^{-1}dM(Q).$$

$dM(Q)$ will here be everywhere positive, and will represent the charge over the rectangular parallelepiped $dx\ dy\ dz$. It will vanish when this parallelepiped contains no point of C_n.

The Green's function of the sphere of radius C about O is

$$G_C(P,Q)=\frac{1}{\overline{PQ}}-\frac{1}{\sqrt{C^2+\dfrac{\overline{OP}^2\,\overline{OQ}^2}{C^2}-2\overline{OP}\ \overline{OQ}\cos(OP,OQ)}}.$$

Let it be noted that

$$G_C(P,Q)\leq 1/\overline{PQ},$$

and that if P and Q both lie at a distance from O not greater than μC $(\mu<1)$,

$$G_C(P,Q)\geq(1-\mu^2)^2/8\overline{PQ}.$$

Hence over a sphere of radius λ^{n-1} about O,

$$\iiint \overline{PQ}^{-1} dM(Q) \geq \iiint G_{\lambda^{n-3/2}}(P, Q) dM(Q)$$
$$\geq \frac{(1-\lambda)^2}{8} \iiint \overline{PQ}^{-1} dM(Q).$$

The function

$$\iiint G_{\lambda^{n-3/2}}(P, Q) dM(Q)$$

is harmonic within our sphere of radius $\lambda^{n-3/2}$ and exterior to C_n It may be verified at once that it vanishes on the sphere. It certainly does not exceed 1 on C_n. Hence

$$F_n(P) \geq \iiint G_{\lambda^{n-3/2}}(P, Q) dM(Q).$$

This can be generalized at once to the case where C_n contains irregular points, by taking a set of regular points C_n' surrounding C_n, assigning boundary values 1 on C_n' and 0 on the sphere, and letting C_n' shrink to C_n.

On a sphere of radius $\lambda^{n+1/2}$ with O as center we shall have

$$F_n(P) \geq \frac{(1-\lambda)^2}{8} \iiint \overline{PQ}^{-1} dM(Q)$$
$$\geq \frac{(1-\lambda)^2}{8} \iiint \tfrac{1}{2}\lambda^{1-n} dM(Q).$$
$$= \frac{(1-\lambda)^2}{16\lambda} \frac{\gamma_n}{\lambda^n}.$$

Now let us consider the potential on a sphere of radius $\lambda^{n+5/2}$ corresponding to a potential of 0 on a sphere of radius $\lambda^{n-3/2}$ about O and a potential 1 on C_n and C_{n+2}. This will be a potential corresponding to a potential at least

$$\frac{(1-\lambda)^2}{16\lambda} \frac{\gamma_n}{\lambda^n}$$

on a sphere of radius $\lambda^{n+1/2}$ about O and 1 on C_{n+2}. Hence it will be at least the sum of

$$\frac{(1-\lambda)^2}{16\lambda} \frac{\gamma_n}{\lambda^n}$$

and the potential due to a boundary potential of 0 on a sphere of radius $\lambda^{n+\frac{1}{2}}$ about O and

$$1-\frac{(1-\lambda)^2}{16\lambda}\frac{\gamma_n}{\lambda^n}$$

on C_{n+2}. In other words, it will at least equal

$$\frac{(1-\lambda)^2}{16\lambda}\frac{\gamma_n}{\lambda^n}+\frac{(1-\lambda)^2}{16\lambda}\frac{\gamma_{n+2}}{\lambda^{n+2}}\left[1-\frac{(1-\lambda)^2}{16\lambda}\frac{\gamma_n}{\lambda^n}\right]$$

$$=1-\left[1-\frac{(1-\lambda)^2}{16\lambda}\frac{\gamma_n}{\lambda^n}\right]\left[1-\frac{(1-\lambda)^2}{16\lambda}\frac{\gamma_{n+2}}{\lambda^{n+2}}\right]$$

on C_{n+4}. In a similar manner, the potential within some sphere about O corresponding to a potential 1 on C_n, C_{n+2}, ... , C_{n+2m} is at least

$$1-\left[1-\frac{(1-\lambda)^2}{16\lambda}\frac{\gamma_n}{\lambda^n}\right]\left[1-\frac{(1-\lambda)^2}{16\lambda}\frac{\gamma_{n+2}}{\lambda^{n+2}}\right]\cdots\left[1-\frac{(1-\lambda)^2}{16\lambda}\frac{\gamma_{n+2m}}{\lambda^{n+2m}}\right]$$

Thus the potential corresponding to a potential 1 on C_n, C_{n+2}, ... , C_{n+2m}, ... assumes continuously the boundary value 1 at O if

$$1-\prod_{m=0}^{\infty}\left[1-\frac{(1-\lambda)^2}{16\lambda}\frac{\gamma_{n+2m}}{\lambda^{n+2m}}\right]=1$$

By a familiar theorem in analysis, this will be the case when and only when

$$\sum_{m=0}^{\infty}\frac{\gamma_{n+2m}}{\lambda^{n+2m}}$$

diverges.

We are now in a position to proceed to a demonstration of the regularity of O. It may be shown immediately by the alternating process, as in my previous paper, that the harmonic function corresponding to boundary values 1 on C_n, C_{n+2}, . . . and 0 on the part of C at a distance at least λ^{n-2} from O is continuous at O. It is easy to show, however, that this latter function is dominated by the harmonic function corresponding to any continuous boundary conditions on C that are non-negative and are 1 over that part of C at a distance from O not exceeding λ^{n-2}

It follows that any harmonic function corresponding to such boundary conditions on C is continuous at O. However, any positive boundary conditions continuous at O and there alone attaining their maximum value 1 can be reduced to boundary conditions of this type by a modification not exceeding the arbitrarily small quantity ϵ. Thus the harmonic function corresponding to these new boundary conditions cannot have an oscillation of more than ϵ at O, and so must assume its boundary values continuously. Since any continuous boundary condition may be represented linearly in terms of two boundary conditions of this sort, O is regular. Hence O is regular if either

$$\frac{\gamma_1}{\lambda}+\frac{\gamma_3}{\lambda^3}+\quad\cdots\quad+\frac{\gamma_{2n+1}}{\lambda^{2n+1}}+\quad\cdots$$

or

$$\frac{\gamma_2}{\lambda^2}+\frac{\gamma_4}{\lambda^4}+\quad\cdots\quad+\frac{\gamma_{2n}}{\lambda^{2n}}+\quad\cdots$$

diverges, and consequently if

$$\frac{\gamma_1}{\lambda}+\frac{\gamma_2}{\lambda^2}+\quad\cdots\quad+\frac{\gamma_n}{\lambda^n}+\quad\cdots$$

diverges.

We now proceed to the converse of this theorem. For this we need a lemma, to the effect that if O is a regular point of the boundary of the domain D, then given any positive numbers η and R, there is a positive number r less than R such that the potential at O corresponding to a boundary potential 0 at infinity and 1 on the set of points exterior to D and lying at a distance from O between r and R inclusive exceeds $1-\eta$. In the first place, the harmonic function $F(P)$ corresponding to boundary values 0 at infinity and 1 on the part of C, the boundary of D, at a distance from O not exceeding R assumes its boundary value tinuously at O, as may be seen by a comparison of this harmonic function with one corresponding to boundary values continuous on C, never exceeding 1, 1 at O, and 0 on the part of C at a distance at least R from O. Hence there is a sphere about O such that in the part of this sphere within D, $F(P)$ exceeds $1-\eta/2$. Furthermore, there is a smaller sphere about O such that the potential on the larger sphere corresponding to a potential

0 at infinity and 1 on the smaller sphere nowhere exceeds $\eta/2$. Let the radius of the smaller sphere be r. Then the potential on the larger sphere corresponding to a boundary potential of 0 at infinity and 1 on the set of points exterior to D at a distance from O between r and R, inclusive, exceeds $1-\eta$. From this fact, and the fact that it corresponds to boundary values 1 over the set of points within the larger sphere, outside the sphere of radius r, and exterior to D, it may readily be concluded that it exceeds $1-\eta$ at O.

Now suppose that

$$\frac{\gamma_1}{\lambda}+\frac{\gamma_2}{\lambda^2}+\ldots+\frac{\gamma_n}{\lambda^n}+\ldots$$

converges. Then given any number ϵ, it is possible to find an n such that

$$\frac{\gamma_n}{\lambda^n}+\frac{\gamma_{n+1}}{\lambda^{n+1}}+\ldots+\frac{\gamma_{n+p}}{\lambda^{n+p}}<\epsilon$$

whatever p may be. The potential at O due to a potential 1 on C_m cannot exceed γ_m/λ^m, since no part of the charge on C_m corresponding to this potential is nearer to O than λ^m. Hence the potential at O corresponding to a potential 1 on $C_n, \ldots, C_{n+p}$ cannot exceed

$$\frac{\gamma_n}{\lambda^n}+\ldots+\frac{\gamma_{n+p}}{\lambda^{n+p}}<\epsilon$$

Since however ϵ is arbitrarily small, the lemma of the last paragraph shows that O is irregular.

We thus have obtained a necessary and sufficient condition for the regularity of O that explicitly involves the boundary. From this all the more special conditions for regularity may be deduced at once. I shall not, however, discuss them all in detail, but shall proceed at once to the theory of monotone cuspidal points of revolution. We shall say that O is a monotone cuspidal point of revolution of the boundary of D if this boundary in the neighborhood of O is a reëntrant surface with equation

$$\phi=f(\rho)$$

in some scheme of spherical coördinates, where $f(\rho)$ is a monotone increasing function of ρ never exceeding π and vanishing for $\rho=0$. Let us write ϕ_m for the value of ϕ corresponding to $\rho=\lambda^m$, and let us define C_m and γ_m as before. Then C_m will be entirely included in a right circular cylinder with base of radius $\lambda^{m-1}\sin\phi_{m-1}$ and altitude λ^{m-1}, and hence in a prolate spheroid with semi-axes $2\lambda^{m-1}\sin\phi_{m-1}$, $2\lambda^{m-1}\sin\phi_{m-1}$, and $2\lambda^{m-1}$. On the other hand, C will ultimately contain a right circular cylinder with base of radius $\lambda^m\sin\phi_m$ and altitude $\dfrac{1-\lambda}{2}\lambda^{m-1}$, and hence a prolate spheroid of semi-axes $\dfrac{\lambda(1-\lambda)}{4}\lambda^{m-1}\sin\phi_m$, $\dfrac{\lambda(1-\lambda)}{4}\lambda^{m-1}\sin\phi_m$, and $\dfrac{\lambda(1-\lambda)}{4}\lambda^{m-1}$. The capacities of these spheroids are respectively[11]

$$\frac{4\lambda^{m-1}\cos\phi_{m-1}}{\log\cot\phi_{m-1}/2}$$

and

$$\frac{\lambda^m(1-\lambda)\cos\phi_m}{2\log\cot\phi_m/2}.$$

Hence

$$\frac{4\cos\phi_{m-1}}{\lambda\log\cot\phi_{m-1}/2}\geq\frac{\gamma_m}{\lambda^m}\geq\frac{(1-\lambda)\cos\phi_m}{2\log\cot\phi_m/2}.$$

It then follows from our general theorem that O *is regular or irregular according as*

$$\sum_1^\infty\frac{\cos\phi_m}{\log\cot\phi_m/2}$$

diverges or converges. This series, however, converges or diverges with

$$\sum_1^\infty\frac{-1}{\log\phi_m}$$

since the ratio between corresponding terms tends to the definite limit 1. In particular, if $f(\rho)>A\rho^n(A>0)$, this series dominates a harmonic progression and diverges. On the other hand, if

[11] Jeans, *Electricity and Magnetism*, p. 248.

$f(\rho) < Ae^{-\frac{n}{\rho}}$, this series is dominated by a geometric progression with ratio between 0 and 1, and converges. From these facts, Lebesgue's results as to cuspidal points with finite and exponential order of contact follow at once.

We can generalize this result considerably with the aid of a few lemmas. To begin with, if we replace the boundary in the neighborhood of the cuspidal point which we have just discussed by a certain sequence of zones of one base, the regularity of O will remain unchanged. These zones are chosen in the following manner: the first consists in those points exterior to D at a distance ρ_1 from O. The second consists in those points exterior to D at a distance $\rho_1 - f(\rho_1)$ from O. The third consists in those points at a distance $\rho_1 - f(\rho_1) - f(\rho_1 - f(\rho_1))$ from O, and so on. Let the set of all points exterior to D and at a distance between λ^m and λ^{m-1}, inclusive, from O be C_m as before. Let the set of all points on the zones just mentioned between a distance of λ^m and λ^{m-1} from O, inclusive be C_m'. It will be noted that to each point of C_m, there can be assigned a zone of C_m' or C_{m-1}' bounded by a circle of radius say α, such that the point in question is remote from the furthest point of the zone by less than 3α. The capacity of the zone will exceed or equal $2\alpha/\pi$. Hence to a potential of 1 on C_m' and C_{m-1}' there will correspond a potential at least $2/3\pi$ on C_m. It may be concluded from this at once by a comparison of the potentials at distant points corresponding respectively to a unit potential on C_m and to a unit potential on C_m' and C_{m-1}' that the capacity of C_m is not greater than $3\pi/2$ times the sum of the capacities of C_m' and C_{m-1}'. From this the regularity of O with respect to the boundary consisting of the zones just defined follows at once.

The next lemma reads as follows: Let $S_1, S_2, \ldots$ be a set of sets of points with capacities $\sigma_1, \sigma_2, \ldots$. Let $T_1, T_2, \ldots$ be a set of sets of points with capacities $\tau_1, \tau_2, \ldots$. Let there be positive numbers a and b such that for every n, $\sigma_n \geq a\,\tau_n$, while for every m and n, the lower bound of the distance between a point of S_m and a point of S_n exceeds b times the upper bound of the distance of a point of T_m from a point of T_n. Then the capacity of the set of points

$$S_1 + S_2 + \ldots$$

is at least as large as the capacity of the set of points

$$T_1+T_2+ \ . \ . \ .$$

multiplied by

$$\frac{ab}{a+b}.$$

The proof of this is simple. Suppose $T_1+T_2+ \ . \ . \ .$ brought to potential 1 by a distribution of charge, and let the charge then found on T_m be ω_m. Transfer this charge to S_m, and it will raise the latter set of points to a potential not exceeding ω_m/a. The remainder of the charge is similarly transferred, and cannot raise any point of S_m to a potential of more than ω_m/b. We thus have a positive distribution of charge over $S_1+S_2+ \ . \ . \ .$ which nowhere on the boundary of this set gives rise to a potential greater than $\Sigma\omega_m(1/a+1/b)$, while the total amount of charge is $\Sigma\omega_m$. A comparison with the charge necessary to bring the whole of $S_1+S_2+ \ . \ . \ .$ to a potential $1/a+1/b$ completes the proof of the theorem.

Combining this theorem with the theorem we have just proved concerning a boundary consisting of spherical caps, and remembering that of all sets of points on the surface of a sphere which have a given area, the spherical cap has the least capacity, we obtain the following theorem: *Let O be a point of the boundary of a region D. Let O' be a monotone cuspidal point of revolution on the boundary of a region D'. Let O' be regular. Let us construct spheres about O and O' respectively as centers with radius r. Let the areas of the parts of D and D' respectively on the surface of these spheres be A and A'. Then if for all sufficiently small values of r we have $A<A'$, O is regular.*

This is the theorem generalizing Lebesgue's result concerning cuspidal points as he generalized Zaremba's result concerning conical points. There is another generalization of the Lebesgue result which resembles rather that of Bouligand concerning flat conical points. Let us return to our sequence of zones of one base terminating in a regular point O. It will be noted that the solid angle subtended by one of these zones will approach 0 as we tend to O, and that consequently these zones will tend in shape to flat discs. Their capacity will hence tend to the capacity of a flat disc bounded by the same circle, and the regularity of O will not be affected if we replace each zone by the corresponding flat disc. If we replace each disc by a disc in the same plane and with the same center, but with a third the radius, we shall have reduced the capacity of each disc to just a third its original value. If we now rotate each disc about its center in such a manner as to bring them all into a single plane passing through O, the least distance between two points of two adjacent discs will be at least a ninth of the greatest distance between two points of the corresponding zones in the original figure. Hence O will still be regular.

It is manifestly true, as Lebesgue has pointed out, that if two boundaries have a point in common, and one boundary is in the neighborhood of this point entirely exterior to another, then if the point is regular for the exterior boundary, it is also regular for the interior boundary. This indeed follows directly from our fundamental theorem. Thus by a comparison on the one hand with the set of discs we have just discussed, and on the other with a monotone cuspidal point of revolution, we get the following criterion for the regularity of a cuspidal point on a lamina: *Let the boundary of the region D in the neighborhood of a point O consist in a plane lamina containing all the points such that*

$$\theta = f(\rho)$$

and no others, in a scheme of polar coördinates in the plain of the lamina with O as origin. Let $f(\rho)$ be a monotone increasing function of ρ never exceeding π and vanishing for $\rho = 0$. Then O is regular or irregular according as

$$\sum_1^\infty \frac{1}{\log f(\lambda^m)}$$

diverges or converges.

We may indeed obtain a still more general theorem concerning flat cuspidal points. To begin with, a unit charge uniformly distributed over any region with a plane projection of area A with respect to its projection can never produce a greater potential at any point than if the region is a circle of area A, and the point where the potential is taken is at the center of the circle. The potential at the center of the circle will then be $2\sqrt{\frac{\pi}{A}}$. This means that the capacity of the region with projection of area A is at least $\frac{1}{2}\sqrt{\frac{A}{\pi}}$, since a charge of this amount will correspond to a potential nowhere exceeding 1. Now the capacity of a circular disc of area A is $\frac{2}{\pi}\sqrt{\frac{A}{\pi}}$. Thus the capacity of any region of projection A is at least $\pi/4$ times as great as the capacity of a circular disc of the same area. Combining this fact with the lemma we have just proved concerning a sequence of discs, we get the following theorem: *Let O be a point of the boundary of a region D. Let O' be a monotone cuspidal point of revolution on the boundary of a region D'. Let O' be regular. Let C be the part of the boundary of D lying at a distance of from r to r', inclusive, from O, and let C' be the part of the boundary of D' similarly situated with reference to O'. Then if for all sufficiently small values of r and r', some projection of C exceeds C' in area, O is regular.*

So much for the three-dimensional case: the theory for the n-dimensional case is closely similar. For $n > 3$, the fundamental theorem reads as follows: *Let O be a point of C, the boundary of an open set of points D, in a space of n dimensions. Let λ be any positive quantity less than 1. Let γ_m be the capacity of the set of all points Q not belonging to D, such that*

$$\lambda^m \leq \overline{OQ} \leq \lambda^{m-1}.$$

Then O is regular or irregular according as

$$\frac{\gamma_1}{\lambda^{n-2}} + \frac{\gamma_2}{\lambda^{2(n-2)}} + \cdots + \frac{\gamma_m}{\lambda^{m(n-2)}} + \cdots$$

diverges or converges.

Here the proof differs in no essential point from that in the three-dimensional case. In the two-dimensional case, a slight complication is introduced by the fact that the potential due to an isolated charge has a logarithmic singularity at infinity instead of vanishing there. This necessitates a recasting of the definition of capacity. The fundamental existence theorem in my previous paper is stated in an incorrect fashion, although the subsequent work of the paper is perfectly correct, and follows immediately from the correct theorem. To begin with, let R be a bounded set of points in the plane, and let it contain a point O in its interior. Form the harmonic function $f(P)$ corresponding to the boundary values $\log \overline{OP}$ for points P on R and finite at infinity. Let this function assume the value $-a$ at infinity. Then we shall say that the function

$$\frac{f(P)-\log \overline{OP}+a}{a}$$

corresponds to boundary values 1 on R and behaves like $\log \overline{OP}$ at infinity. This function will always exist if R contains points properly in its interior and is of sufficiently small linear dimensions. The condition that R should contain points properly in its interior is inessential, if we follow out the spirit of my last paper by regarding the harmonic function corresponding to boundary values 1 on R and behaving like $\log \overline{OP}$ at infinity as the lower bound (which will always exist if R is small enough) of the harmonic functions corresponding to boundary values 1 on contours near to R and containing R and behaving like $\log \overline{OP}$ at infinity. The harmonic function thus obtained may readily be proved unique, and independent of O.

The proper form of the existence theorem for capacity in the two-dimensional case is the legitimate conclusion of the argument used in deriving the erroneous theorem of my previous paper, and reads as follows: *Let R be any bounded set of points in the plane, entirely within a rectangle with opposite corners (a_1, a_2) and (b_1, b_2), and let*

$$|a_1-b_1|<1, \quad |a_2-b_2|<1.$$

Let $u(P)$ behave like $\log \overline{OP}$ at infinity, and correspond to boundary values 1 over R. Then there will be a function $M(P)$ representing the charge on that portion of R consisting of points with both coördinates less than those of P. $M(P)$ will be an increasing function of the coördinates of P. Over every point Q exterior to R we shall have

$$u(Q)=\int_{a_1}^{b_1}\int_{a_2}^{b_2}(-\log \overline{PQ})dM(P)$$

Moreover, if the rectangle with opposite vertices (x_1, x_2) *and* (y_1, y_2) *is entirely exterior to* R, *then*

$$\int_{x_1}^{y_1}\int_{x_2}^{y_2} dM(P)=0,$$

We shall term

$$\int_{a_1}^{b_1}\int_{a_2}^{b_2} dM(P)$$

the capacity of R.

With this definition, our cardinal theorem is the following: *Let* O *be a point of* C, *the boundary of an open set of points* D *in the plane. Let* λ *be any positive quantity less than* 1. *Let* γ_m *be the capacity of the set of all points* Q *not belonging to* D, *such that*

$$\lambda^{2^m} \leq \overline{OQ} \leq \lambda^{2^{m-1}}.$$

Then O *is regular or irregular according as*

$$\gamma_1+2\gamma_2+ \ . \ . \ . \ +2^m\gamma_m+ \ . \ . \ .$$

diverges or converges.

The proof follows that in the three-dimensional case step by step. It is of course necessary to take account of the fact that the logarithm assumes positive as well as negative values, but any difficulty from this source may be avoided by confining our attention to the interior of a circle of radius less than 1/2 about O. The fact that the bounds of $\overline{OQ}$ for points Q on γ_m are λ^{2^m} and $\lambda^{2^{m-1}}$ is accounted for by the fact that the potential at O due to a unit charge at Q will then lie between the numbers $2^m \log \frac{1}{\lambda}$ and $2^{m-1} \log \frac{1}{\lambda}$, which bear to one another a fixed ratio. There is one stage of the proof which perhaps needs a little comment: that in which we determine the potential on a given circle about the origin corresponding to a potential 0 on a circle of larger radius and 1 on γ_m. We shall take the radius of the outer circle as $\lambda^{2^{m-3/2}}$, and that of the smaller circle as $\lambda^{2^{m+1/2}}$. We desire a theorem correlating the potential due to a potential 1 on γ_m and 0 at infinity with the potential in question. As before, we do this by establishing certain inequalities connecting the Green's function of the outer circle with $-\log \overline{PQ}$, the Green's function of the entire plane.

If we write c for the radius of the outer circle, its Green's function will be

$$G(P,Q)=\tfrac{1}{2}\log\left\{\frac{c^2+\dfrac{\overline{OP}^2\,\overline{OQ}^2}{c^2}-2\overline{OP}\,\overline{OQ}\cos(\overline{OP},\overline{OQ})}{\overline{OP}^2+\overline{OQ}^2-2\overline{OP}\,\overline{OQ}\cos(\overline{OP},\overline{OQ})}\right\}$$

$$=\tfrac{1}{2}\log\left\{\frac{1+\left(c-\dfrac{\overline{OP}^2}{c}\right)\left(c-\dfrac{\overline{OQ}^2}{c}\right)}{\overline{PQ}^2}\right\}.$$

We shall always have $G(P,Q)\leq-\log\overline{PQ}$. If $c=\lambda^{2^{m-3/2}}$, while P and Q lie within a circle about O of radius $\lambda^{2^{m-1}}$, we shall have

$$G(P,Q)\geq\tfrac{1}{2}\log\left\{[1-\lambda^{(2-\sqrt{2})2^{m-1}}]^2\frac{\lambda^{2^{m-1/2}}}{\overline{PQ}^2}\right\}$$

$$\geq\tfrac{1}{2}\log\frac{\lambda^{2^{m-1/2}}}{4\overline{PQ}^2}$$

$$\geq(\log\lambda)2^{m-3/2}-\log 4-\log\overline{PQ}$$

for all sufficiently large values of m. Furthermore,

$$-\log\overline{PQ}\geq-\log 2-\log\lambda^{2^{m-1}}$$
$$=-\log 2-\log\lambda^{[2^{m-3/2}]\sqrt{2}}$$
$$=-\log 2-\sqrt{2}\log\lambda^{2^{m-3/2}}$$

Hence for all sufficiently large values of m,

$$(-\log\lambda)2^{m-3/2}+\log 4\leq\frac{-1}{1.4}\log\overline{PQ}$$

It follows that for all sufficiently large values of m, we shall have

$$G(P,Q)\geq-.2\log\overline{PQ}.$$

With this we are in a position to proceed with our two-dimensional theorem in a fashion exactly paralleling that employed in the demonstration of our theorem in three dimensions.

The classical sufficient condition for the solubility of the plane Dirichlet problem is that of Lebesgue[13]. It may be stated as follows: *A point O of the boundary of a region D is regular unless it is possible to construct a circle of arbitrarily small radius about O*

[13] *Rendiconti di Palermo*, loc. cit.

which will lie entirely interior to D. This is completely contained in two conditions, given by Phillips and the author, and the author alone, respectively, which are entirely parallel to the three-dimensional conditions given earlier in this paper. They may readily be deduced from the general condition here given, the series whose divergence is to be investigated reducing to the form

$$A+A+\ .\ .\ .\ +A+\ .\ .$$

A plane problem of particular interest has been discussed by Kellogg in the memoir already cited. He discusses the boundary consisting of all points on the segment (0, 1) which have a representation in ternary fractions not containing the digit 2, and shows that every point of it is regular. If we wish to bring this under our general theorem, we must first show that this set has a non-zero capacity, or in other words that it is possible to distribute over it a positive charge without producing anywhere an infinite potential. This we do as follows: like Kellogg, we form the function $f(x)$ which has the value 0 for $x=0$, 1 for $x=1$, 1/2 for $1/3 \leq x \leq 2/3$, 1/4 for $1/9 \leq x \leq 2/9$, 3/4 for $7/9 \leq x \leq 8/9$, and so on. Then the function

$$-\int_0^1 \log x df(x)$$

is bounded, for it never exceeds

$$\log 6+1/2 \log 18+1/4 \log 54+\ .\ .\ .$$

$$=\log 2(1+1/2+1/4+\ldots)+\log 3\left(1+1+\frac{3}{4}+\frac{4}{8}+\frac{5}{16}+\ldots\right)$$

$$\leq 2 \log 2 + 5 \log 3.$$

Hence *a fortiori* the function

$$-\int_0^1 \log \overline{OP}\, df(O)$$

is bounded.

If the capacity of Kellogg's set of points is at least C, the capacity of each non-null third is at least $\dfrac{C}{1+C \log 3}$, of each non-null ninth $\dfrac{C}{1+C \log 9}$, and so on indefinitely. Moreover if O is a point of Kellogg's set, γ_m will contain at least one non-null

$1/3^{2+k}$th for all sufficiently large values of m, where $1/3^{k-1} \geq \lambda^{2^{m-1}} \geq 1/3^k$. Hence the test-series dominates

$$+\cdots+\frac{2^m}{1-2^{m-1}C\log\lambda+\log 27}+\frac{2^{m+1}}{1-2^m C\log\lambda+\log 27}+\cdots$$

which diverges. Thus Kellogg's case comes under our general theory, even though it can be treated somewhat more elegantly by Kellogg's direct methods[13].

[13] Since this paper has gone to the printer, the author has become more closely acquainted with the very important researches of M. Georges Bouligand. These go back to his Paris thesis of 1914 on the Green and Neumann functions of the cylinder, and culminate in a paper in the *Comptes rendus* for March 24, 1924, yielding conditions of regularity comparable in generality to those here given. In the latter paper are contained very many references to his previous work, which has appeared for the most part in the *Comptes rendus*. His treatment depends upon the discussion of infinite regions, and the use of inversions and Kelvin transformations.

ANALYSE MATHÉMATIQUE. — *Une condition nécessaire et suffisante de possibilité pour le problème de Dirichlet.* Note de M. **Norbert Wiener**, présentée par M. Henri Lebesgue.

Dans un Mémoire récent (¹), M. Lebesgue a montré que le problème de Dirichlet se divise en deux parties dont la première est la détermination d'une fonction harmonique correspondant à certaines conditions sur la frontière, et la seconde est la recherche des propriétés de cette fonction

(¹) *Comptes rendus*, t. 178, 1924, p. 349.

dans le voisinage de cette frontière. Dans un Mémoire [1] paru la même semaine, j'ai fait la même remarque indépendamment, et en plus j'ai donné une définition exacte du sens dans lequel une fonction harmonique dépend des conditions frontières. J'ai démontré aussi que ma méthode assigne une fonction bien déterminée à n'importe quelle condition frontière continue sur un ensemble borné de points dans un nombre quelconque de dimensions. En particulier, j'ai démontré que le potentiel dans l'espace tridimensionnel, correspondant aux valeurs frontières 1 sur un certain ensemble borné de points et correspondant à o à l'infini, détermine dans le sens de ma définition une fonction harmonique engendrée par une distribution de charge étalée sur cet ensemble. La charge ne sera négative dans aucune région, et en général n'aura pas de densité. Le potentiel correspondant à cette charge sera représenté par une intégrale de Stieltjès prise sur l'ensemble sur lequel la charge est étalée. La quantité totale de charge sera dite la capacité de l'ensemble. De cette façon, tout ensemble borné de points, qu'il forme ou non une surface, et en fait qu'il soit mesurable ou non, aura une capacité finie bien déterminée.

Revenant au Mémoire de M. Lebesgue ; comme moi, il montre que la résolution du problème de Dirichlet dans le sens classique, pour une région donnée, ne dépend que des propriétés des éléments de la frontière, et se réduit à la recherche de ce qu'il appelle la régularité des points particuliers. Un point O de la frontière B d'un ensemble D est dit régulier si, toutes les fois que $F(P)$ est une fonction continue sur B et $f(P)$ la fonction harmonique correspondante sur D,

$$\lim_{P \to O} f(P) = F(O).$$

Bien entendu, cette condition doit être remplie pour toute fonction F continue sur B.

En dehors des conditions qui ont été données antérieurement, M. Lebesgue donne une condition qu'on peut formuler de la manière suivante :

S'il existe un segment de droite ou un segment d'une courbe analytique qui se termine en un point O *de la frontière du domaine* D, *si ce segment est entièrement extérieur à* D *et s'il existe des nombres positifs* A *et* B *tels que,* P *étant un point de ce segment, sa distance de* D *est plus grande que* $A\,\overline{OP}^{B}$, O *est régulier. Si, par contre, il existe des nombres positifs* A *et* B, *tels que,*

[1] *Certain problems in Potential Theory* (*Journal of Mathematics and Physics of the Massachusetts Institute of Technology*, t. 3, n° 1, janvier 1924).

lorsque P *est situé sur ce segment*, OP = OQ, *et*

$$\overline{PQ} > A e^{-\frac{B}{\overline{OP}}},$$

Q *est dans* D, O *est irrégulier.*

M. Lebesgue généralise en plus une condition de régularité de M. Zaremba, ayant rapport aux points coniques d'une surface. Il décrit une suite infinie de sphères dont le centre est le point O, indiquant sur chacune la région intérieure à un cône donné dont le sommet est en O, faisant tourner chaque sphère autour de O, et excluant la région indiquée d'un domaine nouveau D′. En outre, il démontre que le point O sera encore régulier comme point de la frontière du domaine D′. Il conclut qu'il n'a pas réussi à étendre ces résultats aux points cuspidaux.

M. Lebesgue déduit tous ses résultats du théorème suivant : *Pour qu'un point* O *de la frontière d'un domaine* D *soit régulier, il faut et il suffit qu'il existe une fonction* $F(x, y, z)$ *continue en* O, *atteignant sa borne inférieure en* O, *et en* O *seulement, et telle que partout, à l'intérieur de* D, $\Delta F \leq 0$. Quoique cette condition soit nécessaire et suffisante, comme elle manque cependant de caractérisation géométrique, on ne peut pas la considérer comme une solution définitive du problème de Dirichlet.

Je viens d'obtenir le théorème suivant, qui donne une caractérisation complète, nécessaire et suffisante des points réguliers. Elle est de nature au moins quasi géométrique, et son énoncé est le suivant : *Soit* O *un point de la frontière* C *d'un ensemble ouvert* D *dans l'espace à trois dimensions. Soit* λ *un nombre positif plus petit que* 1. *Soit* γ_n *la capacité de l'ensemble de tous les points* Q *qui n'appartiennent pas à* D, *tels que*

$$\lambda^n \leq OQ \leq \lambda^{n-1}.$$

O *est alors régulier ou irrégulier selon que*

$$\frac{\gamma_1}{\lambda} + \frac{\gamma_2}{\lambda^2} + \ldots + \frac{\gamma_n}{\lambda^n} + \ldots$$

diverge ou converge.

Voici une application de ce théorème. Nous dirons que O est un point de révolution monotone cuspidal de la frontière de D si cette frontière, dans le voisinage de O, est une surface rentrante donnée par l'équation

$$\varphi = f(\rho)$$

dans un système de coordonnées sphériques, où $f(\rho)$ est une fonction croissante de ρ qui n'excède jamais π et qui s'annule pour $\rho = 0$. *La divergence de*

$$\sum_{1}^{\infty} \frac{1}{\log f(\lambda^m)},$$

λ étant un nombre quelconque positif inférieur à 1, *est donc une condition nécessaire et suffisante pour la régularité de* O.

Un autre corollaire de mon théorème principal est le suivant : *Soit* O *un point de la frontière d'une région* D. *Soit* O′ *un point de révolution monotone cuspidal sur la frontière d'une région* D′. *Supposons en outre* O′ *régulier. Construisons deux sphères de rayon r autour de* O *et de* O′. *Soient* A *et* A′ *les aires des parties de* D *et de* D′ *à l'intérieur de ces sphères. Alors, si pour toutes les valeurs suffisamment petites de r nous avons* $A \leqq A'$, O *est régulier.*

De ce théorème se déduit immédiatement la généralisation du résultat de M. Lebesgue analogue à la généralisation qu'il a faite du théorème de M. Zaremba.

Observations au sujet de la Note de M. N. Wiener, par M. **Henri Lebesgue**.

En janvier dernier, un article de M. G. Bouligand m'a fourni l'occasion de donner quelques résultats obtenus au cours de recherches sur les cas d'impossibilités du problème de Dirichlet, recherches qui remontent à 1913.

M. Wiener parle longuement de ma publication ; ce qu'il ne dit pas et que je tiens à dire, c'est que, dans ces dernières années, le problème de Dirichlet avait été, à mon insu, l'objet de recherches fécondes, dues à divers savants américains et en particulier à M. Wiener. Leurs théorèmes dépassent souvent les miens et ils ont si bien abordé les diverses questions qui m'ont occupé que, si j'avais connu leurs travaux, j'aurais sans doute jugé inutile de revenir sur mes résultats de 1913. J'insisterai sur deux points.

Dès que l'on a étudié la continuité, en un point frontière du domaine, de la fonction fournie par une méthode de résolution du problème de Dirichlet, on a été conduit à formuler des conditions ne faisant pas intervenir la fonction continue donnée sur la frontière étudiée, mais seulement la forme de cette frontière, et dans le seul voisinage du point étudié. Dans ma dernière Note, j'ai surtout insisté sur les notions mathématiques auxquelles conduit la généralisation de l'observation précédente. Or, dans un article : *Certain Notions in potential Theory*, M. Wiener développe, à l'occasion de ce qu'il appelle : *The im kleinen character of the Dirichlet problem*, des considérations à peu près équivalentes aux miennes.

D'autre part, MM. H.-B. Phillips et N. Wiener ont obtenu une condition de régularité très large et comprenant, en particulier, celle que je donnais sous la rubrique A, ainsi que la condition *b*, énoncée par M. Bouligand.

Les travaux des Savants américains avaient aussi échappé à M. Bouligand; il convient d'ajouter que, par contre, l'importante Note publiée par M. Bouligand, en 1919, semble être restée inconnue des Américains. Maintenant que ces efforts parallèles ne s'ignorent plus, on peut espérer qu'ils seront plus fructueux encore ([1]).

([1]) Je renvoie à la Note ci-dessous de M. Bouligand pour les références aux ravaux de MM. Philips et Wiener. Dans leurs publications, on trouvera des indications bibliographiques se rapportant à d'intéressants Mémoires de MM. O.-D. Kellogg, G.-E. Raynor et Gleason.

Comments on [24b, c]

L. Lumer

Wiener's famous criterion for regularity, namely if γ_n is the capacity of that part of the complement of the domain in the intersphere $(\lambda^n, \lambda^{n-1})$, $0 < \lambda < 1$, the regularity is equivalent to the divergence of the series $\Sigma\gamma_n\lambda^{-n}$, is a very deep and difficult result, inspired by a result on analytic functions, which clearly shows the geometric and local (im Kleinen) character of regularity, and its equivalence with a certain "non-smallness" of the complement of the domain in the neighborhood of the point.

This "non-smallness" proved in fact to be the non-thinness of the complement at the point, after Brelot[1] introduced the notion and extended Wiener's criterion to characterize the non-thinness in general (with a simplified proof making use of advanced potential theory).

But Wiener's criterion is really specific of the Euclidian boundary and the Newtonian kernel; for other ideal boundaries or other notions of thinness and in the developments of today's axiomatic theories, it has to be abandoned and replaced by Lebesgue's barrier criterion, which does not have the same geometric character, as it describes the domain close to the boundary through properties of certain superharmonic functions "living" on it, an idea more than once exploited today.

Reference

1. M. Brelot, Bull. Sci. Math., ser. 2, 68(1944), part 1, 12-36.

NOTE ON A PAPER OF O. PERRON

By Norbert Wiener

§1. **The Generalized Dirichlet Problem for Continuous Boundary Values.** In the last few years, the theory of the first boundary value problem of potential has been brought to a very high degree of completeness and definitiveness by the independent work of a considerable number of investigators in Europe and America.[1] In this development there have been three principal stages or moments: (1) the envisagement of the first boundary value problem (which, as Lebesgue has shown, need not always admit a solution) as a special case of a more general linear problem which is always soluble for continuous boundary conditions; (2) the precise determination of the types of boundaries for which the solution of this more general problem for any continuous boundary condition is *ipso facto* a solution of the first boundary-value problem in the classical sense; (3) the extension of the solution to very general cases of discontinuous boundary conditions. Stage (1) has been clearly outlined by Lebesgue, and has been developed in detail by Bouligand and Wiener, who have likewise obtained what may be regarded as a substantially complete theory of stage (2). In stage (3) the furthest developments thus far obtained are those of Evans and of Wiener, who make use of that extremely powerful tool, the Daniell integral.[2]

[1]Cf. H. Lebesgue, *Sur le problème de Dirichlet*, Rend. di Palermo, v. 24, 1907; note in C. R. des séances de la Soc. Math. de France, 1913, p. 17; note in C. R., Jan. 21, 1924 (pp. 349–354); G. Bouligand, *Sur le problème de Dirichlet harmonique*, C. R., Jan. 2, 1924, pp. 55–57; *Domaines infinis et cas d'exception du problème de Dirichlet*, C. R., March 24, 1924, pp. 1054–1057; *Sur les principes de la théorie du potentiel*, Bull. Sci. Math., v. 48, July, 1924, pp. 227–232; O. D. Kellogg, *An Example in Potential Theory*, Proc. Am. Acad. Arts and Sciences, v. 58, No. 14, June, 1923; G. E. Raynor, *Dirichlet's Problem*, Annals of Math., March, 1922; S. Zaremba, Bull. de l'Ac., des Sc. de Cracovie, July, 1909; G. C. Evans, *Problems of Potential Theory*, Proc. Nat. Acad. Sci., v. 7 (1921) pp. 89–98; H. B. Phillips and N. Wiener, *Nets and the Dirichlet Problem*, this Journal, v. 2 (1923) pp. 105–124; N. Wiener; *Certain Notions in Potential Theory*, this Journal, v. 3, No. 1, January, 1924; *The Dirichlet Problem*, this Journal, v. 4, No. 3, April, 1924; *Discontinuous Boundary Conditions and the Dirichlet Problem*, Trans. Am. Math. Soc., v. 25 (1923), pp. 307–314; O. Perron, *Eine neue Behandlung der ersten Randwertaufgabe für* $\Delta u=0$, Math. Ztschr., v. 18, Heft 1/2.

[2]P. J. Daniell, *A General Form of Integral*, Annals of Math., (2) 19, pp. 279–294 (1917–1918).

While the paper of Perron has reached the limit of generality and completeness in none of the three aspects of the Dirichlet problem, and while it is substantially contemporaneous with a much shorter paper of Kellogg in which most of its results are contained, it is distinguished by a peculiar elegance and simplicity. Many of its methods are capable of a development far beyond the confines of the actual paper. It is the purpose of the present note to carry out this development, and to show how it is possible to erect on Perron's foundation a theory substantially equivalent to that of Bouligand and Wiener.

In the course of this paper, however, we shall consider a case of the Dirichlet problem somewhat different from that of Perron. The change is in no way essential, but is introduced to eliminate the more or less accidental complications arising, on the one hand, from the distinction between the different sheets of a Riemann surface of space and, on the other, from the behavior of a logarithmic potential at infinity. Perron discusses a plane-bounded set of points G consisting entirely of interior points. His G may lie on a many-sheeted Riemann surface, but contains no branch point in its interior. We shall consider, on the other hand, a three-dimensional bounded set of points G, consisting entirely of interior points, and nowhere overlapping itself. Again, Perron states his boundary-value problem as follows: *On the boundary R let a bounded function f be given. Let $\underline{f}$ and $\bar{f}$ be its inferior and superior functions.*[3] *To find a function u on the closed point set $G+R$ which fulfils the following conditions:*

I. *On the boundary R, the inferior and superior functions satisfy the inequalities*

$$\underline{f} \leq \underline{u} \leq \bar{u} \leq \bar{f};$$

II. *u is harmonic in G.*

Now this form of putting the Dirichlet problem says at once too much and too little: too much, because it goes beyond the case of continuous functions f; and too little, because for all but the simplest discontinuous boundary conditions, I is not strong

[3] *I.e.*, $\underline{f}(P)$ is the smallest limit which can be approached by $f(P_n)$ as P_n approaches P. Likewise, $\bar{f}(P)$ is the largest limit which can be approached by $f(P_n)$ as P_n approaches P.

enough to secure the uniqueness of u. We shall therefore confine our attention for the present to continuous boundary conditions f, and shall reserve for the end of the paper the more general case.

If K is a sphere which with its interior lies entirely within G, and if v is a function defined over $G+R$, then Perron understands by $M_K v$ a function which coincides with v outside K and on its periphery, but which within K is defined by the Poisson integral generated by the values of v on its periphery. Condition II may then be written

$$u=M_K u.$$

The following definitions are now made:[4] *Every function ϕ continuous on G and satisfying the following conditions is a lower function of the continuous function f:*

$$\overline{\phi}\leq f \text{ on the boundary } R;$$

$$\phi\leq M_K\phi \text{ for every sphere } K.$$

Every function ϕ continuous on G and satisfying the following conditions is an upper function of f:

$$\underline{\phi}\geq f \text{ on } R;$$

$$\phi\geq M_K\phi \text{ for every } K.$$

Perron then proves that the assemblage of all upper functions has a finite lower bound, and that this lower bound is harmonic in G. Similarly the assemblage of all lower functions has a finite upper bound, and this upper bound will be harmonic. Perron shows that under certain assumptions which we shall not stop to consider, but which are substantially identical with those made by Kellogg in the paper which we have cited, these two functions coincide, and assume the assigned boundary values continuously. Even in the case where the Dirichlet problem is not solvable, these harmonic functions may both be regarded as solutions of a generalized Dirichlet problem. However, there is a disadvantage in this: first, they have not been proved to depend linearly on

[4]Perron demands that an " Oberfunktion " or " Unterfunktion " be continuous on $R+G$. In none of his proofs, however, is anything more than continuity on the interior+boundary of K demanded. The Perron definition will not yield the results here desired without the Perron conventions as to Riemann surfaces and spaces, which we expressly do not wish to introduce.

the boundary conditions, and secondly, they are two functions, each of which has an equal right to be considered as a solution of the generalized Dirichlet problem. The theory is much simplified when we can find a single function to play that rôle.

Now, in one of my papers cited above, I have shown that for any continuous f, the Dirichlet problem admits a certain unique solution in a generalized sense. To be more precise, my theorem reads as follows: *Let G, R, and f be defined as above. Then*

(1) *There exists a sequence $\{R_n\}$ of boundaries such that the Dirichlet problem is solvable in the ordinary sense over the interior of each R_n for all continuous boundary conditions; such that R_n always lies in the interior of R_{n+1}; and such that every point of G is interior to some R_n.*

(2) *There exists a function $v(P)$ continuous over $R+G$, and coinciding with $f(P)$ on R.*

(3) *If $u_n(P)$ is the harmonic function determined on the interior of R_n by the boundary values v on R_n, then the sequence of functions u_n converges at any point P of G to a limit $u(P)$. This limit will be a harmonic function of P throughout G, and will be independent of the choice of the sequence $\{R_n\}$ and the function $v(P)$.*

If we agree that our harmonic functions are to vanish at infinity, an exactly similar theorem will hold for the exterior problem.

What I here wish to prove is that *for any continuous f and for any open-bounded G, u will be at once the lower bound of all upper functions and the upper bound of all lower functions.* That is, the apparent difficulties with the Perron generalization of the Dirichlet problem are not actual, and the generalization completely coincides with that which I have given. In addition, Perron's form of generalization is far more elegant than mine, in that no more or less artificial use is made of such auxiliary elements as $v(P)$ or the sequence $\{R_n\}$. It is furthermore better adapted to the direct study of discontinuous boundary conditions.

As a preliminary, let us discuss the simplest possible case. We have given two bounded closed sets of points Σ_1 and Σ_2 with no point in common. Let us now consider the generalization of the Dirichlet problem corresponding to the boundary values 1 on Σ_1 and 0 on Σ_2. As the region exterior to Σ_1 and Σ_2 will not in general be closed, the Dirichlet problem which we are considering is an exterior problem, and it is necessary to assume that the desired harmonic function will vanish at infinity. We shall also assume that $v(P)$ is 1 in the neighborhood of Σ_1 and 0 in the neighborhood of Σ_2. Under these conditions the theorem of my previous paper reads as follows: *Let* $\{R_n\}$ *be a sequence of boundaries, each included in its predecessor, containing in common every point of* Σ_1 *and no other point. Furthermore, let these be so chosen that the exterior Dirichlet problem is solvable for each. Let* $\{S_n\}$ *be a sequence of boundaries satisfying exactly the same conditions as those laid down for* $\{R_n\}$, Σ_2 *being substituted for* Σ_1. *Let* $\phi_{m,n}$ *be the harmonic function vanishing at infinity and corresponding to boundary values* 1 *on* R_m and 0 *on* S_n. $\phi_{m,n}(P)$ *will always exist for sufficiently large values of m and n, and as m and n increase to infinity in any way,*

$$\lim \phi_{m,n}(P)=u(P)$$

exists for all P's not within Σ_1 *nor* Σ_2. *Furthermore, u is harmonic everywhere outside* Σ_1 *and* Σ_2.

Let it be noted that $\phi_{m,n}(P)$ is a monotone non-decreasing function of n and a monotone non-increasing function of m. It is also bounded. Hence

$$\lim_{n\to\infty} \phi_{m,n}(P)=\phi_m(P)$$

and

$$\lim_{m\to\infty} \phi_{m,n}(P)=\phi^{(n)}(P)$$

exist. We shall make the convention that within R_m, $\phi_m(P)=1$, and within S_n, $\phi^{(n)}(P)=0$. There is no difficulty in showing that

$$\lim_{m\to\infty} \phi_m(P)=\lim_{m\to\infty} \phi^{(m)}(P)=u(P)$$

everywhere outside Σ_1 and Σ_2.

I now wish to show that ϕ_m is an upper function and that $\phi^{(m)}$ is a lower function for boundary conditions 1 on Σ_1 and 0 on Σ_2. To begin with, ϕ_m assumes limiting values on Σ_2 which are never negative and is 1 on Σ_1. Hence ϕ_m is never less than the boundary values. In the second place, if K is entirely outside R_m, $\phi_m= M_K\phi_m$. If K cuts R_m, $M_K\,\phi_m$ cannot exceed 1 on the part of R_m

interior to K. Hence over the contour constituted by this part of R_m and the part of K exterior to R_m,

$$\phi_m \geq M_K \phi_m.$$

Therefore this relation must hold within this contour, as the two functions are harmonic. Over the rest of the interior of K,

$$1 = \phi_m \geq M_K \phi_m.$$

That is, ϕ_m is an upper function. The argument in the case of $\phi^{(m)}$ is precisely parallel. Since, however, u is the common limit of the sequences $\{\phi_m\}$ and $\{\phi^{(m)}\}$, and since no lower function anywhere exceeds an upper function, the upper bound of all lower functions coincides with the lower bound of all upper functions, and is u itself.

We may immediately proceed from this to the case where our boundary consists of a finite number of closed, bounded, disconnected pieces, while f assumes a constant value over each piece. Let the pieces be $\Sigma_1, \ldots, \Sigma_n$, and let $f = a_j$ on Σ_j. Let a be the least of the a_j's, and let $f_j = a_j - a$ on Σ_j and $f_j = 0$ on the other Σ_k's. Then

$$f = \Sigma f_j + a.$$

Furthermore, if g_j is an upper function for f_j, then

$$g = \Sigma g_j + a$$

is an upper function for f. Similarly, if h_j is a lower function for f_j, then

$$h = \Sigma h_j + a$$

is a lower function for f. We have, however, proved in substance that g_j and h_j can be chosen as close as we like to one another. Hence we can make the difference between g and h as small as we like, so that the lower bound of all upper functions of f will be at the same time the upper bound of all lower functions of f. Furthermore, as I have shown in a previous paper that my solution of the generalized Dirichlet problem depends linearly on the boundary conditions, it follows, as a moment's consideration will show, that the lower bound of upper functions of f and the upper bound of lower functions of f assume this solution as their common value.

It is now necessary to make the transition to the most general case of continuous boundary conditions. For this purpose, I must recall some definitions and theorems from my previous paper. At a point P the *relative weight* of a point-set Σ_1 to a point-set Σ_2 is the potential corresponding in the sense already explained to a potential 1 on the set consisting of Σ_1 together with its derivative, a potential 0 on the set consisting of Σ_2 together with its derivative, and a potential 0 at infinity. This definition is only directly applicable in case Σ_1 and Σ_2 have no point nor limit-point in common. The *outer weight* at P of a part σ of a bounded point-set Σ is the lower bound of the relative weight of σ to point-sets τ in Σ containing no point nor limit-point in common with σ. It is easy to show that if σ and σ' are two mutually exclusive closed sub-sets of Σ, the outer weight of $\sigma+\sigma'$ is precisely the sum of the outer weights of σ and of σ'. Furthermore, if σ_1 includes σ, the outer weight of σ_1 is at least that of σ.

Now let Σ_ξ consist of those points of Σ with an X-coördinate not exceeding ξ. Considered as a function of ξ, Σ_ξ is monotonic and nondecreasing. It has hence only a denumerable set of discontinuities. If Σ_ξ is continuous for $\xi=\xi_1$, it follows that if ϵ is any positive number, we can choose a positive η so small that

$$\text{outer weight of } [\Sigma_{\xi_1+\eta}-\Sigma_{\xi_1-\eta}]<\epsilon.$$

Hence if σ consists of all the points of Σ with X-coördinates between $\xi_1-\eta/2$ and $\xi_1+\eta/2$, inclusive, it follows from the theorem which we have just proved that

$$\text{outer weight of } \sigma<\epsilon.$$

It consequently follows that for all but a denumerable set of values of ξ, the plane $x=\xi$ may be surrounded by a region of arbitrarily small outer weight. Now let ξ_1 and ξ_2 be the extreme X-coördinates of Σ. The interval (ξ_1, ξ_2) may be subdivided into a finite number of sub-intervals of length less than $\theta/2$ in such a way that no point of subdivision is one of the denumerable set of points of discontinuity of Σ_ξ. Hence the entire set of points of Σ contained in the planes perpendicular to the X-axis and passing through the points of subdivision of the interval (ξ_1, ξ_2) on the X-axis may be surrounded by a region of outer weight less than ϵ. *Mutatis mutandis*, the same thing applies with regard to the Y and Z axes. In other words, given any positive number θ, the closed set Σ may be subdivided into a finite number of mutually exclusive closed compartments with greatest dimension not exceeding θ, plus a set of points contained in three sets each of outer measure less than ϵ, and consequently of outer measure not exceeding 3ϵ.

Let us now consider a continuous function $f(P)$ defined over Σ. Let θ be so chosen that the oscillation of $f(P)$ in a closed region of maximum dimension never exceeds ϵ. Let us form an upper function and a lower function to f in the following manner.

(1) We form a set of compartments in the manner indicated in the last paragraph; we let $\psi_1(P)$ be an upper function of the step-function assuming in each compartment as its value the upper bound of f in that compartment; and we let $\psi_2(P)$ be a lower function of the step-function assuming in each compartment as its value the lower bound of f in that compartment. It follows from what we have said that if χ_1 and χ_2 are respectively these step-functions, we may determine ψ_1 and ψ_1', upper and lower functions to χ_1 such that for a given, P in G

$$\psi_1(P)-\psi_1'(P)<\epsilon.$$

We can similarly determine ψ_2' and ψ_2, upper and lower functions respectively to χ_2, such that for this P

$$\psi_2'(P)-\psi_2(P)<\epsilon.$$

These upper and lower functions may indeed be formed as polynomials in the upper (or lower) functions to the different functions assuming only the values 0 and 1 from which χ_1 and χ_2 are built up as sums. We see at once that

$$\psi_1(P)-\psi_2(P)<3\epsilon.$$

(2) We now consider τ, the set of all points of Σ which are not contained in the compartments of (1). Since the outer weight of τ with respect to Σ does not exceed 3ϵ, we may determine a function $\psi_3(P)$, harmonic except on Σ and within a certain small distance of points of τ, continuous except on Σ, non-negative, equalling 1 in the given neighborhood of τ, and not exceeding 4ϵ at the given P.

(3) We write

$$\Psi_1(P)=\psi_1(P)+\max|f|.\psi_3(P);$$
$$\Psi_2(P)=\psi_2(P)-\max|f|.\psi_3(P).$$

Ψ_1 is clearly an upper function to f, and Ψ_2 a lower function. Furthermore,

$$\Psi_1(P)-\Psi_2(P)<\epsilon(3+8\max|f|).$$

As ϵ is arbitrary, this may be made as small as we like. Hence the upper bound of all lower functions coincides with the lower bound of all upper functions.

(4) In my previous paper, I proved that the generalized solution which I there obtained in the general case was the limit of the harmonic function determined by the boundary values χ_1. This function must lie between ψ_1 and ψ_1', and hence between ψ_1 and ψ_2. *A fortiori* it lies between Ψ_1 and Ψ_2. Hence the func tion which constitutes the solution of the generalized Dirichlet problem of my previous paper is the upper bound of all lower functions of its boundary values and the lower bound of all upper functions.

§2. **Discontinuous Boundary Values.** In the simplest cases of discontinuous boundary values, the Perron generalization of the Dirichlet problem is still satisfactory. For example, let our boundary function be 0 except at a single point of the boundary, where it is 1. It is then possible at a point P interior to the boundary to find an upper function $\phi(P)$ approximating to 0 as closely as we wish. Since, however, 0 is itself a lower function, the upper bound of all lower functions is at the same time the lower bound of all upper functions.

However, we do not need to go much further to find an instance where the Perron generalization breaks down. For example, let G be the interior of a circle with boundary R. Let us establish a system of polar coördinates with the center of G as pole. Let $f(P)$ be 1 for those points P of R with a θ-coördinate rational in terms of 2π, and let $f(P)$ be 0 elsewhere on R. Clearly any upper function of f is at least 1, and no lower function is greater than 0. Hence the Perron method yields us here no unique generalization of the Dirichlet problem.

The difficulty here is of the same character as the difficulty in applying an ordinary integral to the boundary values in question. In this case the Riemann definition breaks down, and we must have recourse to the definition of Lebesgue. In the case of the Dirichlet problem, then, we must develop an analogue to the Lebesgue integral. Now, the fundamental property of the Lebesgue integral is that if $f_1, f_2, \ldots$ are all uniformly bounded and integrable over a closed set of points S, and if $f = \lim f_n$ exists for every point in S, then f is also integrable, and in fact its integral is the limit of the integrals of the f_n's. This suggests the following definitions, which are confined to the case of bounded functions f: *if $f(P)$ is a function defined on R; if $f(P) \leq \lim_{n\to\infty} f_n(P)$ $[f(P) \geq \lim_{n\to\infty} f_n(P)]$ everywhere on R; if every f_n is continuous, and the set of all f_n's is bounded; if $\phi_n(P)$ is an upper* [*lower*] *function of $f_n(P)$; and if $\phi_n(P)$ tends to $\phi(P)$ with increasing n at every point P of G; then $\phi(P)$ will be called a superfunction* [*subfunction*] *to f.*

We now wish to do two things: first to prove that every superfunction of f is at least as large as any subfunction of f; and second to discuss those cases in which the lower bound of all superfunctions coincides with the upper bound of all subfunctions. The common value of these two bounds (which will always be harmonic functions) will be regarded as a solution of the generalized Dirichlet problem corresponding to boundary values f.

Both of our ends can be accomplished with the aid of the Daniell integral and of the methods of a paper of mine entitled *Discontinuous Boundary Conditions and the Dirichlet Problem.* I there pointed out that the potential at a point P of G corresponding to boundary values f on R may be regarded as a type of Daniell integral of f, and in my later papers, I indicated how this result could be extended to cases where the Dirichlet problem was only solvable in the generalized sense. Without going into the assumptions necessary for the application of the Daniell theory, which are fully discussed in my previous paper in so far as they apply to the case in question, the outline of the Daniell theory is the following. We start with a class of functions T_o on a general assemblage, and an operation I on the functions of the class. T_o is here the class of all functions f continuous on R, and $I\{f\}$

is the potential corresponding (in my generalized sense) to the set of boundary values f. We then form the class T_1 of functions obtained as limits of monotone non-descending sequences of functions of T_o. For these functions, the result of applying the operator I is defined as the limit, finite or infinite, of the I's of the functions of the sequence. This definition is always applicable and unambiguous. We next define the upper integral of a function f as the lower bound of the I's of functions of T_1 nowhere less than f, and the lower integral of f as the negative of the upper integral of $-f$. If the upper and lower integrals of f coincide and are finite, f is said to be summable, and the common value of the upper and lower integrals is $I\{f\}$. Daniell proves that if the conditions for the validity of his theory are satisfied, the different stages of the definition of I are compatible; that if f_n tends to f in such a way that there is a summable function exceeding in absolute value every f_n, while every f_n is summable, then f_n is summable, and in fact

$$I\{f\} = \lim I\{f_n\};$$

that if $f \geq g$ for every argument, $I\{f\} \geq I\{g\}$; that I is linear; that the class of summable functions is linear, and contains the absolute value of any one of its members.

It follows from Daniell's definitions and theorems that if we write $I_P\{\psi\}$ for the value at P of the harmonic function determined in accordance with the Daniell generalization process by the boundary values ψ, then if we employ the notation which we used in the definition of superfunctions,

$$\phi(P) = \lim \phi_n(P) \geq \lim I_P\{f_n\} \geq \bar{I}_P\{f\},$$

where the symbol $\bar{I}$ is used to represent the upper integral. Similarly, if ϕ_1 is a subfunction of f, we shall have

$$\phi_1(P) \leq \underline{I}_P\{f\},$$

where $\underline{I}_P\{f\}$ stands of course for the lower Daniell integral of f. Since Daniell proves that

$$\bar{I}_P\{f\} \geq \underline{I}_P\{f\},$$

we see that any superfunction is at least as large as any subfunction. Furthermore, if the upper bound of all subfunctions coincides with the lower bound of all superfunctions, the common value must also be the common value of $\bar{I}_P\{f\}$ and $\underline{I}_P\{f\}$, and must hence be $I_P\{f\}$.

If $f(x)$ is a bounded member of T_1, the upper bound of all its subfunctions and the lower bound of all its superfunctions must coincide. To show this, it is only necessary to point out that we can represent f as $\lim f_n$, where $\{f_n\}$ is an increasing, bounded sequence of continuous functions. We may accordingly choose an upper function $\phi_1(P)$ of $f_1(P)$ less than $\epsilon+$ the lower bound of all upper functions of f_1, an upper function $\phi_2(P)$ of $f_2(P)$ less than $\epsilon/2+$ the lower bound of all upper functions of f_2, and so on. We choose $\bar{\phi}_n(P)$ as a lower function of f_n greater than

$$\text{(upper bound of lower functions of } f_n) - \epsilon/n.$$

Then since $\{I_P\{f_n\}\}$ is a monotonic and hence a convergent sequence, and since $I_P\{f_n\}$ is at once the lower bound of all upper functions and the upper bound of all lower functions of f_n, we have

$$I_P\{f\} = \lim I_P\{f_n\} = \lim \phi_n(P) = \lim \bar{\phi}_n(P).$$

Thus $I_P\{f_n\}$ is at once a superfunction and a subfunction of f.

The transition from T_1 to all bounded summable functions is easy, since it can be shown that any such function may be penned in between two *bounded* functions, one belonging to T_1 and t other the negative of a function belonging to T_1, with " integrals " differing by less than ϵ. Any superfunction to the larger of these two functions is a superfunction to the summable function between, while any subfunction to the smaller of the two confining functions is a subfunction to the function between. From this it may be concluded at once that for any bounded summable function on G, the upper bound of its subfunctions, the lower bound of its superfunctions, and the solution of the generalized Dirichlet problem in the sense of my previous paper, all coincide.

The condition that the summable function be bounded may be eliminated if in the definition of superfunctions and subfunctions, we substitute for the boundedness of the set of f_n's the following condition: *Let $F_n(P)$ be the function always assuming the greatest value of the quantities $|f_1(P)|, \ldots, |f_n(P)|$. Let the set of quantities $I_P\{F_n\}$ be bounded.*

Comments on [25a]

L. Lumer

The resolutivity (that is equality of Perron's envelopes) of every finite continuous function on the boundary is a very fundamental result in the development of the Dirichlet problem by Perron's method of envelopes, usually taken as axiom whenever not true in general (ideal boundaries, as in Brelot, 1956;[1] Gowrisankaran, 1963;[2] Constantinescu and Cornea, 1965[3]).

Here Wiener proves it for the Euclidian boundary (also proving the identity with his generalized solution), and whereas this has been made extremely simple later[4] , Wiener's proof has a degree of generality which made it applicable without modification to other very general ideal boundaries (compact metrizable, Martin's boundary, . . .[5]), even adaptable to nontopological boundaries of harmonic spaces (the present writer, 1969).

The maximum extension of Perron's method to discontinuous boundary functions is only sketched, as Wiener is unexpectedly stopped by a false counterexample (of a discontinuous boundary function summable but thought to be nonresolutive, p. 29). Discovered 15 years later by Brelot in 1939,[6] this lead him to the central result Wiener could have proved too: the equivalence of resolutivity with summability with respect to the harmonic measure. But Wiener's contribution to potential theory will remain, dominated by a perfect mastering of integration theory at its early stage and by the use of arguments valid far beyond the actual situations they were applied to.

References

1. M. Brelot, J. Math. Pures Appl., ser. 9, 35(1956), 297-335.
2. K. N. Gowrisankaran, Ann. Inst. Fourier (Grenoble) 13, no. 2(1963), 307-356.
3. C. Constantinescu and A. Cornea, Nagoya Math. J. 25(1965) 1-57.
4. M. Brelot and G. Choquet, Ann. Inst. Fourier (Grenoble) 3(1951), 199-263.
5. See reference 1.
6. M. Brelot, Acta Sci. Math. (Szeged) 9(1930), 133-153.

IC
Brownian Movement, Wiener Integrals, Ergodic and Chaos Theories, Turbulence and Statistical Mechanics

THE MEAN OF A FUNCTIONAL OF ARBITRARY ELEMENTS.

BY NORBERT WIENER.

1. P. J. Daniell* has recently developed a powerful method whereby if the notion of integration is once defined for a very restricted set of functions of arbitrary elements, it can be extended to a much more comprehensive set of functions. Daniell himself in his second article applied his method to the discussion of integrals in a denumerable infinity of dimensions. Daniell's method, however, leaves the mode of establishing integration over the original restricted set in general undetermined. It is the purpose of this paper to develop a method of setting up a Daniell integral which is applicable to a large group of cases, and in particular to functionals.†

2. **Definitions.** If K be any class, we shall define a *division* of K as a finite set of non-null sub-classes exhausting K at least once. Divisions of K will be represented by Greek letters with the suffix K—thus α_K, β_K, etc. A division depending on a parameter n will be represented by some such symbol as $\alpha_K(n)$. The sub-classes or intervals of a division α_K will be represented by $t_1(\alpha_K)$, $t_2(\alpha_K)$, $\cdots$, $t_m(\alpha_K)$.

A *weighted division* of K will be defined as a division of K to each term of which is assigned a positive number—its *weight*. A weighted division corresponding to α_k will be represented by such a symbol as $\alpha_K{}^U$, $\alpha_K{}^V$, or $\alpha_K{}^W$. The weight of $t_j(\alpha_K)$ will be written $U_\alpha\{t_j(\alpha_K)\}$, $V_\alpha\{t_j(\alpha_K)\}$, or $W_\alpha\{t_j(\alpha_K)\}$, respectively. $U_{\alpha(n)}$, may be abbreviated to U_n if a specific sequence of α's is understood.

A *K-partition*‡ P_K is a sequence of weighted divisions $\alpha_K{}^{U_1}(1)$, $\alpha_K{}^{U_2}(2)$, $\cdots$, $\alpha_K{}^{U_n}(n)$, $\cdots$, such that

(1) Every member of $\alpha_K(n+1)$ is wholly included in one member of $\alpha_K(n)$ and one only.

(2) $U_n\{t_j(\alpha_K(n))\}$ is the sum of all the numbers $U_{n+1}\{t_l(\alpha_K(n+1))\}$ such that $t_l(\alpha_K(n+1))$ is contained in $t_j(\alpha_K(n))$.

(3) If S_n is the 'sum' of a number of intervals from $\alpha_K(n)$, whatever n may be, and if S_{n+1} is always entirely included in S_n, then either there is a K-element common to every S_n, or the sum of the weights of the intervals in S_n approaches 0 as n grows without limit.

* P. J. Daniell, Annals of Mathematics, vol. 19 (1918), vol. 20 (1919).

† Cf. P. Lévy, Comptes Rendus, Aug., 1919.

‡ This notion is related to E. H. Moore's 'development.'

$\alpha_K(n)$, $\alpha_K{}^{U_n}(n)$, and $t_j(\alpha_K(n))$ are said to *belong* to P_k.

A function f defined for all elements of K is said to be a P_K *step-function* if there is an $\alpha_K(n)$ belonging to P_K such that the function is constant for every $t_j(\alpha_K(n))$. We shall say that f has degree n.

The *mean** of a P_K step-function f of degree n, taken over K with respect to P_K, is said to be

$$\frac{\sum_j U_n\{t_j(\alpha_K(n))\}f(x_j)}{\sum_j U_n\{t_j(\alpha_K(n))\}},$$

where x_j is a member of $t_j(\alpha_K(n))$. It will be written briefly $M_{P_K}(f)$.

A function f will be said to *have uniform P_K-continuity* if, given any positive number ϵ, there is an integer n such that if x and y belong to the same $t_j(\alpha_K(n))$, $|f(x) - f(y)| < \epsilon$.†

3. **Application of Daniell's Results.** Daniell's theory of integration is based on the existence of a set T_0 of bounded functions, which shall be closed with respect to multiplication by a constant, the addition of two functions, and the operation of taking the modulus. The class of all P_K step-functions clearly has all these properties. Daniell further postulates a finite functional operation I defined over T_0 and satisfying the conditions

(C) $I(cf) = cI(f)$, if c is any constant,

(A) $I(f_1 + f_2) = I(f_1) + I(f_2)$,

(P) $I(f) \geqq 0$ if $f(p) \geqq 0$ for all p,

(L) If $f_1 \geq f_2 \geq \cdots \geq 0 = \lim f_n$ for every p,

$$\lim I(f_n) = 0.$$

Our operation M_{P_K} satisfies all these conditions. The first three need no proof, being matters merely of elementary algebra. (L) may be proved if we can show that for every positive number a, the total weight of the set of intervals containing the K-elements p for which $f_n(p) \geq a$ approaches 0 as n grows indefinitely. Consider the set S_n of elements p for which $f_n(p) \geq a$. Clearly S_{n+1} is included in S_n. Three conceivable possibilities are open.

* The use of mean instead of integral is found in the posthumous papers of Gateaux (Bull. de la Soc. Math. de France, 1919). This was however unknown to me at the time I wrote this article.

† If a term x is shared by two or more members of some $\alpha(n)$, in determining functions over K, we regard x as a set of different members of K, each consisting of x *qua* member of all of the intervals of a sequence $t_1(\alpha_K(1))$, $t_2(\alpha_K(2))$, $\cdots$, where $t_{n+1}(\alpha_K(n+1))$ is contained in $t_n(\alpha_K(n))$. Each $t_n(\alpha_K(x))$ will contain only this value of x. It may be shown (in the general case, with the aid of Zermelo's axiom) that this change will not affect the validity of the three conditions for a division, and will render every step-function single-valued.

(1) Every S_n contains intervals from some fixed $\alpha_K(m)$. In this case there is some term p common to every S_n, so that for every n, $f_n(p) \geqq a$. This however is contrary to our hypothesis.

(2) (1) is not satisfied, but there is some p common to every S_n. This again violates our hypothesis.

(3) Neither (1) nor (2) is satisfied. Then by (3) of the definition of a partition, the sum of the weights of the intervals in S_n approaches 0 as n grows without limit. In this case it follows from the definition of M_{P_K} that $\lim_{n \doteq \infty} M_{P_K}(f_n) \leqq a$.

Hence (L) is proved, and we are at once in a position to apply Daniell's results. He defines T_1 as the class of all functions f which are the limits of a sequence f_n where every f_n belongs to T_0 and $f_1 \leqq f_2 \leqq \cdots$. $I(f)$ is defined as the limit of $I(f_n)$. Given any function f, $\dot{I}(f)$ is defined as the lower bound of $I(\varphi)$ for all functions φ of class T_1 such that $\varphi \geqq f$, and $\underset{\cdot}{I}(f)$ as $-\dot{I}(-f)$. If $\dot{I}(f) = \underset{\cdot}{I}(f)$ and is finite, f is said to be summable, and it is proved that if $f_1, f_2, \cdots, f_n, \cdots$, is a sequence of summable functions with limit f, and if a summable function φ exists such that $|f_n| \leqq \varphi$ for all n, f is summable, and $\lim I(f_n) = I(f)$. We can translate all this into our language, and in particular we can say that if a function can be obtained as a limit of a set of P_K step-functions all less in modulus than a certain P_K step-function, it is summable and its mean may be determined. A constant is clearly a P_K step-function, and we neither gain nor lose any generality by insisting that all our P_K step-functions be less in modulus than some constant.

Every bounded uniformly P_K-continuous function is summable. For let $f(x)$ be such a function and let $f_n(x)$ be the maximum of $f(x)$ in an interval $t_j(\alpha_K(n))$ that contains x and belongs to P_K. It is clear that by definition f_n is a P_K step-function, and furthermore that $|f_n(x)| < 1 + \max|f(x)|$. Moreover, since f is uniformly P_K-continuous, and since $f_n(x) = f(x_n)$ where x and x_n lie in the same $t_j(\alpha_k(n))$, $\lim_{n \doteq \infty} |f(x) - f_n(x)| = 0$, so that $f(x) = \lim f_n(x)$.

An important generalization of this theorem is the following: *if L_m is the set of all K-elements contained in a finite number of intervals of P_K, if L_m is contained in L_{m+1} for all m and if every K-element is in some L_m, then if f is bounded and uniformly continuous over all the intervals of each L_m, it is summable according to M_{P_K}.* Let $g_m(x) = f(x)$ when x is in L_m and 0 otherwise. Every g_m is summable by the argument of the last paragraph; and the whole set is bounded as no g_m can be larger than f in its largest modulus. As f is the limit of $\{g_m\}$, by Daniell's theorem, f is summable.

4. **Examples.** (*a*) K is the set of all real numbers in the closed interval (a, b). $\alpha_K(n)$ is the set of intervals

$$\left\{\frac{a + h(b - a)}{2^n}, \quad \frac{a + (h + 1)(b - a)}{2^n}\right\} \qquad h < 2^n$$

and each interval of $\alpha_K(n)$ is given the weight $1/2^n$. The mean $M_{P_K}(f)$ is the Lebesgue integral $1/(b - a) \int_a^b f(x)dx$, which is what we should naturally call the mean of $f(x)$ over the interval (a, b). This results immediately from the facts (1) that any continuous function can be obtained as the limit of a bounded sequence of step-functions constant over all the intervals of some $\alpha_K(n)$; (2) that, as Daniell has shown, every Lebesgue summable function is summable in his sense if $I(f)$ be Riemann integration and T be the set of all continuous functions, and hence if $I(f)$ be the Riemann integral confined to some set of functions with respect to which all continuous functions are summable.

(*b*) K is the set of all points $(x_1, x_2, \cdots, x_n)$ in a bounded region V of n-space of 'volume' v. $\alpha_K(l)$ is the set of all intervals

$$a_k/2^l \leq x_k \leq (a_k + 1)/2^l \qquad (1 \leq k \leq n),$$

where the a_k's are integers ranging between bounds not less than the largest coördinate of any point in V. The weight of each interval of $\alpha_K(l)$ is the limit of the sum of the 'volumes' of the intervals of $\alpha_K(m)$ contained in this interval to the sum of the 'volumes' of all intervals of $\alpha_K(m)$, as m grows indefinitely, and V shall be such that this limit always exists. $M_{P_K}(f)$ becomes

$$(1/v) \int\int_V \cdots \int f(x_1, x_2, \cdots, x_n)dV.$$

(*c*) K is the set of all points $(x_1, x_2, \cdots, x_m, \cdots)$ in a region of space of a denumerably infinite number of dimensions such that $a_m \leq x_m \leq b_m$ for all m. $\alpha_K(n)$ is the set of regions

$$\frac{2^{n-m}a_m + h_m(b_m - a_m)}{2^{n-m}} \leq x_m \leq \frac{2^{n-m}a_m + (h_m + 1)(b_m - a_m)}{2^{n-m}}$$

for all $m \leq n$. (Here h_m is some integer between 0 and $2^{n-m} - 1$). The weight of each region is

$$\frac{1}{2^{n(n+1)/2}}.$$

In the case where for every m $a_m = 0$ and $b_m = 1$, $M_{P_K}(f)$ is Daniell's

$$I(f) = \int_0^1 \cdots \int_0^1 \cdots f(p)dx_1dx_2 \cdots dx_n \cdots,$$

and is defined for a class of functions including all continuous functions. This results from the fact that every function belonging to Daniell's T_0, the class of all continuous functions of a finite number of variables, is summable in our system.

(*d*) K is the set of all continuous functions defined between 0 and 1, satisfying a Lipschitz condition and themselves lying between the bounded continuous functions $\varphi(x)$ and $\psi(x)$. The coefficient of irregularity of a function f (written $c(f)$) is the upper bound of the modulus of the slope of a chord of the curve representing the function. $\alpha_K(n)$ is the set of regions each of which consists of all continuous functions f satisfying simultaneously the 2^n pairs of inequalities

$$\frac{2^{2^n-m}\varphi\left(\frac{m}{2^n}\right)+h_m\left\{\psi\left(\frac{m}{2^n}\right)-\varphi\left(\frac{m}{2^n}\right)\right\}}{2^{2^n-m}}\leq f\left(\frac{m}{2^n}\right)$$
$$\leq\frac{2^{2^n-m}\varphi\left(\frac{m}{2^n}\right)+(h_m+1)\left\{\psi\left(\frac{m}{2^n}\right)-\varphi\left(\frac{m}{2^n}\right)\right\}}{2^{2^n-m}}$$

and an inequality either of the form

$$k_m\leq c(f)\leq k_m+1 \qquad \text{or} \qquad c(f)\geq m$$

for all integral values of m not greater than 2^n. Here h_m is an integer between 0 and 2^{2^n-m}, and k_m is an integer between 0 and $m-1$. The coefficient of irregularity of an interval is defined as the smallest integer not less than 2 that is greater than the coefficient of irregularity of some function in the interval. The weighting of $\alpha_K(n)$ is carried on in a progressive manner as follows: when K is divided into intervals of $\alpha_K(1)$ or when any interval of $\alpha_K(n)$ is divided into intervals of $\alpha_K(n+1)$, all intervals with the same coefficient of irregularity are weighted alike. The total weight of all the sub-intervals with a given coefficient of irregularity c greater than that of the original interval we shall then make $w/c!$, where w is the weight of the original interval. The rest of the weight, of course, goes to those sub-intervals with the same coefficient of irregularity as the original interval. If w is the original weight of K, and q the least coefficient of irregularity of any function it contains, the total weight of those intervals in any division whose coefficients of irregularity exceed k may be shown to be no greater than $w(e-2)^k$, if $k>q$.

That is, the total weight of any set of intervals whose coefficient of irregularity is greater than a given number N approaches 0 as N grows without limit. Furthermore, if $S_1, S_2, \cdots S_n, \cdots$ are sets of intervals such that every S_n contains intervals with a coefficient of irregularity not greater than a fixed number N, and if S_{n+1} is always contained in S_n, there is always a continuous function belonging to every S_n. To prove this, in the first place, if we discard from S_n all intervals whose coefficient of irregularity equals or exceeds N, and call the remaining set R_n, R_{n+1} will be contained in R_n, which will always exist, whatever n. Let us divide the intervals of R_1 into those that contain intervals from an infinity of R_n's and those that do not. Clearly there will be at least one interval in the former class. Select one such interval.* This will contain intervals belonging to some R_k. These intervals can again be divided into two classes, and we can again select those that contain sub-intervals of an infinite sequence of orders. In this way an infinite chain $t_1, t_2, \cdots, t_n, \cdots$ can be selected of intervals belonging respectively to $S_1, S_k, \cdots, S_l, \cdots$, and all of a coefficient of irregularity no greater than N. From a certain stage on these t's will contain no function whose coefficient of irregularity is more than $N + 1$. By a theorem of Frechet,† since t_1 is a bounded class of equally continuous functions, it will be compact. It may readily be proved that every t_n is closed and hence extremal.‡ Consequently there will be a continuous function f common to every t_n, and therefore to every S_n. This completes the proof of (3) of the definition of a partition. The satisfaction of (1) and (2) is immediately obvious. We are hence in a position to apply our definition of a mean to functionals of continuous functions, and to give a meaning of $M_{P_{\overline{K}}}\{F\}$.

It should be noted that there is much that is arbitrary in the actual carrying out of this definition. What is really essential is that some coefficient of irregularity be chosen so that every "Einschachtelung" of intervals of less than a given coefficient of irregularity should contain a continuous function, and that then a method of weighting be adopted which shall make the total weight of the set of intervals whose coefficient of irregularity is greater than N a decreasing function of N that approaches 0 as a limit. This method can at once be extended to space-curves, surfaces, and all such entities as are usually made the arguments of functionals.

It clearly follows by the theorem at the end of § 3 that if a functional is bounded and is uniformly P_L-continuous over every P_L that consists of all the $\alpha_K(n)$'s restricted to functions of no more than a given coefficient of irregularity, it is summable with respect to M_{P_K}. Now, the set of all functions lying between two given functions in modulus and of no more than a given coefficient of irregularity is extremal, as may be proved from the fact that it is bounded and equally continuous. Hence by a theorem

* An ordinal arrangement of every $\alpha_K(n)$ can be found which will make this and the following selections perfectly determinate, and will consequently avoid the difficulties of Zermelo's axiom.

† M. Fréchet, Rendiconti del Circolo Matematico di Palermo, vol. 22 (1906), p. 37.

‡ Ibid., p. 7.

of Fréchet, every bounded functional F continuous over the set is uniformly continuous over the set in Fréchet's sense. This means that for every δ there is an ϵ independent of f and g such that if

$$|f(x) - g(x)| < \epsilon$$

for every x,

$$|F(f) - F(g)| < \delta$$

Every functional uniformly continuous in his sense is uniformly continuous in ours, for if two functions f and g have coefficients of irregularity less than or equal to N and if

$$|f(x) - g(x)| < \epsilon$$

for all the points between $x = 0$ and $x = 1$ for which $x = a/2^n$, then

$$|f(x) - g(x)| < \epsilon + \frac{Na}{2^{n-1}}$$

for all points. Since $\epsilon + Na/2^{n-1}$ becomes smaller and smaller as we constrain f and g to lie within smaller and smaller intervals of P_K, the dependence of our definition of uniform continuity on that of Fréchet follows. Hence every bounded continuous functional is summable in accordance with our definition.

MASSACHUSETTS INSTITUTE OF TECHNOLOGY,
November 5, 1919.

PROCEEDINGS

OF THE

NATIONAL ACADEMY OF SCIENCES

Volume 7 SEPTEMBER 15, 1921 Number 9

THE AVERAGE OF AN ANALYTIC FUNCTIONAL[1]

BY NORBERT WIENER

DEPARTMENT OF MATHEMATICS, MASSACHUSETTS INSTITUTE OF TECHNOLOGY

Communicated by A. G. Webster, March 22, 1921

1. Conceive a particle free to wander along the x-axis. Suppose the probability that it wander a given distance independent

(1) of the position from which it starts to wander,

(2) of the time when it starts to wander,

(3) of the direction in which it wanders.

It may be shown[2] that under these circumstances, the probability that after a time, t, it has wandered from the origin to a position lying between $x=x_0$ and $x=x_1$ is

$$\frac{1}{\sqrt{\pi ct}}\int_{x_0}^{x_1} e^{-\frac{x^2}{ct}}\,dx$$

where t is the time and c is a certain constant which we can reduce to 1 by a proper choice of units. This choice we shall make in what follows. The exponential form of this integral needs no comment, while the mode in which t enters results from the fact that

$$\frac{1}{\sqrt{\pi(t_1+t_2)}}\int_{x_0}^{x_1} e^{-\frac{x^2}{t_1+t_2}}\,dx=\frac{1}{\sqrt{\pi t_1}}\int_{-\infty}^{\infty}\left[\frac{1}{\sqrt{\pi t_2}}\int_{x_0}^{x_1} e^{-\frac{(y-x)^2}{t_2}}\,dy\right]e^{-\frac{x^2}{t_1}}\,dx$$

This identity will be presupposed in all that follows.

Let us now consider a particle wandering from the origin for a given period of time, say from $t=0$ to $t=1$. Its position will then be a function of the time, say $x=f(t)$. There are certain quantities—functionals—which may depend on the whole set of values of f from $t=0$ to $t=1$. If we take a large number of particles (i.e. a large number of values of f) at random, it is natural to suppose that the average value of the functional will often approach a limit, which we may call the average value of the functional over its entire range. What will this average be, and how shall we find it?

If $F\{f\}$ is a functional depending on the values of f for only a finite number of values of the argument of f—if $F\{f\}$ is a function[3] of $f(t_1)$, $f(t_2)$,..., $f(t_n)$ of the form $\Phi[f(t_1), \ldots, f(t_n)]$—then it is easy enough to give a natural definition of the average of F, which we shall write $A\{F\}$. We can reasonably say

$$A\{F\} = \frac{1}{\sqrt{\pi^n t_1(t_2-t_1)\ldots(t_n-t_{n-1})}} \int_{-\infty}^{\infty} \cdots \int_{-\infty}^{\infty} \Phi(x_1, \ldots, x_n)$$

$$e^{-\frac{x_1^2}{t_1} - \frac{(x_2-x_1)^2}{t_2-t_1} - \cdots - \frac{(x_n-x_{n-1})^2}{t_n-t_{n-1}}} dx_1 \ldots dx_n.$$

In particular if $F\{f\} = [f(t_1)]^{m_1}[f(t_2)]^{m_2} \ldots [f(t_n)]^{m_n}$,
then

$$A\{F\} = \frac{1}{\sqrt{\pi^n t_1(t_2-t_1)\ldots(t_n-t_{n-1})}} \int_{-\infty}^{\infty} \cdots \int_{-\infty}^{\infty} x_1^{m_1} \ldots x_n^{m_n}$$

$$e^{-\frac{x_1^2}{t_1} - \cdots - \frac{(x_n-x_{n-1})^2}{t_n-t_{n-1}}} dx_1 \ldots dx_n$$

$$= \frac{1}{\sqrt{\pi^n t_1(t_2-t_1)\ldots(t_n-t_{n-1})}} \int_{-\infty}^{\infty} \cdots \int_{-\infty}^{\infty} y_1^{m_1}(y_2+y_1)^{m_2} \ldots (y_n + y_{n-1} + \ldots$$

$$+ y_1)^{m_n} e^{-\frac{y_1^2}{t_1} - \cdots - \frac{y_n^2}{t_n-t_{n-1}}} dy_1 \ldots dy_n$$

$$= \frac{1}{\sqrt{\pi^n}} \int_{-\infty}^{\infty} \cdots \int_{-\infty}^{\infty} (t_1^{1/2} z_1)^{m_1} \left[(t_2-t_1)^{1/2} z_2 + t_1^{1/2} z_1\right]^{m_2} \ldots \left[(t_n-t_{n-1})^{1/2} z_n\right.$$

$$\left. + \ldots + t_1^{1/2} z_1\right]^{m_n} e^{-z_1^2 - \cdots - z_n^2} dz_1 \ldots dz_n.$$

This latter integral is in the form

$$\int_{-\infty}^{\infty} \cdots \int_{-\infty}^{\infty} P(z_1, \ldots, z_n) e^{-z_1^2 - \cdots - z_n^2} dz_1, \ldots dz_n$$

where P is a polynomial, and can be evaluated by means of the well known formulae:

$$\int_{-\infty}^{\infty} e^{-y^2} y^{2n+1} dy = 0,$$

$$\int_{-\infty}^{\infty} e^{-y^2} y^{2n} dy = \sqrt{\pi}\, \frac{1.3.5\ldots(2n-1)}{2^n}.$$

We can thus easily evaluate $A\{F\}$ as a polynomial in $t_1, t_2, \ldots, t_n$, which we shall call $P_{m_1, \ldots, m_n}(t_1, \ldots, t_n)$. It is easy to show that if Σm_k is odd,

$$P_{m_1, \ldots, m_n}(t_1, \ldots, t_n) = 0.$$

To return to the more general functional: there is a large class of so-called analytic functionals,[4] which may be expanded in the form of series such as

$$F\{f\} = a_0 + \int_0^1 f(x)\varphi_1(x)dx + \int_0^1 \int_0^1 f(x)\, f(y)\varphi_2(x,y)dxdy + \ldots$$
$$+ \int_0^1 \ldots \int_0^1 f(x_1) \ldots f(x_n)\varphi_n(x_1, \ldots, x_n)\, dx_1 \ldots dx_n + \ldots$$

and an even wider class of what may be called Stieltjes analyt ic functionals, in which the general term

$$\int_0^1 \ldots \int_0^1 f(x_1) \ldots f(x_n)\varphi_n(x_1, \ldots, x_n)dx_1, \ldots dx_n$$

is replaced by the Stieltjes integral[5]

$$\int_0^1 \ldots \int_0^1 f(x_1) \ldots f(x_n)d\psi_n(x_1, \ldots, x_n).$$

In what follows, we shall confine our discussion to Stieltjes analytic functionals, which we shall call simply analytic. The problem with which we are now concerned is the definition of the average of an analytic functional. Now, the first property which any average ought to fulfil is that the average of the sum of two functionals should equal the sum of their averages. We should expect, therefore, that:

(*a*) Over a wide range of cases, the average of a series should equal the series of the averages of the terms;

(*b*) The average of a Stieltjes integral, single or multiple, of a given functional with respect to such parameters as it may contain, should be equal to the integral of the average;

(*c*) A constant multiplied by the average of a functional should equal the average of a constant times the functional.

In accordance with this, we get the following natural definition of the average of the analytic functional F.

$$A\{F\} = A\left\{a_0 + \int_0^1 f(x)d\psi_1(x) + \int_0^1 \int_0^1 f(x) f(y)d\psi_2(x,y) + \ldots\right\}$$
$$= a_0 + A\left\{\int_0^1 f(x)d\psi_1(x)\right\} + A\left\{\int_0^1 \int_0^1 f(x)\, f(y)d\psi_2(x,y)\right\} + \ldots$$
$$= a_0 + \int_0^1 A\{f(x)\}d\psi_1(x) + \int_0^1 \int_0^1 A\{f(x)f(y)\}d\psi_2(x,y) + \ldots$$

We have already seen how to determine $A\{f(x_1)\ldots f(x_n)\}$ as a polynomial in the x_k's. Hence whenever the above series converges, we have now a way of obtaining a perfectly definite value for $A\{F\}$. It may be noted that every term in the above expression in which the sign of integration is repeated an odd number of times is identically zero.

If $A\,\{F\}$ is to behave as we should expect it to behave, there are certain properties which it must fulfil, at least over a large and important class of cases. Among these are the following:

$$(1)\qquad A\{F_1\} + A\{F_2\} = A\{F_1 + F_2\}$$

$$(2)\qquad cA\{F_1\} = A\{cF_1\}$$

$$(3)\qquad \sum_1^\infty A\{F_n\} = A\left\{\sum_1^\infty F_n\right\}$$

(4) If F_x is a functional depending on the parameter x, and $u(x)$ is a function of limited total variation, then

$$\int_a^b A\{F_x\}du = A\left\{\int_a^b F_x du\right\}$$

(5) Suppose $F_{t_1,\dots,t_n}(x_1, \dots, x_n)$ be defined as $F\left\{f_{t_1,\dots,t_n}^{x_1,\dots,x_n}(t)\right\}$, where $f_{t_1,\dots,t_n}^{x_1,\dots,x_n}(t)$ assumes the value

$$x_k + \frac{(t - t_k)\,(x_{k+1} - x_k)}{t_{k+1} - t_k}$$

for $t_k \leq t \leq t_{k-1}$. Then

$$A\{F\} = \lim \pi^{-\frac{n}{2}} \prod_1^n (t_k - t_{k-1})^{-1/2} \int_{-\infty}^{\infty} \dots \int_{-\infty}^{\infty} F_{t_1,\dots,t_n}(x_1, \dots, x_n)\, e^{-\frac{x_1^2}{t_1} - \sum_2^n \frac{(x_k - x_{k-1})^2}{(t_k - t_{k-1})}}\, dx_1 \dots dx_n,$$

where the limit is taken as the t_k's increase in number in such a manner to divide the interval from 0 to 1 more and more finely.

2. The next task before us is to investigate hypotheses which are sufficient to guarantee the validity of propositions (1)–(5). Propositions (1) and (2) require indeed very little discussion, for they are always satisfied when the series for F_1, F_2, $A\{F_1\}$ and $A\{F_2\}$ converge. In (3), let $F_1, \dots, F_n, \dots$, and the series ΣF_n all possess averages, and let

$$\sum_1^\infty F_n = \sum_1^m F_n + R_m$$

where $A\{R_m\}$ vanishes as m increases without limit. Then

$$A\left\{\sum_1^\infty F_n\right\} = \sum_1^m A\{F_n\} + A\{R_m\}.$$

Therefore

$$\lim_{m=\infty} \left| A\left\{\sum_1^\infty F_n\right\} - \sum_1^m A\left\{F_n\right\}\right| = 0$$

and (3) is proved. If ΣF_n converges and $\lim A\{R_m\} = 0$, we shall say ΣF_n converges *smoothly*.

Proposition (4) reduces to the ordinary inversion of a multiple Stieltjes integral when $F_x\{f\}$ is of the form

$$\int_0^1 \dots \int_0^1 f(x_1) \dots f(x_n) d\psi_n(x_1, \dots, x_n, x)$$

and ψ_n is a function of limited variation in $x_1, \ldots, x_n$. What we wish to prove is that

$$\int_a^b \left\{ \int_0^1 \cdots \int_0^1 A\{f(x_1) \ldots f(x_n)\} d\psi_n(x_1, \ldots, x_n) \right\} du$$
$$= A \left\{ \int_a^b \left\{ \int_0^1 \cdots \int_0^1 f(x_1) \ldots f(x_n)\, d\psi_n\,(x_1, \ldots, x_n) \right\} du \right\}$$

Now

$$\int_a^b \left\{ \int_0^1 \cdots \int_0^1 f(x_1) \ldots f(x_n)\, d\psi_n\,(x_1, \ldots, x_n) \right\} du =$$
$$\int_a^b \left[\lim \sum_{\alpha, \ldots, \vartheta} f(\xi_{1\alpha}) \ldots f\,(\xi_{n\vartheta})\, \Delta_{x_1\alpha \ldots x_{n\vartheta}}^{x_1, \alpha+1 \ldots x_n, \vartheta+1} \psi_n(x_1, \ldots, x_n, x) \right] du.$$

If in this latter expression the total variation of ψ_n is less than a quantity independent of x, we can permute the $\int_a^b$ and the *lim*, and get

$$\lim \sum_{\alpha, \ldots, \vartheta} f(\xi_{1\alpha}) \ldots f(\xi_{n\vartheta})\, \Delta_{x_1\, \alpha \ldots x_{n\vartheta}}^{x_1,\, \alpha+1 \ldots x_n, \vartheta+1} \int_a^b \psi_n\,(x_1, \ldots, x_n, x)\, du,$$

which we may write

$$\int_0^1 \cdots \int_0^1 f(x_1) \ldots f(x_n) d\left[\int_a^b \psi_n(x_1, \ldots, x_n, x)\, du \right].$$

In this we suppose f uniformly continuous. It is easy to show that on our assumptions $\int_a^b \psi_n(x_1, \ldots, x_n, x) du$ is of limited variation. Consequently we obtain

$$A \left\{ \int_a^b \left\{ \int_0^1 \cdots \int_0^1 f(x_1) \ldots f(x_n) d\psi_n(x_1, \ldots, x_n, x) \right\} du \right\}$$
$$= A \left\{ \int_0^1 \cdots \int_0^1 f(x_1) \ldots f(x_n) d \left[\int_a^b \psi_n(x_1, \ldots, x_n, x) du \right] \right\}$$
$$= \int_0^1 \cdots \int_0^1 A\{f(x_1) \ldots f(x_n)\} d \left[\int_a^b \psi_n(x_1, \ldots, x_n, x)\, du \right].$$

A further transformation just like the preceding turns this into

$$\int_a^b \left\{ \int_0^1 \cdots \int_0^1 A\{f(x_1) \ldots f(x_n)\} d\psi_n\,(x_1, \ldots, x_n, x) \right\} du$$

so that we have now a sufficient condition for the validity of our theorem. The extension to non-homogeneous terminating analytic functionals is obvious. The extension to non-terminating analytic functionals may be deduced with the help of (3) and a well-known theorem on the integration of uniformly convergent series, and reads as follows: *let F_x, be an analytic functional of the form*

$$a_0 + \sum_1^\infty \int_0^1 \cdots \int_0^1 f(x_1) \ldots f(x_n)\, d\psi_n(x_1, \ldots, x_n, x)$$

where the total variation of each ψ_n is less than some quantity independent of x, and let each ψ_n be uniformly continuous in x over the interval (a, b). Let $u(x)$ be a function of limited total variation in x over the same interval. Let $A\{\int_a^b F_x du\}$ exist, and let

$$\lim_{m=\infty} \sum_{n=m}^{\infty} \int_0^1 \cdots \int_0^1 A\{f(x_1)\ldots f(x_n)\}\, d\psi_n(x_1, \ldots, x_n, x) = 0$$

uniformly in x. Then

$$\int_a^b A\{F_x\} du = A\left\{\int_a^b F_x\, du\right\}.$$

As to (5), let us begin as above with a functional of the form

$$\Phi\{f\} = \int_0^1 \cdots \int_0^1 f(x_1)\ldots f(x_n)\, d\psi_n(x_1, \ldots, x_n).$$

Consider

$$\mathrm{I} = \int_0^1 \cdots \int_0^1 A\{f(x_1)\ldots f(x_n\}\, d\psi_n(x_1, \ldots, x_n),$$

where A is taken in the original sense as an n-fold integral. By definition

$$\mathrm{I} = \lim \sum_{\alpha, \ldots, \vartheta} A\{f(\xi_{1\alpha})\ldots f(\xi_{n\theta})\}\, \Delta_{x_{1\alpha}\ldots x_{n\vartheta}}^{x_{1,\alpha+1}\ldots x_{n,\vartheta+1}} \psi_n(x_1, \ldots, x_n),$$

where $x_{ko} = 0, x_{k1}, \ldots, x_{k\mu} = 1$ is an increasing sequence of numbers, ξ_{kK} lies between x_{kK} and $x_{k,K+1}$ and *lim* is taken as $max(x_{k,K+1} - x_{kK})$ approaches 0. Let V be the total variation of ψ_n as its arguments range from 0 to 1, and let M stand for $max(x_{k,K+1} - x_{kK})$. Let Q stand for the least upper bound of the variation of $A\{f(x_1)\ldots f(x_n)\}$ as the point $(x_1, \ldots, x_n)$ wanders over an interval $\binom{x_{1,\alpha+1}\ldots x_{n\vartheta+1}}{x_{1\alpha}\ldots x_{n\vartheta}}$. Then

$$\left|\mathrm{I} - A\left\{\sum_{\alpha, \ldots, \vartheta} f(\xi_{1\alpha})\ldots f(\xi_{n\vartheta}) \Delta_{x_{1\alpha}\ldots x_{n\vartheta}}^{x_{1,\alpha+1}\ldots x_{n,\vartheta+1}} \psi_n(x_1, \ldots, x_n)\right\}\right| \leq VQ. \tag{6}$$

Now, let $f_m(x)$ be that function whose graph is the broken line with corners at $(x_{(K)}, f(x_{(K)}))$, where $0 \leq K \leq \mu$, $x_{(o)} = 0$, $x_{(\mu)} = 1$. Then if x lies between $x_{(K)}$ and $x_{(K+1)}$, $f_m(x)$ is of the form $af_m(x_{(K)}) + bf_m(x_{(K)}) \div a+b$. It follows that if $(\xi_{1\alpha}, \ldots, \xi_{n\vartheta})$ lies in the interval $\binom{x_{(\alpha+1)}\ldots x_{(\vartheta+1)}}{x_{(\alpha)}\ldots x_{(\vartheta)}}$, $A\{f_m(\xi_{1\alpha})\ldots f_m(\xi_{n\vartheta})\}$ is of the form

$$\frac{a_1 C_1 + a_2 C_2 + \ldots + a_\rho C_\rho}{a_1 + a_2 + \ldots + a_\rho}$$

where each C_k is the value of some $A\{f(\xi_{1\alpha})\ldots f(\xi_{\eta\vartheta})\}$ such that $(\xi_{1\alpha}, \ldots, \xi_{\eta\vartheta})$ is a corner of the hyperparallelopiped$\binom{x_{(\alpha+1)}\ldots x_{(\vartheta+1)}}{x_{(\alpha)}\ldots x_{(\vartheta)}}$. It readily results from considerations of continuity that $A\{f_m(\xi_1)\ldots f_m(\xi_\mu)\}$ is of the form $A\{f(\eta_{1\alpha})\ldots f(\eta_{n\vartheta})\}$, where $(\eta_{1\alpha}, \ldots, \eta_{n\vartheta})$ also lies in the interval $\binom{x_{(\alpha+1)}\ldots x_{(\vartheta+1)}}{x_{(\alpha)}\ldots x_{(\vartheta)}}$.

Making use of this fact, and of the fact that

$$\frac{1}{\sqrt{\pi^n t_1(t_2-t_1)\dots(t_n-t_{n-1})}}\int_{-\infty}^{\infty}\dots\int_{-\infty}^{\infty}\Phi(x_1,\dots,x_n)\, e^{-\frac{x_1^2}{t_1}-\dots-\frac{(x_n-x_{n-1})^2}{t_n-t_{n-1}}}dx_1\dots dx_n$$

is an increasing functional of $\Phi(x_1, \dots, x_n)$, we can draw the conclusion that

$$A\{\Phi\{f_m\}\} = A\left\{\int_0^1\dots\int_0^1 f_m(x_1)\dots f_m(x_n)\, d\psi_n(x_1,\dots,x_n)\right\}$$

if it exists, lies between the uppermost and lowermost values of

$$\sum_{\alpha,\dots,\vartheta} A\{f_m(\xi_{1\alpha})\dots f_m(\xi_{n\vartheta})\}\Delta_{x(\alpha)\dots x(\vartheta)}^{x(\alpha+1)\dots x(\vartheta+1)}\psi_n(x_1,\dots,x_n)$$

and hence of

$$\sum_{\alpha,\dots,\vartheta} A\{f(\eta_{1\alpha})\dots f(\eta_{n\vartheta})\}\Delta_{x(\alpha)\dots x(\vartheta)}^{x(\alpha+1)\dots x(\vartheta+1)}\psi_n(x_1,\dots,x_n)$$

From this and (6) we can deduce

(7) $$|I - A\{\Phi\{f_m\}\}| \leq 2VQ$$

where Q is taken for $x_{kK} = x_{(K)}$.

This proves our theorem for homogeneous analytic functionals. In precise terms, then, our general theorem will read: *let*

$$F\{f\} = a_0 + \sum_1^{\infty}\int_0^1\dots\int_0^1 f(x_1)\dots f(x_n)\, d\psi_n(x_1,\dots,x_n) = a_0 + \sum_1^{\infty} F_n\{f\}$$

be an analytic functional. Let V_n stand for the total variation of F_n as its arguments range from 0 to 1: Let $(x_{(0)}, \dots, x_{(\mu)})$ be a set of numbers in ascending order from 0 to 1, inclusive. Let M stand for max $(x_{(K+1)} - x_{(K)})$ and Q_n for the upper bound of the variation of $A\{f(x_1)\dots f(x_n)\}$ as the point $(x_1, \dots, x_n)$ wanders over the interval $\binom{x(\alpha+1)\dots x(\vartheta+1)}{x(\alpha)\dots x(\vartheta)}$. Let $f_\mu(x)$ be the function whose graph is the broken line with corners at (0, 0) and $(x_{(K)}, f(x_{(K)}))$. Then if

(*a*) $$\lim_{M=0}\sum_1^{\infty} V_k Q_k = 0,$$

(*b*) $A\{F_n\{f_\mu\}\}$ *exists for every μ and n according to the definition of A as a multiple integral:*

(*c*) *the series for F converges smoothly;*

(*d*) $\lim_{n=\infty} A\left\{\sum_n^{\infty} F_m\{f_\mu\}\right\}$ *exists for every μ, when A is taken as a multiple integral; it follows that*

$$A\{F\} = \lim_{M=0} A\{F\{f_m\}\},$$

where the first A is defined in the sense of the average of an analytic functional, and the second as a multiple integral. A precisely analogous theorem holds when $f_m(x)$ instead of a broken straight line is any broken line with corners at $(0, 0)$ and at $(x_{(K)}, f(x_{(K)}))$, and consists of monotone arcs between these points. This last theorem makes our average of a functional the limit of the average of a function of a discrete set of variables, and justifies our use of the term average.

[1] The problem of the mean of a functional has been attached by Gâteaux (*Bull. Soc. Math. de France*, 1919, pp. 47–70). The idea of using the analytic functional as a basis is there found. The actual definition, however, is essentially different, and does not lend itself readily to the treatment of the Brownian Movement, for which the present method is especially adapted.

[2] Einstein, *Leipzig, Annalen Physik*, **17**, 905.

[3] We here take $t_1, < t_2 < \ldots < t_n$.

[4] Cf. V. Volterra, *Fonctions des Lignes*.

[5] Cf. P. J. Daniell, *Annals of Mathematics*, Sept., 1919, p. 30.

Reprinted from the Proceedings of the NATIONAL ACADEMY OF SCIENCES,
Vol. 7, No. 10, pp. 294–298, October, 1921.

THE AVERAGE OF AN ANALYTIC FUNCTIONAL AND THE BROWNIAN MOVEMENT

By NORBERT WIENER

DEPARTMENT OF MATHEMATICS, MASSACHUSETTS INSTITUTE OF TECHNOLOGY

Communicated by A. G. Webster, March 22, 1921

The simplest example of an average is the arithmetical mean. The arithmetical mean of a number of quantities is their sum, divided by their number. If a result is due to a number of causes whose contribution to the result is simply additive, then the result will remain unchanged if for each of these causes is substituted their mean.

Now, the causes contributing to an effect may be infinite in number and in this case the ordinary definition of the mean breaks down. In this case some sort of measure may be used to replace number, integration to replace summation, and the notion of mean reappears in a generalized form. For instance, the distance form one end of a rod to its center of gravity is the mean of its length with reference to its mass, and may be written in the form

$$\int l dm \div \int dm$$

where l stands for length and m for mass. It is to be noted that l is a function of m, and that the mean we are defining is the mean of a function. Furthermore, the quantity, here the mass, in terms of which the mean is taken, is a necessary part of its definition. We must assume, that is, a normal distribution of some quantity to begin with, in this case of mass.

The mean just discussed is not confined to functions of one variable; it admits of an obvious generalization to functions of several variables. Now, there is a very important generalization of the notion of a function of several variables: the function of a line. For example, the attraction of a charged wire on a unit charge in a given position depends on its shape. The length and area of a curve between two given ordinates depend on its shape. As a curve is essentially a function, these functions of lines may be regarded as functions of functions, and as such are known as functionals. Since a function is determined when its value is known for all arguments, a functional depends on an infinity of numerical determinations, and may hence be regarded as in some wise a function of infinitely many variables.

To determine the average value of a functional, then seems a reasonable problem, provided that we have some convention as to what constitutes a normal distribution of the functions that form its arguments. Two essentially different discussions have been given of this matter: one, by Gâteaux, being a direct generalization of the ordinary mean in n-space;[1] the other, by the author of this paper,[2] involving considerations from the theory of probabilities. The author assumes that the functions $f(t)$ that form the arguments of his functionals have as their arguments the time, and that in any interval of the small length ϵ as many receive increments of value as decrements of equal size. He also assumes that the likelihood that a particle receive a given increment or decrement is independent of its entire antecedent history.

When a particle is acted on by the Brownian movement, it is in a motion due to the impacts of the molecules of the fluid in which it is suspended. While the retardation a particle receives when moving in a fluid is of course due to the action of the individual particles of the fluid, it seems natural to treat the Brownian movement, in a first approximation, as an effect due to two distinguishable causes: (1) a series of impacts received by a particle, dependent only on the time during which the particle is exposed to collisions; (2) a damping effect, dependent on the velocity of the particle. If we consider one component of the total impulse received by a particle under heading (1), we see that it may be considered as a function of the time, and that it will have the sort of distribution which will make our theory of the average of a functional applicable.

It will result directly from the previous paper of the author that if $f(t)$ is the total impulse received by a particle in a given direction when the unit of time is so chosen that the probability that $f(t)$ lie between a and b is

$$\frac{1}{\sqrt{\pi t}}\int_a^b e^{-\frac{x^2}{t}}\,dx$$

then the average value of

$$A + \int_a^b f(t)\,G(t)dt + \int_a^b\int_a^b f(s)\,f(t)\,H(s,t)\,ds\,dt \quad [H(s,t) = H(t,s)]$$

will be

$$A + \int_a^b\int_a^s t\,H(s,t)\,ds\,dt. \tag{1}$$

We now proceed to a more precise and detailed treatment of the question.

2. Einstein[3] has given as the formula for the mean square displacement in a given direction of a spherical particle of radius r in a medium of viscosity η over a time t, under the action of the Brownian movement, the formula

$$\overline{d_t^2} = RTt \div 3\pi r\eta N$$

where R is the gas-constant, T the absolute temperature of the medium, and N the number of molecules per gram-molecule. In the deduction of this formula, Einstein makes two important assumptions. The first is that Stokes' law holds concerning forces of diffusion. Stokes' law states that a force F will carry particle of radius r through a fluid of viscosity η with velocity $F \div 6\pi r\eta$. Einstein's second assumption is that the displacement of a particle in some interval of time small in comparison with those which we can observe is independent, to all intents and purposes, of its entire antecedent history. It is the purpose of this paper to show that even without this assumption, under some very natural further hypotheses, the departure of $\overline{d_t^2}/t$ from constancy will be far too small to observe.

In this connection, it is well to take note of just what the Brownian movement is, and of the precise sense in which Stokes' law holds of particles undergoing a Brownian movement. In the study of the Brownian movement, our attention is first attracted by the enormous discrepancy between the apparent velocity of the particles and that which must animate them if, as seems probable, the mean kinetic energy of each particle is the same as that of a molecule of the gas. This discrepancy is of course due to the fact that the actual path of each particle is of the most extreme sinuosity, so that the observed velocity is almost in no relation to the true velocity. Now, Stokes' law is always applied with reference to movements at least as slow as the microscopically observable motions of a particle. It hence turns out that Stokes' law must be treated as a sort of average effect, or in the words of Perrin,[4] "When a force, constant in magnitude and direction, acts in a fluid on a granule agitated by the Brownian movement, the displacement of the granule, which is perfectly irregular at right angles to the force, takes in the direction of the force a component progressively increasing with the time and in the mean equal to $Ft \div 6\pi\zeta a$, F indicating the force, t the time, ζ the viscosity of the fluid, and a the radius of the granule."

It is a not unnatural interpretation of this statement to suppose that we may assume the validity of Stokes' law for the slower motions which are all that we see directly of the Brownian motion, so that we may regard the Brownian movement as made up (1) of a large number of very brief, independent impulses acting on each particle and (2) of a continual damping action on the resulting velocity in accordance with Stokes' law. It is to be noted that the processes which we treat as impulsive forces need not be the simple results of the collision of individual molecules with the particle, but may be highly complicated processes, involving intricate interactions between the particle and the surrounding molecules. It may readily be shown by a numerical computation that this is the case.

It follows from considerations discussed at the beginning of my paper on *The Average of an Analytic Functional* that after a time the probability that the total momentum acquired by a particle from the impacts of molecules will lie between x_0 and x_1 is of the form

$$\frac{1}{\sqrt{\pi ct}}\int_a^b e^{-\frac{x^2}{ct}}dx.$$

Superimposed on this momentum is that due to the viscosity acting in accordance with Stokes' law, namely $6\pi r\eta V$, where V is the velocity of the particle. We shall write Q for $6\pi r\eta \div M$, where M is the mass of a particle.

Let us write τ for ct. Let the total impulse received by a given particle in time t, neglecting the action of viscosity, be $f(\tau)$. Consider $m(t)$, the actual momentum of the particle, as a function of t. Then

$$m(t+dt) = m(t) + f(ct+cdt) - f(ct) - Qm(ct+c\vartheta dt)dt \quad (o \leq \vartheta \leq 1).$$

We cannot treat this as a differential equation, as we have no reason to suppose that f has a derivative. We can make it into an integral equation, however, which will read

$$m(t) - m(o) = f(ct) - Q\int_o^t m(t)dt.$$

Clearly one solution of this integral equation is

$$m(t) = m(o)e^{-Qt} + f(ct) - Qe^{-Qt}\int_o^t f(ct)e^{Qt}dt,$$

and there is no difficulty in showing that an integral equation of this sort can have only one continuous solution.

Another integration gives for the distance traversed by the particle in time t

$$\begin{aligned} d_t &= \frac{1}{M}\left\{\left[-\frac{m(o)e^{-Qt}}{Q}\right]_o^t + \int_o^t f(ct)dt - \int_o^t Qe^{-Qt}\left[\int_o^t e^{Qt}f(ct)dt\right]dt\right\} \\ &= \frac{1}{M}\left\{\frac{m(o)}{Q}(1-e^{-Qt}) + e^{-Qt}\int_o^t e^{Qt}f(ct)dt\right\} \qquad (2) \\ &= \frac{1}{M}\left\{\frac{m(o)}{Q}(1-e^{-Qt}) + \frac{e^{-Qt}}{c}\int_o^\tau e^{\frac{Q\tau}{c}}f(\tau)d\tau\right\} \end{aligned}$$

Applying the methods of my previous paper, we get for the mean value of d_t^2, in accordance with (1);

$$\begin{aligned} \overline{d_t^2} &= \frac{1}{M^2}\left[\frac{m(o)}{Q}(1-e^{-Qt})\right]^2 + \frac{1}{M^2c^2}e^{-2Qt}\int_o^{ct}\int_o^{c.} A\{f(x)f(y)\}e^{\frac{Q(x+y)}{c}}dydx \\ &= \frac{1}{M^2}\left[\frac{m(o)}{Q}(1-e^{-Qt})\right]^2 + \frac{1}{M^2c^2}e^{-2Qt}\int_o^{ct}\left[\int_o^x ye^{\frac{Q(x+y)}{c}}dy\right]dx \\ &= \left[\frac{m(o)}{MQ}(1-e^{-Qt})\right]^2 + \frac{e^{-2Qt}}{M^2c^2}\int_o^{ct}\left[\frac{cx}{Q}e^{\frac{2Qx}{c}} - \frac{c^2}{Q^2}e^{\frac{2Qx}{c}} + \frac{c^2}{Q^2}e^{\frac{Qx}{c}}\right]dx \qquad (3) \\ &= \left[\frac{m(o)}{MQ}(1-e^{-Qt})\right]^2 + \frac{c}{4M^2Q^3}\left[2Qt - 3 + 4e^{-Qt} - e^{-2Qt}\right] \end{aligned}$$

Therefore

$$\left|\overline{d_i^2}/t - c/(2M^2Q^2)\right| \leq 1/t\left[m(o)(1-e^{-Qt})/(MQ)\right]^2$$
$$+ c(3-e^{-Qt})(1-e^{-Qt})/(4M^2Q^3t)$$
$$\leq \left[m(o)/(MQ)\right]^2 + 3c/(4M^2Q^3)$$

This represents the absolute departure of $\overline{d_i^2}/t$ from constancy. Writing v for $m(o)/M$, the initial velocity of the particle, we get

$$\frac{|\overline{d_i^2}/t - c/(2M^2Q^2)|}{c/(2M^2Q^2)} \leq \frac{v^2/Q^2}{c/2M^2Q^2} + 3/2Q$$

This is a measure of the relative departure of $\overline{d_i^2}/t$ from constancy. v cannot exceed, on the average, the velocity given on the average to the particle on the basis of the equipartition of energy; actually it is much smaller. $c/(2M^2Q^2)$ can be found directly, as it is nearly the observed value of $\overline{d_i^2}/t$. Q can be readily computed from the constants of the particles. Taking as a typical case one of Perrin's experiments on gamboge, Q turns out to be of the order of magnitude of 10^8, $c/2M^2Q^2$ of the order of magnitude of 10^{-8}, and the kinetic energy velocity of the order of magnitude of 10^{-1}. Hence the proportionate error is of the order of magnitude of 10^{-8}.

A proportionate error thus small is quite beyond the reach of our methods of measurement, so that we are compelled to conclude that $\overline{d_i^2}/t$, under the hypotheses we have here formulated, is sensibly constant. There are cases, however, which seem to give a slightly different value of $\overline{d_i^2}/t$ for small values of the time than for larger values. The explanation has been suggested[5] that over small periods the Einstein independence of an interval on previous intervals does not hold. The result of the present paper would be to suggest strongly, if not to demonstrate, that the source of the discrepancy, if, as appears, it is genuine, and not due to experimental error, is in the fact that Stokes' law itself is only a rough approximation, and that the resistance does not vary strictly as the velocity.

[1] *Paris, Bull. Soc. Math. France,* 1919, pp. 47–70.

[2] "The Average of an Analytic Functional," in the last number of these PROCEEDINGS.

[3] *Leipzig, Ann. Physik,* **17,** 1905 (549).

[4] *Ann. Chim. Phys.,* Sept., 1909; tr. by F. Soddy.

[5] Cf. Kleeman, *A Kinetic Theory of Gases and Liquids,* §§ 56, 60.

DIFFERENTIAL-SPACE

By NORBERT WIENER

§1. **Introduction.** The notion of a function or a curve as an element in a space of an infinitude of dimensions is familiar to all mathematicians, and has been since the early work of Volterra on functions of lines. It is worthy of note, however, that the physicist is equally concerned with systems the dimensionality of which, if not infinite, is so large that it invites the use of limit-processes in which it is treated as infinite. These systems are the systems of statistical mechanics, and the fact that we treat their dimensionality as infinite is witnessed by our continual employment of such asymptotic formulae as that of Stirling or the Gaussian probability-distribution.

The physicist has often occasion to consider quantities which are of the nature of functions with arguments ranging over such a space of infinitely many dimensions. The density of a gas, or one of its velocity-components at a point, considered as depending on the coördinates and velocities of its molecules, are cases in point. He therefore is implicitly, if not explicitly, studying the theory of functionals. Moreover, he generally replaces any of these functionals by some kind of average value, which is essentially obtained by an integration in space of infinitely many dimensions.

Now, integration in infinitely many dimensions is a relatively little-studied problem. Apart from certain tentative investigations of Fréchet[1] and E. H. Moore[2], practically all that has been done on it is due to Gâteaux[3], Lévy[4], Daniell[5], and the author of this paper[6]. Of these investigations, perhaps the most complete are those begun by Gâteaux and carried out by Lévy in his *Leçons d'Analyse Fonctionnelle.* In this latter book, the mean value of the functional $U|[x(t)]|$ over the region of function-space

$$\int_0^1 [x(t)]^2 dt \leq 1$$

is considered to be the limit of the mean of the function.

$$U(x_1, \ldots, x_n) = U|[\xi_n(t)]|,$$

$$\left(\text{where} \qquad \xi_n(t) = x_k \quad \text{for} \quad \frac{k-1}{n} \leq t < \frac{k}{n}\right)$$

over the sphere

$$x_1^2 + x_2^2 + \ldots + x_n^2 = n$$

as n increases without limit.

The present paper owes its inception to a conversation which the author had with Professor Lévy in regard to the relation which the two systems of integration in infinitely many dimensions — that of Lévy and that of the author — bear to one another. For this indebtedness the author wishes to give full credit. He also wishes to state that a very considerable part of the substance of the paper has been presented, albeit from a different standpoint and employing different methods, in his previously

1 *Sur l'intégrale d'une fonctionnelle étendue à un ensemble abstrait,* Bull. Soc. Math. de France, Vol. 43, pp. 249–267.

2 Cf. American Mathematical Monthly, Vol. 24 (1917), pp. 31, 333.

3 Two papers, published after his death by Lévy, Bulletin de la Société Mathématique de France, 1919.

4 P. Lévy, *Leçons d'Analyse Fonctionnelle.* Hereinafter to be referred to as " Lévy."

5 P. J. Daniell, *A General Form of Integral,* Annals of Mathematics, Series 2 Vol. 19, pp. 279–294. Hereinafter to be referred to as " Daniell." Also paper in Vol. 20.

6 N. Wiener, *The Average of an Analytic Functional,* Proc. Nat. Acad. of Sci., Vol. 7, pp. 253–26; *The Average of an Analytic Functional and the Brownian Movement, ibid,* 294–298. Also forthcoming paper in Proc. Lond. Math. Soc.

cited publications. It seemed better to repeat a little of what had already been done than to break the continuity of the paper by the constant reference to theorems in other journals and based on treatments sô distinct from the present one that it would need a large amount of explanation to show their relevance.

§2. **The Brownian Movement.** When a suspension of small particles in a liquid is viewed under a microscope, the particles seem animated with a peculiar haphazard motion — the Brownian movement. This motion is of such an irregular nature that Perrin[7] says of it: " One realizes from such examples how near the mathematicians are to the truth in refusing, by a logical instinct, to admit the pretended geometrical demonstrations, which are regarded as experimental evidence for the existence of a tangent at each point of a curve." It hence becomes a matter of interest to the mathematician to discover what are the defining conditions and properties of these particle-paths.

The physical explanation of the Brownian movement is that it is due to the haphazard impulses given to the particles by the collisions of the molecules of the fluid in which the particles are suspended. Of course, by the laws of mechanics, to know the motion of a particle, one must know not only the impulses which it receives over a given time, but the initial velocity with which it is imbued. According, however, to the theory of Einstein,[8] this initial velocity over any ordinary interval of time, is of negligible importance in comparison with the impulses received during the time in question. Accordingly, the displacement of a particle during a given time may be regarded as independent of its entire previous history.

Let us then consider the time-equations of the path of a particle subject to the Brownian movement as of the form $x=x(t)$, $y=y(t)$, $z=z(t)$, t being the time and x, y, and z the coördinates of the particle. Let us limit our attention to the function $x(t)$. Since there is no appreciable carrying over of velocity from one instant to another, the difference between $x(t_1)$ and $x(t)$ $[t_1>t]$ may be regarded as the sum of the displacements incurred by the particle

7 p. 64, *Brownian Movement and Molecular Reality*, tr. by F. Soddy.
8 Perrin, pp. 51–54.

over a set of intervals constituting the interval from t to t_1. In particular, if the constituent intervals are of equal size, then the probability-distribution of the displacements accrued in the different intervals will be the same. Since positive and negative displacements of the same size will, from physical considerations, be equally likely, it will be seen that for intervals of time large with comparison to the intervals between molecular collisions, by dividing them into many equal parts, which are still large with respect to the intervals between molecular collisions, and breaking up the total incurred displacement into the sum of displacements incurred in these intervals, we get very nearly a Gaussian distribution of our total displacement.[9] That is, the probability that $x(t_1)-x(t)$ lie between a and b is very nearly of the form

$$\frac{1}{\sqrt{\pi\phi(t_1-t)}}\int_a^b e^{-\frac{x^2}{\phi(t_1-t)}}\,dx. \tag{1}$$

Since the error incurred over the interval t to t_1 is the sum of the independent errors incurred over the periods from t to t_2 and from t_2 to t_1, we have

$$\frac{1}{\sqrt{\pi\phi(t_1-t)}}\,e^{-\frac{x^2}{\phi(t_1-t)}}$$

$$=\frac{1}{\pi\sqrt{\phi(t_1-t_2)\phi(t_2-t)}}\int_{-\infty}^{\infty} e^{-\frac{y^2}{\phi(t_2-t)}-\frac{(x-y)^2}{\phi(t_1-t_2)}}dy$$

$$=\frac{1}{\pi\sqrt{\phi(t_1-t_2)\phi(t_2-t)}}\int_{-\infty}^{\infty} exp\left\{-\left[y\sqrt{\frac{\phi(t_1-t_2)+\phi(t_2-t)}{\phi(t_1-t_2)\phi(t_2-t)}}\right.\right.$$

$$\left.\left.-\frac{x}{\sqrt{\phi(t_1-t_2)}}\sqrt{\frac{\phi(t_1-t_2)\phi(t_2-t)}{\phi(t_1-t_2)+\phi(t_2-t)}}\right]^2-\frac{x^2}{\phi(t_1-t_2)+\phi(t_2-t)}\right\}dy$$

$$=\frac{e^{-\frac{x^2}{\phi(t_1-t_2)+\phi(t_2-t)}}}{\pi\sqrt{\phi(t_1-t_2)\phi(t_2-t)}}\int_{-\infty}^{\infty} e^{-y^2\left(\frac{\phi(t_1-t_2)+\phi(t_2-t)}{\phi(t_1-t_2)\phi(t_2-t)}\right)}\,dy$$

$$=\frac{e^{-\frac{x^2}{\phi(t_1-t_2)+\phi(t_2-t)}}}{\sqrt{\pi\{\phi(t_1-t_2)+\phi(t_2-t)\}}}. \tag{2}$$

[9] Cf. Poincaré, *Le calcul des probabilités*, Ch. XI.

It is readily verified that this cannot be true unless

$$\phi(t_1-t)=\phi(t_1-t_2)+\phi(t_2-t), \tag{3}$$

whence

$$\phi(u)=Au, \tag{4}$$

and the probability that $x(t_1)-x(t)$ lie between a and b is of the form

$$\frac{1}{\sqrt{\pi A(t_1-t)}}\int_a^b e^{-\frac{x^2}{A(t_1-t)}}\,dx. \tag{5}$$

According to Einstein's theory,[10] $A = \dfrac{RT}{N}\dfrac{a}{3\pi a\xi}$ where R is the constant of a perfect gas, T is the absolute temperature, N is Avogadro's constant, a the radius of the spherical particles subject to the Brownian movement, and ξ the viscosity of the fluid containing the suspension.

§3. **Differential-Space.** In the Brownian movement, it is not the position of a particle at one time that is independent of the position of a particle at another; it is the displacement of a particle over one interval that is independent of the displacement of the particle over another interval. That is, instead of $f\left(\frac{1}{n}\right), \ldots, f\left(\frac{k}{n}\right), \ldots, f(1)$ representing "dimensions" of $f(t)$, the n quantities

$$x_1 = f\left(\frac{1}{n}\right) - f(0)$$

$$x_2 = f\left(\frac{2}{n}\right) - f\left(\frac{1}{n}\right)$$

$$\cdot\quad\cdot\quad\cdot\quad\cdot\quad\cdot$$

$$x_k = f\left(\frac{k}{n}\right) - f\left(\frac{k-1}{n}\right)$$

$$\cdot\quad\cdot\quad\cdot\quad\cdot\quad\cdot$$

$$x_n = f(1) - f\left(\frac{n-1}{n}\right)$$

10 Perrin, p. 53.

are of equal weight, vary independently, and in some degree represent dimensions. It is natural, then, to consider, as Lévy does, the properties of the sphere

$$x_1{}^2+\ .\ .\ .\ +x_n{}^2=r_n{}^2.$$

In particular, let us consider the measure, in terms of the whole sphere, of the region in which $f(a)-f(0)$, assuming it to be representable in terms of the x_s's, lies between the values α and β. Now, we have

$$f(a)-f(0) = \sum_{1}^{na} x_k. \tag{6}$$

Hence, by a change of coördinates, our question becomes: given that

$$\sum_{1}^{n} \xi_k^2 = r_n^2,$$

what is the chance that

$$\alpha \leq \sqrt{na}\,\xi_1 \leq \beta?$$

Letting $\xi_1=\rho \sin \theta$, $\sqrt{\xi_2^2+\ .\ .\ .\ +\xi_n{}^2}=\rho \cos \theta$, this chance becomes

$$\frac{\displaystyle\int_{\sin^{-1}\frac{\alpha}{r_n\sqrt{na}}}^{\sin^{-1}\frac{\beta}{r_n\sqrt{na}}} \cos^n\theta\, d\,\theta}{\displaystyle\int_{-\frac{\pi}{2}}^{\frac{\pi}{2}} \cos^n\theta\, d\,\theta},$$

as has been indicated by Lévy[11]. This may also be written

$$\frac{\displaystyle\int_{\sqrt{n}\sin^{-1}\frac{\alpha}{r_n\sqrt{na}}}^{\sqrt{n}\sin^{-1}\frac{\beta}{r_n\sqrt{na}}} \cos^n \frac{x}{\sqrt{n}}\, dx}{\displaystyle\int_{-\frac{\sqrt{n}\pi}{2}}^{\frac{\sqrt{n\pi}}{2}} \cos^n \frac{x}{\sqrt{n}}\, dx}.$$

[11] Lévy, p. 266.

Now[12], for n large, $\cos^n \frac{x}{\sqrt{n}}$ converges uniformly to $e^{-\frac{x^2}{2}}$ for $|x| < A$. It may be deduced without difficulty from this fact that the integral in the denominator tends to $\int_{-\infty}^{\infty} e^{-\frac{x^2}{2}}\, dx$, which is $\sqrt{2\pi}$. As to the numerator, if the limits of integration approach limiting values, it will also converge. If in particular r_n is a constant, then since

$$\lim_{n\to\infty} \sqrt{n}\ \sin^{-1} \frac{\alpha}{r\sqrt{na}} = \frac{\alpha}{r\sqrt{a}},$$

while a like identity holds for the upper limit, we have for the probability (in the limit) that $f(a)$ lie between α and β.

$$\frac{1}{\sqrt{2\pi}} \int_{\frac{\alpha}{r\sqrt{a}}}^{\frac{\beta}{r\sqrt{a}}} e^{-\frac{x_2}{2}}\, dx = \frac{1}{r\sqrt{2\pi a}} \int_{\alpha}^{\beta} e^{-\frac{u^2}{2ar^2}}\, du. \tag{7}$$

Let it be noted that this expression is of exactly the form (5).

We shall call the space of which the constituent points are the functions $f(t)$, and in which the measure of a region is determined as the limit of a measure in u-space in the way in which formula (7) is obtained from (6), by the name *differential-space*. The appropriateness of this name comes from the fact that it is not the values of $f(t)$, but the small differences, that are uniformly distributed, and act as dimensions.

§4. **The Non-Differentiability Coefficient of a Function.** It thus appears that if we consider the distribution of $f(a) - f(0) = \sum_1^{na} x_k$ in the sphere $\sum_1^{n} x_k^2 = r^2$, we get in the limit a distribution of the values of $f(a) - f(0)$ essentially like the one indicated in (5). Now, we have

$$\sum_1^n x_k^2 = \sum_1^n \left\{ f\left(\frac{k}{n}\right) - f\left(\frac{k-1}{n}\right) \right\}^2$$

[12] Lévy, p. 264.

Our result so far suggests, then, that the probability that $\sum_1^n \left\{ f\left(\frac{k}{n}\right) - f\left(\frac{k-1}{n}\right) \right\}^2$ exceed r^2 becomes increasingly negligible as n increases, or at least the probability that it exceed r^2 by a stated amount. On the hypothesis that $f(t_1)-f(t)$ have the distribution indicated in (5), and that the variation of f over one interval be independent of the variation of f over any preceding interval, let us discuss the distribution of

$$\sum_1^n \left\{ f\left(\frac{k}{n}\right) - f\left(\frac{k-1}{n}\right) \right\}^2.$$

Clearly, the chance that $\sum_1^n \left\{ f\left(\frac{k}{n}\right) - f\left(\frac{k-1}{n}\right) \right\}^2$ lie between α^2 and β^2 is the average value of a function of $x_1, \ldots, x_n$ which is 1 when $x_1^2 + \ldots + x_n^2$ lies between α^2 and β^2, and 0 otherwise, given that the weight of the region

$$\xi_k \leq x_k \leq \eta_k \qquad (k=1, 2, \ldots, n)$$

is

$$\prod_1^n \sqrt{\frac{n}{\pi A}} \int_{\xi_k}^{\eta_k} e^{-\frac{nx^2}{A}} dx$$

$$= \left(\sqrt{\frac{n}{\pi A}}\right)^n \int_{\xi_1}^{\eta_1} dx_1 \ldots \int_{\xi_n}^{\eta_n} dx_n \, e^{-\frac{n}{A}(x_1^2+\ldots+x_n^2)}, \tag{9}$$

as follows from (5). Now, let us put

$$\left.\begin{aligned} x_1 &= \rho \sin\theta_1 \\ x_2 &= \rho \cos\theta_1 \sin\theta_2 \\ x_3 &= \rho \cos\theta_1 \cos\theta_2 \sin\theta_3 \\ &\ldots\ldots\ldots \\ x_{n-1} &= \rho \cos\theta_1 \cos\theta_2 \ldots \cos\theta_{n-2} \sin\theta_{n-1} \\ x_n &= \rho \cos\theta_1 \cos\theta_2 \ldots \cos\theta_{n-1} \end{aligned}\right\} \tag{10}$$

We shall then have

$$\left.\begin{array}{l}\rho^2=x_1^2+\ \ldots\ +x_n^2 \\ dx_1\ldots dx_n=\rho^{n-1}\cos^{n-2}\theta_1\cos^{n-3}\theta_2\ldots\cos\theta_{n-2}\,d\rho\,d\theta_1\ldots d\theta_{n-1}\end{array}\right\}, \tag{11}$$

the last of which formulae may be readily demonstrated by an evaluation of the Jacobian

$$\begin{vmatrix} \frac{\partial x_1}{\partial\rho} & \frac{\partial x_1}{\partial\theta_1} & \cdots & \frac{\partial x_1}{\partial\theta_{n-1}} \\ \frac{\partial x_2}{\partial\rho} & \frac{\partial x_2}{\partial\theta_1} & \cdots & \frac{\partial x_2}{\partial\theta_{n-1}} \\ \cdot & \cdot & \cdot & \cdot \\ \frac{\partial x_n}{\partial\rho} & \frac{\partial x_n}{\partial\theta_1} & \cdots & \frac{\partial x_n}{\partial\theta_{n-1}} \end{vmatrix} \tag{12}$$

Employing a formula of Lévy to the effect that[13]

$$\int_0^{\frac{\pi}{2}}\cos^n\theta d\theta\int_0^{\frac{\pi}{2}}\cos^{n-1}\theta d\theta=\frac{\pi}{2n} \tag{13}$$

we get for the chance that $\sum_1^n\left\{f\left(\frac{k}{n}\right)-f\left(\frac{k-1}{n}\right)\right\}^2$ lie between α^2 and β^2

$$\left(\sqrt{\frac{n}{\pi A}}\right)^n\int_\alpha^\beta\rho^{n-1}e^{-\frac{n\rho^2}{A}}d\rho\prod_{k=1}^{n-2}\int_{-\frac{\pi}{2}}^{\frac{\pi}{2}}\cos^k\theta d\theta$$

$$=\left(\sqrt{\frac{n}{\pi A}}\right)^n 2^{n-2}\frac{\pi}{2(n-2)}\frac{\pi}{2(n-4)}\cdots\frac{\pi}{4}\int_\alpha^\beta\rho^{n-1}e^{-\frac{n\rho^2}{A}}d\rho$$

if n is even

$$=\left(\sqrt{\frac{n}{\pi A}}\right)^n 2^{n-2}\frac{\pi}{2(n-2)}\frac{\pi}{2(n-4)}\cdots\frac{\pi}{6}\int_\alpha^\beta\rho^{n-1}e^{-\frac{n\rho^2}{A}}d\rho$$

if n is odd. (14)

13 Levy, p. 263.

The coefficients of the integral may be reduced in either case to the form

$$\frac{n^{\frac{n}{2}}}{\pi A^{\frac{n}{2}}\,\Gamma\left(\frac{n}{2}\right)},$$

so that the whole expression may be written, by a change of variable,

$$\frac{n^{\frac{n}{2}}}{\pi\Gamma\left(\frac{n}{2}\right)}\int_{\frac{\alpha}{\sqrt{A}}}^{\frac{\beta}{\sqrt{A}}} u^{n-1}e^{-nu^2}\,du. \tag{15}$$

The integrand vanishes for u zero and u infinite. Between those points it attains its maximum for $u^2 = \frac{n-1}{2n}$, which for large values of n is in the neighborhood of $1/2$. It hence becomes interesting to note how much is contributed to our integral by values of u^2 near $1/2$ and how much by values remote from $1/2$.

Let us then evaluate

$$\frac{n^{\frac{n}{2}}}{\pi\Gamma\left(\frac{n}{2}\right)}\int_{u^2=\frac{1}{2}+\epsilon}^{\infty} u^{n-1}\,e^{-nu^2}du,$$

remembering that the integrand is a decreasing function. Let us discuss the ratio

$$\frac{(u+1)^{n-1}\,e^{-n(u+1)^2}}{u^{n-1}\,e^{-nu^2}} = \left(1+\frac{1}{u}\right)^{n-1} e^{-n(2u+1)}$$

$$\leq \frac{3^{n-1}}{e^{2n}} < 1/2. \tag{16}$$

We see at once that it follows from this that

$$\frac{n^{\frac{n}{2}}}{\pi\Gamma\left(\frac{n}{2}\right)}\int_{u^2=\frac{1}{2}+\epsilon}^{\infty} u^{n-1}\, e^{-nu^2}\, du < \frac{n^{\frac{n}{2}}}{\pi\Gamma\left(\frac{n}{2}\right)}\sum_{k=0}^{\infty}\frac{1}{2^k}\,(1/2+\epsilon)^{\frac{n-1}{2}}\, e^{-n(\frac{1}{2}+\epsilon)}$$

$$= \frac{2n^{\frac{n}{2}}}{\pi\Gamma\left(\frac{n}{2}\right)}\,(1/2+\epsilon)^{\frac{n-1}{2}}\, e^{-n(\frac{1}{2}+\epsilon)} \tag{17}$$

For n large, by Stirling's theorem, this latter expression is asymptotically represented by

$$\frac{2n^{\frac{n}{2}}}{\pi\left(\frac{n}{2}-1\right)^{\frac{n}{2}-1} e^{1-\frac{n}{2}}\sqrt{2\pi\left(\frac{n}{2}-1\right)}}\left(\frac{1}{2}+\epsilon\right)^{\frac{n-1}{2}} e^{-n\left(\frac{1}{2}+\epsilon\right)}$$

$$= \frac{2}{\pi e}\sqrt{\frac{n-2}{\pi}}\,\frac{(1+2\epsilon)^{\frac{n-1}{2}}\, e^{-n\epsilon}}{\left(1-\frac{n}{2}\right)^{\frac{n}{2}}}. \tag{18}$$

As n grows larger, this in turn may be represented asymptotically by

$$\frac{2}{\pi(1+2\epsilon)^{\frac{1}{2}}}\sqrt{\frac{n-2}{\pi}}\left(\frac{1+2\epsilon}{e^{2\epsilon}}\right)^{\frac{n}{2}}. \tag{19}$$

Now, $e^{2\epsilon} > 1+2\epsilon$, as may be seen directly from the Maclaurin series for e^x. Hence (19) may be written

$$K\sqrt{n-2}\, C^n,$$

where K is a constant and C is a positive constant less than 1. This, of course, tends to vanish for n large.

We may prove in a similar way that

$$\frac{n^{\frac{n}{2}}}{\pi\Gamma\left(\frac{n}{2}\right)}\int_0^{u^2=\frac{1}{2}-\epsilon} u^{n-1}\, e^{-nu^2}\, du$$

tends to vanish for n large. Clearly, for n sufficiently large, $\frac{n-1}{2n}$ will be greater than $1/2-\epsilon$. Under these circumstances,

$$\frac{n^{\frac{n}{2}}}{\pi\Gamma\left(\frac{n}{2}\right)}\int_0^{u^2=\frac{1}{2}-\epsilon} u^{n-1}\, e^{-nu^2}\, du \leq \frac{n^{\frac{n}{2}}}{2\pi\Gamma\left(\frac{n}{2}\right)}\left(\frac{1}{2}-\epsilon\right)^{\frac{n-1}{2}} e^{-n(\frac{1}{2}-\epsilon)}. \tag{20}$$

This latter quantity is asymptotically represented by

$$\frac{n^{\frac{n}{2}}}{2\pi\left(\frac{n}{2}-1\right)^{\frac{n}{2}-1} e^{1-\frac{n}{2}}\sqrt{2\pi\left(\frac{n}{2}-1\right)}}\left(\frac{1}{2}-\epsilon\right)^{\frac{n-1}{2}} e^{-n(\frac{1}{2}-\epsilon)}$$

$$= \frac{1}{2\pi\epsilon}\sqrt{\frac{n-2}{\pi}}\,\frac{(1-2\epsilon)^{\frac{n}{2}-1}\, e^{n\epsilon}}{\left(1-\frac{2}{n}\right)^{\frac{n}{2}}}. \tag{21}$$

This is in turn asymptotically represented by

$$\frac{1}{2\pi(1-2\epsilon)^{\frac{1}{2}}}\sqrt{\frac{n-2}{\pi}}\left[(1-2\epsilon)\, e^{2\epsilon}\right]^{\frac{n}{2}}. \tag{22}$$

Now, since by Taylor's theorem with remainder, $e^{-2\epsilon}=1-2\epsilon+\frac{\theta^2}{2}$, θ being a positive number less than 2ϵ, we get $(1-2\epsilon)e^{2\epsilon}<1$. Hence expression (22) is again of the form $K\sqrt{n-2}\,C^n$, C being a positive constant less than 1, and approaches zero as n increases.

It will be seen, then, that for n sufficiently large, the chance that $\sum_1^n\left\{f\left(\frac{k}{n}\right)-f\left(\frac{k-1}{n}\right)\right\}^2$ diverge from $A/2$ by more than ϵ is less than an expression of the form $K\sqrt{n-2}\,C^n$, in which K

and C are positive constants independent of n, and $C<1$. It follows that for n sufficiently large, the chance that *any* $\sum_1^N \left\{ f\left(\frac{k}{N}\right) - f\left(\frac{k-1}{N}\right) \right\}^2$ differ from $A/2$ by more than ϵ, for $N \geq n$, is less than

$$K \sum_n^\infty \sqrt{N-2}\, C^N,$$

which is the remainder of a convergent series, and hence vanishes as n becomes infinite. In other words, the chance that

$$\left| \lim_{n\to\infty} \sum_1^n f\left\{ \left(\frac{k}{n}\right) - f\left(\frac{k-1}{n}\right) \right\}^2 - A/2 \right| > \epsilon$$

is less than any assignable positive number, if ϵ is any positive quantity.

Let $f(t)$ be a continuous function of limited total variation T between 0 and 1. Then clearly

$$\sum_1^n \left\{ f\left(\frac{k}{n}\right) - f\left(\frac{k-1}{n}\right) \right\}^2 \leq \max\left| f\left(\frac{k}{n}\right) - f\left(\frac{k-1}{n}\right) \right| \sum_1^n \left| f\left(\frac{k}{n}\right) - f\left(\frac{k-1}{n}\right) \right| \leq T \max\left| f\left(\frac{k}{n}\right) - f\left(\frac{k-1}{n}\right) \right|. \tag{23}$$

Hence

$$\lim_{n\to\infty} \sum_1^n \left\{ f\left(\frac{k}{n}\right) - f\left(\frac{k-1}{n}\right)^2 \right\} = 0. \tag{24}$$

This will in particular be the case when $f(t)$ possesses a derivative bounded over the closed interval (0, 1). Hence it is infinitely improbable, under our distribution of functions $f(t)$, that $f(t)$ be a continuous function of limited total variation, and in particular that it have a bounded derivative. We may regard $\lim_{n\to\infty} \sum_1^n \left\{ f\left(\frac{k}{n}\right) - f\left(\frac{k-1}{n}\right)^2 \right\}$ as in some sort a nondifferentiability coefficient of f.

§5. **The Maximum Gain in Coin-Tossing.** We have now investigated the differentiability of the functions $f(t)$; it behoves us to inquire as to their continuity. To this end we shall discuss the maximum value of $f(t)-f(t_0)$ incurred for $t_0 \leq t \leq t_1$, and shall determine its distribution. As a simple model of this rather complex situation, however, we shall consider a problem in coin-tossing.

A gambler stakes one dollar on each throw of a coin, which is tossed n times, losing the dollar if the throw is heads, and gaining it if the throw is tails. At the beginning he has lost nothing and won nothing; after m throws he will be k_m dollars ahead, k_m being positive or negative. The question here asked is, what is the distribution of the maximum value of k_m for $m \leq n$?

Let n throws be made in all the 2^n possible manners. Let $A_n(p)$ be the number of throw-sequences in which max $(k_m)=p$. Clearly $A_n(-\mu)=0$, if μ is positive, since $k_0=0$. Furthermore, any throw-sequence in which the maximum gain is zero consists of a throw-sequence in which the maximum gain is either 1 or 0 preceded by a throw of heads. That is,

$$A_n(0)=A_{n-1}(0)+A_{n-1}(1). \tag{25}$$

A throw-sequence in which the maximum gain is $p>0$ consists either of a throw-sequence in which the maximum gain is $p-1$ preceded by a throw of tails, or a throw-sequence in which the maximum gain is $p+1$, preceded by a throw of heads. That is,

$$A_n(p)=A_{n-1}(p-1)+A_{n-1}(p+1). \tag{26}$$

Let us tabulate the first few values of $A_n(p)$. We get

$A_n(p)$	$n=1$	2	3	4	5	6
$p=0$	1	2	3	6	10	20
1	1	1	3	4	10	15
2	0	1	1	4	5	15
3	0	0	1	1	5	6
4	0	0	0	1	1	6
5	0	0	0	0	1	1
6	0	0	0	0	0	1

It will be seen that the various numbers in the table are the binomial coefficients, beginning in each column with the middle coefficient for n even and the next coefficient beyond the middle for n odd, and with every coefficient in the expansion $(1+1)^n$ repeated twice, with the exception of the middle coefficient for n even. That is, as far as the table is carried, we have

$$\left.\begin{aligned} A_{2n}(2p) &= A_{2n}(2p-1) = \frac{(2n)!}{(n+p)!\,(n-p)!} \\ A_{2n+1}(2p) &= A_{2n+1}(2p+1) = \frac{(2n+1)!}{(n+p+1)!\,(n-p)!} \end{aligned}\right\}. \tag{27}$$

We may then verify (25) and (26) by direct substitution, thus proving that the values given for $A_m(q)$ in (27) are valid for all values of m and q.

Let us now consider a series of n successive runs of m throws each, for points of $1/\sqrt{m}$ dollars, m being even. The chance that the amount gained in a single run lie between α and β dollars is then

$$1/2^m \sum_{\alpha'\sqrt{m}}^{\beta'\sqrt{m}} B_m(k), \tag{28}$$

where $\alpha'\sqrt{m}$ is the even integer next greater than or equal to $\alpha'\sqrt{m}$, $\beta'\sqrt{m}$ is the even integer next less than $\beta\sqrt{m}$ and $B_m(k)/2^m$ is the chance that of m throws of a coin just k more should be tails than heads — that is,

$$B_m(k) = \frac{m!}{\frac{m-k}{2}!\,\frac{m+k}{2}!}\,. \tag{29}$$

If we begin by assuming α and β positive and less than γ, expression (28) will lie between the values

$$\frac{(\beta'-\alpha')\sqrt{m}}{2^{m+1}}\;\frac{m!}{\left(\frac{m-\beta'\sqrt{m}}{2}\right)!\left(\frac{m+\beta'\sqrt{m}}{2}\right)!} \quad \text{and}$$

$$\frac{\beta('-\alpha')\sqrt{m}}{2^{m+1}}\;\frac{m!}{\left(\frac{m-\alpha'\sqrt{m}}{2}\right)!\left(\frac{m+\alpha'\sqrt{m}}{2}\right)!}\,.$$

Now, by Stirling's theorem, we have uniformly for $\alpha<\gamma$, $\beta<\gamma$

$$\lim_{m\to\infty} \frac{(\beta'-\alpha')\sqrt{m}}{2^{m+1}} \quad \frac{m!}{\left(\frac{m-\beta'\sqrt{m}}{2}\right)!\left(\frac{m+\beta'\sqrt{m}}{2}\right)!}$$

$$= \lim_{m\to\infty} \frac{(\beta-\alpha)\sqrt{m} \quad m^m e^{-m}\sqrt{2\pi m}}{2^{m+1}\left(\frac{m-\beta\sqrt{m}}{2}\right)^{\frac{m-\beta\sqrt{m}}{2}}\left(\frac{m+\beta\sqrt{m}}{2}\right)^{\frac{m+\beta\sqrt{m}}{2}} e^{-m}\pi\sqrt{m^2-\beta^2 m}}$$

$$= \lim_{m\to\infty} \frac{\beta-\alpha}{\sqrt{2\pi}}\left(\frac{m^2}{m^2-\beta^2 m}\right)^{\frac{m}{2}} \left(\frac{m-\beta\sqrt{m}}{m+\beta\sqrt{m}}\right)^{\frac{\beta\sqrt{m}}{2}}$$

$$= \lim_{m\to\infty} \frac{\beta-\alpha}{\sqrt{2\pi}}(1-\beta^2/m)^{\frac{m}{2}}(1-\beta/\sqrt{m})^{\frac{\beta\sqrt{m}}{2}}(1+\beta/\sqrt{m})^{-\frac{\beta\sqrt{m}}{2}}$$

$$= \frac{\beta-\alpha}{\sqrt{2\pi}} e^{-\frac{\beta^2}{2}} \tag{30}$$

We have, then, for m sufficiently large, by combining (30) with an analogous theorem,

$$\frac{\beta-\alpha}{\sqrt{2\pi}} e^{-\frac{\beta^2}{2}} < 1/2^m \sum_{\alpha'\sqrt{m}}^{\beta'\sqrt{m}} \beta_m(k) < \frac{\beta-\alpha}{\sqrt{2\pi}} e^{-\frac{\alpha^2}{2}} \tag{31}$$

Moreover, since (30) holds uniformly, the value of m necessary to make (31) valid depends only on $\beta-\alpha$.

Now, let us divide the interval (α, β) into the p equal parts (α, α_1), (α_1, α_2), . . . , (α_{n-1}, β), and let us choose m sufficiently large for us to have over each interval

$$\frac{\alpha_{k+1}-\alpha_k}{\sqrt{2\pi}} e^{-\frac{\alpha^2_{k+1}}{2}} < 1/2^m \sum_{\alpha'_k\sqrt{m}}^{\alpha''_{k+1}\sqrt{m}} B_m(k) < \frac{\alpha_{k+1}-\alpha_k}{\sqrt{2\pi}} e^{-\frac{\alpha^2_k}{2}},$$

where $\alpha_k'\sqrt{m}$ is $\alpha_k\sqrt{m}$ if this is an even integer, or otherwise the next larger integer, and $\alpha_k''\sqrt{m}=\alpha_k\sqrt{m}-2$. Then we have

$$\frac{1}{\sqrt{2\pi}} \sum_1^n (\alpha_{k+1}-\alpha_k)\, e^{-\frac{\alpha^2_{k+1}}{2}} < 1/2^m \sum_{\alpha'\sqrt{m}}^{\beta'\sqrt{m}} B_m(k)$$

$$< \frac{1}{\sqrt{2\pi}} \sum_1^n (\alpha_{k+1}-\alpha_k)\, e^{-\frac{\alpha^2_k}{2}},$$

where the value of m needed depends only on $\alpha_{k+1}-\alpha_k$. Since $e^{-\frac{x^2}{2}}$ is a decreasing function, we hence have

$$\left|\frac{1}{\sqrt{2\pi}}\int_\alpha^\beta e^{-\frac{x^2}{2}}\,dx-1/2^m\sum_{\alpha'\sqrt{m}}^{\beta'\sqrt{m}}B_m(k)\right|<\frac{\gamma}{\sqrt{2\pi}}\max\left(e^{-\frac{\alpha^2_k}{2}}-e^{-\frac{\alpha^2_{k+1}}{2}}\right)$$

$$<\frac{\gamma}{\sqrt{2\pi e}}(\alpha_{k+1}-\alpha_k), \qquad (32)$$

where a minimum value of m may be determined in terms of $\alpha_{k+1}-\alpha_k$ alone. In other words, $1/2^m\sum_{\alpha'\sqrt{m}}^{\beta'\sqrt{m}}B_m(k)$ converges *uniformly* in α and β to

$$\frac{1}{\sqrt{2\pi}}\int_\alpha^\beta e^{-\frac{x^2}{2}}dx,$$

provided only α and β are positive and smaller than γ. It is obvious that the requirement of positiveness is superfluous, if only $|\alpha|<\gamma$, $|\beta|<\gamma$.

This last restriction can be removed. In the first place

$$B_m(2k)/B_m(k)=\frac{(m-k)!\,(m+k)!}{(m-2k)!\,(m+2k)!}$$

$$=\frac{(m-2k+1)(m-2k+2)\;\ldots\;(m-k)}{(m+k+1)\,(m+k+2)\;\ldots\;(m+2k)}$$

$$<\left(\frac{m-k}{m+2k}\right)^k. \qquad (33)$$

If now $k>l\sqrt{m}$, we have

$$B_m(2k)/B_m(k)<\left(\frac{\sqrt{m}-l}{\sqrt{m}+2l}\right)^{\sqrt{m}}<\left(1-\frac{l}{\sqrt{m}}\right)^{\sqrt{m}}<l^{-l}, \qquad (34)$$

for $k\leq m$. If $k>m$, we have clearly

$$B_m(2k)=B_m(k)=0.$$

Hence in general, for all $k > l\sqrt{m}$,

$$B_m(2^n k) \leq e^{-nl}\, B_m(k). \tag{34}$$

It follows that by making γ sufficiently large, we can make

$$\frac{1}{2^m}\sum_{\gamma'\sqrt{m}}^{\infty} B_m(k) \leq \frac{1}{2^m}\sum_{0}^{\gamma'\sqrt{m}} B_m(k)\ (1+e^{-l}+e^{-2l}+\ .\ .\ .)$$

$$\leq \frac{1}{2^m}\sum_{0}^{\gamma'\sqrt{m}} B_m(k)\ \frac{1}{1-e^{-l}}\ , \tag{35}$$

where $\gamma'\sqrt{m}$ is the integer next larger than $\gamma\sqrt{m}$ and l is independent of m. The result is that

$$\lim_{\gamma\to\infty} \frac{1}{2^m}\sum_{\gamma'\sqrt{m}}^{\infty} B_m(k) = 0 \tag{36}$$

uniformly in m. Hence if we consider

$$\left| \frac{1}{\sqrt{2\pi}} \int_{\alpha} e^{-\frac{x^2}{2}} dx - 1/2^{m+1} \sum_{\alpha'\sqrt{m}}^{\beta'\sqrt{m}} B_m(k) \right| , \tag{37}$$

we may first choose γ so large that the difference in this expression made by replacing α or β by $\pm\gamma$, in case they lie outside $(-\gamma, \gamma)$, is less than $\epsilon/2$, and then choose m so large that within the region $(-\gamma, \gamma)$, the expression is less than $\epsilon/2$. That is, by a choice of m alone, independently of α and β, we can make (37) less than ϵ.

Now let us write

$$\left.\begin{aligned} \frac{1}{2^{m+1}} \sum_{-\infty}^{\beta'\sqrt{m}} B_m(k) &= \phi_m(\beta) \\ \frac{1}{\sqrt{2\pi}} \int_{-\infty}^{\beta} e^{-\frac{x^2}{2}}\, dx &= \phi(\beta) \end{aligned}\right\} . \tag{38}$$

Clearly, the probability that the maximum amount won for any value of k at the end of the first km throws out of mn for $\sqrt{m}$

dollars each lie between 0 and u will be the Stieltjes integral expression

$$\int_{0-0}^{u} d\phi_m(x_1) \int_{-\infty}^{-x_1} d\phi_m(x_2) \int_{-\infty}^{-x_1-x_2} d\phi_m(x_3) \ldots \int_{-\infty}^{-x_1-x_2-\cdots-x_{n-1}} d\phi_m(x_n)$$

$$+\int_{0+0}^{u} d\phi_m(x_1) \int_{0+0}^{u-x_1} d\phi_m(x_2) \int_{-\infty}^{-x_1-x_2} d\phi_m(x_3) \ldots \int_{-\infty}^{-x_1-x_2-\cdots-x_{n-1}} d\phi_m(x_n)$$

$$+\int_{0+0}^{u} d\phi_m(x_1) \int_{-\infty}^{0} d\phi_m(x_2) \int_{x_1+0}^{u} d\phi_m(x_3) \ldots \int_{-\infty}^{-x_1-x_2-\cdots-x_{n-1}} d\phi_m(x_n)$$

$$+\int_{0+0}^{u} d\phi_m(x_1) \int_{0+0}^{u-x_1} d\phi_m(x_2) \int_{x_1+x_2+0}^{u} d\phi_m(x_3) \ldots \int_{-\infty}^{-x_1-x_2-\cdots-x_{n-1}} d\phi_m(x_n)$$

$$+ \quad \cdots\cdots\cdots\cdots\cdots\cdots$$

$$+\int_{0+0}^{u} d\phi_m(x_1) \int_{0+0}^{u-x_1} d\phi_m(x_2) \ldots \int_{0+0}^{u-x_1-x_2-\cdots-x_{n-1}} d\phi_m(x_n), \tag{39}$$

consisting of $1+1+2!+3!+ \ldots +n!$ terms. If in this expression we substitute ϕ for ϕ_m, it results from what we have just said of the uniformity of the convergence of ϕ_m to ϕ that the expression we obtain will be uniformly the limit of (39). This expression will represent the chance that after n independent games, in each of which the chance of winning a sum between α dollars and β dollars is $\frac{1}{\sqrt{2\pi}}\int_\alpha^\beta e^{-\frac{x^2}{2}}dx$, the maximum gain incurred lies between 0 and u. We shall denominate this quantity $G(u, n)$, and expression (39) $G_m(u, n)$.

It is obvious from the definition that $G(u, n)$ is more than

$$1/2^{mn}\sum_{0}^{u'\sqrt{m}} A_{mn}(k) = 1/2^{mn-1}\sum_{k=0}^{k=u'\sqrt{m}} \frac{mn!}{\left(\frac{mn-k}{2}\right)!\left(\frac{mn+k}{2}\right)!}, \tag{40}$$

where $u'\sqrt{m}$ is the integer next larger than $u\sqrt{m}$. By going through an argument precisely analogous to that by which (32) was obtained, we see that

$$\lim_{m\to\infty} 1/2^{mn}\sum_{0}^{u'\sqrt{mn}} A_{mn}(k) = \sqrt{\frac{2}{\pi}}\int_0^{u\sqrt{n}} e^{-\frac{x^2}{2}}\,dx$$
$$= \sqrt{\frac{2}{n\pi}}\int_0^{u} e^{-\frac{x^2}{2n}}\,dx. \tag{41}$$

Hence

$$G\,(u,n) > \sqrt{\frac{2}{n\pi}}\int_0^{u} e^{-\frac{x^2}{2n}}\,dx. \tag{42}$$

§6. **Measure in Differential-Space.** To revert to differential-space, let us consider a functional $F|f|$, f being defined for arguments between zero and one, and let us see if we can frame a definition of its average. To begin with, let us suppose that F only depends on the values of f for the arguments $t_1 < t_2, \ldots < t_n$. F will then be unchanged if we alter f to any other function, assuming the same values for these arguments — in particular, to a step-function $f_\nu(t)$ with steps all of length $1/\nu$, ν being sufficiently large.

Let us now call the difference between the height of the kth step and that of the $(k-1)$st by the name x_k. We shall then have

$$F|f_\nu| = F(f_\nu(t_1), f_\nu(t_2), \ldots, f_\nu(t_n))$$
$$= F(x_1+x_2+\ldots+x_{T_1}, x_1+x_2+\ldots+x_{T_2}, \ldots,$$
$$x_1+x_2+\ldots+x_{T_n}), \tag{43}$$

where T_k is νt_k if this is an integer, and otherwise the next smaller integer. Now, the region of the space $(x_1, x_2, \ldots, x_n)$ which corresponds to differential-space is the interior of the sphere

$$x_1^2+x_2^2+\ldots+x_n^2=r^2.$$

Over the interior of this sphere we may take the average of expression (43), let it approach a limit as ν increases indefinitely, and call this limit the average of F in differential-space.

By the change in coördinates

$$\left.\begin{array}{l} x_1+x_2 \quad + \;.\;.\;.\; +x_{T_1} \quad = \sqrt{T_1}\,\xi_1 \\ x_{T_1+1} \quad + \;.\;.\;.\; +x_{T_2} \quad = \sqrt{T_2-T_1}\,\xi_2 \\ .\quad .\quad .\quad .\quad .\quad .\quad .\quad .\quad .\quad .\quad . \\ x_{T_{n-1}+1} \;+ \;.\;.\;.\;.\; +x_{T_n} = \sqrt{T_n-T_{n-1}}\,\xi_n, \\ \xi_{n+1},\;.\;.\;.\;,\xi_n \text{ represent coördinates orthogonal to } \xi_1,\;.\;.\;.\;,\xi_n \end{array}\right\} \quad (44)$$

we change our sphere into

$$\xi_1^2+\;.\;.\;.\;+\xi_\nu^2=r^2,$$

and have also

$$F|f_\nu| = F(\sqrt{T_1}\xi_1,\ \sqrt{T_2-T_1}\ \xi_2,\ .\ .\ .\ .\ ,\ \sqrt{T_n-T_{n-1}}\ \xi_n).$$

By a transformation like (10), this will become

$$F|f_\nu| = F(\sqrt{T_1}\,\rho \sin\theta_1,\ \sqrt{T_2-T_1}\,\rho\cos\theta_1\sin\theta_2,$$
$$.\ .\ .\ \sqrt{T_n-T_{n-1}}\,\rho\cos\theta_1,\ .\ .\ .\ \cos\theta_{n-1}\sin\theta_n).$$

The average value of this over our sphere is

$$\frac{\left\{\begin{array}{c}\displaystyle\int_0^r d\rho\int_{-\frac{\pi}{2}}^{\frac{\pi}{2}} d\theta_1\ .\ .\ .\int_{-\frac{\pi}{2}}^{\frac{\pi}{2}} d\theta_{\nu-1}\,\rho^{\nu-1}\cos^{\nu-2}\theta_1\cos^{\nu-3}\theta_2\ .\ .\ .\ \cos\theta_{\nu-2} \\ F(\sqrt{T_1}\,\rho\sin\theta_1,\ .\ .\ .\ ,\sqrt{T_n-T_{n-1}}\,\rho\cos\theta_1 .\ .\ .\ \sin\theta_n)\end{array}\right\}}{\displaystyle\int_0^r d\rho\int_{-\frac{\pi}{2}}^{\frac{\pi}{2}} d\theta_1,\ .\ .\ .\int_{-\frac{\pi}{2}}^{\frac{\pi}{2}} d\theta_{\nu-1}\,\rho^{\nu-1}\cos^{\nu-2}\theta_1\ .\ .\ .\ \cos\theta_{\nu-2}}$$

$$=\frac{\left\{\begin{array}{c}\displaystyle\int_0^r d\rho\int_{-\frac{\pi}{2}}^{\frac{\pi}{2}} d\theta_1\ .\ .\ .\int_{-\frac{\pi}{2}}^{\frac{\pi}{2}} d\theta_n\rho^{\nu-1}\cos^{\nu-2}\theta_1\ .\ .\ .\ \cos^{\nu-n-1}\theta_n \\ F\,(\sqrt{T_1}\,\rho\ \sin\theta_1,\ .\ .\ .\ ,\sqrt{T-T_{n-1}}\,\rho\cos\theta_1\ .\ .\ .\ \sin\theta_n)\end{array}\right\}}{\displaystyle\int_0^r d\rho\int_{-\frac{\pi}{2}}^{\frac{\pi}{2}} d\theta_1\ .\ .\ .\int_{-\frac{\pi}{2}}^{\frac{\pi}{2}} d\theta_n\,\rho^{\nu-1}\cos^{\nu-2}\theta_1\ .\ .\ .\ \cos^{\nu-n-1}\theta_n}$$

$$
= \frac{\left\{ \begin{array}{l} \int_0^r d\rho \int_{-\frac{\pi\sqrt{\nu}}{2}}^{\frac{\pi\sqrt{\nu}}{2}} du_1 \ldots \int_{-\frac{\pi\sqrt{\nu}}{2}}^{\frac{\pi\sqrt{\nu}}{2}} du_n \rho^{\nu-1} \cos^{\nu-2} \frac{u_1}{\sqrt{\nu}} \ldots \cos^{\nu-n-1} \frac{u_n}{\sqrt{\nu}} \\ F\left(\sqrt{T}\rho \sin \frac{u_1}{\sqrt{\nu}}, \ldots, \sqrt{T_n - T_{n-1}}\, \rho \cos \frac{u_1}{\sqrt{\nu}}, \ldots \sin \frac{u_n}{\sqrt{\nu}}\right) \end{array} \right\}}{\int_0^r d\rho \int_{-\frac{\pi\sqrt{\nu}}{2}}^{\frac{\pi\sqrt{\nu}}{2}} du_1 \ldots \int_{-\frac{\pi\sqrt{\nu}}{2}}^{\frac{\pi\sqrt{\nu}}{2}} du_n \rho^{\nu-1} \cos^{\nu-2} \frac{u_1}{\sqrt{\nu}} \ldots \cos^{\nu-n-1} \frac{u_n}{\sqrt{\nu}}}. \tag{45}
$$

If we now consider the integrands in the numerator and the denominator for $|u_1| < U, \ldots, |u_n| < U$, and let ν increase indefinitely, we see[14] that

$$
\begin{aligned}
\lim_{\nu \to \infty} \cos^{\nu-2} \frac{u_1}{\sqrt{\nu}} \ldots \cos^{\nu-n-1} \frac{u_n}{\sqrt{\nu}} F\Big(\sqrt{T_1}\, \rho \sin \frac{u_1}{\sqrt{\nu}}, \ldots,& \\
\sqrt{T_n - T_{n-1}}\, \rho \cos \frac{u_1}{\sqrt{\nu}} \ldots \sin \frac{u_n}{\sqrt{\nu}}\Big)& \\
= e^{-\frac{u_1^2}{2} - \cdots - \frac{u_n^2}{2}} F(\rho u_1 \sqrt{t_1}, \ldots, \rho u_n \sqrt{t_n - t_{n-1}}),&
\end{aligned} \tag{46}
$$

and

$$
\lim_{\nu \to \infty} \cos^{\nu-2} \frac{u_1}{\sqrt{\nu}} \ldots \cos^{\nu-n-1} \frac{u_n}{\sqrt{\nu}} = e^{-\frac{u_1^2}{2} - \cdots - \frac{u_n^2}{2}}. \tag{47}
$$

uniformly, provided only F is continuous. If in addition F vanishes for $|u_1| > U, \ldots, |U_n| > U$, we have for the limit of the expression in (45)

$$
\lim_{\nu \to \infty} \frac{\left\{ \begin{array}{l} \int_0^r \rho^{\nu-1} d\rho \int_{-\infty}^{\infty} du_1 \ldots \int_{-\infty}^{\infty} du_n\, e^{-\frac{u_1^2}{2} - \cdots - \frac{u_n^2}{2}} \\ F(\rho u_1 \sqrt{t_1}, \ldots, \rho u_n \sqrt{t_n - t_{n-1}}) \end{array} \right\}}{\int_0^r \rho^{\nu-1} d\rho \int_{-\infty}^{\infty} du_1 \ldots \int_{-\infty}^{\infty} du_n\, e^{-\frac{u_1^2}{2} - \cdots - \frac{u_n^2}{2}}}
$$

14 Cf. note 12.

This will be the case even if F has a set of discontinuities of zero measure, provided F vanishes for $|u_1| > U, \ldots, |U_n| > U$, or even, as may be shown without difficulty by a limit argument, provided merely that $|F|$ is bounded. It follows at once that, under these conditions, the expression in (45) becomes

$$\frac{\int_{-\infty}^{\infty} du_1 \ldots \int_{-\infty}^{\infty} du_n \, e^{-\frac{u_1^2}{2}-\cdots-\frac{u_n^2}{2}} F(ru_1\sqrt{t_1}, \ldots, ru_n\sqrt{t_u-t_{n-1}})}{\int_{-\infty}^{\infty} du_1 \ldots \int_{-\infty}^{\infty} du_n \, e^{-\frac{u_1^2}{2}-\cdots-\frac{u_n^2}{2}}}$$

$$=(2\pi)^{-\frac{n}{2}}\int_{-\infty}^{\infty} du_1 \ldots \int_{-\infty}^{\infty} du_n \, e^{-\frac{u_1^2}{2}-\cdots-\frac{u_n^2}{2}} F(ru_1\sqrt{t_1}, \ldots, ru_n\sqrt{t_n-t_{n-1}})$$

$$=(2\pi)^{-\frac{n}{2}} r^{-n}[t_1(t_2-t_1) \ldots (t_n-t_{n-1})]^{-\frac{1}{2}} \int_{\infty}^{\infty} dy_1 \ldots \int_{-\infty}^{\infty} dy_n \; exp \sum_{t}^{n} \frac{tj_k^2}{2r^2(t_k-t_{k-1})} F(y_1, \ldots y_n), \tag{48}$$

to being zero. In particular, if $F(y_1, \ldots, y_n)$ is 1 when

$$\left.\begin{array}{l} y_{11} < y_1 < y_{12} \\ y_{21} < y_2 < y_{22} \\ \cdots\cdots \\ y_{n1} < y_n < y_{n2} \end{array}\right\}, \tag{49}$$

we get for the average value of F

$$(2\pi)^{-\frac{n}{2}} r^{-n}[t_1(t_2-t_1) \ldots (t_n-t_{n-1})]^{-\frac{1}{2}} \int_{y_{11}}^{y_{12}} dy_1 \ldots \int_{y_{n1}}^{y_{n2}} dy_n \; exp \sum_{1}^{n} \frac{y_k^2}{2r^2(t_k-t_{k-1})}. \tag{50}$$

This we shall term the *measure* or *probability* of region (49).

§7. **Measure and Equal Continuity.** Let us now consider all those functions $f(t)$ $(0 \leq t \leq 1)$ for which $f(0)=0$ and which for some pair of rational arguments, $0 \leq t_1 < t_2 \leq 1$, satisfy the inequality

$$|f(t_2)-f(t_1)| > ar(t_2-t_1)^{\frac{1}{2}-\epsilon} \tag{51}$$

If this inequality is satisfied, we can certainly find an n such that

$$\frac{k-1}{n} \leq t_1 \leq \frac{k}{n} \leq t_2 \leq \frac{k+1}{n}, \quad t_2-t_1 \geq 1/n, \tag{52}$$

while the variation of f over the interval $\left(\frac{k-1}{n}, \frac{k+1}{n}\right)$ will exceed $ar(t_2-t_1)^{\frac{1}{2}-\epsilon}$. That is, the class of functions satisfying the inequality (51) is a sub-class of the class of functions for which the variation over some interval $\left(\frac{k-1}{n}, \frac{k}{n}\right)$ exceeds $\frac{ar}{2} n^{\epsilon-\frac{1}{2}}$. Now, every such function must have either the difference between some *positive* value in the interval and its initial value exceed $\frac{ar}{4} n^{\epsilon-\frac{1}{2}}$, or the difference between its initial value and some *negative* value in the interval must exceed $\frac{ar}{4} n^{\epsilon-\frac{1}{2}}$.

To discuss what we may interpret as the measure of the set of functions (51), we may hence investigate the measure of the set of functions which over an interval of length $1/n$ depart from their initial value (for some rational argument, be it understood) by more than $\frac{ar}{4} n^{\epsilon-\frac{1}{2}}$. Now, if we subdivide the interval of length $1/n$ into m parts, each of length $1/mn$, the probability that in a given one of these sub-intervals the total difference between the initial and the final value of the function lie between α and β is, by (50)

$$\frac{\sqrt{mn}}{r\sqrt{2\pi}} \int_\alpha^\beta e^{-\frac{mnx^2}{2r^2}}\, dx. \tag{53}$$

Hence the probability that the maximum difference between the value of f at the end of one of these sub-intervals and its initial value be greater than $\frac{ar}{4} n^{\epsilon-\frac{1}{2}}$ is by (42) less than

$$\frac{1}{r}\sqrt{\frac{2n}{\pi}} \int_{\frac{ar}{4}n^{\epsilon-\frac{1}{2}}}^{\infty} e^{-\frac{nx^2}{2r^2}} dx$$

$$< \frac{1}{r}\sqrt{\frac{2n}{\pi}} \int_{\frac{ar}{4}n^{\epsilon-\frac{1}{2}}}^{\infty} e^{-\sqrt{\frac{nx^2}{2r^2}}} dx \left(\text{for } \frac{n^{2\epsilon}a^2r^2}{16} \geq 1\right)$$

$$= \frac{1}{r}\sqrt{\frac{2n}{\pi}} \sqrt{\frac{n}{2r^2}}\, e^{-\frac{n^{\epsilon}ar}{4}}$$

$$= \frac{n}{r^2\sqrt{\pi}}\, e^{-\frac{n^{\epsilon}ar}{4}}. \tag{54}$$

This will be true no matter how large m is. Hence those functions f for which the maximum difference between $f\left(\frac{k}{n} + \frac{l}{mn}\right)$ and $f\left(\frac{k}{n}\right)$ is greater than $\frac{ar}{4} n^{\epsilon-\frac{1}{2}}$, l being an integer not greater than m, which is any integer whatever, can be included in a denumerable set of regions such as those contemplated in the last section, of total measure less than $\frac{n}{r^2\sqrt{\pi}} e^{-\frac{n^{\epsilon}ar}{4}}$. If now we let k vary from 1 to n, we shall find that those functions whose positive excursion in one of the intervals $\left(\frac{k}{n}, \frac{k+1}{n}\right)$ for some rational argument exceeds $\frac{ar}{4} n^{\epsilon-\frac{1}{2}}$ will have a total measure less than $\frac{n^2}{r^2\sqrt{\pi}} e^{-\frac{n^{\epsilon}ar}{4}}$, in the sense that they can be included in a denumerable assemblage of sets measurable in accordance with the provisions of the last section, and of total measure less than this amount. If in this last sentence we replace the words " positive excursion " by the words " positive or negative excursion," we get for the total measure of our set of functions a quantity less than $\frac{2n^2}{r^2\sqrt{\pi}} e^{-\frac{n^\epsilon ar}{4}}$. It follows that all the functions satisfying (51) can be included in a denumerable set of regions of total measure less than

$$\frac{2}{r^2\sqrt{\pi}} \sum_1^\infty n^2 e^{-\frac{n^\epsilon ar}{4}} \tag{55}$$

This series converges if a has a sufficiently large value independent of n, as may be seen by comparing it with the series.

$$\frac{2}{r^2\sqrt{\pi}} \sum_1^\infty n^2 (n^\epsilon)^{-e^{\frac{ar}{4}}},$$

which converges for a sufficiently large. Moreover, if (55) converges for $a=a_1$, we may write (55) in the form

$$\frac{2}{r^2\sqrt{\pi}} \sum_1^\infty n^2 e^{-\frac{n^\epsilon ar}{4}} = \frac{2}{r^2\sqrt{\pi}} \sum_1^\infty n^2 e^{-\frac{n^\epsilon a_1 r}{4}} e^{\frac{n^\epsilon (a_1-a)r}{4}}$$

$$< \left[\frac{2}{r^2\sqrt{\pi}} \sum_1^\infty n^2 e^{-\frac{n^\epsilon a_1 r}{4}}\right] e^{\frac{(a_1-a)r}{4}} \tag{56}$$

for $a > a_1$. Hence we have

$$\lim_{a\to\infty} \frac{2}{r^2\sqrt{\pi}} \sum_1^\infty n^2 e^{-\frac{n^\epsilon ar}{4}} = 0. \tag{57}$$

That is, by making a in (51) sufficiently large, the functions satisfying (51) may be included in a denumerable set of such regions as those discussed in §6, of total measure as small as may be desired.

Let us now consider the sphere

$$x_1^2 + \ldots + x_n^2 = r^2,$$

and on this sphere, a sector subtended at the center by a given area of the surface. Let us cut off a portion of this sector by a concentric sphere of radius $r_1 < r$, and let us measure the volume of this new sector in terms of the sphere of radius r. Let us compare the quantity thus obtained with

$$(2\pi)^{-\frac{n}{2}} r^{-n} n^{\frac{n}{2}} \int dx_1 \ldots \int dx_n\, e^{-\frac{(x_1^2+\ldots+x_n^2)n}{2r^2}}, \tag{58}$$

the integral being taken over the same sector. The ratio between these two quantities may readily be shown to be

$$\frac{\dfrac{r^{-n}\left(\dfrac{n}{2}\right)!}{\pi^{\frac{n}{2}}}\displaystyle\int_0^{r_1} \rho^{n-1}\,d\rho}{(2\pi)^{-\frac{n}{2}} r^{-n} n^{\frac{n}{2}} \displaystyle\int_0^{r_1} \rho^{n-1} e^{-\frac{\rho^2 n}{2r^2}}\,d\rho}$$

$$= \frac{\left(\dfrac{n}{2}\right)! \displaystyle\int_0^{\frac{r_1\sqrt{n}}{r}} u^{n-1} du}{\left(\dfrac{n}{2}\right)^{\frac{n}{2}} \displaystyle\int_0^{\frac{r_1\sqrt{n}}{r}} u^{n-1} e^{-\frac{n^2}{2}}\,du} \tag{59}$$

Since $e^{-\frac{u^2}{2}}$ is a decreasing function, we have

$$\frac{\left(\dfrac{n}{2}\right)! \displaystyle\int_0^{\frac{r_1\sqrt{n}}{r}} u^{n-1}\,du}{\left(\dfrac{n}{2}\right)^{\frac{n}{2}} \displaystyle\int_0^{\frac{r_1\sqrt{n}}{r}} u^{n-1} e^{-\frac{u^2}{2}}\,du}$$

$$> \frac{\left(\dfrac{n}{2}\right)! \displaystyle\int_0^{\sqrt{n}} u^{n-1} du}{\left(\dfrac{n}{2}\right)^{\frac{n}{2}} \displaystyle\int_0^{\sqrt{n}} u^{n-1} e^{-\frac{u^2}{2}}\,du}$$

$$> \frac{\left(\dfrac{n}{2}\right)!\, n^{\frac{n}{2}-1}}{\left(\dfrac{n}{2}\right)^{\frac{n}{2}} \displaystyle\int_0^{\infty} u^{n-1} e^{-\frac{u^2}{2}}\,du}$$

$$= \frac{\left(\dfrac{n}{2}\right)!\, n^{\frac{n}{2}-1}}{\left(\dfrac{n}{2}\right)^{\frac{n}{2}} \left(\dfrac{n}{2}-1\right)!\, 2^{\frac{n}{2}-1}} = 1. \tag{60}$$

Hence if we consider a region of the sphere $\Sigma x_k^2 = r^2$ bounded by radii and the sphere $\Sigma x_k^2 = r_1^2$, its measure in terms of the volume of the sphere $\Sigma x_k^2 = r^2$ is greater than its measure as given by an expression such as (58). In this statement we may replace the sphere $\Sigma x_k^2 = r_1^2$ by any region such that if it contains a point $(x_1, \ldots, x_n)$ it contains $(\theta x_1, \ldots, \theta x_n)$ $(0 \leq \theta \leq 1)$, since such a region may be approximated to within any desired degree of accuracy by a sum of sectors. It hence results that the measure of all the points in $\Sigma x_k^2 = r^2$ but *without* such a region is less than the measure after the fashion of (58) of all the points without such a region. We may conclude from this fact and (57) that in the space $\Sigma x_k^2 = r^2$, the measure of all of the points $(x_1, \ldots, x_n)$ such that for some k and $l > k$,

$$\left| \sum_{k+1}^{l} x_i \right| > a\, r \left(\frac{l-k}{n} \right)^{\frac{1}{2}-\epsilon}. \tag{61}$$

vanishes uniformly in n as a increases.

§8. **The average of a Bounded, Uniformly Continuous Functional.** Let $F|f|$ be a functional, defined in the first instance for all step-functions constant over each of the intervals $\left(\frac{k}{n}, \frac{k+1}{n}\right)$ for some n, and having the following properties:

(1) $F|f| < A$ for all f,

(2) There is a function $\phi(x)$ such that

$$\lim_{x \to 0} \phi(x) = 0,$$

and

$$\left| F|f| - F|f+g| \right| < \phi(\max |g|).$$

We shall speak of the function $F(x_1, \ldots, x_n)$, which is $F|f_n|$, where

$$f_n(t) = \sum_{l}^{k} x_i \quad \text{for} \quad \frac{k-1}{n} < t \leq \frac{k}{n},$$

as the *nth section* of $F|f|$. I say that for any functional $F|f|$

satisfying (1) and (2), the average of the nth section over the sphere $\sum_1^n x_k^2 = r^2$ approaches a limit as n increases indefinitely, which limit we shall term the *average of* $F\,|f|$.

To begin with, let us have given a positive number η; we shall construct a value of n such that the difference between the averages of the nth and the $(n+p)$th sections of F is always less than 2η. First choose a so great that the measure of the points in $\Sigma x_k^2 \leq r^2$ satisfying (61) is for all n less than $\eta/2A$. Next find a number ξ so small that

$$\phi(ar\xi^{1/2-\epsilon}) < \eta/4. \tag{62}$$

Divide the interval (0, 1) into portions of width no greater than ξ, and let $t_1, \ldots, t_m$ be the boundaries of these intervals. Let

$$\psi(t) = f(t_n) \qquad (t_n \leq t \leq t_{n+1}),$$

and let

$$G|f| = F|\psi|.$$

Form the average of G as in §6, and let n be so large that this average differs from the average of

$$G(x_1+x_2+ \ldots +x_{T_1}, x_1+x_2+ \ldots +x_{T_2}, \ldots, x_1+x_2+ \ldots +x_{T_m})^{15}$$

by less than $\eta/4$. Then it is immediately obvious that the difference between the averages of the nth and the $(n+p)$th sections of F is less than 2ϵ. In other words, the average of a functional satisfying (1) and (2) necessarily exists.

An example of such a functional is

$$\frac{1}{1+\int_0^1 [f(x)]^2 dx}.$$

Another method of defining the average of a bounded, uniformly continuous functional F is as

$$\lim_{n\to\infty} (2\pi)^{-\frac{n}{2}} r^{-n} n^{\frac{n}{2}} \int_{-\infty}^{\infty} dx_1 \ldots \int_{-\infty}^{\infty} dx_n\, e^{-\frac{(x_1^2+\ldots+x_n^2)n}{2r^2}} F(x_1, \ldots, x_n). \tag{63}$$

[15] Cf. (43).

There is no particular difficulty in showing the equivalence of the two definitions with the aid of (57). A fuller discussion of this definition of the average of a functional is to be found in the papers of the author which have been already cited. It is easy to demonstrate, as the author has done, that the division into parts that are exactly equal plays no essential rôle in definition (63), and can be replaced by a much more general type of division.

For the average of $F\,|f|$ as here defined, we shall write

$$A_r\{F\}. \tag{64}$$

§9. **The Average of an Analytic Functional.** Let $F|f|$ be a functional such that:

(1) Given any positive number B, there is an increasing function $A(B)$, such that if

$$\max |f(t)| \leq B,$$

then

$$|F|f|| < A(B).$$

(2) Given any positive number B, there is a function $\phi(x)$ such that

$$\lim_{x \to 0} \phi(x) = 0,$$

while if

$$\max |f(t)| \leq B, \quad \max |g(t)| \leq B$$

then

$$|F|f| - F|g|| < \phi(\max |f-g|).$$

Let us write

$$\left. \begin{array}{ll} F_H|f| = F|f| & \text{for } \max |f(t)| \leq H \\ F_H|f| = F\left|\dfrac{f(H)}{\max |f(t)|}\right| & \text{for } \max |f(t)| > H \end{array} \right\}. \tag{65}$$

$F_H|f|$ is clearly bounded and uniformly continuous, and as such comes under the class of those functionals which have averages in the sense of the last paragraph.

Let us define for any functional $G|f|$ the function $G(x_1, \ldots, x_n)$ as in the last paragraph, and let us write $A_r^n\{F\}$ for the average of $F(x_1, \ldots, x_k)$ over $\Sigma x_k^2 = r^2$. Let H and $K > H$ be any two positive numbers. Clearly by (42) and the argument which follows (60), the set of functions for which max $|f(t)| \geq H$ can be enclosed over $\Sigma x_k^2 = r^2$ in a region of measure

$$\frac{2}{r}\sqrt{\frac{2}{\pi}}\int_H^\infty e^{-\frac{x^2}{2r^2}}\,dx.$$

Hence

$$|A_r^n\{F_H\} - A_r^n\{F_K\}| < \frac{4}{r}\sqrt{\frac{2}{\pi}}\int_H^\infty e^{-\frac{x^2}{2r^2}}\,dx\; A(K). \qquad (66)$$

In particular, if $\theta \leq 1$

$$|A_r^n\{F_H\} - A_r^n\{F_{H+\theta}\}| < \frac{4}{r}\sqrt{\frac{2}{\pi}}\int_H^\infty e^{-\frac{x^2}{2r^2}}\,dx\; A(H+1).$$

Hence if $K - H < m$, m being an integer, we have

$$|A_r^n\{F_H\} - A_r^n\{F_K\}| < \frac{4}{r}\sqrt{\frac{2}{\pi}}\sum_1^m \int_{H+i}^\infty e^{-\frac{x^2}{2r^2}}\,dx\; A(H+i+1). \qquad (67)$$

It follows that if

$$\lim_{p\to\infty} \frac{A(p+1)\int_p^\infty e^{-\frac{x^2}{2r^2}}\,dx}{A(p)\int_{p-1}^\infty e^{-\frac{x^2}{2r^2}}\,dx} < 1, \qquad (68)$$

then

$$\lim_{H\to\infty} |A_r^n\{F_H\} - A_r^n\{F_K\}| = 0 \qquad (69)$$

uniformly in n and K. This will be the case, for example, if $A(p)$ is a polynomial in p, or is of the form a^p. Under these circumstances

$$\lim_{n\to\infty} A_r^n\{F\} = \lim_{n\to\infty}\lim_{H\to\infty} A_r^n\{F_H\}$$

$$= \lim_{H\to\infty}\lim_{n\to\infty} A_r^n\{F_H\} = \lim_{H\to\infty} A_r\{F_H\} \qquad (70)$$

exists. We shall call this quantity $A_r\{F\}$. There is no difficulty in showing that formula (63) holds in this case also.

Now let us have an $F|f|$ of the form

$$F|f| = a_0 + \int_0^1 K_1(t)f(t)dt + \ldots + \int_0^1 dt_1 \ldots \int_0^1 dt_n K_n(t_1, \ldots, t_n) f(t_1) \ldots f(t_n) + \ldots, \tag{71}$$

given that

$$A(u) = a_0 + u \int_0^1 |K_1(t)|dt + \ldots + u^n \int_0^1 dt_1 \ldots \int_0^1 dt_n |K_n(t_1, \ldots, t_n)| + \ldots \tag{72}$$

exists for all u and satisfies (68). Then if

$$\max f|(t)| \leq B,$$

we have as an obvious result that (1) at the beginning of this section is satisfied. Moreover

$$\begin{aligned} |F|f| - F|g|\,| &\leq \Big| \sum_1^\infty \int_0^1 dt_1 \ldots \\ &\int_0^1 dt_n K_n(t_1, \ldots, t_n) [f(t_1) \ldots f(t_n) - g(t_1) \ldots g(t_n)] \\ &\leq \sum_1^\infty \int_0^1 dt_1 \ldots \int_0^1 dt_n |K_n(t_1, \ldots, t_n)| \\ &\times \{ [\max|G(x)| + \max|f(x) - g(x)|]^n - [\max|g(x)|]^n \}. \end{aligned} \tag{73}$$

If $\max|f(x)| < B$, and $\max |f(x)-g(x)| < \epsilon$,

$$|F|f|-F|g|| \leq \sum_{1}^{\infty} [(B+\epsilon)^n - B^n] \int_0^1 dt_1, \ldots \int_0^1 dt_n |K_n(t_1, \ldots, t_n)|$$

$$= \sum_{1}^{\infty} \int_B^{B+\epsilon} ny^{n-1} \int_0^1 dt_1 \ldots \int_0^1 dt_n |K_n(t_1, \ldots, t_n)|$$

$$\leq \sum_{1}^{\infty} \epsilon n(B+\epsilon)^{n-1} \int_0^1 dt_1 \ldots \int_0^1 dt_n | K_n(t_1, \ldots, t_n) |. \tag{74}$$

Series (74) converges for all ϵ, since series (72) has a convergent derivative series. Hence condition (2) may be proved to be satisfied. In other words, $F|f|$ has an average in the sense of this paper.

This result may be much generalized without any great difficulty. It may be extended to functionals containing Stieltjes integrals such as

$$a_1 + \int_0^1 f(t) dQ_1(t) + \ . \ . \ . \ .$$

$$+ \int_0^1 \ldots \ldots \int_0^1 f(t_1) \ . \ . \ . \ f(t_n)\, d^n Q_n(t_1, \ . \ . \ . \ . \ , t_n) + \ . \ . \ . \ . \ , \tag{75}$$

provided

$$A(u) = a_0 + \sum_{1}^{\infty} u^n \int_0^1 \ldots \int_0^1 |d^n Q(t_1, \ . \ . \ . \ t_n)| \tag{76}$$ [16]

exists for all u and satisfies (68). This class of functionals includes such expressions as

$$\int_0^1 \int_0^1 \int_0^1 [f(t_1)]^m [f(t_2)]^n [f(t_3)]^p K(t_1, t_2, t_3)\, dt_1\, dt_2\, dt_3.$$

16 Cf. Daniell, *Functions of Limited Variation in an Infinite Number of Dimension*, Annals of Mathematics, Series 2, Vol. 21, pp. 30–38.

On the other hand, $A(u)$ in (72) may be replaced by

$$a_0+\sqrt{u}\sqrt{\int_0^1 [K_1(t)]^2\,dt} + \ldots$$

$$+\ u^{\frac{n}{2}}\sqrt{\int_0^1 \ldots \int_0^1 dt_1 \ldots dt_n\,[K_n(t_1, \ldots, t_n)]^2}, \tag{77}$$

the K's being all summable and of summable square, and by the use of the Schwarz inequality, (1) and (2) may be deduced. Again, we may under the proper conditions concerning ϕ show that

$$\phi(F_1|f|, \ldots, F_n|f|) \tag{78}$$

has an average in the sense of this paper, if $F_1, \ldots, F_n$ are such functionals as we have already described.

I here wish to discuss only such a functional as

$$F|f| = \int_0^1 \ldots \int_0^1 [f(t_1)]^{\mu_1} \ldots [f(t_\nu)]^{\mu_\nu} K(t_1, \ldots t_\nu)\,dt \ldots dt_\nu.$$

The average of this functional will be the limit as n increases indefinitely of

$$\sum_{k_1=1}^{n} \cdots \sum_{k_\nu=1}^{n} (2\pi)^{-\frac{n}{2}} r^{-n} n^{-\frac{n}{2}} \int_{-\infty}^{\infty} dx_1 \ldots \int_{-\infty}^{\infty} dx_n\, e^{-\frac{(x_1^2+\cdots+x_n^2)n}{2r^2}}$$

$$(x_1+\ldots+x_{k_1})^{\mu_1} \ldots (x_1+\ldots+x_{k_\nu})^{\mu_\nu}$$

$$\int_{\frac{k_1-1}{n}}^{\frac{k_1}{n}} dt_1 \ldots \int_{\frac{k_\nu-1}{n}}^{\frac{k_\nu}{n}} dt_\nu K(t_1, \ldots, t_\nu)$$

$$= \sum_{k_1=1}^{n} \cdots \sum_{k_\nu=1}^{n} \frac{(2\pi)^{-\frac{\nu}{2}} r^{-\nu} n^{\frac{\nu}{2}}}{[k_1(k_2-k_1)\ldots(k_\nu-k_{\nu-1})]^{\frac{1}{2}}}$$

$$\int_{-\infty}^{\infty} d\xi_1 \ldots \int_{-\infty}^{\infty} d\xi_\nu\, e^{-\frac{\xi_1^2 n}{2r^2 k_1} - \cdots \frac{\xi_\nu^2 n}{2r^2(k_\nu-k_{\nu-1})^{\mu_1}}}$$

$$\xi_1(\xi_1+\xi_2)^{\mu_2} \ldots (\xi_1+\ldots+\xi_\nu)^{\mu_2}$$

$$\int_{\frac{k_1-1}{n}}^{\frac{k_1}{n}} dt_1 \ldots \int_{\frac{k_\nu-1}{n}}^{\frac{k_\nu}{n}} dt_n K(t_1, \ldots, t_\nu), \tag{80}$$

because of (63). Expression (80) clearly approaches the limit

$$\int_0^1 dt_1 \ldots \int_0^1 dt_n K(t_1, \ldots, t_n) \frac{(2\pi)^{-\frac{\nu}{2}} r^{-\nu}}{[t_1(t_2-t_1) \ldots (t_\nu - t_{\nu-1})]^{\frac{1}{2}}}$$

$$\int_{-\infty}^{\infty} d\xi_1 \ldots \int_{-\infty}^{\infty} d\xi_\nu \, exp \left(\frac{-\xi_1^2}{2r^2 t_1} - \ldots - \frac{\xi_\nu^2}{2r^2(t_\nu - t_{\nu-1})} \right)$$

$$\xi_1^{\mu_1}(\xi_1+\xi_2)^{\mu_2} \ldots (\xi_1 + \ldots + \xi_\nu)^{\mu_\nu}, \qquad (81)$$

provided that K is of limited total variation. This agrees with a definition already obtained by the author.[17]

§10. **The Average of a Functional as a Daniell Integral.** Daniell[18] has discussed a generalized definition of an integral in the following manner: he starts with a set of functions $f(p)$ of general elements p. He assumes a class T_0 of such functions which is closed with respect to the operations, multiplication by a constant, addition, and taking the modulus. He also assumes that to each f of class T_0 there corresponds a number K, independent of p, such that

$$|f(p)| \leq K,$$

and that to each f there corresponds a finite "integral" $U(f)$ having the properties

$$(C) \qquad Uc(f) = c\,U(f),$$

$$(A) \qquad U(f_1+f_2) = U(f_1) + U(f_2),$$

$$(L) \quad \text{If} \quad f_1 \geq f_2 \geq \ldots \geq 0 = \lim f_n, \text{ then } \lim U(f_n) = 0,$$

$$(P) \qquad U(f) \geq 0 \text{ if } f \geq 0.$$

There is no difficulty in showing that if T_0 be taken as the set of all functionals F such as those defined by (1) and (2) of §8, and the operator A_r is taken as U, all these conditions are ful-

17 *The Average of an Analytic Functional*, p. 256.
18 Daniell, p. 280.

filled save possibly (L). I here wish to discuss the question as to whether (L) is fulfilled.

Let it be noted that (L) involves the knowledge of those entities which form the arguments to the members of T_0. Up to this point we have regarded the arguments of our functionals F as step-functions. However, if $F|f|$ is a bounded, uniformly continuous functional of all step-functions constant over every interval $\left(\frac{k-1}{n}, \frac{k}{n}\right)$, it may readily be shown that if f is the uniform limit of a sequence of such step-functions f_n, then $\lim F|f_n|$ exists, is independent of the particular sequence f_n chosen, and is bounded and uniformly continuous. We shall call this limit $F|f|$. This extension of the arguments of F does not vitiate the validity of any of the Daniell conditions. In fact, no condition save possibly (L) is vitiated if we then restrict the arguments of our functionals of T_0 to continuous functions $f(t)$ such that $f(0)=0$.

We wish, then, to show that if $F_1 \geq F_2 \geq \ldots \geq 0 = \lim F_n$, for every f that is continuous, then

$$\lim A_r\{F_n\} = 0.$$

Let us notice that it follows from (50), (55) and (63) that

$$A_r\{F_n\} \leq [\max F_1]\frac{2}{r^2\pi}\sum_1^\infty n^2 e^{-\frac{n^\epsilon ar}{4}} + \max_{S_a} F_n, \tag{82}$$

where S_a is the set of all functions f for which for every t_1 and t_2 between 0 and 1,

$$|f(t_2) - f(t_1)| \leq ar\,(t_2 - t_1)^{1/2-\epsilon}.$$

By (57), the first term in (82) can be made as small as we please by taking a large enough. Accordingly, we can prove (L) if we can show that for every a, the set of functions Σ_n consisting in all the functions f in S_a for which $F_n|f| \geq \eta$ can be made to become null by making n large enough, whatever η may be. We shall prove this by a reductio ad absurdum.

Σ_n is in the language of Fréchet an extremal set, in that it is closed, equicontinuous, and uniformly bounded.[19] Hence either all the Σ_n's from a certain stage are null, or there is an element common to every Σ_n.[20] Let this element be $f(t)$. Then for all n, we have

$$F_n|f| \geq \eta \tag{83}$$

This however contradicts our hypothesis.

Our operation of taking the average is hence a Daniell integration, and as such capable of the extensions which Daniell develops. Daniell first proves that if $f_1 \leq f_2 \leq \ldots$ is a sequence from T_0, then the sequence $U(f_n)$ is an increasing sequence, and hence either becomes positively infinite or has a limit. If then $f = \lim f_n$ exists, Daniell defines $U(f)$ as $\lim U(f_n)$, and says that f belongs to T_1. Our T_1 will contain all the functions discussed in §9, and it admits of an easy proof that the definition of $A_r\{F\}$ in §9 accords with the definition arising from the Daniel extension of A_r whenever the former definition is applicable. Daniell then defines for any function f the upper semi-integral $\dot{U}(f)$ as the lower bound of $U(g)$ for g in T_1 and $g > f$. He defines $\underset{\cdot}{U}(f)$ as $-\dot{U}(-f)$, and $U(f)$ as $\dot{U}(f)$ if $\dot{U}(f) = \underset{\cdot}{U}(f) =$ finite, f being then called summable. All these extensions are applicable to our average operator A_r, as is also Daniell's theorem to the effect that if $f_1, \ldots, f_n, \ldots$ is a sequence of summable functions with limit f, and if a summable function ϕ exists such that $|f_n| \leq \phi$ for all n, f is summable, $\lim U(f_n)$ exists and $= U(f)$.

It may be shown from theorems of Daniell that if the measure of a set of functions be held to be the average of a functional 1 over the set and zero elsewhere, and the outer measure the upper semi-average of a functional 1 over the set and zero elsewhere, then the definition in §6 will coincide with this whenever it is applicable. §4 may be interpreted as saying that the measure of the set of functions for which the non-differentiability coefficient differs from r by more than ϵ is zero, and §7 as saying that the outer measure of the functions satisfying (51) is less than (55).

19 *Sur quelques points du calcul fonctionnel*, Rend. Cir. Math. di Palermo Vol. 22, pp. 7, 37.
20 Ibid, p. 7.

§11. **Independent Linear Functionals.** Let us consider a functional

$$F|f| = \phi\left(\int_0^1 K(x)f(x)dx, \int_0^1 Q(x)f(x)dx\right),$$

K and Q being summable of summable square, and ϕ being bounded and uniformly continuous for all arguments from $-\infty$ to ∞. We then have

$$F(x_1, \ldots, x_n) = \phi\left(\sum_1^n x_k \int_{\frac{k-1}{n}}^1 K(x)dx, \sum_1^n x_k \int_{\frac{k-1}{n}}^1 Q(x)dx\right).$$

$$= \phi\left\{\sum_1^n x_k \int_{\frac{k-1}{n}}^1 K(x)dx, \frac{\sum_1^n \int_{\frac{k-1}{n}}^1 K(x)dx \int_{\frac{k-1}{n}}^1 Q(x)dx}{\sum_1^n \left[\int_{\frac{k-1}{n}}^1 K(x)dx\right]^2} \times \sum_1^n x_k \int_{\frac{k-1}{n}}^1 K(x)dx \right.$$

$$\left. + \sum_1^n x_k \left[\int_{\frac{k-1}{n}}^1 Q(x)dx - \frac{\sum_1^n \int_{\frac{j-1}{n}}^1 K(x)dx \int_{\frac{j-1}{n}}^1 Q(x)dx}{\sum_1^n \left[\int_{\frac{j-1}{n}}^1 K(x)dx\right]^2} \times \int_{\frac{k-1}{n}}^1 K(x)dx\right]\right\}$$

$$= \phi\left\{\xi\sqrt{\sum_1^n \left[\int_{\frac{k-1}{n}}^1 K(x)dx\right]^2}, \xi \frac{\sum_1^n \int_{\frac{k-1}{n}}^1 K(x)dx \int_{\frac{k-1}{n}}^1 Q(x)dx}{\sqrt{\sum_1^n \left[\int_{\frac{k-1}{n}}^1 K(x)dx\right]^2}}\right.$$

$$+\eta\sqrt{\sum_1^n\Big[\int_{\frac{k-1}{n}}^1 Q(x)dx\Big]^2-\frac{\Big[\sum_1^n\int_{\frac{k-1}{n}}^1 K(x)dx\int_{\frac{k-1}{n}}^1 Q(x)dx\Big]^2}{\sum_1^n\Big[\int_{\frac{k-1}{n}}^1 K(x)dx\Big]^2}}\Bigg\},$$

ξ and η being two orthogonal unit functions. As η increases, the arguments of ϕ approach uniformly

$$\xi\sqrt{\int_0^1\Big[\int_x^1 K(x)dx\Big]^2dx}$$

and

$$\xi\frac{\int_0^1\Big[\int_x^1 K(x)dx\int_x^1 Q(x)dx\Big]dx}{\sqrt{\int_0^1\Big[\int_x^1 K(x)dx\Big]^2dx}}$$

$$+\eta\sqrt{\int_0^1\Big[\int_x^1 Q(x)dx\Big]^2dx-\frac{\Big[\int_0^1\Big[\int_x^1 K(x)dx\int_x^1 Q(x)dx\Big]dx\Big]^2}{\int_0^1\Big[\int_x^1 K(x)dx\Big]^2dx}}.$$

If in particular $\int_x^1 K(x)dx$ and $\int_x^1 Q(x)dx$ are normal and orthogonal,

$$A_r\{F\}=\lim_{n\to\infty}\frac{\int_{-\frac{\pi}{2}}^{\frac{\pi}{2}}\int_{-\frac{\pi}{2}}^{\frac{\pi}{2}}\cos^{n-1}\theta_1\cos^{n-2}\theta_2\,\phi(\sin\theta_1,\sin\theta_2)\,d\theta_1\,d\theta_2}{\int_{-\frac{\pi}{2}}^{\frac{\pi}{2}}\int_{-\frac{\pi}{2}}^{\frac{\pi}{2}}\cos^{n-1}\theta_1\cos^{n-2}\theta_2\,d\theta_1\,d\theta_2}$$

$$=\frac{1}{2\pi r^2}\int_{-\infty}^{\infty}dx_1\int_{-\infty}^{\infty}dx_2\,e^{-\left(\frac{x_1^2+x_2^2}{2r}\right)}\phi(x_1,x_2). \tag{84}$$

That is, the average of F with respect to f can be obtained by first forming the average with respect to f of

$$\phi\left(\int_0^1 K(x)f(x)dx,\ \int_0^1 Q(x)g(x)dx\right),$$

then forming the average of this average with respect to g, and finally putting $f=g$. A similar theorem can be shown by the same means to hold of

$$\phi\left(\int_0^1 K_1(x)f(x)dx,\ .\ .\ .\ ,\ \int_0^1 K_n(x)f(x)dx\right),$$

in case the set of functions $\left[\int_x^1 K_k(x)dx\right]$ is normal and orthogonal.

If now ϕ is not a bounded function, but is merely uniformly continuous over any finite ranges of its arguments, and

$$\frac{1}{(2\pi r^2)^{\frac{n}{2}}}\int_{\infty}^{\infty} dx_1\ .\ .\ .\ \int_{-\infty}^{\infty} dx_n e^{-\frac{\sum_1^n x_k^2}{2r^2}}\phi(x_1,\ .\ .\ .\ ,\ x_n) \tag{85}$$

exists, it may be proved by a simple limit argument that (85) represents the average of

$$\phi\left(\int_0^1 K_1(x)f(x)dx,\ .\ .\ .\ ,\ \int_0^1 K_n(x)f(x)dx\right).$$

§12. **Fourier Coefficients and the Average of a Functional.** The functions

$$-\sqrt{2}\sin n\pi x = \int_x^1 n\pi\sqrt{2}\cos n\pi x$$

form a normal and orthogonal set. Accordingly if we have a functional

$$\phi(a_1,\ .\ .\ .\ ,\ a_n)$$

of the function

$$f(x) = a_0 + a_1\sqrt{2}\cos\sqrt{\pi x} +\ .\ .\ .\ + a_n\sqrt{2}\cos n\pi x +,\ .\ .\ .$$

its average will be

$$\frac{1}{(2\pi r^2)^{\frac{n}{2}}}\int_{-\infty}^{\infty} dx_1 \; . \; . \; . \int_{-\infty}^{\infty} dx_n \, e^{-\frac{\sum_1^n x_k^2}{2r^2}} \phi\left(\frac{x_1}{\pi}, \; . \; . \; . \; , \frac{x_n}{\pi n}\right) \quad (86)$$

provided this exists and ϕ is continuous, or even provided ϕ is a sum of step-functions. In particular if $F\,|f(x)|=1$ if $a_m^2 + \; . \; . \; . + a_n^2 + \; . \; . \; . \geq a^2$ and zero otherwise, its average will be less than the average of a functional which is 1 if for some k between m and n included,

$$a_k^2 > \frac{a^2 k^{-\frac{3}{2}}}{\sum_1^{\infty} j^{-\frac{3}{2}}}$$

and zero otherwise. The upper average of this latter functional is by (86) not greater than

$$\sum_m^{\infty} \frac{2}{(2\pi r^2)^{\frac{1}{2}}} \int_{\frac{a k^{\frac{1}{4}} \pi}{\left(\sum_1^{\infty} j^{-\frac{3}{2}}\right)^{\frac{1}{2}}}}^{\infty} e^{-\frac{x^2}{2}} dx$$

$$< \frac{2}{(2\pi r^2)^{\frac{1}{2}}} \sum_m^{\infty} L^{-k^{\frac{1}{4}}}, \quad (87)$$

for m sufficiently large, where

$$L = e^{\frac{a\pi}{\left(\sum_1^{\infty} j^{-\frac{3}{2}}\right)^{\frac{1}{2}}}} > 1.$$

Series (87) converges. Hence as m increases, the measure of the set of functions for which $a_m^2 + \; . \; . \; . + a_n^2 + \; . \; . \; . > a^2$ approaches zero.

In this demonstration, we have made use of several theorems of Daniell which it did not seem worth while to enumerate in detail. They may all be found in his discussion of measure and integration.[21]

[21] Daniell, loc. cit.

Let us now consider a bounded functional $F|f|$ that is uniformly continuous in the more restrictive sense that for any positive η, there is a positive θ such that

$$|F|f|-F|g||<\eta$$

whenever

$$\int_0^1 [f(t)-g(t)]^2dt<\theta.$$

Let $F|f|$ be invariant, moreover, when a constant is added to f. It will then be possible to write $F|f|$ in the form

$$F|f|=\phi(a_1, \ldots, a_n, \ldots),$$

where

$$|\phi(a_1, \ldots, a_n, 0, 0 \ldots)-\phi(a_1, \ldots, a_n, a_{n+1}, \ldots)|<\eta$$

whenever

$$\sum_{n+1}^{\infty} a_k^2<\theta.$$

Hence

$$\left| A_r\{F\} - \frac{1}{(2\pi r^2)^{\frac{n}{2}}}\int_{-\infty}^{\infty} dx_1, \ldots \int_{-\infty}^{\infty} dx_n e^{-\frac{\sum_1^n x_k^2}{2r^2}} \right.$$

$$\left. \times\phi\left(\frac{x_1}{\pi}, \ldots, \frac{x_n}{\pi_n}, 0, 0, \ldots\right)\right| < \eta+2 \max |F|M(\theta). \quad (87)$$

where $M(\theta)$ is the outer measure of all the functions for which $\sum_{n+1}^{\infty} a_k^2\geq\theta$. N can, as we have seen, be made arbitrarily small by making n sufficient y large. It can hence be shown that

$$A_r\{F\} = \lim_{n\to\infty} \frac{1}{(2\pi r^2)^{\frac{n}{2}}}\int_{-\infty}^{\infty} dx_1 \ldots \int_{-\infty}^{\infty} dx_n e^{-\frac{\sum_1^n x_k^2}{2r^2}}$$

$$\times\phi\left(\frac{x_1}{\pi}, \ldots, \frac{x_n}{\pi n}, 0, 0, \ldots\right). \quad (88)$$

This theorem may be established for the more general case of f functional F which is uniformly continuous for all functions a for which

$$\int_0^1 [f(x)]^2\, dx < \theta,$$

whatever θ may be, and for which there is an increasing $\psi(u)$ such that

$$F|f| \leq \psi\left[\int_0^1 [f(x)]^2 dx\right],$$

while

$$\sum_1^\infty \psi(n+1)[M(n+1)-M(n)]$$

converges.

Now let us consider a functional of the form

$$F|f| = H_{i_1}\left(\frac{x_1\pi}{r\sqrt{2}}\right) \;.\;.\;.\; H_{i_n}\left(\frac{nx_n\pi}{r\sqrt{2}}\right), \tag{89}$$

where the H's are any set of Hermite polynomials corresponding to normalized Hermite functions. If $G|f|$ is another such function, it is easy to show that

$$A_r\{F|f|G|f|\} = 0, \tag{90}$$

and to compute

$$A_r\{F|f|\}^2. \tag{91}$$

Now let us arrange all these functionals in a progression $\{F_n\}$ It follows from theorems analogous to those familiar in the ordinary theory of orthogonal functions that if $G|f|$ is a continuous function ϕ of $a_1, \ldots, a_n$ for which expression (86) exists, then

$$\sum_1^\infty F_k|f| \frac{A_r\{F_k|f|G|f|\}}{\sqrt{\{A_r F_k|f|\}^2}} \tag{92}$$

converges in the mean to $G|f|$ in the sense that if $G_n|f|$ is its nth partial sum,

$$\lim_{n\to\infty} A_r\{G|f| - G_n|f|\}^2 = 0. \tag{93}$$

With the aid of generalizations of familiar theorems concerning orthogonal functions, it may be shown that this result remains valid if $\{G|f|\}^2$ fulfils the conditions laid down for F in (88), and G does likewise. The functionals (89) are then in a certain sense a complete set of normal and orthogonal functions.

One final remark. By methods which exactly duplicate §4, it can be shown that the set of functions f for which it is not true that

$$\left| \frac{r^2}{\pi^2} - \lim_{n \to \infty} \sum_1^n \frac{k^2 a_k^2}{n} \right| < \epsilon \tag{94}$$

is of zero measure, whatever ϵ. This suggests interesting questions relating to the connection between the coefficient of non-differentiability of a function and $\lim_{n \to \infty} \sum_1^n \frac{k^2 a_k^2}{n}$.

THE AVERAGE VALUE OF A FUNCTIONAL*

By NORBERT WIENER.

[Received February 27th, 1922.—Read March 9th, 1922.]

1. The notion of a mean or average of a quantity is most familiar to us in those cases where the quantity averaged is regarded as a function of an independent variable, the range of which consists either of a discrete set of values or of a manifold of a finite number of dimensions. Such general theories of the average or the integral as have been formulated have, however, been founded on a much wider basis, and have concerned themselves with an independent variable subject to very few restrictions. Papers by Fréchet,† Daniell,‡ and the author§ may be mentioned in this connection.

Among the simplest ranges more complex than space of a finite number of dimensions are space of a denumerable infinity of dimension and function-space. Daniell has discussed averages in space of the former type, while there is a very elegant theory of the average of a function of a line due to Gâteaux¶ and Lévy.** E. H. Moore†† has also done some work which should be mentioned in this connection. These theories, however, while they give very natural definitions of averages in function-space, by no means exhaust the field. Other definitions are possible, and may in some cases be useful. In three previous papers‡‡

* Presented to the American Mathematical Society, December 1920.

† M. Fréchet, *Bull. Soc. Math. de France*, Vol. 43 (1915), p. 248.

‡ P. J. Daniell, "A General Form of Integral", *Annals of Mathematics*, Vol. 19 (1918), p. 279.

§ N. Wiener, *Annals of Mathematics*, Vol. 21 (1920), p. 66.

|| P. J. Daniell, *Annals of Mathematics*, Vol. 21 (1919), p. 281.

¶ R. Gâteaux, *Bull. Soc. Math. de France*, Vol. 47 (1919), p. 47.

** P. Lévy, *Leçons d'analyse fonctionnelle*, Paris, 1922.

†† Cf. T. H. Hildebrandt, *Bull. American Math. Soc.*, Vol. 24 (1918), p. 201.

‡‡ N. Wiener, "The Average of an Analytic Functional", *Proc. Nat. Acad. Sci., Washington*, Vol. 7, pp. 253–260; N. Wiener, "The Average of an Analytic Functional and the Brownian Movement", *ibid.*, pp. 294–298; N. Wiener, "Differential-Space", *Jour. Math. and Phys. Mass. Inst. Technology*, Vol. 2, pp. 131–174.

the author has developed the properties of a type of function-space essentially different from the familiar type, and has given a partial discussion of the theory of integration in this space. It is the purpose of the present paper to give a self-contained discussion of this type of integration which shall be in many respects more complete than those which he has yet developed. In order that the content of this paper be intelligible by itself, a certain amount of repetition from the other papers has been unavoidable, as is also a certain amount of preliminary discussion which treats of well-known points in classical probability theory.

2. Let us consider a particle free to wander along the X-axis. Let the probability that it wander a given distance in a given time be independent

(1) of the position from which it starts to wander,

(2) of the time when it starts to wander,

(3) of the direction in which it wanders.

It may then be shown* that the probability that after a time t it has wandered from the origin to a position lying between $x = x_1$ and $x = x_2$ is

$$(1) \qquad \frac{1}{\sqrt{\{\pi c t\}}} \int_{x_1}^{x_2} e^{-(x^2/ct)}\, dx,$$

where c is a certain constant which we may reduce to 1 by a proper choice of units. This choice we shall make. The exponential form of the integral needs no comment, while the mode in which t enters results from the fact that

$$(2) \qquad \frac{1}{\sqrt{\{\pi(t_1+t_2)\}}} \int_{x_1}^{x_2} e^{-\{x^2/(t_1+t_2)\}}\, dx$$

$$= \frac{1}{\sqrt{\{\pi t_1\}}} \int_{-\infty}^{\infty} \left[\frac{1}{\sqrt{\{\pi t_2\}}} \int_{x_1}^{x_2} e^{-\{(y-x)^2/t_2\}}\, dy\right] e^{-(x^2/t_1)}\, dx.$$

This identity is tantamount to the statement that the probability that after a period t_1+t_2 the particle has wandered a distance between x_1 and x_2 is the total compound probability that it be anywhere at time t_1 and that it then wander to a position between x_1 and x_2 in a subsequent period of length t_2. This fact lies at the bottom of all that follows.

* A. Einstein, *Annalen der Physik*, Vol. 17, p. 905.

3. We shall say that the particle of § 2 has its displacements normally distributed. Let us represent the history of such a particle by an equation $x = \Phi(t)$, Φ being a continuous function. If we suppose the particle to start from rest, we shall have $\Phi(0) = 0$. We shall consider in what follows particle-histories or time-paths in which the range of t is from 0 to 1.

There are certain assemblages of time-paths to which we can immediately assign a measure, a probability. These assemblages are obtained by restricting the position of the particle at certain specified times, finite in number, to certain specified finite or infinite intervals. An example is the set of all time-paths $x = \Phi(t)$, such that

$$x_{11} \leqslant \Phi(t_1) \leqslant x_{12};$$

$$x_{21} \leqslant \Phi(t_2) \leqslant x_{22};$$

$$\cdots \qquad \cdots \qquad \cdots$$

$$x_{n1} \leqslant \Phi(t_n) \leqslant x_{n2}.$$

$$(0 \leqslant t_1 \leqslant t_2 \leqslant \ldots \leqslant t_n \leqslant 1.)$$

This assemblage of time-paths may be considered as possessing the compound probability that after a time t_1 the particle shall occupy a position between x_{11} and x_{12}, that by time t_2 it shall have wandered from whatever position it shall have occupied at time t_1 to a position between x_{21} and x_{22}, and so on. By (1), this compound probability is

$$(3) \qquad \frac{1}{\sqrt{\{\pi^n t_1(t_2-t_1)(t_3-t_2)\ldots(t_n-t_{n-1})\}}}$$

$$\times \int_{x_{11}}^{x_{12}}\int_{x_{21}}^{x_{22}}\cdots\int_{x_{n1}}^{x_{n2}} e^{-[(\xi_1^2/t_1)+\{(\xi_2-\xi_1)^2/(t_2-t_1)\}+\ldots+\{(\xi_n-\xi_{n-1})^2/(t_n-t_{n-1})\}]}\,d\xi_1\,d\xi_2\ldots d\xi_n.$$

We shall call the assemblage of time-paths which we have just specified an *interval* and shall term (3) its *weight*. We shall allow any x_{k1} to be $-\infty$ and any x_{k2} to be $+\infty$. It is clear that the weight of any interval consisting of the logical sum of a finite number of mutually exclusive intervals is the sum of the weights of its component intervals; and that if *every* x_{k1} is $-\infty$, and *every* x_{k2} is $+\infty$, the weight of the interval is 1.

4. Under certain circumstances, if an assemblage can be divided into sub-assemblages with assignable weights, it is possible to define the mean of a function ranging over the assemblage in question. These circumstances have been set forth by the author in an earlier paper already noted. The mean thus defined, which was a Daniell integral, presupposed

that the assemblage could be divided into successive collections I_n, each of a finite number of weighted intervals, satisfying the following conditions:

(*a*) Every interval of I_{n+1} forms part of one interval I_n and of one only;

(*b*) The weight of an interval of I_n is the sum of the weights of the component intervals in I_{n+1};

(*c*) If S_n consists of all the elements in a number of intervals of I_n and S_{n+1} is contained in S_n, then either there is an element common to every S_n, or the total weight of S_n approaches 0 with $1/n$.

For the purposes of the present paper, we shall define the intervals of I_n in the following somewhat arbitrary manner: we shall take for our intervals intervals in the sense of § 3, letting

$$t_h = h/2^n \quad (1 \leqslant h \leqslant 2^n),$$

$$x_{1h} = \tan (k_h \pi/2^n),$$

$$x_{2h} = \tan \left((k_h + 1) \pi/2^n \right),$$

k_h being some positive or negative integer. It will be seen that there are 2^n such segments (x_{1h}, x_{2h}), running from $-\infty$ to $+\infty$, that every segment correlated with I_n is made up of a finite number of segments correlated with I_{n+1}, and that if H is any finite positive quantity, and ϵ is any positive quantity, then there is an m such that $n > m$, the segments correlated with I_n between $-H$ and H are all less in length than ϵ. The accompanying figure shows an interval of I_3. The interval is indicated by the vertical black lines.

The black horizontal segments form an interval of I_3 containing the curve drawn. The line *OPQRS* satisfies any condition of equal continuity fulfilled by curves of the interval.

I_n itself is now defined as the assemblage of $(2^n)^{2^n}$ intervals which are obtained by associating a k_h with each h from 1 to 2^n inclusive. The weight of the intervals of I_n is defined as in § 3. Conditions (*a*) and (*b*) will clearly be satisfied.

Condition (*c*) requires a consideration somewhat more careful. In this connection, let us make use of the function

$$\psi(u) = 2^{1-k/4}/(1-2^{-\frac{1}{4}}),$$

where k is the largest integer such that $1/2^k > u$. Clearly

$$\lim_{u \to 0} \psi(u) = 0. \tag{4}$$

Now consider the functions $f(t)$ such that $f(0) = 0$ and that for all values of t_1 and t_2 between 0 and 1, inclusive,

$$|f(t_1)-f(t_2)| \leqslant h\psi(|t_1-t_2|), \tag{5}$$

where h is a given positive integer. These functions are manifestly what Ascoli* calls equally continuous, and are bounded as a set. They therefore

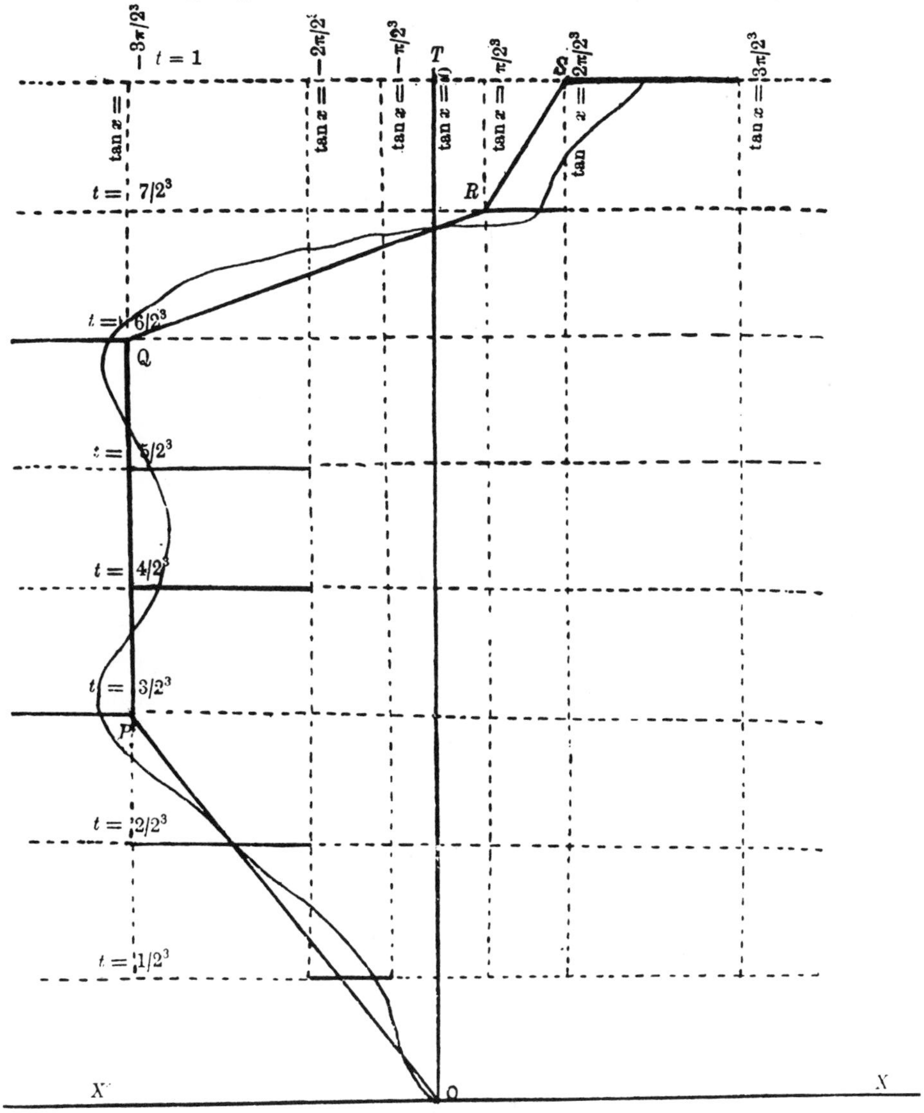

form what Fréchet† calls a compact set, and there is no difficulty in

* Ascoli, "Le curve limiti di una varietà data di curve", *Lincei*, Vol. 18, pp. 521–586.

† M. Fréchet, "Sur quelques points du calcul fonctionnel", *Rend. Cir. Mat. Pal.*, Vol. 22, pp. 6, 37.

showing that every limit-function of the set satisfies the same condition of equal continuity as the members of the set, and hence belongs to the set,* which is closed. Now an interval is closed, and hence those functions in an interval satisfying our condition of equal continuity form a closed set. Hence, if every S_n contains intervals which contain functions satisfying the given condition of equal continuity, there is a function common to every S_n.† Then (*c*) is satisfied.

On the other hand, suppose that some S_k contains no function satisfying the given condition of equal continuity. That is to say, given any function $f(t)$ from S_k, suppose that there are two numbers t_1 and t_2 between 0 and 1, inclusive, such that

$$|f(t_1)-f(t_2)| > h\psi(|t_1-t_2|). \tag{6}$$

We shall show that by choosing h sufficiently large, the total weight possible for S_k may be made arbitrarily small. This will be sufficient to complete the proof of (*c*).

To show this, let us represent t_1 as the binary fraction

$$0 \cdot a_1 a_2 \ldots a_n \ldots$$

and t_2 as the binary fraction

$$0 \cdot b_1 b_2 \ldots b_n \ldots.$$

Let t_3 be a number whose binary expansion may be made to agree with that of t_1 up to and including a_j and with that of t_2 up to and including b_j. Choose t_3 in such a manner that j may be as large as possible.‡ In general t_3 will admit of a terminating expression. The interval from t_1 to t_3 will then be expressible in the form

$$0 \cdot 00 \ldots 0 c_{j+1} c_{j+2} \ldots,$$

where there are j consecutive 0's and every c is 0 or 1. The interval from t_3 to t_2 may be expressed in a similar manner. In other words, the interval from t_1 to t_2 may be reduced to the sum of a denumerable set of intervals from terminating binaries to adjacent terminating binaries of the same number of figures, such that there are not more than two inter-

* *Ibid.*, p. 6.

† *Ibid.*, p. 7.

‡ This may necessitate the use of an expression for t_3 ending in 111... to agree with the smaller of the quantities t_1 and t_2, and of a terminating expression for t_3 to agree with the larger.

vals in the set of magnitude $1/2^{j+k}$, where k is any positive integer, and such that every interval is of one of these sizes.

Now clearly

$$|t_2-t_1| \geqslant 2^{-1-j}. \tag{7}$$

Hence, by the definition of ψ,

$$|f(t_1)-f(t_2)| > h\,.\,2^{1-(j-1)/4}/(1-2^{-1/4}), \tag{8}$$

which we may write in the form

$$|f(t_1)-f(t_2)| > 2h \sum_{j+1}^{\infty} 2^{-n/4}. \tag{9}$$

If we make use of the fact that f is continuous, and correlate the terms of (9) with the digits of the steps from t_3 to t_1 and t_2, we obtain the result that for some interval from $m/2^{j+k}$ to $(m+1)/2^{j+k}$, where m and k are integers and $0 \leqslant m < 2^{j+k}$, we shall have

$$\left|f\left(\frac{m}{2^{j+k}}\right)-f\left(\frac{m+1}{2^{j+k}}\right)\right| > h\,.\,2^{-(j+k)/4} \tag{10}$$

Writing i for $j+k$, this becomes

$$\left|f(m2^{-i})-f\left((m+1)2^{-i}\right)\right| > h\,.\,2^{-i/4}. \tag{11}$$

Every function satisfying (6) satisfies (11) for some m and i. We may go further and say that any interval consisting entirely of functions satisfying (6) has all its functions satisfy (11) for the same value of i. This follows immediately from the fact that, as shown in the diagram, every interval possesses what we may term a most continuous time-path, which fulfils any condition of equal continuity fulfilled by any time-path in the interval.

We shall now show that all functions fulfilling (11) for any one of a finite number of values of i may be enclosed in a finite set of intervals the total weight of which may be made to vanish as h becomes infinite. To begin with, let i and m be fixed, and let n be so great that the segments to which $f(m2^{-i})$ and $f\left((m+1)2^{-i}\right)$ are confined are all less in magnitude than ϵ, over a range running from $-E$ to $+E$. That this can be done is clear from the definition of I_n. If we divide all intervals containing functions satisfying (11) into those going beyond $\pm E$ for $t = m2^{-i}$ or $t = (m+1)2^{-i}$ and all others, we get for the total weight of these intervals a quantity not exceeding

$$(12) \quad P = \frac{2}{\sqrt{(\pi m 2^{-i})}} \int_E^\infty e^{(-2^i x^2)/m}\, dx + \frac{2}{\sqrt{\{\pi(m+1)2^{-i}\}}} \int_E^\infty e^{(-2^i x^2)/(m+1)}\, dx$$

$$+ \frac{2}{\sqrt{(\pi 2^{-i})}} \int_{(h\,.\,2^{i/4}-2\epsilon)}^\infty e^{-2^i x^2}\, dx$$

$$= \frac{2}{\sqrt{\pi}} \left\{ \int_{E\sqrt{(2^i/m)}}^\infty e^{-x^2}\, dx + \int_{E\sqrt{\{(2^i)/(m+1)\}}}^\infty e^{-x^2}\, dx + \int_{(h\,.\,2^{i/4}-2^{i/2}\,.\,2\epsilon)}^\infty e^{-x^2}\, dx \right\}.$$

We may prove without difficulty that given any positive η, by choosing n sufficiently great, and hence ϵ sufficiently small and E sufficiently large,

$$(13) \qquad P < \eta + \frac{2}{\sqrt{\pi}} \int_{h\,.\,2^{i/4}}^\infty e^{-x^2}\, dx < \eta + \frac{2}{\sqrt{\pi}} \int_{h\,.\,2^{i/4}}^\infty e^{-x}\, dx$$

if $h > 1$. Hence

$$(14) \qquad P < \eta + \frac{2}{\sqrt{\pi}}\, e^{-h\,.\,2^{i/4}}.$$

It may readily be shown that η may be chosen once for all values of m. Hence if we let m range from 1 to 2^i-1, the total probability that (11) should be satisfied for a given i is less than

$$(15) \qquad P_i = 2^i \eta + \frac{2^{i+1}}{\sqrt{\pi}}\, e^{-h\,.\,2^{i/4}}.$$

Let us have $\eta < 1/(h\,.\,2^{2i})$, and choose an appropriate n. Then

$$(16) \qquad P_i < \frac{1}{h\,.\,2^i} + \frac{2^{i+1}}{\sqrt{\pi}}\, e^{-h\,.\,2^{i/4}}.$$

Since, for a sufficiently large x, $e^x > x^p$, if we take $p = 8$ and let i be sufficiently large,

$$(17) \qquad P_i < \frac{1}{h\,.\,2^i} + \frac{2}{\sqrt{\pi}} \cdot \frac{1}{h^8\,.\,2^i}.$$

It is thus clear that ΣP_i converges to a limit which vanishes as h becomes infinite.

Now, it follows from the definition of P_i that the weight of S_k is less than ΣP_i if every function in S_k satisfies (6). This completes the proof of (c), and allows us to apply at once the theorems contained in the author's earlier paper, and hence those due to Daniell.

Before we proceed further, it may be remarked that none of the results already obtained are affected if we discard from consideration all functions satisfying equation (11) for all values of h. There is no interval which consists exclusively of such functions. We shall, accordingly, consider them discarded in all that follows.

We shall designate as a *step-functional* a functional* F which is constant over every interval of some I_n. A step-functional is allowed to be multiple-valued for arguments common to several intervals. If i_k be used to represent the general interval from I_n, we shall say that the *average value* of F is the quantity

$$A\{F\} = \sum_k F\{f\} p(i_k), \tag{18}$$

where f is a function belonging to i_k and $p(i_k)$ is the weight of i_k. It follows from the results of my previous paper that A is what Daniell terms an I-integral.†

5. A brief résumé of Daniell's main definitions and results in so far as they apply to the operation A is now in order. The class of all step-functionals plays the rôle of what he calls T_0. The class which is the T_1 of his terminology consists of all those functionals F which are limits of non-decreasing sequences of step-functionals

$$F_1 \leqq F_2 \leqq \dots \leqq F_n \leqq \dots .$$

We then make the definition

$$A\{F\} = \lim A\{F_n\}. \tag{19}$$

For *any* functional G, $\dot{A}\{G\}$ is defined as the lower bound of $A\{F\}$ for all functionals F belonging to T_1 and such that $F \geqq G$ for all arguments. $\underset{\cdot}{A}\{G\}$ is defined by

$$\underset{\cdot}{A}\{G\} = -\dot{A}\{-G\}. \tag{20}$$

If

$$\dot{A}\{G\} = \underset{\cdot}{A}\{G\} = \text{finite}, \tag{21}$$

their common value is by definition $A\{G\}$, and we call G summable. Clearly every function in T_0 is summable.

The most important theorem which Daniell proves reads as follows:

If $\{F_n\}$ is a sequence of summable functions with the limit F, and if a summable functional G exists such that for all n

$$|F_n| \leqq G,$$

* Cf. Fréchet, *loc. cit.*, p. 4.

† *A General Form of Integral*, p. 280.

then F is summable, and

$$A\{F\} = \lim A\{F_n\}. \tag{22}$$

In addition to this, A is a linear positive operator.

6. Let us form as in § 4 sets of intervals σ_h of total weight not exceeding ΣP_i containing all the continuous functions satisfying (11) for specified values of h. Let $\tau_1 = \sigma_1$, let τ_2 consist of those parts of intervals of σ_2 that lie in intervals of τ_1, and in general let τ_n consist of those parts of intervals of σ_n that lie in intervals of τ_{n-1}. Let G be a functional that differs from a step-functional only in the fact that it is 0 for every function in τ_n. I say that it is summable. For let us assume that the original step-functional F is constant over the intervals of I_k. Arrange the intervals of τ_n, which form a denumerable set, in a progression, and let F_m be the step-functional which differs from F by being 0 over the first m of these. Clearly $F_m \leqslant |F|$ and $\lim F_m = G$, so that G is summable.

Next let H be a functional which differs from a bounded continuous* functional only in that it vanishes for all functions in τ_n. Then H is summable. For let F_m be that functional which is constant over each interval of I_m, excluding functions in τ_n (for which it vanishes), which assumes in each interval of I_m the upper bound of H for an f in that interval and not in τ_n. Since an interval is closed and since the set of all time-paths not in τ_n is extremal,† the region to which f is here restricted is extremal, and the upper bound of H is actually the largest value assumed by H in the region in question.‡ Since for later and later I_m's the portions of intervals not in τ_n come to have smaller and smaller maximum difference between any two functions they contain,

$$\lim F_m = H. \tag{23}$$

As $F_m \leqslant |H|$, and as H is bounded, it follows that H is summable.

Finally—and this is our cardinal theorem—*any bounded continuous functional defined over all time-paths is summable.* Let F be such a functional. Consider the sequence $\{F_n\}$, where F_n differs from F only by being 0 over τ_n. F_n is summable by the last theorem. $F = \lim F_n$, as there is no function of the region which we consider which is common to every τ_n. Furthermore, every F_n is not greater than the constant step-functional $\max |F|$. Thus our theorem follows at once.

---

* Fréchet, *loc. cit.*, p. 7.
† Fréchet, *loc. cit.*, p. 37.
‡ Fréchet, *loc. cit.*, p. 29.

7. We shall now search for a more explicit expression for $A\{F\}$. In the first place, by definition (3), the weight of an interval J of I_n is

$$(24)\quad \pi^{-2^n-1}\,2^{n.2^{n-1}}\int_J\!\!\cdots\!\int \exp\left\{-2^n\left[x_1^2+\sum_{j=2}^{2^n}(x_j-x_{j-1})^2\right]\right\}dx_1\ldots dx_{2^n},$$

and this weight is also by definition $A\{G\}$, where G is 1 over J and 0 elsewhere. It follows at once that if F is any step-functional assuming constant values over intervals of I_n,

$$(25)\quad A\{F\} = \pi^{-2^n-1}\,2^{n.2^{n-1}}\int_{-\infty}^{\infty}\!\!\cdots\!\int_{-\infty}^{\infty}F\{f_{x_1,\ldots,x_{2^n}}\}$$

$$\times\exp\left\{-2^n\left[x_1^2+\sum_{j=2}^{2^n}(x_j-x_{j-1})^2\right]\right\}dx_1\ldots dx_{2^n},$$

where $f_{x_1,\ldots,x_{2^n}}(t)$ represents the broken line with corners successive at $(0, 0)$, $(1/2^n, x_1)$, $(2/2^n, x_2)$, $\ldots$, $(1, x_{2^n})$.

By a theorem of Daniell which we have already given, this formula may be extended to all bounded functionals $F\{f\}$ that are only dependent on the values of f for arguments of the form $k/2^n$, and that are limits of uniformly bounded sequences of step-functionals constant over the intervals of I_n. It is in particular true of all bounded continuous functionals for which $F\{f\} = F\{f_{x_1,\ldots,x_{2^n}}\}$.

Under this supposition we may elide, as it were, any x_k in (25) such that a change in x_k does not alter $F\{f_{x_1,\ldots,x_{2^n}}\}$. If we write

$$(26)\quad \{\Psi(t_1,\ldots,t_n)(x_1,\ldots,x_n)\}(t)$$

for that broken line with corners successively at $(0, 0)$, (t_1, x_1), $\ldots$, (t_n, x_n), then

$$(27)\ A\{F\} = \pi^{-n/2}\prod_1^n(t_k-t_{k-1})^{-\frac{1}{2}}\int_{-\infty}^{\infty}\!\!\cdots\!\int_{-\infty}^{\infty}F\left\{\{\Psi(t_1,\ldots,t_n)(x_1,\ldots,x_n)\}(t)\right\}$$

$$\times\exp\left\{-\frac{x_1^2}{t_1}-\frac{(x_2-x_1)^2}{t_2-t_1}-\cdots-\frac{(x_n-x_{n-1})^2}{t_n-t_{n-1})}\right\}dx_1\ldots dx_n,$$

if $t_1, \ldots, t_n = 1$ are terminating positive binaries, arranged in order of magnitude, and $F\{f\}$ is any continuous, bounded functional dependent only on $f(t_1), \ldots, f(t_n)$.

The restriction that the binaries $t_1, \ldots, t_n$ terminate is inessential. Let U_h be the set of all time-paths satisfying (5). Let

$$(28)\quad F'\{f\} = F\left\{(t_1,\ldots,t_n)\Big(f(t_1),\ldots,f(t_{k-1}),f(t_k+\delta),f(t_{k+1}),\ldots,f(t_n)\Big)\right\}.$$

In the first place there is a $\phi(\delta, h)$ such that $\lim_{|\delta| \to 0} \phi(\delta, h) = 0$, and that

(29)

$$\Psi(t_1, \ldots, t_n)\Big(f(t_1), \ldots, f(t_n)\Big) - \Psi(t_1, \ldots, t_n)\Big(f(t_1), \ldots, f(t_k+\delta), \ldots, f(t_n)\Big) \leqslant \phi(\delta, h)$$

for every f in U_h. In the second place, U_h is bounded and extremal,* so that F is uniformly continuous over U_h.† Hence there is a Φ such that $\lim_{|\delta| \to 0} \Phi(\delta, h) = 0$ and that

(30) $$|F'\{f\} - F\{f\}| \leqslant \Phi(\delta, h)$$

over U. Now, we may write

(31) $$|F'\{f\} - F\{f\}| = Q_1\{f\} + Q_2\{f\},$$

where $Q_1\{f\}$ equals $|F'\{f\} - F\{f\}|$ over U_h and 0 elsewhere, while Q_2 equals 0 over U_h and $|F' - F|$ elsewhere. Hence

(32) $$\begin{aligned} |A\{F'\} - A\{F\}| &= |A\{F' - F\}| \\ &\leqslant A\{|F' - F|\} \\ &\leqslant \dot{A}\{Q_1\} - \dot{A}\{Q_2\} \\ &\leqslant \dot{A}\{Q_1\} + 2 \max|F| \,.\, A\{R\}, \end{aligned}$$

where $R = 1$ over σ_h and $R = 0$ elsewhere. By definition, σ_h is the sum of a denumerable set of intervals, and hence R is the limit of a bounded set of step-functionals R_n and is summable to the limit $A\{R_n\}$.‡ By the discussion following (17), it may be shown that this limit is no greater than ΣP_i, which vanishes as h becomes infinite. Choose h so that

(33) $$2 \max|F| \,.\, A\{R\} < \epsilon/2,$$

and then choose δ so that $\Phi(\delta, h) < \epsilon/2$. Then by (32)

(34) $$|A\{F'\} - A\{F\}| < \epsilon.$$

It follows that $|A\{F'\} - A\{F\}|$ approaches 0 with δ.

* Fréchet, *loc. cit.*, p. 37.

† Fréchet, *loc. cit.*, p. 29.

‡ Daniell, *A General Form of Integral*, 7 (7).

Now let $0 \leqslant t_1 \leqslant \ldots \leqslant t_n = 1$, and let every t but t_k be a terminating binary. Let $t_k + \delta$ terminate. Let $t'_j = t_j$ if $j \neq k$ and let $t'_k = t_k + \delta$. By (27) and (28),

(35)

$$A\{F'\} = \pi^{-(n/2)} \prod_1^n (t'_j - t'_{j+1})^{-\frac{1}{2}} \int_{-\infty}^{\infty} \cdots \int_{-\infty}^{\infty} F'\{\{\Psi(t'_1, \ldots, t'_n)(x_1, \ldots, x_n)\}(t)\}$$

$$\times \exp\left\{-\frac{x_1^2}{t'_1} - \ldots - \frac{(x_n - x_{n-1})^2}{t'_n - t'_{n-1}}\right\} dx_1 \ldots dx_n$$

$$= \pi^{-(n/2)} \prod_1^n (t'_j - t'_{j+1})^{-\frac{1}{2}} \int_{-\infty}^{\infty} \cdots \int_{-\infty}^{\infty} F\{\{\Psi(t_1, \ldots, t_n)(x_1, \ldots, x_n)\}(t)\}$$

$$\times \exp\left\{-\frac{x_1^2}{t'_1} - \ldots - \frac{(x_n - x_{n-1})^2}{t'_n - t'_{n-1}}\right\} dx_1 \ldots dx_n.$$

Hence

(36)

$$\Bigg| A\{F'\} - \pi^{-(n/2)} \prod_1^n (t_j - t_{j+1})^{-\frac{1}{2}} \int_{-\infty}^{\infty} \cdots \int_{-\infty}^{\infty} F\{\{\Psi(t_1, \ldots, t_n)(x_1, \ldots, x_n)\}(t)\}$$

$$\times \exp\left\{-\frac{x_1^2}{t_1} - \ldots - \frac{(x_n - x_{n-1})^2}{t_n - t_{n-1}}\right\} dx_1 \ldots dx_n \Bigg|$$

$$\leqslant \max |F| \Bigg| \Delta \Bigg\{ \pi^{-(n/2)} \prod (t_j - t_{j+1})^{-\frac{1}{2}}$$

$$\times \int_{-\infty}^{\infty} \cdots \int_{-\infty}^{\infty} \exp\left\{-\frac{x_1^2}{t_1} - \ldots - \frac{(x_n - x_{n-1})^2}{t_n - t_{n-1}}\right\} dx_1 \ldots dx_n \Bigg\} \Bigg|$$

where the Δ is taken for a change of δ in t_k. As the expression under the Δ is a continuous function of the t's (so long as no two are equal), the left-hand side of (36) vanishes with δ. Combining this fact with (34), we see that (27) is valid for all continuous bounded functionals of f that depend only on $f(t_1), \ldots, f(t_n)$. By a continuation of this process we can make (27) hold when none of the t_j's need be terminating binaries.

Now let F be *any* bounded continuous functional of time-paths. Let us write

(37) $$F_{t_1, \ldots, t_n}\{f\} = F\{\Psi(t_1, \ldots, t_n)\big((t_1), \ldots, f(t_n)\big)\}.$$

Let Φ equal 0 on σ_h and F elsewhere, and let $\Phi_{t_1, \ldots, t_n}$ equal to 0 on σ_h and $F_{t_1, \ldots, t_n}$ elsewhere. Each of these functionals is summable by a theorem of Daniell,* being the "logical product" of a summable functional such as

* *A General Form of Integral*, p. 280.

F and the limit of a denumerable set of step-functionals—*i.e.* the functional which is 0 on σ_h and $\max|F|$ elsewhere.* Furthermore

$$\lim_{\max(t_{i+1}-t_i)\to 0} \Phi_{t_1,\ldots,t_n} = \Phi. \tag{38}$$

Hence, as the $\Phi_{t_1,\ldots,t_n}$'s form a bounded set,†

$$\lim_{\max(t_{i+1}-t_i)\to 0} A\{\Phi_{t_1,\ldots,t_n}\} = A\{\Phi\}. \tag{39}$$

Moreover, $|A\{F-\Phi\}|$ and $|A\{F_{t_1,\ldots,t_n}-\Phi_{t_1,\ldots,t_n}\}|$ both exist, and are both less than or equal to $\max|F|.A\{R\}$ {Cf. 32)}. Hence both these quantities vanish as h becomes infinite.

From these facts it may be concluded that if F be a bounded continuous functional,

$$A\{F\} = \lim_{\max(t_{i+1}-t_i)\to 0} \pi^{-(n/2)} \prod_1^n (t_k-t_{k-1})^{-\frac{1}{2}} \tag{40}$$

$$\times \int_{-\infty}^{\infty} \cdots \int_{-\infty}^{\infty} F\{\{\Psi(t_1,\ldots,t_n)(x_1,\ldots,x_n)\}(t)\}$$

$$\times \exp\left\{-\frac{x_1^2}{t_1}-\cdots-\frac{(x_n-x_{n-1})^2}{t_n-t_{n-1}}\right\} dx_1 \ldots dx_n.$$

MASSACHUSETTS INSTITUTE OF TECHNOLOGY.

* *Ibid.*, 7 (5).
† *Ibid.*, 7 (7).

2 H 2

Commentary on [20f], [21c, d], [23d], [24d]

K. Ito

These five papers aimed mainly at the rigorous proof of the existence of the probability law governing the Brownian motion process, which now also bears the name of Wiener process.

In the first paper [20f] Wiener gave a condition for a sequence of weighted subdivisions of a given abstract space compatible with each other to determine a σ-additive probability measure or an average operation. This is merely a version of Daniell's theory of abstract integral in terms of measures. However, it is more convenient in defining concrete measures and is a prototype of the so-called extension theorem of a finitely additive measure to a σ-additive one. The usage of this theorem is explained by three examples. In the last example he discussed a certain average operator on the Hölder continuous functions. Although nothing is mentioned here about Brownian motion, we can imagine by his further development that he aimed at it in this paper.

In the second paper [21c] Wiener explained by the central limit theorem (in a classical form) that it is natural to assume the displacement of the Brownian particle in any time interval to be Gauss distributed. Starting with this consideration he obtained the joint distribution $\mu_{x_1 \ldots x_n}$ of the positions of the particle at time points $x_1 \ldots x_n$. The average $A(F_n)$ of a functional F_n of the following type, called an *analytic functional of order n*,

$$F_n(f) = \int_0^1 \cdots \int_0^1 f(x_1) \cdots f(x_n)\, d\psi_n(x_1 \cdots x_n),$$

is defined by

$$A(F_n) = \int_0^1 \cdots \int_0^1 \int_{-\infty}^{\infty} \cdots \int_{-\infty}^{\infty} y_1 \cdots y_n\, d\mu_{x_1 \ldots x_n}(y_1 \cdots y_n)\, d\psi_n(x_1 \cdots x_n).$$

This can be extended to the average $A(F)$ of an analytic functional $F = \Sigma_n F_n$ by $A(F) = \Sigma_n A(F_n)$ under certain conditions. $A(F)$ is clearly the integral of F with respect to the Wiener measure whose existence was proved later in the fourth paper [23d]. The present paper is an intermediate step to reach his goal.

The purpose of the third paper [21d] is to make use of the result of paper [21c] to justify the experimental fact that the square mean of the displacement of a particle in a fluid is almost proportional to the time length. Wiener assumed the increment of the momentum of the particle to be obtained by the superposition of the random impacts caused by the molecular motion of the fluid and the friction caused by the viscosity. This idea had been known among physicists as Langevin's model which was given in 1908, though Wiener did not mention it. Because random impacts were supposed to act independently at different instants, Wiener equated their effect during the infinitesimal interval to the increment of the Brownian motion introduced in paper [21c]. Then he proved that the square of the displacement of the particle up to time t is an analytic functional of order 2 of the Brownian motion. Applying the formula in paper [21c] to this functional, it was proved that its average is almost proportional to t. The moment process appearing in this paper is now known in the name of Ornstein-Uhrenbeck process.

The fourth paper "Differential space" [23d] is the most important contribution in the theory of stochastic processes. The idea of establishing probability theory on the basis of measure and integration goes back to E. Borel who proved the strong law of large numbers (in a classical form) in terms of Lebesgue measure in 1909. However, in order to discuss stochastic processes we need the theory of measure and integration on function space. It was so difficult in the 1920s that all mathematicians but Wiener discussed only the finite joint distributions. To obtain the average of a functional depending on the whole trajectory of a stochastic process they defined it to be the average for a sequence of approximating functionals depending on a finite number of time points without any justification. Wiener gave a solid foundation to the theory of Brownian motion by defining the average as a Daniell integral. Although he dealt with the Brownian motion in this paper, his method served as a model for the modern theory of stochastic processes.

Let us review the content of this paper. It is written in a heuristic way so that we can see how he approached the rigorous definition of Wiener measure.

Brownian motion is introduced in the same way as in the second paper [21c]. Being influenced by P. Lévy's work, Wiener introduced the notion of *differential space,* which is now known as *Wiener space.* The differential space is the space C of all continuous functions on [0,1] vanishing at 0 endowed with a certain ideal limit of the measures μ_n on C defined as follows. The μ_n-measure of the set of $f \in C$ such that $a_i < f(i/n) - f((i-1)/n) < b_i$, $i = 1, 2, \ldots, n$, is the normalized spherical measure of the set

$a_i < x_i < b_i$ on the sphere $\sum_{i=1}^{n} x_i^2 = r^2$ (r = positive constant). The ideal limit of μ_n as $n \to \infty$ is the *Wiener measure* whose rigorous definition will be given later. It is explained that the Brownian motion and the differential space are essentially the same. It is obvious that the μ_n-measure of the set of f such that

$$\sum_i \left[f\left(\frac{i}{n}\right) - f\left(\frac{i-1}{n}\right) \right]^2 = r^2$$

is 1. It is essentially proved that the left-hand side of this equation tends to r^2 as $n \to \infty$ almost everywhere on the differential space. Wiener claimed that this would be a mathematical justification of the nondifferentiable character of the trajectory of the Brownian particle stated by the famous physicist J. Perrin. The rigorous proof of this fact and an even stronger result was given in Wiener's joint paper with Paley and Zygmund [33a].

It is essentially proved in Section 7 that almost all functions in the differential space are Hölder continuous of order less than 1/2. To do this, he proved that the measure of $\{f \in C: \max_{s \leq t \leq s+\delta} (f(s) - f(r)) > C\}$ is twice the measure of $\{f \in C: f(r+\delta) - f(s) > C\}$ by first proving a similar fact for the random walk to obtain the fact for Brownian motion by grouping. In this proof we can find a prototype of the useful invariance principle of Kac. A more satisfactory proof of Hölder continuity was given in Wiener's joint paper [33a] with Paley and Zygmund cited above.

In Sections 8 and 9 the average of an analytic functional is defined in the same way as in the second paper [21c].

Having investigated the differential space from various directions, Wiener defines the Wiener measure as a σ-additive probability measure by means of Daniell's theory of integral. Once this is done, the heuristic observations made above are transformed into rigorous proofs.

In Section 10 he proves that if $\{\psi_k(x) \equiv \int_x^1 \varphi_k(y)\, dy,\ k = 1,2,\ldots,n\}$ is an orthonormal system on $L^2[0,1]$, then $\int_0^1 \varphi_k(x) f(x)\, dx,\ k = 1,2,\ldots,n,$ are independently and identically Gauss distributed. By formal application of integration by parts, we have

$$\int_0^1 \varphi_k(x) f(x)\, dx = \int_0^1 \psi_k(x)\, df(x).$$

The integral on the right-hand side, called the *Wiener integral* (of the first order), for $\psi_k \in L^2[0,1]$ was defined in Wiener's later work and played an important role in the theory of Brownian motion. Setting $\varphi_k(x) = \cos(k\pi x)$, Wiener obtained the Fourier expansion

$$f(x) = a_0 + \sum_{n=1}^{\infty} a_n \cos(n\pi x),$$

$a_0, a_1, \ldots$ being independently and identically Gauss distributed. This expansion was used in his later work on Brownian motion. The Hermite expansion of a functional on the differential space is mentioned in this connection. This expansion was used by S. Kakutani in his proof of the σ-Lebesgue property of the flow of Brownian motion.

In spite of such an abundant content, this paper was not easy to read because of the heuristic nature of the presentation. In the fifth paper [24d] Wiener gave a proof of the existence of Wiener measure which is more satisfactory from the logical point of view.

A word should be mentioned as to an application of Brownian motion which was discussed in his later paper on generalized harmonic analysis [30a]. Let f be a complex-valued measurable function on $(-\infty,\infty)$ for which

$$\varphi(x) = \lim_{T\to\infty} \frac{1}{2T} \int_{-T}^{T} f(t+x)\overline{f(t)}\, dt$$

exists for every such x and is continuous at $x = 0$. Then $\varphi(x)$ is written as the Fourier transform of a finite measure on $(-\infty,\infty)$:

$$\varphi(x) = \int_{-\infty}^{\infty} e^{ix\lambda}\, dF(\lambda),$$

which dF is called the spectral intensity of f. If f is almost periodic, then $\varphi(x)$

exists and the spectral intensity dF is purely discontinuous. Wiener also constructed a function f for which dF is absolutely continuous. Let $\{a_n(w), n = 0, \pm 1, \pm 2, \ldots\}$ be a sequence of independent random variables, each being governed by the law

$$P(a_n = \pm 1) = 1/2.$$

Define a stochastic process $f(t,w)$ by

$$f(t,w) = a_n \quad \text{on} \quad [n, n+1), \quad n = 0, \pm 1, \pm 2, \ldots .$$

Then the spectral intensity $dF(\lambda,w)$ for almost every sample function of $f(t,w)$ is given by

$$dF(\lambda,w) = \frac{1}{\pi} \frac{1-\cos\lambda}{\lambda^2} d\lambda.$$

Let $\xi(t,w)$ be a Brownian motion with $\xi(0,w) \equiv 0$ and $\vartheta(t)$ be quadratically summable such that $\vartheta(t)\sqrt{1+t^2}$ is of bounded variation over $(-\infty,\infty)$. Set

$$f(t,w) = \int_{-\infty}^{\infty} \xi(t+\tau, w)\, d\vartheta(\tau);$$

$f(t,w)$ is also expressed by

$$f(t,w) = \int_{-\infty}^{\infty} \vartheta(t+l)\, d\xi(t,w)$$

in terms of the Wiener integral defined by Wiener later. Writing $\psi(\lambda)$ for the Fourier transform of $\vartheta(t)$, Wiener proved that the spectral measure $dF(\lambda,w)$ for $f(t,w)$ is given by

$$dF(\lambda, w) = (1/2)|\psi(u)|^2\, d\lambda$$

for almost all sample functions of $f(t,w)$. It is now known that for an arbitrary finite measure dF on $(-\infty,\infty)$ one can make use of the same idea to construct a complex Gaussian stationary process for which almost every sample function has the spectral intensity equal to dF.

Wiener's success in his rigorous definition of Brownian motion lies mainly in his remark that Wiener measure is concentrated in a compact family of continuous functions with arbitrarily high percentage. This idea, now called *tightness,* has become a basis of the modern theory of probability measures on function space initiated by Prohorow and developed by W. Sazonov, L. Gross, I. M. Gelfand, R. A. Minlos, and others.

Wiener measure is one particular probability measure in function space. Nevertheless Wiener's work on this measure is extremely important. We divide its merit into three categories.

First, it contains several techniques that are powerful in the theory of stochastic processes. We have alluded above to this point several times.

Second, Wiener proved striking properties of Brownian motion, such as nondifferentiability and Hölder continuity, that sounded even pathological at that time and drew the attention of many mathematicians later. The attempt to strengthen his results has produced many interesting works due to P. Lévy, A. Dvoretzky, P. Erdös, S. Kakutani, and others.

Third, Brownian motion is useful for many purposes in the theory of stochastic processes and even in analysis in general.

Wiener considered the increments of Brownian motion as elementary random effects, called *pure chaos* in his paper "Homogeneous chaos" [38a], and tried to express a general *homogeneous chaos* as an analytic function of pure chaos. Following the same idea he used pure chaos to express the random noise that occurred in communication theory. This led him to the theory of prediction, filtering, and cybernetics.

K. Ito proved that the sample function of the Kolmogorov diffusion is realized on the Wiener space by solving a certain stochastic differential equation. The sample function of the regular one-dimensional diffusion is obtained from the Brownian motion by a certain stochastic time change, as was shown by K. Ito and H. P. McKean, Jr. R. H. Cameron and W. T. Martin discussed the linear transformation on Wiener space and gave new light to the theory of functional analysis.

The extension of the notion of Brownian motion to several dimensions is obvious. Several-dimensional Brownian motion has some interesting properties other than the mere extension of those for the one-dimensional case because of dimensionality: for example, multiple points, winding numbers, and so on. These were discussed by P. Lévy, S. Kakutani, A. Dvoretzky, F. Spitzer, and others. What is more important in this field is the discovery of a close relation between several-dimensional Brownian motion and potential theory. This was initiated by P. Lévy and S. Kakutani and fully developed by J. L. Doob and others. According to this theory the solution of Dirichlet's boundary value problem for harmonic functions is given in terms of Brownian motion. Doob gave the boundary properties of harmonic functions a nice probabilistic interpretation. M. Kac expressed the solution of the heat equation with the absorption term (or the potential term) by means of Brownian motion.

It is astonishing that all such developments stand on the basis given by Wiener's work on Brownian motion.

O szeregach $\sum_1^\infty(\pm 1/n)$. — *Note on the series* $\sum_1^\infty(\pm 1/n)$.

Note

de M. ***N. WIENER,***

présentée dans la séance du 3 décembre 1923 par M. W. Sierpiński m. t.

In volume IV of Fundamenta Mathematicae there appeared a paper by H. Steinhaus entitled, „Les probabilités dénombrables et leur rapport à la théorie de la mesure". Among other things, Steinhaus showed that if the sequence of the sings of the terms in the series

$$\pm 1 \pm 1/2 \pm 1/3 \pm \ldots \pm 1/n \pm \ldots$$

is allowed to vary throuh all possible values, the probability that it converges — that is. the average value of a function of the sequence which is 1 when it converges and 0 otherwise — is 1. The author of this note had at the time of publication of this note arrived at the same conclusion independently by means of a method involving the extremely useful generalized intégrale of P. J. Daniell[1]). It may perhaps still be of interest to develop these methods.

We shall represent sign-sequences by Greek letters. If $\varphi(\alpha)$ is a bounded function of the first n signs in α, we can define its average readily enough, as the arithmetical average of φ for its 2^n essentially distinct arguments. Let the class of such functions be termed T_0. Let it be noted that if f and g belong to T_0, so do $f+g$, cf and $|f|$. Now let us write $I(\varphi)$ for the average value of φ. We then find that

[1]) *A General Form of Integral*, Annals of Mathematics vol. 19 (1917—18), pp. 279—295.

(C) $I(c\varphi) = cI(\varphi)$ if c is any constant,

(A) $I(\varphi_1 + \varphi_2) = I(\varphi_1) + I(\varphi_2)$.

(P) $I(\varphi) \geqslant 0$ if $\varphi \geqslant 0$ for all sign-sequences.

These conditions constitute all but one of the conditions given by Daniell for the existence of a generalized integral with certain important properties. The remaining condition is

(L) If $\varphi_1 \geqslant \varphi_2 \geqslant \ldots \geqslant 0 = \lim \varphi_n(\alpha)$ for all α, then

$$\lim I(\varphi_n) = 0.$$

To prove this, it is enough to show that

$$\lim_{n \to \infty} \max_{\alpha} \varphi_n(\alpha) = 0. \tag{2}$$

Suppose this were false. Then there would be an ε such that for any n, there is an α_n such that

$$\varphi_n(\alpha_n) \geqslant \varepsilon. \tag{3}$$

Now let E_n be the class of α_n's for which (3) holds. Clearly E_{n+1} is contained in E_n. Let s_1 stand for the sign $+$ if every E_n contains an α beginning with a $+$, and otherwise for $-$. Let the terms in E_n beginning with the sign s_1, be represented by E_n'. Clearly either E_n' exists for every n or else all E_n's from some n on vanish. Moreover E_{n+1} is contained in E_n'. Now let s_2 stand for $+$ if every E_n' contains an α with its second term a $+$, and for $-$ otherwise. Let the terms in E_n beginning s_1, s_2 be represented by E_n^2. This process may be extended indefinitely, determining a sequence E_n^k, and a correlated sequence

$$s_1, s_2, \ldots, s_k \ldots \tag{4}$$

If sequence (4) terminates, it may be shown that every E_n from some stage on fails to exist. On the other hand, since each E_n is conditioned only by the sings of some n of the terms in the α's it contains, it may be shown that (4), if it does not terminate, is common to every E_n.

By (3), the sequence $\{E_n'\}$ does not terminate. Hence if we represent sequence (4) by α, we find that for every n,

$$\varphi_n(\alpha) \geqslant \varepsilon. \tag{5}$$

This contradicts the hypothesis of (L).

We thus see that the operation I is a Daniell integration, and is capable to the extension which Daniell has given. It is in particular true that I can be extended to any function f which is less than a sum of a denumerable number of functions belonging to T_0, whose averages total up to an arbitrarily small quantity, and that for such an f,

$$I(f) = 0. \tag{6}$$

I shall prove that this is actually the case when $f(\alpha) = 1$ if (1) diverges and $f(\alpha) = 0$ otherwise.

Let us group the terms of (1) as follows:

$$\pm 1 + \left(\pm \frac{1}{2} \pm \frac{1}{3}\right) + \left(\pm \frac{1}{4} \pm \frac{1}{5} \pm \frac{1}{6}\right) + \dots + \tag{7}$$
$$+ \left[\pm \frac{1}{n(n+1)/2} \pm \frac{1}{n(n+1)/2+1} \pm \dots \pm \frac{1}{(n+1)(n+2)/2-1}\right] + \dots$$

Let us consider the chance that at any stage the sum of the terme in the general bracket exceed in absolute value a quantity A_n. It will be seen immediately thet any partial sum of the terms in the general bracket differs from the corresponding sum of

$$\pm \frac{2}{n(n+1)} \pm \frac{2}{n(n+1)} \pm \dots \pm \frac{2}{n(n+1)} \qquad (n+1 \text{ times}) \tag{8}$$

by less than

$$\sum_1^n \left| \frac{1}{\frac{n(n+1)}{2} + k} - \frac{1}{\frac{n(n+2)}{2}} \right| < \sum_1^n \frac{k}{\left[\frac{n(n+1)}{1}\right]^2} = \frac{2}{n(n+1)}.$$

Hence the chance that any partial sum in the bracket exceed A_n is less than the chance that the expression (8) exceed $A_n - \frac{2}{n(n+1)} = B$. The chance that expression (8) have some partial sum exceeding B in absolute value is the same as the chance that

$$\pm 1 \pm 1 \pm \dots \pm 1 \qquad (n+1 \text{ times}) \tag{9}$$

have some partial sum exceeding $\frac{n(n+1)}{2}B$ in absolute value. This chance is clearly less than twice the chance that series (9) have some positive partial sum exceeding $\frac{n(n+1)}{2}B$ in value, and is hence less than a quantity which I have elsewhere[1]) shown to be asymptotically represented by

$$(10)\qquad \frac{2}{\sqrt{\Pi(n+1)}}\int_{\frac{n(n+1)}{2}B}^{\infty} e^{-\frac{x^2}{n+1}}\,dx = \sqrt{\frac{2}{\Pi}}\int_{\frac{nB\sqrt{n+1}}{2}}^{\infty} e^{-y^2}\,dy.$$

Now let us choose A_n the value $n^{-5/4}$. For expression (10) we shall get

$$(11)\qquad \sqrt{\frac{2}{\Pi}}\int_{\frac{(n+1)^{1/2}}{2n^{1/4}}-\frac{1}{2(n+1)^{1/2}}}^{\infty} e^{-y^2}\,dy.$$

When n is sufficiently large, this will be less than

$$(12)\qquad \sqrt{\frac{2}{\Pi}}\int_{\frac{n^{1/2}}{2}}^{\infty} e^{-y^2}\,dy < e^{-\frac{n^{1/4}}{2}}.$$

It may be shown that the series $\sum\limits_{n=1}^{\infty} e^{-\frac{n^{1/4}}{2}}$ converges. Hence if p_n is the probability that some partial sum in the general bracket in (7) exceeds in absolute value $n^{-5/4}$, then Σp_n converges. It will be noted that the function which is 1 if every partial sum in the bracket in (7) is less in absolute value then $n^{-5/4}$ and 0 otherwise belongs to T_0, und has p_n as its average value. Now, if every bracket in (7) from some stage on has all its partial sums less in absolute value than $n^{-5/4}$, since $\Sigma n^{-5/4}$ converges, it follows at once that (1) converges. The probability that this is not true is clearly less than Σp_n from the corresponding value of n on. This however is arbitrarly small, so that the probability that (P) does not converge is reduced to something less than any assigned positive quantity, and is hence zero.

It is obvious that the methods of this paper are of general applicability, and could be used to determine more precisely the average degree of convergence of (1).

[1]) *Differential-Space*, Journal of Mathematics and Physisc of the Massachusetts Institute of Tehnology, vol. II, n° 3.

Comments on [23e]

J. -P. Kahane

The paper contains (a) an application of the Daniell integral in order to define the set $\{-,+\}^{\mathbb{N}}$ as a probability space; (b) a proof that the random series

$$\pm 1 \pm \frac{1}{2} \cdots \pm \frac{1}{n} \pm \cdots \quad (1)$$

converges a.s. In the Steinhaus approach the probability space is always the interval [0,1] of the real line provided with the Lebesgue measure. The idea of using other probability spaces occurs here as in many places in Wiener's work.

Today, the a.s. convergence of the series (1) is seen to be a particular case of more general results of Rademacher, Khintchin and Kolmogorov, about sums of independent random variables.[1-3]

Errata

p. 87, line 10: write (1) before the formula

p. 87, line 8: write signs instead of sings

References

1. H. Rademacher, *Einige Sätze über Reihen von allgemeinen Orthogonal Functionen,* Math. Ann. 87(1922), 112-138.

2. A. Khintchin and A. N. Kolmogorov, *Über Konvergenz von Reihen deren Glieder durch den Zufall bestimmt werden,* Mat. Sb. 32(1925), 668-677.

3. A. N. Kolmogorov, *Über die Summen durch den Zufall bestimmter unabhangiger Grossen,* Math. Ann. 102(1929), 484-488.

UN PROBLÈME DE PROBABILITÉS DÉNOMBRABLES;

PAR M. NORBERT WIENER [1].

L'étude des lois de probabilité concernant les systèmes d'une infinité dénombrable de variables est due principalement à M. E. Borel [2]. M. H. Steinhaus [3] a obtenu des résultats très intéressants sur cette question, particulièrement au sujet de la convergence des séries telles que

$$\sum_{1}^{\infty} \pm \frac{1}{n},$$

où chaque signe est choisi indépendamment des autres, les signes + et — ayant à chaque choix des probabilités égales. J'ai également écrit sur ce sujet [4]. En outre la partie des *Leçons d'Analyse fonctionnelle* de M. Paul Lévy, qui traite de la mesure dans l'espace fonctionnel du point de vue d'une infinité dénombrable de coordonnées, se rattache au même ordre d'idées.

Dans plusieurs articles [5] j'ai développé la théorie d'un type d'espace fonctionnel qui diffère par plusieurs points essentiels du type classique, et que j'appelle espace différentiel. Cet espace est caractérisé par le fait que l'on considère comme des variables indépendantes, non les différentes valeurs de la fonction $f(x)$

---

(1) J'adresse ici mes remerciements à M. Paul Lévy, qui a bien voulu m'aider dans la rédaction de ce Mémoire en français.

(2) E. BOREL, *Les probabilités dénombrables et leurs applications arithmétiques* (*Rendiconti del Circ. Mat. di Palermo*, t. XXVII, 1er semestre 1909, p. 247 à 271).

(3) H. STEINHAUS, *Les probabilités dénombrables et leur rapport à la théorie de la mesure* (*Fund. Math.*, t. IV, 1923).

(4) NORBERT WIENER, *Notes on the Series* $\Sigma \pm \frac{1}{n}$ (*Bulletin de l'Académie des Sciences de Pologne*, 1923).

(5) *The Average of an Analytic functional* (*Proc. Nat. Acad. Sc.*, Washington, t. VII, p. 253-260); *The Average of an Analytic functional and the Brownian Movement* (*Ibid.*, p. 294-298); *Differential-Space* (*Journ. Math. and Phys., Mass. Inst. Techn.*, t. II, p. 131-174); *The Average Value of a functional* (*Proc. London Math. Soc.*, 2e série, t. XXII, Part 6, p. 454-467).

que représente un point de cet espace, pour des valeurs de x très nombreuses et en progression arithmétique, mais les accroissements de $f(x)$ entre deux valeurs consécutives de x. Comme dans l'espace fonctionnel ordinaire, on peut remplacer l'infinité continue de coordonnées qui détermine un point de cet espace par une infinité dénombrable de coordonnées indépendantes. Pour préciser, si

$$(1) \qquad f(x) \sim a_0 + a_1\sqrt{2}\cos\pi x + \ldots + a_n\sqrt{2}\cos n\pi x + \ldots \quad (^1)$$

est une fonction de l'espace différentiel définie dans l'intervalle (o, 1), les quantités

$$(2) \qquad \pi a_1, \quad 2\pi a_2, \quad \ldots, \quad n\pi a_n, \quad \ldots$$

constituent une infinité dénombrable de coordonnées ayant toutes des poids égaux. Si

$$\varphi(a_1, \ldots, a_n)$$

est une fonction bornée uniformément continue de $a_1, \ldots, a_n$, j'ai démontré (²) que sa valeur moyenne dans un domaine que j'appelle sphère de rayon r est

$$\frac{1}{(2\pi r^2)^{\frac{n}{2}}}\int_{-\infty}^{+\infty} dx_1 \ldots \int_{-\infty}^{+\infty} dx_n\, e^{-\frac{1}{2r^2}\sum_1^n x_i^2}\, \varphi\left(\frac{x_1}{\pi}, \ldots, \frac{x_n}{\pi n}\right)$$

et l'on obtient la même moyenne, si toutefois elle existe, lorsque φ, quoique non borné, est uniformément continu dans tout domaine où les a_n sont bornés.

On verra que nous supposons essentiellement que πa_1, $2\pi a_2$, $\ldots$, $n\pi a_n$, $\ldots$ sont des variables indépendantes, obéissant à la loi de Gauss avec le même paramètre. C'est dans cette hypothèse que nous trouvons pour la moyenne d'une fonctionnelle la valeur obtenue d'autre part dans l'espace différentiel. Nous sommes ainsi conduits à la question : est-il possible d'appliquer cette transformation de coordonnées à une classe plus étendue de fonctionnelles? En d'autres termes, est-il possible de développer toute la théorie

(¹) Le signe $\sim$ est mis à la place du signe $=$ parce que la série considérée peut n'être que convergente en moyenne.

(²) *Differential Space*, p. 171.

de l'espace différentiel en se plaçant au point de vue d'une infinité dénombrable de variables?

L'objet du présent travail est de donner à cette question une réponse affirmative. Dans ce but, il est nécessaire de rappeler brièvement quelques points de mon précédent Mémoire; le procédé d'intégration développé dans ce Mémoire était un cas particulier de l'intégration dans les ensembles abstraits au sens de Daniell (¹). M. Daniell considère un ensemble T_0 de fonctions bornées $f(p)$ d'éléments quelconques p. Cet ensemble est fermé relativement aux opérations suivantes : multiplier par une constante; ajouter deux fonctions; et prendre les modules. M. Daniell appelle « intégrale » une fonctionnelle $U(f)$ ayant les propriétés suivantes :

(C) $$U(cf) = c\,U(f),$$

(A) $$U(f_1+f_2) = U(f_1) + U(f_2),$$

(L) Si $f_1 \geqq f_2 \geqq \ldots \geqq 0$ et $\lim f_n = 0$ alors $\lim U(f_n) = 0$,

(P) $U(f) \geqq 0$ si f est toujours $\geqq 0$.

Il étend autant que possible le domaine dans lequel il définit cette opération, comme M. Lebesgue l'a fait en partant de l'intégration au sens de Riemann. Il prouve d'abord que, si $f_1 \leqq f_2 \leqq \ldots$ est une suite de fonctions appartenant à T_0, la suite des $U(f_n)$ est non décroissante, et par suite devient infinie par valeurs positives ou a une limite. Si alors f_n a une limite f, M. Daniell définit $U(f)$ comme limite de $U(f_n)$, et désigne par T_1 l'ensemble des fonctions telles que f. Il définit ensuite pour toute fonction f la semi-intégrale supérieure $\dot{U}(f)$ comme la borne inférieure de $U(g)$ pour toute fonction g appartenant à T_1 et $\geqq f$. Il définit $\underset{\cdot}{U}(f)$ comme $-\dot{U}(-f)$. Si $\dot{U}(f) = \underset{\cdot}{U}(f)$, ces quantités étant finies, il les désigne par $U(f)$, et la fonction f est dite sommable. Un des principaux théorèmes de Daniell est le suivant : si $f_1, \ldots, f_n, \ldots$ est une suite de fonctions sommables ayant pour limite f, et s'il existe une fonction φ sommable telle que $|f_n| \leqq \varphi$ pour toutes les valeurs de n, alors f est sommable et $U(f_n)$ a pour limite $U(f)$. Il faut aussi, pour la suite, rappeler les résultats

(¹) P. J. Daniell, *A General Form of Integral* (*Annals of Mathematics*, 2ᵉ série, t. XIX, p. 279-294).

suivants : si f est sommable, cf et $|f|$ sont sommables et $U(cf) = cU(f)$; de plus, si f et g sont sommables $f+g$ est sommable et $U(f+g) = U(f) + U(g)$, et si $f \geq g$, $U(f) \geq U(g)$.

Enfin, si $f(x) \geq g(x) \geq h(x)$, $f(x)$ et $h(x)$ étant sommables et ayant même intégrale U, $g(x)$ est également sommable.

Dans mon précédent article, les éléments p de l'espace auquel j'appliquais la théorie de l'intégration ou de la moyenne étaient les fonctions continues $f(x)$ s'annulant avec x et définies dans l'intervalle $(0, 1)$. L'ensemble T_0 était celui des fonctionnelles bornées et telles qu'il existe une fonction $\varphi(x)$, s'annulant pour $x = 0$, pour laquelle

$$|F[f] - F[f+g]| < \varphi(\max|g|).$$

Nous désignions par $F(x_1, x_2, \ldots, x_n)$ la valeur de $F[f_n]$, où

$$f_n(t) = \sum_1^k x_i, \qquad \left[\frac{k-1}{n} < t \leq \frac{k}{n}\right],$$

et nous l'appelions $n^{\text{ième}}$ section de $F[f]$. Nous avons établi que pour toute fonctionnelle de l'ensemble T_0, la moyenne de la $n^{\text{ième}}$ section dans la sphère $\sum_1^n x_n^2 = r^2$ a, pour n infini, une limite que nous appelons moyenne de $F[f]$ dans la sphère de rayon r et désignons par $A_r\{F\}$. Nous avons montré que les résultats de Daniell s'appliquent dans ces conditions, et comme cas particulier nous avons considéré la moyenne de fonctionnelles telles que $\varphi(a_1, a_2, \ldots, a_n)$, les a_n ayant la même signification que plus haut, formule (2).

Nous allons maintenant, indépendamment de la théorie précédente, définir une intégrale de Daniell représentant la moyenne d'une fonction des a_n. Nous considérerons des éléments p d'une nature plus générale : toute suite de nombres réels a_1, a_2, ..., a_n, ... constitue un argument pour nos fonctionnelles, que la série (1) converge en moyenne ou non. Notre ensemble T_0 sera celui des fonctions $\varphi(a_1, a_2, \ldots, a_n)$ (n étant fini), bornées, uniformément continues, et nulles pour les valeurs suffisamment grandes des variables, et notre opération U sera définie par la formule

$$U_r \{\varphi(a_1, \ldots, a_n)\}$$
$$= \frac{1}{(2\pi r^2)^{\frac{n}{2}}} \int_{-\infty}^{+\infty} dx_1 \ldots \int_{-\infty}^{+\infty} dx_n \, e^{-\frac{1}{2r^2}\sum_1^n x_i^2} \varphi\left(\frac{x_1}{\pi}, \ldots, \frac{x_n}{\pi n}\right).$$

Il est évident que T_0 vérifie les conditions de Daniell, et que l'opération (U) vérifie les conditions (C), (A) et (P). Elle vérifie aussi la condition (L). On peut l'établir par la méthode employée par Daniell dans un autre Mémoire, en utilisant le fait que U_r, qui est une moyenne, ne peut pas dépasser le module maximum de son argument. La méthode de Daniell s'applique donc pour généraliser l'opération U_r.

L'opération U_r, appliquée à φ considéré comme fonction des a_n, coïncide avec l'opération A_r, appliquée à φ considéré comme fonctionnelle de f, la fonction f et les coefficients a_n étant liés par la relation (1). Alors, d'après le théorème de Daniell rappelé plus haut, pour toute fonction de l'ensemble T_1 liée à l'opération U_r, l'opération A_r s'applique et conduit à la même valeur. Il en résulte, d'après un autre théorème de Daniell également rappelé plus haut, que les extensions de ces deux opérations coïncident; en ce sens que dans tout domaine où U_r peut être défini, A_r peut être défini et U_r et A_r ont la même signification.

L'objet du présent travail est de montrer que le domaine où U_r est défini coïncide avec celui où A_r est défini, et que presque tous les systèmes de coefficients a_n (dans le sens qui résulte de la loi de probabilité indiquée plus haut), correspondent à des fonctions $f(x)$ continues. Dans mon précédent article (¹), j'ai démontré un théorème revenant à dire que la fonction égale à l'unité si $a_m^2 + \ldots + a_n^2 + \ldots > a^2$ et nulle dans le cas contraire a une semi-intégrale supérieure tendant vers zéro pour m infini. J'ai montré ensuite que, si $F(f)$ est une fonctionnelle bornée et uniformément continue au sens restreint, c'est-à-dire si à tout η positif correspond un θ positif tel que

$$\int_0^1 [f(t) - g(t)]^2 \, dt < \theta$$

(¹) *Differential-Space*, p. 171-172.

entraîne

$$|F[f]-F[g]|<\eta,$$

F est sommable A_r; et la méthode employée permet de voir que F, considéré comme fonction des a_n, est sommable U_r.

Établissons maintenant la sommabilité U_1 de fonctionnelles d'un type particulier. Si $f(x)$ est la fonction de carré sommable correspondant à la série des cosinus

$$a_0+a_1\sqrt{2}\cos\pi x+\ldots+a_n\sqrt{2}\cos n\pi x+\ldots,$$

la fonction

$$\int_0^1 f(\pi y)\frac{1-\rho^2}{1-2\rho\cos\pi(x-y)+\rho^2}dy,$$

qui correspond à la série

$$a_0+a_1\sqrt{2}\rho\cos\pi x+\ldots+a_n\sqrt{2}\rho^n\cos n\pi x+\ldots,$$

est continue par rapport à x pour tout $\rho<1$. Choisissons maintenant pour a_0 une fonction de ρ telle que notre nouvelle fonction s'annule pour $x=0$, c'est-à-dire que cette fonction sera

$$f_\rho(x)=\int_0^1 f(\pi y)(1-\rho^2)$$
$$\times\left[\frac{1}{1-2\rho\cos\pi(x-y)+\rho^2}-\frac{1}{1-2\rho\cos y+\rho^2}\right]dy.$$

Considérons alors une fonctionnelle de la forme

$$\varphi[f_\rho(x_1),\ \ldots,\ f_\rho(x_n)],$$

où φ est une fonction bornée et continue de ses arguments. On voit aisément que φ sera une fonctionnelle bornée de a_1, a_2, ..., continue au sens restreint indiqué tout à l'heure. Elle sera sommable A_1 et sommable U_1, et les deux opérations conduiront à la même valeur.

Il n'y a aucune difficulté à étendre un peu cette classe de fonctionnelles; φ peut être une fonction de n variables ne prenant que les valeurs 1 et 0, prenant la première de ces valeurs dans un ensemble mesurable borné. Une telle fonction peut en effet être considérée comme limite d'une suite décroissante de fonctions continues, et différentes de zéro seulement dans un domaine borné. D'ailleurs, d'après un autre théorème de Daniell, si deux fonctionnelles sont sommables, la fonctionnelle toujours égale à la plus grande des deux (ou à la plus petite) est sommable. On en conclut que, si une fonctionnelle de f est égale à 1 ou 0 suivant que des inégalités en nombre fini entre un nombre fini d'expressions $f_\rho(x)$ (les ρ et les x ayant des valeurs comprises entre 0 et 1) sont vérifiées ou non, cette fonctionnelle est sommable, pourvu que ces inégalités permettent de borner supérieurement les $f_\rho(x)$. Comme d'ailleurs la limite d'une suite décroissante de fonctionnelles sommables est elle-même sommable, on peut augmenter indéfiniment le nombre des valeurs considérées des ρ et des x, et obtenir par exemple pour les x l'ensemble des nombres rationnels de l'intervalle (0, 1), limites comprises, et pour les ρ un ensemble de nombres inférieurs à 1 mais approchant indéfiniment de cette valeur. Ainsi la fonctionnelle égale à 1 quand les $f_\rho(x)$ (pour les valeurs de ρ considérées) vérifient une condition de Lipschitz donnée pour toutes les valeurs rationnelles de x, et à 0 dans le cas contraire, est sommable U_1 et sommable A_1, les deux opérations donnant la même valeur; bien entendu, les fonctions $f_\rho(x)$ étant continues, la restriction que l'on ne considère que les valeurs rationnelles de x est sans importance, et peut être supprimée.

On peut remarquer que, si $f(x)$ vérifie une condition de Lipschitz, les $f_\rho(x)$ la vérifient également, car $f_\rho(x+h)-f_\rho(x)$ est une moyenne des valeurs de $f(x+h)-f(x)$, calculée avec des poids convenables. Par suite la probabilité [1] que tous les $f_\rho(x)$ vérifient une condition de Lipschitz est au moins égale à la probabilité que $f(x)$ la vérifie; et cela est vrai aussi bien en définissant la probabilité par les moyennes U_r ou A_r. Or j'ai démontré [2] que la probabilité qu'il existe deux nombres t_1 et t_2 compris entre 0 et 1 tels que

$$(3) \qquad |f(t_2)-f(t_1)| \leq ar\,|t_2-t_1|^{\frac{1}{2}-\varepsilon}$$

tend vers 1 quand a augmente indéfiniment, la probabilité étant

(1) La probabilité ou mesure d'un ensemble de fonctions est par définition la moyenne d'une fonctionnelle égale à 1 pour ces fonctions et 0 pour les autres.

(2) *Differential-Space*, p. 166.

liée à la moyenne A_r. Donc, qu'il s'agisse de moyenne A_r ou U_r, la probabilité que, pour tous les ρ considérés,

$$(4) \qquad |f_\rho(t_2) - f_\rho(t_1)| \leq ar\,|t_2 - t_1|^{\frac{1}{2}-\varepsilon},$$

tend vers 1 pour a infini.

Quel que soit a, si cette dernière égalité est vraie pour toutes les valeurs considérées de ρ, les $f_\rho(x)$ sont également continus. Il est donc possible de choisir parmi ces fonctions une suite convergeant uniformément vers une limite. Les $f_\rho(x)$ convergent d'ailleurs en moyenne vers f et ne peuvent converger uniformément vers une autre limite. La fonction $f(x)$ est donc la limite d'une suite uniformément convergente de $f_\rho(x)$, et vérifie toute condition d'égale continuité que les $f_\rho(x)$ vérifient. Donc, non seulement (4) résulte de (3), mais (3) résulte de (4). Donc l'ensemble des fonctions vérifiant l'inégalité (3) a une mesure U_1 aussi bien qu'une mesure A_1, et ces mesures coïncident. En particulier, la probabilité A_1 de l'inégalité (3) tendant vers 1 pour a infini, sa probabilité U_1 tend aussi vers 1. De ce résultat, et des théorèmes principaux de Daniell, nous pouvons conclure que l'ensemble des suites a_n ne correspondant à aucune fonction $f(x)$, ou correspondant à des fonctions pour lesquelles on ne puisse pas trouver a tel que pour tout système de valeurs de t_1 et t_2,

$$|f(t_2) - f(t_1)| \leq ar\,|t_2 - t_1|^{\frac{1}{2}-\varepsilon},$$

a une mesure nulle. Donc, à l'exception de suites constituant un ensemble de mesure nulle, toute suite de a_n représente une fonction continue.

Considérons maintenant les fonctionnelles définies, si f est continu, par la formule

$$F[f] = \varphi[f(x_1), \ldots, f(x_n)],$$

où φ est borné et continu, et si f est discontinu (1) par la formule

$$F[f] = 0.$$

(1) Nous comprenons aussi dans ce cas celui où f est une notation purement symbolique représentant une suite de coefficients a_n.

Nous considérons de même la fonctionnelle définie, si f est continu, par la formule

$$F_\rho[f] = \varphi[f_\rho(x_1), \ldots, f_\rho(x_n)]$$

et si f est discontinu par la formule

$$F_\rho[f] = 0.$$

Il est évident que, si f est continu,

$$\begin{aligned}\lim_{\rho \to 1} F_\rho[f] &= \lim_{\rho \to 1} \varphi[f_\rho(x_1), \ldots, f_\rho(x_n)] \\ &= \varphi\left[\lim_{\rho \to 1} f_\rho(x_1), \ldots, \lim_{\rho \to 1} f_\rho(x_n)\right] \\ &= \varphi[f(x_1), \ldots, f(x_n)] = F[f],\end{aligned}$$

et que, dans le cas contraire,

$$\lim_{\rho \to 1} F_\rho[f] = F[f].$$

En outre l'ensemble des F_ρ est uniformément borné, et chaque F_ρ est sommable comme étant toujours égal à la plus petite de deux fonctionnelles sommables. Alors, d'après les théorèmes de Daniell, $F[f]$ est sommable U_1.

On voit aisément que ce résultat subsiste si φ est une fonction simple (step-function), car une telle fonction peut être obtenue à la limite en partant de fonctions continues.

Cet ensemble de fonctionnelles est précisément l'ensemble T_0 d'un de mes précédents Mémoires ([1]), où j'étudiais la même opération A_r que dans mon Mémoire sur l'espace différentiel. Il en résulte que toute fonctionnelle sommable A_r au sens du présent Mémoire est sommable U_r, les deux opérations conduisant à la même moyenne, pourvu qu'elle s'annule pour toute fonction discontinue. Cette dernière restriction, qui ne s'applique qu'à des fonctions formant un ensemble de mesure nulle, peut d'ailleurs être omise. Les deux opérations U_1 et A_1 sont donc complètement identiques, ayant la même extension et conduisant à la même valeur.

Tous les résultats de mon article sur l'espace différentiel peuvent

---

([1]) *The Average Value of a Functional.*

donc être considérés comme des résultats sur les probabilités dénombrables ([1]).

([1]) Je rectifie ici quelques erreurs d'impression commises dans cet article.

Première formule non numérotée après la formule (78), page 164, lire, pour le champ d'intégration,

$$\int_0^1 \int_{t_1}^1 \cdots \int_{t_{\nu-1}}^1 .$$

Même correction pour la première formule de la page 165; dans cette formule, remplacer aussi t_n par t_ν.

Même page, formule (C), lire

$$U(cf) = cU(f).$$

Comments on [24e]

J. -P. Kahane

This paper discusses "Differential spaces," but contains a new introduction to Brownian motion, using the Fourier-Wiener series (1) (p. 570, where the a_n are random variables) as a definition.

Notes on random functions.

By

R. E. A. C. Paley †, N. Wiener and A. Zygmund.

---

1. The introduction of the notion of the random into analysis is in the first instance the work of Borel[1]). His theory of *probabilités dénombrables* concerns itself with quantities depending upon an infinite sequence of choices and with their average values. In the simplest case, the choices in question are between two alternatives, which may be taken as the signs + and —. Thus questions as to the probability of convergence of the series

$$\sum_{n=0}^{\infty} \pm C_n \tag{1.01}$$

and of the distribution of its sums belong to this order of ideas. Such questions have been much discussed in the pages of Fundamenta Mathematicae and Studia Mathematica, and are associated with the names of Steinhaus and Rademacher, among others. To Steinhaus in particular is due the reduction of such questions to questions concerning the Lebesgue integral.

The next step forward in introducing the notion of probability into analysis consists in replacing numbers dependent on an infinite sequence of choices by functions dependent on an infinite sequence of choices. This step was taken by Paley and Zygmund[2]), who have developed the theory of the analytic properties shared by "almost all" functions of the form

$$\sum_{n=0}^{\infty} \pm C_n f_n(x), \tag{1.02}$$

and the closely related problem of the properties shared by "almost all" functions of the form

$$\sum_{n=0}^{\infty} C_n e^{2\pi i \vartheta_n} f_n(x), \tag{1.03}$$

---

[1]) Borel.

[2]) Paley and Zygmund (1), (2), (3) hereinafter referred to as PZ 1, 2, 3.

where the numbers ϑ_n vary independently in the interval (0, 1). In this they were systematically developing a method of Littlewood for the production of *Gegenbeispiele*. From this point of view, random functions present themselves as a somewhat sophisticated analytical tool, and may seem to belong rather to the pathology than to the natural history of functions of the real variable.

There is however a way in which functions with an element of randomness present themselves to the physicist. The Brownian movement, the motion of a suspended particle in a fluid in response to the irregularities of the molecular collisions to which it is subjected, is of so strikingly random a character that it does not even seem to possess a velocity. The physicist Perrin[3]) has commented on the striking way in which this motion suggests the non-differentiable continuous functions of Weierstrass, and Borel[4]) has commented on this in connection with his theory of denumerable probabilities. The components of the displacement of such a particle, regarded as functions of the time, are random functions in a sense differing considerably from that of Zygmund and Paley. Wiener[5]) has reduced the theory of such random functions to a mathematically definite form, also depending upon a Lebesgue integration. For the purposes of the present paper, we shall represent one of the components of the displacement of the moving particle in the Wiener theory by $\chi(\alpha, t)$, where t is the time and α the parameter on which Lebesgue integration is performd for the purpose of averaging over all functions and determining probabilities. Wiener's $\chi(\alpha, t)$ is then, as he shows, a continuous function of t for almost all values of α, defined for $0 \leqq \alpha \leqq 1$, $0 \leqq t \leqq 1$, and vanishing for $t = 0$.

The purpose of this paper is to bridge the gap between the PZ and the W theories, by proving for the W random functions theorems analogous to those proved in the PZ papers. The analogy in question takes two different forms, which are in a manner of speaking Fourier duals of one another. The simplest and most obvious correspondence between the two theories is that $\chi(\alpha, t)$ determines what is really the limit of a large number of choices of sign of small quantities, each choice being whether over a small period of time a particle subject to the Brownian motion will wander to the right or to the left. Thus formally

$$(1.04) \qquad \int_0^1 f(x, t)\, c(t)\, d_t \chi(\alpha, t)$$

[3]) E. Perrin (1)

[4]) E. Borel (1), (2).

[5]) N. Wiener (1).

is an analogue to (1.02). On the other hand, it may be established that if $\{\gamma_n(t)\}$ is a set of normal and orthogonal functions (0, 1), the quantities

$$\int_0^1 \gamma_n(t)\, d_t \chi(\alpha, t) \tag{1.05}$$

appropriately defined all have a Gaussian distribution with the same modulus, and are completely independent of one another. Thus

$$\sum_{n=0}^{\infty} c_n f_n(x) \int_0^1 \gamma_n(t)\, d_t \chi(\alpha, t) \tag{1.06}$$

is an analogue of (1.02), and like it depends on an infinite sequence of like independent probabilities, with the difference that the probabilities of (1.02) are all-or-none probabilities such as arise in the tossing of a coin, whereas the probabilities of (1.06) are continuously distributed like those of the errors in shooting at a target.

It will turn out that by far the greater part of the PZ theory has a precise W analogue in both possible senses. The W theory is a little less adapted to the construction of Gegenbeispiele than the PZ theory, in that the absolute magnitude of the coefficients in (1.02) is determined in advance, while that in (1.06) itself depends on probability considerations. On the other hand, many averages concerning which the PZ theory only yields inequalities are expressible in closed form in the *W* theory.

If the finite range of t in $\chi(\alpha, t)$ is replaced by an infinite range, as is easily possible, the theory undergoes but little change. On the other hand, the Fourier series is replaced by the Fourier integral, and the symmetry of the latter makes the two analogues of the PZ theory reduce essentially to one. Furthermore, a new type of functional arises over the infinite range — the mean as contrasted with the integral, and the evaluation of the distribution of such means introduces new considerations. Under certain circumstances, such means will almost always assume a certain fixed value.

2. By W 1, if $t_0 = 0 \leqq t_1 \leqq t_2 \leqq \dots \leqq t_n \leqq 1$,

$$\int_0^1 F(\chi(\alpha, t_1), \chi(\alpha, t_2) - \chi(\alpha, t_1), \dots, \chi(\alpha, t_n) - \chi(\alpha, t_{n-1}))\, d\alpha \tag{2.01}$$

$$= \pi^{-n/2} \prod_1^n (t_k - t_{k-1})^{-1/2} \int_{-\infty}^{\infty} du_1 \dots \int_{-\infty}^{\infty} du_n$$

$$\exp.\left[-\sum_1^n \frac{u_k^2}{t_k - t_{k-1}}\right] F(u_1, \dots, u_n).$$

From this it is easy to show that if $t_2 \geqq t_1$,

$$(2.02) \qquad \int_0^1 [\chi(\alpha, t_2) - \chi(\alpha, t_1)]^{2n} d\alpha = \frac{1 \cdot 3 \cdot 5 \ldots 2n-1}{2^n} (t_2 - t_1)^n;$$

$$\int_0^1 [\chi(\alpha, t_2) - \chi(\alpha, t_1)]^{2n+1} d\alpha = 0;$$

and that if none of the intervals $(t_{1,p}, t_{2,p})$ overlap,

$$(2.03) \qquad \int_0^1 \prod_{p=1}^{\nu} [\chi(\alpha, t_{2,p}) - \chi(\alpha, t_{1,p})]^{n_p} d\alpha$$

$$= \prod_{p=1}^{\nu} \int_0^1 [\chi(\alpha, t_{2,p}) - \chi(\alpha, t_{1,p})]^{n_p} d\alpha.$$

Since $1 \cdot 3 \cdot 5 \ldots (2n-1)$ represents the number of distinct ways of parcelling $2n$ objects into pairs, it follows from (2.02) and (2.03) that the integral in (2.03) may be evaluated by parcelling the product in the integrand into pairs in all possible ways, integrating the product of each pair, multiplying the integrals for all the pairs in each parcelling, and adding these products over all parcellings. From this it results immediately that

$$(2.04) \qquad \int_0^1 \chi(\alpha, t_1) \chi(\alpha, t_2) \ldots \chi(\alpha, t_{2n}) d\alpha$$

$$= \int_0^1 \chi(\alpha, t_1) \chi(\alpha, t_2) d\alpha \int_0^1 \chi(\alpha, t_3) \chi(\alpha, t_4) d\alpha \ldots \int_0^1 \chi(\alpha_1, t_{2n-1}) \chi(\alpha, t_{2n}) d\alpha$$

+ all similar sums for all other parcellings of $t_1, \ldots, t_{2n}$ into pairs;

$$\int_0^1 \chi(\alpha, t_1) \chi(\alpha, t_2) \ldots \chi(\alpha, t_{2n+1}) d\alpha = 0.$$

This enables us to establish:

Lemma 1. *Let $\varrho_1(t)$, $\varrho_2(t)$, ... $\varrho_{2n}(t)$ be functions of limited total variation over* (0. 1). *Then*

$$(2.05) \qquad \int_0^1 d\alpha \int_0^1 \chi(\alpha, t_1) d\varrho_1(t_1) \int_0^1 \chi(\alpha, t_2) d\varrho_2(t_2) \ldots \int_0^1 \chi(\alpha, t_{2n}) d\varrho_{2n}(t_{2n})$$

$$= \int_0^1 d\varrho_1(t_1) \ldots \int_0^1 d\varrho_{2n}(t_{2n}) \int_0^1 d\alpha_1 \ldots \int_0^1 d\alpha_n \{\chi(\alpha_1, t_1) \chi(\alpha_1 t_2) \chi(\alpha_2, t_3) \chi(\alpha_2, t_4) \ldots$$

$$\chi(\alpha_n, t_{2n-1}) \chi(\alpha_n, t_{2n}) + \text{etc.}\},$$

where the "etc." stands for all other distributions of the $2n$ t_k's over the n α_k's, two to each, permutations of the α_k's among themselves not being counted as distinct.

Another very easy result to establish is that if ϱ_1 and ϱ_2 are of limited total variation and $\varphi_1(1) = \varphi_2(1) = 0$,

$$\text{(2.06)} \qquad \int_0^1 d\alpha \int_0^1 \chi(\alpha, t_1)\, d\varrho_1(t_1) \int_0^1 \chi(\alpha, t_2)\, d\varrho_2(t_2)$$

$$= \tfrac{1}{2} \int_0^1 t_1\, d\varrho_1(t_1) \int_{t_1}^1 d\varrho_2(t_2) + \tfrac{1}{2} \int_0^1 t_2\, d\varrho_2(t_2) \int_{t_2}^1 d\varrho_1(t_1)$$

$$= -\tfrac{1}{2} \int_0^1 t \{\varrho_2(t)\, d\varrho_1(t) + \varrho_1(t)\, d\varrho_2(t)\} = \tfrac{1}{2} \int_0^1 \varrho_1(t)\, \varrho_2(t)\, dt.$$

Thus we may give to Lemma 1 the somewhat compacter form:

Lemma 1′. *Let* $\varrho_1(t), \ldots \varrho_{2n}(t)$ *be functions of limited total variation over* $(0, 1)$. *Then*

$$\text{(2.07)} \qquad \int_0^1 d\alpha \int_0^1 \chi(\alpha, t_1)\, d\varrho_1(t_1) \ldots \int_0^1 \chi(\alpha, t_{2n})\, d\varrho_{2n}(t_{2n})$$

$$= 2^{-n} \int_0^1 \varrho_1(t)\, \varrho_2(t)\, dt \int_0^1 \varrho_3(t)\, \varrho_4(t)\, dt \ldots \int_0^1 \varrho_{2n-1}(t)\, \varrho_{2n}(t)\, dt + \text{etc.}$$

In particular, if $\int_0^1 [\varrho(t)]^2 = 1$, and $\varrho(t)$ is of limited total variation,

$$\text{(2.08)} \qquad \int_0^1 d\alpha \left[\int_0^1 \chi(\alpha, t)\, d\varrho(t) \right]^{2n} = \frac{1 \cdot 3 \cdot 5 \ldots 2n - 1}{2^n}.$$

From this we may readily prove that if $P(w)$ is any polynomial in w,

$$\text{(2.09)} \qquad \int_0^1 d\alpha\, P\left[\int_0^1 \chi(\alpha, t)\, d\varrho(t) \right] = \frac{1}{\sqrt{\pi}} \int_{-\infty}^{\infty} P(u)\, e^{-u^2}\, du.$$

We may readily extend this result to:

Lemma 2. *Let* $\varrho(t)$ *be a function of limited total variation, and let* $\int_0^1 [\varrho(t)]^2\, dt = 1$. *Let*

$$\text{(2.10)} \qquad I = \frac{1}{\sqrt{\pi}} \int_{-\infty}^{\infty} F(u)\, e^{-u^2}\, du$$

exist as a Lebesgue integral. Then

$$\text{(2.11)} \qquad \int_0^1 d\alpha\, F\left[\int_0^1 \chi(d, t)\, d\varrho(t) \right] = I.$$

This may be expressed in words by saying that $\int_0^1 \chi(\alpha, t)\, d\varrho(t)$ has a Gaussian distribution with modulus dependent only on $\int_0^1 [\varrho(t)]^2\, dt$.

A reasoning of the same sort will establish:

Lemma 3. *Let $\varrho_1(t), \ldots, \varrho_n(t)$ be functions of limited total variation for which*

$$\int_0^1 \varrho_k(t)\,\varrho_2(t)\,dt = \begin{cases} 0; & [k \neq l] \\ 1; & [k = l] \end{cases} \tag{2.12}$$

and let

$$\pi^{-n/2} \int_{-\infty}^{\infty} du_1 \ldots \int_{-\infty}^{\infty} du_n F(u_1, \ldots, u_n)\, e^{-\sum_1^n u_k^2} \tag{2.13}$$

exist as a Lebesgue integral. Then

$$\int_0^1 d\alpha\, F\left[\int_0^1 \chi(\alpha, t)\, d\varrho_1(t), \ldots, \int_0^1 \chi(\alpha, t)\, d\varrho_n(t)\right] \tag{2.14}$$

exists and has the same value.

Let it be noted that we have hereby established that the set of functions

$$g_{\lambda_1, \lambda_2, \ldots, \lambda_n}(\alpha) = \int_0^1 \chi(\alpha, t)\, d\varrho_{\lambda_1}(t) \ldots \int_0^1 \chi(\alpha, t)\, d\varrho_{\lambda_n}(t) \qquad [\lambda_j \neq \lambda_k \text{ if } j \neq k] \tag{2.15}$$

is normal and orthogonal. They are analogous to a set of normal and orthogonal functions given by Walsh, and assuming only the values ± 1. Unlike the Walsh functions, they are continuous.

We may put Lemma 3 more strikingly by saying that the quantities $\int_0^1 \chi(\alpha, t)\, d\varrho_u(t)$ are completely independent — not merely linearly independent — of one another, and that they all have a Gaussian distribution with the same modulus.

3. Let us now introduce:

Definition 1. *If $\varrho(t)$ is a function of limited total variation over $(0, 1)$, we shall write*

$$\int_0^1 \varrho(t)\, d\chi(\alpha, t) = \varrho(1)\,\chi(\alpha, 1) - \int_0^1 \chi(\alpha, t)\, d\varrho(t) \tag{3.01}$$

As before, let $\{\varrho_n(t)\}$ be a complete set of normal and orthogonal functions of limited total variation. Let $\sum_1^\infty a_n^2$ converge. We wish to consider the behaviour of

$$\int_0^1 \left[\sum_1^n a_k \varrho_k(t)\right] d\chi(\alpha, t) \tag{3.02}$$

as $n \to \infty$. By Lemma 3, this question may be answered in terms of the distribution of $\sum_1^n a_k u_k$ when the individual variables u_k have independent Gaussian distributions with the same modulus. This again is formally the same question as that of the distribution of $\chi(\alpha, a_1^2 + a_2^2 + \ldots + a_n^2)$. As has been said, Wiener has shown that for almost all values of α, $\chi(\alpha, t)$ is continuous in t. Thus almost always

$$(3.03) \qquad \lim_{n \to \infty} \chi(\alpha, \sum_1^n a_k^2) = \chi(\alpha, \sum_1^\infty a_k^2)$$

from which it follows that $\sum_1^\infty a_k u_k$ almost always converges, and that

$$(3.04) \qquad \lim_{n \to \infty} \int_0^1 \left[\sum_1^n a_k \varrho_k(t)\right] d\chi(\alpha, t)$$

almost always exists. We now introduce:

Definition 2. *Let $\varrho(t)$ belong to L_2, and let*

$$(3.05) \qquad \varrho(t) \sim \sum_1^\infty a_k \varrho_k(t).$$

Then we shall put

$$(3.06) \qquad \int_0^1 \varrho(t)\, d\chi(\alpha, t) = \lim_{n \to \infty} \int_0^1 \left[\sum_1^n a_k \varrho_k(t)\right] d\chi(\alpha, t).$$

We now wish to show that Definitions 1 and 2 are consistent, in the sense that they almost always lead to the same value of $\int_0^1 \varrho(t)\, d\chi(\alpha, t)$ when they are both applicable. Let us then adhere to Definition 1, and let us suppose that

$$(3.07) \qquad \int_0^1 [\varrho(t)]^2 dt - \sum_1^n a_k^2 < \varepsilon^2.$$

It will follow from Lemma 2 that

$$(3.08) \qquad \left| \int_0^1 \varrho(t)\, d\chi(\alpha, t) - \int_0^1 \left[\sum_1^n a_k \varrho_k(t)\right] d\chi(\alpha, t) \right| < \varepsilon$$

except over a set of values of α of measure not exceeding

$$(3.09) \qquad \frac{1}{\varepsilon \sqrt{\pi}} \int_{-\infty}^{\infty} (\varepsilon u)^2 e^{-u^2} du = \varepsilon/2.$$

Thus (3.08) cannot tend almost everywhere to any limit other than 0, and we have established the consistency of the two definitions. A precisely similar argument will show that Definition 2 is independent of the particular choice of the set $\{\varrho_k(t)\}$.

Let it be noted that by (2.08)

$$(3.10) \qquad \int_0^1 d\alpha \left[\sum_1^n a_k \int_0^1 \varrho_k(t)\, d\chi(\alpha, t)\right]^{2m} \leq \frac{1 \cdot 3 \cdot 5 \ldots 2m-1}{2^m} \left[\int_0^1 [\varrho(t)]^2 dt\right]^m.$$

42*

Let it also be noted that

$$\text{(3.11)} \qquad \lim_{n\to\infty} \int_0^1 d\alpha \left\{ \int_0^1 \left[\sum_1^n a_k \varrho_k(t)\, d\chi(\alpha, t) - \int_0^1 \varrho(t)\, d\chi(\alpha, t) \right\}^2 = 0.$$

This results from (3.06) and the fact that by (2.08)

$$\text{(3.111)} \qquad \lim_{m,n\to\infty} \int_0^1 d\alpha \left\{ \int_0^1 \sum_m^n a_k \varrho_k(t) \right] d\chi(\alpha, t) \right\}^2 = 0.$$

From (3.11) and (3.10) it readily follows that

$$\text{(3.12)} \qquad \int_0^1 d\alpha \left[\int_0^1 \varrho(t)\, d\chi(\alpha, t) \right]^{2m} \leqq \frac{1 \cdot 3 \cdot 5 \ldots 2m-1}{2^m} \left[\int_0^1 [\varrho(t)]^2\, dt \right]^m.$$

Applying this to

$$\text{(3.13)} \qquad \varrho_*(t) = \varrho(t) - \sum_1^n a_k \varrho_k(t),$$

we get

$$\text{(3.14)} \qquad \lim_{n\to\infty} \int_0^1 d\alpha \left[\int_0^1 \varrho_*(t)\, d\chi(\alpha, t) \right]^{2m} = 0.$$

This together with (2.08) yields

$$\text{(3.15)} \qquad \int_0^1 d\alpha \left[\int_0^1 \varrho(t)\, d\chi(\alpha, t) \right]^{2m} = \frac{1 \cdot 3 \cdot 5 \ldots 2m-1}{2^m} \left[\int_0^1 [\varrho(t)]^2\, dt \right]^m.$$

We have thus established:

Lemma 2′. *Let*

$$\text{(2.10)} \qquad I = \frac{1}{\sqrt{\pi}} \int_{-\infty}^{\infty} F(u)\, e^{-u^2}\, du$$

exist as a Lebesgue integral, and let $\int_0^1 [\varrho(t)]^2\, dt = 1$. *Then*

$$\text{(3.16)} \qquad \int_0^1 d\alpha\, F\left[\int_0^1 \varrho(t)\, d\chi(\alpha, t) \right] = I.$$

Lemma 3 also possesses an extension of the same sort to functions $\varrho_k(t)$ not of limited total variation. Once these lemmas are established, it follows that Definition 2 will yield the same result even if the ϱ_k's are not functions of limited total variation. Of course, in this form it would be circular and of no value as a definition.

4. Let us now consider a function of t and the parameter τ of the form

$$\text{(4.01)} \qquad \Phi_n(\tau, t) = \sum_1^n \sigma_k(\tau)\, \varrho_k(t)$$

where the functions $\varrho_k(t)$ are normal and orthogonal over (0, 1) and

$$\text{(4.02)} \qquad \sum_1^n \int_0^1 [\sigma_k(\tau)]^2\, d\tau = S < \infty.$$

Then except for a set of values of τ of zero measure,

$$(4.03)\qquad \int_0^1 d\alpha\left[\int_0^1 \Phi_n(\tau,t)\,d\chi(\alpha,t)\right]^{2m} = \frac{1\cdot 3\cdot 5\ldots 2m-1}{2^m}\left[\sum_1^n [\sigma_k(\tau)]^2\right]^m$$

$$= \frac{1\cdot 3\cdot 5\ldots 2m-1}{2^m}\left[\int_0^1 [\Phi_n(\tau,t)]\,dt\right]^m.$$

It follows that

$$(4.04)\qquad \int_0^1 d\alpha \int_0^1 d\tau\left[\int_0^1 \Phi_n(\tau,t)\,d\chi(\alpha,t)\right]^{2m}$$

$$= \frac{1\cdot 3\cdot 5\ldots 2m-1}{2^m}\int_0^1\left[\sum_1^n [\sigma(\tau)]^2\right]^m d\tau.$$

Thus if $\Phi(\tau,t)$ is any function belonging to L_2 as a function of t and τ simultaneously, and

$$(4.05)\qquad \Phi_n(\tau,t) = \sum_1^n \varrho_k(t)\int_0^1 \Phi(\tau,s)\,\varrho_k(s)\,ds,$$

an argument precisely similar to that of the last paragraph will show that for almost all values of α

$$(4.06)\qquad \underset{n\to\infty}{\text{l. i. m}} \int_0^1 \Phi_n(\tau,t)\,d\chi(\alpha,t)$$

exists, and that as before we may *define* $\int_0^1 \Phi(\tau,t)\,d\chi(\alpha,t)$ by

$$(4.07)\qquad \int_0^1 \Phi(\tau,t)\,d\chi(\alpha,t) = \underset{n\to\infty}{\text{l. i. m.}} \int_0^1 \Phi_0(\tau,t)\,d\chi(\alpha,t).$$

The consistency of this definition with previous definitions and its independence of the choice of ϱ_n's involve no new discussion. If we reflect that the finiteness of

$$(4.08)\qquad \int_0^1 d\tau\left[\int_0^1 [\Phi(\tau,t)]^2\,dt\right]^m$$

carries with it the boundedness of

$$(4.09)\qquad \int_0^1 d\tau\left[\int_0^1 [\Phi_n(\tau,t)]^2\,dt\right]^m$$

for all n, and the fact that

$$(4.10)\qquad \lim_{m,n\to\infty} \int_0^1 d\tau\left[\int_0^1 [\Phi_m(\tau,t)-\Phi_n(\tau,t)]^2\,dt\right]^m = 0,$$

then the methods by which we have established Lemma 3′ are adequate to establish:

Theorem I. *If*

$$\int_0^1 d\tau \left[\int_0^1 [\Phi(\tau, t)]^2 \, dt\right]^m < \infty, \tag{4.11}$$

then

$$\int_0^1 d\alpha \int_0^1 d\tau \left[\int_0^1 \Phi(\tau, t)\, d\chi(\alpha, t)\right]^{2m} \tag{4.12}$$

$$= \frac{1 \cdot 3 \cdot 5 \ldots 2m-1}{2^m} \int_0^1 d\tau \left[\int_0^1 [\Phi(\tau, t)]^2 \, dt\right]^m,$$

and

$$\int_0^1 d\tau \left[\int_0^1 \Phi(\tau, t)\, d\chi(\alpha, t)\right]^{2m} \tag{4.13}$$

exists for almost all values of α.

Particular cases for which (4, 11) is valid in case $q = 2m$ are: (a)

$$\Phi(\tau, t) = \beta(t) f(\tau, t), \tag{4.14}$$

where $\beta(t)$ belongs to L_2 and

$$\int_0^1 |f(\tau, t)|^q \, d\tau < B \tag{4.15}$$

and B is an absolute constant; (b)

$$\Phi(\tau, t) \sim \sum_1^\infty b_n f_n(\tau) \varrho_n(t), \tag{4.16}$$

where

$$\sum_1^\infty b_n^2 < \infty; \tag{4.17}$$

$$\int_0^1 \varrho_k(t) \varrho_l(t)\, dt = \begin{cases} 1; & [k = l] \\ 0; & [k \neq l] \end{cases}, \tag{4.18}$$

$$\int_0^1 |f_n(\tau)|^q \, d\tau < B; \tag{4.19}$$

and B is an absolute constant. These two cases form the two different analogues by the two methods we have mentioned, of Theorem II of PZ 1. A more precise analogue of this is:

Theorem II. *If hypothesis* (a) *or hypothesis* (b) *holds, and* $q \geqq 2$, *without necessarily being an integer,*

$$\int_0^1 d\tau \left|\int_0^1 \Phi(\tau, t)\, d\chi(\alpha, t)\right|^q \tag{4.20}$$

exists for almost all values of α.

Here the proof is exactly alike in the two cases. We have in case (a) for $2m-2 < q \leqq 2m$

$$(4.21)\quad \left\{\int_0^1 d\alpha \left|\int_0^1 \Phi(\tau,t)\,d\chi(\alpha,t)\right|^q\right\}^{1/q} \leqq \left\{\int_0^1 d\alpha\left[\int_0^1 \Phi(\tau,t)\,d\chi(\alpha,t)\right]^{2m}\right\}^{1/2m}$$

$$= \left(\frac{1\cdot 3\cdot 5\ldots 2m-1}{2^m}\right)^{1/2m}\left[\int_0^1 [\beta(t)]^2[f(\tau,t)]^2\,dt\right]^{1/2}$$

$$\leqq C_q\left[\int_0^1 [\beta(t)]^2\,|f(\tau,t)|^q\,dt\right]^{1/q}\left[\int_0^1 [\beta(t)]^2\,dt\right]^{q-2/2q}$$

where C_q depends only on q.

Raising both sides to the qth power, and integrating with respect to τ, we get:

$$(4.22)\quad \int_0^1 d\alpha\int_0^1 d\tau\left|\int_0^1 \Phi(\tau,t)\,d\chi(\alpha,t)\right|^q$$

$$\leqq C_q^q\left[\int_0^1 [\beta(t)]^2\,dt\right]^{q-2/2}\int_0^1 d\tau\int_0^1 [\beta(t)]^2\,|f(\tau,t)|^q\,dt$$

$$\leqq C_q^q\left[\int_0^1 [\beta(t)]^2\,dt\right]^{q/2} B,$$

from which Theorem II follows at once.

It follows at once from Theorem I that:

Theorem III. *Let*

$$(4.23)\quad \int_0^1 [\Phi(\tau,t)]^2\,dt \leqq w^2$$

where w is independent of τ. Then if

$$(4.24)\quad G(u) = \sum_0^\infty a_n u^n$$

has only positive coefficients in its Maclaurin series, and

$$(4.25)\quad \frac{1}{\sqrt{\pi}}\int_{-\infty}^{\infty} G(uw)\,e^{-u^2}\,du < \infty$$

then for almost all α

$$(4.26)\quad \int_0^1 d\tau\, G\left[\int_0^1 \Phi(\tau,t)\,d\chi(\alpha,t)\right] < \infty.$$

In particular, almost always

$$(4.27)\quad \int_0^1 d\tau \exp\lambda\left[\int_0^1 \Phi(\tau,t)\,d\chi(\alpha,t)\right]^2 < \infty$$

where

$$(428)\quad \lambda < 1/w^2.$$

In the particular case where $\Phi(\tau, t)$ is of form (4.16), where (4.17) and (4.18) hold, and where (4.19) is replaced by the condition that $f_n(\tau)$ is bounded for all values of n and τ, the methods of Theorem III of PZ 1 will enable us to show that condition (4.28) is unnecessary.

5. Now that we have obtained the formulae for the mean values of expressions of the form

$$\int_0^1 d\tau \left[\int_0^1 \Phi(\tau, t)\, d\chi(\alpha, t) \right]^{2m} \tag{5.01}$$

let us investigate the mean squares of the departures from these means. We have formally:

$$\begin{aligned} &\int_0^1 d\alpha \left\{ \int_0^1 d\tau \left[\int_0^1 \Phi(\tau, t)\, d\chi(\alpha, t) \right]^{2m} \right. \\ &\qquad \left. - \frac{1 \cdot 3 \cdot 5 \ldots 2m-1}{2^m} \int_0^1 d\tau \left[\int_0^1 [\Phi(\tau, t)]^2\, dt \right]^m \right\}^2 \\ &= \int_0^1 d\alpha \int_0^1 d\sigma \int_0^1 d\tau \left[\int_0^1 \Phi(\tau, t)\, d\chi(\alpha, t) \right]^{2m} \int_0^1 \Phi(\sigma, t)\, d\chi(\alpha, t) \Big]^{2m} \\ &\qquad - \left\{ \frac{1 \cdot 3 \cdot 5 \ldots 2m-1}{2^m} \int_0^1 d\tau \left[\int_0^1 [\Phi(\tau, t)]^2\, dt \right]^m \right\}^2. \end{aligned} \tag{5.02}$$

If we evaluate (5.02) in accordance with Lemma 1′, we get for the case $m = 1$

$$\frac{1}{2} \int_0^1 d\sigma \int_0^1 d\tau \left[\int_0^1 \Phi(\sigma, t)\, \Phi(\tau, t)\, dt \right]^2; \tag{5.03}$$

for the case $m = 2$,

$$\begin{aligned} &\frac{9}{8} \int_0^1 d\sigma \int_0^1 d\tau \int_0^1 \Phi(\sigma, t)\, \Phi(\tau, t)\, dt \int_0^1 [\Phi(\sigma, t)]^2\, dt \int_0^1 [\Phi(\tau, t)]^2\, dt \\ &\qquad + \frac{3}{4} \int_0^1 d\sigma \int_0^1 d\tau \left[\int_0^1 \Phi(\sigma, t)\, \Phi(\tau, t)\, dt \right]^3; \end{aligned} \tag{5.04}$$

and for the case $m = 3$,

$$\begin{aligned} &\frac{9}{2} \int_0^1 d\sigma \int_0^1 d\tau \left[\int_0^1 \Phi(\sigma, t)\, \Phi(\tau, t)\, dt \right]^2 \int_0^1 [\Phi(\sigma, t)]^2\, dt \int_0^1 [\Phi(\tau, t)]^2\, dt \\ &\qquad + \frac{3}{2} \int_0^1 d\sigma \int_0^1 d\tau \left[\int_0^1 \Phi(\sigma, t)\, \Phi(\tau, t) \right]^4. \end{aligned} \tag{5.05}$$

By the Schwarz inequality, each of these expressions will be finite with the appropriate

$$\int_0^1 d\tau \left[\int_0^1 [\Phi(\tau, t)]^2\, dt \right]^m. \tag{5.06}$$

In the first instance, we have only established (5.03), (5.04), and (5.05) in the case of functions $\Phi(\sigma, t)$ of limited total variation in t, but there is no difficulty in extending them by processes now familiar to the general case for which (5.06) is finite.

If the mean square departure of a quantity from the value A is B^2, the quantity cannot differ from A by an amount exceeding B except over a set of values of measure not exceeding B^2/C_u^2. We thus obtain:

Theorem IV. *Let*

$$\frac{1}{2}\int_0^1 d\tau \int_0^1 dt\,[\Phi_n(\tau, t)]^2 = A \qquad [n = 1, 2, \ldots] \tag{5.07}$$

and let

$$\sum_{n=1}^{\infty}\left\{\frac{1}{2}\int_0^1 d\sigma \int_0^1 d\tau\left[\int_0^1 \Phi_n(\sigma, t)\,\Phi_n(\tau, t)\,dt\right]^2\right\}^{1/2} \tag{5.08}$$

converge. Then except for a set of values of α of zero measure,

$$\lim_{n\to\infty}\int_0^1 d\tau\left[\int_0^1 \Phi_n(\tau, t)\,d\chi(\alpha, t)\right]^2 = A. \tag{5.09}$$

Similar theorems hold for cases (5.04), (5.05), etc.

In particular, let

$$\Phi_n(\tau, t) = n^2 \text{ if } \tau - \frac{1}{n^4} < t < \tau,\ a + \frac{1}{n^4} < \tau < t; \tag{5.10}$$

$$\Phi_n(\tau, t) = 0 \text{ otherwise.}$$

Then

$$\int_0^1 \Phi_n(\sigma, t)\,\Phi_n(\tau, t)\,dt = O(1) \text{ if } |\sigma - \tau| < n^{-4}; \tag{5.11}$$

$$\int_0^1 \Phi_n(\sigma, t)\,\Phi_n(\tau, t)\,dt = O \text{ otherwise.}$$

Thus the series of (5.08) reduces to

$$\sum_{n=1}^{\infty} O(n^{-2}) < \infty, \tag{5.12}$$

Formula (5.09) becomes

$$\lim_{n\to\infty} n^4 \int_{a+\frac{1}{n^4}}^{b}\left[\chi(\alpha, \tau) - \chi\left(\alpha, \tau - \frac{1}{n^4}\right)\right]^2 d\tau = \frac{1}{2}(b - a). \tag{5.13}$$

We have already proved, however (W 1) that except for a set of values of α of zero measure, there is an A such that

$$[\chi(\alpha, \tau) - \chi(\alpha, \sigma)]^2 < A\,|\tau - \sigma|^{1/2} \tag{5.14}$$

for all τ and σ. Without substantial change, the argument will show that there is almost always an A such that

$$[\chi(\alpha, \tau) - \chi(\alpha, \sigma)]^2 < A\,|\tau - \sigma|^{9/10} \tag{5.15}$$

for all τ and σ. Thus if $(n+1)^{-4} \leqq \varepsilon \leqq n^{-4}$,

$$(5.16) \qquad n^4 \int_{a+n^{-4}}^{b} [\chi(\alpha, \tau-\varepsilon) - \chi(\alpha, \tau - n^{-4})]^2 \, d\tau = O(n^{-2})$$

uniformly in ε, so that by (5.13) we have almost always

$$(5.17) \qquad \lim_{n\to\infty} n^4 \int_{a+n^{-4}}^{b} [\chi(\alpha,\tau) - \chi(\alpha, \tau-\varepsilon)]^2 \, d\tau = \frac{1}{2}(b-a)$$

and hence

$$(5.18) \qquad \lim_{\varepsilon\to 0} \frac{1}{\varepsilon} \int_{a+\varepsilon}^{b} [\chi(\alpha,\tau) - \chi(\alpha,\tau-\varepsilon)]^2 \, d\tau = \frac{1}{2}(b-a)$$

as $\varepsilon \to 0$ in any way whatever. This is almost always, true for every rational a and b, and it may consequently be proved to hold almost always for *every* a and b in $(0,1)$

6. We now come to the consideration of random functions over an infinite range. If we put:

$$(6.01) \qquad \xi(\alpha, t) = \sqrt{\pi} \int_{1/2}^{\frac{1}{\pi}\cot^{-1}(-t)} \chi(\alpha, u) \, d \csc \pi u - \chi\left(\alpha \frac{1}{\pi} \cot^{-1}(-t)\right) \sqrt{1+t^2} + \chi(\alpha, \tfrac{1}{2})$$

we shall find that $\xi(\alpha, 0) = 0$, and that if $-\infty \leqq t_1 \leqq t_2 \leqq \dots \leqq t_n$

$$(6.02) \qquad \int_0^1 F(\xi(\alpha,t_2) - \xi(\alpha,t_1), \; \xi(\alpha,t_3) - \xi(\alpha,t_2), \; \dots, \; \xi(\alpha,t_n) - \xi(\alpha,t_{n-1})) \, d\alpha$$

$$= \pi^{-\frac{n-1}{2}} \prod_2^n (t_k - t_{k-1}) \int_{-\infty}^{\infty} du_1 \dots \int_{-\infty}^{\infty} du_n \exp\left[-\sum_2^n \frac{u_k^2}{t_k - t_{k-1}}\right] F(u_2, \dots, u_n).$$

Formally,

$$(6.03) \qquad \int_{-\infty}^{\infty} \xi(\alpha, t) \, d\varrho(t) = \sqrt{\pi} \int_0^1 \chi(\alpha, t) \, d\,[\csc \pi t \, \varrho(-\cot \pi t)]$$

so that if $\varrho(t)$ belongs to L_2, and $\varrho(t)\sqrt{1+t^2}$ is of limited total variation over $(-\infty, \infty)$ we may define

$$(6.04) \qquad \int_{-\infty}^{\infty} \varrho(t) \, d\xi(\alpha, t) = -\int_{-\infty}^{\infty} \xi(\alpha, t) \, d\varrho(t).$$

Up to Definition 2, the theory of random functions over the infinite interval is precisely parallel to that over the finite interval. In Definition 2, we may now generalize $\int_{-\infty}^{\infty} \varrho(t) \, d\xi(\alpha, t)$ so as to eliminate not merely the condition that $\varrho(t)$ should be of limited total variation, but all conditions concerning its order of magnitude, and as before, we now merely require that $\varrho(t)$ shall belong to L_2, this time over $(-\infty, \infty)$. Up to

Theorem III, no change is made by introducing an infinite range of both t and τ, and even in Theorem III the changes that are introduced over a finite range of τ are trivial and obvious. In particular, the reasoning of Theorem IV is still valid, and all that part of W 1 which applies to the proof that the spectral density of $\int_{-\infty}^{\infty} \varrho(t)\, d\,\xi(d, t)$ is almost always

$$(6.05) \qquad \frac{1}{4\pi}\left|\underset{M\to\infty}{\mathrm{l.\,i.\,m.}} \int_{-M}^{M} \varrho(t)\, e^{-iut}\, dt\right|^2$$

will remain valid when $\varrho(t)$ merely belongs to L_2, and

$$(6.06) \qquad \int_{-\infty}^{\infty} du \left[\int_{-\infty}^{\infty} \varrho(t)\, \varrho(t+u)\, dt\right]^2 < \infty$$

If we put

$$(6.07) \qquad \varrho(t) = \frac{1}{\sqrt{2\pi}} \underset{M\to\infty}{\mathrm{l.\,i.\,m.}} \int_{-M}^{M} \psi(u)\, e^{iut}\, du,$$

these conditions reduce themselves to the condition that $\psi(u)$ should belong both to L_2 and L_4.

7. We now restrict ourselves to the case (4. 14) and prove a theorem analogous to Theorem VIII of PZ 1. We have first to generalize Lemma 6 of that paper. This may be done in the following way. Let

$$(7.01) \qquad 0 = t_0 < t_1 < t_2 \ldots < t_N = 1$$

divide the interval (0, 1) into N portions. We write

$$(7.02) \qquad F_x(\alpha) = \sum_{n=0}^{N-1} \varphi_n(x) \int_{t_n}^{t_{n+1}} \beta(t)\, \alpha\, \chi(\alpha, t)$$

where $\varphi_n(x)$ denotes Rademacher's function, defined in PZ 1. We denote by $S_x(\alpha)$ the maximum

$$(7.03) \qquad S_x(\alpha) = \underset{0 \leqq m \leqq N-1}{\mathrm{Max}} \left| \sum_{n=0}^{m} \varphi_n(x) \int_{t_n}^{t_{n+1}} \beta(t)\, d\chi(\alpha, t)\right|.$$

We wish to investigate the behaviour of the average

$$(7.04) \qquad \int_0^1 S_0^q(\alpha)\, d\alpha.$$

We observe first that, if we write,

$$(7.05) \qquad \nu(t) = n \quad (t_n \leqq t < t_{n+1},\ 0 \leqq n \leqq N-1)$$

then, for a given value of x, we may write

$$(7.06) \qquad \varphi_{\nu(t)}(x)\, d\chi(\alpha, t) = d\chi(\alpha' t),$$

since the left hand side of (7.06) clearly represents an arbitrary Brownian motion. The correspondance between α and α' is $1-1$ except in a set of measure zero, and measure (in space α) is preseired. It follows that

$$S_x(\alpha) = S_0(\alpha'), \tag{7.07}$$

and that

$$\int_0^1 S_x^q(\alpha)\,d\alpha = \int_0^1 S_0^q(\alpha)\,d\alpha. \tag{7.08}$$

Applying Lemma 6 of PZ 1, we have

$$\begin{aligned} \int_0^1 S_0^q(\alpha)\,d\alpha &= \int_0^1 dx \int_0^1 S_x^q(\alpha)\,d\alpha \\ &= \int_0^1 d\alpha \int_0^1 S_x^q(\alpha)\,dx \\ &\leqq C_q \int_0^1 d\alpha \int_0^1 |F_x(\alpha)|^q\,dx \\ &= C_q \int_0^1 dx \int_0^1 |F_x(\alpha)|^q\,d\alpha \\ &\leqq C_q \left(\int_0^1 \beta^2(t)\,dt\right)^{\frac{1}{2}q}, \end{aligned} \tag{7.09}$$

where the constants C_q depend only on q. Now denote by $S^*(\alpha)$ the upper bound

$$S^*(\alpha) = \sup_{0 \leqq t' \leqq 1} \left| \int_0^{t'} \beta(t)\,d\chi(\alpha, t) \right|. \tag{7.10}$$

By proceeding to the limit in (7.09) we obtain

$$\int_0^1 S^{*q}(\alpha)\,d\alpha \leqq C_q \left(\int_0^1 \beta^2(t)\,dt\right)^{\frac{1}{2}q}. \tag{7.11}$$

The above result is analogous to Lemma 6. We may now prove the following result, analogous to Theorem VIII.

Theorem V. *Let $S^*(\tau, \alpha)$ denote*

$$\sup_{0 \leqq t' \leqq 1} \left| \int_0^{t'} \beta(t) f(\tau, t)\,d\chi(\alpha, t) \right|, \tag{7.12}$$

where

$$\int_0^1 \beta^2(t)\,dt < \infty, \tag{7.13}$$

$$\int_0^1 |f(\tau, t)|^q\,d\tau < B, \tag{7.14}$$

and B is an absolute constant. Then, for almost all α, we have

$$(7.15)\qquad \int_0^1 S^{*q}(\tau, \alpha)\, d\tau < \infty .$$

Suppose that $\beta(t)$ is a function of bounded variation for which

$$(7.16)\qquad \int_0^1 \beta^2(t)\, dt = \infty .$$

It may be shown*) that, for almost all α, the integral

$$(7.17)\qquad \int_0^1 \beta(t)\, d\chi(\alpha, t)$$

does not exist is the sense (3.01). We may now prove the following theorem analogous to Theorem IX of PZ 1.

Theorem VI. *Suppose that, corresponding to every set E of positive measure in the interval* (0.1), *we can find a set* $G = G(E)$, *such that*

$$(7.18)\qquad \int_G \beta^2(t)\, dt = \infty ,$$

$$(7.19)\qquad \int_E f^2(\tau, t)\, d\tau > \delta_E \qquad (t < G)$$

where δ_E *is a positive constant, which may depend on E. Then, for almost all α, the integral*

$$(7.20)\qquad \int_0^1 \beta(t)\, f(\tau, t)\, d\chi(\alpha, t)$$

exists in the sense (3.01) *for almost no* τ.

8. It was proved in PZ 1 that almost all the functions

$$(8.01)\qquad f(z) = \sum_{n=0}^{\infty} a_n z^n \varphi_n(t) \qquad (z = r e^{i\theta}),$$

where Σa_n^2 converges, satisfy the inequality

$$(8.02)\qquad \max_{0 \leq \theta \leq 2\pi} |f(r e^{i\theta})| = O\left(\sqrt{\log \frac{1}{1-r}}\right).$$

A *gegenbeispiel* was given to show that (8.02) was best possible. This *gegenbeispiel* was a lacunary trigonometric series, and therefore, being of a somewhat special type, may not be regarded as altogether sufficient evidence that almost all functions are unbounded. Leater, in PZ. 3, it was shown that the function (8.01) was unbounded for almost all t subject only to the divergence of the series

$$(8.03)\qquad |a_0| + \sum_{m=0}^{\infty} \left(\sum_{n=2^m}^{2^{m+1}} a_n^2 \right)^{1/2} .$$

*) The proof is almost identical with that of Theorem B in the paper Zygmund (1).

Here we give a very simple *gegenbeispiel* with smooth coefficients, which illustrates the phenomenon in a rather different way. We write

$$F_y^{(\nu)}(r,\Theta) = \sqrt{\nu}\sum_{n=1}^{[k\nu]}\frac{2\sin\frac{n}{\nu}}{n}\varrho^{-n}r^n\cos n\Theta\, y_x, \tag{8.04}$$

where ν is an even integer, k is a number which remains to be chosen, and $y_1, y_2, \ldots, y_n$ represent independent Gaussian distributions of modulus 1, and $\varrho = \varrho_\nu$ is defined by the equation

$$\varrho^{-\nu} = e. \tag{8.05}$$

Clearly the average value of

$$\int_0^{2\pi}(F_y^{(\nu)}(1,\Theta))^2\,d\Theta \tag{8.05$_1$}$$

is less than on absolute constant independent of ν (but depending on k). Now consider

$$\begin{aligned} F_y^{(\nu)}(\varrho,\Theta) &= \sqrt{\nu}\sum_{n=1}^{(k\nu)}\frac{2\sin\frac{n}{\nu}}{n}\cos n\Theta\, y_n \\ &= \sqrt{\nu}\left(\sum_{n=1}^{\infty}\frac{\sin n\left(\Theta+\frac{1}{\nu}\right)}{n}y_n - \sum_{n=1}^{\infty}\frac{\sin n\left(\Theta-\frac{1}{\nu}\right)}{n}y_n\right) \\ &\quad - \sqrt{\nu}\left(\sum_{n=[k\nu]+1}^{\infty}\frac{\sin n\left(\Theta+\frac{1}{\nu}\right)}{n}y_n - \sum_{n=[k\nu]+1}^{\infty}\frac{\sin n\left(\Theta-\frac{1}{\nu}\right)}{n}y_n\right) \\ &= F_y^*(\varrho,\Theta) + R_y(\Theta). \end{aligned} \tag{8.06}$$

Wiener has shown*) that

$$\sqrt{\left(\frac{2}{\pi}\right)}\sum_{n=1}^{\infty}\frac{\sin n\Theta}{n}y_n \tag{8.07}$$

represents an arbitrary Brownian motion $\chi(\alpha,\Theta)$, and thus

$$F_y^*\left(\varrho,\frac{2\lambda}{\nu}\right) \qquad (\lambda = 0, 1, \ldots, \tfrac{1}{2}\nu - 1) \tag{8.08}$$

represent independent Gaussian distributions

$$\sqrt{\nu}\left(\chi\left(\alpha,\frac{2\lambda+1}{\nu}\right) - \chi\left(\alpha,\frac{2\lambda-1}{\nu}\right)\right), \tag{8.09}$$

whose modulus is independent of ν. Also

$$\begin{aligned} \text{Average}\max_{0\le\Theta\le2\pi}\left|F_y^*(\varrho,\Theta)\right| &\geqq \text{Average}\max_{0\le\lambda\le{}^1\!/_2\nu-1}\left|F_y^*\left(\varrho\,\frac{2\lambda}{\nu}\right)\right| \\ &= \int_0^1\sqrt{\nu}\max_{0\le\lambda\le{}^1\!/_2\nu-1}\left|\chi\left(\alpha,\frac{2\lambda+1}{\nu}\right)-\chi\left(\alpha,\frac{2\lambda-1}{\nu}\right)\right|d\alpha. \end{aligned} \tag{8.10}$$

*) Wiener (2).

The last integral may be evaluated by means of (2.01), and exceeds a fixed multiple of $\sqrt{\log \nu}$. It follows that

$$\text{(8.11)} \qquad \underset{0 \leqq \Theta \leqq 2\pi}{\text{Average Max}} \; |F_y^*(\varrho, \Theta)| \geqq B \sqrt{\log\left(\frac{1}{1-\varrho}\right)}.$$

Insuring now to $R_y(\Theta)$, we observe that, since

$$\text{(8.12)} \qquad \text{Average} \int_0^{2\pi} R_y^2(\Theta)\, d\Theta = O\left(\frac{1}{k}\right),$$

we have

$$\text{(8.13)} \qquad \underset{0 \leqq \Theta \leqq 2\pi}{\text{Average Max}} \left| \sqrt{\nu} \sum_{[k\nu]+1}^{\infty} \frac{2 \sin \frac{n}{\nu} \cos n\Theta}{n} \bar{\varrho}^n y_n \right| = O\left\{ k^{-1/2} \sqrt{\log\left(\frac{1}{1-\bar{\varrho}}\right)} \right\},$$

$\bar{\varrho}$ being a number which remains to be chosen. Also since

$$\text{(8.14)} \qquad \text{Average} \int_0^{2\pi} \left[\frac{\partial}{\partial r} \left\{ \sqrt{\nu} \sum_{[k\nu]+1}^{\infty} \frac{2 \sin \frac{n}{\nu} \cos u\Theta}{n} r^{1/2\, n} y_n \right\} \right]^2 d\Theta = O\left(\frac{\nu}{(1-r)}\right),$$

it fellows that

$$\text{(8,15)} \qquad \underset{0 \leqq \Theta \leqq 2\pi}{\text{Average Max}} \left| \frac{\partial}{\partial r} \left\{ \sqrt{\nu} \sum_{[k\nu]+1}^{\infty} \frac{2 \sin \frac{n}{\nu} \cos n\Theta}{n} r^n y_n \right\} \right| = O\left(\frac{\nu \log\left(\frac{1}{1-r}\right)}{1-r} \right)^{1/2}$$

Integrating the result (8.15) from $r = \bar{\varrho}$ to $r = 1$, we get

$$\text{(8.16)} \qquad \underset{0 \leqq \Theta \leqq 2\pi}{\text{Average Max}} \left| \sqrt{\nu} \sum_{[k\nu]+1}^{\infty} \frac{2 \sin \frac{n}{\nu} \cos n\Theta}{n} (1-\bar{\varrho}^n)\, y_n \right| = O\left\{ \nu (1-\bar{\varrho}) \log\left(\frac{1}{1-\bar{\varrho}}\right) \right\}^{1/2}$$

If we now choose $\bar{\varrho}$ so that $k\nu(1-\bar{\varrho}) = 1$, then, combining (8.13) and (8.16), we have

$$\text{(8.17)} \qquad \underset{0 \leqq \Theta \leqq 2\pi}{\text{Average Max}} \left| \sqrt{\nu} \sum_{[k\nu]+1}^{\infty} \frac{2 \sin \frac{n}{\nu} \cos n\Theta}{n} y_n \right| = O\left(\frac{\log(k\nu)}{k}\right)^{1/2}.$$

Finally by choosing k sufficiently large, and combining (8.11) with (8.17), we obtain the desired result

$$\text{(8.18)} \qquad \underset{0 \leqq \Theta \leqq 2\pi}{\text{Average Max}} \; |F_y^{(\nu)}(\varrho, \Theta)| \geqq \tfrac{1}{2} B \sqrt{\log\left(\frac{1}{1-\varrho}\right)}.$$

It is not difficult to combine functions of the form $F_y^{(\nu)}$ so as to form a function $F_y(r, \Theta)$ for which

$$\text{(8.19)} \qquad \underset{0 \leqq \Theta \leqq 2\pi}{\text{Max}} |F_y(r, \Theta)| = \Omega\left\{ \varepsilon (1-r) \sqrt{\log\left(\frac{1}{1-r}\right)} \right\},$$

for almost all y, $\varepsilon(1-r)$ being an arbitrary function which tends to zero with $1-r$.

Suppose that $\vartheta_n(m)$ denote a set of functions for which

(8.20) $$\sum_{m=1}^{N} \vartheta_n^2(m) = 1;$$

(8.21) $$\sum_{m=1}^{N} \vartheta_n(m)\,\vartheta_{n'}(m) = 0.$$

We may show, by means of the argument already given, that, if y_1, $y_2, \ldots, y_N$ denote independent Gaussian distributions of modulus 1, then

(8.22) $$\text{Average} \max_{1 \leq m \leq N} \left| \sum_{n=1}^{N} \vartheta_n(m)\, y_n \right| \geq B\sqrt{(N \log N)},$$

where B is an absolute constant.

A further property of $\chi(\alpha, t)$, which has attracted the attention of Perrin as an experimentally observable fact, is that for almost all values of α, it nowhere possesses a derivative with respect to t. We shall in fact prove the more comprehensive.

Theorem VII. *The values of* α *for which there exists at such that*

(9.01) $$\lim_{\varepsilon \to 0} (\chi(\alpha, t+\varepsilon) - \chi(\alpha, t))/\varepsilon^{\lambda} < \infty \qquad [\lambda > \tfrac{1}{2}]$$

form a set of zero measure.

To prove this, let us subdivide the interval $(0, 1)$ binarily. If $(m\,2^{-q}$, $(m+1)\,2^{-q})$ is a subdivision of the qth order, its middle point will be $(2m+1)\,2^{-q-1}$, and the differential determining the distribution of the pair of quantities $\chi(\alpha, (2m+1)\,2^{-q-1}) - \chi(\alpha, m\,2^{-q}) = u_1$, and $\chi(\alpha, (m+1)\,2^{-q}) - \chi(\alpha, (2m+1)\,2^{-q-1}) = u_2$, will be

(9.02) $$\frac{1}{\pi\, 2^{-q-1}} \exp.\left(-\frac{u_1^2 + u_2^2}{2^{-q-1}}\right) d\,u_1\, d\,u_2.$$

If we put $u_1 + u_2 = u$, this becomes

(9.03) $$\frac{1}{\pi\, 2^{-q-1}} \exp.\left(-\frac{u_1^2 + (u-u_1)^2}{2^{-q-1}}\right) d\,u, d\,u$$

$$= \frac{1}{\pi\, 2^{-q-1}} \exp. \frac{-\left(u_1 - \frac{u}{2}\right)^2 - \frac{u^2}{4}}{2^{-q-1}} d\,u_1\, d\,u.$$

From this we may readily conclude that the probability that either u_1 or u_2 is less than A_q in modulus does not exceed

(9.04) $$\frac{4}{\sqrt{\pi\, 2^{-q-2}}} \int_{-A}^{A} e^{-\frac{u^2}{2^{-q-2}}}\, d\,u \leq \frac{16}{\sqrt{\pi}} A\, 2^{q/2}.$$

The probability that of the 2^{q+1} quantities $|\chi(\alpha, (\mu+1)\,2^{-q-1}) - \chi(\alpha, \mu\, 2^{-q-1})|$, at least N_q do not exceed A_q, whatever the distribution of the 2^q analogous quantities of the proceeding series, thus does not exceed

$$(9.05)\qquad \sum_{\nu=N}^{2^{q+1}} \frac{(2^{q+1})!}{\nu!\,(2^{q+1}-\nu)!}\left(1-\frac{16}{\sqrt{\pi}}A\,2^{q/2}\right)^{2^{q+1}-\nu}\left(\frac{16}{\sqrt{\pi}}A\,2^{q/2}\right)^{\nu} < \int\limits_{\frac{N-\frac{16}{\sqrt{\pi}}A\,2^{\frac{3q}{2}+1}}{\sqrt{\frac{16}{\sqrt{\pi}}A\,2^{\frac{3q}{2}+1}}}}^{\infty} \frac{e^{-u^2}\,du}{\sqrt{\pi}}.$$

Let us now put

$$(9.06)\qquad A_q = 2^{-\lambda q}a,\ N_q = \frac{16}{\sqrt{\pi}}\,2^{q\left(\frac{3}{2}-\mu\right)} \qquad [0 \leqq \mu < \lambda].$$

Then the dominant expression in (9.05) becomes:

$$(9,07)\qquad \int\limits_{\left(2^{(\lambda-\mu)q}-2a\right)2^{q\left(\frac{3}{4}-\frac{\lambda}{2}\right)}}^{\infty} \frac{e^{-u^2}\,du}{\sqrt{\pi}} = T_q$$

and we have

$$(9.08)\qquad \sum_1^{\infty} T_q < \infty.$$

Thus it is infinitely improbable that more than a finite number of q's can be found for which at least N_q of the 2^{q+1} quantities $|\chi(\alpha,(\mu+1)2^{-q-1}) - \chi(\alpha,\mu\,2^{-q-1})|$ do not exceed A_q, N_q and A_q being defined as in (9.06). On the other hand, the probability that the N_q intervals of the qth stage have any point in common for all values of q from Q_1+1 to Q_2, does not exceed

$$(9.09)\qquad 2^{Q_2+1}\prod_{Q_1+1}^{Q_2}\frac{16}{\sqrt{\pi}}\,\frac{2^{q\left(\frac{3}{2}-\mu\right)}}{2^{q+1}} \leqq C_1\,C_2^{Q_2}\,2^{\left(\frac{1}{2}-\mu\right)Q_2^2}$$

where C_1 and C_2 are constants. If $\mu > \frac{1}{2}$, this tends to 0 as $Q_2 \to \infty$. Thus it is infinitely improbable that there is any point common to S_q for all q's from some stage on, S_q consisting of all the points in the N_q intervals of the qth stage for which

$$(9.10)\qquad |\chi(\alpha,(\mu+1)2^{-q-1}) - \chi(\alpha,\mu\,2^{-q-1})| \leqq 2^{-\lambda q}a.$$

It is thus, infinitely improbable that there exist any point common to an infinite sequence of intervals satisfying (9.10), and *a fortiori* infinitely improbable that any point lie in intervals of the binary subdivision, all of which from some stage on satisfy (9.10). This is in the first instance true for one value of a, but since the sum of a denumerable number of sets of zero measure is itself of zero measure, we may extend it to the case in which (9.10) is supposed to hold for *any* value of a. From this, Theorem VII follows at once.

We have already noted that if $\lambda < \frac{1}{2}$, then except for a set of cases of zero probability,

$$\lim_{\varepsilon \to 0} (\chi(\alpha, t+\varepsilon) - \chi(\alpha, t))/\varepsilon^{\lambda} = 0 \tag{9.11}$$

uniformly for all values of t.

This was proved in W_1 for the case $\lambda = 1/4$, and the methods are substantially unchanged. The study of precisely what happens in the case $\lambda = \frac{1}{2}$ demands the use of more refined methods, and is not treated here.

Bibliography

E. Borel (1): Les probabilités dénombrables et leurs applications arithmétiques, Rend. di Palermo **27** (1904), p. 247—271.

Borel (2): The book of the opening of the Rice Institute. Houston (Texas) 1917.

R. E. A. C. Paley and A. Zygmund (1), (2), (3): On some series of functions (1), (2), Proc. Camb. Phil. Soc. **26** (1930), p. 337—357, 458—474, (3) to appear soon.

E. Perrin (1): Les atomes, 4e éd. Paris, 1914.

Perrin (2): Brownian movement and molecular reality. Tr. by F. Soddy, London, 1910.

H. Rademacher (1): Einige Sätze über Reihen von allgemeinen Orthogonalfunktionen. Math. Annalen **87** (1922), S. 112—138.

N. Wiener (1): Generalized harmonic analysis, Acta Mathematica **55** (1930), p. 117—258.

Wiener (2): Differential-space. Journal of Mathematics and Physics **2** (1923), p. 131—174.

A. Zygmund (1): On the convergence of lacunary trigonometric series, Fundamenta Math. **16**, (1930), p. 90—107.

(Eingegangen am 24. April 1932.)

Comments on [33a]

J.-P. Kahane

The main part of this paper is redeveloped in essence in chapter 9 of the book [34d] of Paley and Wiener, *Fourier transforms in the complex domain.* We shall not therefore give it a systematic review. Some information on the present state of the subject of series of functions with random independent coefficients can be found in reference 1. In particular, the techniques of references 2 and 3 can be used in order to derive the main properties of Brownian motion from the Fourier-Wiener series. More information on sample functions of Brownian motion can be found in reference 3a.

We now turn to some specific comments on the paper.

A. In connection with the reference to the work of Borel and of Paley-Zygmund (PZ, see p. 647 of [33a]), it should be noted that the idea of considering random functions also goes back to Borel;[4] he stated, without proof, that the circle of convergence of a random Taylor series is a natural boundary. The precise statement and the first proof were given by Steinhaus[5] (see also reference 6 and reference 1, chap. 4).

B. Regarding the use of the random for the construction of Gegenbeispiele, mentioned on pp. 648-649, the simplest example (PZ, I theorem 9) is: *If* $\Sigma c_n^2 = \infty$, *almost all series* $\Sigma \pm c_n^2 \cos nt$ *are not Fourier-Lebesgue series.* Gaussian Fourier series are used in order to provide examples in many parts of harmonic analysis, for example, Malliavin theorem on spectral synthesis, Salem sets, Rudin sets.[7]

C. Regarding the Gaussian distribution of the integrals (1.05) (p. 649), it should be mentioned that the following related fundamental property of Brownian motion has been adopted as a definition: given an isometric mapping from $L^2(0,1)$ to a Hilbert space of Gaussian variables, the Brownian motion function $\chi(t)$ is the image of the function $1_{[0,t]}$ $(0 < t < 1)$.[8]

D. In regard to the reference to Theorems 2 and 3 of PZ, I on pp. 656, 658, it is conveneint to rerender them in the following form:

Theorem: If $\Sigma c_n^2 < \infty$ *and* $\sup_n \|f_n\|_{L^q(0,1)} < \infty$, $q > 2$, (*resp.* $\sup_n \|f_n\|_{L^\infty(0,1)} < \infty$), *and* $s = \Sigma \pm c_n f_n$, *then* $s \in L^q$ (*resp.* $\exp(\lambda s^2) \in L^1$ *for all* $\lambda > 0$) a. s. ;

E. Regarding the dispensibility of (4.28) alluded to on p. 658, we claim that to get rid of the condition (4.28) it suffices to remove a number of terms in (4.01).

F. It should be noted that in Theorem IV, p. 659, the Φ_n do not have the same meaning as in (4.01).

G. As regards the (5.15) (p.659), the best estimate is

$$|\chi(\alpha,\tau) - \chi(\alpha,\sigma)|^2 < A|\tau - \sigma| \log \frac{1}{|\tau - \sigma|} \, .$$

H. In regard to the Gegenbeispiele on pp. 664-665, let us recall the counter-example of Paley:[9] if $\lim_{\rho\to 0} \epsilon(\rho) = 0$, there exists a square-summable sequence c_n such that all series

$$F(z) = \sum_0^\infty \pm c_n z^n \text{ are not } O\left(\epsilon(1-r)\sqrt{\log\frac{1}{1-r}}\right), \quad (r \to 1).$$

The construction given on pp. 664-665 yields the same result, except that the signs ± are replaced by independent normalized Gaussian variables.

Concerning the remark at the end of p. 665, it is not difficult to show that

$$\sup_t |F(r,t)| \neq O\left(\epsilon(1-r)\sqrt{\log\frac{1}{1-r}}\right), \text{ a.s.}$$

but it is *not possible* to obtain

$$\lim_{r\to 1} \frac{\sup_t |F(r,t)|}{\epsilon(1-r)\sqrt{\log\frac{1}{1-r}}} = \infty \text{ p.s.} \qquad (\ddagger)$$

instead of (8.19).

In order to explain this point, let us consider

$$F(r,t) = \sum_{n=0}^\infty a_n r^n Y_n \cos nt$$

$$M(r) = \sup_t |F(r,t)| \quad (0 < r < 1),$$

where $a_n \geqslant 0$, $\sum_0^\infty a_n^2 < \infty$ and the Y_n are independent real random variables, such that $E(\exp \lambda Y_n) \leqslant \exp \lambda^2/2$ (E denotes the expectation, exp the exponential). We write

$$F_m(r,t) = \sum_{n=m+1}^{\infty} a_n r^n Y_n \cos nt$$

$$M_m(r) = \sup_t |F_m(r,t)|.$$

Since $E(\Sigma a_n^2 Y_n^2) = \Sigma a_n^2 < \infty$, we have $\Sigma a_n^2 Y_n^2 < \infty$ a. s., and Schwarz's inequality gives $|F(r,t) - F_m(r,t)| < C\sqrt{m}$ (C is a positive random variable), then

$$M(r) < C\sqrt{m} + M_m(r). \tag{α}$$

Moreover

$$\frac{dF_m}{dt}(r,t) < \sum_{n=m+1}^{\infty} |a_n n r^n Y_n| \leqslant \frac{C}{(1-r)^2}$$

(C is always a positive random variable, not the same on each line). Therefore, if $r = 1-1/N$, we have

$$M_m(r) < M_m^*(r) + C \tag{β}$$

$$M_m^*(r) = \max_{1 \leqslant k \leqslant N^2} |F_m(r, 2k\pi/N^2)|.$$

Since the Y_n are independent we have

$$E(\exp(\lambda F_m(r,t))) = \prod_{n=m+1}^{\infty} E \exp(\lambda a_n r^n Y_n \cos nt)$$

$$\leqslant \exp \frac{\lambda^2}{2} \eta_m(r)$$

with

$$\eta_m(r) = \sum_{n=m+1}^{\infty} a_n^2 r^{2n}.$$

Moreover

$$E(\exp(\lambda M_m^*(r))) \leqslant \sum_{k=1}^{N^2} E(\exp(\lambda F_m(r,t_k)) + \exp(-\lambda F_m(r,t_k))),$$

with $t_k = 2k\pi/N^2$. Therefore

$$E(\exp(\lambda M_m^*(r) - \frac{\lambda^2}{2}\eta_m(r) - 3\log N)) \leqslant \frac{2}{N}.$$

Let us choose $\lambda = \lambda_r = \sqrt{\eta_m^{-1}(r)\log N}$, and consider a sequence $r_j = 1 - 1/N_j$ tending to 0 in such a way that $\Sigma 1/N_j < \infty$, and a sequence of integers m_j. We obtain

$$\sum_{j=1}^{\infty} E(\exp \lambda_{r_j}(M_{m_j}^*(r_j) - 4\sqrt{\eta_{m_j}(r_j)\log N_j})) < \infty,$$

therefore

$$M_{m_j}^*(r_j) < 4\sqrt{\eta_{m_j}(r_j)\log N_j} \quad (j \text{ large});$$

therefore, using (α) and (β),

$$M(r_j) < C\sqrt{m_j} + 4\sqrt{\eta_{m_j}(r_j)\log N_j}. \qquad (\gamma)$$

From (γ) and the convexity of $\log M(e^\sigma)$ results immediately

$$M(r) = o\left(\sqrt{\log\frac{1}{1-r}}\right), \quad (r \nearrow 1),$$

that is theorem 4 of PZ, I. Moreover we have

$$M^2(r_j) < C(m_j + \eta_{m_j}(r_j)\log N_j)$$

and

$$\eta_{m_j}(r_j) \leqslant \sum_{n=m_j+1}^{jN_j} a_n^2 + e^{-j} \qquad (j \text{ large}).$$

Let us choose $m_{j+1} = jN_j$ and $N_{j+1} = \exp(j^2 m_{j+1}) = \exp(j^3 N_j)$. Then we obtain

$$\sum_{j=1}^{\infty} \frac{M^2(r_j)}{\log N_j} < \infty.$$

Therefore, if the function $\epsilon(1-r)$ satisfies $\Sigma_1^\infty \epsilon^2(1-r_j) = \infty$, we cannot have (‡).

I. More material relating to the statement on p. 663 regarding PZ, 3 and the role of the series (8.03), will be found in references 10.

J. In connection with the assertion (9.11) p. 668, it should be pointed out that actually

$$\overline{\lim_{\epsilon\to 0}} \frac{|\chi(t+\epsilon) - \chi(t)|}{|\epsilon|^{1/2}} > 0 \quad \text{for all } t$$

holds a. s. See reference 11 for a combinatorial proof, and reference 12 for a proof using the Fourier-Wiener series.

Errata

p. 647,
last line: write

$$\sum_{n=0}^{\infty} C_n e^{2\pi i\theta_n} f_n(x)$$

p. 652,
line 15: write "have a continuous distribution" instead of "are continuous"
p. 655,
last line: write lemma 2′ instead of lemma 3′
p. 659,
line 6: write C instead of B

line 7: write B^2/C instead of B^2/C_u
p. 664,
line 3: write y_n instead of y_x
p. 665,
equation (4.07): write $\Phi(\tau, t)$ instead of $\Phi_0(\tau, t)$
p. 666,
line 10: write "9." before "A further property. . ."
line 14: write "a t such" instead of "at such"
line 5: write $\overline{\lim}$ instead of lim.

References

1. J. -P. Kahane, Some Random Series of Functions, Heath, Lexington, Mass., 1968.

2. R. Salem and A. Zygumnd, *Some properties of trigonometric series whose terms have random signs,* Acta Math. 91(1954), 245-301.

3. J. -P. Kahane, *Propriétés locales des fonctions à séries de Fourier aléatoires,* Studia Math. 19(1960), 1-25.

3a. R. M. Dudley, *Sample functions of the Gaussian process,* Ann. Prob. 1(1973), 66-103.

4. E. Borel, *Sur les séries de Taylor,* C. R. Acad. Sci. Paris, 123(1896), 1051-1052.

5. H. Steinhaus, *Über Die Wahrseheinlichkeit dafür, dass der Konvergenzkreis einer Potenzreihe ihre naturlich Grenze ist,* Math. Z. 31(1930), 408-416.

6. A. Zygmund, *On continuability of power series,* Acta Sci. Math. (Szeged) 6, 1933, 80-84.

7. See reference 1.

8. S. Kakutani, *On the brownian motion in n-spaces,* Proc. Imp. Acad. Tokyo 20(1944), 648-652.

9. R. E. A. C. Paley, *On some problems connected with the Weierstrass's non differentiable function,* Proc. London Math. Soc. 31(1930), 301-328.

10. P. Billard, *Séries de Fourier aléatoirement bornées, continues, uniformément convergentes,* Studia Math. 22(1963), 309-329. See also references 3 and 1, chap. 3.

11. A. Dvoretski, *On the oscillation of the brownian motion process,* Israel J. Math. 1(1963), 212-214.

12. See reference 1, chap. 8.

Reprinted from JOURNAL OF MATHEMATICS AND PHYSICS
Vol. XIV, No. 1, March, 1935

RANDOM FUNCTIONS

BY N. WIENER

By a random function we understand a function which, in a certain way to be specified in what follows, depends upon a formally expressed variable and a usually suppressed parameter of distribution. Such functions occur throughout statistical mechanics. In statistical mechanics a certain possible state of the universe is expressed by a function of one or more variables which give the geometrical coordinates and the time, together with a parameter which singles out the universe considered as one among all possible universes and determines its probability. As a more specific example let us take the path of a particle subject to a Brownian motion and let us consider one coordinate of the particle (let us say the X-coordinate) as a function of the time t. Then for any particular Brownian motion or motion of a particle impelled by the collision of neighboring molecules subject to a thermal agitation, x will be a well-defined function of t. If however, instead of considering the actual traces by a specified particle, we consider all possible courses traced by all possible particles, in addition to the variable t, x will have an argument which singles out the specific Brownian motions in question from all possible Brownian motions. This variable is introduced for purposes of integration; that is, a certain range of this variable measures by its length the probability of the set of Brownian motion it represents, and an integration with respect to this variable yields a probability average of the quantity integrated. Naturally this probability theory is not at all simple and needs to be specified in detail.

It will be seen that the theory of random functions is in essentials a theory of integration in function space. As such it may be subsumed under the general theories of integration of Radon, Daniell, and Wiener. A previous attempt at a theory of integration in function space was due to Gâteaux, but this earlier theory is not a special case of the Daniell-Radon integral. The earlier theory attempted to treat every value $F(t_0)$ of a function $F(t)$ as an independent variable. In such a scheme we find that a succession of regions of positive measure may include one another successively in such a way that there is no element common to all of them. This violates one of the most essential canons of the Daniell theory. In order to eliminate it, it is necessary that the class of functions over which we are integrating should be in some sense or other compact; that is, that any sequence of functions should contain a subsequence with a limit function. If this is not strictly true with the original class with which we are dealing, it must at any rate be possible to make it true by removing or adjoining a set of functions of arbitrarily small measure. Compactness in the ordinary sense is equivalent to uniform boundedness and equi-continuity. Now, it is a priori obvious that Gâteaux's type of integration must yield us functions which almost never are continuous, and completely breaks down on this point.

The Brownian motion points us a way out of the difficulty. In this motion it is not the position of a particle which is independent of the position at another time, but the travel of a particle between two stated times which is independent of the position at the first time. In crude language, the differentials of the function $x(t)$ correspond to the independent coordinates of a point in space of a finite number of dimensions, and not the values of the function. Physically it is at least reasonable that the Brownian motion of a particle is continuous, and we shall in fact show that the Einstein theory of the Brownian motion permits us to say that in fact it is almost always continuous. We shall show even more, that it is subject to a condition of equi-continuity except for a set of cases whose measure, or what is equivalent, whose probability, we can reduce to as small a value as we want.

The theory of random functions always makes the impression of a much greater degree of artificiality than corresponds to the facts. The reason is that in the theory of Daniell or Lebesgue integration of the sum of a denumerable set of null sets is itself null, while the sum of a set of null sets of the power of the continuum need not be null. This fact makes it extremely desirable to characterize a random function by a denumerable set of conditions, while the characterization of such a function by its value for all its arguments yields a number of characterizations of the power of continuum. If then we attempt to characterize a random function by a set of its values, it is almost necessary for the purposes of our technical development that the original set of values which we take should be denumerable. For example, we may define such a function by its values at the rational or the binary points of the line. On the other hand, we may give up entirely the characterization of such a function by its values and define it by its Fourier coefficients or its coefficients in some other scheme of development. In any case it will turn out that the function which we have so defined will, except in a set of cases of probability zero (or measure zero, which is the same thing), either be continuous or in some readily specified sense be equivalent to a continuous function. Once this is done, we may define the value of this continuous function uniquely over the whole of its interval of definition. We thus have generalized our initial function defined by a denumerable set of parameters to a function apparently defined by a larger number of parameters. In doing so we have made use of a method which masks the true invariance of the result we have obtained. It is, however, a perfectly legitimate mathematical method to obtain our result in a non-invariantive way and to establish its proper invariance later by specific theorems. This mode of procedure unluckily has the disadvantage of being most non-heuristic and of demanding from the reader the patience to take on faith the necessity of a large amount of material whose justification is only given after the completion of the argument. It will ultimately turn out that the random functions we deal with and the measure that properly belongs to them determine a function-space distinct from that of Hilbert but covariant with it under all unitary transformations which leave Hilbert space invariant. While such a theory may be built up for real Hilbert space, and while this is the case in my previous writings, the present theory of differential space applies to one covariant with complex Hilbert space.

It will be seen that although this note is apparently devoted to integration in a continuous infinity of dimensions, it really forms a chapter in what E. Borel has designated the theory of "denumerable probabilities."[1] The first close approach to the specific problems which concern us here was made by Steinhaus.[2] He considers the series

$$\sum_{1}^{\infty} \pm \frac{1}{n}$$

where the signs $\pm$ represent independent choices, and shows that this series converges in almost every case. The methods of Steinhaus were made into a powerful analytic tool by Paley and Zygmund,[3] more espe-

[1] E. Borel, Les probabilités dénombrables et leurs applications arithmetiques, Rendiconti del Circolo Matematico di Palermo, vol. 27 (1909), pp. 247–271.

[2] Cf. Sur la probabilité de la convergence de series, Studia Mathematica, vol. 2 (1930), pp. 21–39, and earlier papers there referred to.

[3] On some series of functions, Proceedings of the Cambridge Philosophical Society, vol. 26 (1930), pp. 337–357, 458–474; vol. 28, pp. 190–205. Cf. also Notes

cially in the formation of Gegenbeispiele, and the final theory has been employed with great effect by Bohr and Jessen,[4] in the study of the Riemann zeta function and of almost periodic function in the complex domain.

Wiener[5] has developed a theory of random functions in many respects parallel to those discussed by the authors already mentioned, but not identical with them. Since the Wiener theory of random functions is derived from considerations in the theory of the Brownian motion and of statistical mechanics, in which Gaussian distributions play an important rôle, they also play an important rôle in his theory. In this respect, Wiener's theory resembles the Einstein theory of the Brownian motion, to which we shall later show it to be equivalent.[6] Indeed a sharp distinction may be made between Wiener's theory and all others so far mentioned in that, while in both cases a denumerable set of terms are assigned coefficients with a certain distribution, in all other theories these coefficients either have at random the value ± 1, or are distributed at random around the unit circle in the complex plane. In addition to its applicability to statistical mechanics, Wiener's theory possesses a larger degree of symmetry then the alternative theories, or what is the same thing in other words, it has a more extensive group of transformations under which it is invariant. The reason for this is that if a number of terms have independent Gaussian distributions, any linear combination of these terms has itself a Gaussian distribution. On the other hand, if we start with a distribution of the original terms over the values ± 1, or around the unit circle, the distribution of a linear combination of the original terms becomes extremely complex and unmanageable. This property of Gaussian distribution is thus intimately allied with the covariance which the theory of random functions developed here shows with Hilbert space.

By a change of scale, a real Gaussian distribution may be reduced to the distribution of a real parameter uniformly over the interval (0, 1). In the same way, a complex Gaussian distribution may be reduced to

---

on random functions, Paley, Wiener, and Zygmund, Mathematische Zeitschrift, vol. 37 (1933), pp. 647–688, on which the present chapter is largely based.

[4] H. Bohr and B. Jessen, Über die Werteverteilung der Riemannschen Zetafunktion, Acta Mathematica, vol. 54 (1930), pp. 1-35; vol. 58 (1932), pp. 1–55.

[5] Cf. N. Wiener, Generalized harmonic analysis, Acta Mathematica, vol. 55, pp. 214 ff., and the references there given.

[6] A. Einstein, Annalen der Physik, vol. 17 (1905), p. 549 ff.; vol. 19 (1906), p. 371 ff.

the simultaneous and independent uniform distribution of two parameters over an interval of that sort. By this artifice our integration in function space may be carried back to the integration of a function of a denumerable infinity of variables over a cube in its proper space. Such an integration has been discussed by Daniell, Jessen and others.[7] Since however it is probably not familiar to the reader of this note, we shall derive it from the ordinary integration of a function of one variable over an interval of a line by process of mapping. This process of mapping is merely an elaboration of the process by which a square may be mapped on a line in such a way that planar measures are mapped into equal linear measure.

The points of Wiener's space correspond to the formal Fourier series

$$\sum_{-\infty}^{\infty} a_n \, e^{inx} \tag{1}$$

where each a_n has real and imaginary components all independent and all subject to a Gaussian distribution with the same modulus. We may determine the distribution of all the coefficients a_n by a single parameter, distributed uniformly. Such a distribution may be defined by a Daniell measure. The series (1) almost never converges in the mean, but the series arising from them by formal integration: namely

$$\psi(x, \alpha) = a_0 \, x + \sum_{1}^{\infty} \left(\frac{a_n \, e^{inx}}{in} + \frac{a_{-n} \, e^{-inx}}{-in} \right)$$

almost always converges in the mean. It determines a function which has been shown to be almost always continuous, and indeed almost always to satisfy uniform Lipschitz conditions of every order below $\frac{1}{2}$. The probability that there exists any point for which it satisfies a Lipschitz condition of order exceeding $\frac{1}{2}$ is zero. Accordingly, almost all such functions share the properties of the Weierstrass non-differentiable functions.

If $a < b < c < d$, the correlation between the excursus

$$\psi(a, \alpha) - \psi(b, \alpha)$$

and that $\psi(c, \alpha) - \psi(d, \alpha)$ is zero. Indeed, these quantities are not merely linearly, but completely independent in their distributions. If

[7] P. J. Daniell, Integrals in an infinite number of dimensions, Annals of Mathematics (2), vol. 20 (1919), pp. 281–288. B. Jessen, Bidrag til Integratteorien for Funktioner af uendelig mange Variable, Copenhagen, 1930.

t be taken as the time, $R\psi(t, \alpha)$ be plotted along the x-axis, and $I\psi(t, \alpha)$ along the Y-axis, the function $\psi(t, \alpha)$ will represent a typical Brownian motion of a particle in a fluid, impelled by the molecular agitation of the fluid.

The formal series (1) is transformed by the unitary transformation T into the formal series

$$\sum_{-\infty}^{\infty} a_n \, Te^{inx},$$

where Te^{inx} is any closed set of normal and orthogonal functions. This may be rearranged into a series

$$\sum_{-\infty}^{\infty} b_n \, e^{inx}$$

by expanding the functions Te^{inx} in Fourier series. It may be shown that the new coefficients b_n all have completely independent Gaussian distribution for their real and imaginary parts with the same moduli as those of the a_n's.

Let T be any unitary transformation, and let T^n be its nth iterate. If for every functions $f(x)$ and $g(x)$ of L_2,

$$\lim_{n\to\infty} \int_{-\pi}^{\pi} (T^n f(x)) \, \bar{g}(x) \, dx = 0$$

then for almost all α,

$$\lim_{N\to\infty} \frac{1}{N} \sum_{0}^{N-1} \varphi\left(\int_{-\pi}^{\pi} T^n f(x) \, d\psi(x, \alpha)\right) = \int_0^1 \varphi\left(\int_{-\pi}^{\pi} f_n(x) \, d\psi(x, \alpha)\right) d\alpha \, .$$

Here we use Birkhoff's ergodic theorem.

We may easily define the random function $\psi(x, \alpha)$ over $(-\infty, \infty)$ by a change in variable. We may construct the analytic function

$$\int_{-\infty}^{\infty} f(z + x) \, d\psi(x, \alpha)$$

dependent on the parameter z. The problem of the distribution of the zeros of this function is an analytic problem quite analogous to Dr. Jessen's problem of the zeros of an almost periodic function in the complex domain. We show under certain very general restrictions that if $N_A(y_0, y_1)$ is the number of zeros in the rectangle

$$| Rz | < A, \qquad y_0 < Iz < y_1,$$

and

$$\varphi(u) = \underset{A\to\infty}{\text{l.i.m.}} \int_{-A}^{A} f(x+iy)\, e^{-in(x+iy)}\, dx\,,$$

then

$$N_A(y_0, y_1) \sim \frac{A}{2\pi}\left\{\frac{\int_{-\infty}^{\infty} 2u\,|\,\varphi(u)\,|^2\, e^{-2uy_0}\, du}{\int_{-\infty}^{\infty} |\,\varphi(u)\,|^2\, e^{-2uy_0}\, du} - \frac{\int_{-\infty}^{\infty} 2u\,|\,\varphi(u)\,|^2\, e^{-2uy_1}\, du}{\int_{-\infty}^{\infty} |\,\varphi(u)\,|^2\, e^{-2uy_1}\, du}\right\}.$$

I wish to emphasize the essential similarity of the problems we discuss and those treated by Bohr and Jessen. The coefficients of Bohr and Jessen are distributed around the unit circle, whereas mine have complex Gaussian distributions. So long as we adhere to one special system of orthogonal functions, one distribution is as natural as the other. However, the distribution which we give is, as we have seen, not tied to any particular set of normal and orthogonal functions.[8]

[8] Cf. Paley and Wiener, Fourier Transforms in the Complex Domain, New York, 1934, Chapters IX and X.

Comments on [34a]

J. -P. Kahane

This paper summarizes previous results and some developed in Chapters IX and X of the book [34d] by Paley and Wiener. In fact, pp. 17-20 and the first paragraph on p. 21 are verbatim copies of the material on pp. 140-143 of the book.

THE HOMOGENEOUS CHAOS.

By NORBERT WIENER.

1. Introduction. Physical need for theory. Statistical mechanics may be defined as the application of the concepts of Lebesgue integration to mechanics. Historically, this is perhaps putting the cart before the horse. Statistical mechanics developed through the entire latter half of the nineteenth century before the Lebesgue integral was discovered. Nevertheless, it developed without an adequate armory of concepts and mathematical technique, which is only now in the process of development at the hands of the modern school of students of integral theory.

In the more primitive forms of statistical mechanics, the integration or summation was taken over the manifold particles of a single homogeneous dynamical system, as in the case of the perfect gas. In its more mature form, due to Gibbs, the integration is performed over a parameter of distribution, numerical or not, serving to label the constituent systems of a dynamical ensemble, evolving under identical laws of force, but differing in their initial conditions. Nevertheless, the study of the mode in which this parameter of distribution enters into the individual systems of the ensemble does not seem to have received much explicit study. The parameter of distribution is essentially a parameter of integration only. As such, questions of dimensionality are indifferent to it, and it may be replaced by a numerical variable α with the range $(0, 1)$. Any transformation leaving invariant the probability properties of the ensemble as a whole is then represented by a measure-preserving transformation of the interval $(0, 1)$ into itself.

Among the simplest and most important ensembles of physics are those which have a spatially homogeneous character. Among these are the homogeneous gas, the homogeneous liquid, the homogeneous state of turbulence. In these, while the individual systems may not be invariant under a change of origin, or, in other words, under the translation of space by a vector, the ensemble as a whole is invariant, and the individual systems are merely permuted without change of probability. From what we have said, the translations of space thus generate an Abelian group of equi-measure transformations of the parameter of distribution.

One-dimensional groups of equi-measure transformations have become well known to the mathematicians during the past decade, as they lie at the root of Birkhoff's famous ergodic theorem.[1] This theorem asserts that if we have given a set S of finite measure, an integrable function $f(P)$ on S, and a one-parameter Abelian group T^λ of equi-measure transformations of S into itself, such that

$$(1) \qquad T^\lambda \cdot T^\mu = T^{\lambda+\mu} \qquad (-\infty < \lambda < \infty, \ -\infty < \mu < \infty),$$

then for all points P on S except those of a set of zero measure, and provided certain conditions of measurability are satisfied,

$$(2) \qquad \lim_{A\to\infty} \frac{1}{A} \int_0^A f(T^\lambda P)\, d\lambda$$

will exist. Under certain more stringent conditions, known as metric transitivity, we shall have

$$(3) \qquad \lim_{A\to\infty} \frac{1}{A} \int_0^A f(T^\lambda P)\, d\lambda = \int_S f(P)\, dV_P$$

almost everywhere. The ergodic theorem thus translates averages over an infinite range, taken with respect to λ, into averages over the set S of finite measure. Even without metric transitivity, the ergodic theorem translates the distribution theory of λ averages into the theory of S averages.

In the most familiar applications of the ergodic theorem, S is taken to be a spatial set, and the parameter λ is identified with the time. The theorem thus becomes a way of translating time averages into space averages, in a

[1] Cf. Eberhard Hopf, "Ergodentheorie," *Ergebnisse der Mathematik und ihrer Grenzgebiete*, vol. 5. See particularly § 14, where further references to the literature are given.

manner which was postulated by Gibbs without rigorous justification, and which forms the entire basis of his methods. Strictly speaking, the space averages are generally in phase-space rather than in the ordinary geometrical space of three dimensions. There is no reason, however, why the parameter λ should be confined to one taking on values of the time for its arguments, nor even why it should be a one-dimensional variable. We are thus driven to formulate and prove a multidimensional analogue of the classical Birkhoff theorem.

In the ordinary Birkhoff theorem, the transformations T^λ are taken to be one-one point transformations. Now, the ergodic theorem belongs fundamentally to the abstract theory of the Lebesgue integral, and in this theory, individual points play no rôle. In the study of chaos, individual values of the parameter of integration are equally unnatural as an object of study, and it becomes desirable to recast the ergodic theorem into a true Lebesgue form. This we do in paragraph 2.

Of all the forms of chaos occurring in physics, there is only one class which has been studied with anything approaching completeness. This is the class of types of chaos connected with the theory of the Brownian motion. In this one-dimensional theory, there is a simple and powerful algorithm of phase averages, which the ergodic theorem readily converts into a theory of averages over the transformation group. This theory is easily generalized to spaces of a higher dimensionality, without any very fundamental alterations. We shall show that there is a certain sense in which these types of chaos are central in the theory, and allow us to approximate to all types.

Physical theories of chaos, such as that of turbulence, or of the statistical theory of a gas or a liquid, may or may not be theories of equilibrium. In the general case, the statistical state of a chaotic system, subject to the laws of dynamics, will be a function of the time. The laws of dynamics produce a continuous transformation group, in which the chaos remains a chaos, but changes its character. This is at least the case in those systems which can continue to exist indefinitely in time without some catastrophe which essentially changes their dynamic character. The study, for example, of the development of a state of turbulence, depends on an existence theory which avoids the possibility of such a catastrophe. We shall close this paper by certain very general considerations concerning the demands of an existence theory of this sort.

2. Definition. Types of Chaos. A *continuous homogeneous chaos* in n dimensions is a scalar or vector valued measurable function $\rho(x_1, \cdots, x_n; \alpha)$ of $x_1, \cdots, x_n; \alpha$, in which $x_1, \cdots, x_n$ assume all real values, while α ranges over $(0, 1)$; and in which the set of values of α for which

(4) $$\rho(x_1 + y_1, \cdots, x_n + y_n; \alpha) \text{ belongs to } S,$$

if it has a measure for any set of values of $y_1, \cdots, y_n$, has the same measure for any other set of values. In this paper, we shall confine our attention to scalar chaoses. A continuous homogeneous chaos is said to be *metrically transitive*, if whenever the sets of values of α for which $\rho(x_1, \cdots, x_n; \alpha)$ belongs to S and to S_1, respectively, have measures M and M_1, the set of values of α for which simultaneously

$$\rho(x_1, \cdots, x_n; \alpha) \text{ belongs to } S$$

and

$$\rho(x_1 + y_1, \cdots, x_n + y_n; \alpha) \text{ belongs to } S_1$$

has a measure which tends to M_1 as $y_1^2 + \cdots + y_n^2 \to \infty$.

If ρ is integrable, it determines the additive set-function

(5) $$\mathfrak{F}(\Sigma; \alpha) = \int \cdots_{\Sigma} \int \rho(x_1, \cdots, x_n; \alpha) \, dx_1 \cdots dx_n.$$

On the other hand, not every additive set-function may be so defined. This suggests a more general definition of a *homogeneous chaos*, in which the chaos is defined to be a function $\mathfrak{F}(\Sigma; \alpha)$, where α ranges over $(0, 1)$ and Σ belongs to some additively closed set Ξ of measurable sets of points in n-space. We suppose that if Σ and Σ_1 do not overlap,

(6) $$\mathfrak{F}(\Sigma + \Sigma_1; \alpha) = \mathfrak{F}(\Sigma; \alpha) + \mathfrak{F}(\Sigma_1; \alpha).$$

We now define the new point-set $\Sigma(y_1, \cdots, y_n)$ by the assertion

(7) $\Sigma(y_1, \cdots, y_n)$ contains $x_1 + y_1, \cdots, x_n + y_n$ when and only when Σ contains $x_1, \cdots, x_n$.

This leads to the definition of the additive set-function $\mathfrak{F}_{y_1, \ldots, y_n}(\Sigma; \alpha)$ by

(8) $$\mathfrak{F}_{y_1, \ldots, y_n}(\Sigma; \alpha) = \mathfrak{F}(\Sigma(y_1, \cdots, y_n); \alpha).$$

If, then, for all classes S of real numbers,

(9) Measure of set of α's for which $\mathfrak{F}_{y_1, \ldots, y_n}(\Sigma; \alpha)$ belongs to class S

is independent of $y_1, \cdots, y_n$, in the sense that if it exists for one set of these numbers, it exists for all sets, and has the same value, and if it is measurable in $y_1, \cdots, y_n$, we shall call $\mathfrak{F}$ a homogeneous chaos. The notion of metrical transitivity is generalized in the obvious way, replacing $\rho(x_1, \cdots, x_n; \alpha)$ by $\mathfrak{F}(\Sigma; \alpha)$, and $\rho(x_1 + y_1, \cdots, x_n + y_n; \alpha)$ by $\mathfrak{F}_{y_1, \ldots, y_n}(\Sigma; \alpha)$.

The theorem which we wish to prove is the following:

THEOREM I. *Let $\mathfrak{F}(\Sigma;\alpha)$ be a homogeneous chaos. Let the functional*

$$\Phi\{\mathfrak{F}(\Sigma;\alpha)\} = g(\alpha) \tag{10}$$

be a measurable function of α, such that $\int_0^1 |\, g(\alpha) \log^+ |\, g(\alpha)|\,|\, d\alpha$ is finite. Then for almost all values of α,

$$\lim_{r\to\infty} \frac{1}{V(r)} \int \cdots_R \int \Phi\{\mathfrak{F}_{y_1,\ldots,y_n}(\Sigma;\alpha)\} dy_1 \cdots dy_n \tag{11}$$

exists, where R is the interior of the sphere

$$y_1^2 + y_2^2 + \cdots + y_n^2 = r^2 \tag{12}$$

and $V(r)$ is its volume. If in addition, $\mathfrak{F}(\Sigma;\alpha)$ is metrically transitive,

$$\lim_{r\to\infty} \frac{1}{V(r)} \int \cdots_R \int \Phi\{\mathfrak{F}_{y_1,\ldots,y_n}(\Sigma;\alpha)\} dy_1 \cdots dy_n = \int_0^1 \Phi\{\mathfrak{F}(\Sigma;\beta)\} d\beta \tag{13}$$

for almost all values of α.

3. Classical ergodic theorem. Lebesgue form. Theorem I is manifestly a theorem of the ergodic type. Let it be noticed, however, that we nowhere assume that the transformation of α given by $\beta = T\alpha$ when

$$\mathfrak{F}_{y_1,\ldots,y_n}(\Sigma;\beta) = \mathfrak{F}(\Sigma;\alpha) \tag{14}$$

is one-one. This should not be surprising, as the ergodic theorem is fundamentally one concerning the Lebesgue integral, and in the theory of the Lebesgue integral, individual points play no rôle.

Nevertheless, in the usual formulation of the ergodic theorem, the expression $f(T^\lambda P)$ enters in an essential way. Can we give this a meaning without introducing the individual transform of an individual point?

The clue to this lies in the definition of the Lebesgue integral itself. If $f(P)$ is to be integrated over a region S, we divide S into the regions $S_{a,b}(f)$, defined by the condition that over such a region,

$$a < f(P) \leq b. \tag{15}$$

We now write

$$\int_S f(P)\, dV_P = \lim_{\epsilon\to 0} \sum_{-\infty}^{\infty} n\,\epsilon\, m(S_{(n-1)\epsilon,\, n\epsilon}(f)). \tag{16}$$

The condition that $f(P)$ be integrable thus implies the condition that it be measurable, or that all the sets $S_{a,b}(f)$ be measurable.

Now, if T is a measure-preserving transformation on S, the sets $TS_{a,b}(f)$

will all be measurable, and will have, respectively, the same measures as the sets $S_{a,b}(f)$. We shall define the function $f(TP) = g(P)$ by the conditions

$$TS_{a,b}(f) = S_{a,b}(g). \tag{17}$$

If T conserves relations of inclusion of sets, up to sets of zero measure, this function will clearly be defined up to a set of values of P of zero measure, and we shall have

$$\int_S f(TP)\,dV_P = \int_S f(P)\,dV_P. \tag{18}$$

We may thus formulate the original or discrete case of the Birkhoff ergodic theorem, as follows: *Let S be a set of points of finite measure. Let T be a transformation of all measurable sub-sets of S into measurable sub-sets of S, which conserves measure, and the relation between two sets, that one contains the other except at most for a set of zero measure. Then except for a set of points P of zero measure,*

$$\lim_{N\to\infty} \frac{1}{N+1} \sum_0^N f(T^n P) \tag{19}$$

will exist.

The continuous analogue of this theorem needs to be formulated in a somewhat more restricted form, owing to the need of providing for the integrability of the functions concerned. It reads: *Let S be a set of points of finite measure. Let T^λ be a group of transformations fulfilling the conditions we have laid down for T in the discrete case just mentioned. Let $T^\lambda P$ be measurable in the product space of λ and of P. Then, except for a set of points P of zero measure,*

$$\lim_{A\to\infty} \frac{1}{A} \int_0^A f(T^\lambda P)\,d\lambda \tag{20}$$

will exist.

In the proofs of Birkhoff's ergodic theorem, as given by Khintchine and Hopf, no actual use is made of the fact that the transformation T is one-one, and the proofs extend to our theorem as stated here, without any change. The restriction of measurability, or something to take its place, is really necessary for the correct formulation of Khintchine's statement of the ergodic theorem, as Rademacher, von Neumann, and others have already pointed out.[2]

With the aid of the proper Lebesgue formulation of the ergodic theorem, the one-dimensional case of Theorem I follows at once. Actually more follows, as it is only necessary that g belong to L, instead of to the logarithmic class

[2] Cf. J. v. Neumann, *Annals of Mathematics*, 2, vol. 33, p. 589, note 11.

with which we replace it. To prove Theorem I in its full generality, we must establish a multidimensional ergodic theorem.

4. Dominated ergodic theorem. Multidimensional ergodic theorem. As a lemma to the multidimensional ergodic theorem, we first wish to establish the fact that if the function $f(P)$ in the ergodic theorem satisfies the condition

$$\int_S |f(P)| \log^+ |f(P)| \, dV_P < \infty, \tag{21}$$

then the expressions (19) and (20) not only exist, but the limits in question will be approached under the domination of a summable function of P. We shall prove this in the discrete case, for the sake of simplicity, but the result goes over without difficulty to the continuous case.

Let T be an equimeasure transformation of the set S of finite measure into itself, in the generalized sense of the last paragraph, and let W be a measurable sub-set of S, with the characteristic function $W(P)$. Let U be the set of all points P for which some $T^{-j}P$ belongs to W. Let $i(P)$, when P belongs to W, be defined as the smallest positive number n such that $T^{-n}P$ belongs to W, and let us call it the *index* of P. Every point of W, except for a set of measure 0, will have a finite index, since if we write W_∞ for the set of points without a finite index, no two sets $T^m W_\infty$ and $T^n W_\infty$ can overlap, while they all have the same measure and their sum has a finite measure. Thus except for a set of zero measure, we may divide W into the sets W_p, each consisting of the points of W of index p. It is easy to show that if $1 \leq p < \infty$, $1 \leq p' < \infty$, $0 \leq j < p$, $0 \leq j' < p'$, the sets $T^{-j}W_p$ and $T^{-j'}W_{p'}$ can not overlap over a set of positive measure unless $j = j'$, $p = p'$. Similarly, the sets $T^{-p}W_p$ and $T^{-p'}W_{p'}$ can not overlap over sets of positive measure, unless $p = p'$, and represent a dissection of W, except for a set of zero measure. Let us put W_{pq} for the logical product of W_p and $T^p W_q$; W_{pqr} for the logical product of W_{pq} and $T^{p+q}W_r$; and so on. Then if k is fixed, the sets $T^{-j}W_{p_1p_2\ldots p_k}$ cover U (except for sets of zero measure) once as j goes from 0 to p_1, once as it goes from p_1 to $p_1 + p_2$, and so on; making just k times between 0, inclusive, and $p_1 + p_2 + \cdots + p_k$, exclusive. Thus if S_K is the set of all the points in all the $W_{p_1p_2\ldots p_k}$ for which $p_1 + p_2 + \cdots + p_k = K$, the total measure of all the S_K's for $2^{N+1} \geq K \geq 2^N$ can not exceed $km(S)/2^N$.

Now let P lie in $T^{-j}W_{p_1p_2\ldots p_k}$, where $0 \leq j < p_1 + p_2 + \cdots + p_k$. Let us consider the sequence of numbers a_j, where if $p_1 + \cdots + p_l \leq j < p_1 + \cdots + p_{l+1}$, a_j is the greatest of the numbers

$$\frac{1}{j - p_1 - \cdots - p_l + 1}, \quad \frac{2}{j - p_1 - \cdots - p_{l-1} + 1},$$
$$\frac{3}{j - p_1 - \cdots - p_{l-2} + 1}, \ldots, \frac{l+1}{j+1}.$$

Then a_j will be the largest of the numbers

$$W(P), \frac{W(P)+W(TP)}{2}, \cdots, \frac{W(P)+W(TP)+\cdots+W(T^jP)}{j+1}.$$

The sum $\sum_0^k a_j$, for a fixed K, will have its maximum value when $p_1 = p_2 = \cdots = p_{k-1} = 1$; $p_k = K+1-k$, when it will be $k + \sum_{k+1}^{k} k/j \leq k(1+\log K/k)$. In this case the sequence of the a_j's will be

$$\underbrace{1, 1, \cdots, 1}_{k \text{ times}}, \frac{k}{k+1}, \frac{k}{k+2}, \cdots, \frac{k}{K}.$$

This remark will be an easy consequence of the following fact: let us consider the sequence

$$\cdots, \frac{\lambda}{n-1}, \frac{\lambda}{n}, 1, \frac{1}{2}, \cdots, \frac{1}{\left[\frac{n}{\lambda}\right]}, \frac{\lambda+1}{\left[\frac{n}{\lambda}\right]+n+1}, \cdots \tag{22}$$

and the modified sequence

$$\cdots, \frac{\lambda}{n-1}, 1, \frac{1}{2}, \cdots, \frac{1}{\left[\frac{n-1}{\lambda}\right]}, \frac{\lambda+1}{\left[\frac{n-1}{\lambda}\right]+n}, \frac{\lambda+1}{\left[\frac{n-1}{\lambda}\right]+n+1}, \cdots \tag{23}$$

where of course

$$\frac{\lambda+1}{n+\left[\frac{n}{\lambda}\right]} \leq \frac{1}{\left[\frac{n}{\lambda}\right]}; \quad \frac{\lambda+1}{n+\left[\frac{n}{\lambda}\right]+1} > \frac{1}{\left[\frac{n}{\lambda}\right]+1};$$

$$\frac{\lambda+1}{n-1+\left[\frac{n-1}{\lambda}\right]} \leq \frac{1}{\left[\frac{n-1}{\lambda}\right]}; \quad \frac{\lambda+1}{n+\left[\frac{n-1}{\lambda}\right]} > \frac{1}{\left[\frac{n-1}{\lambda}\right]+1}.$$

Apart from the arrangement, the terms of (22) will be the same as the terms of (23), except that in (23), $\dfrac{\lambda+1}{\left[\frac{n-1}{\lambda}\right]+n}$ replaces λ/n. Now,

$$\frac{\lambda+1}{\left[\frac{n-1}{\lambda}\right]+n} > \frac{\lambda}{n},$$

so that the sum of the terms in (23) is greater than that in (22).

Since the transforms of a given set have the same measure as the set, and the sets $T^{-j}S_K$ cover U exactly k times, we have

$$\frac{1}{k}\int_S \Big\{ W(P) + \max\left(W(P), \frac{W(P)+W(TP)}{2}\right) + \max\left(W(P), \frac{W(P)+W(TP)}{2}, \frac{W(P)+W(TP)+W(T^2P)}{3}\right) + \cdots + \max\left(W(P), \frac{W(P)+W(TP)}{2}, \cdots, \frac{W(P)+\cdots+W(T^{k-1}P)}{k}\right)\Big\} dV_P$$

$$\leq \sum_{K=0}^{\infty} m(S_K)\left(1+\log\frac{K}{k}\right)$$

$$\leq m(W)\left(1+\log\frac{2^N}{k}\right) + \sum_{j=1}^{\infty}\frac{km(S)}{2^{N+j}}\left(1+\log\frac{2^{N+j}}{k}\right)$$

$$\leq \left(1+\log\frac{2^N}{k}\right)\left\{ m(W) + \text{const.}\,\frac{km(S)}{2^N}\right\},$$

the constant being absolute. If we now put

$$N = \left[\log\frac{km(S)}{m(W)} / \log 2\right],$$

this last expression is dominated by

$$\text{const.}\, m(W)\left(1+\log\frac{m(S)}{m(W)}\right).$$

Since the constant is independent of k, we see that the Cesàro average of

$$\max\left(W(P), \frac{W(P)+W(TP)}{2}, \cdots, \frac{W(P)+\cdots+W(T^{k-1}P)}{k}\right)$$

is dominated by the same term. Thus

$$\int_S \max\left(W(P), \cdots, \frac{W(P)+\cdots+W(T^{k-1}P)}{k}\right) dV_P \leq \text{const.}\, m(W)\left(1+\log\frac{m(S)}{m(W)}\right),$$

and it follows by monotone convergence that there exists a function $W^*(P)$ such that

$$\int_S W^*(P)\,dV_P \leq \text{const.}\, m(W)\left(1+\log\frac{m(S)}{m(W)}\right), \tag{24}$$

and for all positive m,

$$\frac{W(P)+\cdots+W(T^mP)}{m+1} \leq W^*(P). \tag{25}$$

Now let $f(P)$ be a function such that

$$\int_S |f(P)| \log^+ |f(P)|\, dV_P < \infty. \tag{21}$$

Let $W^{(N)}$ be the set of points such that

$$2^N \leq |f(P)| \leq 2^{N+1}, \tag{26}$$

and let $f^{(N)}(P)$ be the function equal to $f(P)$ over this set of points, and 0 elsewhere. Then there exists a function $f^{(N)*}(P)$, such that

$$\frac{f^{(N)}(P) + \cdots + f^{(N)}(T^m P)}{m+1} \leq f^{(N)*}(P) \qquad (m = 0, 1, 2, \cdots), \tag{27}$$

and

$$\int_S f^{(N)*}(P)\, dV_P \leq \text{const. } m(W^{(N)}) \left(1 + \log^+ \frac{m(S)}{m(W^{(N)})}\right) 2^N. \tag{28}$$

Hence if

$$\sum_{-\infty}^{\infty} m(W^{(N)}) \left(1 + \log^+ \frac{m(S)}{m(W^{(N)})}\right) 2^N < \infty \tag{29}$$

and

$$f^*(P) = \sum_{-\infty}^{\infty} f^{(N)*}(P), \tag{30}$$

then

$$\frac{f(P) + \cdots + f(T^m P)}{m+1} \leq f^*(P) \qquad (m = 0, 1, 2, \cdots), \tag{31}$$

and

$$\int_S f^*(P)\, dV_P \leq \text{const. } \sum m(W^{(N)}) \left(1 + \log^+ \frac{m(S)}{m(W^{(N)})}\right) 2^N. \tag{32}$$

However,

$$\begin{aligned} \sum_{-\infty}^{\infty} m(W^{(N)}) &\left(1 + \log^+ \frac{m(W^{(N)})}{m(S)}\right) 2^N \\ &\leq \sum_{-\infty}^{\infty} \int_{W^{(N)}} |f(P)|\, dV_P \left(1 + \log^+ \frac{2^{N+1} m(S)}{\int_S |f(P)|\, dV_P}\right) \\ &\leq \int_S |f(P) \left(1 + \log^+ \frac{2\,|f(P)|\, m(S)}{\int_S |f(Q)\, dV_Q}\right) dV_P \\ &\leq \int_S |f(P)|\, dV_P \left(1 + \log^+ \frac{2m(S)}{\int_S |f(P)|\, dV_P}\right) \\ &\quad + \int_S |f(P)| \log^+ |f(P)|\, dV_P. \end{aligned} \tag{33}$$

Thus $\int_S f^*(P)dV_P$ has an upper bound which is less than a function of $\int_S |f(P)|\,dV_P$ and $\int_S |f(P)|\log^+|f(P)|\,dV_P$, tending to 0 as they both tend to 0. This establishes our theorem of the existence of a uniform dominant.

There is a sense in which (21) *is a best possible condition.* That is, if

$$\psi(x) = o(\log^+ x), \tag{34}$$

the condition

$$\int_S (\psi(|f(P)|)\,|f(P)|\,dV_P < \infty \tag{35}$$

is not sufficient for the existence of a uniform dominant. For let S be a set of measure 1, subdivided into mutually exclusive sets S_n, of measures respectively 2^{-n}. Let S_n be divided into mutually exclusive sets $S_{n,1}, \cdots, S_{n,\nu_n}$, all of equal measures. Let T transform $S_{n,k}$ into $S_{n,(k+1)}$ $(k < \nu^n)$, and S_{n,ν_n} into $S_{n,1}$. Let $f(P)$ be defined by

$$f(P) = a_n > 0 \text{ on } S_{n_1}, \quad (n = 1, 2, \cdots); \quad f(P) = 0 \text{ elsewhere.} \tag{36}$$

Then the smallest possible uniform dominant of

$$\frac{1}{N+1}\sum_0^N f(T^nP) \tag{37}$$

is

$$f^*(P) = \frac{a_n}{k} \text{ on } S_{n,k}, \tag{38}$$

and we have

$$\int_S f^*(P)dV_P = \sum \frac{a_n\Omega(\log \nu_n)2^{-n}}{\nu_n}. \tag{39}$$

Thus if

$$\nu_n = 2^{2n}a_n, \qquad a_n = \Omega(n!), \tag{40}$$

the function $f^*(P)$ will belong to L if and only if

$$\infty > \sum \frac{a_n\Omega(\log a_n)2^{-n}}{\nu_n} > \text{const.} \int_S f(P)\log^+ f(P)dV_P. \tag{41}$$

While we have proved the dominated ergodic theorem merely as a lemma for the multidimensional ergodic theorem, the theorem, and more particularly, the method by which we have proved it, have very considerable independent interest. We may use these methods to deduce von Neumann's mean ergodic theorem from the Birkhoff theorem; or vice versa, we may deduce the Birkhoff theorem, at least in the case of a function satisfying (21), from the von

Neumann theorem. These facts however are not relevant to the frame of the present paper, and will be published elsewhere.

We shall now proceed to the proof of the multidimensional ergodic theorem, which we shall establish in the two-dimensional case, although the method is independent of the number of dimensions. Let $T_1^\lambda T_2^\mu$ be a two-dimensional Abelian group of transformations of the set S (of measure 1) into itself, in the sense in which we have used this term in paragraph 3. Let $T_1^\lambda T_2^\mu P$ be measurable in λ, μ, and P. We now introduce a new variable x, ranging over $(0, 1)$, and form the product space Σ of P and x. We introduce the one-parameter group of transformations of this space, T^ρ, by putting

$$T^\rho(P, x) = (T_1^{\rho \cos 2\pi x} T_2^{\rho \sin 2\pi x} P, x) \tag{42}$$

The expressions $T^\rho(P, x)$ will be measurable in ρ and (P, x), and the transformations T^ρ will all preserve measure on Σ. Thus by the ergodic theorem, for almost all points (P, x) of Σ, if $f(P) = f(P, x)$ belongs to L, the limit

$$\lim_{A \to \infty} \frac{1}{A} \int_0^A f(T^\rho(P, x))\, d\rho = \lim_{A \to \infty} \frac{1}{A} \int_0^A f(T_1^{\rho \cos 2\pi x} T_2^{\rho \sin 2\pi x} P)\, d\rho \tag{43}$$

will exist. If condition (26) is satisfied, it will follow by dominated convergence that the limit

$$\lim_{A \to \infty} \frac{1}{A} \int_0^A d\rho \int_0^1 f(T_1^{\rho \cos 2\pi x} T_2^{\rho \sin 2\pi x} P)\, dx \tag{44}$$

will exist for almost all points (P, x), and hence for almost all points P.

For the moment, let us assume that $f(P)$ is non-negative. Then there is a Tauberian theorem, due to the author,[3] which establishes that the expression (44) is equivalent to

$$\lim_{A \to \infty} \frac{1}{\pi A^2} \int_0^A \rho\, d\rho \int_0^{2\pi} f(T_1^{\rho \cos \theta} T_2^{\rho \sin \theta} P)\, d\theta. \tag{45}$$

The only point of importance which we must establish in order to justify this Tauberian theorem is that

$$\int_0^1 \rho^{1+iu} d\rho = \frac{1}{2 + iu} \neq 0 \tag{46}$$

for real values of u. Since every function $f(P)$ satisfying (21) is the difference of two non-negative functions satisfying this condition, (45) is established in the general case.

It will be observed that we have established our multidimensional ergodic

[3] N. Wiener, "Tauberian theorems," *Annals of Mathematics*, 2, vol. 33, p. 28.

theorem on the basis of assumption (21), and not on that of the weaker assumption that $f(P)$ belongs to L. What the actual state of affairs may be, we do not know. At any rate, all attempts to arrive at a direct analogue of the Khintchine proof for one dimension have broken down. The one-dimensional proof makes essential use of the fact that the difference of two intervals is always an interval, while the difference between two spheres is not always a sphere.

The precise statement of the multidimensional ergodic theorem is the following: *Let S be a set of points of measure 1, and let $T_1^{\lambda_1}T_2^{\lambda_2}\cdots T_n^{\lambda_n}$ be an Abelian group of equimeasure transformations of S into itself, in the sense of paragraph 3. Let $T_1^{\lambda_1}\cdots T_n^{\lambda_n}P$ be measurable in $\lambda_1,\cdots,\lambda_n$; P. Let R be the set of values of $\lambda_1,\cdots,\lambda_n$ for which*

$$\lambda_1^2+\lambda_2^2+\cdots+\lambda_n^2\leq r^2 \tag{47}$$

and let $V(r)$ be its volume. Let $f(P)$ satisfy the condition (21). *Then for almost all values of P,*

$$\lim_{r\to\infty}\frac{1}{V(r)}\int\cdots_R\int f(T_1^{\lambda_1}\cdots T_n^{\lambda_n}P)\,d\lambda_1\cdots d\lambda_n \tag{48}$$

exists.

That part of Theorem I which does not concern metric transitivity is an immediate corollary.

5. Metric transitivity. Space and Phase Averages in a Chaos. If the function $f(P)$ is positive, clearly

$$\begin{aligned}\int\cdots\int_{\lambda_1^2+\cdots+\lambda_n^2\leq r^2-\mu_1^2-\cdots-\mu_n^2} & f(T_1^{\lambda}\cdots T_n^{\lambda}P)\,d\lambda_1\cdots d\lambda_n \\ &\leq \int\cdots\int_{\lambda_1^2+\cdots+\lambda_n^2\leq r^2} f(T_1^{\lambda_1}\cdots T_n^{\lambda_n}T_1^{\mu_1}\cdots T_n^{\mu_n}P)\,d\lambda_1\cdots d\lambda_n \\ &\leq \int\cdots\int_{\lambda_1^2+\cdots+\lambda_n^2\leq r^2+\mu_1^2+\cdots+\mu_n^2} f(T_1^{\lambda_1}\cdots T_n^{\lambda_n}P)\,d\lambda_1\cdots d\lambda_n.\end{aligned} \tag{49}$$

Hence

$$\begin{aligned}\lim_{r\to\infty}\frac{1}{V(r)}\int\cdots_R\int & f(T_1^{\lambda_1}\cdots T_n^{\lambda_n}T_1^{\mu_1}\cdots T_n^{\mu_n}P)\,d\lambda_1\cdots d\lambda_n \\ &=\lim_{r\to\infty}\frac{1}{V(r)}\int\cdots_R\int f(T_1^{\lambda_1}\cdots T_n^{\lambda_n}P)\,d\lambda_1\cdots d\lambda_n,\end{aligned} \tag{50}$$

and expression (48) has the same value for P and all its transforms under the group $T_1^{\lambda_1}\cdots T_n^{\lambda_n}$. The condition of positivity is clearly superfluous. Thus in case expression (48) does not almost everywhere assume a single value, there will be two classes S_1 and S_2 of elements of S, each of positive measure, and each invariant under all the transformations $T_1^{\lambda_1} \cdots T_n^{\lambda_n}$.

A condition which will manifestly exclude such a contingency is that if S_1 and S_2 are two sub-sets of S of positive measure, and ϵ is a positive quantity, there always exists a transformation $T = T_1^{\lambda_1} \cdots T_n^{\lambda_n}$, such that

$$\left| \frac{mS_1(TS_2)}{mS_2} - mS_1 \right| < \epsilon. \tag{51}$$

From this it will immediately follow that if a chaos is metrically transitive in the sense of paragraph 3, the group of transformations of the α space generated by translations of the chaos will have the property we have just stated, and under the assumptions of Theorem I,

$$\lim_{r\to\infty} \frac{1}{V(r)} \int \cdots_R \int \Phi\{\mathfrak{F}_{y_1, \ldots, y_n}(\Sigma ; \alpha)\} dy_1 \cdots dy_n \tag{52}$$

will exist and have the same value for almost all values of α.

If almost everywhere

$$\lim_{r\to\infty} \frac{1}{V(r)} \int \cdots_R \int \Phi\{\mathfrak{F}_{y_1, \ldots, y_n}(\Sigma ; \alpha)\} dy_1 \cdots dy_n = A, \tag{53}$$

and

$$\frac{1}{V(r)} \int \cdots_R \int \Phi\{\mathfrak{F}_{y_1, \ldots, y_n}(\Sigma ; \alpha)\} dy_1 \cdots dy_n < g(\alpha) \tag{54}$$

where $g(\alpha)$ belongs to L, then by dominated convergence,

$$\begin{aligned} A &= \lim_{r\to\infty} \frac{1}{V(r)} \int \cdots_R \int dy_1 \cdots dy_n \int_0^1 d\alpha \Phi\{\mathfrak{F}_{y_1, \ldots, y_n}(\Sigma ; \alpha)\} \\ &= \int_0^1 \Phi\{\mathfrak{F}(\Sigma ; \alpha)\} d\alpha. \end{aligned} \tag{55}$$

That is, the average of $\Phi\{\mathfrak{F}(\Sigma ; \alpha)\}$, taken over the finite phase space of α, is almost everywhere the same as the average of $\Phi\{\mathfrak{F}_{y_1, \ldots, y_n}(\Sigma ; \alpha)\}$ taken over the infinite group space of points $y_1, \cdots, y_n$. This completes the establishment of Theorem I, and gives us a real basis for the study of the homogeneous chaos.[4]

6. Pure one-dimensional chaos. The simplest type of pure chaos is that which has already been treated by the author in connection with the Brownian motion. However, as we wish to generalize this theory to a multi-

[4] The material of this chapter, in the one-dimensional case, has been discussed by the author with Professor Eberhard Hopf several years ago, and he wishes to thank Professor Hopf for suggestions which have contributed to his present point of view.

plicity of dimensions, instead of referring to existing articles on the subject, we shall present it in a form which emphasizes its essential independence of dimensionality.

The type of chaos which we shall consider is that in which the expression $\mathfrak{F}(\Sigma;\alpha)$ has a distribution in α dependent only on the measure of the set Σ; and in which, if Σ_1 and Σ_2 do not overlap, the distributions of $\mathfrak{F}(\Sigma_1,\alpha)$ and $\mathfrak{F}(\Sigma_2,\alpha)$ are independent, in the sense that if $\phi(x,y)$ is a measurable function, and either side of the equation has a sense,

$$(56)\qquad \int_0^1\int_0^1 \phi(\mathfrak{F}(\Sigma_1;\alpha),\mathfrak{F}(\Sigma_2,\beta))\,d\alpha d\beta = \int_0^1 \phi(\mathfrak{F}(\Sigma_1,\alpha),\mathfrak{F}(\Sigma_2,\alpha))\,d\alpha.$$

We assume a similar independence when n non-overlapping sets $\Sigma_1,\cdots,\Sigma_n$ are concerned. It is by no means intuitively certain that such a type of chaos exists. In establishing its existence, we encounter a difficulty belonging to many branches of the theory of the Lebesgue integral. The fundamental theorem of Lebesgue assures us of the possibility of adding the measures of a denumerable assemblage of measurable sets, to get the measure of their sum, if they do not overlap. Accordingly, behind any effective realization of the theory of Lebesgue integration, there is always a certain denumerable family of sets in the background, such that all measurable sets may be approximated by denumerable combinations of these. This family is not unique, but without the possibility of finding it, there is no Lebesgue theory.

On the other hand, a theory of measure suitable for the description of a chaos must yield the measure of any assemblage of functions arising from a given measurable assemblage by a translational change of origin. This set of assemblages is essentially non-denumerable. Any attempt to introduce the notion of measure in a way which is invariant under translational changes of origin, without the introduction of some more restricted set of measurable sets, which does not possess this invariance, will fail to establish those essential postulates of the Lebesgue integral which deal with denumerable sets of points. There is no way of avoiding the introduction of constructional devices which seem to restrict the invariance of the theory, although once the theory is obtained it may be established in its full invariance.

Accordingly, we shall start our theory of randomness with a division of space, whether of one dimension or of more, into a denumerable assemblage of sub-sets. In one dimension, this division may be that into those intervals whose coördinates are terminating binary numbers, and in more dimensions, into those parallelepipeds with edges parallel to the axes and with terminating binary coördinates for the corner points. We then wish to find a self-consistent distribution-function for the mass in such a region, dependent only on the volume, and independent for non-overlapping regions.

This problem does not admit of a unique solution, although the solution becomes essentially unique if we adjoin suitable auxiliary conditions. Among these conditions, for example, is the hypothesis that the distribution is symmetric, as between positive and negative values, has a finite mean square, and that the measure of the set of α's for which $\mathfrak{F}(S;\alpha) > A$ is a continuous function of A.[5] Without going into such considerations, we shall assume directly that the measure of the set of instances in which the value of $\mathfrak{F}(S;\alpha)$ in a region S of measure M lies between a and $b > a$, is

$$\frac{1}{\sqrt{2\pi M}}\int_a^b \exp\left(-\frac{u^2}{2M}\right)du. \tag{57}$$

The formula

$$\frac{1}{\sqrt{2\pi M_1 M_2}}\int_{-\infty}^{\infty} \exp\left(-\frac{u^2}{2M_1}-\frac{(v-u)^2}{2M_2}\right)du = \frac{1}{\sqrt{M_1+M_2}}\exp\left(-\frac{v^2}{2(M_1+M_2)}\right) \tag{58}$$

shows the consistency of this assumption.

The distributions of mass among the sets of our denumerable assemblage may be mapped on the line segment $0 \leq \alpha \leq 1$, in such a way that the measure of the set of instances in which a certain contingency holds will go into a set of values of α of the same measure. This statement needs a certain amount of elucidation. To begin with, the only sets of instances whose measures we know are those determined by

$$\begin{array}{c} a_1 \leq \mathfrak{F}(S_1;\alpha) \leq b_1 \\ a_2 \leq \mathfrak{F}(S_2;\alpha) \leq b_2 \\ \cdots\cdots\cdots \\ a_n \leq \mathfrak{F}(S_n;\alpha) \leq b_n; \end{array} \tag{59}$$

where $S_1, S_2, \cdots, S_n$ are to be found among our denumerable set of subdivisions of space. However, once we have established a correspondence between the measures of these specific sets of contingencies and their corresponding sets of values of α, we may use the measure of any measurable set of values of α to define the measure of its corresponding set of contingencies.

The correspondence between sets of contingencies and points on the line $(0, 1)$ is made by determining a hierarchy of sets of contingencies

$$\begin{array}{c} a_1^{(m,n)} \leq \mathfrak{F}(S_1^{(m,n)};\alpha) \leq b_1^{(m,n)} \\ \cdots\cdots\cdots \\ a_\nu^{(m,n)} \leq \mathfrak{F}(S_\nu^{(m,n)};\alpha) \leq b_\nu^{(m,n)}. \end{array} \tag{60}$$

---

[5] Cf. the recent investigations of Cramér and P. Lévy.

Let us call such a contingency $C_{m,n}$. If m is fixed, let all the contingencies $C_{m,n}$ $(n = 1, 2, \cdots)$ be mutually exclusive, and let them be finite in number. Let us be able to write

$$(61) \qquad C_{m,n} = \sum_{k=1}^{N} C_{m+1,n_k}.$$

If S' is one of our denumerable sets of regions of space, let every $C_{m,n}$ with a sufficiently large index m be included in a class determined by a set of conditions concerning the mass on S' alone, and restricting it to a set of values lying in an interval (c, d), corresponding to an integral of form (57) and of arbitrarily small value. (Here d may be ∞, or c may be $-\infty$.) Let us put $C_{1,1}$ for the entire class of all possible contingencies, and let us represent it by the entire interval $(0, 1)$. Let us assume that $C_{m,n}$ has been mapped into an interval of length corresponding to its probability, in accordance with (57), and let this interval be divided in order of the sequence of their n_k's into intervals corresponding respectively to the component C_{m+1,n_k}'s, and of the same measure. Except for a set of points of zero measure, every point of the segment $(0, 1)$ of α will then be determined uniquely by the sequence of the intervals containing it and corresponding to the contingencies $C_{m,n}$ for successive values of m. This sequence will then determine uniquely (except in a set of cases corresponding to a set of values of α of zero measure) the value of $\mathfrak{F}(S_n; \alpha)$ for every one of our original denumerable set of sets S_n.

So far, everything that we have said has been independent of dimensionality. We now proceed to something belonging specifically to the one-dimensional case. If the original sets S_n are the sets of intervals with binary end-points, of such a form that they may be written in the binary scale

$$(62) \qquad (d_1 d_2 \cdots d_k \cdot d_{k+1} \cdots d_l, d_1 \cdots d_k \cdot d_{k+1} \cdots d_l + 2^{-l-1})$$

where $d_1, \cdots, d_l$ are digits which are either 0 or 1, then any interval whatever of length not exceeding $2^{-\mu}$ and lying in (a, b) (where a and b are integers) may be written as the sum of not more than two of the $(b-a)2^{\mu+2}$ intervals of form (62) lying in (a, b) and of length $2^{-\mu-1}$, not more than two of the $(b-a)2^{\mu+2}$ intervals of length $2^{-\mu-2}$, and so on. The probability that the value of $|\mathfrak{F}(S_n; \alpha)|$ should exceed A, or in other words, the measure of the set of α's for which it exceeds A, is

$$(63) \qquad \frac{2}{\sqrt{2\pi m(S_n)}} \int_A^\infty \exp\left(-\frac{u^2}{2m(S_n)}\right) du = o\{\exp(-Am(S_n)^{-\frac{1}{2}})\}.$$

Now let us consider the total probability that the value of $|\mathfrak{F}(S_n; \alpha)|$ should exceed $2^{-(\mu+1)(\frac{1}{2}-\epsilon)}$ for any one of the $(b-a)2^{\mu+1}$ intervals of length

$2^{-\mu-1}$, or $2^{-(\mu+2)(\frac{1}{2}-\epsilon)}$ for any one of the $(b-a)2^{\mu+2}$ intervals of length $2^{-\mu-2}$, or so on. This probability can not exceed

$$(64) \qquad \sum_{k=1}^{\infty}(b-a)2^{\mu+k}\,o(\exp(-2^{(\mu+k)\epsilon}))=o(2^{\mu}\exp(-2^{(\mu+1)\epsilon})).$$

On the other hand, the sum of $|\mathfrak{F}(S_n;\alpha)|$ for all the $2+2+\cdots$ intervals must in any other case be equal to or less than

$$(65) \qquad 2\sum_{k=1}^{\infty}2^{-(\mu+k)(\frac{1}{2}-\epsilon)}=O(2^{-\mu(\frac{1}{2}-\epsilon)}).$$

Thus there is a certain sense in which over a finite interval, and except for a set of values of α of arbitrarily small positive measure, the total mass in a sub-interval of length $\leq 2^{-\mu}$ tends uniformly to 0 with $2^{-\mu}$. On this basis, we may extend the functional $\mathfrak{F}(S;\alpha)$ to all intervals S. It is already defined for all intervals with terminating binary end-points. If (c,d) is any interval whatever, let $c_1, c_2, \cdots$ be a sequence of terminating binary numbers approaching c, and let $d_1, d_2, \cdots$ be a similar sequence approaching d. Then except for a fixed set of values of α of arbitrarily small measure,

$$(66) \qquad \lim_{m,n\to\infty}|\mathfrak{F}((c_n,d_n)+\alpha)-\mathfrak{F}((c_m,d_m):\alpha)|=0,$$

and we may put

$$(67) \qquad \mathfrak{F}((c,d);\alpha)=\lim_{m\to\infty}\mathfrak{F}((c_m,d_m):\alpha).$$

Formula (57), and the fact that $\mathfrak{F}(S_1;\alpha)$ and $\mathfrak{F}(S_2;\alpha)$ vary independently for non-overlapping intervals S_1 and S_2, will be left untouched by this extension.

Thus if Σ is an interval and T^{λ} a translation through an amount λ, we can define $\mathfrak{F}(T^{\lambda}\Sigma:\alpha)$, and it will be equally continuous in λ over any finite range of λ except for a set of values of α of arbitrarily small measure. From this it follows at once that it is measurable in λ and α together. Furthermore, we shall have

(68) Measure of set of α's for which $\mathfrak{F}(T^{\lambda}\Sigma;\alpha)$ belongs to C = Measure of set of α's for which $\mathfrak{F}(\Sigma;\alpha)$ belongs to C.

Thus $\mathfrak{F}(\Sigma;\alpha)$ is a homogeneous chaos. We shall call it *the pure chaos.*

If $\Phi\{\mathfrak{F}(\Sigma;\alpha)\}$ is a functional dependent on the values of $\mathfrak{F}(\Sigma;\alpha)$ for a finite number of intervals Σ_n, then if λ is so great that none of the intervals Σ_n overlaps any translated interval $T^{\lambda}\Sigma_m$, $\Phi\{\mathfrak{F}(\Sigma;\alpha)\}$ will have a distribution entirely independent of $\Phi\{\mathfrak{F}(T^{\lambda}\Sigma;\alpha)\}$. As every measurable functional may be approached in the L sense by such a functional, we see at once that $\mathfrak{F}(\Sigma:\alpha)$ is metrically transitive.[6]

[6] Except that the method of treatment has been adapted to the needs of § 7, the

7. Pure multidimensional chaos. In order to avoid notational complexity, we shall not treat the general multidimensional case explicitly, but shall treat the two-dimensional case by a method which will go over directly to the most general multidimensional case. If our initial sets S_n are the rectangles with terminating binary coördinates for their corners and sides parallel to the axes, and we replace (a, b) by the square with opposite vertices (p, q) and $(p+r, q+r)$, an argument of exactly the same sort as that which we have used in the last paragraph will show that except in a set of cases of total probability not exceeding

$$(69) \qquad \text{const.} \sum_{k=1}^{\infty} \sum_{l=1}^{\infty} 2^{\mu+k} 2^{l}\, o(\exp(-2^{(\mu+k+l)\epsilon})) = o(2^{\mu} \exp(-2^{(\mu+1)\epsilon}))$$

the sum of $|\mathfrak{F}(S_n;\alpha)|$ for a denumerable set of binary rectangles with base $\leq 2^{-\mu}$, of the form (62), and adding up to make a vertical interval lying in the square (p, q), $(p+r, q+r)$, must be equal to or less than

$$(65) \qquad 2\sum_{k=1}^{\infty} 2^{-(\mu+k)(\frac{1}{2}-\epsilon)} = O(2^{-\mu(\frac{1}{2}-\epsilon)}).$$

If we now add this expression up for all the base intervals of type (62) necessary to exhaust a horizontal interval of magnitude not exceeding $2^{-\mu}$, we shall again obtain an expression of the form (65). It hence follows that if we take the total mass on the coördinate rectangles within a given square, this will tend to zero uniformly with their area, except for a set of values of α of arbitrarily small measure. From this point the two-dimensional argument, and indeed the general multidimensional argument, follows exactly the same lines as the one-dimensional argument. It is only necessary to note that if

$$a_n \to a, \; b_n \to b, \; c_n \to c, \; d_n \to d$$

then the rectangles (a_m, b_m), (c_m, d_m) and (a_n, b_n), (c_n, d_n) differ at most by four rectangles of small area.[7]

From this point on, we shall write $\mathfrak{P}(S;\alpha)$ for a pure chaos, whether in one or in more dimensions.

8. Phase averages in a pure chaos. If $f(P)$ is a measurable step-function, the definition of

$$(70) \qquad \int f(P) d_P \mathfrak{P}(S;\alpha)$$

results of this section have previously been demonstrated by the author. (*Proceedings of the London Mathematical Society*, 2, vol. 22 (1924), pp. 454-467).

[7] Here we represent a rectangle by giving two opposite corners.

is obvious, for it reduces to the finite sum

$$\sum_{1}^{N} f_n \mathfrak{P}(S_n; \alpha) \tag{71}$$

where f_n are the N values assumed by $f(P)$, and S_n respectively are the sets over which these values are assumed. Let us notice that

$$\begin{aligned} \int_0^1 d\alpha \left| \int f(P)\,d_P\mathfrak{P}(S;\alpha) \right|^2 &= \sum_{m=1}^{N}\sum_{n=1}^{N} \int_0^1 d\alpha f_m \bar{f}_n \mathfrak{P}(S_m;\alpha)\,\mathfrak{P}(S_n;\alpha) \\ &= \sum_1^N |f_n|^2 \int_0^1 (\mathfrak{P}(S_n;\alpha))^2 d\alpha \\ &= \sum_1^N |f_n|^2 \frac{1}{\sqrt{2\pi m(S_n)}} \int_{-\infty}^{\infty} u^2 \exp\left(-\frac{u^2}{2m(S_n)}\right) du \\ &= \sum_1^N |f_n|^2 m(S_n) \frac{1}{\sqrt{2\pi}} \int_{-\infty}^{\infty} u^2 e^{-(u^2/2)} du \\ &= \sum_1^N |f_n|^2 m(S_n) = \int |f(P)|^2 dV_P, \end{aligned} \tag{72}$$

the integral being taken over the whole of space. In other words, the transformation from $f(P)$ as a function of P, to $\int f(P)\,d_P\mathfrak{P}(S;\alpha)$ as a function of α, retains distance in Hilbert space.[8] Such a transformation, by virtue of the Riesz-Fischer theorem, may always be extended by making limits in the mean correspond to limits in the mean. Thus both in the one-dimensional and in the many-dimensional case, we may *define*

$$\int f(P)\,d_P\mathfrak{P}(S;\alpha) = \underset{n\to\infty}{\text{l. i. m.}} \int f_n(P)\,d_P\mathfrak{P}(S;\alpha) \tag{73}$$

where $f(P)$ is a function belonging to L^2, and the sequence $f_1(P), f_2(P), \cdots$ is a sequence of step-functions converging in the mean to $f(P)$ over the whole of space. The definition will be unambiguous, except for a set of values of α of zero measure.

If S is any measurable set, we have

$$\begin{aligned} \int_0^1 d\alpha\{\mathfrak{P}(S;\alpha)\}^n &= \frac{1}{\sqrt{2\pi m(S)}} \int_{-\infty}^{\infty} u^n \exp\left(-\frac{u^2}{2m(S)}\right) du \\ &= (m(S))^{n/2} \frac{1}{\sqrt{2\pi}} \int_{-\infty}^{\infty} u^n e^{-(u^2/2)} du \\ &\begin{cases} = 0 \text{ if } n \text{ is odd} \\ = (m(S))^{n/2}(n-1)(n-3)\cdots 1 \text{ if } n \text{ is even.} \end{cases} \end{aligned} \tag{74}$$

[8] Cf. Paley, Wiener, and Zygmund, *Mathematische Zeitschrift*, vol. 37 (1933), pp. 647-668.

This represents $(m(S))^{n/2}$, multiplied by the number of distinct ways of representing n objects as a set of pairs. Remembering that if $S_1, S_2, \cdots, S_{2n}$ are non-overlapping, their distributions are independent, we see that if the sets $\Sigma_1, \Sigma_2, \cdots, \Sigma_{2n}$ are either totally non-overlapping, or else such that when two overlap, they coincide, we have [9]

$$(75) \qquad \int_0^1 \mathfrak{P}(\Sigma_1;\alpha)\cdots\mathfrak{P}(\Sigma_n;\alpha)\,d\alpha = \Sigma\Pi\int_0^1 \mathfrak{P}(\Sigma_j;\alpha)\,\mathfrak{P}(\Sigma_k;\alpha)\,d\alpha,$$

where the product sign indicates that the $2n$ terms are divided into n sets of pairs, j and k, and that these factors are multiplied together, while the addition is over all the partitions of $1,\cdots,2n$ into pairs. If $2n$ is replaced by $2n+1$, the integral in (75) of course vanishes.

Since $\mathfrak{P}(S;\alpha)$ is a linear functional of sets of points, and since both sides of (75) are linear with respect to each $\mathfrak{P}(\Sigma_k;\alpha)$ separately, (75) still holds when $\Sigma_1, \Sigma_2, \cdots, \Sigma_{2n}$ can be reduced to sums of sets which either coincide or do not overlap, and hence holds for all measurable sets.

Now let $f(P_1,\cdots,P_n)$ be a measurable step-function: that is, a function taking only a finite set of finite values, each over a set of values $P_1,\cdots,P_n$ which is a product-set of measurable sets in each variable P_k. Clearly we may define

$$(76) \qquad \int\cdots\int f(P_1,\cdots,P_n)\,d_{P_1}\mathfrak{P}(S;\alpha)\cdots d_{P_n}\mathfrak{P}(S;\alpha)$$

in a way quite analogous to that in which we have defined (70), and we shall have

$$(77) \qquad \int_0^1 d\alpha\int\cdots\int f(P_1,\cdots,P_n)\,d_{P_1}\mathfrak{P}(S;\alpha)\cdots d_{P_n}\mathfrak{P}(S;\alpha)$$
$$= \Sigma\int\cdots\int f(P_1,P_1,P_2,P_2,\cdots,P_n,P_n)\,dV_{P_1}\cdots dV_{P_n}$$

where the summation is carried out for all possible divisions of the $2n$ P's into pairs. Similarly in the odd case

$$(78) \qquad \int_0^1 d\alpha\int\cdots\int f(p_1,\cdots,P_{2n+1})\,d_{P_1}\mathfrak{P}(S;\alpha)\cdots d_{P_n}\mathfrak{P}(S;\alpha) = 0.$$

We may apply (77) to give a meaning to

$$(79) \qquad \int_0^1 d\alpha\left|\int\cdots\int f(P_1,\cdots,P_n)\,d_{P_1}\mathfrak{P}(S;\alpha)\cdots d_{P_n}\mathfrak{P}(S;\alpha)\right|^2.$$

If $f(\mathrm{P}_1,\cdots,P_n)$ is a measurable step-function, and

[9] Cf. Paley, Wiener, and Zygmund, *loc. cit.*, formula (2.05).

$$(80)\qquad |f(P_1,\cdots,P_n)| \leq |f_1(P_1)\cdots f_n(P_n)|\,;$$
$$\int |f_k(P)|^2 dV_P \leq A \qquad (k=1,2,\cdots)$$

we shall have

$$(81)\qquad \int_0^1 d\alpha \left|\int\cdots\int f(P_1,\cdots P_n)\,d_{P_1}\mathfrak{P}(S;\alpha)\cdots d_{P_n}\mathfrak{P}(S;\alpha)\right|^2 \leq A^n(2n-1)(2n-3)\cdots 1.$$

If now $f(P_1,\cdots,P_n)$ is an integrable function satisfying (80), but not necessarily a step-function, let

$$(82)\qquad f(\nu;P_1,\cdots,P_n)=\frac{1}{\nu}\operatorname{sgn} f(P_1,\cdots,P_n)\,[\nu f(P_1,\cdots,P_n)\operatorname{sgn} f(P_1,\cdots,P_n)].$$

Clearly almost everywhere

$$(83)\qquad \int\cdots\int f(\nu;P_1,\cdots,P_n)\,d_{P_1}\mathfrak{P}(S;\alpha)\cdots d_{P_n}\mathfrak{P}(S;\alpha) \leq \prod_1^n \int f_k(P)\,d_P\mathfrak{P}(S;\alpha)$$

and

$$(84)\qquad \overline{\lim_{\mu,\nu\to\infty}} \left|\int\cdots\int f(\mu;P_1,\cdots,P_n)\,d_{P_1}\mathfrak{P}(S;\alpha)\cdots d_{P_n}\mathfrak{P}(S;\alpha) - \int\cdots\int f(\nu;P_1,\cdots,P_n)\,d_{P_1}\mathfrak{P}(S;\alpha)\,d_{P_2}\mathfrak{P}(S;\alpha)\right|$$
$$\leq \prod_1^n \int f_k(P)\,d_P\mathfrak{P}(S;\alpha)\left[\epsilon+\sum_1^n\left\{\frac{\int_R f_k(P)\,d_P\mathfrak{P}(S;\alpha)}{\int f_k(P)\,d_P\mathfrak{P}(S;\alpha)}\right\}\right]$$

where R represents the exterior of a sphere of arbitrarily large volume. Let it be noted that both the numerator and the denominator of this fraction have Gaussian distributions, but that the mean square value of the numerator is arbitrarily small. Thus except for a set of values of α of arbitrarily small measure, the right side of expression (84) is arbitrarily small, so that we may write

$$(85)\qquad \lim_{\mu,\nu\to\infty}\left\{\int\cdots\int f(\mu;P_1,\cdots,P_n)\,d_{P_1}\mathfrak{P}(S;\alpha)\cdots d_{P_n}\mathfrak{P}(S;\alpha) - \int\cdots\int f(\nu;P_1,\cdots,P_n)\,d_{P_1}\mathfrak{P}(S;\alpha)\cdots d_{P_n}\mathfrak{P}(S;\alpha)\right\}=0.$$

Thus by dominated convergence,

$$(86)\qquad \lim_{\mu\to\infty}\int\cdots\int f(\mu;P_1,\cdots,P_n)\,d_{P_1}\mathfrak{P}(S;\alpha)\cdots d_{P_n}\mathfrak{P}(S;\alpha)$$

exists for almost all values of α, and we may write it by definition

$$(87) \qquad \int \cdots \int f(P_1, \cdots, P_n) d_{P_1} \mathfrak{P}(S; \alpha) \cdots d_{P_n} \mathfrak{P}(S; \alpha).$$

This will clearly be unique, except for a set of values of α of zero measure. There will then be no difficulty in checking (77), (78), and (81).

9. Forms of chaos derivable from a pure chaos. Let us assume that $f(P)$ belongs to L^2, or that $f(P_1, \cdots, P_n)$ is a measurable function satisfying (80). Let us write $\widehat{PQ}$ for the vector in n-space connecting the points P and Q. Then the function

$$(88) \quad \int \cdots \int f(\widehat{PP_1}, \cdots, \widehat{PP_n}) d_{P_1} \mathfrak{P}(S; \alpha) \cdots d_{P_n} \mathfrak{P}(S; \alpha) = F(P; \alpha)$$

is a metrically transitive differentiable chaos. This results from the fact that $\mathfrak{P}(S; \alpha)$ is a metrically transitive chaos, and that a translation of P generates a similar translation of all the points P_k. The sum of a finite number of functions of the type (88) is also a metrically transitive differentiable chaos. To show that $F(P; \alpha)$ is measurable in P and α simultaneously, we merely repeat the argument of (83)–(86) with both P and α as variables.

We shall call a chaos such as (88) a *polynomial chaos homogeneously of the n-th degree*, and a sum of such chaoses a *polynomial chaos* of the degree of its highest term. In this connection, we shall treat a constant as a chaos homogeneously of degree zero.

By the multidimensional ergodic theorem, if Φ is a functional such that

$$(89) \qquad \int_0^1 | \Phi(F(P; \alpha)) | \log^+ | \Phi(F(P; \alpha)) | \, d\alpha < \infty$$

we shall have

$$(90) \qquad \lim_{r \to \infty} \frac{1}{V(r)} \int_R \Phi(F(P; \alpha)) dV_P = \int_0^1 \Phi(F(P; \alpha)) d\alpha$$

for almost all values of α. Since the distribution of $F(P; \alpha)$ is dominated by the product of a finite number of independent Gaussian distributions, we even have

$$(91) \qquad \int_0^1 | F(P; \alpha) |^n d\alpha < \infty$$

for all positive integral values of n. In a wide class of cases this enables us to establish a relation of the type of (89).

In formula (90), we have an algorithm for the computation of the right-hand side. For example, if

$$F(P;\alpha) = \int f(\widehat{PP_1})\, d_{P_1}\mathfrak{P}(S;\alpha), \tag{92}$$

and $P + Q$ is the vector sum of P and Q, we have for almost all α,

$$\lim_{r\to\infty} \frac{1}{V(r)} \int_R F(P+Q;\alpha)\bar{F}(Q;\alpha)\,dV_Q = \int f(P+Q)\bar{f}(Q)\,dV_Q, \tag{93}$$

the integral being taken over the whole of space; if

$$F(P;\alpha) = \int \cdots \int f(\widehat{PP_1}, \widehat{PP_2})\, d_{P_1}\mathfrak{P}(S;\alpha)\, d_{P_2}\mathfrak{P}(S;\alpha), \tag{94}$$

we have almost always

$$\begin{aligned} \lim_{r\to\infty} \frac{1}{V(r)} \int_R & F(P+Q;\alpha)\bar{F}(Q;\alpha)\,dV_Q \\ &= \left| \int f(Q,Q)\,dV_Q \right|^2 + \int\int f(P+Q,P+M)\bar{f}(Q,M)\,dV_Q dV_M \\ &\quad + \int\int f(P+Q,P+M)\bar{f}(M,Q)\,dV_Q dV_M; \end{aligned} \tag{95}$$

and if

$$F(P;\alpha) = \int\int\int f(\widehat{PP_1}, \widehat{PP_2}, \widehat{PP_3})\, d_{P_1}\mathfrak{P}(S;\alpha)\, d_{P_2}\mathfrak{P}(S;\alpha)\, d_{P_3}\mathfrak{P}(S;\alpha), \tag{96}$$

we have almost everywhere

$$\begin{aligned} \lim_{r\to\infty} \frac{1}{V(r)} \int_R & F(P+Q;\alpha)\bar{F}(Q;\alpha)\,dV_Q \\ = \int\int\int \{ & f(Q,Q,P+M)\bar{f}(M,S,S) \\ & + f(Q,Q,P+M)\bar{f}(S,M,S) + f(Q,Q,P+M)\bar{f}(S,S,M) \\ & + f(Q,P+M,Q)\bar{f}(M,S,S) + f(Q,P+M,Q)\bar{f}(S,M,S) \\ & + f(Q,P+M,Q)\bar{f}(S,S,M) + f(P+M,Q,Q)\bar{f}(M,S,S) \\ & + f(P+M,Q,Q)\bar{f}(S,M,S) + f(P+M,Q,Q)\bar{f}(S,S,M) \\ & + f(P+Q,P+M,P+S)\bar{f}(Q,M,S) \\ & + f(P+Q,P+M,P+S)\bar{f}(Q,S,M) \\ & + f(P+Q,P+M,P+S)\bar{f}(M,Q,S) \\ & + f(P+Q,P+M,P+S)\bar{f}(M,S,Q) \\ & + f(P+Q,P+M,P+S)\bar{f}(S,Q,M) \\ & + f(P+Q,P+M,P+S)\bar{f}(S,M,Q) \}\,dV_Q dV_M dV_S. \end{aligned} \tag{97}$$

We have similar results in the non-homogeneous case. Thus if

$$\begin{aligned} F(P;\alpha) = A + \int & f(\widehat{PP_1})\, d_{P_1}\mathfrak{P}(S;\alpha) \\ & + \int\int g(\widehat{PP_1}, \widehat{PP_2})\, d_{P_1}\mathfrak{P}(S;\alpha)\, d_{P_2}\mathfrak{P}(S;\alpha), \end{aligned} \tag{98}$$

we have almost everywhere

$$\lim_{r\to\infty}\frac{1}{V(r)}\int_R F(P+Q;\alpha)\bar{F}(Q;\alpha)dV_Q \tag{99}$$
$$= A\int \bar{g}(Q,Q)dV_Q + \bar{A}\int g(Q,Q)dV_Q$$
$$+\int f(P+Q)\bar{f}(Q)dV_Q + \left|\int g(Q,Q)dV_Q\right|^2$$
$$+\int\int g(P+Q,P+M)\bar{g}(Q,M)dV_Q dV_M$$
$$+\int\int g(P+Q,P+M)\bar{g}(M,Q)dV_Q dV_M.$$

10. Chaos theory and spectra.[10] The function

$$\lim_{r\to\infty}\frac{1}{V(r)}\int_R F(P+Q)\bar{F}(Q)dV_Q = G(P) \tag{100}$$

occupies a central position in the theory of harmonic analysis. If it exists and is continuous for every value of P, the function $F(P)$ is said to have an n-dimensional spectrum. To define this spectrum, we put

$$F_r(P) = \begin{cases} F(P) & \text{on } R; \\ 0 & \text{elsewhere.}\end{cases} \tag{101}$$

It is then easy to show by an argument involving considerations like those of (49) that if

$$G_r(P) = \frac{1}{V(r)}\int_\infty F_r(P+Q)\bar{F}_r(Q)dV_Q, \tag{102}$$

the integral being taken over the whole of space, then we have

$$G(P) = \lim_{r\to\infty} G_r(P). \tag{103}$$

Since, if O is the point with zero coördinates, by the Schwarz inequality,

$$|G(P)| \leq G(O), \tag{104}$$

the limit in (103) is approached boundedly.

If now we put

$$\phi_r(U) = (2\pi)^{-(n/2)}V(r)^{-\frac{1}{2}}\,\underset{s\to\infty}{\text{l. i. m.}}\int_S F_r(P)e^{iU\cdot P}dV_P \tag{105}$$

where S is the interior of a sphere of radius s about the origin, the n-fold Parseval theorem will give us

[10] Cf. N. Wiener, "Generalized harmonic analysis," *Acta Mathematica*, vol. 55 (1930).

$$(106) \qquad |\phi_r(U)|^2 = (2\pi)^{-n} \int_\infty G_r(P) e^{iU \cdot P} dV_P.$$

If $M(U)$ is a function with an absolutely integrable Fourier transform, we shall have

$$(107) \quad \int_{-\infty}^{\infty} |\phi_r(U)|^2 M(U) dV_U = (2\pi)^{-n} \int_\infty G_r(P) dV_P \int_\infty M(U) e^{iU \cdot P} dV_U,$$

and hence

$$(108) \quad \lim_{r\to\infty} \int_{-\infty}^{\infty} |\phi_r(U)|^2 M(U) dV_U = (2\pi)^{-n} \int G(P) dV_P \int_\infty M(U) e^{iU \cdot P} dV_U$$

which will always exist. Let us put

$$(109) \qquad \mathfrak{H}\{M(U)\} = (2\pi)^{-n} \int_\infty G(P) dV_P \int_\infty M(U) e^{iU \cdot P} dV_U.$$

If S is any set of points of finite measure, and $S(P)$ is its characteristic function, let us put

$$(110) \qquad \overline{\mathfrak{H}}(S) = \underset{M(U) \geq S(U)}{\text{l. u. b.}} \mathfrak{H}(M(U)),$$

and

$$(111) \qquad \underline{\mathfrak{H}}(S) = \underset{M(U) \leq S(U)}{\text{g. l. b.}} \mathfrak{H}(M(U)).$$

If $\underline{\mathfrak{H}}(S)$ and $\overline{\mathfrak{H}}(S)$ have the same value, we shall write it $\mathfrak{H}(S)$, and shall call it the *spectral mass* of F on S. It will be a non-negative additive set-function of S, and may be regarded as determining the spectrum of F.

If $f(P_1, \cdots, P_n)$ satisfies (80) and $F(P;\alpha)$ is defined as in (88), we know that for any given P,

$$(112) \quad G(P;\alpha) = \lim_{r\to\infty} \frac{1}{V(r)} \int_R F(P+Q;\alpha) \bar{F}(Q;\alpha) = \int_0^1 F(P,\beta) \bar{F}(0,\beta) d\beta$$

for almost all values of α. This alone is not enough to assure that $F(\mathrm{P},\alpha)$ has a spectrum for almost all values of α, as the sum of a non-denumerable set of sets of zero measure is not necessarily of zero measure. On the other hand, except for a set of values of α of zero measure, $G(P,\alpha)$ exists for all points P with rational coördinates.

We may even extend this result, and assert that if

$$(113) \qquad F_\theta(P;\alpha) = \frac{1}{V(\theta)} \int_{\text{length of } \overline{PS} \leq \theta} F(S;\alpha) dV_S$$

and

$$(114) \qquad G_\theta(P:\alpha) = \lim_{r\to\infty} \frac{1}{V(r)} \int_R F_\theta(P+Q;\alpha) \bar{F}_\theta(Q;\alpha) dV_Q,$$

then except for a set of values of α of zero measure, $G_\theta(P;\alpha)$ exists for all points P with rational coördinates and all rational parameters θ, and it is easily proved that for almost all values of α, as θ tends to 0 through rational values,

$$\lim_{\theta\to 0}\lim_{r\to\infty}\frac{1}{V(r)}\int_R |F_\theta(Q;\alpha)-F(Q;\alpha)|^2\,dV_Q=0. \tag{115}$$

Now, by the Schwarz inequality,

$$\begin{aligned}(116)\quad &\Big|\frac{1}{V(r)}\int_R F_\theta(P+Q;\alpha)\bar{F}_\theta(Q;\alpha)dV_Q\\ &-\frac{1}{V(r)}\int_R F_\theta(P_1+Q;\alpha)\bar{F}_\theta(Q;\alpha)dV_Q\Big|\\ &\le\left\{G_\theta(O;\alpha)\left(\frac{1}{V(r)}\int_R |F_\theta(P+Q;\alpha)-F_\theta(P_1+Q;\alpha)|^2\,dV_Q\right)\right\}^{\frac12}\\ &\le\left\{G_\theta(O;\alpha)\left(\frac{1}{V(r)}\int_R dV_Q\left(\frac{1}{V(\theta)}\left[\int_{|\widehat{P+Q,S}|\le\theta}-\int_{|\widehat{P+Q,S}|\le\theta}\right]|F(S;\alpha|^2\,dV_S\right)\right.\right.\\ &\quad\times\left.\left.\left(\frac{1}{V(\theta)}\left[\int_{|\widehat{P+Q,S}|\le\theta}-\int_{|\widehat{P_1+Q,S}|\le\theta}\right]dV_S\right)\right\}^{\frac12}\\ &\le G_\theta(O;\alpha)O(|\widehat{PP_1}|^{\frac12}).\end{aligned}$$

It thus follows that if (114) exists for a given θ and all P's with rational coördinates, it exists for that θ and all real P's whatever. We may readily show that

$$G_\theta(O;\alpha)\le G(O;\alpha). \tag{117}$$

By another use of the Schwarz inequality,

$$\begin{aligned}(118)\quad &\Big|\frac{1}{V(r)}\int_R F_\theta(P+Q;\alpha)\bar{F}_\theta(Q;\alpha)dV_Q\\ &\qquad-\frac{1}{V(r)}\int_R F(P+Q;\alpha)\bar{F}(Q;\alpha)dV_Q\Big|\\ &\qquad\le\frac{1}{V(r)}\int_R |F_\theta(P+Q;\alpha)-F(P+Q;\alpha)|\;|F_\theta(Q;\alpha)|\,dV_Q\\ &\qquad\quad+\frac{1}{V(r)}\int_R |F(P+Q;\alpha)|\;|F_\theta(Q;\alpha)-F(Q;\alpha)|\,dV_Q\\ &\qquad\le\{G(O;\alpha)\frac{1}{V(r)}\int_R |F_\theta(P+Q;\alpha)-F(P+Q;\alpha)|^2\,dV_Q\}^{\frac12}\\ &\qquad\quad+\{G(O;\alpha)\frac{1}{V(r)}\int_R |F_\theta(Q;\alpha)-F(Q;\alpha)|^2\,dV_Q\}^{\frac12}.\end{aligned}$$

Combining (115) and (118), we see that except for a set of values of α of zero measure, we have for all P,

$$G(P;\alpha)=\lim_{\theta\to 0}G_\theta(P;\alpha). \tag{119}$$

We thus have an adequate basis for spectrum theory. This will extend, not merely to functions $F(P,\alpha)$ defined as in (88), but to finite sums of such functions. It will even extend to the case of any differentiable chaos $F(P;\alpha)$, for which $F(P+Q;\alpha)\bar{F}(Q;\alpha)$ is an integrable function of α, and for which (115) holds. For a metrically transitive chaos, this latter will be true if

$$(120) \qquad \lim_{\vartheta\to 0}\int_0^1 |F_\vartheta(P;\alpha)-F(P;\alpha)|^2\,d\alpha = 0.$$

Under this assumption, we have proved that $F(P)$ has a spectrum, and the same spectrum, for all values of α.

This enables us to answer a question which has been put several times, as to whether there is any relation between the spectrum of a chaos and the distribution of its values. There is no unique relation of the sort. The function

$$(121) \qquad \int g(P+Q)\bar{g}(Q)\,dV_Q,$$

where g belongs to L^2, may be so chosen as to represent any Fourier transform of a positive function of L, and if $f(P,Q,M)$ is a bounded step-function, the right-hand side of (97) will clearly be the Fourier transform of a positive function of L. In particular, let $f(P_1,P_2,P_3)=f(P_1)f(P_2)f(P_3)$,

$$(122) \qquad F_1(P;\alpha)=\int f_1(\widehat{PP_1})\,d_{P_1}\mathfrak{P}(S;\alpha)$$

and choose $f_1(Q)$ in such a way that

$$(123) \qquad \int f_1(P+Q)\bar{f}_1(Q)\,dV_Q = \text{right-hand side of (97)}.$$

Then

$$(124) \qquad \int_0^1 (F_1(P;\alpha))^{2n}d\alpha=\left(\int_\infty |f_1(P)|^2\,dV_P\right)^n(2n-1)(2n-3)\cdots 1$$

and if $F(P,\alpha)$ is defined as in (96),

$$(125) \qquad \int_0^1 (F(P;\alpha))^{2n}d\alpha=\left(\int |f(P)|^2dV_P\right)^{3n}(6n-1)(6n-3)\cdots 1)$$

so that for all but at most one value of n,

$$(126) \qquad \int_0^1 (F_1(P;\alpha)^{2n}d\alpha \neq \int_0^1 (F(P;\alpha))^{2n}d\alpha$$

and we obtain in F and F_1 two chaoses with identical spectra but different distribution functions. On the other hand, if

$$(127) \qquad \int_\infty |f_1(P)|^2\,dV_P=\int_\infty |f_2(P)|^2\,dV_P,$$

the chaoses

$$\int f_1(\widehat{PP_1})\, d_{P_1}\mathfrak{P}(S;\alpha) \tag{128}$$

and

$$\int f_2(\widehat{PP_1})\, d_{P_1}\mathfrak{P}(S;\alpha) \tag{129}$$

will have the same distribution functions, but may have very different spectra.

11. The discrete chaos.[11] Let us now divide the whole of Euclidean n-space dichotomously into sets $S_{m,n}$, such that every two sets S_{m_1,n_1} and S_{m_2,n_2} have the same measure, and that each $S_{m,n}$ is made up of exactly two non-overlapping sets $S_{m+1,k}$. Let us divide all these sets into two categories, "occupied," and "empty." Let us require that the probability that a set be empty depend only on its measure, and that the probability that two non-overlapping sets be empty be the product of the probabilities that each be empty. Let us assume that both empty and occupied sets exist. Let every set contained in an empty set be empty, while if a set be occupied, let at least one-half always be occupied. We thus get an infinite class of schedules of emptiness and occupiedness, and methods analogous to those of paragraph 6 may be used to map the class of these schedules in an almost everywhere one-one way on the line $(0,1)$ of the variable α, in such a way that the set of schedules for which a given finite number of regions are empty or occupied will have a probability equal to the measure of the corresponding set of values of α.

By the independence assumption, the probability that a given set $S_{m,n}$ be empty must be of the form $e^{-Am(S_{m,n})}$. If $S_{m,n}$ is divided into the 2^ν intervals $S_{m+\nu,n_1}, \cdots, S_{m+\nu,n_{2^\nu}}$ at the ν-th stage of sub-division, the probability that just one is occupied and the rest are empty is

$$2^\nu(1-\exp(-Am(S_{m,n})/2^\nu)\exp\left(-\frac{2^\nu-1}{2^\nu}Am(S_{m,n})\right). \tag{130}$$

This contingency at the $\nu+1$-st stage is a sub-case of this contingency at the ν-th stage. If we interpret probability to mean the same thing as the measure of the corresponding set of α's, then by monotone convergence, the probability that at every stage, all but one of the subdivisions of $S_{m,n}$ are empty, while the remaining one is occupied, will be the limit of (130), or

$$Am(S_{m,n})\exp(-Am(S_{m,n})). \tag{131}$$

[11] The ideas of this paragraph are related to discussions the author has had with Professor von Neumann, and the main theorem is equivalent to one enunciated by the latter.

10

Such a series of stages of subdivision will have as its occupied regions exactly those which contain a given point.

The probability that the occupied regions are exactly those which contain two points is the probability that each half of $S_{m,n}$ contain exactly one point, plus the probability that one-half is empty, and that in the occupied half, each quarter will contain exactly one point, plus and so on. This will be

$$(132)\quad \left\{ \frac{Am(S_{m,n})}{2} \exp\left(-\frac{Am(S_{m,n})}{2}\right)\right\}^2 + 2\exp\left(-\frac{Am(S_{m,n})}{2}\right)\left\{\frac{Am(S_{m,n})}{2}\exp\left(-\frac{Am(S_{m,n})}{4}\right)^2 + \cdots = \exp(-Am(S_{m,n})m(S_{m,n}))^2(\tfrac{1}{4}+\tfrac{1}{8}+\cdots) = \frac{(Am(S_{m,n}))^2}{2}\exp(-Am(S_{m,n})).$$

If the probability that the occupied regions are exactly those containing $k-1$ points is

$$\frac{(Am(S_{m,n}))^{k-1}}{(k-1)!}\exp(-Am(S_{m,n}))$$

then a similar argument will show that the probability that the occupied regions are exactly those containing k points will be

$$(133)\quad \sum_{j=1}^{k-1}\frac{1}{j!}\frac{1}{(k-j)!}(Am(S_{m,n}))^k\exp(-Am(S_{m,n}))\left(1+\frac{1}{2^{k-1}}+\frac{1}{4^{k-1}}+\cdots\right) = \frac{1}{k!}(2^k-2)\left(\frac{1}{1-\frac{1}{2^{k-1}}}\right)(Am(S_{m,n}))^k\exp(-Am(S_{m,n})) = \frac{1}{k!}(Am(S_{m,n}))^k\exp(-Am(S_{m,n})).$$

Thus by mathematical induction, the probability that the occupied regions are exactly those containing k points will be

$$\frac{1}{k!}(Am(S_{m,n}))^k\exp(-Am(S_{m,n}))$$

and the sum of this for all values of k will be

$$(134)\quad \sum_0^\infty \frac{1}{k!}(Am(S_{m,n}))^k\exp(-Am(S_{m,n}))=1.$$

In other words, except for a set of contingencies of probability zero, the occupied regions will be exactly those containing a given finite number of points.

We may proceed at once from the fact that the probability that a set S_1 contains exactly k points is

$$\frac{1}{k!}(Am(S_1))^k e^{-Am(S_1)}$$

while the probability that the non-overlapping set S_2 contains exactly k points is

$$\frac{1}{k!}(Am(S_2))^k e^{-Am(S_2)}$$

to the fact that the probability that the set $S_1 + S_2$ contains exactly k points is

$$\begin{aligned}(135)\qquad & \sum_0^k \frac{1}{j!}\frac{1}{(k-j)!}(Am(S_1))^j(Am(S_2))^{k-j}e^{-Am(S_1+S_2)} \\ & = \frac{1}{k!}(Am(S_1+S_2))^k e^{-Am(S_1+S_2)}.\end{aligned}$$

From this, by monotone convergence, it follows at once that the probability that any set S which is the sum of a denumerable set of our fundamental regions $S_{m,n}$ should contain exactly k points is

$$\frac{1}{k!}(Am(S))^k e^{-Am(S)}.$$

It is then easy to prove this for all measurable sets S.

We are now in a position to prove that the additive functional $\mathfrak{D}(S;\alpha)$, consisting in the number of points in the region S on the basis of the schedule corresponding to α, is a homogeneous metrically transitive chaos. The rôle which continuity filled in paragraphs 6 and 7, of allowing us to show that $\mathfrak{F}_{y_1,\ldots,y_n}(S;\alpha)$ was measurable in $y_1,\cdots,y_n$ and α, is now filled by the fact that the probability that any of the points in a region lie within a very small distance of the boundary, is for any Jordan region the probability that a small region be occupied, and is small. The metric transitivity of the chaos results as before from the independence of the distribution in non-overlapping regions.

The discrete or Poisson chaos which we have thus defined is the chaos of an infinite random shot pattern, or the chaos of the gas molecules in a perfect gas in statistical equilibrium according to the old Maxwell statistical mechanics. It also has important applications to the study of polycrystalline aggregates, and to similar physical problems.

Two important formulae are

$$(136)\qquad \int_0^1 \mathfrak{D}(S;\alpha)\,d\alpha = e^{-Am(S)}\sum_1^\infty \frac{k}{k!}(Am(S))^k = Am(S).$$

and

$$(137)\quad \int_0^1 (\mathfrak{D}(S;\alpha))^2 d\alpha = e^{-Am(S)} \sum_1^\infty \frac{k^2}{k!} (Am(S))^k = (Am(S))^2 + Am(S).$$

Let it be noted that if we define

$$(138)\quad \int f(P)\, d_P \mathfrak{D}(S;\alpha)$$

for a measurable step-function $f(P)$ as in (70), by

$$(139)\quad \sum_1^N f_n \mathfrak{D}(S_n;\alpha),$$

(72) is replaced by

$$(140)\quad \int_0^1 d\alpha \left| \int f(P)\, d_P \mathfrak{D}(S;\alpha) - A \int_\infty f(Q)\, dV_Q \right|^2$$
$$= \sum_{m=1}^N \sum_{n=1}^N \int_0^1 d\alpha\, f_m \bar{f}_n \mathfrak{D}(S_m;\alpha) \mathfrak{D}(S_n;\alpha)$$
$$- 2\Re \{ \bar{A} \sum_{m=1}^\infty \int_0^1 d\alpha\, f_m \mathfrak{D}(S_m;\alpha) \int_\infty f(Q)\, dV_Q + | A \int_\infty f(Q)\, dV_Q |^2 \}$$
$$= \sum_{m=1}^N | f_m |^2 Am(S_m)$$
$$= A \int_\infty | f(P) |^2 dV_P.$$

Thus the transformation from $f(P)$ as a function of P, to

$$(141)\quad \int f(P)\, d_P \mathfrak{D}(S;\alpha) - A \int_\infty f(Q)\, dV_Q$$

as a function of α, retains distance in Hilbert space, apart from a constant factor, and if $f(P)$ belongs to L and L^2 simultaneously, and $\{f_n(P)\}$ is a sequence of step-functions converging in the mean both in the L sense and in the L^2 sense to $f(P)$, we may *define*

$$(142)\quad \int f(P)\, d_P \mathfrak{D}(S;\alpha) = A \int_\infty f(Q)\, dV_Q$$
$$+ \underset{n\to\infty}{\text{l. i. m.}} \left(\int f_n(P)\, d_P \mathfrak{D}(S;\alpha) - A \int_\infty f_n(Q)\, dV_Q \right).$$

As in the case of (73), this definition is substantially unique. We may prove the analogue of (93) in exactly the same way as (93) itself, and shall obtain

$$(143)\quad \lim_{r\to\infty} \frac{1}{V(r)} \int_R \left\{ \int_\infty (f(P + \widehat{Q)M};\alpha)\, d_M \mathfrak{D}(S;\alpha) - A \int_\infty f(M)\, dV_M \right\}$$
$$\times \left\{ \int_\infty \bar{f}(\widehat{QM};\alpha)\, d_M \mathfrak{D}(S;\alpha) - A \int_\infty \bar{f}(M)\, dV_M \right\} dV_Q$$
$$= A^2 \int_\infty f(P+Q) \bar{f}(Q)\, dV_Q.$$

As we may see by appealing to the theory of spectra, one interpretation of this in the one-dimensional case is the following: *If a linear resonator be set into motion by a haphazard series of impulses forming a Poisson chaos, the effect, apart from that of a constant uniform stream of impulses, will have the same power spectrum as the energy spectrum of the response of the resonator to a single impulse.*

12. The weak approximation theorem for the polynomial chaos. We wish to show that the chaoses of paragraph 9 are in some sense everywhere dense in the class of all metrically transitive homogeneous chaoses. We shall show that if $\mathfrak{F}(S;\alpha)$ is any homogeneous chaos in n dimensions, there is a sequence $\mathfrak{F}_k(S;\alpha)$ of polynomial chaoses as defined in paragraph 9, such that if $S_1,\cdots,S_\nu$ is any finite assemblage of bounded measurable sets in n-space selected from among a denumerable set, and

$$(144)\qquad \int_0^1 |\mathfrak{F}(S_\lambda;\alpha)|^\mu\,d\alpha<\infty \qquad (\lambda=1,2,\cdots,\nu)$$

is finite, then

$$(145)\qquad \int_0^1 \mathfrak{F}(S_1;\alpha)\cdots\mathfrak{F}(S_\nu;\alpha)\,d\alpha=\lim_{n\to\infty}\int_0^1 \mathfrak{F}_n(S_1;\alpha)\cdots\mathfrak{F}_n(S_\nu;\alpha)\,d\alpha.$$

We first make use of the fact that if the probability that a quantity u be greater in absolute value than A, be less than

$$(146)\qquad \frac{2}{\sqrt{2\pi B}}\int_A^\infty e^{-(u^2/2B)}\,du,$$

then if $\psi(u)$ is any even measurable function bounded over $(-\infty,\infty)$, we may find a polynomial $\psi_\epsilon(u)$, such that the mean value of

$$(147)\qquad |\psi(u)-\psi_\epsilon(u)|^n,$$

which will be

$$(148)\qquad \frac{1}{\sqrt{2\pi B}}\int_{-\infty}^\infty |\psi(u)-\psi_\epsilon(u)|^n\,e^{-(u^2/2B)}\,du,$$

is less than ϵ. Since it is well known that if $\phi(u)$ is a continuous function vanishing outside a finite interval, and

$$(149)\qquad \sum_1^\infty A_nH_n(u)\,e^{-(u^2/2)}$$

is the series for $\phi(u)$ in Hermite functions, then we have uniformly

(150) $$\phi(u) = \lim_{t\to 1-0} \sum_1^\infty A_n t^n H_n(u) e^{-(u^2/2)},$$

to establish the existence of $\psi_\epsilon(u)$, we need only prove it in the case in which

(151) $$\psi(u) = u^k e^{-Cu^2}$$

for an arbitrarily small value of C: as for example for $C = 1/4n_1B$. We shall then have

(152) $$\left|\psi(u) - u^k \sum_0^N \frac{(cu^2)^k}{k!}\right| \leq |u|^k \sum_0^\infty \frac{(cu^2)^k}{k!} = |u|^k e^{u^2/4n_1B},$$

so that by dominated convergence, and if we take N large enough, we may make

(153) $$\frac{1}{\sqrt{2\pi B}}\int_{-\infty}^{\infty} |\psi(u) - \psi_\epsilon(u)|^n e^{-(u^2/2B)} du < \epsilon \qquad (n \leq n_1).$$

Now let

(154) $$\psi_K(u) = \begin{cases} 0 & (|u| < K); \\ 1 & (|u| \geq K); \end{cases}$$

and let us put

(155) $$\mathfrak{G}(P;\alpha) = \frac{1}{V(r)}\mathfrak{P}(S;\alpha) \quad (S = \text{interior of } |\widehat{QP}| \leq r).$$

The chaos

(156) $$\mathfrak{L}(P;\alpha) = \psi_K(\mathfrak{G}(P;\alpha))$$

may then be approximated by polynomial chaoses in such a way as to approximate simultaneously to all polynomials in $\mathfrak{L}(P;\alpha)$ by corresponding polynomials in the approximating chaoses. Since the distribution of the values of $\mathfrak{G}(P;\alpha)$ will be Gaussian, with a root mean square value proportional to a power of r, and $\mathfrak{G}(P;\alpha)$ will be independent in spheres of radius η about two points P_1 and P_2 more remote from each other than $2r + 2\eta$, it follows that if we take K to be large enough, we may make the probability that $\mathfrak{L}(P;\alpha)$ differs from 0 between two spheres of radii respectively $r + \eta$ and H about a given point where it differs from 0, as small as we wish.

We now form the new chaos

(157) $$\int_{|\widehat{PQ}| < r} \mathfrak{L}(Q;\alpha)\, dV_Q,$$

which we may also approximate, with all its polynomial functionals, by a sequence of polynomial chaoses. The use of polynomial approximations tending boundedly to a step function over a finite range will show us that this is also true of the chaos determined by

$$\psi_\gamma\left(\int_{|\widehat{PQ}|<x} \mathcal{L}(Q;\alpha)\,dV_Q = \mathfrak{M}(P;\alpha).\right. \tag{158}$$

By a proper choice of the parameters, this can be made to have arbitrarily nearly all its mass uniformly distributed over regions arbitrarily near to arbitrarily small spheres, all arbitrarily remote from one another, except in an arbitrarily small fraction of the cases. We then form

$$\frac{1}{(2\pi k)^{n/2}}\int_\infty (\mathfrak{M}(Q;\alpha) + \delta)\exp\left(-\frac{|\widehat{PQ}|^2}{2k}\right)dV_Q = \mathfrak{N}(P;\alpha) \tag{159}$$

where δ is taken to be very small. This chaos again, as far as all its polynomial functionals are concerned, will be approximable by polynomial chaoses. Since it is bounded away from 0 and ∞, and since over such a range the function $1/x$ may be approximated uniformly by polynomials, it follows that in our sense,

$$1/\mathfrak{N}(P;\alpha) \tag{160}$$

is approximable by polynomial chaoses.

If $\varpi(P)$ is any measurable function for which arbitrarily high moments are always finite, it is easy to show that

$$\frac{1}{(2\pi k)^{n/2}}\int_\infty \varpi(Q)(\mathfrak{M}(Q;\alpha) + \delta)\exp\left(-\frac{|\widehat{PQ}|^2}{2k}\right)dV_Q = \mathfrak{W}(P;\alpha) \tag{161}$$

is approximable by polynomial chaoses. Multiplying expressions (160) and (161), it follows that

$$\mathfrak{W}(P;\alpha)/\mathfrak{N}(P;\alpha) = \mathfrak{U}(P;\alpha) \tag{162}$$

is approximable by polynomial chaoses.

If A is a large enough constant, depending on the choice of the constant ϵ, we have

$$\begin{aligned}\frac{1}{(2\pi k)^{n/2}}\int_{|P|>A} \exp\left(-\frac{|P|^2}{2k}\right)dV_P &= \int_A^\infty x^n e^{-(x^2/2k)}\,dx \Big/ \int_0^\infty x^n e^{-(x^2/2k)}\,dx \\ &< \frac{1}{(2\pi k)^{n/2}}\exp\left(-\frac{(A-\epsilon)^2}{2k}\right).\end{aligned} \tag{163}$$

Thus by the proper choice of the parameters of $\mathfrak{M}(P;\alpha)$, if we take k small enough and then δ small enough, the chaos (162) will consist as nearly as we wish, from the distribution standpoint, of an infinite assemblage of convex cells of great minimum dimension, in each of which the function $\varpi(P)$ is repeated, with the origin moved to some point remote from the boundary.

Now let $\mathfrak{F}(S;\alpha)$ be a metrically transitive homogeneous chaos. Let us form

$$(164)\qquad \mathfrak{F}(r;S;\alpha)=\frac{1}{V(r)}\int_R \mathfrak{F}_{x_1,\cdots,x_n}(S;\alpha)\,dx_1\cdots dx_n.$$

Clearly by the fundamental theorem of the calculus, over any finite region in $(x_1,\cdots,x_n)$, we shall have for almost all points and almost all values of α,

$$(165)\qquad \mathfrak{F}(S;\alpha)=\lim_{r\to 0}\mathfrak{F}(r;S;\alpha);$$

and if (144) holds, it is easy to show that

$$(166)\qquad \int_0^1 |\mathfrak{F}(r;S;\alpha)|^n d\alpha < \text{const.}$$

From this it follows that

$$(167)\qquad \lim_{r\to 0}\int_0^1 |\mathfrak{F}(r;S;\alpha)-\mathfrak{F}(S;\alpha)|^n d\alpha=0$$

and by the ergodic theorem, except for a set of values of α of zero measure, as r tends to 0 through a denumerable set of values,

$$(168)\qquad \lim_{r\to 0}\frac{1}{V(r)}\int_R |\mathfrak{F}_{x_1,\cdots,x_n}(r;S;\alpha) - \mathfrak{F}_{x_1,\cdots,x_n}(S;\alpha)|^n dx_1\cdots dx_n=0.$$

With this result as an aid, enabling us to show that the distribution of $\mathfrak{F}(S;\alpha)$ is only slightly affected by averaging within a small sphere with a given radius, or even within any small region near enough to a small sphere with a given radius, we may proceed as in (161) and (162) and form the chaos

$$(169)\qquad \mathfrak{F}_k(S;\alpha)=\frac{1}{\mathfrak{N}(\mathfrak{P};\alpha)(2\pi k)^{n/2}}\int_\infty \mathfrak{F}_{x_1,\cdots,x_n}(S;\beta) \times(\mathfrak{M}(x_1,\cdots,x_n;\alpha)+\delta)\exp\frac{\left(-\sum_1^n x_j^2\right)}{2k}dx_1\cdots dx_n.$$

For almost all β, in each of the large cells of this chaos, (169) will have as nearly as we wish the same distribution as some $\mathfrak{F}_{x_1,\ldots,x_n}(S;\alpha)$, where $(x_1,\cdots,x_n)$ lies in the interior of the cell, remote from the boundary. These cells may so be determined that except for those filling an arbitrarily small proportion of space, all are convex regions with a minimum dimension greater than some given quantity.

To establish (143), it only remains to show that the average of a quantity depending on a chaos over a large cell tends to the same limit as its average over a large sphere. To show this, we only need to duplicate the argument of paragraph 4, where we prove the multidimensional ergodic theorem, for large pyramids with the origin as a corner, instead of for large spheres about the origin. We may take the shapes and orientations of these pyramids to form a denumerable assemblage, from which we may pick a finite assemblage which will allow us to approach as closely as we want to any cell for which the ratio of the maximum to the minimum distance from the origin within it does not exceed a given amount. It is possible to show that by discarding cells whose measure is an arbitrarily small fraction of the measure of all space, the remaining cells will have this property.

13. The physical problem. The transformation of a chaos. The statistical theory of a homogeneous medium, such as a gas or liquid, or a field of turbulence, deals with the problem, given the statistical configuration and velocity distribution of the medium at a given initial time, and the dynamical laws to which it is subject, to determine the configuration at any future time, with respect to its statistical parameters. This of course is not a problem in the first instance of the history of the individual system, but of the entire ensemble, although in proper cases it is possible to show that almost all systems of the ensemble do actually share the same history, as far as certain specified statistical parameters are concerned.

The dynamical transformations of a homogeneous system have the very important properties, that they are independent of any choice of origin in time or in space. Leaving the time variable out of it, for the moment, the simplest space transformations of a homogeneous chaos $\mathfrak{F}(S;\alpha)$ which have this property are the polynomial transformations which turn it into

$$\begin{aligned}(170)\qquad & K_0+\int K_1(x_1-y_1,x_2-y_2,\cdots,x_n-y_n)\mathfrak{F}_{y_1,\ldots,y_n}(S;\alpha)\,dy_1\cdots dy_n \\ & +\cdots \\ & +\int\cdots\int K_\nu(x_1-y_1^{(1)},\cdots,x_n-y_n^{(1)},\cdots,x_1-y_1^{(\nu)},\cdots, \\ & \quad x_n-y_n^{(\nu)})\mathfrak{F}_{y_1^{(1)},\ldots,y_n^{(1)}}(S;\alpha)\cdots \\ & \qquad \mathfrak{F}_{y_1^{(\nu)},\ldots,y_n^{(\nu)}}(S;\alpha)\,dy_1^{(1)}\cdots dy_n^{(\nu)}.\end{aligned}$$

These are a sub-class of the general class of polynomial transformations

$$(171)\quad \begin{aligned} &K_0 + \int K_1(x_1, \cdots, x_n; y_1, \cdots, y_n) \mathfrak{F}_{y_1, \ldots, y_n}(S; \alpha)\, dy_1 \cdots dy_n \\ &+ \cdots \\ &+ \int \cdots \int K_\nu(x_1, \cdots, x_n; y_1^{(1)}, \cdots, y_n^{(1)}; \cdots; y_1^{(\nu)}, \cdots, y_n^{(\nu)}) \\ &\quad \times \mathfrak{F}_{y_1^{(1)}, \ldots, y_n^{(1)}}(S; \alpha) \cdots \mathfrak{F}_{y_1^{(\nu)}, \ldots, y_n^{(\nu)}}(S; \alpha)\, dy_1^{(1)} \cdots dy_n^{(\nu)}. \end{aligned}$$

If a transformation of type (171) is invariant with respect to position in space, it must belong to class (170). On the other hand, in space of a finite number of dimensions and in any of the ordinary spaces of an infinite number of dimensions, polynomials are a closed set of functions, and hence every transformation may be approximated by a transformation of type (171).

A polynomial transformation such as (170) of a polynomial chaos yields a polynomial chaos. If then we can approximate to the state of a dynamical system at time 0 by a polynomial chaos, and approximate to the transformation which yields its status at time t by a polynomial transformation, we shall obtain for its state at time t, the approximation of another polynomial chaos. The theory of approximation developed in the last section will enable us to show this.

On the other hand, the transformation of a dynamical system induced by its own development is infinitely subdivisible in the time, and except in the case of linear transformations, this is not a property of polynomial transformations. Furthermore, when these transformations are non-linear, they are quite commonly not infinitely continuable in time. For example, let us consider the differential equation

$$(172)\qquad \frac{\partial u}{\partial t} + u \frac{\partial u}{\partial x} = 0.$$

This corresponds to the history of a space-distribution of velocity transferred by particles moving with that velocity. Its solutions are determined by the equation

$$(173)\qquad u(x, t) = u(x - tu(x, t), 0),$$

or if ψ is the inverse function of $u(x, 0)$,

$$(174)\qquad x - tu(x, t) = \psi(u(x, t)).$$

Manifestly, if two particles with different velocities are allowed to move long enough to allow their space-time paths to cross, $u(x, t)$ will cease to exist as a single-valued function. This will always be the case for *some* value of t if $u(x, 0)$ is not constant, and for almost all values of t and α if it is a polynomial chaos.

By Lagrange's formula, (174) may be inverted into

$$u(x, t) = \sum_{0}^{\infty} \frac{(-t)^n}{n!} \frac{\partial^n}{\partial x^n} \{(u(x, 0))^n\}. \tag{175}$$

In a somewhat generalized sense, the partial sums of this formally represent polynomial transformations of the initial conditions. However, it is only for a very special sort of bounded initial function, and for a finite value of the time, that they converge. It is only in this restricted sense that the polynomial transformation represents a true approximation to that given by the differential equation.

It will be seen that the useful application of the theory of chaos to the study of particular dynamical chaoses involves a very careful study of the existence theories of the particular problems. In many cases, such as that of turbulence, the demands of chaos theory go considerably beyond the best knowledge of the present day. The difficulty is often both mathematical and physical. The mathematical theory may lead inevitably to a catastrophe beyond which there is no continuation, either because it is not the adequate presentation of the physical facts; or because after the catastrophe the physical system continues to develop in a manner not adequately provided for in a mathematical formulation which is adequate up to the occurrence of the catastrophe; or lastly, because the catastrophe does really occur physically, and the system really has no subsequent history. The hydrodynamical investigations required in the case of turbulence are directly in the spirit of the work of Oseen and Leray, but must be carried much further.

The study of the history of a mechanical chaos will then proceed as follows: we first determine the transformation of the initial conditions generated by the dynamics of the ensemble. We then determine under what assumptions the initial conditions admit of this transformation for either a finite or an infinite interval of time. Then we approximate to the transformation for a given range of values of the time by a polynomial transformation. Then, having regard to a definition of distance between two functions determined by the transformation, we approximate to the initial chaos by a polynomial chaos. Next we apply the polynomial transformation to the polynomial chaos, and obtain an approximating polynomial chaos at time t. Finally, we apply our algorithm of the pure chaos to determine the averages of the statistical parameters of this chaos, and express these as functions of the time.

The results of such an investigation belong to a little-studied branch of statistical mechanics: the statistical mechanics of systems not in equilibrium. To study the classical, equilibrium theory of statistical mechanics by the methods of chaos theory is not easy. As yet we lack a method of representing all forms of homogeneous chaos, which will tell us by inspection when two differ merely by an equimeasure transformation of the parameter of distribution. In certain cases, in which the equilibrium is stable, the study of the history of a system with an arbitrary initial chaos will yield us for large values of t an approximation to equilibrium, but this will often fail to be so, particularly in the case of differentiable chaoses, or the only equilibrium may be that in which the chaos reduces to a constant.

MASSACHUSETTS INSTITUTE OF TECHNOLOGY.

Comments on [38a]

L. Gross

The mathematical content of the paper may be divided into two parts. The first part, sections 2 to 5, deals with an extension of the Birkhoff ergodic theorem to the case where the usual one-parameter group of measure-preserving transformations is replaced by an n-parameter commutative group of measure-preserving transformations. This portion of the paper was later amplified by Wiener in his work "The ergodic theorem," [39a].

The second part of the paper, dealing with various kinds of chaos has had extensive development in diverse directions and is presently having an influence on quantum field theory. By a chaos Wiener meant an additive function defined on a ring of subsets of a given set whose values are random variables on a probability space. Sections 6 to 10 are greatly clarified, extended and largely superceded by the work of Kakutani,[1] Ito,[2] and Segal.[3] The basic theorem, in Segal's terminology, is this. Let H be a real Hilbert space, H_c its complexification, $S(H_c)$ the Hilbert space of all symmetric tensors over H_c and d the isotropic normal distribution over H with variance parameter one. Then there is a unitary transformation W from $S(H_c)$ onto $L^2(H,d)$ such that for any orthogonal transformation T on H the unitary operators naturally induced by T on the above two spaces are unitarily equivalent via W. Kakutani's cited paper was the first one stating a theorem of this type. It stated essentially this theorem for $H =$ real $L^2(-\infty, \infty)$ and for T in the group of orthogonal transformations induced by translation on the line. The cited paper of Ito describes W in terms of stochastic integrals when $H =$ real $L^2(S,m)$ for some measure space (S,m). Wiener did not, in his paper, construct W, but rather he constructed a related map (section 8) with less useful properties. Prior to Kakutani's work Cameron and Martin[4] constructed an orthonormal system in L^2 (Wiener space) which is closely related to the operator W and could easily be used to construct W. They did not consider the above described intertwining property. Influenced, perhaps, by Cameron and Martin's work, Wiener later (*Nonlinear problems in random theory,* [58i]) sketched a construction of W, apparently unaware of the earlier work of Kakutani, Ito, and Segal.

Segal's cited paper also relates W (called the duality transform by Segal) to operators that occur in quantum field theory and forms a point of contact of Wiener's paper with quantum field theory.

Section 12 of Wiener's paper was clarified and extended by Nisio.[5]

Ergodicity of transformation groups in the above context has been discussed by Segal.[6]

We refer the reader to Kakutani[7] for a survey of topics immediately related to the unitary operator *W*. For a more detailed and more recent survey and exposition of this subject and other properties of *W* we refer the reader to the notes of Jacques Neveu.[8] These notes also contain an extensive bibliography of papers dealing with other aspects of integration over infinte-dimensional spaces. For a recent account of the important role of the duality transform in constructive quantum field theory and some related problems we refer the reader to Simon[9] (note, for example, §I.3).

Editor's Note. Vis-à-vis Professor Gross's remarks on the construction of the unitary operator *W* in [58i, ch. 2-4], it should be said that Wiener had discussed this subject with Professors Kakutani and Segal, and was aware of their work in a general way. The absence of the Refs. 1, 3 from [58i] was thus an unfortunate oversight. For the purposes of [58i], however, his own approach based on multiple stochastic integrals appears to be appropriate, cf. Commentary by Drs. McMillan and Deem on pp. 654-671. For more on the nexus between the treatments of Cameron and Martin, Kakutani, and Wiener the reader is referred to my paper: *Wiener's contributions to generalized harmonic analysis, prediction theory and filter theory,* Bull. Amer. Math. Soc. 77 (1966), 73-125, §28.

References

1. S. Kakutani, Proc. Nat. Acad. Sci. U.S.A., 36(1950), 319-323.
2. K. Ito, J. Math. Soc. Japan 3(1951), 157-169.
3. I. E. Segal, Trans. Amer. Math. Soc. 81(1956), 106-134.
4. R. H. Cameron and W. T. Martin, Ann. Math. (2) 48(1947), 385-392.
5. M. Nisio, J. Math. Soc. Japan 12(1960), 207-226.
6. I. E. Segal, Ann. Math. (2), 66(1957), 297-303.
7. S. Kakutani, Proc. Fourth Berkeley Symposium on Mathematical Statistics and Probability, vol. 2(1961), 239-247.
8. J. Neveu, Processus Aléatoires Gaussiens, University of Montreal Press, Montreal, 1968, 224 pages.
9. B. Simon, The $P(\phi)_2$ Euclidean (Quantum) Field Theory, Princeton University Press, Princeton, N. J., 1974, 392 pages.

THE DISCRETE CHAOS.*

By Norbert Wiener and Aurel Wintner.

Introduction

A characteristic difficulty of problems requiring the introduction of a measure with infinitely many dimensions is that ratios which are normally finite become either infinite or zero in the majority of cases. Such difficulties are of a trivial nature in case of complete independence of infinitely many random variables, since in this case the proper measure is simply the product measure. However, most of the physical theories presuppose measure which belongs to infinitely many dimensions but not to the trivial case of a product measure.

In classical mechanics, the Maxwellian picture is that of a large number of individual particles, and averages are taken when this number becomes infinite; while in the picture of Boltzmann and Gibbs, a system with a fixed finite degree of freedom is studied as a statistical function of its initial phase determination. The standard case of a Maxwellian description is that in which the particles have a vanishingly small coupling; so that the measure which determines probabilities reduces to a product measure in the limiting case. On the other hand, the standard formulation of the Boltzmann-Gibbs statistics involves densities which cease to exist when the degree of freedom is *actually* infinite. Nevertheless, notions like " number of distinct phases " or statements like that of the " almost certain increase of entropy " are meaningless or untrue with reference to the probability measure of a system with a *fixed* finite degree of freedom.

Thus there arises the question as to the existence of a definite probability measure in terms of which it is possible to unite the advantages, and eliminate the mathematical inadequacies, of the models of Maxwell and Boltzmann-Gibbs.

Although the present paper does not deal with statistical mechanics, the statistical problems to be considered are characterized by difficulties of the type just described. Of course, the problem will not be the postulation of a suitable measure, without which statistical statements are meaningless, but rather an existence proof, which must necessarily be based on an explicit construction in terms of a finite number of random variables.

* Received September 24, 1941.

PART I

The Distribution Functions

1. Let $\mathcal{E}$ be a field of sets E which is closed with respect to finite or enumerable logical addition and multiplication. Let $\mathcal{E}^{(n)}$ denote the field which is the closure (with respect to addition and multiplication) of the product $\mathcal{E}^n = \mathcal{E}^{n-1} \times \mathcal{E}$, where $\mathcal{E}^1 = \mathcal{E}$. It will be convenient to let $\mathcal{E}^{(0)}$ (and $\mathcal{E}^0$) denote the field consisting of the empty set. The logical sum of all sets E will be denoted by S, provided S is an E.

Let there be given, for every n, a non-negative, additive function Ψ_n of the sets which constitute $\mathcal{E}^{(n)}$. For those sets in $\mathcal{E}^{(n)}$ which are in $\mathcal{E}^n$, i. e., which are of the form $E_1 \times \cdots \times E_n$, the function Ψ_n can be written as a function $\Psi_n = \Psi_n(E_1, \cdots, E_n)$ of n sets E_j each of which is in $\mathcal{E}$. It will be assumed that $\Psi_n(E_1, \cdots, E_n)$ is a symmetric function of the n variables E_j if $n \geqq 1$, while $\Psi_0 = 1$.

Let $E_1, \cdots, E_k$ be mutually disjoint sets of $\mathcal{E}$. For every integer $k \geqq 0$ and for every k-uple, $(n_1, \cdots, n_k)$, of integers $n_j \geqq 0$, let

$$(1) \qquad \Phi_{n_1 \ldots n_k}(E_1, \cdots, E_k) = \Psi_n(E_1, \cdots, E_1, \cdots, E_k, \cdots, E_k), \quad \text{where } n = n_1 + \cdots + n_k;$$

it being understood that every E_j occurs n_j times in Ψ_n and that sets with different indices are disjoint. In particular

$$(1 \text{ bis}) \qquad \Psi_k(E_1, \cdots, E_k) = \Phi_{1 \ldots 1}(E_1, \cdots, E_k), \text{ where } 1 + \cdots + 1 = k.$$

Hence, the assignment of the functions Ψ_n is equivalent to the assignment of the functions $\Phi_{n_1 \ldots n_k}$. Since $\Psi_0 = 1$, the Φ which has no subscript ($k = 0$) is 1. It is important that (1) is undefined unless $E_1, \cdots, E_k$ are mutually disjoint.

It will be convenient to assume that, for every k and arbitrary E_j,

$$(2) \qquad \Phi_{n_1 \ldots n_k}(E_1, \cdots, E_k) < c^{n_1 + \cdots + n_k}, \quad \text{where} \quad c = c(E_1, \cdots, E_k),$$

holds for a sufficiently large c which is independent of the n_j. This is the case if and only if

$$(3) \quad f_{E_1 \ldots E_k}(z_1, \cdots, z_k) = \sum_{n_1=0}^{\infty} \cdots \sum_{n_k=0}^{\infty} \frac{(z_1 - 1)^{n_1} \cdots (z_k - 1)^{n_k}}{n_1! \cdots n_k!} \Phi_{n_1 \ldots n_k}(E_1, \cdots, E_k)$$

is an entire function of exponential type in the k complex variables z_j, where k and $E_1, \cdots, E_k$ are arbitrary. It is clear from (1) and from the symmetry of $\Psi_n(E_1 \cdots, E_n)$ in the E_j, that (3) is invariant under any simultaneous permutation of the z_j and the E_j. The f which has no subscript ($k = 0$) is the constant 1. Furthermore,

$$(4) \qquad f_{E_1 \ldots E_k}(1, \cdots, 1) = 1.$$

More generally, (3) and (1) imply that

(5) $$f_{E_1\ldots E_{k-1}}(z_1,\cdots,z_{k-1}) = f_{E_1\ldots E_k}(z_1,\cdots,z_{k-1},1),$$

since the Φ which has no index is 1. It is similarly verified that, in virtue of the binomial theorem and the additivity of the set functions Ψ, the relation

(6) $$f_{E_0E_1E_2\ldots E_k}(z_1,z_1,z_2,\cdots,z_k) = f_{E_0+E_1\,E_2\ldots E_k}(z_1,z_2,\cdots,z_k)$$

holds for any pair of disjoint sets E_0, E_1 of $\mathcal{E}$.

2. Reorder (3) according to the powers of the z_j, obtaining

(7) $$f_{E_1\ldots E_k}(z_1,\cdots,z_k) = \sum_{m_1=0}^{\infty}\cdots\sum_{m_k=0}^{\infty}\phi_{m_1\ldots m_k}(E_1,\cdots,E_k)z_1^{m_1}\cdots z_k^{m_k}$$

as the definition of the ϕ's. It is clear from the corresponding properties of the coefficients of (3), that $\phi_{m_1\ldots m_k}(E_1,\cdots,E_k)$ is invariant under any simultaneous permutation of the m_j and the E_j, and that the ϕ which has no subscript ($k=0$) is 1. It also is clear from (7) that (4) can be written in the form

(8) $$\sum_{m_1=0}^{\infty}\cdots\sum_{m_k=0}^{\infty}\phi_{m_1\ldots m_k}(E_1,\cdots,E_k) = 1;$$

that (5) is equivalent to

(9) $$\sum_{m_k=0}^{\infty}\phi_{m_1\ldots m_k}(E_1,\cdots,E_k) = \phi_{m_1\ldots m_{k-1}}(E_1,\cdots,E_{k-1});$$

finally that, according to (6),

(10) $$\sum_{h=0}^{\infty}\phi_{h\;m_1-h\;m_2\ldots m_k}(E_0,E_1,E_2,\cdots,E_k) = \phi_{m_1\ldots m_k}(E_0+E_1,E_2,\cdots,E_k)$$

holds for any pair of disjoint sets E_0, E_1 of $\mathcal{E}$. In particular, if $k=1$,

(10 bis) $$\sum_{h=0}^{m}\phi_{h\;m-h}(E_0,E_1) = \phi_m(E_0+E_1), \quad \text{where} \quad E_0E_1 = O,$$

O denoting the empty set.

In addition,

(11) $$\Phi_{n_1\ldots n_k}(E_1,\cdots,E_k) = \sum_{m_1=0}^{\infty}\cdots\sum_{m_k=0}^{\infty}\frac{(m_1+n_1)!\cdots(m_k+n_k)!}{m_1!\cdots m_k!}$$
$$\text{times } \phi_{n_1+m_1\ldots n_k+m_k}(E_1,\cdots,E_k),$$

as is seen by comparing the partial derivatives of (3) and (7) at $(z_1,\cdots,z_k) = (1,\cdots,1)$. If the same is done at $(z_1,\cdots,z_k)=(0,\cdots,0)$, the result is

$$(12)\qquad \phi_{m_1\ldots m_k}(E_1,\cdots,E_k)=\frac{1}{m_1!\cdots m_k!}\sum_{n_1=0}^{\infty}\cdots\sum_{n_k=0}^{\infty}\frac{(-1)^{n_1+\cdots+n_k}}{n_1!\cdots n_k!}$$

$$\text{times}\quad \Phi_{m_1+n_1\ldots m_k+n_k}(E_1,\cdots,E_k).$$

The infinite (and, in general, non-recursive) linear substitutions (11), (12) are reciprocal (it being understood that the multiple series involved are absolutely convergent in virtue of (2)). According to (11) and (1 bis),

$$(11\text{ bis})\qquad \Psi_k(E_1,\cdots,E_k)=\sum_{n_1=0}^{\infty}\cdots\sum_{n_k=0}^{\infty}(m_1+1)\cdots(m_k+1)$$

$$\text{times}\quad \phi_{m_1+1\ldots m_k+1}(E_1,\cdots,E_k).$$

3. It is clear from (1) and (1 bis) that

$$(13)\qquad \Phi_{n_1\ldots n_k}\geqq 0$$

is equivalent to

$$(13\text{ bis})\qquad \Psi_n\geqq 0.$$

The assumption (13 bis), made in **1**, has not been used thus far. Actually, what will be needed is that

$$(14)\qquad \phi_{m_1\ldots m_k}\geqq 0.$$

It is obvious from (11) and (11 bis) that (14) is sufficient but not necessary for (13 bis), that is, for (13).

Suppose that not only (13) but also (14) is satisfied. Then (8) shows that, if k and the disjoint sets $E_1,\cdots,E_k$ of $\mathcal{E}$ are arbitrarily fixed, the k-fold sequences $\phi_{m_1\ldots m_k}(E_1,\cdots,E_k)$, where $m_j=0,1,2,\cdots$; $j=1,\cdots,k$, can be interpreted as representing a discrete probability distribution of events which are characterized by the variable subscript $(m_1,\cdots,m_k)$. More specifically, this probability distribution will be thought of as realized in terms of a repartition of "points" P on the sets E of $\mathcal{E}$. In other words, $\phi_{m_1\ldots m_k}(E_1,\cdots,E_k)$ will be interpreted as the probability that there be exactly m_j points P in E_j for all k values of j simultaneously, where the k sets E_j of $\mathcal{E}$ can not overlap. This interpretation is consistent for mutually k disjoint sets E_j, since

(i) as has been seen after (7), the value $\phi_{m_1\ldots m_k}(E_1,\cdots,E_k)$ is invariant under any simultaneous permutation of the m_j and the E_j;

(ii) according to (9), the probability that $k-1$ of k specified sets contain specified numbers of points and the k-th set contain another specified number of points, is such as to supply, when summed over all possible values $(=0,1,2,\cdots)$ of this last specified number, precisely the probability that the $k-1$ sets contain the specified number of points;

(iii) according to (10), the probabilities of disjoint sets E_0, E_1 are logically additive for arbitrarily fixed $E_2, \cdots, E_k$.

According to (11), the expected value of the number of arrays consisting of an ordered n_1-uple of points P in $E_1, \cdots$, of an ordered n_k-uple of points P in E_k is $\Phi_{n_1 \ldots n_k}(E_1, \cdots, E_k)$. This is also clear from (1 bis), since (11 bis) shows that $\Psi_n(E_1, \cdots, E_n)$ represents the expected value of the number of ordered n-uples of points P such that the j-th point is in E_j for all n values of j simultaneously. It is understood that the E_j cannot overlap in these interpretations of the Φ and Ψ.

It is now clear why (13) or, equivalently, (13 bis) is necessary but not sufficient for (14). In fact, the expected values of the "distribution" which represent average values of non-negative integers can be non-negative if some of the "probabilities" are negative.

4. Suppose that the logical sum, S, of all sets E contained in the field $\mathcal{E}$ is a set E contained in $\mathcal{E}$, and that the whole space S (which can, but need not, be Euclidean) contains only a finite number of points P. This means that there exists a positive integer, ν, such that

$$(15_1) \quad \Phi_{n_1 \ldots n_k}(E_{n_1}, \cdots, E_{n_k}) \equiv 0 \text{ for every } n > \nu, \text{ where } n = n_1 + \cdots + n_k.$$

Suppose further that

$$(15_2) \quad \Phi_{n_1 \ldots n_k}(E_{n_1}, \cdots, E_{n_k}) = \frac{1}{(\nu - n)!} \Phi_{n_1 \ldots n_k \, \nu - n}(E_{n_1}, \cdots, E_{n_k}, E)$$
$$\text{for every } n \leqq \nu, \text{ where } n = n_1 + \cdots + n_k, \; E = S - (E_1 + \cdots + E_k).$$

This is a condition of consistency, if the number of points P in S is exactly ν.

It will be shown that, in this case of a finite number of points P in S, the requirement (14) is automatically satisfied in virtue of (13). In fact, it will be shown that, if (15_1) and (15_2) are satisfied, then

$$(16_1) \quad \phi_{n_1 \ldots n_k}(E_1, \cdots, E_k) \equiv 0 \text{ if } n > \nu, \text{ where } n = n_1 + \cdots + n_k,$$

while

$$(16_2) \quad \phi_{n_1 \ldots n_k}(E_1, \cdots, E_k) = \frac{1}{n_1! \cdots n_k! (\nu - n)!} \Phi_{n_1 \ldots n_k \, \nu - n}(E_1, \cdots, E_k, E)$$
$$\text{if } n \leqq \nu, \text{ where } n = n_1 + \cdots + n_k,$$
$$E = S - (E_1 + \cdots + E_k); \; E_i E_j = O (i \neq j),$$

O denoting the empty set.

First, (16_1) is clear from (15_1) and (12). In order to verify (16_2), note that, under the assumptions of (16_2),

$$S^\nu = E_1^{n_1} \times \cdots \times E_k^{n_k} \times E^{\nu-n}$$

$$= \sum_{m_1+\cdots+m_k+m_{k+1}=\nu-n} \frac{(\nu-n)!}{m_1!\cdots m_k!\, m_{k+1}!} \text{ permutations of}$$

$$E_1^{m_1+n_1} \times \cdots \times E_k^{m_k+n_k} \times E^{m_{k+1}}$$

is a logical identity, where A^l denotes $A^{l-1} \times A$ or A according as $l > 1$ or $l = 1$, while $A^0 \times B = B$. Since the set functions Ψ_n are additive, it follows that

$$\Phi_{n_1\ldots n_k}(E_1,\cdots,E_k) = \sum_{m_1+\cdots+m_k+m_{k+1}=\nu-n} \frac{1}{m_1!\cdots m_k!\, m_{k+1}!}$$

$$\text{times } \Phi_{m_1+n_1\ldots m_k+n_k\, m_{k+1}}(E_1,\cdots,E_k,E)$$

in virtue of (1) and the assumptions (15_1), (15_2). This, when compared with the uniqueness of the reciprocal relations (11), (12), completes the proof of (16_2).

Note that, while the uniqueness of the correspondences expressed by the relations (11), (12) presupposes a restrictive condition of the type (2), this condition is trivially satisfied in the present case, (15_1). Thus (14) is a consequence of (13) in the case of (15_1), (15_2).

5. The sufficient condition of **4** for (14) implies a sufficient condition for (14) even if the number of points P in the space S is infinite. In fact, suppose that there are assigned, for every ν, functions $\Phi^{(\nu)}{}_{n_1\ldots n_k}$ such that (15_1) and (15_2) are satisfied by $\Phi = \Phi^{(\nu)}$, and let $\phi^{(\nu)}{}_{n_1\ldots n_k}$ denote the corresponding functions $\phi_{n_1\ldots n_k}$. Thus $\phi^{(\nu)}{}_{n_1\ldots n_k} \geqq 0$, by (16_1)-(16_2). It follows that, if the functions Φ are representable as averages,

$$\Phi_{n_1\ldots n_k}(E_1,\cdots,E_n) = \sum_{\nu=0}^{\infty} \lambda_\nu \Phi^{(\nu)}{}_{n_1\ldots n_k}(E_1,\cdots,E_k),$$

of the respective functions $\Phi^{(\nu)}$, where the weight factors λ_ν depend only on ν and satisfy

$$\lambda_\nu \geqq 0 \qquad \text{and} \qquad \sum_{\nu=0}^{\infty} \lambda_\nu = 1,$$

then the $\phi_{n_1\ldots n_k}$ which belong to the $\Phi_{n_1\ldots n_k}$ satisfy (14). In fact, it is easily verified that

$$\phi_{n_1\ldots n_k}(E_1,\cdots,E_k) = \sum_{\nu=0}^{\infty} \lambda_\nu \phi^{(\nu)}{}_{n_1\ldots n_k}(E_1,\cdots,E_k).$$

6. A rather particular case of a system of functions $\Phi_{n_1\ldots n_k}$ which is of quite another type than the case considered in **4** is represented by the assumption of quasi-independence and corresponds, therefore, to an infinite chain:

$$\Phi_{n_1\ldots n_k}(E_1,\cdots,E_k)=\prod_{j=1}^{k}\Phi_{n_j}(E_j).$$

It is understood that every Φ on the right has a single subscript; so that all $\phi_{n_1\ldots n_k}(E_1,\cdots,E_k)$ are determined by the $\Phi_n(E)$ alone. For a fundamental characterization of this case of quasi-independence, cf. the end of **13.**

It turns out that, in the present case, (14) may but need not be satisfied.

First, it is easily verified from (12) that

$$\phi_{m_1\ldots m_k}(E_1,\cdots,E_k)=\prod_{j=1}^{k}\phi_{m_j}(E_j);$$

so that (14) is equivalent to $\phi_m \geqq 0$, where m is arbitrary. Furthermore, (8) is equivalent to

$$\sum_{m=0}^{\infty}\phi_m(E)=1,$$

while (10 bis) reduces to the addition rule

$$\sum_{h=0}^{m}\phi_h(E_0)\phi_{m-h}(E_1)=\phi_m(E_0+E_1),\text{ where } E_0E_1=O.$$

Since (12) implies that

$$\phi_m(E)=\frac{1}{m!}\sum_{n=0}^{\infty}\frac{(-1)^n}{n!}\Phi_{n+m}(E),$$

it is clear that the present formulation, $\phi_m(E)\geqq 0$, of (14) may, but need not, be satisfied, if the present formulation, $\Phi_n(E)\geqq 0$, of (13) is satisfied. It also is seen from the last series that a sufficient condition for $\phi_m(E)\geqq 0$ consists in

$$\Phi_0(E)\geqq\Phi_1(E)\geqq\Phi_2(E)\geqq\cdots(\geqq 0),$$

where m and E are arbitrary.

Actually, it is sufficient to assume that every E be decomposable into a finite number of mutually disjoint subsets E in such a way that this monotony condition is satisfied for each of these subsets. In fact, $\phi_m(E)\geqq 0$ then holds for every m and for each of these subsets E and so, by the addition rule given before, for every m and for an arbitrary set E.

6 bis. Consider, finally, the case where the assumption of **6** concerning quasi-independence is replaced by the assumption of complete independence; so that $\Phi_n(E)=[\Phi_1(E)]^n$ and

$$\Phi_{n_1 \ldots n_k}(E_1, \cdots, E_k) = \prod_{j=1}^{k} [\Phi_1(E_j)]^{n_j}.$$

Then, according to (12),

$$\phi_{m_1 \ldots m_k}(E_1, \cdots, E_k) = \prod_{j=1}^{k} \left(\frac{[\Phi_1(E_j)]^{m_j}}{m_j!} \sum_{n=0}^{\infty} \frac{[-\Phi_1(E_j)]^n}{n!} \right) = \prod_{j=1}^{k} \phi_{m_j}(E_j).$$

Thus

$$\phi_m(E) = \frac{a^m e^{-a}}{m!}, \text{ where } a = \Phi_1(E), \qquad (m = 0, 1, \cdots);$$

so that the distribution (belonging to $k = 1$) is the Poisson distribution of variance $a = \Phi_1(E)$.

In particular, (14) is implied by the assumption of complete independence.

Notice that the monotony condition, mentioned in **6**, reduces to

$$1 \geqq \Phi_1(E) \geqq [\Phi_1(E)]^2 \geqq [\Phi_1(E)]^3 \geqq \cdots$$

and is, therefore, satisfied only if E is so "small" that $\Phi_1(E) \leqq 1$.

PART II

The Space of Contingencies and its Lebesgue Measure

7. In what follows, the assumptions will be that the functions (1) satisfy (2) and (14), and that the field $\mathcal{E}$ of sets E has a finite or enumerable basis (cf. the beginning of **1**). This basis will be thought of as fixed. A set E which occurs in this basis will be denoted by E^*.

By a fundamental contingency, ω, will be meant a consistent allotment, assigning to every E^* a non-negative integer $m = m(E^*)$ which can be interpreted as follows: The set E^* of the basis of $\mathcal{E}$ contains exactly $m(E^*)$ points P. By the consistency of the allotment is meant, of course, that $m(E^*) = m(E^*_1) + m(E^*_2)$ whenever E^*_1 and E^*_2 are disjoint and $E^*_1 + E^*_2 = E^*$.

By a fundamental contingency set will be meant, of course, a set of fundamental contingencies ω; the ω which constitute the set being thought of as "points" of the set. It is understood that the basis of $\mathcal{E}$ which occurs in the assignment $m = m(E^*)$ of one ω is chosen to be the same for all points ω of the set of ω's.

By a primitive contingency set of order k will be meant a set consisting of all those fundamental contingencies for which $E^*_1, \cdots, E^*_k$ contain $m_1, \cdots, m_k$ points respectively, where the k non-negative integers m_j and the k sets E^*_j in the (fixed) basis of the field $\mathcal{E}$ are arbitrarily fixed but disjoint.

Consider the Borel field generated by the collection of all primitive contingency sets belonging to the fixed basis of $\mathcal{E}$. Let Γ denote an arbitrary set in this field, and Ω the particular Γ which is the logical sum of all sets Γ.

If Γ is a primitive contingency set, determined by k; $m_1, \cdots, m_k$ and $E^*_1, \cdots, E^*_k$, put

$$\mu(\Gamma) = \phi_{m_1 \ldots m_k}(E^*_1, \cdots, E^*_k) .$$

It is then clear from (8), (9), (10) and (14), that the requirement of complete additivity extends this function μ of the primitive contingency sets Γ to a unique Lebesgue measure $\mu = \mu(\Gamma)$, defined for all sets contained in the Borel field generated by the primitive contingency set Γ. Obviously,

$$\mu(\Omega) = 1.$$

In particular, Birkhoff's ergodic theorem is applicable on Ω, if there is given on the Borel sets Γ of Ω a μ-preserving transformation of Ω into itself.

7 bis. Let $\mathcal{E}^n$ and $\Psi_n(E_1, \cdots, E_n)$ be defined as at the beginning of **1**. Let $E^{(n)}$ denote a set in the Borel field generated by $\mathcal{E}^n$, and let $\Psi^{(n)}(E^{(n)})$ be the completely additive set function which reduces to $\Psi_n(E_1, \cdots, E_n)$ if $E^{(n)} = E_1 \times \cdots \times E_n$. It is seen from the definitions of **7**, that such an $E^{(n)}$ can be thought of as representing a set Γ.

In order to illustrate this remark, suppose first that there exists a ν for which the assumptions (15_1), (15_2) of **4** are satisfied. Then, according to (16_1), (16_2), (1) and the definition of μ in **7**,

$$\mu(\Gamma) = \frac{1}{\nu!} \Psi^{(\nu)}(\Gamma)$$

for every Γ of the type just mentioned. Furthermore, the assumption (14) of **7** is now automatically satisfied, by **4**.

It follows that, under the more general assumptions of **5**,

$$\mu(\Gamma) = \sum_{\nu=0}^{\infty} \frac{\lambda_\nu}{\nu!} \Psi^{(\nu)}(\Gamma^{(\nu)}),$$

where $\Gamma^{(\nu)}$ denotes the projection of Γ on the space $\Omega^{(\nu)}$ of exactly ν points P. Again, (14) is automatically satisfied, by **5**.

Still more general assumptions can be obtained by considering limits of systems of the type described in **5**. However, it remains problematic, in what sense, if any, is such an approximation possible in case of a given μ on a given Ω.

It is now seen that the measure μ introduced in **7** serves the purpose of supplying a substitute for the explicitly given measure of a system of the elementary type, considered in **4** or **5**, in the case of a general system; a case in which the measure μ must be constructed, instead of being available *a priori.* Of course, (14) is not a consequence of (13) in this general case.

8. Suppose, for simplicity, that the sets E of the fixed $\mathcal{E}$ are situated in a Euclidean space, S, of given dimension number, d; so that every E is a set of points x, where x is a vector with d components. Let $\mathcal{E}$ be the field of all Borel sets E of S. As an enumerable basis of $\mathcal{E}$, choose the collection of all those parallelepipeds which are parallel to the coördinate axes and have rational numbers for all d coördinates of each of their 2^d corners. In accordance with the notations of **7**, E^* will denote one of these rational parallelepipeds.

Thus, a basis contingency is given by a consistent allotment to every E^* of a non-negative integer $m = m(E^*)$ which can be interpreted as the number of points P contained in E^*. Since the allotment is consistent, $m(E^*) = m(E^*_1) + m(E^*_2) + \cdots$ for every decomposition of a basis parallelepiped E^* into mutually disjoint basis parallelepipeds $E^*_1, E^*_2, \cdots$.

For a fixed E^*, consider a sequence of such decompositions, having the property that the $(l+1)$-th decomposition is obtained by a subdivision of the l-th and the maximum diameter of the parallelepipeds occurring in the l-th decomposition tends to 0 as $l \rightarrow \infty$. For every fixed l, not more than $m(E^*)$ of the parallelepipeds of the l-th decomposition contain at least one point P. Since $m(E^*)$ is independent of l, it follows that in the sequence of successive decompositions of E^* not more than $m(E^*)$ nested sequences of parallelepipeds contain points P. Let $\bar{m}(E^*)$ denote the number ($\geqq 0$) of these non-empty nested sequences of parallelepipeds. Thus $\bar{m}(E^*) \leqq m(E^*)$, where $\bar{m}(E^*) < m(E^*)$ cannot be excluded, since some of the points P can be multiple. On the other hand, $\bar{m}(E^*) \geqq 1$ unless $m(E^*) = 0$.

9. If x is one of the $\bar{m}(E^*)$ points of E^* to which the $\bar{m}(E^*)$ nested sequences of parallelepipeds tend as $l \rightarrow \infty$, there exists a non-negative integer $i = i(x)$ such that the l-th parallelepiped in the sequence of nested parallelepipeds about x contains exactly $i(x)$ points P for every sufficiently large x. Furthermore, the sum of the $\bar{m}(E^*)$ integers $i(x)$ is $m(E^*)$, if suitable precaution is taken for the faces of E^*, i. e., for the boundary of the set E^*. (It is possible to take such a precaution, since E^* can be chosen arbitrarily in the basis of the Euclidean space S.) If x is any point which lies in the interior of E^* and is distinct from the $\bar{m}(E^*)$ points x for which $i(x)$ was just defined, put $i(x) = 0$.

Since E^* can be chosen arbitrarily, there is now defined a non-negative integer $i = i(x)$ as a function of the position x on the Euclidean space S.

Since there exist in every $E*$ at most a finite number ($\leqq m(E*)$, possibly 0) of points x for which $i(x) \neq 0$, and since the set of all $E*$ is enumerable, there exists in S an at most enumerable set of distinct points X which have no finite cluster point in S and are characterized by the fact that the index $i(x)$ of a point x of S is zero or a positive integer according as x is or is not an X. Let $\{X\}$ denote the sequence of all points X of S. It is understood that the set $\{X\}$ can be finite.

10. It is clear from **8** that a basis contingency is equivalent to a consistent allotment of a non-negative integer $\bar{m}(E*)$ and of $\bar{m}(E*)$ points of $E*$, along with the multiplicities of these points, for each of the parallelepipeds $E*$ which form a basis of the field $\mathcal{E}$ of the Borel sets E of S. It follows, therefore, from **9**, that a basis contingency is equivalent to the assignment of an arbitrary sequence $\{X\}$ of distinct points X of S which have no finite cluster point, and of an arbitrary positive integer $i(X)$ as a function of the position X on $\{X\}$. In fact, a basis contingency then is defined by assigning that every basis set $E*$ contain exactly $m(E*) = \Sigma\, i(X)$ points P, where the summation runs through those points X of $\{X\}$ which are in $E*$.

While only *fundamental contingencies*, and not contingencies (unqualified) have been defined so far, it is now possible to define *contingencies*, as follows: A contingency is given by a consistent allotment to every bounded Borel set E in S of a non-negative integer, $m(E)$, which represents the number of points P contained in E and is extended, either as a non-negative integer $m(E)$ or as $m(E) = \infty$, to every (not necessarily bounded) Borel set E of S by the requirement of complete additivity. In fact, this definition of a contingency is equivalent to the assignment of an arbitrary positive integer $i(X)$ as a function of the position X on an arbitrary sequence $\{X\}$ of points of the Cartesian space S which have no finite cluster point. It is understood that the number ($\geqq 0, \leqq \infty$) of points P contained in an arbitrary Borel set E of S is then given by $m(E) = \Sigma\, i(X)$, where the summation runs through those points X of $\{X\}$ which are in E.

11. In **7**, the symbol ω was used to denote an arbitrary fundamental contingency. Let ω now denote an arbitrary contingency. While the sets which in **7** were denoted by Γ were there introduced by a Borel extension of the collection of all primitive contingency sets (instead of as sets consisting of contingencies as points), it is now possible to consider every set Γ as a Borel set of points ω. In particular, the set of all points ω is the particular Γ which in **7** was denoted by Ω. Thus Ω is now a space consisting of all contingencies ω as points, and $\mu(\Gamma)$ is a measure, defined on the field of sets Γ which are subsets of the contingency space Ω.

The definition of $\bar{m}(E^*)$ in **8** is extended from basis sets E^* to arbitrary Borel sets E of S, if $\bar{m}(E)$ denotes the number of those points X of $\{X\}$ which are in E. Since $i(X)$ is a positive integer, $\bar{m}(E) \leqq m(E)$. The sign of equality holds for every E if and only if $i(X) = 1$ for every X. Since if x is an arbitrary point of S, then $i(x) = 0$ or $i(x) \geqq 1$ according as x is or is not an X, it is clear that $\bar{m}(E) = m(E)$ for every E if and only if $i(x)$ is either 0 or 1 for every x.

It follows that, if $\bar{m}(E) = m(E)$ for every E, then a contingency is equivalent to an arbitrary allotment of a sequence $\{X\}$ of points X of S which do not have a finite cluster point. In fact, the allotment of the function $i(X)$ on $\{X\}$, as required by **10**, is then given by $i(X) \equiv 1$.

In what follows, there will be delimited cases in which $\bar{m}(E) \equiv m(E)$, i. e. $i(X) \equiv 1$, is true, though not necessarily for all, at least for almost all, points ω of the contingency space Ω. By this is meant that, in the cases to be considered, the exceptional contingencies are of measure 0 with respect to the μ-measure on Ω (cf. **7** and **10**). Since this measure $\mu(\Gamma)$ on Ω was defined (**7**) in terms of the system of set functions $\phi_{m_1\ldots m_k}(E_1,\cdots,E_k)$, it depends, of course, on the choice of this system of set functions whether $i(X) \equiv 1$ is or is not true for almost all contingencies on Ω; so that what must be delimited is a class of suitable systems of set functions $\phi_{m_1\ldots m_k}$.

12. In view of (11), (12), (1) and the assumption (1), it will be sufficient to consider classes of suitable systems of set functions $\Psi_n(E_1,\cdots,E_n)$. Actually, it turns out that, in order to obtain a useful sufficient criterion for the validity of $i(X) \equiv 1$ almost everywhere on Ω, it is not necessary to consider the full sequence $\Psi_n(E_1,\cdots,E_n)$, where $n = 1, 2, \cdots$, but only Ψ_1 and Ψ_2.

In fact, suppose that there exists for every point x of the Euclidean space S and for every $\epsilon > 0$ a parallelepiped $E^*_x(\epsilon)$ about x in such a way that

$$(*) \quad \Psi_2(E, E) \leqq \epsilon\, \Psi_1(E) \text{ whenever } E \text{ is in } E^*_x(\epsilon)$$

(it being understood that E denotes a Borel set in S). It will be shown that, if this condition is satisfied, the set of those points ω of the contingency space Ω for which $0 \leqq i(x) \leqq 1$ does not hold for all points x of S is of μ-measure 0.

Let E^* be a fixed parallelepiped in the enumerable basis of S (cf. **8**). Let $\bar{m}(E^*)$ denote the same integer as in **8**. It is then clear from **9-10** that the set of those fundamental contingencies for which E^* contains exactly $\bar{m}(E^*)$ distinct points P is a primitive contingency set in the sense of **7**. Consider these point groups as *ordered* point groups, and let E^* contain the point P.

For every point x of the parallelepiped $E*$ and for a given $\epsilon > 0$, choose the parallelepiped $E*_x(\epsilon)$ in accordance with $(*)$. Then, if $E*$ is closed and $\epsilon > 0$ is fixed, the Heine-Borel theorem assures that there is a finite number of points x in $E*$ such that the $E*_x(\epsilon)$ which belong to these x cover the whole of $E*$. Hence, there exists a finite number of mutually disjoint parallelepipeds $E*_l$ such that $\sum_l E*_l = E*$ and, for every l,

$$\Phi_2(E*_l) = \Psi_2(E*_l, E*_l) \leqq \epsilon\Psi_1(E*_l).$$

Thus

$$\sum_l \Phi_2(E*_l) = \sum_l \Psi_2(E*_l, E*_l) \leqq \epsilon\Psi_1(E*),$$

since $\Psi_1(E)$ is additive (**1**). In view of the interpretation of Φ_n in **3**, the sum on the left of the last inequality cannot be less than the expected value of the number of those points X in E at which $i(X) > 1$. It follows, therefore, from the definition of μ in **7**, that, since $\epsilon > 0$ is arbitrarily small, the set of those contingencies for which the *first* of the ordered group of $\bar{m}(E*)$ points X in $E*$ satisfies $i(X) > 1$ is of μ-measure zero. Since the sum of a sequence of contingency sets of μ-measure 0 is of μ-measure 0, the proof is complete.

13. Suppose that

(i) $$\Psi_k(E_1, E_2, \cdots, E_k) \leqq c^k \Psi_1(E_1)\Psi_1(E_2)\cdots\Psi_1(E_k)$$

holds for a sufficiently large constant c, and that

(ii) $$\Psi_1(E) \to 0 \text{ as } E \to x$$

(that is, that $\Psi_1(E)$ tends to 0 whenever a nested sequence of Borel sets E shrinks to any fixed point x of the Euclidean space S). It is clear that (ii) implies, in virtue of (i), the restriction $(*)$ of **12**. Furthermore, (2) can be replaced by the sharper estimate

(i bis) $$\Phi_{n_1\ldots n_k}(E_1, \cdots, E_k) \leqq c^{n_1+\cdots+n_k}[\Phi_1(E_1)]^{n_1}[\Phi_1(E_2)]^{n_2}\cdots[\Phi_1(E_k)]^{n_k},$$

which is obvious from (1) by the assumption (i).

Since Φ_1 is additive and non-negative (**1**), it is clear that $\Phi_1(E)$ can be thought of as defining a Lebesgue measure on the d-dimensional Euclidean space S. Then $\Phi_1(E_1)\Phi_1(E_2)\cdots\Phi_1(E_k)$ defines a product measure on the product space $S^k = S^{k-1} \times S$, where $S^1 = S$.

It is easily seen from (i) and (ii) that the set function defined by the Lebesgue extension of the set function (1 bis), **1** on S^k is absolutely continuous with respect to the product measure define by the corresponding extension of

$\Phi_1(E_1)\Phi_1(E_2)\cdots\Phi_1(E_k)$. Accordingly, there exists on the product space S^k a function δ_k of the position $(x_1,\cdots,x_k)$ such that

$$\Psi_k(E_1,\cdots,E_k)=\int_{E_1}\cdots\int_{E_k}\delta_k(x_1,\cdots,x_k)d_{x_1}\Phi_1(x_1)\cdots d_{x_k}\Phi_1(x_k),$$

where every x_j represents a point of the Euclidean space $S^1=S$ of given dimension number d. Needless to say, the L-integrable point function δ_k is determined by the set function Ψ_k almost everywhere (that is, up to an $(x_1,\cdots,x_k)$-set of vanishing product measure), and is non-negative (almost everywhere), since $\Psi_k\geqq 0$ (**1**). Thus, from (i),

$$0\leqq\delta_k(x_1,\cdots,x_k)\leqq c^k$$

(almost everywhere)

It has been seen in **3** that $\Psi_k(E_1,\cdots,E_k)$ represents the expected value of the number of ordered k-uples of points the j-th of which is in E_j for all k values of j simultaneously. It follows, therefore, from the result of **12** and from the preceding integral representation of Ψ_k, that $\delta_k(x_1,\cdots,x_k)$ is the density of probability of the ordered k-uples at $(x_1,\cdots,x_k)$.

Obviously, the case of quasi-independence (**6**) results if each of the densities (with respect to product measure!), $\delta_k(x_1,\cdots,x_k)$, is independent of the position $(x_1,\cdots,x_k)$ on S^k.

PART III

Functions of the Contingencies

14. In what follows, the assumption will be that the conditions of **7**, **8** and **12** (but not necessarily those of **13**) are satisfied. In particular, a contingency (unqualified) is defined, by **11**, as a point ω of space Ω which carries a Lebesgue measure μ such that $\mu(\Omega)=1$.

For k given Borel sets E_j of the underlying d-dimensional Euclidean space S and for k given non-negative integers m_j, define a function

$$\phi=\phi(\omega)\equiv\phi^{\omega}{}_{m_1\ldots m_k}(E_1,\cdots,E_k)$$

of the position ω on Ω as follows: The point function $\phi^{\omega}{}_{m_1\ldots m_k}(E_1,\cdots,E_k)$ of ω is the characteristic function of that set

$$\Gamma=\Gamma_{m_1\ldots m_k}(E_1,\cdots,E_k)$$

of contingencies in which E_j contains exactly m_j points P. Since this $\Gamma_{m_1\ldots m_k}(E_1,\cdots,E_k)$ obviously is a μ-measurable subset of Ω, its characteristic function is integrable over Ω. Furthermore, if the k subscripts of this characteristic function are varied, a straightforward counting verifies the orthogonality relation

$$(*)\quad \int_\Omega \phi^\omega{}_{m_1\ldots m_k}(E_1,\cdots,E_k)\phi^\omega{}_{l_1\ldots l_k}(E_1,\cdots,E_k)\mu(\delta\Omega)$$

$$= \begin{cases} 0 \text{ if } (m_1,\cdots,m_k)\neq(l_1,\cdots,l_k), \\ \phi_{m_1\ldots m_k}(E_1,\cdots,E_k) \text{ if } (m_1,\cdots,m_k)=(l_1,\cdots,l_k), \end{cases}$$

where $E_1,\cdots,E_k$ are fixed and $\phi_{m_1\ldots m_k}(E_1,\cdots,E_k)$ denotes the same set function as in **2-3**.

15. Let a non-negative integer n and n Borel sets $E_1,\cdots,E_n$ of S be given. Define a non-negative integer

$$\Psi=\Psi(\omega)=\Psi_n{}^\omega(E_1,\cdots,E_n)$$

as a function of the position ω on Ω as follows: $\Psi_n{}^\omega(E_1,\cdots,E_n)$ is the number of ordered n-uples $(P_1\cdots P_n)$ which correspond to ω and to the restriction that the point P_j is in E_j for $j=1,\cdots,n$, where ω is a given contingency.

Corresponding to (1), define a function

$$\Phi=\Phi(\omega)\equiv\Phi^\omega{}_{n_1\ldots n_k}(E_1,\cdots,E_k)$$

of the position ω on Ω by placing

$$(\text{I})\quad \Phi^\omega{}_{n_1\ldots n_k}(E_1,\cdots,E_k)=\Psi_n{}^\omega(E_1,\cdots,E_1,\cdots,E_k,\cdots,E_k),$$

$$\text{where } n=n_1+\cdots+n_k,$$

it being understood that every E_j occurs exactly n_j times in $\Psi_n{}^\omega$. In particular

$$(\text{I bis})\quad \Psi_k{}^\omega(E_1,\cdots,E_k)=\Phi^\omega{}_{1\ldots 1}(E_1,\cdots,E_k),\text{ where } 1+\cdots+1=n.$$

It is easily seen from the definition of $\phi^\omega{}_{m_1\ldots m_k}(E_1,\cdots,E_k)$ in **14** that, corresponding to (11),

$$(\text{II})\quad \Phi^\omega{}_{n_1\ldots n_k}(E_1,\cdots,E_k)=\sum_{m_1=0}^{\infty}\cdots\sum_{m_k=0}^{\infty}\frac{(m_1+n_1)!\cdots(m_k+n_k)!}{m_1!\cdots m_k!}$$

$$\text{times } \phi^\omega{}_{n_1+m_1\ldots n_k+m_k}(E_1,\cdots,E_k);$$

in fact, all but one of the terms of the non-negative series (II) vanish. Similarly, corresponding to (12),

$$(\text{III})\quad \Phi^\omega{}_{m_1\ldots m_k}(E_1,\cdots,E_k)=\frac{1}{m_1!\cdots m_k!}\sum_{n_1=0}^{\infty}\cdots\sum_{n_k=0}^{\infty}\frac{(-1)^{n_1+\cdots+n_k}}{n_1!\cdots n_k!}$$

$$\text{times } \Phi^\omega{}_{m_1+n_1\ldots m_k+n_k}(E_1,\cdots,E_k);$$

7

it being clear that all but a finite number of the terms of the series (III) vanish.

16. While (II) and (III) are obvious identities for every fixed ω, it is by no means obvious that both of these expansions are valid in the (L^2)-mean of the μ-measure on Ω also. It will now be shown that such happens to be the case.

In other words, it will be shown that the quadratic mean error of the partial sums of the series, an error represented for (III) by

$$\text{(IV)} \quad \int_\Omega [\phi^\omega_{m_1\ldots m_k}(E_1,\cdots,E_k) - \frac{1}{m_1!\cdots m_k!}\sum_{n_1=0}^{N_1}\cdots\sum_{n_k=0}^{N_k}\frac{(-1)^{n_1+\cdots+n_k}}{n_1!\cdots n_k!}$$
$$\text{times } \Phi^\omega_{m_1+n_1\ldots m_k+n_k}(E_1,\cdots,E_k)]^2\mu(d\Omega),$$

tends to zero as $N_1\to\infty,\cdots,N_k\to\infty$, if k and $E_1,\cdots,E_k$ are arbitrarily fixed; and that the corresponding relation holds for (II) also.

In view of the orthogonality relation, mentioned at the end of **14**, the validity of the expansions (II), (III) in the mean can be interpreted as expressing the completeness of the orthogonal system at hand. Actually, it is precisely this completeness property that will be needed in the sequel.

16 bis. The proof proceeds as follows:

Since the series (II) cannot have more than one non-vanishing term for a fixed ω, it is easily verified from the orthogonality relations, mentioned at the end of **14**, that

$$\int_\Omega \phi^\omega_{m_1\ldots m_k}(E_1\cdots,E_k)\Phi^\omega_{n_1\ldots n_k}(E_1,\cdots,E_k)\mu(d\Omega)$$
$$=\begin{cases}\dfrac{m_1!\cdots m_k!}{(m_1-n_1)!\cdots(m_k-n_k)!}\phi_{m_1\ldots m_k}(E_1,\cdots,E_k) \text{ if } m_1>n_1,\cdots,m_k>n_k;\\ 0 \text{ otherwise}\end{cases}$$

and

$$\int_\Omega \Phi^\omega_{m_1\ldots m_k}(E_1\cdots,E_k)\Phi^\omega_{n_1\ldots n_k}(E_1,\cdots,E_k)\mu(d\Omega)$$
$$=\sum_{j_1=\min(m_1,n_1)}^{\infty}\cdots\sum_{j_k=\min(m_k,n_k)}^{\infty}\frac{(j_1!\cdots j_k!)^2\,\phi_{j_1\ldots j_k}(E_1,\cdots,E_k)}{(j_1-n_1)!\cdots(j_k-n_k)!(j_1-m_1)!\cdots(j_k-m_k)!}$$

Hence, the mean square error, represented by the integral (IV), is identical with

$$-\phi_{m_1\ldots m_k}(E_1,\cdots,E_k)-\frac{2}{m_1!\cdots m_k!}\sum_{n_1=\min(m_1,N_1)}^{\infty}\cdots\sum_{n_k=\min(m_k,N_k)}^{\infty}$$

$$\frac{(-1)^{n_1+\cdots+n_k}m_1!\cdots m_k!}{n_1!\cdots n_k!(m_1-n_1)!\cdots(m_k-n_k)!}\phi_{m_1\ldots m_k}(E_1,\cdots,E_k)$$

$$+\frac{1}{(m_1!\cdots m_k!)^2}\sum_{n_1=0}^{N_1}\cdots\sum_{n_k=0}^{N_k}\sum_{l_1=0}^{N_1}\cdots\sum_{l_k=0}^{N_k}\frac{(-1)^{n_1+\cdots+n_k+l_1+\cdots+l_k}}{n_1!\cdots n_k!\,l_1!\cdots l_k!}\sum_{j_1=m_1+\min(n_1,l_1)}^{\infty}\cdots\sum_{j_k=m_k+\min(n_k,l_k)}^{\infty}$$

$$\frac{(j_1!\cdots j_k!)^2\,\phi_{j_1\ldots j_k}(E_1,\cdots,E_k).}{(j_1-m_1-n_1)!\cdots(j_k-m_k-n_k)!(j_1-m_1-l_1)!\cdots(j_k-m_k-l_k)!},$$

and reduces therefore, if $N_1\geqq m_1,\cdots,N_k\geqq m_k$, to

$$-\phi_{m_1\ldots m_k}(E_1,\cdots,E_k)+\sum_{j_1=m_1}^{\infty}\cdots\sum_{j_k=m_k}^{\infty}\phi_{j_1\ldots j_k}(E_1,\cdots,E_k)(j_1!\cdots j_k!)^2\text{ times}$$

$$\sum_{n_1=\min(N_1,j_1-m_1)}^{\infty}\cdots\sum_{n_k=\min(N_k,j_k-m_k)}^{\infty}\;\sum_{l_1=\min(N_1,j_1-m_1)}^{\infty}\cdots\sum_{l_k=\min(N_k,j_k-m_k)}^{\infty}$$

$$\frac{(-1)^{n_1+\cdots+n_k+l_1+\cdots+l_k}}{n_1!\cdots n_k!(j_1-m_1-n_1)!\cdots(j_k-m_k-n_k)!(j_1-m_1-l_1)!\cdots(j_k-m_k-l_k)!}$$

or simply to

$$-\phi_{m_1\ldots m_k}(E_1,\cdots,E_k)+\sum_{j_1=m_1}^{\infty}\cdots\sum_{j_k=m_k}^{\infty}\phi_{j_1\ldots j_k}(E_1,\cdots,E_k)(j_1!\cdots j_k!)^2\text{ times}$$

$$\left(\sum_{n_1=\min(N_1,j_1-m_1)}^{\infty}\cdots\sum_{n_k=\min(N_k,j_k-m_k)}^{\infty}\frac{(-1)^{n_1+\cdots+n_k}}{n_1!\cdots n_k!(j_1-m_1-n_1)!\cdots(j_k-m_k-n_k)!}\right)^2.$$

Since k and $m_1,\cdots,m_k$ are fixed, it follows that, as $N_1\to\infty,\cdots,N_k\to\infty$, the integral (IV) is majorized by a constant multiple of

$$\sum_{j_1=N_1+m_1}^{\infty}\cdots\sum_{j_k=N_k+m_k}^{\infty}\phi_{j_1\ldots j_k}(E_1,\cdots,E_k)\left(\frac{j_1!\cdots j_k!\,2^{j_1+\cdots+j_k}}{(j_1-m_1)!\cdots(j_k-m_k)!}\right)^2.$$

In order to estimate the coefficients of ϕ in this series, notice that, by (2), the function (3) is an entire function of the exponential type in the z_j-1; hence, it is an entire function of the exponential type in the z_j also. This means, in view of (7), that

$$|\phi_{m_1\ldots m_k}(E_1,\cdots,E_k)|<\frac{C^{m_1+\cdots+m_k}}{m_1!\cdots m_k!}$$

holds for a sufficiently large C which is independent of $(m_1,\cdots,m_k)$. Hence, the dominant series of (IV), found before, is majorized by a constant multiple of

$$\sum_{j_1=N_1+m_1}^{\infty}\cdots\sum_{j_k=N_k+m_k}^{\infty}\frac{C^{j_1+\cdots+j_k}}{j_1!\cdots j_k!}\left(\frac{j_1!\cdots j_k!\,2^{j_1+\cdots+j_k}}{(j_1-m_1)!\cdots(j_k-m_k)!}\right)^2.$$

This series can be written in the form

$$\sum_{j_1=N_1}^{\infty}\cdots\sum_{j_k=N_k}^{\infty}\frac{(4C)^{j_1+\cdots+j_k}}{j_1!\cdots j_k!}\,\frac{(j_1+m_1)!\cdots(j_k+m_k)!}{j_1!\cdots j_k!}$$

and tends, therefore, to zero as $N_1\to\infty,\cdots,N_k\to\infty$, the integers k and $m_1,\cdots,m_k$ being fixed. This proves that (IV) tends to zero as $N_1\to\infty,\cdots,N_k\to\infty$.

In other words, the expansion (III) is valid in the mean. The proof of the corresponding statement of **16** concerning the expansion (II) is similar.

17. According to **11**, a contingency ω can be thought of as consisting of certain assignments of definite points P in the Euclidean space S; while **12** assures that these points are all distinct. Thus, it is clear from the definition of $\Psi_n^{\omega}(E_1,\cdots,E_n)$ in **15** that, for every fixed n,

$$\Psi_n^{\omega}(E_1,\cdots,E_n)=\Sigma\cdots\Sigma\,\psi^{\omega}{}_{E_1\ldots E_n}(P_1,\cdots,P_n),$$

where the n summation sign $\Sigma\cdots\Sigma$ represents summation over all n-uples of distinct points $P_1,\cdots,P_n$ corresponding to the contingency ω, and $\psi^{\omega}{}_{E_1\ldots E_n}(P_1,\cdots,P_n)$ denotes the characteristic function of the product set $E_1\times\cdots\times E_n$.

Correspondingly, the completeness theorem proved in **16** bis (cf. the end of **16**), when applied to the ψ_n^{ω} insteal of the $\Phi^{\omega}{}_{n_1\ldots n_k}$, can be expressed by saying that the system

$$1,\quad \Sigma\,\psi_E^{\omega}(P),\quad \Sigma\Sigma\,\psi^{\omega}{}_{E_1E_2}(P_1,P_2),\quad \Sigma\Sigma\Sigma\,\psi^{\omega}{}_{E_1E_2E_3}(P_1,P_2,P_3),\cdots$$

is a closed (L^2) system of functions on the contingency space. Since these functions are independent, they could be ortho-normalized.

18. The preceding completeness theory of contingencies is relevant for an ergodic theory of contingencies. This will now be illustrated by the simplest case, that of the translation group.

Suppose that the functions (1 bis), **1** are given so as to remain invariant if the sets $E_1,\cdots,E_k$ of the d-dimensional Euclidean space S are subject to the same translation, which is allowed to be arbitrary. In other words, let, for every k and for arbitrary $E_1,\cdots,E_k$,

$$\Psi_k(E_1+s,\cdots,E_k+s)=\Psi_k(E_1,\cdots,E_k),$$

where s is an arbitrary vector with d components and $E+s$ denotes the set of all those vectorial sums $e+s$ for which the vector $x=e$ represents a point of the Borel set E of S. Using the assumptions and notations of **13**, one can also say that, on the one hand,

$$\delta_k(x_1+s,\cdots,x_k+s)=\delta_k(x_1,\cdots,x_k),$$

and, on the other hand, $d_x\Psi_1$ is proportional to the d-dimensional Euclidean volume element, dx. Accordingly, the representation of $\Psi_k(E_1,\cdots,E_k)$ in **13** reduces to

$$\text{(i)}\qquad \Psi_k(E_1,\cdots,E_k)=\int_{E_1}\cdots\int_{E_k}\rho_k(x_1,\cdots,x_k)\,dx_1\cdots dx_k,$$

where $\rho_k(x_1,\cdots,x_k)$ is proportional to the density $\delta_k(x_1,\cdots,x_k)$,

$$\rho_k(x_1,\cdots,x_k)=(\text{const.})^k\,\delta_k(x_1,\cdots,x_k),\qquad(\delta_0\equiv 1);$$

so that

$$\text{(ii)}\qquad \rho_k(x_1+s,\cdots,x_k+s)=\rho_k(x_1,\cdots,x_k)$$

and, according to **13**,

$$\text{(iii)}\qquad 0\leqq\rho_k(x_1,\cdots,x_k)\leqq(\text{Const.})^k.$$

It is clear from **1-12** that addition of an arbitrary constant vector, s, to every point X of a contingency $\omega=\{X\}$ generates a μ-measure-preserving transformation, say τ_s, of the contingency space Ω into itself. All transformations τ_s together, which belong to translations s of the d-dimensional Euclidean space S, constitute a group. Clearly, the multi-dimensional ergodic theorem is applicable to this group. Accordingly, if $F(\omega)$ is a function of class (L) on Ω, and if the point ω of Ω is fixed not on a certain subset of μ-measure 0, the s-average of $F\ (\tau_s\omega)$ over the interior of an s-sphere about the unit of the group τ_s tends to a limit when this sphere increases indefinitely.

19. In view of the physical applications alluded to in the Introduction, it is of fundamental importance to have a criterion which assures that the "flow" τ_s on Ω makes almost all "paths" on the configurations statistically independent and such as to correspond to the case of asymptotic equidistribution.

It will be shown that a simple sufficient criterion to this effect is represented by the approximate independence of remote regions of S; that is, by the assumption that

$$(*)\qquad \rho_k(x_1,\cdots,x_l,x_{l+1}+s,\cdots,x_k+s)\rightarrow\rho_l(x_1,\cdots,x_l)\rho_{k-l}(x_{l+1},\cdots,x_k)$$

as $|s|\rightarrow\infty$, where k and l are arbitrary. In fact, it will be shown that, if $(*)$ is satisfied, τ_s is a mixture.

In particular, $(*)$ is sufficient for the metrical transitivity of τ_s, that is, for the situation in which the s-average of $F(\tau_s\omega)$ is, for almost all points ω of Ω, independent of ω and, therefore equals

$$\int_\Omega F(\omega)\mu(d_\omega\Omega),$$

where $F(\omega)$ is any fixed function of class (L) on Ω.

19 bis. The proof proceeds as follows:

First, every $p_k(x_1,\cdots,x_k)$ is bounded, by (iii), **18.** Hence, it is clear from (i), **18** and (*), **19** that

$$\Psi_k(E_1,\cdots,E_l,E_{l+1}+s,\cdots,E_k+s)\to\Psi_l(E_1,\cdots,E_l)\Psi_{k-l}(E_{l+1},\cdots,E_k)$$

as $|s|\to\infty$. It follows, therefore, from (2), **1** and from the connections between the set functions Ψ, Φ, ϕ, that

$$\phi_{m_1\ldots m_k}(E_1,\cdots,E_l,E_{l+1}+s,\cdots,E_k+s)\to\phi_{m_1\ldots m_l}(E_1,\cdots,E_l)\phi_{m_{l+1}\ldots m_k}(E_{l+1},\cdots,E$$

Cf. **1-3** and **16** bis.

Since a basis of the measure μ on Ω can be obtained by assigning the measure $\phi_{m_1\ldots m_k}(E_1,\cdots,E_k)$ to that set of contingencies ω for which E_j contains exactly m_j points P for $j=1,\cdots,k$, it is now seen from **17** that, if F is the characteristic function of any $\mu\times\mu$-measurable set on the product space $\Omega\times\Omega$, then

$$\int_\Omega F(\Omega,\tau_s\Omega)\mu(d_\omega\Omega)\to\int_\Omega\int_\Omega F(\omega,\bar\omega)\mu(d_\omega\Omega)\mu(d_{\bar\omega}\Omega)$$

as $|s|\to\infty$. Thus τ_s is a mixture on Ω, i.e., the product flow, $\tau_s\times\tau_s$, is metrically transitive on $\Omega\times\Omega$.

MASSACHUSETTS INSTITUTE OF TECHNOLOGY,
THE JOHNS HOPKINS UNIVERSITY.

Comments on [43a]

J. Feldman

The authors are mainly engaged in axiomatizing stochastic processes which assign, in a random way, a "number of particles" N_A to each set $A \in \mathcal{E}$, where $\mathcal{E}$ is a σ-field of subsets of a basic "configuration space" S. This can most easily be done through the joint distributions $\phi_{n_1}, \ldots, {}_{n_k} (A_1, \ldots, A_k)$ $= \text{Prob} \{N_{A_1} = n_1, \ldots, N_{A_k} = n_k\}$, by applying the Kolmogorov existence theorem. However, sections 8-11 are devoted to performing the construction of the stochastic process without recourse to that theorem. More interesting is the attempt to characterize the process through the "moments" $\theta_{n_1, \ldots, n_k} (A_1, \ldots, A_k) = E \{(N_{A_1})_{n_1} \ldots (N_{A_k})_{n_k}\}$, where $(j)_m = j(j-1) \ldots (j-n+1)$, so that $\phi_{n_1}, \ldots, {}_{n_k} (A_1, \ldots, A_k)$ = expected number of ordered arrays of n_1 particles from $A_1, \ldots, n_k$ particles from A_k. Under certain boundedness assumptions (2) on these moments, it is possible to establish formulas giving the Φ in terms of the ϕ and conversely. However, positivity of the Φ does not imply that of the ϕ, and since the latter is necessary for the construction of the process, some effort is made (sections 4, 5) to establish this under various assumptions. It is my belief that these arguments are incorrect.

After making assumptions to guarantee that no two particles can be in the same place, the authors establish a representation of $\mathcal{L}_2$ (Prob) analagous to the representation of the $\mathcal{L}_2$-space of a Gaussian process as a direct sum of spaces of polynomials of various degrees in the process which occurs in the paper "Homogeneous chaos" [38a].

Finally, in the case where S is R^n, and the process is stationary under translation, an assumption of "asymptotic independence" is used to give metric transitivity of the action of the translation group.

Unlike its sister paper "Homogeneous chaos" [38a], this paper has had little direct influence; the one paper I know which refers to it is Shale and Stinespring, "Wiener Processes II" (J. Functional Analysis, vol. 5, (1970), pp. 334-353). However, the paper is very much in the spirit of certain quite recent developments in statistical mechanics, and its essential contents have been reinvented. The basic space of contingencies is precisely that in Ruelle (J. Math. Phys. vol. 8 (1967), pp. 1657-1668). In this context, the technical assumption (2) turns out to be satisfied for systems of particles in equilibrium and having "reasonable" interactions (Ruelle, Comm. Math. Phys. vol. 18

(1970), pp. 727-759). And the "asymptotic independence" assumption is related to "cluster decomposition" properties. Also in this general direction is recent work of Dobrushin, Minlos, and Sinai.

References

1. R. Dobrushin, *Gibbsian random fields for lattice systems with pairwise interactions.* J. Functional Analysis, 2, no. 4 (1968), 431-443. Gibbsian random fields: general case.

2. Ibid., 3, no. 1 (1969), 27-35.

3. G. Gallavotti and S. Miracle-Sole, *A variational principle for the equilibrium of hard sphere systems,* Ann. Inst. H. Poincaré, 8, no. 3 (1968), 287-299.

4. D. Ruelle, Statistical Mechanics: Rigorous Results, Benjamin, New York, 1969; see section 7.1.2.

5. O. Lanford, *Classical mechanics of one-dimensional systems of infinitely many particles,* Comm. Math. Phys. 11(1969), 257-292.

Comments on [43a]

B. McMillan

It may be clarifying to recognize that in this paper a "discrete chaos" is a random process based on random variables $\{\mathcal{D}(A), A \in \mathsf{A}\}$. Here A is a ring of bounded Borel subsets of a Euclidean space X, containing an enumerable basis for the field E of the paper. The sample (set) functions $\mathcal{D}(\mathsf{A})$ are, with probability one, countably additive; with probability one, $\mathcal{D}(\cdot)$ possesses a carrier $\gamma \leqslant X$ that is a (random) discrete subset of X to each point of which $\mathcal{D}(\cdot)$ assigns unit measure. A specific realization for such a process is discussed in the commentary in this volume on Wiener's work in statistical physics.[1]

Motivated by [39h] and by the work cited in reference 1, the authors consider those processes for which moments exist and are sufficiently small that the expectation $E\{\exp[\alpha\mathcal{D}(A)]\}$ is, for any $A \in \mathsf{A}$, an entire function of α of exponential type. Such a limitation does not rule out models of physical systems since typically in gases and fluids the number of particles in a bounded set $A \in \mathsf{A}$ is itself bounded by a constant times the volume of A. The authors seek conditions on a pregiven family of set functions $\Phi_{n_1,\ldots,n_k}(A_1,\ldots,A_k)$ defined for collections $\{A_1,\ldots,A_k\}$ of pairwise disjoint sets in A, such that a discrete chaos exists with

$$\Phi_{n_1,\ldots,n_k}(A_1,\ldots,A_k) = E\left\{[N(A_1)]^{[n_1]} \ldots [N(A_k)]^{[n_k]}\right\},$$

where $[x]^{[n]}$ denotes the factorial product $x(x-1)\ldots(x-n+1)$. They conclude that such a chaos exists if the (putative) probabilities of all "cylinder sets" (events) $\{\mathcal{D}(A_1) = n_1 \;\&\ldots\&\; \mathcal{D}(A_k) = n_k\}$ are nonnegative. This rather weak result can be derived simply from Bochner[2] by considering the (putative) expectations of stochastic integrals ("polynomial chaoses") with respect to $\mathcal{D}(\cdot)$. More complete results can be obtained in a more general setting from Bochner[3]. The authors' results on ergodicity are of interest in applications and have not been improved upon elsewhere.

A definitive solution to the problem considered by Wiener and Wintner has not been published. Partial results were announced in reference 4; stronger results will be stated below, for their possible interest.

Given any function $E(A)$, defined for sets A in the ring A introduced above, given $k \geqslant 0$ and a collection $A_0, A_1, \ldots, A_k$ of elements of A, pairwise disjoint, we define (reference 5, paragraph 2) the quantity

$$\Delta^k E(A_0 | A_1, \ldots, A_k)$$

recursively from

$$\Delta^0 E(A_0) = E(A_0),$$

$$\Delta^{n+1} E(A_0 | A_1, \ldots, A_{n+1}) = \Delta^n E(A_0 \cup A_{n+1} | A_1, \ldots, A_n) - \Delta^n E(A_0 | A_1, \ldots, A_n).$$

A function $E(A)$, $A \in \mathsf{A}$ is called completely monotone if, for all $k \geqslant 0$ and all sets $A_0, \ldots, A_k$ of pairwise disjoint elements of A

$$(-1)^k \Delta^k E(A_0 | A_1, \ldots, A_k) \geqslant 0.$$

Theorem: Given a discrete chaos, the function

$$E(A) \equiv P\{\mathcal{D}(A) = 0\}, \quad A \in \mathsf{A},$$

has the following properties:

(a) $E(\phi) = 1$;
(b) $E(A)$ is completely monotone;
(c) If $A_1 \geqslant A_2 \geqslant \ldots$ are sets in A such that

$$A = \bigcap_{i=1}^{\infty} A_i \in a,$$

then

$$\lim_{i \to \infty} E(A_i) = E(A).$$

Conversely, given a set function $E(A)$ satisfying (a), (b), and (c), there exists a subset $X_0 \leqslant X$, and a ring A_0, namely

$$\mathsf{A}_0 = \{A \cap X_0 | A \in \mathsf{A}\},$$

and there exists a discrete chaos $\{\mathcal{D}(A) | A \in \mathsf{A}\}$ such that if $A \in \mathsf{A}_0$ then

$$E(A) = P\{\mathcal{D}(A) = 0\} . \blacksquare \tag{1}$$

In the converse here, X_0 is that part of X in which, with probability one, $\mathcal{D}(\cdot)$ has a "discrete" carrier γ. The theorem does not, and cannot, assert at this level of generality that $X_0 \neq \phi$.

To insure that indeed $X_0 = X$, one must adjoin to (a), (b), and (c) above some condition that guarantees that for each $A \in \mathsf{A}$, $\mathcal{D}(A) < \infty$ with probability one. In the spirit of the theorem as stated, the natural condition must be stated in terms defined in reference 5, paragraph 4, that we will not take the space to redefine here. The condition is

(d) For each given $B \in \mathsf{A}$, the function $E(B - A)$, as a function of $A \in \mathsf{A}$, $A \leqslant B$, is analytic absolutely monotone. (*(b)* implies that it is already absolutely monotone.)

The theorem obtained by adjoining (d) is then that (a), (b), (c), and (d) are necessary and sufficient that there exists a discrete chaos such that (1) holds for all $A \in \mathsf{A}$.

One can take another tack that introduces a moment condition: this approach also shows the relation of the theorems above to the formalism used by Wiener and Wintner.

Perfectly generally, given a discrete chaos, define

$$E(\theta;A) \equiv E\{(1-\theta)^{\mathcal{D}(A)}\} . \tag{2}$$

It is easy to see that this function is regular for $|1 - \theta| < 1$, and right continuous at $\theta = 0$. Also, to relate it to $E(A)$ above, note that

$$P\{\mathcal{D}(A) = 0\} = E(1;A).$$

The theorem below considers a special case, more general than that of Wiener and Wintner.

Theorem: Suppose that a discrete chaos is given, such that for each $A \in \mathsf{A}$ the function (2) is regular for $|1 - \theta| \leqslant 1$. Then the chaos is such that, given $A \in \mathsf{A}$, there exists a $\delta(A) > 0$ such that

$$E\ \{(1 + \delta(A))^{\mathcal{D}(A)}\} < \infty. \tag{3}$$

Furthermore, the function $E(\theta;A)$ also has the following properties:

(e) $E(\theta;\phi)\equiv 1,\ |1-\theta|\leqslant 1$;

(f) For each θ in $0\leqslant\theta\leqslant 1$, $E(\theta;A)$ is a completely monotone function of $A\in\mathsf{A}$;

(g) If $A_1\geqslant A_2\geqslant\ldots$ is a sequence of sets in A such that

$$\bigcap_{i=1}^{\infty} A_i = A\in\mathsf{A},$$

then for each fixed θ in $|\theta|\leqslant 1$

$$\lim_{i\to\infty} E(\theta;A_i) = E(\theta;A).$$

Conversely, suppose that a function $E(\theta;A)$ is given that, for each $A\in\mathsf{A}$, is regular in $|1-\theta|\leqslant 1$ and satisfies (e), (f), and (g). Then there exists a discrete chaos such that (2) holds. This chaos is such that, for each $A\in\mathsf{A}$, a $\delta(A)>0$ exists such that (3) holds. ∎

The chaoses to which this last theorem applies all meet the conditions sufficient for the calculus theorem, cited in reference 1, to hold.

The positivity conditions of Wiener and Wintner are contained in the completely monotone conditions (b) and (f). In fact, $E(\theta;A)$ is also completely monotone in θ, a fact also equivalent to positivity of the probabilities.

The numerous consistency conditions stated by Wiener and Wintner for the quantities Φ are redundant in the formulation here; they follow automatically from the fact that the Φ's can be derived by differencing operations from $E(\theta;A)$, and therefore from the quantities $\Phi_n(A)\equiv\mu_n(A^n)$ that define the Maclaurin's coefficients for $E(\theta;A)$ about $\theta=0$. For example if $A\cap B=\phi$,

$$2\mu_2(A\times B)=\mu_2\left((A\cup B)^2\right)-\mu_2(A^2)-\mu_2(B^2).$$

so that only the $\mu_n(A^n)$, hence only $E(\theta;A)$, are needed to define all the joint factorial moments Φ. This particular kind of formalism is exploited in extenso in reference 5, for quite other purposes.

References

1. B. McMillan and G. S. Deem, The Wiener program in statistical physics (Commentary on [38a], [39b, h], [40d], [43a]; These works, vol. 1 p. 654).

2. S. Bochner, *Additive set functions on groups,* Ann. Math. 40(1939), pp. 769-799.

3. S. Bochner, *Stochastic processes,* Ann. Math. 48(1942), pp. 1014-1061.

4. B. McMillan, *Random point patterns,* Abstract 49-11-265, Bull. Amer. Math. Soc., 49(1943), 865.

5. B. McMillan, *Absolutely montone functions,* Ann. Math. 60, no. 3(1954), 467-501, paragraph 2.

THE USE OF STATISTICAL THEORY IN THE STUDY OF TURBULENCE

by

NORBERT WIENER

It has been realized since the beginning that the problem of turbulence is a statistical problem; that is, a problem in which we study instead of the motion of a given system, the distribution of motions in a family of systems. For example, the measurements which TAYLOR and others have made in turbulence theory are measures of correlation coefficients between velocity in different directions at the same points or at different points. It has not, however, been adequately realized just what has to be assumed in a statistical theory of turbulence. In general, people have studied a small number of statistical coefficients and have attempted in some way to tie these up with the laws of hydrodynamics. The most thorough-going attempt of this sort is that of Professor von KARMAN. Von KARMAN arrives at an infinite sequence of non-linear equations, none of which can be solved rigorously without heavy assumptions on the character of the solution of all the subsequent equations. The purpose of this report is to indicate a systematic attack on problems of statistics in the theory of turbulence which, it is hoped, can be used practically for the purpose of computation in further work.

Behind all statistical mechanics lies GIBBS' notion of the ensemble. GIBBS studies the individual dynamical system with given initial conditions for a whole family of mechanical systems possessing the same assigned laws of force but with some sort of random distribution of the initial conditions; that is, he takes all possible sets of initial conditions, assigns to them a label which we shall call α in the present discussion, and discusses what percentage of these systems will have some assigned property. In other words, this label α is tied with a notion of probability or measure. In the first instance the parameter α may have a very varied character. It may be one-dimensional, two-dimensional or even of infinitely many dimensions but since we are concerned only about α from the point of view of measure, it is possible by a well-known mathematical theorem to represent α as a one-dimensional variable running over the values between zero and one.

Among the simplest systems which occur in the theory of turbulence are those having the property of spatial homogeneity. These have been extensively studied by TAYLOR, von KARMAN and others, and are characterized by the fact that from the standpoint of distribution any one point is exactly like any other. In strict mathematical language these systems have the property that a translation in space leads to a new labeling of variable α with conservation of measure properties. As the mathematician would put it, the group of translations in space generates a group of equi-measure transformations of α into itself.

The most important set of theorems concerning equi-measure transformations is the set of ergodic theorems. The original ergodic theorem of BIRKHOFF states that if T is an equi-measure transformation of the set values of α into itself and the function $f(\alpha)$ is integrable then for almost all values of α the average of the sequence $f(\alpha)$, $F(T\alpha)$,..., $f(T^n\alpha)$,... exists. For the study of turbulence we need a similar theorem replacing a single transformation T by the whole group of transformations corresponding to all translations of space. The proof of this theorem was not possible on the basis of the Birkhoff theorem, but this summer I have been able to give a proof that for at least for a very large class of functions f the space average of $f(T\alpha)$, where T ranges over equi-measure transformations corresponding to the translation group in space, exists for almost every α.

The importance of this theorem is that it permits us to translate averages over all systems of an ensemble into averages over all points of space for a given system in the ensemble. In a very wide class of cases known as the metrically transitive class these two averages will be identical. This fact may be used in both directions. On the one hand the averages which have been physically measured by Professor G.I. TAYLOR and his school, although they have been observed in an individual system, throw light on the statistical distribution of these same averages on all systems. On the other hand, theoretical results computed over all systems will tell us the actual averages in the individual system.

After we have built up a theory of the very general properties on such ensembles as occur in statistical mechanics, it is important to have a repertory of specific ensembles which we can describe in detail and which every average can be computed. This is done by introducing the notions (A) of the pure chaos; and(B) of a chaos not in itself pure but derived from a pure chaos. The pure chaos represents a mass distribution over the whole of space with the following properties:

1. The mass in a given region has a Gaussian distribution dependent only on the volume of the region.

2. The mass distribution in non-overlapping regions is independent. We shall represent the pure chaos independently in the number of dimensions by the symbol $P(S;\alpha)$, where S represents the set of points over which we have examined the distribution and α the label of the particular system which we have been examining. $P(S;\alpha)$ then will represent the amount of mass on the set S in the case α The distribution represented by P will in almost no case have a mass density at any point, but it is nevertheless possible to integrate with respect to α as if it had a

mass density.

It is possible to prove the following formula under appropriate definitions: if $\Sigma_1, \ldots, \Sigma_n$ do not overlap,

$$\int_0^1 P(\Sigma_1;\alpha)\ldots.P(\Sigma_{2n};\alpha)d\alpha = \Sigma\Pi\int_0^1 P(\Sigma_j;\alpha)P(\Sigma_k;\alpha)d\alpha \ ,$$

where the product sign indicates that the $2n$ terms are divided into n sets of pairs and that these factors are multiplied together, while the addition is over all the partitions of $1,\ldots,2n$ into pairs. We now define a homogeneous differentiable polynomial chaos of the nth degree in the following manner:

$$F(P;\alpha) = \int\ldots.\int f(\overline{PP}_1,\ldots.,\overline{PP}_n)d_{P1}P(S;\alpha)\ldots.d_{Pn}P(S;\alpha) \ .$$

where the formal Stieltjes integrations reduce to finite sums in the case where f is a step-function, and are extended to a more general case by an obvious mathematical artifice. It then will be possible by the formulae already given to determine the averages of all quantities depending on a homogeneous polynomial chaos or on the sum of a finite number of chaoses of this sort. These averages can be computed not merely for the whole ensemble but for almost all the individual systems in the ensemble. What gives importance to this fact is the further theorem that in a certain very definite sense we may approximate to any chaos whatever by a sum of homogeneous polynomial chaoses.

This is to a certain extent an analogue of Weierstrass' theorem in ordinary analysis.

The statistical theory of a homogeneous medium, such as a gas or liquid, or a field of turbulence, deals with the problem, given the statistical configuration and velocity distribution of the medium at a given initial time, and the dynamical laws to which it is subject, to determine the configuration at any future time, with respect to its statistical parameters. This of course is not a problem in the first instance of the history of the individual system, but of the entire ensemble, although in proper cases it is possible to show that almost all systems of the ensemble do actually share the same history, as far as certain specified statistical parameters are concerned.

The dynamical transformation of a homogeneous system have the very important properties, that they are independent of any choice of origin in time or in space. Leaving the time variable out of it, for the moment, the simplest space transformations of a homogeneous chaos $F(S;\alpha)$ which have this property are the polynomial transformations which turn it into

-6-

$$K_0 + \int K_1(x_1-y_1,\ldots x_n-y_n)F_{y_1\ldots y_n}(S;\alpha)dy_1\ldots dy_n + \ldots$$

$$\ldots + \int_{\nu}\!\!\ldots\int K_\nu(x_1-y_1^{(1)}\ldots x_n-y_n^{(1)};x_1-y_1^{(2)},\ldots x_n-y_n^{(2)};\ldots;x_1-y_1^{(\nu)}\ldots x_n-y_n^{(\nu)})\times$$

$$\times F_{y_1^{(1)}\ldots y_n^{(1)}}(S;\alpha)\ldots F_{y_1^{(\nu)}\ldots y_n^{(\nu)}}(S;\alpha)dy_1^{(1)}\ldots dy_n^{(\nu)} \qquad (1)$$

These are a sub-class of the general class of polynomial transformations

$$K_0 + \int K_1(x_1\ldots x_n;y_1\ldots y_n)F_{y_1\ldots y_n}(S;\alpha)dy_1\ldots dy_n + \ldots$$

$$\ldots + \int_{\nu}\!\!\ldots\int K_\nu(x_1\ldots x_n;y_1^{(1)}\ldots y_n^{(1)};\ldots;y_1^{(\nu)}\ldots y_n^{(\nu)})F_{y_1^{(1)}\ldots y_n^{(1)}}(S;\alpha)\ldots$$

$$\ldots F_{y_1^{(\nu)}\ldots y_n^{(\nu)}}(S;\alpha)dy_1^{(1)}\ldots dy_n^{(\nu)} . \qquad (2)$$

If a transformation of type (2) is invariant with respect to position in space, it must belong to class (1). On the other hand, in space of a finite number of dimensions and any of the ordinary spaces of an infinite number of dimensions, polynomials are a closed set of functions, and hence every transformation may be approximated by a transformation of type (2).

A polynomial transformation such as (1) of a polynomial chaos yields a polynomial chaos. If then we can approximate to the state of a dynamical system at time 0 by a polynomial chaos, and approximate to the transformation which leads to its state at time t by a polynomial transformation, we shall obtain for its state at time t, the approximation by another polynomial chaos.

On the other hand, the transformation of a dynamical system induced by its own development is infinitely subdivisible in the time, and except for linear transformations, this is not a property of polynomial transformations. Furthermore, when these transformations are non-linear, they are quite commonly not infinitely continuable in time. For example, let us consider the differential equation

$$\frac{\partial u}{\partial t} + u\frac{\partial u}{\partial x} = 0 \ . \tag{3}$$

This corresponds to the history of a space-distribution of velocity transferred by particles moving with that velocity. Its solutions are determined by the equation

$$u(x,t) = u(x - tu(x,t),0) \tag{4}$$

or if ψ is the inverse function of

$$x - tu(x,t) = \psi\left[u(x,t)\right] \ . \tag{5}$$

Manifestly, if two particles with different velocities are allowed to move long enough to allow their space-time paths to cross, $u(x,t)$ will cease to exist as a single-valued function. This will always be the case for some value of t if $u(x,0)$ is not constant, and for almost all values of t and α if it is a polynomial chaos.

By Lagrange's formula, (5) may be inverted into

$$u(x,t) = \sum_{n=0}^{\infty} \frac{(-t)^n}{n!} \frac{\partial^n}{\partial x^n}\left\{\left[u(x,0)\right]^n\right\} . \tag{6}$$

In a somewhat generalized sense, the partial sums of this represent formally polynomial transformations of the initial conditions. However, it is only for a special sort of bounded initial function, and for a finite value of the time, that they converge. It is only in this restricted case that the polynomial transformation represents a true approximation to that given by the differential equation.

It will be seen that the useful application of the theory of chaos to the study of particular dynamical chaoses involves a very careful knowledge of the existence theories of the particular problems. In many cases, such as in that of turbulence, the demands of chaos theory go considerably beyond the best knowledge of the present day. The difficulty is often both mathematical and physical. The mathematical theory may lead inevitably to a catastrophe beyond which there is no continuation, either because it is not an adequate presentation of the physical facts, or because after the catastrophe the physical system continues to develop in a manner not adequately provided for in a mathematical formulation which is adequate up to the occurrence of the catastrophe, or lastly, the catastrophe does really occur physically, and the system really has no subsequent history. The hydrodynamical investigations required in the case of turbulence are directly in the spirit of the work of OSEEN and LERAY, but must be carried much further.

The study of the history of a mechanical chaos will then procede as follows: we first determine the transformation of initial conditions generated by the dynamics of the ensemble. We then determine under what assumptions the initial conditions admit of this transformation for either a finite or an infinite interval of time. Then we approximate to the transformation for a given range of values of the time by a polynomial transformation. Then, having regard to a definition of distance between two functions determined by this transformation, we approximate to the initial chaos by a polynomial chaos. Next we apply the polynomial transformation to the polynomial chaos, and obtain an approximating polynomial chaos at time t . Finally, we apply our algorithm of the pure chaos to determine the averages of the statistical parameters of this chaos, and express them as functions of the time.

The results of such an investigation belong to a little-studied branch of statistical mechanics: - the statistical mechanics of systems not in equilibrium.

To study the classical, equilibrium theory of statistical mechanics by the methods of chaos theory is not easy. As yet we lack a method of representing all forms of homogeneous chaos, which will tell us by inspection when two differ merely by an equimeasure transformation of the parameter of distribution. In certain cases, in which the equilibrium is stable, the study of the history of a system with an arbitrary initial chaos will yield us for large values of t an approximation to equilibrium, but this will often fail to be so, particularly in the case of differentiable chaoses, or the only equilibrium may be that in which the chaos reduces to a constant.

133. Norbert Wiener and Brockway McMillan: *A new method in statistical mechanics.*

The calculus of the discrete homogeneous chaos is developed and extended to permit applications in the field of statistical mechanics. In particular, as a model for other applications, a discrete homogeneous chaos $M(P; \alpha)$, representing a distribution of phase points in a six-dimensional phase space, is used as initial conditions for the dynamical equations of an infinite monatomic gas with intermolecular forces. The transformation in time of the chaos in accordance with the laws of dynamics enables one to calculate the time derivatives of linear, quadratic, and higher degree functionals of the chaos. These are at time $t=0$ simply quadratic and higher degree functionals of $M(P; \alpha)$, and this permits a large class of statistical parameters of the gas to be expanded, at least formally, in power series in the time. The ergodic theorem enables one to calculate these statistical parameters for "almost all" dynamical systems in the ensemble in question by a simple averaging over the α of the initial chaos. Methods of extending these statistical parameters to their equilibrium values at $t=\infty$ will then give information about the physical gas. This extension is made for the mechanical analogue of temperature. (Received January 26, 1939.)

133. Norbert Wiener: *A canonical series for symmetric functions in statistical mechanics.*

A symmetric function of a set of particles can under certain conditions be expanded in a series of which the nth term is an n-tuple sum over the particles. This method is used to express symmetric functions of the positions of a set of molecules of a monatomic gas at time t in terms of the positions at time 0. The distribution function of the particles is deduced. (Received November 22, 1939.)

The Wiener Program in Statistical Physics. Commentary on [38a], [39b,h], [40d], [43a]

B. McMillan and G. S. Deem

Introduction

It is clear in the content of his work, from 1923 ("Differential space" [23d]) until its close, that Norbert Wiener was intensely interested in statistical problems deriving from physics. From his writings, his lectures, and his personal conversation, one saw that this interest went beyond the mathematical structures he examined, to the physics that these structures reflected or modeled. Like his nineteenth century predecessors Maxwell and Boltzmann, he was impressed by the statistical regularities exhibited by a chamber of gas containing of the order of 10^{23} molecules. He saw these as regularities imposed by inescapable mathematical laws, whose working out could be thwarted only by almost universal comspiracy among the molecules.* He forged novel mathematical tools for modeling such phenomena and made repeated efforts to use these tools to probe the physical consequences of a variety of statistical systems.

This essay is based on the period 1938-1941 of Wiener's work. During that time "The homogeneous chaos" [38a] appeared, and the work was done that resulted in the "Ergodic theorem" [39a] and the "Discrete chaos" [43a]. Our aim is to show how the work of these papers, and subsequent related work, bears on the study of the statistical phenomena Wiener was seeking to understand. Our approach is to suggest motivation, by anecdote and by reference to specific writings, to describe Wiener's own efforts at application, with which one of us was associated, and to indicate applications that have been made by others including, independently, each of us.

"The homogeneous chaos" is the primary document of this period. Applying the somewhat idiosyncratic methods of "Generalized harmonic analysis" [30a] to the problem of defining a stochastic process, this paper constructs what is now called a "multidimensional Wiener process" and develops the calculus of multiple stochastic integrals with respect to that process. This paper clearly reflects preoccupations already evident in 1923, in "Differential space" [23a], ideas that are now generalized and whose usefulness is significantly expanded by adjunction of the "weak approximation theorem."

*As, indeed, occurs in some quantum systems at low temperatures, leading to such bizarre phenomena as superconductivity and superfluidity.

The weak approximation theorem gives conditions on an arbitrary random process which insure that it can be arbitrarily closely approximated by a finite sum of multiple stochastic integrals. This theorem is central to the concerns of our essay because it provides the mathematical basis for representing more general random processes in terms of well-defined functionals of the mathematically more tractable multidimensional Wiener process. Explicit in paragraph 13 of "Homogeneous chaos" is Wiener's own belief that by these means one might solve problems of fluid mechanics, of the statistical theory of turbulence, and of statistical mechanics.

In the following, we divide this "Wiener program" into two lines of inquiry—one concerned with the theory of turbulence, in which the Wiener process and its functional calculus are basic, and the other directed toward statistical mechanics in which a simpler process, the discrete chaos, plays an analogous role.

The Homogeneous Chaos and Turbulence Theory

Wiener's interest in the problem of fluid dynamical turbulence spanned over forty years of his life, preceding even his first major mathematical contribution on Brownian motion. His long association with the turbulence problem recalls that portion of his autobiography [56g, p. 42] where he states,

> If the career of a mathematician is to be anything but an anti-climactic one, he must devote [his] brief springtime of top creative ability to the discovery of new fields and new problems, of such richness and compelling character that he can scarcely exhaust them in his lifetime. It has been my good fortune that the problems which excited me as a youth, and which I did a considerable amount to initiate, still do not seem to have lost their power to make maximum demands on me in my sixtieth year.

Wiener's impact on the theory of fluid dynamical turbulence can be judged only in the context of the rich and continuing impetus he has passed along to later investigators. Wiener's only formal contributions in this area are found in two papers, "The homogeneous chaos" [38a] and "The use of statistical theory in the study of turbulence" [39b], and in his book *Nonlinear problems in random theory* [58i]. Even in these works, references to the problem of turbulence are in the form of proposals, involving application of multiple stochastic integrals; specific dynamical systems are not considered. Only in the years following Wiener's death in 1964 has any significant headway been made in understanding and developing some of the consequences of his ideas.

Wiener's interest in the theory of turbulence began in 1921 when he read G. I. Taylor's now classical paper on turbulent diffusion entitled "Diffusion by continuous movements."[1] In that period of his life, after his arrival at MIT in 1919, Wiener had developed an awareness that

> . . . it was within nature itself that I must seek the language and the problems of my mathematical investigations. [56g, p. 33]

Moreover, the random trajectories of turbulently diffusing air particles described in Taylor's paper reminded Wiener of his own current interests on functional integration with respect to families of curves:

> . . . the problem of turbulence was too complicated for immediate attack, but there was a related problem which I found to be just right for the theoretical considerations of the field which I had chosen for myself. This was the problem of the Brownian motion, and it was to provide the subject of my first major mathematical work. [56g, p. 37]

Wiener's interests in turbulence reemerged in his paper "The homogenous chaos" [38a]. The importance of this paper, with regard to the theory of turbulence, is the development of the pure multidimensional chaos, a generalization of the one-dimensional Brownian motion random process (Wiener process), and the construction of multiple stochastic integrals with respect to the pure chaos.

Wiener's concept of a homogeneous chaos corresponds to what is now called a *stationary random measure.*[2] A functional calculus which readily generalizes to functions on higher dimensional spaces, as required for multidimensional turbulence theory,[3] can be constructed as follows: let $C[0, 1]$ denote the space of continuous functions $\xi(\cdot)$ on $[0, 1]$ satisfying $\xi(0) = 0$, let $x_1, \ldots, x_m$ be an increasing sequence of points in $[0, 1]$, and let E be a Lebesgue measurable subset of $[0, 1]^m$. The corresponding interval in C is defined as the set of $\xi(\cdot)$ which assume values in E according to

$$I(m; x_1, \ldots, x_m; E) = \{\xi(\cdot): [\xi(x_1), \ldots, \xi(x_n)] \in E\}. \tag{1}$$

The intervals constitute a field of sets in C and generate a smallest σ-field B. With x a point in $[0, 1]$, define

$$p(\xi; x) = (2\pi x)^{-1/2} \exp\left[-\frac{\xi^2}{2x}\right]. \tag{2}$$

The measure of an interval is given by the following conditional probability that $[\xi(x_1), \ldots, \xi(x_m)]$ belongs to E:

$W[I(m; x_1, \ldots, x_m; E)]$

$$= \int_E p(\xi_1; x_1) p(\xi_2 - \xi_1; x_2 - x_1) \ldots p(\xi_m - \xi_{m-1}; x_m - x_{m-1}) d\xi_1 \ldots d\xi_m . \tag{3}$$

W is countably additive on the field of intervals (Wiener), so it possesses a unique extension to B. W is *Wiener measure* for (C, B). The above construction can be generalized to $C(-\infty, \infty)$ and to higher dimensions. In the following C will denote the space of continuous functions defined on $x \in R^n$.

The probability space (C, B, W), or less precisely the random process $\{\xi(\cdot)\}$, induces random variables $\{\xi(x), \text{fixed } x\}$. These satisfy

(i) for fixed points x_1 and x_2, $\{\xi(x_1)\}$ and $\{\xi(x_2)\}$ are independent random variables,

(ii) for fixed x, $\{\xi(x)\}$ is a Gaussian random variable with mean zero and variance equal to the volume of the n-dimensional cube with vertices at 0 and x.

Existence of the measure space (C, B, W) with intervals (1) is sufficient to define a linear integral operator with respect to a class of Wiener summable functionals $F[\xi(\cdot)]$ on C. This *Wiener integral* is constructed, using (3), from the characteristic functionals with respect to intervals (the characteristic functional for interval I is unity if $\xi(\cdot)$ lies in I, zero otherwise). The Wiener integral of a Wiener summable functional F taken over all C defines the expectation of F:

$$< F[\xi(\cdot)] > \equiv \int_C^W F[\xi] \, d_W \xi . \tag{4}$$

Properties (i) and (ii) permit the construction of integral, polynomial functionals on C. The simplest is the linear stochastic integral

$$\int \phi^{(1)}(\mathrm{x}_1) \, d\xi(x_1) \tag{5}$$

where $\phi^{(1)}(\mathrm{x}) \in L_2(R^n)$. It is easily seen[4] that (5) is a Gaussian random process with respect to Wiener measure.

Higher order, non-Gaussian, homogeneous polynomial functionals are similarly defined. For example let $\phi^{(m)}(\mathrm{x}_1, \ldots, \mathrm{x}_m) \in L_2(R^{m \times n})$, and denote the *homogeneous polynomial functionals*

$$\int \phi^{(m)}(x_1, \ldots, x_m)\, d\xi(x_1) \ldots d\xi(x_m). \quad (6)$$

An important conclusion given in "The homogeneous chaos" (Sec. 12) is that the homogeneous polynomial functionals (6) are everywhere dense in the space of Wiener summable functionals (weak approximation theorem). More precisely, any Wiener square-summable functional may be approximated arbitrarily closely, in the mean-square sense with respect to expectation, by a finite sum of homogeneous polynomial functionals.[5]

In the final section of "The homogeneous chaos," Wiener proposed that at least for bounded time intervals, suitable sums of homogeneous polynomial functionals might be used to approximate the dynamical evolution of more general random processes, for example, the velocity fields $\{\mathbf{u}\}$ evolving from random initial conditions according to the incompressible Navier-Stokes equations.

Wiener restated this proposal before the International Congress of Mathematicians (1938) in his paper entitled "The use of statistical theory in the study of turbulence" [39b]. He drew only one small response in a review article[6] of the following year:

> Behind the seeming innocence of this title there lies in this paper the beginning of a more general and systematic method of attack on the important problem of turbulence.

The theory of fluid turbulence was not sufficiently developed at that time for the application of Wiener's ideas. And indeed, the use of the ordinary polynomial chaos was operationally complicated and inappropriate for the study of high Reynolds number turbulence. The major developments in the theory until 1958 are contained in Wiener's book *Nonlinear problems in random theory* [58i]. The work of Cameron and Martin[7] and Kakutani[8] (after a suggestion of J. von Neumann) led to the introduction of the Wiener-Hermite[9] polynomial development of Wiener square summable functionals in terms of sums of statistically *orthogonal* polynomial functionals.

To discuss the orthogonal development of multidimensional, *vector-valued* random processes, which occur in turbulence theory, it is convenient to extend[10] the above construction of $\{\xi(\cdot)\}$ to a vector-valued random process $\{\boldsymbol{\xi}(\cdot)\}$. Properties (i) and (ii) are preserved, with the exception that for fixed x and distinct vector indices $i \neq j$, $\{\xi_i(x)\}$ and $\{\xi_j(x)\}$ are uncorrelated random variables. It is also operationally convenient to introduce the vector, white-noise Gaussian random process $\{\mathbf{H}^{(1)}(\cdot)\}$ formally[11] according to

$$\xi(x) = \int_0^x H^{(1)}(x')d^n(x') . \tag{7}$$

It satisfies, cf. (4),

$$\langle H_i^{(1)}(\mathrm{x}_1)\rangle = 0 , \tag{8}$$

$$\langle H_i^{(1)}(\mathrm{x}_1)H_j^{(1)}(\mathrm{x}_2)\rangle = \delta_{ij}\,\delta(\mathrm{x}_1 - \mathrm{x}_2) , \tag{9}$$

with higher order moments determined by Gaussianity. In (9), $\delta(\mathbf{x})$ denotes the Dirac distribution. The statistical properties of the higher-order polynomial functionals are recovered by replacing $d\xi(x) \to \mathbf{H}^{(1)}(x)dx$ in the generalized version of (6), and using the above properties of $\mathbf{H}^{(1)}$.

To construct the Wiener-Hermite polynomial functionals on C, the space of continuous vector-valued functions of $\mathbf{x}$, define $H^{(0)} \equiv 1$ and determine $H^{(m)}_{i_1 \ldots i_m}(\mathrm{x}_1, \ldots, \mathrm{x}_m)$ as an m-degree polynomial in $H^{(l)}_{i_1 \ldots i_l}(\mathrm{x}_1, \ldots, \mathrm{x}_l)$, $l \leqslant m$, requiring statistical orthogonality of distinct $H^{(m)}_{i_1 \ldots i_m}$ in the sense that

$$\langle H^{(l)}_{i_1 \ldots i_l}\; H^{(m)}_{j_1 \ldots j_m}\rangle = 0 , \quad l \neq m . \tag{10}$$

For example

$$H^{(2)}_{ij}(\mathrm{x},\mathrm{x}') = H^{(1)}_i(\mathrm{x})H^{(1)}_j(\mathrm{x}') - \delta_{ij}\,\delta(\mathrm{x} - \mathrm{x}') . \tag{11}$$

For a function ϕ of m variables, define

$$G^{(m)}_{j_1 \ldots j_m}[\phi, C]$$

$$= \int \phi^{(m)}(\mathrm{x}_1, \ldots, \mathrm{x}_m) \cdot H^{(m)}_{j_1 \ldots j_m}(\mathrm{x}_1, \ldots, \mathrm{x}_m)\, d\mathrm{x}_1 \ldots d\mathrm{x}_m . \tag{12}$$

The denseness of the ordinary homogeneous polynomial functionals in the space of Wiener square-summable functionals, and the statistical orthogonality of the Wiener-Hermite polynomial functionals imply the following representation theorem:

Let $\mathrm{F}[\xi(\cdot)]$ *be a vector-valued Wiener square-summable functional on C. Except for a set of* $\xi(\cdot)$ *having zero Wiener measure, there exist unique*

Lebesgue square-integrable functions $\phi^{(m)}_{i|j_1\ldots j_m}(\mathbf{x}_1,\ldots,\mathbf{x}_m)$, *symmetric in their m arguments, such that*

$$\lim_{N\to\infty}\left\langle\left|F_i[\xi(\cdot)] - \sum_{m=0}^{N} G^{(m)}_{j_1\ldots j_m}\left[\phi^{(m)}_{i|j_1\ldots j_m}, C\right]\right|^2\right\rangle = 0 \tag{13}$$

(summation convention over repeated subscripts). *For N finite, the lower order* $\phi^{(m)}$ *which minimize the above expression remain unchanged as N is increased.*

If F is dependent on additional parameters, such as $\mathbf{x}$ or a time t, as for an ensemble $\{\mathbf{u}(\mathbf{x},t)\}$ of turbulent velocity fields satisfying the Navier-Stokes equations, these can be included as additional arguments in the basis functions ϕ, [38a].[12] In particular, if $\{\mathbf{u}(\mathbf{x},t)\}$ is statistically homogeneous in $\mathbf{x}$, its Wiener-Hermite expansion (13) can be written[13]

$$u_i(\mathbf{x},t) = \sum_{n=0}^{\infty} G^{(n)}_{j_1\ldots j_n}\left[\phi^{(n)}_{i|j_1\ldots j_n}, T^{\mathbf{x}}C\right] . \tag{14}$$

The $\phi^{(n)}_{i|j_1\ldots j_n}(\mathbf{x}_1,\ldots,\mathbf{x}_n;t)$ are spatially square integrable kernel functions, determined so that the individual $\mathbf{u}$ satisfy the dynamical equations of motion. T^x is the Wiener measure preserving transformation on C which carries functions $\xi(x')$ into $\xi(x'-x)$, where $x'-x$ is the translation of points x' by x.

The first application of Wiener's ideas to fluid dynamical turbulence appears to have been made by Poduska (1962).[14] This work was carried out using the original homogeneous polynomial functionals, for which there is no convergence theorem analogous to (13).

In a series of papers beginning in 1964, Meecham and his coworkers applied (14) first to the one-dimensional Burger's equation,[15] and then to the full[16,17] Navier-Stokes equations. The demand that individual realizations $\mathbf{u}$ in (14) satisfy the Navier-Stokes equations leads to a closure problem for the $\phi^{(n)}$. In all applications of the theory, it has been necessary to assume that (14) is rapidly convergent about the $G^{(1)}$, Gaussian term.

Orszag and Bissonnette[18] and Crow and Canavan[19] have shown that at high Reynolds numbers, the expansion (14) is likely to be slowly convergent for times much longer than the initial time. This possibility was actually anticipated by Wiener in 1958, when he wrote

As time goes on, the relation of the parameters of the changing system to a fixed [$\{\xi(\mathbf{x})\}$] becomes more and more remote. To describe the dynamics of the system in these terms is to refer its randomness to an epoch that is continually escaping us in the past. Such a description is not too well suited to the employment of differential-equation theory, where the discussion of each moment in the system is complete in itself and we are not perpetually hovering between two instants. [58i, p. 122]

More recently, work by Bodner,[20] Canavan,[21, 22] and Doi and Imamura[23] has attempted to resolve the above difficulty by allowing the ensemble $\{\xi(\mathbf{x})\}$ itself to evolve according to a time-dependent, Wiener measure-preserving transformation S^t of C onto itself:

$$u_i(\mathbf{x},t) = \sum_{n=0}^{\infty} G^{(n)}_{j_1 \ldots j_n} \left[\phi^{(n)}_{i|j_1 \ldots j_n}, S^t T^{\mathbf{x}} C\right]. \tag{15}$$

In Bodner's work, for example, the transformation S^t is chosen so that exactly Gaussian, inviscid, equipartition solutions[24] to the Navier-Stokes equations can be described by the $n = 1$, Gaussian term in (15). The expectation here is that expansion (15) is more rapidly convergent, at least for large Reynolds number turbulence. It must be emphasized that the choice of S^t is somewhat arbitrary, as emphasized by the differences in the theories in references 20 through 23.

Saffman (1969)[25] has applied a version of (15) to describe the diffusion of a passive scalar convected in a turbulent flow, the original problem discussed by Taylor[26] which introduced Wiener to the problem of turbulence. It is interesting that Saffman is able to arrive at many of Taylor's results using expansion (15) truncated after only the $n = 2$ terms.

The use of Wiener's statistical theory in the problem of turbulence is still far from complete. The convergence problems mentioned above and the general usefulness of expansion (15) will most likely be answered only by further advances in computer technology, needed to treat the complex dynamical equations for the kernel functions $\phi^{(n)}$. The advantages of Wiener's approach over other theories continue to offer promise for its eventual success:

1. It is "honest"[27] in the sense that physically based, intuitive approximations are not built into the mathematical foundations;
2. The convergence of the expansions (15) is assured, giving a rational means for approximating actual turbulence ensembles $\{\mathbf{u}\}$ satisfying the Navier-Stokes equations;

3. The expansion (15) is a representation of the random process $\{u\}$ itself; it yields information concerning the probability distribution function for $\{u\}$ and not merely a few statistical moments. The truncated version of (15) is a realizable random process whose statistics approximate those of $\{u\}$.

The Discrete Chaos and Statistical Mechanics

Section ll of "Homogeneous chaos" introduces the concept of the "discrete chaos," or more specifically, the Poisson chaos. Apparently Wiener hoped to use the Poisson chaos as a basic random process, in terms of which more general discrete random processes, such as those in statistical mechanics, could be expressed, in analogy to the use of the Wiener process above. The idea was not developed, and later work has taken a different turn, as we shall see.

We shall first define a discrete chaos. Quite generally, let X be a space and let $\mathcal{A}$ be a countable ring of subsets of X with property that, given any point $x \in X$, there exists a sequence $A_1 \supseteq A_2 \ldots$ of sets from $\mathcal{A}$ such that

$$x = \bigcap_{i=1}^{\infty} A_i \, .$$

A subset $\gamma \subseteq X$ will be called a "point pattern" if, for each $A \in \mathcal{A}$, the set $\gamma \cap A$ has a finite, possibly zero, number of points in it. Let Γ be the class of all point patterns γ. Given $\gamma \in \Gamma$, define the set function $\mathcal{D}(A;\gamma)$, a function of sets $A \in \mathcal{A}$, by

$$\mathcal{D}(A;\gamma) = \text{the number of points in } \gamma \cap A.$$

An "interval" in Γ is a subset of Γ defined by an integer $k > 0$, a set of nonnegative integers $n_1, \ldots, n_k$, and a set $\{A_1, \ldots, A_k\}$ of pairwise disjoint elements of $\mathcal{A}$. The corresponding interval is the set of all $\gamma \in \Gamma$ such that

$$\mathcal{D}(A_i;\gamma) = n_i \, , \; 1 \leqslant i \leqslant k \, .$$

A discrete chaos can be defined as a probability measure on Γ such that each interval is measurable. In this formulation, for every $\gamma \in \Gamma$, $\mathcal{D}(\cdot;\gamma)$ can be extended to be a countably additive, nonnegative set function defined (though possibly infinite) for all subsets $\Sigma \subseteq X$; in particular, it is defined for all sets Σ in the Borel field $\mathcal{B}$ determined by $\mathcal{A}$.

Thus defined, a discrete chaos is a stochastic process with probability label space Γ, based on the jointly distributed random variables $\{\mathcal{D}(A;\cdot)|A \in \mathcal{A}\}$. For $\gamma \in \Gamma$, the sample function $\mathcal{D}(\cdot;\gamma)$ is a countably additive set function of a particularly simple form. As in Section 2, when considering the sample functions $\mathcal{D}(A;\gamma)$ we usually drop reference to the label γ.

In the "Homogeneous chaos", X is a Euclidean space of finite dimension and $\mathcal{A}$ can be taken to be the ring generated by the closed bounded parallelepipeds with faces parallel to the coordinate planes, having corners whose coordinates are terminating binary numbers.

Wiener starts with this geometry, but fixes attention on sets $S_{m,n} \in \mathcal{A}$ that make up a family of coverings of X. He considers an abstract space Γ^* of "schedules of occupiedness and emptiness" which contains Γ. He postulates

(W1) If $S_{m,n}$ and $S_{p,q}$ are disjoint, then the events $\gamma \cap S_{m,n} = \phi$ and $\gamma \cap S_{p,q} = \phi$ are independent.

(W2) The probability of the event $\gamma \cap S_{m,n} = \phi$ is a function only of the Lebesgue measure, $\mu(S_{m,n})$, of $S_{m,n}$.

He then shows from these postulates that there exists a probability measure on Γ^*, such that the outer measure

$$\overline{P}\{\Gamma^* - \Gamma\} = 0\,,$$

and such that P, restricted to Γ, defines a discrete chaos. Moreover there exists a constant $a > 0$ such that for any $B \in \mathcal{B}$, and $k = 0, 1, 2, \ldots,$

$$P\{\mathcal{D}(B) = k\} = \frac{[a\mu(B)]^k}{k!}\, e^{-a\mu(B)}\,.$$

Hence the name "Poisson chaos."

Wiener's theorem that (W1) and (W2) imply the existence of a Poisson chaos is an example of a general fact. A discrete chaos is determined by the events $\gamma \cap A = \phi$, $A \in \mathcal{A}$, and their probabilities. Alternatively, it is characterized* by the set function

*McMillan[28] gives this for the case where $E\{z^{\mathcal{D}(A)}\} < \infty$ for $A \in \mathcal{A}$.

$$E(A) = P\{\mathcal{D}(A) = 0\}$$

defined for sets $A \in \mathcal{A}$.

Because a discrete chaos $\{\mathcal{D}(B)\}$ is nonnegative and countably additive, the calculus of stochastic integrals such as

$$\int f_1(x)\ \mathcal{D}(dx)$$

$$\iint f_2(x_1, x_2)\ \mathcal{D}(x_1)\ \mathcal{D}(dx_2)$$

is vastly simpler than in the case of the Wiener process. In general, a calculus exists for any discrete chaos such that, for each $A \in \mathcal{A}$, all of the moments of $\mathcal{D}(A)$ are finite.* Given such a chaos, one finds that there exists a sequence of nonnegative measures $\{\mu_n\}$, μ_n being a measure on the Borel field $\mathcal{B}^n$ of subsets of X^n determined by the Cartesian products $B_1 \otimes, \ldots, \otimes B_n$, $B_i \in \mathcal{B}$, $1 \leqslant i \leqslant n$ (or simply $B_i \in \mathcal{A}$). These measures have the further property that

$$\mu_n(Z_n) = 0,\ \ n = 1, 2, \ldots,$$

where $Z_1 = \phi$ and, for $n \geqslant 2$, Z_n is that subset of X consisting of all points $\{x_1, \ldots, x_n\}$ such that, for some $1 \leqslant i < j \leqslant n$,

$$x_i = x_j\,.$$

The calculus is then completely determined by the following *Theorem:* Suppose that $f_n(x_1, \ldots, x_n)$ is defined on X^n; that if $\{x_1, \ldots, x_n\} \in Z_n$, then $f_n(x_1, \ldots, x_n) = 0$; and that

$$\int \cdots \int\ |f_n(x_1, \ldots, x_n)|\ dx_1 \ldots dx_n$$

exists, where dx is Lebesgue measure in X. Then with probability one

$$\int \cdots \int f_n(x_1, \ldots, x_n)\ \mathcal{D}(dx_1) \ldots \mathcal{D}(dx_n) \qquad (16)$$

exists and is equal to the random variable

$$F(\gamma) = \sum_{x_1 \in \gamma} \ldots \sum_{x_n \in \gamma} f_n(x_1, \ldots, x_n)\,.$$

*McMillan[29] covers the Poisson case.

The series defining $F(\gamma)$ converges absolutely with probability one.

Furthermore, the expectation is

$$E\{F(\gamma)\} = \int \cdots \int f_n(x_1, \ldots, x_n)\; dx_1 \ldots dx_n \; . \quad \blacksquare \tag{17}$$

It follows as a simple corollary that, for $A \in \mathsf{A}$, the nth factorial moment

$$E\{\mathcal{D}(A)[\mathcal{D}(A)-1] \cdots [\mathcal{D}(A)-n+1]\} = \mu_n(A^n)\,. \tag{18}$$

This identifies the μ_n.

To our knowledge, this calculus has not been published. For the case in which the factorial moment generating function

$$E(\theta;A) = E\{(1-\theta)^{\mathcal{D}(A)}\} = \sum_{n=0}^{\infty} \frac{(-\theta)^n}{n!}\; \mu_n(A^n)$$

is entire in θ for each $A \in \mathsf{A}$, it was certainly known to Wiener; the knowledge is implicit in "The discrete chaos" [43a] and was tacit in much of the work reported below. Reference 28 applies to the more general case in which $E(\theta;A)$ is regular for $|\theta| \leqslant 1$. It follows rather easily in this case that the polynomial chaoses, i.e., those random variables of which (16) is the prototype homogeneously of degree n, are dense in L^p, $p \geqslant 1$. This is the analogue of the weak representation theorem for the Wiener process, here true for a board class of underlying (discrete, nonnegative) processes.

In the spring of 1938 one of us (McM), having, as he believed, mastered "Differential space" [23d] and "Generalized harmonic analysis" [30a], sought and received permission from Wiener to do a thesis under him. Wiener spoke briefly about the possibility of work on the "shot effect" in electron tubes, or in statistical mechanics, and suggested preparatory reading from a reference on the shot effect that is now lost to memory. This is our first evidence of Wiener's interest, specifically, in applying the discrete chaos to physical problems.* With it was initiated the work on statistical mechanics outlined below.

From the fall of 1938 through July of 1939, McMillan and Wiener† at-

*See also the footnote on page 925 of the "Homogeneous chaos" [38a], referring to conversations with J. V. Neumann.

†At times during June and July, both E. R. VanKampen and Manuel Vallarta were interested, albeit skeptical observers.

tempted to exploit the calculus of stochastic integrals and their averages, as just outlined, in the first instance simply for the Poisson case, to develop formulas in the equilibrium statistical mechanics of a gas filling an infinite space. For this purpose, the space X above is replaced by the Euclidean phase space $X \otimes U$, where X is position space and U a Euclidean velocity space of the same dimensionality as X. In place of Lebesgue measure, one uses the measure which, for the product set $S \otimes T$, $S \subseteq X$, $T \subseteq U$, takes the value

$$\lambda\theta^{p/2}\,\mu(S)\int_T \exp\left\{-\pi\theta u\cdot u\right\}\,du\,.$$

Here $\lambda \geqslant 0$ is a parameter of the ensemble, namely, its particle density in X, $\theta \geqslant 0$ is another (specifically, $\pi\theta = (kT)^{-1}$, where T is the temperature and k is Boltzman's constant), and p is the dimensionality of U-space, so that

$$\theta^{p/2}\int_U \exp\left\{-\pi\theta u\cdot u\right\}\,du = 1\;;$$

here $(u \cdot u)^{1/2}$ is the Euclidean length of the vector from the origin in U to the point u.

Physically, a system of particles is imagined whose initial states are drawn from a given Poisson chaos in $X \otimes U$, described by the parameters λ, θ above. That is, at time $t = 0$ each sample point pattern $\gamma \subseteq X \otimes U$ represents a possible collection of particles with their respective phase points at the $x_n \otimes u_n \in \gamma$. Particles at $x_1 \otimes u_1$ and $x_2 \otimes u_2$ influence each other through a force field defined by a two-body potential $\phi(x_1, x_2)$ depending only on the spatial coordinates x_1, x_2. It is assumed that $\phi(x_1, x_2)$ is spherically symmetric and translation-invariant in X; i.e.,

$$\phi(x_1, x_2) = \phi(|x_1 - x_2|) = \phi(x_2, x_1)\,.$$

From time $t = 0$ on, the force field applies, and the pattern γ evolves through configurations, $\gamma(t) \in X \otimes U$ for each t, determined by the laws of dynamics.

The first calculations employed direct continuation by power series in t. Consider, for example, the simplest polynomial chaos, the random variable

$$F\Big(\gamma(t)\Big) = \sum_{x\otimes u\in\gamma(t)} f(x, u)\,.$$

Suppose that f vanishes identically outside of a bounded sphere in $X \otimes U$. Formally the laws of dynamics imply that

$$\frac{dF}{dt} = \sum_{x \otimes u \in \gamma(t)} u \cdot \nabla_x f(x, u)$$

$$+ \sum_{\substack{x_1 \otimes u_1 \in \gamma(t) \\ x_1 \otimes u_1 \neq x_2 \otimes u_2}} \sum_{x_2 \otimes u_2 \in \gamma(t)} K(x_1 - x_2) \cdot \nabla_u f(x, u) .$$

Here

$$K(x_1 - x_2) = - \nabla_{x_1} \phi(x_1, x_2)$$

is the interparticle force and ∇_x, ∇_u are gradient operators in X, U respectively.

At time $t = 0$, $F(\gamma)$ is homogeneous of degree one in $\mathcal{D}(\cdot)$, and dF/dt of degree two. Continuing, $d^n F/dt^n$ is of degree $n + 1$ in $\mathcal{D}$; f appears only to degree one, and K (hence ϕ) appears to degree n.

One formally has

$$F\left(\gamma(t)\right) = F\left(\gamma(0)\right) + \sum_{n=1}^{\infty} \frac{t^n}{n!} \left\{ \frac{d^n}{dt^n} F\left(\gamma(t)\right) \right\}_{t=0} . \tag{19}$$

The nth term on the right is a polynomial in $\mathcal{D}$ of degree $n + 1$, and involves kernels or integrands that are derivatives of ϕ and f of degree one in f. The calculus of averages gives a formula for its expectation. The result, after suitable integrations by parts (applications of Green's formula) is an expression linear in f (without derivatives), and of degree n in ϕ (without derivatives) having coefficients that are polynomials in θ and λ, of degree $n + 1$ in the latter. The expressions involve multiple integrals of ϕ. They resemble expressions used by others in applying the classical Gibbs approach to an infinite gas.

The abstract [39h] dates from about this stage of the calculations. No real results were in fact known, and the paper, which was given only orally, simply amounted to an elaboration of what is given in the last few paragraphs above, cf. [39h].

At about this time, an obvious fact became evident to both collaborators, namely that the result of taking the expectation of (19), which is, as we have already noted, of the form

$$\int dx \int du \;\; \rho_1(\lambda;\theta;t;x,u) f(x,u) \,, \tag{20}$$

is at least formally the expectation of $F(\gamma(t))$ with respect to a discrete chaos—that at time t—whose μ_1 has a density $\rho_1(\lambda;\theta;t;x;u)$ with respect to $dx\, du$ in $X \otimes \mathsf{U}$. A similar interpretation applies to higher order polynomials. These observations brought about a distinct change in point of view. First, the functions such as f in the example above ceased to be of physical importance themselves, but became, as elements of a conjugate space, merely tools for studying an operator by means of its adjoint. The densities

$$\rho_n(\lambda;\theta;t;x_1,u_1;\ldots;x_n,u_n) \,, \tag{21}$$

that define the μ_n of the chaos at time t, became the true quantities of physical interest. Second, the Poisson chaos remained a tool but was not necessarily a unique tool; in particular, one no longer sought to define random variables at time t in terms always of stochastic integrals over a Poisson chaos, one simply considered them as quantities measurable on the chaos at time t, and then sought to characterize that chaos by means of its densities (21).

This change in point of view brought forth a new, general, mathematical problem, distinct from that of calculating the densities (21) which describe a physical gas, namely, the problem of characterizing those sequences of functions (21) that validly define a discrete chaos. This can be regarded as an extended version of the moment problem, to find conditions on a sequence of measures μ_n, respectively on X^n, such that there exists a discrete chaos for which (18) holds for all $A \in \mathsf{A}$. This problem is the subject of the later Wiener and Wintner paper "Discrete chaos" [43a[, discussed elsewhere in this Volume, and was the genesis of the abstract (reference 28) cited earlier.

It was also observed at this time that the densities (21) describing a physical system, not necessarily in equilibrium, satisfy a sequence of equations generalizing Liouville's equation:

$$\frac{\partial \rho_n}{\partial t} + \sum_{i=1}^{n} u_i \cdot \nabla_{x_i} \rho_n + \sum_{i=1}^{n} \sum_{\substack{j=1 \\ j \neq i}}^{n} K(x_i - x_j) \cdot \nabla_{u_i} \rho_n$$

$$+\sum_{i=1}^{n} \int_X dx_{n+1} \int_U du_{n+1}\, K(x_i - x_{n+1}) \cdot \nabla_{u_i} \rho_{n+1} = 0. \qquad (22)$$

Here the ρ_n and ρ_{n+1} are functions as described by (21).

The collaborators recognized that, could they find time-invariant solutions to (22) that met whatever (then unknown) consistency conditions required to define a chaos, their problem would be solved. These ideas were set aside at the time, however, and calculations continued, generally along the original plan.

Using a device suggested by Wiener, too elaborate to include here, it was possible to group the expectation of (19) to an infinite sum of explicit functions of t, in each of which one could in fact take the limit as $t \to \infty$. Each ρ_n in (21) then turned out to be represented by a double power series in λ and θ with coefficients that were intricate combinations of multiple integrals on ϕ. Physicists call some of these quantities "viral coefficients" and the multiple integrals "cluster integrals." Efforts were made to sum the double series in λ and θ in various ways without success. Both parties became discouraged and the collaboration stopped at the end of July in 1939.

There are only three further bits of evidence relating to Wiener's efforts in this particular direction. One dates from 1939. Independently, both collaborators returned briefly to their calculations in the fall of that year, and both were able to sum on n the terms in $\lambda\theta^n$ for one quantity of interest. The result is that near $\lambda = 0$

$$\rho_2(\lambda;\theta;\infty;x_1,u_1;x_2,u_2) = \lambda \exp\left\{-\pi\theta\, [u_1 \cdot u_1 + u_2 \cdot u_2 + \phi(x_1,x_2)]\right\} + O(\lambda^2).$$

Here the ∞ symbolizes that this applies to the equilibrium ensemble. Wiener pointed out that this result is well known to physicists.

A second item of evidence is the fact that Wiener submitted a manuscript to the *Journal of Chemical Physics* in 1940 or 1941. The existence of this unpublished manuscript is noted by Mayer and Montroll.[30] In conversation with one of us, Montroll recalls that this paper explained the use of the densities (21), derived (22), and outlined formulas for the ρ_n that would solve (22) in the equilibrium case. At that time, the paper of Mayer and Montroll was already in press, giving explicit formulas for the equilibrium $\{\rho_n\}$.

It seems a likely conjecture that during the refereeing of his manuscript for the *Journal of Chemical Physics,* Wiener for the first time learned of the extent of work then going on elsewhere on these problems. Indeed, in 1938 Born and Fuchs[31] had already proposed the use of the densities (21) and had stated the equivalent of equations (22). Their paper refers to earlier work of Mayer, which in turn identifies the formalism used as a "well known device." The first publication, to our knowledge, of formal time-invariant solutions to (22) is that of Mayer and Montroll. Much work has since been done; results include fairly general theorems[32] on the convergence of the expansions for the equilibrium $\{\rho_n\}$. Extensive references can be found in Frisch and Lebowitz.[33] McMillan has recently announced[34] formulas and converence theorems for nonequilibrium solutions to (22). To our knowledge, however, no adequate existence theory for these equations is yet available.

None of this last mentioned work makes any direct use of the calculations reported above, and none of it, apart from reference 34 makes any use of chaos theory, per se. Many authors have recognized that there must be consistency relations of some kind, among the ρ_n if they are to be densities defining a chaos. The only published treatment of this latter problem is the "Discrete chaos" paper [43a], reviewed elsewhere in this Volume. Accordingly, it appears that the impact of Wiener's extensive computational efforts has been slight. The impact of the more mathematical aspects of his theory of the discrete chaos, that is actually visible in published papers, is evident in the "Discrete chaos" and in Lectures 14 and 15 of [58i]. This latter material represents the last evidence of Wiener's continuing interest in statistical mechanics. The approach outlined there reflects the point of view developed during the period 1938-1940, that one should consider the chaos as defined by densities that evolve with time. The text exhibits formulas similar to those encountered during the computations of 1938-1940.

References

1. G. I. Taylor, Proc. London Math. Soc. 20(1921), 196.

2. P. Masani, Bull. Amer. Math. Soc. 72, no. 1, pt. II(1966), p. 113.

3. G. S. Deems, *A nearly-normal theory of zero-mean turbulence,* Ph.D. thesis, New York University, 1969.

4. Ibid.

5. R. H. Cameron and W. T. Martin, Ann. Math. 48(1947), 385.

6. Nature 144(1939), 728.

7. See reference 5.

8. S. Kakutani, Proc. Nat. Acad. Sci. U.S.A. 36(1950), 319.

9. W. C. Meecham and A. Siegal, Phys. Fluids 7(1964), 1178.

10. See reference 3.

11. T. Imamura, W. C. Meecham, and A. Siegal, J. Math. Phys. 6(1965), 95.

12. See reference 9.

13. See references 2, 11, and S.E. Bodner, Phys. Fluids 12(1969), 33.

14. J. W. Poduska, *Random theory of turbulence,* Sc.D. thesis, Massachusetts Institute of Technology, 1962.

15. See reference 9.

16. W. C. Meecham and D. Jeng, J. Fluid Mech. 32(1968), 225.

17. W. C. Meecham, J. Fluid Mech. 41(1970), 179.

18. S. A. Orszag and L. R. Bissonnette, Phys. Fluids 10(1967), 2603.

19. S. C. Crow and G. H. Canavan, J. Fluid Mech. 41(1970), 387.

20. S. E. Bodner, Phys. Fluids 12(1969), 33.

21. G. H. Canavan and C. E. Leith, Phys. Fluids 11(1968), 2759.

22. G. H. Canavan, J. Fluid Mech. 41(1970), 405.

23. M. Doi and T. Imamura, Progr. Theoret. Phys. 41(1969), 358.

24. See reference 18.

25. P. G. Saffman, Phys. Fluids 12(1969), 1786.

26. See reference 1.

27. J. B. Keller, Proc. Symp. Appl. Math. 13(1962), 227.

28. B. McMillan, *Random point patterns,* Bull. Amer. Math. Soc. 49(1943), abstract 49-11-265.

29. B. McMillan, *The calculus of the discrete homogeneous chaos,* Ph.D. thesis, Massachusetts Institute of Technology, 1939.

30. J. E. Mayer and E. W. Montroll, J. Chem. Phys. 9(1941), 2.

31. M. Born and K. Fuchs, Proc. Roy. Soc. London Ser. A 166(1938), 391.

32. See, for example, D. Ruelle, Ann. Phys. 25(1963), 100.

33. H. L. Frisch and J. L. Lebowitz, editors, Equilibrium Theory of Classical Fluids, Benjamin, New York, 1964.

34. B. McMillan, *A rigorous statistical mechanics for a classical fluid not in equilibrium,* presented at the SIAM Annual Meeting, 13 October 1970 (to be published).

THE ERGODIC THEOREM

By Norbert Wiener

1. Ergodic theory has its roots in statistical mechanics. Both in the older Maxwell theory and in the later theory of Gibbs, it is necessary to make some sort of logical transition between the average behavior of all dynamical systems of a given family or ensemble, and the historical average behavior of a single system. This transition was not carried out with any rigor until the theory of Lebesgue measure had been developed. The fundamental theorems are those due to Koopman, von Neumann, and Carleman, on the one hand, and to Birkhoff, on the other. A careful account of them and their proofs is to be found in Eberhard Hopf's *Ergodentheorie* (Berlin, 1937) in the series *Ergebnisse der Mathematik und ihrer Grenzgebiete.* The bibliography of that monograph is so complete that it relieves me from all need of furnishing one on my own account. It is important to point out, however, that Birkhoff's first paper (Proc. Nat. Acad. Sci. (1931)) contains Theorem IV of this paper.

The fundamental theorems of ergodic theory, which emerged in the epoch 1931–32, are recognized as theorems pertaining in the first instance to the abstract theory of the Lebesgue integral. Much of this paper will be devoted to new proofs of these known results and to their proper orientation mutually and with respect to the rest of analysis. They have several variant forms, but perhaps the simplest form of the von Neumann theorem reads:

Theorem I. *Let S be a measurable set of points of finite measure. Let T be a transformation of S into itself, which transforms every measurable subset of S into a set of equal measure, and whose inverse has the same property. Let $f(P)$ be a function defined over S and of Lebesgue class L^2. Then there exists a function $f_1(P)$, also belonging to L^2 and such that*

$$\lim_{N\to\infty} \int_S \left| f_1(P) - \frac{1}{N+1} \sum_{n=0}^{N} f(T^n P) \right|^2 dV_P = 0. \tag{1.01}$$

Birkhoff's theorem reads:

Theorem II. *Let S and T be as in Theorem* I. *Let $f(P)$ be a function defined over S and of Lebesgue class L. Then, except for a set of points P of zero measure,*

$$f_1(P) = \lim_{N\to\infty} \frac{1}{N+1} \sum_{n=0}^{N} f(T^n P) \tag{1.02}$$

will exist and belong to L.

These theorems have continuous analogues. These are, respectively:

THEOREM I′. *Let S be as in the hypothesis of Theorem* I. *Let* T^λ *satisfy the conditions given for T in the hypothesis of Theorem* I, *for every real value of* λ, *and let*

$$T^\lambda(T^\mu P) = T^{\lambda+\mu}P. \tag{1.03}$$

Let $f(P)$ *be a function defined over S and of Lebesgue class* L^2, *and let* $f(T^\lambda P)$ *be measurable in the product space of* λ *and P. Then there exists a function* $f_1(P)$, *also belonging to* L^2 *and such that*

$$\lim_{N\to\infty} \int_S \left| f_1(P) - \frac{1}{N}\int_0^N f(T^\lambda P)\,d\lambda \right|^2 dV_P = 0. \tag{1.04}$$

THEOREM II′. *Let S and* T^λ *be as in the hypothesis of Theorem* I′. *Let* $f(P)$ *be a function defined over S and of Lebesgue class L, and let* $f(T^\lambda P)$ *be measurable in the product space of* λ *and P. Then, except for a set of points P of zero measure,*

$$f_1(P) = \lim_{N\to\infty} \frac{1}{N}\int_0^N f(T^\lambda P)\,d\lambda \tag{1.05}$$

will exist as a function in L.

Theorem II′ has a very close formal analogy to the fundamental theorem of the calculus. One form of the latter reads:

THEOREM III. *Let* $f(x)$ *belong to L. Then, for all values of x with the exception of a set of zero measure,*

$$\lim_{\epsilon\to 0} \frac{1}{\epsilon}\int_0^\epsilon f(x+\lambda)\,d\lambda = f(x). \tag{1.06}$$

This analogy becomes even closer if we substitute for Theorem III

THEOREM III′. *On the hypothesis of Theorem* II′, *except for a set of points P of zero measure,*

$$f(P) = \lim_{\epsilon\to 0} \frac{1}{\epsilon}\int_0^\epsilon f(T^\lambda P)\,d\lambda. \tag{1.07}$$

It will be seen that the only difference between the pattern of Theorem II′ and Theorem III′ is that in the first case $N \to \infty$, while in the second case $\epsilon \to 0$. This suggests the existence of a central theorem in which we suppose neither the one nor the other, but deal with $\frac{1}{A}\int_0^A f(T^\lambda P)\,d\lambda$ for all values of A. Birkhoff in fact has proved a theorem of this type.[1] A sharper form of it is the following:

THEOREM IV. *On the hypothesis of Theorem* II, *let* $f(P) \geqq 0$ *on S and let*

$$f^*(P) = \underset{0<A<\infty}{\text{l.u.b.}} \frac{1}{A+1} \sum_{n=0}^{A} f(T^n P); \tag{1.08}$$

[1] Cf. G. D. Birkhoff, Proc. Nat. Acad. Sci., vol. 17(1931), pp. 650–660; also N. Wiener, *The homogeneous chaos*, Amer. Jour. Math., vol. 60(1938), p. 907.

or on the hypothesis of Theorem II′, *let*

$$f^*(P) = \underset{0<A<\infty}{\text{l.u.b.}} \frac{1}{A} \int_0^A f(T^\lambda P)\, d\lambda. \tag{1.09}$$

In either case, if $\alpha > 0$, *the measure of the set of points* P *for which* $f^*(P) \geqq \alpha$ *does not exceed*

$$\frac{1}{\alpha} \int_S f(P)\, dV_P. \tag{1.10}$$

It also does not exceed

$$\frac{2}{\alpha} \int_{f(P) \geqq \frac{1}{2}\alpha} f(P)\, dV_P. \tag{1.11}$$

The importance of Theorem IV has been much neglected in the subsequent literature and can be properly appreciated only by a direct reference to Birkhoff's own work. As a corollary of Theorem IV, we have

THEOREM V. *If we define* $f^*(P)$ *as in Theorem* IV, *then if* $f(P)$ *belongs to* L^p $(p > 1)$, *so does* $f^*(P)$; *while if*

$$\int_S f(P) \log^+ f(P) dV_P < \infty, \tag{1.12}$$

then $f^*(P)$ *belongs to* L.

This last theorem is intimately connected with an inequality of Hardy and Littlewood.[2]

All of the theorems so far quoted have analogues in which the group T^λ is replaced by an Abelian group with more than one generator. Theorem I becomes

THEOREM I″. *Let* S *be a measurable set of points of finite measure. Let* $T_1^{\lambda_1} T_2^{\lambda_2} \cdots T_n^{\lambda_n}$ *satisfy the conditions given for* T *in the hypothesis of Theorem* I, *for every set of real values of* $(\lambda_1, \cdots, \lambda_n)$, *and let*

$$T_1^{\lambda_1} T_2^{\lambda_2} \cdots T_n^{\lambda_n} (T_1^{\mu_1} \cdots T_n^{\mu_n} P) = T_1^{\lambda_1+\mu_1} \cdots T_n^{\lambda_n+\mu_n} P. \tag{1.13}$$

Let $f(P)$ *be a function defined over* S *and of Lebesgue class* L^2, *and let* $f(T_1^{\lambda_1} \cdots T_n^{\lambda_n} P)$ *be measurable in the product space of* $(\lambda_1, \cdots, \lambda_n)$ *and* P. *Then there exists a function* $f_1(P)$, *also belonging to* L^2 *and such that*[3]

$$\lim_{\Lambda\to\infty} \int_S \left| f_1(P) - \frac{1}{V(\Lambda)} \int \cdots \int_{\lambda_1^2+\cdots+\lambda_n^2 \leq \Lambda^2} f(T_1^{\lambda_1} \cdots T_n^{\lambda_n} P)\, d\lambda_1 \cdots d\lambda_n \right|^2 dV_P = 0. \tag{1.14}$$

[2] Cf. Hardy, Littlewood, and Pólya, *Inequalities*, Cambridge, 1934.

[3] Here we use $V(r)$ for the volume of a sphere of radius r in n-space.

Theorem II becomes

THEOREM II″. *Let S and* $T_1^{\lambda_1} \cdots T_n^{\lambda_n}$ *be as in the hypothesis of Theorem* I″, *and let f(P) be a function defined over S and of Lebesgue class L. Let* $f(T_1^{\lambda_1} \cdots T_n^{\lambda_n}P)$ *be measurable in the product space of* $(\lambda_1, \cdots, \lambda_n)$ *and P. Then, except for a set of points P of zero measure,*

$$f_1(P) = \lim_{\Lambda \to \infty} \frac{1}{V(\Lambda)} \int \cdots \int_{\lambda_1^2 + \cdots + \lambda_n^2 \leqq \Lambda^2} f(T_1^{\lambda_1} \cdots T_n^{\lambda_n}P) \, d\lambda_1 \cdots d\lambda_n \tag{1.15}$$

will exist.

Theorem III′ will become

THEOREM III″. *On the hypothesis of Theorem* III′, *except for a set of points P of zero measure,*

$$f(P) = \lim_{\Lambda \to 0} \frac{1}{V(\Lambda)} \int \cdots \int_{\lambda_1^2 + \cdots + \lambda_n^2 \leqq \Lambda^2} f(T_1^{\lambda_1} \cdots T_n^{\lambda_n}P) \, d\lambda_1 \cdots d\lambda_n . \tag{1.16}$$

Theorem IV will become

THEOREM IV′. *On the hypothesis of Theorem* II″, *let*

$$f^*(P) = \underset{0 < \Lambda < \infty}{\text{l.u.b.}} \frac{1}{V(\Lambda)} \int \cdots \int_{\lambda_1^2 + \cdots + \lambda_n^2 \leqq \Lambda^2} f(T_1^{\lambda_1} \cdots T_n^{\lambda_n}P) \, d\lambda_1 \cdots d\lambda_n . \tag{1.17}$$

If $\alpha > 0$, *the measure of the set of points P for which* $f^*(P) \geqq \alpha$ *does not exceed*

$$\frac{B_n}{\alpha} \int_S f(P) \, dV_P , \tag{1.18}$$

where B_n *depends only on the number n of dimensions. It also does not exceed*

$$\frac{2B_n}{\alpha} \int_{f(P) \geqq \frac{1}{2}\alpha} f(P) dV_P . \tag{1.19}$$

The generalization of Theorem V has verbally the same statement as Theorem V itself, with the exception that the definition of $f^*(P)$ is made as in Theorem IV′ instead of as in Theorem IV. We shall call this theorem *Theorem* V′.

The order of proof will be the following: we shall first establish the theorems of the I type.[4] Then we give a completely autonomous proof of the theorems of the IV type. Theorems of the III and the V type result from theorems of the IV type alone, while theorems of the II type result from the combination of theorems of the I and the IV type.

[4] Theorem I has been established in a very direct way in a recent unpublished paper by F. Riesz, and Theorem I″ by Dunford.

2. In the proof of theorems of the I type, the following lemmas are of great service:

LEMMA A. *Let $\phi_k(x)$ belong to L^2 for $1 \leqq k \leqq n$. Let*

$$\int |\phi_k(x)|^2 \, dx = A \tag{2.01}$$

for all admissible values of k. Then

$$\int \left| \frac{1}{n} \sum_1^n \phi_k(x) \right|^2 dx \leqq A. \tag{2.02}$$

LEMMA A′. *Let $\phi(x, y)$ be measurable in x and y, and let*

$$\int |\phi(x, y)|^2 \, dx = A \qquad \textit{for all } y. \tag{2.03}$$

Let $\psi(y)$ belong to L. Then $\int \phi(x, y)\psi(y) \, dy$ belongs to L^2, and

$$\int \left| \int \phi(x, y)\psi(y) \, dy \right|^2 dx \leqq \left[\int |\psi(y)| \, dy \right]^2 A. \tag{2.04}$$

LEMMA B. *Let $\phi_k(x)$ be a set of functions of L^2 for $1 \leqq k \leqq n$. Then there exists a value of k, say k_1, such that*

$$\int \left| \phi_{k_1}(x) - \frac{1}{n} \sum_{k=1}^n \phi_k(x) \right|^2 dx \leqq \int |\phi_{k_1}(x)|^2 \, dx - \int \left| \frac{1}{n} \sum_{k=1}^n \phi_k(x) \right|^2 dx. \tag{2.05}$$

LEMMA B′. *Let $\phi(x, y)$ be measurable in x and y, and let $\int |\phi(x, y)|^2 \, dx$ be uniformly bounded in y. Let $\psi(y)$ be a non-negative function of class L such that $\int \psi(y) \, dy = 1$. Then there exists a value of y, say y_1, such that*

$$\begin{aligned} \int \left| \phi(x, y_1) - \int \phi(x, y)\psi(y) \, dy \right|^2 dx &\leqq \int |\phi(x, y_1)|^2 \, dx \\ &\quad - \int \left| \int \phi(x, y)\psi(y) \, dy \right|^2 dx. \end{aligned} \tag{2.06}$$

In all these lemmas, the limits of integration when left blank may be given any suitable values.

To establish Lemma A, we notice that

$$\begin{aligned} \int \left| \frac{1}{n} \sum_1^n \phi_k(x) \right|^2 dx &= \frac{1}{n^2} \sum_{j=1}^n \sum_{k=1}^n \int \phi_j(x)\overline{\phi_k(x)} \, dx \\ &\leqq \frac{1}{n^2} \sum_{j=1}^n \sum_{k=1}^n \left\{ \int |\phi_j(x)|^2 \, dx \int |\phi_k(x)|^2 \, dx \right\}^{\frac{1}{2}} = A. \end{aligned} \tag{2.07}$$

As to Lemma B, we may obviously choose k_1 so that

$$R \int \overline{\phi_{k_1}(x)} \left(\frac{1}{n} \sum_{k=1}^{n} \phi_k(x) \right) dx \geqq \int \left| \frac{1}{n} \sum_{k=1}^{n} \phi_k(x) \right|^2 dx. \tag{2.08}$$

Otherwise we should have

$$R \frac{1}{n} \sum_{j=1}^{n} \int \overline{\phi_j(x)} \left(\frac{1}{n} \sum_{k=1}^{n} \phi_k(x) \right) dx < \int \left| \frac{1}{n} \sum_{k=1}^{n} \phi_k(x) \right|^2 dx, \tag{2.09}$$

and this is a contradiction. Then

$$\begin{aligned} \int \left| \phi_{k_1}(x) - \frac{1}{n} \sum_{k=1}^{n} \phi_k(x) \right|^2 dx & \\ = \int | \phi_{k_1}(x) |^2 dx - 2R \int & \overline{\phi_{k_1}(x)} \left(\frac{1}{n} \sum_{k=1}^{n} \phi_k(x) \right) dx \\ & + \int \left| \frac{1}{n} \sum_{k=1}^{n} \phi_k(x) \right|^2 dx \\ \leqq \int | \phi_{k_1}(x) |^2 dx - \int & \left| \frac{1}{n} \sum_{k=1}^{n} \phi_k(x) \right|^2 dx. \end{aligned} \tag{2.10}$$

Lemmas A′ and B′ are proved in an exactly analogous manner.

We now turn to the proof of Theorem I. Let us put

$$f_{(k)}(P) = \frac{1}{2^k} \sum_{j=0}^{2^k-1} f(T^j P). \tag{2.11}$$

We then have for $k > m$

$$f_{(k)}(P) = \frac{1}{2^{k-m}} \sum_{j=0}^{2^{k-m}-1} f_{(m)}(T^{2^m j} P). \tag{2.12}$$

Thus by Lemma A, the numbers $\int_S | f_{(k)}(P) |^2 dV_P$ form a decreasing sequence of positive terms, so that

$$F = \lim_{k \to \infty} \int_S | f_{(k)}(P) |^2 dV_P \tag{2.13}$$

exists and is non-negative. By Lemma B, if $k > m > 0$, there exists an integer j in the range $0, 2^{k-m} - 1$ such that

$$\begin{aligned} \int_S | f_{(m)}(T^{2^m j} P) - f_{(k)}(P) |^2 dV_P &\leqq \int_S | f_{(m)}(T^{2^m j} P) |^2 dV_P - \int_S | f_{(k)}(P) |^2 dV_P \\ &\leqq \int_S | f_{(m)}(P) |^2 dV_P - F, \end{aligned} \tag{2.14}$$

since T preserves measure. Also

$$\int_S |f_{(m)}(T^\nu P) - f_{(k)}(T^{\nu-2^m j} P)|^2 dV_P \leqq \int_S |f_{(m)}(P)|^2 dV_P - F. \tag{2.15}$$

Now let m_μ be so large that

$$\int_S |f_{(m_\mu)}(P)|^2 dV_P < 2^{-\mu} + F. \tag{2.16}$$

It will follow that there exists a sequence of numbers ν_μ such that

$$\int_S |f_{(m_\mu)}(T^{\nu_\mu} P) - f_{(m_{\mu+1})}(T^{\nu_{\mu+1}} P)|^2 dV_P < 2^{1-\mu}. \tag{2.17}$$

Then if $\mu > \lambda$,

$$\int_S |f_{(m_\mu)}(T^{\nu_\mu} P) - f_{(m_\lambda)}(T^{\nu_\lambda} P)|^2 dV_P \leqq \frac{2^{\frac{1}{2}(1-\lambda)}}{2^{\frac{1}{2}} - 1}. \tag{2.18}$$

Thus, by the Riesz-Fischer theorem, there exists a function $f_1(P)$ such that

$$\lim_{\mu\to\infty} \int_S |f_1(P) - f_{(m_\mu)}(T^{\nu_\mu} P)|^2 dV_P = 0. \tag{2.19}$$

Again,

$$\lim_{\mu\to\infty} \int_S |f_1(TP) - f_{(m_\mu)}(T^{\nu_\mu+1} P)|^2 dV_P = 0. \tag{2.20}$$

Now,

$$\begin{aligned} &\int_S |f_{(m_\mu)}(T^{\nu_\mu} P) - f_{(m_\mu)}(T^{\nu_\mu+1} P)|^2 dV_P \\ &\qquad = \int_S |2^{-m_\mu} f(T^{\nu_\mu} P) - 2^{-m_\mu} f(T^{\nu_\mu+2^{m_\mu}} P)|^2 dV_P \\ &\qquad \leqq 2^{2-2m_\mu} \int_S |f(P)|^2 dV_P. \end{aligned} \tag{2.21}$$

Thus

$$\lim_{\mu\to\infty} \int_S |f_{(m_\mu)}(T^{\nu_\mu} P) - f_{(m_\mu)}(T^{\nu_\mu+1} P)|^2 dV_P = 0, \tag{2.22}$$

and

$$\int_S |f_1(P) - f_1(TP)|^2 dV_P = 0. \tag{2.23}$$

Hence for all n

$$f_1(P) = f_1(TP) = f_1(T^n P). \tag{2.24}$$

Now let us put

$$f_2(P) = f(P) - f_1(P). \tag{2.25}$$

Then there exists a sequence $\{m_\mu\}$ such that

$$\lim_{\mu\to\infty}\int_S |f_{2(m_\mu)}(P)|^2\,dV_P = 0. \tag{2.26}$$

However, as in (2.13),

$$\lim_{k\to\infty}\int_S |f_{2(k)}(P)|^2\,dV_P \tag{2.27}$$

exists. Thus

$$\lim_{k\to\infty}\int_S |f_{(k)}(P) - f_1(P)|^2\,dV_P = 0. \tag{2.28}$$

Now, let the integer N have the development

$$N = 2^\nu + \sum_{\mu=0}^{\nu-1} a_\mu 2^\mu \qquad (a_\mu = 0 \text{ or } 1;\, a_\nu = 1) \tag{2.29}$$

in the binary scale. Then

$$\begin{aligned}&\int_S \left| f_1(P) - \frac{1}{N}\sum_{k=0}^{N-1} f(T^k P)\right|^2 dV_P\\ &\qquad \leqq \left\{\sum_{\mu=0}^{\nu} \frac{a_\mu 2^\mu}{N}\left\{\int |f_1(P) - f_{(\mu)}(P)|^2\,dV_P\right\}^{\frac12}\right\}^2.\end{aligned} \tag{2.30}$$

This establishes (1.01) and Theorem I.

Theorem I′ may be established in a like manner. We shall proceed to the more general case of Theorem I″. Here we shall prove in the first instance that there exists a function $f_1(P)$ of L^2 such that

$$\begin{aligned}\lim_{\Lambda\to\infty}\int_S \Bigg| f_1(P) &- \frac{1}{(2\pi\Lambda)^{\frac12 n}}\int_{-\infty}^{\infty}\cdots\int_{-\infty}^{\infty} f(T_1^{\lambda_1}\cdots T_n^{\lambda_n}P)\\ &\cdot\exp\left(-\sum_1^n \frac{\lambda_k^2}{2\Lambda}\right) d\lambda_1\cdots d\lambda_n\Bigg|^2 dV_P = 0.\end{aligned} \tag{2.31}$$

Let us notice that

$$\begin{aligned}&\frac{1}{[2\pi(\Lambda_1+\Lambda_2)]^{\frac12 n}}\int_{-\infty}^{\infty}\cdots\int_{-\infty}^{\infty} f(T_1^{\lambda_1}\cdots T_n^{\lambda_n}P)\exp\left(-\sum_1^n \frac{\lambda_k^2}{2(\Lambda_1+\Lambda_2)}\right)d\lambda_1\\ &\cdots d\lambda_n = \frac{1}{(2\pi\Lambda_1)^{\frac12 n}}\int_{-\infty}^{\infty}\cdots\int_{-\infty}^{\infty}\exp\left(-\sum_1^n\frac{\lambda_k^2}{2\Lambda_1}\right)d\lambda_1\cdots d\lambda_n\,\frac{1}{(2\pi\Lambda_2)^{\frac12 n}}\\ &\cdot\int_{-\infty}^{\infty}\cdots\int_{-\infty}^{\infty} f(T_1^{\lambda_1+\mu_1}\cdots T_n^{\lambda_n+\mu_n}P)\exp\left(-\sum_1^n\frac{\mu_k^2}{2\Lambda_2}\right)d\mu_1\cdots d\mu_n.\end{aligned} \tag{2.32}$$

By Lemma A′,

$$\int_S\left|\frac{1}{(2\pi\Lambda)^{\frac12 n}}\int_{-\infty}^{\infty}\cdots\int_{-\infty}^{\infty} f(T_1^{\lambda_1}\cdots T_n^{\lambda_n}P)\exp\left(-\sum_1^n\frac{\lambda_k^2}{2\Lambda}\right)d\lambda_1\cdots d\lambda_n\right|^2 dV_P \tag{2.33}$$

is monotone decreasing in Λ and has a non-negative limit F. Thus by Lemma B′, if $\Lambda > \Lambda_1 > 0$, there exists a set $(\mu_1, \cdots, \mu_n)$ such that

$$\begin{aligned}\int_S \Bigg| &\frac{1}{(2\pi\Lambda)^{\frac{1}{2}n}} \int_{-\infty}^{\infty} \cdots \int_{-\infty}^{\infty} f(T_1^{\lambda_1} \cdots T_n^{\lambda_n} P) \exp\left(-\sum_1^n \frac{\lambda_k^2}{2\Lambda}\right) d\lambda_1 \cdots d\lambda_n \\ &- \frac{1}{(2\pi\Lambda_1)^{\frac{1}{2}n}} \int_{-\infty}^{\infty} \cdots \int_{-\infty}^{\infty} f(T_1^{\lambda_1+\mu_1} \cdots T_n^{\lambda_n+\mu_n} P) \exp\left(-\sum_1^n \frac{\lambda_k^2}{2\Lambda_1}\right) d\lambda_1 \\ &\cdots d\lambda_n \Bigg|^2 dV_P \leqq \int_S \left| \frac{1}{(2\pi\Lambda_1)^{\frac{1}{2}n}} \int_{-\infty}^{\infty} \cdots \int_{-\infty}^{\infty} f(T_1^{\lambda_1} \cdots T_n^{\lambda_n} P) \right. \\ &\left. \cdot \exp\left(-\sum_1^n \frac{\lambda_k^2}{2\Lambda_1}\right) d\lambda_1 \cdots d\lambda_n \right|^2 dV_P - B.\end{aligned} \tag{2.34}$$

As in the proof of (2.19), we may choose the vectors $(\mu_{\nu_1}, \cdots, \mu_{\nu_n})$ and the numbers Λ_ν in such a way that the sequence

$$\frac{1}{(2\pi\Lambda_\nu)^{\frac{1}{2}n}} \int_{-\infty}^{\infty} \cdots \int_{-\infty}^{\infty} f(T_1^{\lambda_1+\mu_{\nu_1}} \cdots T_n^{\lambda_n+\mu_{\nu_n}} P) \exp\left(-\sum_1^n \frac{\lambda_k^2}{2\Lambda_\nu}\right) d\lambda_1 \cdots d\lambda_n \tag{2.35}$$

converges in the mean in the L^2 sense to a function $f_1(P)$.

The argument corresponding to (2.21)–(2.23) proceeds as follows. We have

$$\begin{aligned}\lim_{\Lambda_\nu\to\infty} \frac{1}{(2\pi\Lambda_\nu)^{\frac{1}{2}n}} \int_{-\infty}^{\infty} \cdots \int_{-\infty}^{\infty} \Bigg| &\exp\left(-\sum_1^n \frac{(\lambda_k + \mu_{\nu_k} + \mu_k)^2}{2\Lambda_\nu}\right) \\ &- \exp\left(-\sum_1^n \frac{(\lambda_k + \mu_{\nu_k})^2}{2\Lambda_\nu}\right)\Bigg| d\lambda_1 \cdots d\lambda_n = 0.\end{aligned} \tag{2.36}$$

Again,

$$\begin{aligned}&\frac{1}{(2\pi\Lambda_\nu)^{\frac{1}{2}n}} \int_{-\infty}^{\infty} \cdots \int_{-\infty}^{\infty} \{f(T_1^{\lambda_1+\mu_{\nu_1}+\mu_1} \cdots T_n^{\lambda_n+\mu_{\nu_n}+\mu_n} P) \\ &\qquad - f(T_1^{\lambda_1+\mu_{\nu_1}} \cdots T_n^{\lambda_n+\mu_{\nu_n}} P)\} \exp\left(-\sum_1^n \frac{\lambda_k^2}{2\Lambda_\nu}\right) d\lambda_1 \cdots d\lambda_n \\ &= \frac{1}{(2\pi\Lambda_\nu)^{\frac{1}{2}n}} \int_{-\infty}^{\infty} \cdots \int_{-\infty}^{\infty} f(T_1^{\lambda_1} \cdots T_n^{\lambda_n} P) \left\{\exp\left(-\sum_1^n \frac{(\lambda_k + \mu_{\nu_k} + \mu_k)^2}{2\Lambda_\nu}\right)\right. \\ &\qquad \left. - \exp\left(-\sum_1^n \frac{(\lambda_k + \mu_{\nu_k})^2}{2\Lambda_\nu}\right)\right\} d\lambda_1 \cdots d\lambda_n.\end{aligned} \tag{2.37}$$

Thus by Lemma A′,

$$\begin{aligned}\lim_{\Lambda_\nu\to\infty} \int_S \Bigg| &\frac{1}{(2\pi\Lambda_\nu)^{\frac{1}{2}n}} \int_{-\infty}^{\infty} \cdots \int_{-\infty}^{\infty} f(T_1^{\lambda_1+\mu_{\nu_1}+\mu_1} \cdots T_n^{\lambda_n+\mu_{\nu_n}+\mu_n} P) \\ &\qquad \cdot \exp\left(-\sum_1^n \frac{\lambda_k^2}{2\Lambda_\nu}\right) d\lambda_1 \cdots d\lambda_n \\ &- \frac{1}{(2\pi\Lambda_\nu)^{\frac{1}{2}n}} \int_{-\infty}^{\infty} \cdots \int_{-\infty}^{\infty} f(T_1^{\lambda_1+\mu_{\nu_1}} \cdots T_n^{\lambda_n+\mu_{\nu_n}} P) \\ &\qquad \cdot \exp\left(-\sum_1^n \frac{\lambda_k^2}{2\Lambda_\nu}\right) d\lambda_1 \cdots d\lambda_n \Bigg|^2 dV_P = 0.\end{aligned} \tag{2.38}$$

The argument of (2.23)–(2.28) is duplicated without substantial change.

To proceed from (2.31) to (1.14), let us notice that if $S_\Lambda(\lambda_1, \cdots, \lambda_n)$ is the characteristic function of the sphere $\lambda_1^2 + \lambda_2^2 + \cdots + \lambda_n^2 \leqq \Lambda^2$, we have

$$\lim_{\Lambda\to\infty} \frac{1}{V(\Lambda)} \int_{-\infty}^{\infty} \cdots \int_{-\infty}^{\infty} \Big| S_\Lambda(\lambda_1, \cdots, \lambda_n) - \frac{1}{(2\pi\Lambda_1)^{\frac{1}{2}n}} \int_{-\infty}^{\infty} \cdots \int_{-\infty}^{\infty} S_\Lambda(\lambda_1 + \mu_1, \cdots, \lambda_n + \mu_n) \cdot \exp\left(-\sum_1^n \frac{\mu_k^2}{2\Lambda_1}\right) d\mu_1 \cdots d\mu_n \Big|^2 d\lambda_1 \cdots d\lambda_n = 0. \tag{2.39}$$

Thus, by Lemma A′,

$$\lim_{\Lambda\to\infty} \int_S dV_P \Big| \int_{-\infty}^{\infty} \cdots \int_{-\infty}^{\infty} \{f(T_1^{\lambda_1} \cdots T_n^{\lambda_n} P) - f_1(T_1^{\lambda_1} \cdots T_n^{\lambda_n} P)\} \frac{1}{V(\Lambda)} \Big\{ S_\Lambda(\lambda_1, \cdots, \lambda_n) - \frac{1}{(2\pi\Lambda_1)^{\frac{1}{2}n}} \int_{-\infty}^{\infty} \cdots \int_{-\infty}^{\infty} S_\Lambda(\lambda_1 + \mu_1, \cdots, \lambda_n + \mu_n) \cdot \exp\left(-\sum_1^n \frac{\mu_k^2}{2\Lambda_1}\right) d\mu_1 \cdots d\mu_n \Big\} d\lambda_1 \cdots d\lambda_n \Big|^2 = 0. \tag{2.40}$$

Hence by (2.31)

$$\overline{\lim_{\Lambda\to\infty}} \int_S \Big| f_1(P) - \frac{1}{V(\Lambda)} \int \underset{\lambda_1^2+\cdots+\lambda_n^2\leqq\Lambda^2}{\cdots} \int f(T_1^{\lambda_1} \cdots T_n^{\lambda_n} P)\, d\lambda_1 \cdots d\lambda_n \Big|^2 dV_P \leqq \overline{\lim_{\Lambda_1\to\infty}} \int_S \Big| f_1(P) - \frac{1}{(2\pi\Lambda_1)^{\frac{1}{2}n}} \int_{-\infty}^{\infty} \cdots \int_{-\infty}^{\infty} f(T_1^{\lambda_1} \cdots T_n^{\lambda_n} P) \cdot \exp\left(-\sum_1^n \frac{\lambda_k^2}{2\Lambda_1}\right) d\lambda_1 \cdots d\lambda_n \Big|^2 dV_P = 0. \tag{2.41}$$

Theorems I, I′, and I″ deal with functions of the class L^2. From these, similar theorems arise concerning functions of the class L. If in the hypothesis of any of these theorems we assume $f(P)$ only to belong to L, we shall show that (1.01) may be replaced by

$$\lim_{N\to\infty} \int_S \Big| f_1(P) - \frac{1}{N+1} \sum_{n=0}^{N} f(T^n P) \Big| dV_P = 0; \tag{2.42}$$

(1.04) by

$$\lim_{N\to\infty} \int_S \Big| f_1(P) - \frac{1}{N} \int_0^N f(T^\lambda P)\, d\lambda \Big| dV_P = 0; \tag{2.43}$$

and (1.14) by

$$(2.44)\quad \lim_{\Lambda\to\infty} \int_S \left| f_1(P) - \frac{1}{V(\Lambda)} \int \cdots \int_{\lambda_1^2+\cdots+\lambda_n^2 \leqq \Lambda^2} f(T_1^{\lambda_1} \cdots T_n^{\lambda_n} P)\, d\lambda_1 \cdots d\lambda_n \right| dV_P = 0.$$

As all the proofs are exactly alike, we shall establish (2.42). We may approximate in the L sense to $f(P)$ with any desired degree of accuracy by a bounded function $g(P)$, which obviously belongs to L^2. Next we form $g_1(P)$ as we have formed $f_1(P)$ in the L^2 case. By the Schwarz inequality,

$$(2.45)\quad \lim_{N\to\infty} \int_S \left| g_1(P) - \frac{1}{N+1} \sum_{n=0}^{N} g(T^n P) \right| dV_P = 0.$$

Again

$$(2.46)\quad \int_S \left| \frac{1}{N+1} \sum_0^N f(T^n P) - \frac{1}{N+1} \sum_0^N g(T^n P) \right| dV_P \leqq \int_S |f(P) - g(P)|\, dV_P.$$

Thus

$$(2.47)\quad \begin{aligned} \overline{\lim_{M,N\to\infty}} \int_S \left| \frac{1}{N+1} \sum_0^N f(T^n P) - \frac{1}{M+1} \sum_0^N f(T^n P) \right| dV_P \\ \leqq 2 \int_S |f(P) - g(P)|\, dV_P. \end{aligned}$$

Since $g(P)$ converges in the mean in the L sense to $f(P)$, the right side of (2.47) tends to 0, so that we may replace it by 0. If we now use the L form of the Riesz-Fischer theorem, (2.42) follows at once.

3. We now turn to the proofs of Theorems IV and IV′. For Theorem IV′ we need a lemma of the Vitali type:[5]

LEMMA C′. *Let S be a set of points of finite outer measure in n-space. Let each point P of S be the center of a sphere $\Sigma(P)$. Then there exists a finite set of spheres $\Sigma(P)$, non-overlapping, and of total volume exceeding $A_n m_e(S)$, where A_n is a positive absolute constant depending only on the number n of dimensions.*

To establish this lemma, let us notice that it is trivially true unless the radii of the spheres $\Sigma(P)$ have a finite upper bound. If they have such a bound, let it be B. Let S_ν be that subset of S for which the spheres $\Sigma(P)$ have radii $\leqq 2^{1-\nu}B$ and $> 2^{-\nu}B$. Let $S_1' = S_1$, and let us define by mathematical induction S_ν' as that part of S_ν which is more remote than $5B$ from any point of S_1', more remote than $5B \cdot 2^{-1}$ from any point of S_2', and so on, being more remote than $5B \cdot 2^{-\nu+2}$ from any point of $S_{\nu-1}'$.

Let us lay upon S_1 a mesh of cubes with side $5B$, and if any cube of this mesh contain a point of S_1, let us pick out one of these points, P_m. Then the

[5] Cf. E. W. Hobson, *The Theory of Functions of a Real Variable*, vol. 1, Cambridge, 1927, §136. Cf. also Titchmarsh, *Theory of Functions*, §11.41.

sum of the volumes of the spheres $\Sigma(P_m)$ will be more than a fixed fraction of the sum of the meshes of the cubes corresponding to them, and hence more than a fixed fraction of the volumes of these cubes, together with all adjacent cubes. None of these spheres can overlap a sphere whose center is not in an adjacent cube, and hence each can overlap at most a fixed finite number of spheres. Thus if we enumerate these spheres in any order, and discard in turn every sphere overlapping a sphere which we have not already discarded, we shall ultimately obtain a finite set of non-overlapping spheres whose total volume will exceed a fixed submultiple A'_n of the volume of all the cubes containing points of S_1, together with all adjacent cubes. These cubes contain every point of S_1, together with all the points which are in any S_ν, but fail to be included in S'_ν by virtue of their vicinity to points of S_1.

We now cover S'_2 with a mesh of half the linear magnitude, and proceed as above. We shall find a finite set of spheres with centers at points of S'_2, obviously not overlapping any sphere of the previous set, not overlapping one another, and with a total volume exceeding A'_n multiplied by the total volume of all the cubes containing S'_2, together with all adjacent cubes. These cubes together with the similar set obtained from S_1 contain S_1, S_2, and all the points in later S_ν, but are excluded from the corresponding S'_ν on account of their propinquity to S_1 or S'_2.

By continuing this process, we obtain a denumerable set of spheres $\Sigma(P)$, exceeding in volume $A'_n m_e(S)$ and non-overlapping. We may clearly take a finite number of these with total volume exceeding $A_n m_e(S)$, if $A_n < A'_n$.

In the one-dimensional case, Lemma C′ is replaced by the somewhat more precise

LEMMA C. *Let S be a set of points of finite measure on the line. Let every point P of S be the right-hand terminus of an interval $I(P)$. Then there exists a finite subset of intervals $I(P)$, non-overlapping, and of total length exceeding $(1 - \epsilon)m_e(S)$, where ϵ is any positive number.*

This lemma is due to Sierpinski (see footnote 5) and is completely elementary. From Lemma C and Lemma C′, respectively, we may deduce

LEMMA D. *Let $f(x)$ belong to L on the infinite line. Then the set S_α of values of x for which*

$$\operatorname*{l.u.b.}_{0<\Lambda<\infty} \frac{1}{\Lambda} \int_{x-\Lambda}^{x} |f(\xi)|\, d\xi > \alpha > 0 \tag{3.01}$$

does not exceed $\frac{1}{\alpha} \int_{-\infty}^{\infty} |f(\xi)|\, d\xi$ *in measure.*

LEMMA D′. *Let $f(x_1, \cdots, x_n)$ belong to L in infinite n-space. Then the set S_α of points $(y_1, \cdots, y_n)$ for which*

$$\operatorname*{l.u.b.}_{0<\Lambda<\infty} \frac{1}{V(\Lambda)} \int \cdots \int_{x_1^2+\cdots+x_n^2 \leq \Lambda^2} |f(x_1 + y_1, \cdots, x_n + y_n)|\, dx_1 \cdots dx_n > \alpha > 0 \tag{3.02}$$

does not exceed

$$\frac{1}{\alpha A_n}\int_{-\infty}^{\infty}\cdots\int_{-\infty}^{\infty}|f(x_1,\cdots,x_n)|\,dx_1\cdots dx_n \tag{3.03}$$

in measure.

The proof is the same in both cases. In Lemma D, we can pick out a finite set of non-overlapping intervals, of total length exceeding $(1-\epsilon)m(S_\alpha)$ and such that the integral of $f(x)$ over each interval exceeds α multiplied by its length. Thus

$$\alpha(1-\epsilon)m(S_\alpha) \leqq \int_\Sigma |f(\xi)|\,d\xi, \tag{3.04}$$

where Σ is a finite set of non-overlapping intervals.

The discrete analogue of Lemma D is

LEMMA E. *Let* $\sum_{-\infty}^{\infty}|A_\nu| < \infty$. *Then the number of values of* ν *for which*

$$\underset{0<\mu<\infty}{\text{l.u.b.}}\ \frac{1}{\mu}\sum_{k=\nu-\mu+1}^{k=\nu}|A_k| > \alpha > 0 \tag{3.05}$$

does not exceed $\frac{1}{\alpha}\sum_{-\infty}^{\infty}|A_k|$.

Here the proof goes exactly as in Lemma D. The discrete equivalent of Lemma C is trivial.

We proceed to the proof of Theorem IV. It is clear that, for a particular P, the number of values of ν on $(0, N)$ for which

$$\underset{0<\mu\leqq\nu}{\text{l.u.b.}}\ \frac{1}{\mu}\sum_{k=\nu-\mu+1}^{k=\nu}f(T^{-k}P) > \alpha > 0 \tag{3.06}$$

does not exceed

$$\frac{1}{\alpha}\sum_{0}^{N}f(T^{-k}P). \tag{3.07}$$

If we now integrate over S with respect to P, we see that the sum of the measures of the sets of values of P for which

$$\underset{0<\mu\leqq\nu}{\text{l.u.b.}}\ \frac{1}{\mu}\sum_{k=\nu-\mu+1}^{k=\nu}f(T^{-k}P) > \alpha > 0 \qquad (\nu = 1, 2, \cdots, N) \tag{3.08}$$

will not exceed

$$\frac{N+1}{\alpha}\int_S f(P)\,dV_P. \tag{3.09}$$

Since $T^{-\nu}$ is an equimeasure transformation, we get

$$\frac{1}{N}\sum_{\nu=1}^{N}\left(\text{measure of set of values of } P \text{ for which } \underset{0<\mu\leqq\nu}{\text{l.u.b.}}\ \frac{1}{\mu}\sum_{k=0}^{k=\mu-1}f(T^{k}P) > \alpha > 0\right) \leqq \frac{1}{\alpha}\int_S f(P)\,dV_P. \tag{3.10}$$

We now use the simple lemma that if B_ν is an increasing sequence for which

$$\frac{1}{N+1}\sum_{\nu=1}^{N} B_\nu \leqq B \qquad \text{for all } N, \tag{3.11}$$

then

$$B_\nu \leqq B \qquad \text{for all } \nu. \tag{3.12}$$

Thus for all values of ν,

$$\left(\text{measure of set of values of } P \text{ for which } \operatorname*{l.u.b.}_{0<\mu\leqq\nu} \frac{1}{\mu}\sum_{k=0}^{\mu-1} f(T^k P) > \alpha > 0\right) \leqq \frac{1}{\alpha}\int_S f(P)\, dV_P, \tag{3.13}$$

which gives us

$$\left(\text{measure of set of all values of } P \text{ for which } \operatorname*{l.u.b.}_{0<\mu<\infty} \frac{1}{\mu}\sum_{k=0}^{\mu-1} f(T^k P) > \alpha > 0\right) \leqq \frac{1}{\alpha}\int_S f(P)\, dV_P. \tag{3.14}$$

Let us now turn to Theorem IV′. In Lemma D′, let

$$f(x_1, \cdots, x_n) = \begin{cases} f(T_1^{x_1} \cdots T_n^{x_n} P) & \text{if } x_1^2 + \cdots + x_n^2 \leqq \Lambda^2; \\ 0 & \text{otherwise.} \end{cases} \tag{3.15}$$

Then if we integrate with respect to P over S, we see that in the product space of $(\mu_1, \cdots, \mu_n)$ and P, the measure of the set of points for which

$$\operatorname*{l.u.b.}_{0<\Lambda<\infty} \frac{1}{V(\Lambda)} \underset{\substack{\lambda_1^2+\cdots+\lambda_n^2\leqq\Lambda^2 \\ (\lambda_1+\mu_1)^2+\cdots+(\lambda_n+\mu_n)^2\leqq\Lambda^2}}{\int\cdots\int} f(T_1^{\lambda_1+\mu_1} \cdots T_n^{\lambda_n+\mu_n} P)\, d\lambda_1 \cdots d\lambda_n > \alpha > 0 \tag{3.16}$$

does not exceed

$$\frac{1}{\alpha A_n}\int_S dV_P \underset{\lambda_1^2+\cdots+\lambda_n^2\leqq\Lambda^2}{\int\cdots\int} f(T_1^{\lambda_1} \cdots T_n^{\lambda_n} P)\, d\lambda_1 \cdots d\lambda_n = \frac{V(\Lambda)}{\alpha A_n}\int_S f(P)\, dV_P. \tag{3.17}$$

Since $T_1^{\mu_1} \cdots T_n^{\mu_n}$ is an equimeasure transformation, we may replace the measure of the set for which (3.16) holds by

$$\int\cdots\int d\mu_1 \cdots d\mu_n \left\{\text{measure of the set of values of } P \text{ for which } \operatorname*{l.u.b.}_{0<\Lambda<\infty} \frac{1}{V(\Lambda)} \underset{\substack{\lambda_1^2+\cdots+\lambda_n^2\leqq\Lambda^2 \\ (\lambda_1+\mu_1)^2+\cdots+(\lambda_n+\mu_n)^2\leqq\Lambda^2}}{\int\cdots\int} f(T_1^{\lambda_1} \cdots T_n^{\lambda_n} P)\, d\lambda_1 \cdots d\lambda_n > \alpha > 0\right\} \tag{3.18}$$

A fortiori,

$$\int_{\mu_1^2+\cdots+\mu_n^2\leq\frac{1}{4}\Lambda^2}\cdots\int d\mu_1\cdots d\mu_n \Big\{\text{measure of set of values of } P \text{ for which}$$

$$\text{(3.19)}\qquad \underset{0<\Lambda\leqq\frac{1}{2}\Lambda}{\text{l.u.b.}}\ \frac{1}{V(\Lambda)}\int_{\lambda_1^2+\cdots+\lambda_n^2\leqq\Lambda^2}\cdots\int f(T_1^{\lambda_1}\cdots T_n^{\lambda_n}P)\,d\lambda_1\cdots d\lambda_n > \alpha > 0\Big\}$$

$$\leqq \frac{V(\Lambda)}{\alpha A_n}\int_S f(P)\,dV_P.$$

Thus

$$\Big\{\text{measure of set of values of } P \text{ for which}$$

$$\text{(3.20)}\qquad \underset{0<\Lambda\leqq\frac{1}{2}\Lambda}{\text{l.u.b.}}\ \frac{1}{V(\Lambda)}\int_{\lambda_1^2+\cdots+\lambda_n^2\leqq\Lambda^2}\cdots\int f(T_1^{\lambda_1}\cdots T_n^{\lambda_n}P)\,d\lambda_1\cdots d\lambda_n > \alpha > 0\Big\}$$

$$\leqq \frac{2^n}{\alpha A_n}\int_S f(P)\,dV_P.$$

If we now put $B_n = A_n 2^{-n}$, Theorem IV′ follows at once.

Theorems IV and IV′ have the secondary conclusions (1.11) and (1.19), respectively. To obtain these, let us introduce the function $h(P)$ which is $f(P)$ when this exceeds $\frac{1}{2}\alpha$ and is 0 otherwise. Clearly

$$\text{(3.21)}\qquad f^*(P) \leqq h^*(P) + \tfrac{1}{2}\alpha.$$

Thus in Theorem IV

$$\text{(3.22)}\qquad \begin{aligned} &\{\text{measure of set of values of } P \text{ for which } f^*(P) > \alpha\}\\ &\quad\leqq \{\text{measure of set of values of } P \text{ for which } h^*(P) > \tfrac{1}{2}\alpha\}\\ &\quad\leqq \frac{2}{\alpha}\int_S h(P)\,dV_P = \frac{2}{\alpha}\int_{f(P)\geqq\frac{1}{2}\alpha} f(P)\,dV_P.\end{aligned}$$

This yields (1.11), and (1.19) may be obtained in the identical manner. These are similar to certain results of Hardy and Littlewood, but are somewhat weaker.

Let us now turn to the proof of Theorem II. By (2.42), we may choose N so large that

$$\text{(3.23)}\qquad \int_S \left| f_1(P) - \frac{1}{N+1}\sum_{n=0}^{N} f(T^n P)\right| dV_P < \epsilon^2.$$

Then by Theorem IV, except for a set of values of P of measure not exceeding ϵ, we shall have

$$\text{(3.24)}\qquad \underset{0<A<\infty}{\text{l.u.b.}}\left|\frac{1}{A+1}\sum_{n=0}^{A}\left[f_1(T^n P) - \frac{1}{N+1}\sum_{m=0}^{N} f(T^{m+n}P)\right]\right| < \epsilon.$$

We have, for large A,

$$\text{(3.25)}\quad \begin{aligned}\frac{1}{A+1}\sum_{N+1}^{A} f(T^n P) &\leqq \frac{1}{A+1}\sum_{n=0}^{A}\frac{1}{N+1}\sum_{m=0}^{N} f(T^{m+n}P)\\ &\leqq \frac{1}{A+1}\sum_{0}^{A+N} f(T^n P).\end{aligned}$$

Thus, except for a set of values of P of measure not exceeding ϵ, we have

$$\text{(3.26)}\quad \overline{\lim_{A\to\infty}}\left| f_1(P) - \frac{1}{A+1}\sum_{n=0}^{A} f(T^n P)\right| < \epsilon.$$

Since ϵ can be given consecutively values from a convergent series, we see that, except over a set whose measure does not exceed the remainder of this series, we have uniformly

$$\text{(3.27)}\quad f_1(P) = \lim_{A\to\infty}\frac{1}{A+1}\sum_{n=0}^{A} f(T^n P).$$

This establishes the Birkhoff ergodic theorem. The parallel Theorems II′ and II″ are proved in exactly the same way.

Theorem III and its n-dimensional analogue, which is the particular case of Theorem III″ which we obtain when the group reduces to the n-dimensional translation group, may be proved on the basis of Theorems IV and IV′, or even on the basis of the Lemmas D and D′. Every function of class L is the sum of a continuous bounded function and a function of small L-norm. Let $f(x)$ be the sum of $\phi(x)$ and $\psi(x)$, where $\phi(x)$ is continuous and $\int_{-\infty}^{\infty} |\psi(x)|\,dx < \frac{1}{2}\epsilon^2$. Then

$$\text{(3.28)}\quad \begin{aligned}\overline{\lim_{\eta\to 0}}\left| f(x) - \frac{1}{\eta}\int_x^{x+\eta} f(\xi)\,d\xi\right| &= \overline{\lim_{\eta\to 0}}\left|\psi(x) - \frac{1}{\eta}\int_x^{x+\eta}\psi(\xi)\,d\xi\right|\\ &\leqq |\psi(x)| + \psi^*(x),\end{aligned}$$

where

$$\text{(3.29)}\quad \psi^*(x) = \underset{0<\eta<\infty}{\text{l.u.b.}}\ \frac{1}{\eta}\int_x^{x+\eta} |\psi(\xi)|\,d\xi.$$

By Lemma D, the set of points for which $\psi^*(x) > \epsilon$ has a measure not exceeding $\frac{1}{2}\epsilon$, and the same is clearly true of $|\psi(x)|$. Thus, except for a set of points of measure not exceeding ϵ, we have

$$\text{(3.30)}\quad \overline{\lim_{\eta\to 0}}\left| f(x) - \frac{1}{\eta}\int_x^{x+\eta} f(\xi)\,d\xi\right| \leqq \epsilon.$$

The proof now proceeds as in Theorem II. Theorem III″, as far as the translation group is concerned, is proved in the same way.

To prove Theorem III′ and the general case of Theorem III″, let us note that, for a fixed P, the set of values of λ for which we fail to have

$$f(T^\lambda P) = \lim_{\eta \to 0} \frac{1}{\eta} \int_\lambda^{\lambda+\eta} f(T^\xi P)\, d\xi, \tag{3.31}$$

or the set of values of $(\lambda_1, \cdots, \lambda_n)$ for which we fail to have

$$f(T_1^{\lambda_1} \cdots T_n^{\lambda_n} P) = \lim_{\Lambda \to 0} \frac{1}{V(\Lambda)} \underset{\mu_1^2+\cdots+\mu_n^2 \leqq \Lambda^2}{\int \cdots \int} f(T_1^{\lambda_1+\mu_1} \cdots T_n^{\lambda_n+\mu_n} P)\, d\mu_1 \cdots d\mu_n \tag{3.32}$$

are sets of zero measure. Thus in the product space of these spaces and P, the corresponding sets of values are of zero measure. That is, for almost all values of λ or $(\lambda_1, \cdots, \lambda_n)$, the set of points P for which we fail to have (3.31) or (3.32) valid is of zero measure. Therefore, for at least one of these values, the set is of zero measure. The result follows by an equimeasure transformation.

We finally come to Theorem V. Let $M(x)$ be the measure of the set of points over which $f(P) > x$, and let $M^*(x)$ be the measure of the set of points over which $f^*(P) \geqq x$. We shall have

$$\begin{aligned} \int_S s(f(P))\, dV_P &= -\int_0^\infty s(x)\, dM(x); \\ \int_S s(f^*(P))\, dV_P &= -\int_0^\infty s(x)\, dM^*(x). \end{aligned} \tag{3.33}$$

By Theorem IV, we have formally for $\nu > 0$

$$\begin{aligned} \int_0^\infty m^*(x) x^\nu\, dx &\leqq -\text{const.} \int_0^\infty x^{\nu-1}\, dx \int_x^\infty y dM(y) \\ &= -\text{const.} \int_0^\infty y dm(y) \int_0^y x^{\nu-1}\, dx \\ &= -\text{const.} \int_0^\infty y^{\nu+1}\, dM(y). \end{aligned} \tag{3.34}$$

Thus if $\int_0^\infty y^{\nu+1}\, dM(y)$ is bounded, it will follow that

$$0 = \lim_{\xi \to \infty} \int_\xi^{2\xi} m^*(x) x^\nu\, dx = \lim_{\xi \to \infty} M^*(2\xi)\xi^{\nu+1}. \tag{3.35}$$

We may hence integrate by parts in (3.34) and obtain

$$-\int_0^\infty x^{\nu+1}\, dm^*(x) \leqq -\text{const.} \int_0^\infty y^{\nu+1}\, dM(y). \tag{3.36}$$

Similarly,

$$\int_1^\infty m^*(x)\,dx \leqq -\text{const.} \int_1^\infty y\,dM(y) \int_1^y \frac{dx}{x} \tag{3.37}$$
$$= -\text{const.} \int_{1}^\infty y \log y\,dM(y).$$

As before, if $\int f(P) \log^+ f(P)\, dV_P$ is bounded,

$$-\int_1^\infty x\,dm^*(x) \leqq -\text{const.} \int_1^\infty y \log y\,dM(y). \tag{3.38}$$

These theorems, apart from the less accurate constants, are the same as theorems given by Hardy, Littlewood, and Pólya (loc. cit.).

Comments on [39a]

E. M. Stein

Background

Let $P \to TP$ be a measure-preserving transformation of a measure space, whose points are $P, \ldots,$ and consider the induced mapping $f(P) \to f(T(P))$. The ergodic theorems of von Neumann and Birkhoff (1931-1932) deal with the existence of the limit

$$\lim_{N\to\infty} \frac{1}{N+1} \sum_{n=0}^{N} f(T^n P) \tag{A}$$

either in the norm or almost everywhere. Of particular interest here is the almost everywhere result, whose continuous analogue contains a very well-known classical case, namely

$$\lim_{t\to 0} \frac{1}{2t} \int_{-t}^{t} f(x-s)\, ds = f(x), \quad \text{a.e.,}$$

for f integrable on the line. Hardy and Littlewood had somewhat earlier (1930) introduced the associated "maximal function"

$$m_f(x) = \sup_{t>0} \frac{1}{2t} \int_{-t}^{t} |f(x-s)|\, ds$$

in this case, and proved for it the appropriate inequalities. They also showed how this maximal function is controlling in several other situations.

Soon thereafter Riesz[1] simplified the argument and in effect isolated the "weak-type inequality" for L^1, namely

$$|\{x: m_f(x) > \alpha\}| \leqslant \frac{A}{\alpha} \|f\|_1 . \tag{B}$$

Wiener's paper and later developments

Wiener attacks these problems anew, unifies the above ideas, and goes considerably further. For the purposes of discussion we set out three main strands in his argument.

1. A covering lemma: (p.11).

This states, in effect, that if a set of finite measure in Euclidean space is suitably covered by a collection of balls, then there is a disjoint subcollection whose total volume is of least a positive fraction of the measure of the original set. This lemma, a variant of a classical lemma of Vitali, is the key step for the weak-type inequality for the maximal function. A closely related form of the covering lemma (dealing with rectangles instead of balls) was found at about the same time by Marcinkiewicz and Zygmund[2] in their study of multiple Poisson integrals. Since that time the lemma has been greatly generalized and abstracted so that it would apply, for example, to the context of appropriate metric spaces with application to harmonic functions by K. T. Smith,[3] differentiation theorems for groups by Edwards and Hewitt,[4] and differentiability with respect to measures in Euclidean space other than Lebesgue measure by Besicovitch.[5]

2. Maximal functions and weak-type inequalities

If f is a locally integrable function on Euclidean n-space, Wiener defines

$$\sup_{0<\Lambda} \frac{1}{V(\Lambda)} \int_{|y|\leqslant \Lambda} |f(x+y)|\, dy = m_f(x) \tag{C}$$

and shows that the mapping $f \to m_f$ is of weak-type (1,1) (it satisfies (B) above). From this he deduces the L^p inequality

$$\|m_f\|_p \leqslant A_p \|f\|_p\,, \quad 1 < p \leqslant \infty,$$

generalizing the results which were known in the one-dimensional case. The argument by which he obtains the L^p inequalities from the weak-type (1,1) result can with hindsight be interpreted as a basic special case of the Marcinkiewicz interpolation theorem (see Zygmund[6]). Incidentally, the occurence of weak-type inequalities and maximal functions are by now a commonplace in several branches of analysis.

3. Passage to the measure-preserving case

To complete his chain of reasoning Wiener develops a simple argument to pass from inequalities for the Euclidean maximal function (given by (C)) to its general measure-preserving analogue

$$f^*(P) = \sup_{0<\Lambda} \frac{1}{V(\Lambda)} \int_{|\lambda|\leqslant \Lambda} |f(T^{\lambda}P)|\, d\lambda, \tag{D}$$

where $\{T^\lambda\}$ is an n-parameter Abelian family of measure-preserving transformations; he also shows how to pass to discrete analogues corresponding to what arises in (A). Part of the argument here was made more direct by Yosida and Kakutani,[7] and later by F. Riesz.[8] Cotlar[9] takes the matter up again, with the additional feature that he is then able to give a unified treatment of the ergodic theorem, differentiation theorems, and the existence of the Hilbert transform. In some other development (originating with work by E. Hopf) the operators $f \to f(TP)$ are replaced by more general operators not arising from point transformation. (For some of these results see Dunford-Schwartz.[10])

Historical Note

The reader may find it instructive to observe for himself to what extent some of the ideas of the present paper are already present (albeit in rather implicit form) in Wiener's paper "The homogeneous chaos" [38a], in particular in section 4 of that paper.

References

1. F. Riesz, *Sur un théorème de maximum de M. M. Hardy et Littlewood,* J. London Math. Soc. 7(1932), 10-13.

2. J. Marcinkiewicz and A. Zygmund, *On the summability of double Fourier series,* Fund. Math. 32(1939), 122-132.

3. K. T. Smith, *A generalization of an inequality of Hardy and Littlewood,* Canad. J. Math. 8(1956), 157-170.

4. R. Edwards and E. Hewitt, *Pointwise limits for sequences of convolution operators,* Acta Math. 113(1965), 181-218.

5. A. Besicovitch, *A general form of the covering principle and relative differentiation of additive functions,* Proc. Cambridge Philos. Soc. 41(1945), 103-110; 42(1946), 1-10

6. A. Zygmund, Trigonometric Series, Cambridge, 1959, chap. 12.

7. K. Yosida and S. Kakutani, *Birkhoff's ergodic theorem and the maximal ergodic theorem,* Proc. Imp. Acad. Japan 15(1939).

8. F. Riesz, *Sur la théorie ergodique,* Comm. Math. Helv. 17(1945), 221-239.

9. M. Cotlar, *A unified theory of Hilbert transforms and ergodic theory,* Rev. Mat. Cuyana 1(1955), 105-167.

10. N. Dunford and J. T. Schwartz, *Convergence almost everywhere of operator averages,* J. Rational Mech. 5(1956), 129-178.

HARMONIC ANALYSIS AND ERGODIC THEORY.*

By Norbert Wiener and Aurel Wintner.

Introduction. Let a space S of points P carry a measure which supplies for S a finite non-vanishing value, and let τ be a measure-preserving transformation of S into itself (so that Birkhoff's ergodic theorem[1] is applicable). Then, if $f(P)$ is any L-integrable function of position P on S, the j-average $M_j\{f(\tau^j P)\}$, where

$$(1) \qquad M_j\{c_j\} = \lim_{l\to\infty} \frac{1}{2l+1} \sum_{j=-l}^{l} c_j,$$

exists for almost all P.

Let C be the boundary of a circle, ρ a *fixed* rotation of C, and let the measure on C be the ordinary Lebesgue measure. Consider on the product space of S and C the product flow $\tau \times \rho$ and the corresponding product measure, and apply the ergodic theorem to this product model. It is then clear from Fubini's theorem that Birkhoff's ergodic theorem implies the following fact as a trivial consequence:[2] If $f(P)$ is any L-integrable function on S and if λ is any fixed real number, then the set of those points P of S for which $M_j\{e^{ij\lambda} f(\tau^j P)\}$ does not exist is a set of measure 0. Hence, it is clear that if Λ is any given enumerable set of real numbers λ, then there exists on S a set which is independent of the points λ of Λ and is such that if P does not belong to that 0-set, then $M_j\{e^{ij\lambda} f(\tau^j P)\}$ exists for every λ contained in Λ.

On the other hand, it is by no means clear that there exists on S a 0-set which contains all exceptional points P belonging to all points λ of Λ simultaneously, if Λ is, for instance, a Cantor set of frequencies λ. Thus there arises the question as to the existence of all Fourier averages of the function $f(\tau^j P)$ of j for almost all points P of S, if the 0-set of points excluded is required to be independent of λ, where λ is any real number.

In this direction it was recently shown[3] that, in the case of the problem

* Received November 8, 1940.

[1] G. D. Birkhoff, "Proof of a recurrence theorem for strongly transitive systems," *Proceedings of the National Academy of Sciences*, vol. 17 (1931), pp. 650-655; cf. N. Wiener, "The ergodic theorem," *Duke Mathematical Journal*, vol. 5 (1939), pp. 1-20.

[2] E. Hopf, "Ergodentheorie," *Ergebnisse der Mathematik und ihrer Grenzgebiete*, vol. 5, no. 2 (1937), pp. 55-56, where further references are given.

[3] A. Wintner, "On an ergodic analysis of the remainder term of mean motions," *Proceedings of the National Academy of Sciences*, vol. 26 (1940), pp. 126-128.

of the remainder term of the classical problem of astronomical mean motions, there exists such an ergodic harmonic analysis, i. e. that the 0-set can be chosen independent of λ $(\neq 0)$. Although in this case the ergodic flow is of a rather particular structure, it suggests the possibility that the result has nothing to do with the torus character of the underlying model (Kronecker-Weyl). We shall prove in the metrically transitive case that such is actually the case:

The Fourier average

$$M_j\{e^{ij\lambda}f(\tau^j P)\}, \quad \textit{where} -\infty < \lambda < \infty,$$

exists for every λ *and for almost all* P, *where the* 0-*set excluded is independent of* λ.

By an application of a well known procedure which amounts to the introduction of the "irreducible integral surfaces" of the isoenergetic system (cf. J. von Neumann, *loc. cit.*[9]) the case of an arbitrary measure preserving τ could, for instance, be reduced to the case of a metrically transitive τ, if suitable local-topological assumptions are made. However, only the metrically transitive case will be considered in what follows.

The proof will be based on a Lemma (§ 4) the rôle of which is that of connecting the spectrum in the sense of Wiener [4] with that part of the spectrum in the sense of Wintner [5] which is called the point spectrum. The existence of such a connection is indicated by the fact that the two terminologies indeed correspond, from the point of view of optics, to intensity and frequency distributions, respectively. The proof of this connection will require an adaptation of a proof which has been used in Wiener's harmonic analysis from the beginning.[6] In particular, use will be made of both the Abelian and the Tauberian parts of the following theorem: [7]

If a sequence

$$\cdots, c_{-2}, c_{-1}, c_0, c_1, c_2, \cdots$$

has an average, (1), *and if*

---

[4] N. Wiener, *The Fourier Integral and Certain of Its Applications*, Cambridge 1933, p. 163.

[5] A. Wintner, *Spektraltheorie der unendlichen Matrizen*, Leipzig, 1929, pp. 267-271; cf. B. Jessen and A. Wintner, "Distribution functions and the Riemann zeta function," *Transactions of the American Mathematical Society*, vol. 38 (1935), pp. 48-88 (second footnote on p. 51).

[6] N. Wiener, "The spectrum of an array," *Journal of Mathematics and Physics*, vol. 6 (1927), pp. 145-157.

[7] N. Wiener, *loc. cit.*[4], § 20; cf. *loc. cit.*[6], pp. 152-153.

$$\sum_{j=-l}^{l} |c_j|^2 = O(l) \text{ as } l \to \infty,$$

then the series

$$(2) \qquad \frac{2\epsilon}{\pi} \sum_{j=-\infty}^{\infty} c_j \left(\frac{\sin j\epsilon}{j\epsilon}\right)^2, \quad \text{where} \quad \left(\frac{\sin j\epsilon}{j\epsilon}\right)_{j=0} \equiv 1,$$

is convergent for $\epsilon > 0$ *and tends to the limit* (1) *as* $\epsilon \to 0$; *whilst if it is only assumed that the series* (2) *is convergent for* $\epsilon > 0$ *and tends to a limit as* $\epsilon \to 0$, *then the Tauberian condition* $|c_j| < const.$ *is sufficient to ensure the existence of the limit* (1).

After the general Lemma (§ 4) in harmonic analysis is established (§ 5), we combine it (§ 7) with a fact (§ 6) observed by Wiener several years ago, but not published by him.[8] In this manner the theorem italicized above follows for the case where the transformation τ of S is metrically transitive. The passage from metrically transitive transformations to arbitrary volume-preserving transformations under suitable topological restrictions for S and τ could then be treated by an adaptation of von Neumann's decomposition [9] of an arbitrary measure-preserving flow into metrically transitive components.

In § 8, the result will be applied to that case of statistical independence [10] in which τ is a mixture. It then follows, in particular, that Kolmogoroff's theorem of large numbers [11] for the case of independent random variables which have a common distribution function can be refined in the same way as the ergodic theorem of Birkhoff is refined by the theorem italicized above. In fact, Kolmogoroff's theorem is a particular case of the mixture case of Birkhoff's theorem.

It should finally be mentioned that, after obvious modifications, the results to be obtained can be transferred from sequences (sums) to functions (integrals).

1. In order to avoid an interruption of the following considerations, it will first be shown that if

$$(3) \qquad \cdots, a_{-2}, a_{-1}, a_0, a_1, a_2, \cdots$$

is any sequence of real or complex numbers for which

$$(4) \qquad \sum_j{}' \frac{|a_j|^2}{j^2} < \infty,$$

---

[8] Cf. E. Hopf, *loc. cit.*[2], p. 55.

[9] J. v. Neumann, "Zur Operatorenmethode in der klassischen Mechanik," *Annals of Mathematics*, vol. 33 (1932), pp. 587-642 and pp. 789-791.

[10] In this connection, cf. P. Hartman and A. Wintner, "Statistical independence and statistical equilibrium," *American Journal of Mathematics*, vol. 62 (1940), pp. 646-654.

[11] A. Kolmogoroff, "Grundbegriffe der Wahrscheinlichkeitsrechnung," *Ergebnisse der Mathematik und ihrer Grenzgebiete*, vol. 2, no. 3 (1933), p. 59.

where

$$\sum_j{}' = \sum_{j=-\infty}^{-1} + \sum_{j=1}^{\infty}, \tag{5}$$

then

$$\frac{1}{\epsilon}\sum_{j=-\infty}^{\infty}\left(a_j\bar{a}_{j-k}\frac{\sin j\epsilon}{j}\,\frac{\sin(j-k)\epsilon}{j-k} - a_j\bar{a}_{j-k}\frac{\sin^2 j\epsilon}{j^2}\right)\to 0 \text{ as } \epsilon\to 0 \tag{6}$$

holds for every fixed k $(=0, \pm 1, \pm 2, \cdots)$, where it is understood that

$$\frac{\sin h\epsilon}{h} = \epsilon \text{ if } h = 0. \tag{7}$$

In fact, if k is fixed and $j \to \pm\infty$, then, uniformly for $0 < \epsilon < 1$,

$$\left|\frac{\sin(j-k)\epsilon}{(j-k)\epsilon} - \frac{j\epsilon}{\sin j\epsilon}\right| = \left|\int_{j\epsilon}^{(j-k)\epsilon}\frac{d}{dx}\frac{\sin x}{x}\,dx\right| < \text{const.}\left|\int_{j\epsilon}^{(j-k)\epsilon}\frac{dx}{x}\right|,$$

while

$$\left|\int_{j\epsilon}^{(j-k)\epsilon}\frac{dx}{x}\right| < \frac{|(j-k)\epsilon - j\epsilon|}{\operatorname{Min}(|j-k|\,\epsilon,\ |j|\,\epsilon)} = \frac{|k|}{\operatorname{Min}(|j-k|,\ |j|)} < \frac{\text{const.}}{|j|}.$$

Hence, if k is fixed and $j \to \pm\infty$, then

$$\left(\frac{\sin(j-k)\epsilon}{(j-k)\epsilon} - \frac{\sin j\epsilon}{j\epsilon}\right)^2 = O\left(\frac{1}{j^2}\right)$$

holds uniformly for $0 < \epsilon < 1$. Since (4) and (7) imply that

$$\sum_{j=-\infty}^{\infty} |a_j|^2\frac{\sin^2 j\epsilon}{j^2} \to 0 \text{ as } \epsilon \to 0,$$

it follows that

$$\sum_{j=-\infty}^{\infty} |a_j|^2 \sin^2 j\epsilon\left(\frac{\sin(j-k)\epsilon}{(j-k)\epsilon} - \frac{\sin j\epsilon}{j\epsilon}\right)^2 \to 0 \text{ as } \epsilon \to 0,$$

where k is arbitrarily fixed. It follows, therefore, from (4), (5) and from the Schwarz inequality that, for every fixed k,

$$\sum_{j=-\infty}^{\infty} a_j\bar{a}_{j-k}\frac{\sin j\epsilon}{j}\left(\frac{\sin(j-k)\epsilon}{(j-k)\epsilon} - \frac{\sin j\epsilon}{j\epsilon}\right) \to 0 \text{ as } \epsilon \to 0.$$

This completes the proof of (6).

2. It follows that if (3) is any sequence satisfying (4), and if

$$\phi(x) \sim a_0 x + \sum_j{}' \frac{a_j}{ij}e^{ijx} \tag{8}$$

[cf. (5)], then, for every fixed k,

$$(9)\quad \frac{1}{8\pi\epsilon}\int_0^{2\pi} e^{-ikx}\,|\,\phi(x+\epsilon)-\phi(x-\epsilon)\,|^2\,dx - \sum_{j=-\infty}^{\infty} a_j\bar{a}_{j-k}\frac{\sin^2 j\epsilon}{j^2\epsilon} \rightarrow 0 \text{ as } \epsilon\rightarrow 0.$$

First, it is clear from the Fischer-Riesz theorem and from the assumption (4), that (8) defines $\phi(x) - a_0x$ as a function of class (L^2) and of period 2π. It is also clear from (8) that, if ϵ is arbitrarily fixed,

$$\phi(x+\epsilon)-\phi(x-\epsilon) \sim 2a_0\epsilon + \sum_j{}' a_j\frac{e^{ij\epsilon}-e^{-ij\epsilon}}{ij}e^{ijx}.$$

According to (5) and (7), this can be written in the form

$$\phi(x+\epsilon)-\phi(x-\epsilon) \sim 2\sum_{j=-\infty}^{\infty} a_j\frac{\sin j\epsilon}{j}e^{ijx}.$$

Hence, if k is any integer,

$$e^{ikx}\{\phi(x+\epsilon)-\phi(x-\epsilon)\} \sim 2\sum_{j=-\infty}^{\infty} a_{j-k}\frac{\sin(j-k)\epsilon}{j-k}e^{ijx}.$$

But application of the polarized form of the Parseval relation to the last two Fourier series (L^2) gives

$$\frac{1}{2\pi}\int_0^{2\pi} e^{-ikx}\,|\,\phi(x+\epsilon)-\phi(x-\epsilon)\,|^2\,dx = 4\sum_{j=-\infty}^{\infty} a_j\bar{a}_{j-k}\frac{\sin j\epsilon}{j}\,\frac{\sin(j-k)\epsilon}{j-k}.$$

Hence, (9) follows from (6).

3. Let (3) be a sequence for which the sequence of auto-correlations,

$$(10)\qquad \cdots, \mu_{-2}, \mu_{-1}, \mu_0, \mu_1, \mu_2, \cdots,$$

exists. By this is meant that the average (1) exists for the product $c_j = a_j\bar{a}_{j-k}$, where k is arbitrary, and that μ_k is the resulting function of k; so that

$$(11)\qquad \mu_k = M_j\{a_j\bar{a}_{j-k}\}; \qquad (k = 0, \pm 1, \pm 2, \cdots).$$

It is easy to see that

$$\mu_{-k} = \bar{\mu}_k \quad\text{and}\quad |\,\mu_k\,| \leqq \mu_0.$$

Since $\mu_0 < \infty$ by assumption, it is clear from (11) and (1) that

$$\sum_{j=-l}^{l} |\,a_j\,|^2 = O(l) \text{ as } l\rightarrow\infty;$$

hence, partial summation shows that (4) is necessarily satisfied.

It is known that every sequence of auto-correlations is a sequence of trigonometric momenta.[12] In other words, if all averages (10) exist for (3), then there exists a monotone function

[12] Cf. N. Wiener, *loc. cit.*[6]

$$\sigma(x); \qquad 0 \leqq x \leqq 2\pi,$$

for which

$$(12) \qquad M_j\{a_j\bar{a}_{j-k}\} = \frac{1}{2\pi}\int_0^{2\pi} e^{ikx} d\sigma(x); \qquad (k = 0, \pm 1, \pm 2, \cdots).$$

[In order to see this, it is sufficient to verify that

$$(12\text{ bis}) \qquad \sum_{m=0}^{N} \sum_{m=0}^{N} M_j\{a_j\bar{a}_{j-m+n}\} y_m \bar{y}_n \geqq 0$$

for arbitrary $y_0, y_1, \cdots, y_N$ and for every N. In fact, the representation (12) of (11) then follows from the Herglotz criterion for the Toeplitz-Carathéodory problem in the theory of the logarithmic potential. But (12 bis) may be verified by approximating the given sequence (3) by periodic sequences (3), for which sequences the existence of (11) and the representation (12) are trivial. The passage from the σ of the approximating sequences (3) to the σ of the given sequence (3) follows directly from Helly's theory of monotone functions.] [13]

It may be mentioned that also the converse of the theorem quoted is true.[14] In other words, there exists for every given σ a sequence (3) for which all averages (11) exist and satisfy (12). Needless to say, (3) is not uniquely determined by σ.

4. In view of the Weierstrass approximation theorem, the trigonometric momenta of an angular distribution σ determine σ uniquely, if the monotone function $\sigma(x)$, $0 \leqq x \leqq 2\pi$, is thought of as normalized, for instance, by

$$\sigma(x - 0) = \sigma(x), \qquad 0 < x \leqq 2\pi \quad \text{and} \quad \sigma(0) = 0.$$

In this sense, σ is, in view of (12), uniquely determined by the sequence (3), and will be called the spectral function of (3). On the other hand, the spectrum of (3) may be defined as the set of those points x_0 of the interval $0 \leqq x \leqq 2\pi$ for which

$$\sigma(x') < \sigma(x'') \quad \text{whenever} \quad 0 \leqq x' < x_0 < x'' \leqq 2\pi,$$

---

[13] Cf. A. Wintner, "On the distribution function of almost-periodic angular variables," *American Journal of Mathematics*, vol. 55 (1933), pp. 606-610, where further references are given.

[14] Cf. N. Wiener and A. Wintner, "On singular distributions," *Journal of Mathematics and Physics* (M. I. T.), vol. 17 (1938), pp. 233-246, § 5.

It should be mentioned as a correction of a statement made incidentally but erroneously in the introduction of the paper just mentioned, (p. 234, top), that § 1, *loc. cit.* does not lead to a monotone function which is constant almost everywhere and has Fourier-Stieltjes constants which tend to 0 as rapidly as a power of $1/n$.

where it is understood that $x = 0$ is taken to be identical with $x = 2\pi$. It is clear that every point which is a discontinuity point (mod 2π) of σ is in the spectrum of σ, but not conversely. The set of those points of the interval $0 \leqq x \leqq 2\pi$ at which σ is discontinuous (mod 2π) will be called the point spectrum of (3). Accordingly, the point spectrum is either vacuous or forms a finite or infinite (possibly dense) sequence of points on the interval $0 \leqq x \leqq 2\pi$.

The lemma indicated in the Introduction can now be formulated as follows:

LEMMA. *If a sequence* (3) *is bounded, and if its sequence* (10) *of auto-correlations* (11) *exists, then its Fourier average* $M_j\{a_j e^{ij\lambda}\}$ *exists and vanishes for every real* λ *which, when reduced* mod 2π, *is not contained in the point spectrum of* (3).

Since the point spectrum consists of the discontinuity points of the monotone function σ, it follows, in particular, that those wave lengths λ for which the amplitude $M_j\{a_j e^{ij\lambda}\}$ either does not exist or exists but does not vanish form a set which is at most enumerable. Actually, this exceptional set, if any, is stated to be contained in the set of those points each of which carries a non-vanishing fraction of the total optical intensity. In particular, the amplitude exists and vanishes for every wave length whenever the spectral function has no discontinuities. It may be mentioned that the Lemma could be refined by restricting the square sum of the absolute values of the upper averages by an inequality of the Bessel type.

It will be clear from the proof that the restriction $| a_j | <$ const. can be replaced by more general Tauberian assumptions for (3). However, it remains undecided whether or not the Lemma remains correct if (3) is not restricted by *some* Tauberian assumption.

5. The proof of the Lemma proceeds as follows:

Since $| a_j | <$ const., the series (2) is convergent for every $\epsilon > 0$ if $c_j = a_j \bar{a}_{j-k}$, where k is arbitrarily fixed. In addition, the averages (11) are supposed to exist. Hence, the Abelian theorem quoted in connection with (2) is applicable to $c_j = a_j \bar{a}_{j-k}$. Accordingly, if k is arbitrarily fixed,

$$\lim_{\epsilon \to 0} \frac{2\epsilon}{\pi} \sum_{j=-\infty}^{\infty} a_j \bar{a}_{j-k} \left(\frac{\sin j\epsilon}{j\epsilon} \right)^2 = M_j\{a_j \bar{a}_{j-k}\}, \quad \text{where} \quad \left(\frac{\sin j\epsilon}{j\epsilon} \right)_{j=0} \equiv 1.$$

Hence, from (9),

$$\lim_{\epsilon \to 0} \epsilon^{-1} \int_0^{2\pi} e^{-ikx} \, | \phi(x+\epsilon) - \phi(x-\epsilon) |^2 \, dx = 4\pi^2 M_j\{a_j \bar{a}_{j-k}\}.$$

According to (12), this can be written in the form

$$\lim_{\epsilon\to 0} \epsilon^{-1} \int_0^{2\pi} f(x) \mid \phi(x+\epsilon) - \phi(x-\epsilon) \mid^2 dx = 2\pi \int_0^{2\pi} f(x) d\sigma(x),$$

where $f(x) = e^{-ikx}$. Since $k = 0, \pm 1, \pm 2, \cdots$, it follows that the last relation holds for every $f(x)$ which is a trigonometric polynomial, and so for every x which is continuous and of period 2π. Hence, the last relation holds for any $f(x)$ which is the characteristic function of a subinterval $\alpha \leqq x \leqq \beta$ of $0 \leqq x \leqq 2\pi$, provided that $x = \alpha$ and $x = \beta$ are continuity points of $\sigma(x)$. Consequently,

$$\limsup_{\epsilon\to 0} \epsilon^{-1} \int_\alpha^\beta \mid \phi(x+\epsilon) - \phi(x-\epsilon) \mid^2 dx \leqq 2\pi \int_{\alpha-0}^{\beta+0} d\sigma(x)$$

holds whether $x = \alpha$ and $x = \beta$ are or are not continuity points of $\sigma(x)$. Accordingly, if λ is any point between α and β,

$$\limsup_{\epsilon\to 0} \epsilon^{-1} \int_{\lambda-\epsilon}^{\lambda+\epsilon} \mid \phi(x+\epsilon) - \phi(x-\epsilon) \mid^2 dx \leqq 2\pi \int_{\alpha-0}^{\beta+0} d\sigma(x).$$

Let λ be arbitrarily fixed in the interval $\alpha < \lambda < \beta$, and let $\alpha \to \lambda - 0$ and $\beta \to \lambda + 0$. Then the integral on the right tends to the saltus of $\sigma(x)$ at $x = \lambda$. On the other hand, the Schwarz inequality shows that the integral on the left is a majorant of

$$\frac{1}{2\epsilon} \left| \int_{\lambda-\epsilon}^{\lambda+\epsilon} \{\phi(x+\epsilon) - \phi(x-\epsilon)\} dx \right|^2, \quad \text{since} \quad \frac{1}{2\epsilon} = 1/\int_{\lambda-\epsilon}^{\lambda+\epsilon} 1^2\, dx.$$

Consequently,

$$\limsup_{\epsilon\to 0} \left| \int_{\lambda-\epsilon}^{\lambda+\epsilon} \frac{\phi(x+\epsilon) - \phi(x-\epsilon)}{\epsilon} dx \right|^2 \leqq 4\pi\{\sigma(\lambda+0) - \sigma(\lambda-0)\}$$

holds for every λ.

Finally, it is easily verified from (8) that

$$\int_{\lambda-\epsilon}^{\lambda+\epsilon} \{\phi(x+\epsilon) - \phi(x-\epsilon)\} dx = 2 \sum_j{}' a_j \frac{\sin j\epsilon}{j} \frac{e^{ij\lambda}}{ij} (e^{ij\epsilon} - e^{-ij\epsilon}) + 4a_0\epsilon^2;$$

cf. (5). According to (7), this can be written in the form

$$\int_{\lambda-\epsilon}^{\lambda+\epsilon} \frac{\phi(x+\epsilon) - \phi(x-\epsilon)}{\epsilon} dx = 4\epsilon \sum_{j=-\infty}^{\infty} a_j e^{ij\lambda} \left(\frac{\sin j\epsilon}{j\epsilon}\right)^2.$$

Hence, the last inequality is equivalent to

$$\limsup_{\epsilon\to 0}\left|\frac{2\epsilon}{\pi}\sum_{j=-\infty}^{\infty} a_j e^{ij\lambda}\left(\frac{\sin j\epsilon}{j\epsilon}\right)^2\right|^2 \leqq \frac{\sigma(\lambda+0)-\sigma(\lambda-0)}{\pi}.$$

Now suppose that λ is not in the point spectrum (cf. § 4). Then $\sigma(\lambda+0)=\sigma(\lambda-0)$, and so the preceding inequality reduces to

$$\lim_{\epsilon\to 0}\frac{2\epsilon}{\pi}\sum_{j=-\infty}^{\infty} a_j e^{ij\lambda}\left(\frac{\sin j\epsilon}{j\epsilon}\right)^2 = 0.$$

This means that if $c_j = a_j e^{ij\lambda}$, then (2) tends to 0 as $\epsilon\to 0$. Furthermore, these c_j remain bounded as $j\to\pm\infty$, since $|a_j| <$ const. by assumption. It follows, therefore, from the Tauberian theorem quoted in connection with (2), that $M_j\{c_j\}$ exists and is 0 for $c_j = a_j e^{ij\lambda}$.

This completes the proof of the Lemma announced in § 4.

6. Let a space S of points P carry a Lebesgue measure, and let the measure of S be finite. Let τ be an essentially one-to-one and measure-preserving transformation of S into itself. Finally, let $f(P)$ be any given function of class (L) on S. Then, according to Birkhoff's ergodic theorem, there exists an average $M_j\{f(\tau^j P)\}$ for almost all points P of S.

Suppose that the given function $f(P)$ is of class (L^2). Then, if k is any fixed integer, $g_k(P) = f(P)\bar{f}(\tau^{-k}P)$ is a function of class (L) on S. Hence there exists a finite average $M_j\{g_k(\tau^j P)\}$ for almost all points P of S, if k is fixed. Since $k = 0, \pm 1, \pm 2, \cdots$, and since the sum of a sequence of zero sets is a zero set, it follows from $g_k(P) = f(P)\bar{f}(\tau^{-k}P)$ that there exists on S a set of measure zero such that if P is any point not in that zero set, the sequence of auto-correlations,

$$(13)\qquad \mu_k \equiv \mu_k(P) = M_j\{f(\tau^j P)\bar{f}(\tau^{j-k}P)\}, \qquad (k = 0, \pm 1, \pm 2, \cdots),$$

exists for the sequence of numbers,

$$(14)\qquad a_j \equiv a_j(P) = f(\tau^j P), \qquad (j = 0, \pm 1, \pm 2, \cdots);$$

it being understood that (3) and (11) are represented by (14) and (13) respectively, where P is fixed. Correspondingly, (12) becomes

$$(15)\quad M_j\{f(\tau^j P)\bar{f}(\tau^{j-k}P)\} = \frac{1}{2\pi}\int_0^{2\pi} e^{ikx} d\sigma_P(x);\quad (k = 0, \pm 1, \pm 2, \cdots),$$

where $\sigma_P(x)$ denotes, for almost all P, the spectral function of (14), if P is fixed on the complement of a zero set.

It is clear from this proof for the existence of the averages (13), that the sequence (13) is independent of P for almost all P whenever the transformation τ is metrically transitive on S.

7. It is now easy to show that if $f(P)$ is any function of class (L) on a space S on which τ is metrically transitive, then all Fourier averages,

$$M_j\{e^{ij\lambda}f(P)\}, \qquad -\infty < \lambda < \infty,$$

exist for almost all P, where the zero set excluded is independent of λ.

It will be sufficient to prove this on the assumption that $f(P)$ is bounded. In fact, suppose that the proof has been completed for the case of bounded functions. For the given function $f(P)$ of class (L), choose a sequence of bounded functions $f_1(P), f_2(P), \cdots$ such that

$$\int_S |f(P) - f_n(P)|\, dS \to 0 \text{ as } n \to \infty.$$

By assumption, the limit

$$\lim_{l\to\infty} \frac{1}{2l+1} \sum_{j=-l}^{l} e^{ij\lambda} f_n(\tau^j P) = M_j\{e^{ij\lambda} f_n(\tau^j P)\}$$

exists for almost every P and for every λ, if n is fixed. Since the sum of a sequence of zero sets is a zero set, it follows that the zero set excluded can be chosen as independent of both λ and n. Hence, in order to show that $M_j\{e^{ij\lambda}f(\tau^j P)\}$ exists for almost all P and for every λ, it is sufficient to ascertain that

$$\limsup_{l\to\infty} \left| \frac{1}{2l+1} \sum_{j=-l}^{l} e^{ij\lambda} f(\tau^j P) - \frac{1}{2l+1} \sum_{j=-l}^{l} e^{ij\lambda} f_n(\tau^j P) \right|$$

tends with $1/n$ to 0 for almost all P and for every λ. But this upper limit is, for fixed n, fixed P, and for every λ, majorized by the upper limit of

$$\frac{1}{2l+1} \sum_{j=-l}^{l} |f(\tau^j P) - f_n(\tau^j P)|$$

as $l \to \infty$; while the latter upper limit is, for fixed n and for almost all P, identical with

$$\lim_{l\to\infty} \frac{1}{2l+1} \sum_{j=-l}^{l} |f(\tau^j P) - f_n(\tau^j P)| = \frac{1}{\text{meas } S} \int_S |f(P) - f_n(P)|\, dS,$$

since τ is metrically transitive.

This proves that $f(P)$ can be assumed to be bounded without loss of generality. In particular, $f(P)$ is now of class (L^2), and so § 6 is applicable.

Hence, the sequence of auto-correlations, (13), exists for almost all choices of P in (14). In addition, the sequence (13) is independent of P for almost all P, since τ is metrically transitive (cf. the end of § 6). It follows, therefore, from (15) and from the uniqueness theorem of the trigonometric momentum problem, that the spectral function, $\sigma_P(x)$, of (14) is independent of P for almost all P. Let

$$\lambda_1, \lambda_2, \cdots, \lambda_m, \cdots$$

denote the sequence of those real numbers which are congruent (mod 2π) with values x contained in the point spectrum of this common monotone function $\sigma \equiv \sigma_P$; it being understood that a zero set of points P is excluded, and that the λ_m exist only if σ has at least one saltus.

Since $f(P)$ is a bounded function, (14) is a bounded sequence for every fixed P. It follows, therefore, from the Lemma of § 4, that if P is chosen arbitrarily but not on a zero set, then $M_j\{e^{i\lambda j}f(\tau^jP)\}$ exists (and, incidentally, vanishes) for every λ which is distinct from every λ_m. Since $M_j\{e^{i\lambda_m j}f(\tau^jP)\}$ exists for almost every P and for every m (cf. the second paragraph of the Introduction), the proof is complete.

8. As an application, suppose that τ is a mixture. This implies that τ, besides being metrically transitive, is such that the condition of statistical independence [10]

$$M_j\{f(\tau^jP)g(\tau^jQ)\} = M_j\{f(\tau^jP)\}M_j\{g(\tau^jQ)\}$$

is satisfied for almost all points (P, Q) of the product space S with itself, if f and g are of class (L^2) on S. In particular

$$M_j\{f(\tau^jP)\bar{f}(\tau^{j-k}P)\} = M_j\{f(\tau^jP)\}M_j\{\bar{f}(\tau^{j-k}P)\}$$

holds, by Fubini's theorem, for almost all points P of S, if k is fixed; hence also if k varies. But, if meas $S = 1$ for simplicity, then

$$M_j\{f(\tau^jP)\} = \int_S f(P)\,dS \quad \text{and} \quad M_j\{\bar{f}(\tau^{j-k}P)\} = \int_S \bar{f}(P)\,dS,$$

since τ is metrically transitive. Hence, (15) reduces to

$$\left| \int_S f(P)\,dS \right|^2 = \frac{1}{2\pi}\int_0^{2\pi} e^{ikx}\,d\sigma(x),$$

where $\sigma \equiv \sigma_P$ for almost all P. Thus the k-th trigonometric moment of σ is independent of k. A σ for which this is true has at $x = 0$ a saltus and is

13

constant for $0 < x < 2\pi$. Hence, by the uniqueness of the trigonometric momentum problem, the set of those real numbers which are congruent (mod 2π) with values contained in the point spectrum consists of the multiples of 2π.

It follows, therefore, from the parenthetical remark in the last paragraph of § 7, that if τ is a mixture and if $f(P)$ is any function of class (L), then, for almost all P,

$$(16) \qquad M_j\{e^{ij\lambda}f(\tau^j P)\} = \begin{cases} 0 \text{ if } \lambda \not\equiv 0 \pmod{2\pi}, \\ \int_S f dS : \int_S dS \text{ if } \lambda \equiv 0 \pmod{2\pi}, \end{cases}$$

where the zero set excluded is independent of λ.

This implies an harmonic law of large numbers, which may be formulated as follows:

If

$$\cdots, x_{-2}(t), x_{-1}(t), x_0(t), x_1(t), x_2(t), \cdots,$$

where $0 \leqq t \leqq 1$, *is a sequence of independent functions* [15] *which are of class* (L) *and have one and the same distribution function of vanishing first moment (expectation), then* $M_j\{e^{ij\lambda}x_j(t)\}$ *exists and vanishes for almost all* t *and for every real* λ, *where the zero set excluded is independent of* λ.

Kolmogoroff's law of large numbers in case of a common distribution function [11] concerns the case $\lambda = 0$. The possibility of simultaneously including all wave lengths λ expresses a high degree of chaotic homogeneity [16] in the distribution of the values of the function $x_j(t)$ of j, where t is arbitrarily fixed in the complement of a t-set of measure 0.

MASSACHUSETTS INSTITUTE OF TECHNOLOGY,
THE JOHNS HOPKINS UNIVERSITY.

---

[15] By this is meant that any finite number of these functions forms a set of functions which are independent; cf. A. Kolmogoroff, *loc. cit.*[11], p. 50.

[16] In this connection, cf. P. Erdös and A. Wintner, "Additive functions and almost periodicity (B^2)," *American Journal of Mathematics*, vol. 62 (1940), pp. 635-645, end of the second paragraph of § 1 (p. 636).

ON THE ERGODIC DYNAMICS OF ALMOST PERIODIC SYSTEMS.*

By Norbert Wiener and Aurel Wintner.**

Introduction. The classical "trigonometric" developments in celestial mechanics,[1] that is, the developments to which the whole of the second volume of the *Méthodes Nouvelles* is devoted, are purely formal in nature. It was shown by Birkhoff[2] that the *successive stages* of the process defining these expansions are subject on *any finite time range* to certain estimates of the order of magnitude; estimates which, being exactly of the same type as those underlying his theory of non-linear difference equations, can be adapted to local proofs of existence of periodic motions. Correspondingly, the considerations of the astronomers and the investigations of Poincaré and of Birkhoff do not attempt to prove that there exist trigonometric expansions, say

$$F(t) \sim \sum_k \alpha_k e^{i\lambda_k t}, \qquad (\lambda_k \neq \lambda_j \text{ for } k \neq j),$$

which are *Fourier series*, representing for the function $F(t)$ an anharmonic analysis in a sense suggested by Poincaré himself;[3] namely, in the sense that

$$\lim_{T\to\infty} \frac{1}{2T} \int_{-T}^{T} e^{-i\mu t} F(t)\,dt = \begin{cases} 0 & \text{if } \mu \neq \lambda_k \text{ for every } k, \\ \alpha_k & \text{if } \mu = \lambda_k \text{ for some } k, \end{cases}$$

where μ is any real number.

By a recondite connection of Diophantine intricacy, the problem must somehow depend on the celebrated small divisors in celestial mechanics.[4] On the other hand, it turned out that, by virtue of a property of uniformly almost periodic angular variables, the Diophantine small divisors prove to be

* Received March 28, 1941.

** Fellow of the Guggenheim Foundation.

[1] The classical literature of the subject is collected in R. Marcolongo's bibliography, *Il problema dei tre corpi* (Manuali Hoepli, nos. 403-405), 1919, pp. 61-79. For more recent references cf. A. Wintner, *The Analytical Foundations of Celestial Mechanics*, Princeton, 1941, § 523.

[2] For a short presentation, cf. G. D. Birkhoff, *Dynamical Systems*, New York, 1927, Chap. III–Chap. IV, where references are given to the original papers.

[3] H. Poincaré, "Sur la convergence des séries trigonométriques," *Bulletin Astronomique*, vol. 1 (1884), pp. 319-327.

[4] Cf. *loc. cit.*[1], § 523–§ 529.

harmless, not only in the integrable case of Liouville systems,[5] but also in such cases (by no means trivial, though relatively simple) as Hill's theory of the perigee[6] or Adams' theory of the lunar node.[7] This suggested that the treatment of the general problem might be attempted if Bohr's theory of uniformly almost periodic functions, which deals indeed with a situation quite degenerate from the dynamical (or, equivalently, topological) point of view, were replaced by the only result of real generality established for the solutions of dynamical systems; namely, by the ergodic theorem of Birkhoff.[8]

In this direction, it was shown, first in the case of Lagrange's problem of mean motions[9] and then,[10] via the Fourier integral theory of auto-correlations,[11] for any metrically transitive system, that Birkhoff's ergodic theorem actually leads to the existence of an anharmonic analysis, as defined by the above pair of formulae. The case of an arbitrary measure-preserving flow can be reduced to the metrically transitive case, if use is made of von Neumann's transfinite decomposition of the flow into its irreducible components.[12] In this sense, the existence problem for an anharmonic analysis, as formulated above, can be considered as solved in the general case.

However, this general result does not supply the answer to the more specific dynamical question which concerns not merely the existence of an anharmonic analysis but also the problem whether the resulting Fourier series of the function $F(t)$ does or does not converge to $F(t)$ in the mean $(-\infty < t < \infty)$. Obviously, the answer to this specific question of *completeness* cannot be in the affirmative in the general case. For instance, if a surface of constant negative curvature is of finite connectivity and of finite area, then its geodesic flow is, according to Hedlund,[13] a mixture, and so it is easy to see that the anharmonic analysis cannot satisfy the condition of completeness.

---

[5] *Ibid.*, § 194–§ 198. [6] *Ibid.*, § 520–§ 522. [7] *Ibid.*, § 484–§ 487.

[8] G. D. Birkhoff, "Proof of a recurrence theorem for strongly transitive systems," "Proof of the ergodic theorem," *Proceedings of the National Academy of Sciences*, vol. 17 (1931), pp. 650-655, 656-660.

[9] A. Wintner, "On an ergodic analysis of the remainder term of mean motions," *Ibid.*, vol. 26 (1940), pp. 126-128.

[10] N. Wiener and A. Wintner, "Harmonic analysis and ergodic theory," *American Journal of Mathematics*, vol. 63 (1941), pp. 415-426.

[11] N. Wiener, *The Fourier Integral and Certain of its Applications*, Cambridge, 1933, § 20–§ 23.

[12] J. v. Neumann, "Zur Operatorentheorie in der klassischen Mechanik," *Annals of Mathematics*, vol. 33 (1932), pp. 587-642 and 789-791.

[13] G. A. Hedlund, "Fuchsian groups and mixtures," *ibid.*, vol. 40 (1939), pp. 370-383.

The object of the present paper is to delimit the class of those flows on which the condition of completeness is satisfied, i. e., on which the anharmonic analysis (in the sense defined above) becomes a Fourier analysis (in the usual sense of the word).

1. In what follows, the existence of an average, $M_t\{g(t)\}$, of a function $g(t)$, $-\infty < t < \infty$, will be meant in the following sense: $g(t)$ is of class (L) on every bounded t-interval and there exists a finite limit

$$(1) \qquad M\{g(t)\} = \lim_{\substack{-A\to\infty \\ B\to\infty}} \frac{1}{B-A}\int_A^B g(t)\,dt$$

(in other words,

$$\frac{1}{B}\int_0^B g(t)\,dt \quad\text{and}\quad \frac{1}{B}\int_{-B}^0 g(t)\,dt$$

tend to a common finite limit as $B\to\infty$). It will be very important that the limit process occurring in the definition of $M_t\{g(t)\}$ is the one given under (1), and not the limit process

$$(1\text{ bis}) \qquad B - A \to \infty$$

which is compatible with $B\to-\infty$ or $A\to\infty$. In fact, while Birkhoff's ergodic theorem [8] holds (for almost all points P of the phase space) if its statement is formulated in terms of the average definition (1), it does not,[14] in general, hold (for almost all P) if the average is referred to the unrestricted limit process (1 bis). On the other hand, it is well-known that the mean ergodic theorem,[15] resulting by integration with respect to P over the phase space, is sufficiently rough to hold even if the t-average involved belongs to the unrestricted limit process (1 bis). The method of proof in § 8 will depend precisely on this discrepancy between the two ergodic theorems.

2. Let $f(t)$, $-\infty < t < \infty$, be of class (L) on every bounded t-interval. For a fixed real number x, the x-th Fourier constant of f is defined by

$$(2) \qquad a(x) = M_t\{e^{-ixt}f(t)\},$$

provided that this average exists. If $a(x)$ exists, $a(-x)$ need not exist, since it is not assumed that $\bar{f} = f$. If (2) exists for every x, the function $f(t)$

[14] Examples to this effect can be constructed from a perusal of the proof of the dominated ergodic theorem of N. Wiener, "The ergodic theorem," *Duke Mathematical Journal*, vol. 5 (1939), pp. 1-18.

[15] Cf. G. Birkhoff, "The mean ergodic theorem," *ibid.*, pp. 19-20.

will be said to possess an amplitude function, $a(x)$. The amplitude function $a(x)$, $-\infty < x < \infty$, if it exists, is a rather discontinuous function in the relevant cases.

Let $f(t)$, $-\infty < t < \infty$, be of class (L^2) on every bounded t-interval. If the average

(3) $$c(s) = M_t\{f(t+s)\bar{f}(t)\}$$

exists for every real number s, and if $c(s)$ is a continuous function, $f(t)$ is said to possess a correlation function, $c(s)$.

It is known [11] that every function $c(s)$, $-\infty < s < \infty$, which is a correlation function (for a suitable f) is the Fourier-Stieltjes transform,

(4) $$c(s) = \int_{-\infty}^{\infty} e^{isx} d\phi(x),$$

of a non-decreasing bounded function $\phi(x)$, $-\infty < x < \infty$, the "periodogram" [16] of $f(t)$. If the monotone function ϕ is thought of as normalized, for instance by

(4 bis) $$\phi(-\infty) = 0 \quad \text{and} \quad \phi(x-0) = \phi(x), \quad (-\infty < x < \infty),$$

then, according to the uniqueness theorem of the Fourier-Stieltjes transform, the periodogram ϕ is uniquely determined by the function (4) and therefore, via (3), by the function f (on the other hand, the function (3) does not, of course, determine the function f). It may be mentioned that there exists to every non-decreasing bounded function ϕ a function f possessing the given ϕ as periodogram.[17]

It is well-known [18] that the Fourier-Stieltjes transform, (4), of any non-decreasing bounded function, ϕ, has the amplitude function

(5) $$M_s\{e^{-ixs}c(s)\} = \phi(x+0) - \phi(x-0)$$

(so that, in particular,

(5 bis) $$M_s\{e^{-ixs}c(s)\} \geqq 0$$

---

[16] This nomenclature, which we now propose in order to eliminate the confusion that both of us had a definition of "spectrum" (cf. *loc. cit.*[10], the fourth and fifth footnotes), differs from Schuster's terminology in the theory of hidden periodicities only insofar as Schuster, having been interested only in cases where $\phi(x)$ is a step function, has called the jump ($\geqq 0$) of ϕ at x (that is, the "histogram" of ϕ), and not ϕ itself, the periodogram of f.

[17] N. Wiener and A. Wintner, "On singular distributions," *Journal of Mathematics and Physics* (M. I. T.), vol. 17 (1938), pp. 233-246, § 5.

[18] This useful elementary lemma, found by P. Lévy in his *Calcul des probabilités*, Paris, 1925, pp. 169-172, has its historical origin in physical optics.

for $-\infty < x < \infty$). Clearly, (5) remains unchanged if the function $\phi(x)$ occurring in (4) is replaced by its purely discontinuous component [which can be $\equiv 0$]; that is, by the non-decreasing bounded function defined for $-\infty < x < \infty$ by

(6) $$\sum_{x_n < x} (\phi(x_n + 0) - \phi(x_n - 0)),$$

where $x_1, x_2, \cdots$ is any sequence containing all discontinuity points of ϕ.

$\phi(x)$ is called a step function if it is identical with the function (6). In this case, and only in this case, the bounded, uniformly continuous function (4), which is almost periodic (B^2) even if $\phi(x)$ has a (continuous) singular and/or absolutely continuous component, is a uniformly almost periodic function.

3. Let $f(t)$, $-\infty < t < \infty$, be any function which has a correlation function, (3), and is such that the x-th Fourier constant, (2), exists for a *fixed* x.

Under these assumptions, it is easy to verify that the function of t defined (for the fixed value of x) by

(7) $$f^{(x)}(t) = f(t) - a(x)e^{ixt}$$

possesses a correlation function,

(8) $$c^{(x)}(s) = M_t\{f^{(x)}(t+s)\bar{f}^{(x)}(t)\},$$

which turns out to be the function

(9) $$c^{(x)}(s) = c(s) - |a(x)|^2 e^{ixs};$$

and that the x-th Fourier constant of this function of s exists and is represented by

(10) $$M_s\{e^{-ixs}c^{(x)}(s)\} = \phi(x+0) - \phi(x-0) - |a(x)|^2,$$

where ϕ denotes the periodogram of $f(t)$.

In fact, since (3) exists for every s, substitution of (7) into (8) gives

$$c^{(x)}(s) = c(s) - a(x)e^{ixs}M_t\{\bar{f}(t)e^{ixt}\} - \bar{a}(x)M_t\{f(t+s)e^{-ixt}\} + |a(x)|^2 e^{ixs},$$

provided that the averages

$$M_t\{\bar{f}(t)e^{ixt}\} \quad \text{and} \quad M_t\{f(t+s)e^{-ixt}\} \equiv M_t\{f(t)e^{-ixt}e^{ixs}\}$$

exist. But they exist, and are represented by $\bar{a}(x)$ and $a(x)e^{isx}$ respectively, since (2) is supposed to exist for the given value of x. Accordingly,

$$c^{(x)}(s) = c(s) - a(x)e^{ixs}\bar{a}(x) - \bar{a}(x)a(x)e^{isx} + |a(x)|^2 e^{ixs}.$$

Since this reduces to (9), and since (9) and (5) imply (10), the proof is complete.

Its trivial character notwithstanding, the result of this simple calculation will be fundamental in the sequel, for the following reason: (10) *can be interpreted as a commutation rule, expressing the deviation which results by subjecting the function* $e^{-ixs}\bar{f}(t)f(t+s)$ *of the two variables* s, t *(where* x *is fixed) to the two iterated average operators* M_tM_s, M_sM_t; *the error committed by the interchange of the two limit processes* M_t, M_s *having precisely the value* (10).

In fact, from (2), where x is fixed,

$$|a(x)|^2 = M_t\{e^{ixt}\bar{f}(t)\}M_s\{e^{-ixs}f(s)\} \equiv M_t\{e^{ixt}\bar{f}(t)M_s\{e^{-ixs}f(s)\}\};$$

so that, since $M_s\{e^{-ixs}f(s)\} = e^{-itx}M_s\{e^{-ixs}f(t+s)\}$ for every t,

$$|a(x)|^2 = M_t\{\bar{f}(t)M_s\{e^{-ixs}f(t+s)\}\} \equiv M_t\{M_s\{e^{-ixs}f(t+s)\bar{f}(t)\}\}.$$

On the other hand, substitution of (3) into (5) gives

$$\begin{aligned}\phi(x+0)-\phi(x-0) &= M_s\{e^{-ixs}M_t\{f(t+s)\bar{f}(t)\}\}\\ &\equiv M_s\{M_t\{e^{-ixs}f(t+s)\bar{f}(t)\}\}.\end{aligned}$$

Hence, (10) can be written as a commutator,

$$\text{(10 bis)}\qquad M_s\{e^{-ixs}c^{(x)}(s)\} = [M_sM_t - M_tM_s]e^{-ixs}f(t+s)\bar{f}(t).$$

In particular

$$(11_1)\qquad M_s\{M_t\{e^{-ixs}f(t+s)\bar{f}(t)\}\} = M_t\{M_s\{e^{-ixs}f(t+s)\bar{f}(t)\}\}$$

holds, for the given value of x, if and only if

$$(11_2)\qquad M_s\{e^{-ixs}c^{(x)}(s)\} = 0$$

for the same x.

The above results can be restated as follows:

THEOREM 1. *If a function* $f(t)$, $-\infty < t < \infty$, *has a correlation function,* (3), *and if the* x*-th Fourier constant,* (2), *of* $f(t)$ *exists for a fixed* x, *then the periodogram,* ϕ, *of* $f(t)$ *is subject, at the given point* x, *to the inequality*

$$(12)\qquad |a(x)|^2 \leqq \phi(x+0) - \phi(x-0),$$

where the sign of equality holds if and only if the commutability condition (11_1) *is satisfied for the given* x.

In fact, since (7) has a correlation function, (8), it has a periodogram, say $\phi^{(x)} = \phi^{(x)}(y)$, $-\infty < y < \infty$; so that, corresponding to (4),

$$c^{(x)}(s) = \int_{-\infty}^{\infty} e^{iys} d\phi^{(x)}(y).$$

On applying the corollary (5 bis) of (5) to this Fourier-Stieltjes transform instead of to (4), one obtains the inequality

$$M_s\{e^{-iys}c^{(x)}(s)\} \geqq 0$$

for $-\infty < y < \infty$. The particular case $y = x$ of this inequality, when combined with the identity (10), implies Theorem 1, since (11_2) has been seen to be equivalent to (11_1).

4. The proof of Theorem 3 (concerning almost periodic (B^2) functions; cf. § 5 below) will depend (§ 6) on Theorem 1 alone. On the other hand, before Theorem 3 becomes applicable to the ergodic problem, it will (in § 9) be necessary to use, besides Theorem 3, a Tauberian counterpart of Theorem 1, namely the following result:[19]

THEOREM 2. *If a bounded function $f(t)$, $-\infty < t < \infty$, has a correlation function, (3)-(4), then the x-th Fourier constant, (2), of $f(t)$ exists and vanishes for all those values x which are continuity points of the periodogram, $\phi(x)$, of $f(t)$.*

The Tauberian element is represented by the assumption $|f(t)| <$ const. of Theorem 2.

Loc. cit.[10], the proof of Theorem 2 was given for sequences, instead of for functions, but it was pointed out there that the proof applies without change to the case of functions as well. It is clear also from the proof that the assumption $|f(t)| <$ const. of Theorem 2 can be replaced by other, more general, Tauberian conditions; for instance, by the restriction $f(t) \geqq 0$, if $f(t)$ is real-valued. However, it is undecided whether or not the assertion of Theorem 2 is true for *every* function possessing a correlation function.

5. Let the almost periodicity (B^2) of a function $f(t)$, $-\infty < t < \infty$, be meant in the sense that the trigonometric approximability in quadratic mean refers to the definition (1) of the average.

It will be shown that, in terms of the notions considered in § 2 and § 3, the almost periodicity (B^2) of a function $f(t)$ can be characterized, without

[19] N. Wiener and A. Wintner, *loc. cit.*[10], Lemma.

any explicit reference to the notions of approximation, translation or compactness, as follows:

THEOREM 3. *A function* $f(t)$, $-\infty < t < \infty$, *is almost periodic* (B^2) *if and only if it has*

(I) *an amplitude function;*
(II) *a correlation function;*
(III) *a periodogram which is a step function;*
(IV) *the commutability property* (11_1) *for every* x.

As an illustration, let $f(t) = \sin |t^{\frac{1}{2}}|$. Then it is easily verified that $M_t\{|f(t)|^2\} \neq 0$, and that the functions (2), (3) exist and reduce to constants: $a(x) \equiv 0$, $c(s) \equiv M_t\{|f(t)|^2\}$. Since $c(s) = \text{const.}$ means, by (4), that $\phi(x)$ is a constant multiple of the step function $\operatorname{sgn} x$, it follows that (I), (II), (III) are satisfied. But $f(t)$ is not almost periodic (B^2), since $M_t\{|f(t)|^2\} \neq 0$ and the Parseval relation are at variance with $a(x) \equiv 0$.

Accordingly, the conditions (I), (II), (III) together are not sufficient for almost periodicity (B^2); that they are necessary, is quite straightforward (cf. the beginning of § 6 below). Thus the emphasis in Theorem 2 is on the form of the additional condition, (IV).

That (IV) becomes, in virtue of (I), (II), (III), equivalent to the definition of almost periodicity (B^2), is not as surprising as appears at first glance. In fact, the validity of (11_1) for every x is equivalent to the assumption that (11_2) is an identity in x. But (11_2) can be thought of as an orthogonality relation (for every fixed x). Then the criterion represented by the set of the four conditions (I)–(IV) corresponds to the definition of the almost periodic class (B^2) (that is, to the condition of trigonometric approximability in quadratic mean) in the same way that the notion of a *closed* sequence of orthogonal functions relates to the equivalent notion of a *complete* sequence of orthogonal functions in the Hilbert space of the functions $f(t)$, $0 \leqq t \leqq 1$, of class (L^2). Of course, the situation is more delicate in the present case, since the integrals become replaced by averages and, correspondingly, the index x cannot be restricted to an enumerable set which is independent of f.

6. In order to prove Theorem 3, suppose first that $f(t)$ is almost periodic (B^2). Then (2) exists for every x, and vanishes except when x belongs to a set which is enumerable (at most). If $x_1, x_2, \cdots$ is a sequence of distinct numbers containing this set, then, by the Parseval relation,

$$(13) \qquad \sum_n |a(x_n)|^2 < \infty.$$

In these notations, the Fourier expansion (B^2) of $f(t)$ is

$$f(t) \sim \sum_n a(x_n) e^{ix_n t}.$$

Furthermore, if s is arbitrarily fixed, the translated function, $f(t+s)$, of t is almost periodic (B^2), and

$$f(t+s) \sim \sum_n a(x_n) e^{ix_n s} e^{ix_n t}.$$

Hence, by the polarized form of the Parseval relation,

$$M_t\{f(t+s)\bar{f}(t)\} = \sum_n |a(x_n)|^2 e^{ix_n s}.$$

This function of s is, by (12), continuous (and, as a matter of fact, uniformly almost periodic), and appears in the Fourier-Stieltjes form (4) if one puts

$$\phi(x) = \sum_{x_n < x} |a(x_n)|^2; \tag{14}$$

it being assured by (13) that (14) defines a non-decreasing bounded function for $-\infty < x < \infty$. Furthermore, (14) can be written in the form (6), where

$$\phi(x_n + 0) - \phi(x_n - 0) = |a(x_n)|^2; \tag{15}$$

so that $\phi(x)$ is purely discontinuous.

This completes the proof of the necessity of the conditions (I), (II), (III), if $f(t)$ is almost periodic (B^2).

Next, suppose that a given $f(t)$, which need not be almost periodic (B^2), satisfies the three conditions (I), (II), (III). The two conditions (I), (II) together can be expressed by saying that the assumptions of Theorem 1 are satisfied not for a fixed x but for every x; while condition (III) means that there exists a sequence of distinct numbers $x_1, x_2, \cdots$ by means of which the non-decreasing bounded function $\phi(x)$, $-\infty < x < \infty$, implicitly defined by (3), (4) and (4 bis), can be represented in the form (6).

Under these assumptions, define, for every positive integer N, a function $f_N(t)$, $-\infty < t < \infty$, by placing

$$f_N(t) = f(t) - \sum_{n \leq N} a(x_n) e^{ix_n t}. \tag{16}$$

Since (7) has a correlation function (8) which reduces to (9), it follows from the orthogonality relation

$$M_t\{e^{ixt} e^{-iyt}\} = \begin{cases} 0, & x \neq y \\ 1, & x = y \end{cases} \tag{17}$$

by induction from $N-1$ to N, that (16) has a correlation function

(18) $$c_N(s) = M_t\{f_N(t+s)\bar{f}_N(t)\}$$

which reduces to

(19) $$c_N(s) = c(s) - \sum_{n \leqq N} |a(x_n)|^2 e^{ix_n s}.$$

Since the assumptions of Theorem 1 are satisfied for every x, the inequality (12) holds for $-\infty < x < \infty$; so that

(20) $$|a(x_n)|^2 \leqq \phi(x_n + 0) - \phi(x_n - 0)$$

for every n. This implies (13), since

(21) $$\sum_n (\phi(x_n + 0) - \phi(x_n - 0)) < \infty,$$

$\phi(x)$ being a non-decreasing bounded function for $-\infty < x < \infty$. Furthermore, since $x_n \neq x_m$ for $n \neq m$, it is clear from (20) and (21) that the function

(22) $$\phi_N(x) = \sum_{\substack{x_n < x \\ n \leqq N}} (\phi(x_n + 0) - \phi(x_n - 0) - |a(x_n)|^2) + \sum_{\substack{x_n < x \\ n > N}} (\phi(x_n + 0) - \phi(x_n - 0))$$

is non-decreasing and bounded for $-\infty < x < \infty$. Since $\phi(x)$ is supposed to be the sum (6), one sees from (22), (19) and (4) that

(23) $$c_N(s) = \int_{-\infty}^{\infty} e^{isx} d\phi_N(x);$$

so that, since (18) is the correlation function of (16), the function (22) is the periodogram of (16). In particular

(24) $$M_t\{|f_N(t)|^2\} = \phi_N(\infty) - \phi_N(-\infty),$$

as is seen by placing $s = 0$ in (18) and (23).

All of this was deduced under the assumption that $f(t)$ is a function which satisfies (I), (II), (III). In order to complete the proof of Theorem 3, it remains to be shown that, if (I), (II), (III) are satisfied, $f(t)$ is or is not almost periodic (B^2) according as it does or does not satisfy (IV).

By definition, $f(t)$ is almost periodic (B^2) if and only if the functions (16) satisfy the requirement

$$M_t\{|f_N(t)|^2\} \to 0, \qquad (N \to \infty).$$

According to (24), this can be written in the form

$$\phi_N(\infty) - \phi_N(-\infty) \to 0, \qquad (N \to \infty).$$

This condition is reduced by (20), (21) and (22) to the equivalent requirement

$$\sum_n (\phi(x_n + 0) - \phi(x_n - 0) - |a(x_n)|^2)^2 = 0.$$

According to (20), this requirement is satisfied if and only if (15) holds for every n. Since it is not assumed that the Fourier constant $a(x_n)$ be distinct from 0 for $n = 1, 2, \cdots$, one can adjoin every given real number x to the sequence $x_1, x_2, \cdots$. In this sense, the requirement that (15) holds for every n is equivalent to the condition that the function (10) of x should vanish identically. But this is precisely the condition (IV), since (11_1) is equivalent to (11_2).

This completes the proof of Theorem 3.

7. The results of § 3–§ 6, which refer to a *single* function $f(t)$, will now be applied to the set of almost all functions in Birkhoff's ergodic theorem. To this end, it will first be necessary to collect those straightforward consequences of this theorem which involve the notions discussed in § 2.

Let a space S of points P carry a Lebesgue measure for which S is of finite measure, and let T^t, $-\infty < t < \infty$, be a cyclic group of transformations which map S on itself, are measure-preserving, and such that the functions T^tP of (t, P) satisfies the usual measurability condition. Then Birkhoff's ergodic theorem states that, for every function $f(P)$ of class (L) on S, the average $M_t\{f(T^tP)\}$ exists for almost all P, and represents a function $f^*(P)$ of class (L) on S.

Let R be a *fixed* rotation of a circle C on which the arc length is assigned as Lebesgue measure. Consider on the product space $S \times C$ the product flow $T^t \times R$ and the corresponding product measure, and apply Birkhoff's ergodic theorem to this product model. It is then clear from Fubini's theorem that, if x is any *fixed* real number, the x-th Fourier constant,

$$a_P{}^f(x) = M_t\{e^{-ixt}f(T^tP)\}, \tag{25}$$

of $f(t) = f(T^tP)$ exists, for every given $f(P)$ of class (L), for almost all P.

The proof of this well-known trivial consequence of Birkhoff's ergodic theorem has been given here only because its analogue corresponding to the mean ergodic theorem will also be needed. Let $f(P)$ be of class (L^2) on S. It has been pointed out by Wiener [20] that application of Birkhoff's ergodic theorem to a product model, when combined, as before, with Fubini's theorem, assures the existence of the correlation function

$$(26) \qquad c_P{}^f(s) = M_t\{f(T^{t+s}P)\bar{f}(T^tP)\}$$

of $f(t) = f(T^tP)$ for almost all P. The corresponding fact supplied by the mean ergodic theorem (in the mean of the function space $(L) = (L^1)$ on S) is that

$$(27) \qquad \lim_{B-A\to\infty} \int_S \left| c_P{}^f(s) - \frac{1}{B-A}\int_A^B f(T^{t+s}P)\bar{f}(T^tP)\,dt \right| d_PS = 0$$

holds for $-\infty < s < \infty$ (it is understood that d_PS denotes the volume element on S).

Let

$$(28) \qquad \phi_P{}^f = \phi_P{}^f(x), \qquad -\infty < x < \infty,$$

denote the periodogram of $f(t) = f(T^tP)$, $-\infty < t < \infty$, (for almost all P); so that, corresponding to (4),

$$(29) \qquad c_P{}^f(s) = \int_{-\infty}^{\infty} e^{isx} d_x\phi_P{}^f(x), \qquad -\infty < s < \infty,$$

and, by (5)

$$(30) \qquad M_s\{e^{-ixs}c_P{}^f(s)\} = \phi_P{}^f(x+0) - \phi_P{}^f(x-0) \text{ for every } x.$$

It turns out that, if x is any *fixed* real number, then the averages (25) and (26), which exist for almost all P, satisfy the relation

$$(31) \qquad M_s\{e^{-ixs}c_P{}^f(s)\} = |a_P{}^f(x)|^2$$

for almost all P. It is precisely (31) that will make applicable Theorems 1 and 3. Corresponding to this central rôle of (31), the proof of (31) is somewhat lengthy; it will occupy the whole of § 8.

[20] Cf. *loc. cit.*[10], the eighth footnote.

8. Let $f(P)$ be of class (L^2) on S. According to the mean ergodic theorem, there exists on S a function $f^*(P)$ of class (L^2) such that, *in the sense of* (1 bis),

$$\lim_{B-A\to\infty} \int_S | f^*(P) - \frac{1}{B-A} \int_A^B f(T^tP)\,dt\,|^2\, d_PS = 0.$$

Hence, if x is any real number, it follows, by a repetition of the product argument which led from the Birkhoff ergodic theorem to (25), that there exists for the given x a function $f^*_x(P)$ of class (L^2) on S such that

$$\lim_{B-A\to\infty} \int_S | f^*_x(P) - \frac{1}{B-A} \int_A^B e^{-ixt} f(T^tP)\,dt\,|^2\, d_PS = 0.$$

Since this means convergence in the mean (L^2) of S, it implies the existence of a subsequence which is convergent almost everywhere on S. Thus there exist two sequences, say $A_1, A_2, \cdots$ and $B_1, B_2, \cdots$, such that

$$A_n \to -\infty \quad \text{and} \quad B_n \to \infty \quad \text{as} \quad n \to \infty$$

and, for almost all P,

$$\frac{1}{B_n - A_n} \int_{A_n}^{B_n} e^{-ixs} f(T^tP)\,dt \to f^*_x(P) \quad \text{as} \quad n \to \infty,$$

where x is fixed. Since (25) exists for almost all P, it follows that

$$f^*_x(P) = a_P{}^f(x).$$

On substituting this into the definition of $f^*_x(P)$, one sees that

$$\lim_{B-A\to\infty} \int_S | a_P{}^f(x) - \frac{1}{B-A} \int_A^B e^{-ixt} f(T^tP)\,dt\,|^2\, d_PS = 0.$$

The complex conjugate of this (real) relation is

$$\lim_{D-C\to\infty} \int_S | \bar{a}_P{}^f(x) - \frac{1}{D-C} \int_C^D e^{ixs} \bar{f}(T^sP)\,ds\,|^2\, d_PS = 0,$$

where s, C, D are written in place of t, A, B. Since the quadratic mean on S has the properties of a distance (Schwarz, Minkowski), the last two relations imply that

$$\lim_{\substack{B-A\to\infty \\ D-C\to\infty}} \int_S \| a_P{}^f(x)|^2 - \frac{1}{B-A}\int_A^B e^{-ixt} f(T^tP)\,dt \frac{1}{D-C}\int_C^D e^{ixs}\bar{f}(T^sP)\,ds \mid d_PS = 0,$$

where the limit sign refers to a *double* limit, (A, B) and (C, D) being independent of one another.

Since the time ranges, $A \leqq t \leqq B$ and $C \leqq s \leqq D$, are chosen independently, and since the length $D - C$ remains unchanged if the range $C \leqq s \leqq D$ is replaced by $C + t \leqq s \leqq D + t$, where t is arbitrary, the last relation can be written in the form

$$\lim_{\substack{B-A\to\infty \\ D-C\to\infty}} \int_S \| a_P{}^f(x)|^2 - \frac{1}{B-A}\,\frac{1}{D-C}\int_A^B e^{-ixt} f(T^tP)\,dt \int_{C+t}^{D+t} e^{ixs}\bar{f}(T^sP)\,ds \mid d_PS = 0$$

But

$$\int_{C+t}^{D+t} e^{ixs}\bar{f}(T^sP)\,ds = e^{ixt}\int_C^D \bar{f}(T^{t+s}P)\,ds;$$

so that, by Fubini's theorem,

$$\int_A^B e^{-ixt} f(T^tP)\,dt \int_{C+t}^{D+t} e^{ixs}\bar{f}(T^sP)\,ds = \int_C^D e^{ixs}ds \int_A^B \bar{f}(T^{t+s}P) f(T^tP)\,dt.$$

Hence, the complex conjugate of the preceding limit relation can be written in the form

$$\lim_{\substack{B-A\to\infty \\ D-C\to\infty}} \int_S \| a_P{}^f(x)|^2 - \frac{1}{B-A}\,\frac{1}{D-C}\int_C^D e^{-ixs}\Big[\int_A^B f(T^{t+s}P)\bar{f}(T^tP)\,dt\Big]ds \mid d_PS = 0$$

Since this relation expresses convergence in the mean of $(L) = (L^1)$ on S, it implies the existence of a suitable subsequence which is convergent almost everywhere on S. Thus there exist, for the fixed value of x, two pairs of sequences of numbers, say

$$A_1, A_2, \cdots, B_1, B_2, \cdots \quad \text{and} \quad C_1, C_2, \cdots, D_1, D_2, \cdots,$$

such that

$$A_n \to -\infty,\ B_n \to \infty \text{ as } n \to \infty \text{ and } C_m \to -\infty,\ D_m \to \infty \text{ as } m \to \infty$$

and, for almost all P,

$$\lim_{\substack{n\to\infty\\ m\to\infty}} \frac{1}{B_n - A_n} \frac{1}{D_m - C_m} \int_{C_m}^{D_m} e^{-ixs} [\int_{A_n}^{B_n} f(T^{t+s}P)\bar{f}(T^tP)\,dt]\,ds = |\, a_P{}^f(x)\,|^2.$$

Since the limit on the left is a double limit, it can be written as an iterated limit, so that

$$\lim_{m\to\infty} \frac{1}{D_m - C_m} \lim_{n\to\infty} \int_{C_m}^{D_m} \frac{e^{-ixs}}{B_n - A_n} \int_{A_n}^{B_n} f(T^{t+s}P)\bar{f}(T^tP)\,dt\,ds = |\, a_P{}^f(x)\,|^2$$

for almost all P.

Since $f(P)$ is of class (L^2) on S, there exists, by Birkhoff's ergodic theorem, a finite limit

$$\lim_{\substack{-A\to\infty\\ B\to\infty}} \frac{1}{B-A} \int_A^B |\, f(T^tP)\,|^2\,dt = M_t\{|\, f(T^tP)\,|^2\}$$

for almost all P. Hence, it is easily seen from the integrability properties of f and T^tP, that, if two numbers, say C and $D(>C)$, are arbitrarily fixed, and if the point P is fixed in the complement of a set of measure zero, then the function

$$\frac{1}{B-A} \int_A^B |\, f(T^{t+s}P)\,|^2\,dt$$

of s remains for $C \leqq s \leqq D$ under a fixed bound[21] (depending on C, D, and P), as $A \to -\infty$, $B \to \infty$. Thus it is clear from the Schwarz inequality and from the assumption

$$A_n \to -\infty, \quad B_n \to \infty, \qquad (n \to \infty),$$

that, if m and P are fixed, then, unless P belongs to a set of measure zero, the functions

$$\frac{e^{-ixs}}{B_n - A_n} \int_{A_n}^{B_n} f(T^{t+s}P)\bar{f}(T^t)\,dt$$

of s, where $n = 1, 2, \cdots$, are uniformly bounded on the interval $C_m \leqq s \leqq D_m$. Furthermore, since the average (26), defined by (1), exists for $-\infty < s < \infty$ unless P belongs to a set of measure zero, the expression in the last formula

[21] For the details of the rather elementary proof, cf. N. Wiener, *loc. cit.*[11], p. 155.

line tends for $C_m \leqq s \leqq D_m$ and for almost all P to the limit $c_P{}^f(s)$, as $n \to \infty$. It follows, therefore, from Lebesgue's theorem on term-by-term integration, that

$$\lim_{n\to\infty} \int_{C_m}^{D_m} \frac{e^{-ixs}}{B_n - A_n} \int_{A_n}^{B_n} f(T^{t+s}P)\bar{f}(T^tP)\,dt\,ds = \int_{C_m}^{D_m} \quad {}_P{}^f(s)\,ds$$

for almost all P and for every m.

Accordingly, the relation containing the iterated limit reduces to

$$\lim_{m\to\infty} \frac{1}{D_m - C_m} \int_{C_m}^{D_m} e^{-ixs} c_P{}^f(s)\,ds = |\,a_P{}^f(x)\,|^2$$

for almost all P. But the average on the left of (30) exists whenever the correlation function (29) exists; so that, since the latter exists, b for almost all P,

$$\lim_{m\to\infty} \frac{1}{D_m - C_m} \int_{C_m}^{D_m} e^{-ixs} c_P{}^f(s)\,ds = \lim_{\substack{-C\to\infty \\ D\to\infty}} \frac{1}{D-C} \int_C^D e^{-ixs} c_P{}^f(s)\,ds \equiv M_s\{e^{-ixs} c_P{}^f(s)\}$$

holds, in view of $\lim_{m\to\infty} C_m = -\infty$, $\lim_{n\to\infty} D_m = \infty$, for almost all P.

The last two formula lines show that the proof of (31) is now complete; it being understood that a zero set of points P is excluded in (31) for every fixed x.

9. It is now easy to prove a theorem which in its simplest case may be formulated as follows:

If $f(P)$ is a bounded, measurable function on a space S of finite measure, and if the flow T^t is metrically transitive on S, then the function $f(T^tP)$, $-\infty < t < \infty$, is or is not almost periodic (B^2) for almost all P according as its periodogram (which, by § 7, exists for almost all P) is or is not a step function for almost all P.

In fact, the metrical transitivity of T^t means that $M_t\{g(T^tP)\}$ is constant almost everywhere on S for every function $g(P)$ of class (L). But then it is clear from the proof of the existence of the correlation functions (26), that (26) is independent of P for almost all P. It follows, therefore, from (29) and from the uniqueness theorem of the Fourier-Stieltjes transform,

9

that the monotone function (28) of x is independent of P for almost all P. Hence, the preceding italicized statement is implied by the following theorem:

THEOREM 4. *Let $f(P)$ be a bounded, measurable function on a space S of finite measure, and let T^tP, $-\infty < t \infty$, be a measure-preserving, (t, P)-measurable transformation group of S into itself; so that, by § 7, the function $f(T^tP)$, $-\infty < t < \infty$, has a periodogram $\phi_P{}^f(x)$, $-\infty < x < \infty$, for almost all P. Suppose that there exists an enumerable set of points x which contains the set $X_P{}^f$ for almost all P, where $X_P{}^f$ denotes, for almost all P, the sequence of those points x at which the monotone function $\phi_P{}^f$ is discontinuous. Then the function $f(T^tP)$, $-\infty < t < \infty$, is almost periodic (B^2) for almost all P if and only if the function $\phi_P{}^f(x)$ of x is a step function of x for almost all P.*

In the proof of Theorem 4, the assumption that $f(P)$ is a bounded function will be used only via Theorem 2. Hence, if the answer to the question formulated at the end of § 4 were in the affirmative, Theorem 4 would follow for every function $f(P)$ of class (L^2). Incidentally, it will follow from Theorem 6 below, that every function of class (L^2) can be admitted in the particular case italicized before Theorem 4.

In order to prove Theorem 4, let $f(P)$ be a fixed, bounded, measurable function on S. According to § 7, there exists, for every real number x, a zero set, say Z_x, such that the x-th Fourier constant, $a_P{}^f(x)$, of $f(T^tP)$ exists for every point P of $S - Z_x$. Since $f(T^tP)$ has, by § 7, a correlation function, $c_P{}^f$, and therefore a periodogram, $\phi_P{}^f$, for almost all P, the zero set Z_x can so be chosen that not only the x-th Fourier constant but also the periodogram of $f(T^tP)$ exists for every point P of $S - Z_x$. Finally, since (31) holds for almost all P for every fixed x, the zero set Z_x can so be chosen that (31) is satisfied by every point P of $S - Z_x$.

According to the hypothesis of Theorem 4, there exists a zero set, say Z^*, and a sequence, say $x_1, x_2, \cdots$, such that, if P is any point of $S - Z^*$, then the monotone function $\phi_P{}^f(x)$, $-\infty < x < \infty$, is defined (i. e., $f(T^tP)$ has a correlation function) and has no discontinuity points x distinct from every x_n. It follows, therefore, from Theorem 2, that, if P is any point of $S - Z^*$, the x-th Fourier constant, $a_P{}^f(x)$, of $f(T^tP)$ exists and vanishes for every x distinct from every x_n. Thus, if P is any point of $S - Z^*$, then

$$\phi_P{}^f(x+0) - \phi_P{}^f(x-0) = 0 = |a_P{}^f(x)|^2$$

for every x distinct from every x_n. Hence, it is clear from (30), that, if P is any point of $S - Z^*$, then (31) is satisfied for every x distinct from every x_n.

This, when compared with the above definition of the zero sets Z_x, $-\infty < x < \infty$, implies that, if the point P is in none of the zero sets

$$Z^*;\ Z_{x_1}, Z_{x_2}, \cdots,$$

then the x-th Fourier constant of $f(T^tP)$ exists for every x, the correlation function of $f(T^tP)$ exists, and (31) is satisfied for every x. In view of (30) and of the equivalence of the two relations (11_1), (11_2), one can express this by saying that conditions (I), (II), (IV) of Theorem 3 are satisfied by $f(t) = f(T^tP)$ for every point P not contained in the zero set

$$Z^* + Z_{x_1} + Z_{x_2} + \cdots.$$

Since Theorem 3 implies that, in case (I), (II), (IV) are satisfied, (III) is necessary and sufficient for the almost periodicity (B^2) of $f(t)$, the proof of Theorem 4 is complete.

It is clear from this proof that the assertion of Theorem 4 supplying a *necessary* condition for the almost periodicity (B^2) of almost all function $f(T^tP)$ certainly holds for arbitrary, and not only for bounded, functions $f(P)$ of class (L^2), and is, in addition, independent of the hypothesis of an enumerable set $x_1, x_2, \cdots$.

10. The assumptions of Theorem 4 are so specific that if its assertion is true for *certain* bounded, measurable functions $f(P)$ on S and for a given flow, its assertion need not be true for *every* bounded, measurable function $f(P)$ on S and for the same flow. Thus there arises the question as to a criterion which corresponds to that supplied by Theorem 4 but involves only the flow on S, without involving a particular function $f(P)$ on S. In order to deduce from Theorem 4 such a criterion, *it will from now on be assumed that there exists on S a complete orthogonal system.* This assumption is made because it will be necessary to establish the connection between the periodograms $\phi_P{}^f$, belonging to individual points P of S, on the one hand, and the considerations of Carleman [22] and Koopman,[23] concerning integrals of $f(T^tP)$ over S, on the other hand.

---

[22] T. Carleman, "Application de la théorie des équations intégrales linéaires aux équations différentielles de la dynamique," *Arkiv för Mat., Astr. och Fys.*, vol. 22 (1931), no. 7.

[23] B. O. Koopman, "Hamiltonian systems and linear transformations in Hilbert space," *Proceedings of the National Academy of Sciences*, vol. 17 (1931), pp. 315-318.

The latter considerations center around the remark that, if the transition from a function $f_0 \equiv f(P)$ of class (L^2) on S to the function $f_t \equiv f(T^tP)$, where t is fixed, is thought of as a transformation, $f_t = U^t f_0$, of the Hilbert space (L^2) of S into itself, then the measure-preserving property of T^t on S can be expressed by saying that U^t is unitary.[24] Correspondingly, the group property of T^t and the (P,t)-measurability of T^tP mean that U^t, $-\infty < t < \infty$, is a cyclic group such that, in the sense of strong convergence, $U^s \rightarrow U^t$ whenever $s \rightarrow t$.

Let $f.g$ denote the scalar product,

$$(32) \qquad f.g = \int_S f(P)\bar{g}(P)\,d_PS,$$

of two vectors of the Hilbert space (L^2) on S, and let

$$(33) \qquad \bar{f}.\Phi(x)f, \qquad -\infty < x < \infty,$$

be the spectral form of the group of unitary forms,

$$(34) \qquad \bar{f}.U^tf, \qquad -\infty < t < \infty.$$

Thus, if x is fixed, (33) is a bounded Hermitian form in f (boundedness being meant in the sense of Hilbert); while if f is fixed, (33) is a non-negative, non-decreasing, bounded function of x and can be chosen as normalized, corresponding to (4 bis), by

$$\bar{f}.\Phi(-\infty)f = 0 \text{ and } \bar{f}.\Phi(x-0)f = \bar{f}.\Phi(x)f; \text{ so that } \bar{f}.\Phi(\infty)f = \bar{f}.f.$$

Then U^t, $-\infty < t < \infty$, determines, according to Stone,[25] exactly one $\Phi(x)$, $-\infty < x < \infty$, such that (34) is representable by means of (33) in the form

[24] Since every unitary operator is bounded, and since every bounded operator in Hilbert space is, according to Toeplitz, a bounded matrix, the spectral theory of unitary matrices (A. Wintner, *loc. cit.*[25]) has *not* been generalized by replacing it by the corresponding theory of unitary operators. In this connection, cf. A. Wintner, "Dynamische Systeme und unitäre Matrizen," *Mathematische Zeitschrift*, vol. 36 (1933), pp. 630-637.

[25] M. H. Stone, "Linear transformations in Hilbert space," *Proceedings of the National Academy of Sciences*, vol. 16 (1930), pp. 137-139. Previously, the theory of the discrete, instead of continuous, unitary cyclic groups in Hilbert space, was developed by A. Wintner, "Zur Theorie der beschränkten Bilinearformen," *Mathematische Zeitschrift*, vol. 30 (1929), pp. 228-282. It was shown by S. Bochner ("Spektralzerlegung linearer Scharen unitärer Operatoren," *Sitzungsberichte der Preussischen Akademie der Wissenschaften*, 1933, p. 371), that the method applied there in the discrete case can easily be transcribed to the continuous case, and this approach appears to be the simplest among the known proofs of Stone's result.

$$(35) \qquad \bar{f}.U^t f = \int_{-\infty}^{\infty} e^{itx} d_x(\bar{f}.\Phi(x)f).$$

Hence, Φ is uniquely determined by the flow T^t, $-\infty < t < \infty$.

Correspondingly, the point spectrum of the flow can be defined to be the point spectrum in the sense of Hilbert, that is, the sequence

$$(36) \qquad X: \qquad x_1, x_2, \cdots$$

of those points x_n at which the monotone function (33) of x is discontinuous for at least one f. In particular, the flow will be said to have a *pure* point spectrum if the spectral form, (33), of the flow reduces to

$$(37) \qquad \bar{f}.\Phi(x)f = \sum_{x_n < x} \bar{f}.(\Phi(x_n + 0) - \Phi(x_n - 0))f,$$

where x and f are arbitrary and x_n ranges over the point spectrum, X.

A connection between these notions, which refer to the whole of the flow, and notions involving individual stream lines in the flow can easily be established, as follows:

THEOREM 5. *If (33) denotes the spectral form of a measure-preserving, (P, t)-measurable flow T^t on a space S of finite measure, then*

$$(38) \qquad \int_S \phi_P{}^f(x)\, d_P S = \bar{f}.\Phi(x)f, \qquad -\infty < x < \infty,$$

for every function $f(P)$ of class (L^2) on S, where $\phi_P{}^f$ denotes, for almost all P, the periodogram of $f(T^tP)$, $-\infty < t < \infty$.

COROLLARY. *A real number, x_0, is a discontinuity point of the monotone function $\phi_P{}^f$ of x for at least one fixed f and for a P-set of positive measure, if and only if x_0 is in the point spectrum, X, of the flow; furthermore, the flow has a pure point spectrum if and only if $\phi_P{}^f$ is, for every fixed f and for almost all P, a step function of x.*

In order to prove Theorem 5, let $f(P)$ be a function of class (L^2) and let s, A, B be real numbers. Then, according to Fubini's theorem,

$$\int_S \int_A^B f(T^{s+t}P)\bar{f}(T^tP)\, dt\, d_P S = \int_A^B \int_S f(T^{s+t}P)\bar{f}(T^tP)\, d_P S\, dt.$$

Hence, on replacing in the inner integral on the right the integration variable P by T^tP, where t is fixed, one sees from

$$\int_A^B \int_S f(T^sP)\bar{f}(P)\,d_PS\,dt = (B-A)\int_S f(T^sP)\bar{f}(P)\,d_PS$$

that

$$\frac{1}{B-A}\int_S \int_A^B f(T^{s+t}P)\bar{f}(P)\,dt\,d_PS = \int_S f(T^sP)\bar{f}(P)\,d_PS.$$

It follows, therefore, from (27) that

$$\int_S c_P{}^f(s)\,d_PS = \int_S f(T^sP)\bar{f}(P)\,d_PS.$$

On applying (29) on the left and (32) on the right, one obtains

$$\int_S \int_{-\infty}^{\infty} e^{isx} d_x\phi_P{}^f(x)\,d_PS = \bar{f}\,.\,U^sf,$$

since $U^sf = f(T^sP)$, by the definition of U^s. Hence, on writing t for s, one sees, by applying Fubini's theorem on the left and (35) on the right, that

$$\int_{-\infty}^{\infty} e^{itx} d_x\left(\int_S \phi_P{}^f(x)\,d_PS\right) = \int_{-\infty}^{\infty} e^{itx} d_x(\bar{f}\,.\,\Phi(x)f),$$

where $-\infty < t < \infty$; whence (38) follows from the uniqueness theorem of the Fourier-Stieltjes transform.

11. The flows which are almost periodic in the sense indicated at the beginning of § 10 can now be characterized as follows:

THEOREM 6. *Let T^t be a measure-preserving, (P,t)-measurable flow on a space S of finite measure, and let $q \geqq 1$. Then $f(T^tP)$, $-\infty < t < \infty$, is, for every function $f(P)$ of class (L^q) on S and for almost all P, almost periodic (B^q) if and only if the flow has a pure point spectrum.*

In other words, *the flow transforms every Lebesgue integrable function on S into a Besicovitch almost periodic function of t for almost all P if and only if the flow is Bohr almost periodic.* It is understood that by the Bohr almost periodicity of the flow is meant the Bohr almost periodicity of the scalar function (35) of t, where $f = f(P)$ is any function of class (L^2) on S. Then the equivalence of the last italicized statement with Theorem 6 is clear from the remark made at the end of § 2.

In order to prove Theorem 6, associate with every $f(P)$ on S the sequence of functions $f_1(P)$, $f_2(P)$, $\cdots$ on S which are defined by

$$(39) \qquad f_n(P) = \begin{cases} f(P), & \text{if } |f(P)| \leq n; \\ 0, & \text{if } |f(P)| > n. \end{cases}$$

It is then clear, either from Birkhoff's proof of his ergodic theorem or from the proof of Wiener's dominated ergodic theorem, that

$$(39 \text{ bis}) \qquad M_t\{|f(T^tP) - f_n(T^tP)|^q\} \to 0 \quad \text{as} \quad n \to \infty$$

holds for almost all P, if $f(P)$ is any function of class (L^q) on S. Hence, if $f_n(T^tP)$ is almost periodic (B^q) for every n and for a fixed P which is not in a certain set of measure zero, then $f(T^tP)$ is almost periodic (B^q) for the same P. But every $f_n(T^tP)$ is, by (39), a bounded function of t and is, therefore, either almost periodic (B^q) for every $q \geqq 1$ or not almost periodic (B^q) for any $q \geqq 1$. Hence, Theorem 6 is equivalent to the following statement: For every measurable, bounded function $f(P)$ on S, the function $f(T^tP)$ of t is almost periodic (B^2) for almost all P if and only if the flow has a pure point spectrum. Since the truth of this statement is clear from Theorem 4 and from the Corollary of Theorem 5, the proof of Theorem 6 is complete.

In order to attempt a dynamical understanding of the actual content of Theorem 6, use will be made of Koopman's interpretation [26] of the point spectrum in terms of the "first integrals" of the flow T^t; an interpretation which holds also when the point spectrum is not pure. In fact, whether the point spectrum, (36), does or does not satisfy the condition, (37), of purity, a real number, x, is in the point spectrum if and only if e^{ixt} is a characteristic number of the unitary form (35) for every t, i. e., if and only if there exists on S a function $F = F(P)$ which is not 0 (almost everywhere on S), is of class (L^2), and has the property that

$$(40) \qquad e^{ixt}F(P) = F(T^tP), \quad \text{where} \quad -\infty < t < \infty,$$

is an identity (for almost all P). A corresponding statement holds concerning the multiplicities of characteristic numbers. But if $x = 0$, then (40) reduces to

$$F(P) = F(T^tP), \qquad -\infty < t < \infty,$$

which means that $F(P)$ is a "first integral" of the flow T^t (provided that a subset of

[26] B. O. Koopman, *loc. cit.*[23], p. 318.

S of measure zero is disregarded), unless $F(P)$ is constant on S (almost everywhere). If, on the other hand, $x \neq 0$, then, on assuming that $F(P)$ is regular enough to make

$$F^*(P) = \arg F(P), \text{ i.e., } \exp iF^*(P) = F(P)/|F(P)|,$$

a meaningful definition of a function F^* of the position P on S (almost everywhere), one sees that (40) can be written in the form

(40 bis) $$F^*(T^tP) \equiv xt + F^*(P) \pmod{2\pi}, \quad \text{where} \quad -\infty < t < \infty.$$

But (40 bis) means that also $F^*(P)$ corresponds to a "first integral," namely, to one belonging to an "ignorable coördinate" leading to an "angular variable." This is the interpretation of Koopman, referred to before.

Suppose now that the flow on S has a *pure* point spectrum. Then, and only in this case, the characteristic numbers x and the corresponding characteristic functions $F(P)$ determine all unitary invariants of the flow on the Hilbert space (L^2) of S. Accordingly, the flows of pure point spectrum can be thought of as characterized by the fact that their complete system of unitary invariants is expressible in terms of "first integrals" alone. But then Theorem 6 means that the "first integrals" of exactly those flows have this completeness property which determine "conditionally periodic" paths as "general (or, rather, generic) solutions."

This seems to agree with certain statements of physicists, which have never been motivated in a mathematically reasonable direction; actually, most of the problems in question are not even meaningful without definite topological assumptions in the large. Correspondingly, the curious fact expressed by the following theorem is likely to have topological implications in the dynamical case, where S is a manifold and T^t is a continuous flow on S.

THEOREM 7. *If a measure-preserving, $(P.t)$-measurable flow T^t on a space S of finite measure has the property that $f(T^tP)$, $-\infty < t < \infty$, is, for every function $f(P)$ of class (L^2) on S and for almost all P, almost periodic (B^2), then all the Fourier exponents of $f(T^tP)$ are contained, for almost all P, in the point spectrum, X, of the flow, and so in a single enumerable set which is independent of P (and also of f).*

Theorem 6 and the hypothesis of Theorem 7 imply that the flow has a pure point spectrum. Hence, Theorem 7 follows from Theorem 2 and from the Corollary of Theorem 5. In fact, a real number x is called a Fourier exponent of an almost periodic function $f(t)$, $-\infty < t < \infty$, if the x-th Fourier constant of $f(t)$ does not vanish.

12. The results obtained will now be applied to the case of metrical transitivity. In order to simplify some of the formulae belonging to this case, it will be assumed that meas $S = 1$. This is only a normalization, involving no loss of generality, since the case meas $S = \infty$ was always excluded, while the case meas $S = 0$ is always trivial. Since meas $S = 1$, the flow T^t on S is metrically transitive if and only if

$$M_t\{f(T^tP)\} = \int_S f(P)\, d_PS$$

holds for almost all P, whenever $f(P)$ is of class $(L) = (L^1)$ on S.

THEOREM 8. *If the flow T^t on S, where meas $S = 1$, is metrically transitive, there exists for every function $f(P)$ of class (L) on S a subset, Z, of S which is of measure zero and such that, unless P is in $S - Z$,*

(i) *the x-th Fourier constant, $a_P{}^f(x)$, of the function $f(T^tP)$, $-\infty < t < \infty$, (which need not be almost periodic (B) in this case) exists for every x and determines an "intensity," $|a_P{}^f(x)|$, which is independent of P (so that only the "phase," arg $a_P{}^f(x)$, of the x-th "amplitude," $a_P{}^f(x)$, can vary with P), where Z is independent of x;*

(ii) *the point spectrum, X, of the flow (which may or may not be a pure point spectrum) contains every Fourier exponent ("frequency") of $f(T^tP)$, $-\infty < t < \infty$, i. e., every real number x satisfying $a_P{}^f(x) \neq 0$;*

(iii) *in the particular case where the function f on S is of class (L^2), the correlation and intensity functions of $f(T^tP)$, $-\infty < t < \infty$, are explicitly given by*

$$\text{(I)} \qquad \phi_P{}^f(x) = \bar{f}.\Phi(x)f, \qquad -\infty < x < \infty$$

and

$$\text{(II)} \qquad |a_P{}^f(x)|^2 = \bar{f}.(\Phi(x+0) - \Phi(x-0))f, \qquad -\infty < x < \infty,$$

where $\bar{f}.\Phi(x)f$ denotes the spectral form of the flow.

In order to prove this theorem, it will be sufficient to establish the first part of (iii), i. e. (I), for every f of class (L^2), but (i), (ii) and the second part of (iii), i. e. (II), only under the assumption that the given function on S, instead of being of class (L) and of class (L^2) respectively, is bounded and measurable. In fact, if f is of class (L) and f_n denotes the function (39), then an obvious adaptation of (39 bis) shows that, when x is fixed,

$$\lim_{n\to\infty} a_P{}^{f_n}(x) = a_P{}^f(x) \text{ for almost all } P;$$

so that the statements (i), (ii) follow for the given function f of class (L), if they are granted for every f_n. If in addition f is of class (L^2), then, by (39),

$$\lim_{n\to\infty} |\bar{f}.f - \bar{f}_n.f_n| = 0;$$

so that (II) follows for the given f, if it is granted for every f_n.

Next, let f be of class (L^2). Then, as shown in the proof of the italicized statement at the beginning of § 9 (in the paragraph preceding Theorem 4), the periodogram $\phi_P{}^f$ (exists and) is independent of P for almost all P. Hence, (I) follows from (38), where meas $S = 1$, for every f of class (L^2).

Accordingly, the proof of Theorem 8 will be complete if one verifies (i), (ii) and (II) for every bounded $f(P)$. But (II) implies the whole of (ii) and it also implies the second part of (i), i.e., the statement that only the phase of $a_P{}^f(x)$ can vary with P. Finally, (I) reduces (II) to the relation

$$|a_P{}^f(x)|^2 = \phi_P{}^f(x+0) - \phi_P{}^f(x-0).$$

Hence, in order to complete the proof of Theorem (8), it is sufficient to prove that there exists for every bounded measurable function, f, a zero set, Z, such that $a_P{}^f(x)$ exists for $-\infty < x < \infty$ and satisfies the last relation, unless P is in $S - Z$. But the existence of such a Z follows from (31) and from Theorems 2 and 3 in the same way as in the proof of Theorem 8; in fact, (I) implies that the enumerability hypothesis of Theorem 8 is satisfied.

The simplest illustration of Theorem 8 is supplied by the "strongly mixing" flows, considered by Koopman and von Neumann.[27] In fact, these flows can be characterized as those metrically transitive flows for which the point spectrum, X, consists of a single point, $x = 0$; so that the criterion of Theorem 6 for almost periodicity (B) is violated in its full extent. The other extreme case, that is, the case of a metrically transitive flow with a pure point spectrum, is represented by any Kronecker-Weyl flow on a torus.

Appendix.

As mentioned at the beginning of § 5, the almost periodicity (B^2) of a measurable function $f(t)$, $-\infty < t < \infty$, has thus far been meant in the sense that there exists for $f(t)$ a sequence of trigonometric polynomials, say

(41) $$r_1(t), r_2(t), \cdots, r_n(t), \cdots,$$

such that

(42) $$M_t\{|f(t) - r_n(t)|^2\} \text{ exists for } n = 1, 2, \cdots$$

and

(43) $$M_t\{|f(t) - r_n(t)|^2\} \to 0 \quad \text{as} \quad n \to \infty,$$

where $M_t\{\cdots\}$ is meant in the following sense:

(44) $$M_t\{g(t)\} = \lim_{\substack{-A\to\infty \\ B\to\infty}} \frac{1}{B-A}\int_A^B g(t)\,dt, \text{ i.e.,} = \lim_{A\to\pm\infty} \frac{1}{A}\int_0^A g(t)\,dt.$$

Almost periodicity (B^2) in this sense is a sharper property than almost periodicity (B^2) according to the original definition of Besicovitch.[28] In fact, he requires the existence of a sequence (41) of trigonometric polynomials

[27] B. O. Koopman and J. v. Neumann, "Dynamical systems of continuous spectra," *Proceedings of the National Academy of Sciences*, vol. 18 (1932), pp. 255-263.

[28] Cf. A. S. Besicovitch, *Almost Periodic Functions*, Cambridge, 1932.

satisfying (42) and (43) when $M_t\{\cdots\}$ is defined not by (44) but as a principal limit, i. e., by

$$M_t\{g(t)\} = \lim_{A\to\infty} \frac{1}{2A} \int_{-A}^{A} g(t)\,dt. \tag{45}$$

The replacement of (44) by (45) is essential, for instance, in the theory of the Riemann zeta-function,[29] since those theorems on ordinary Dirichlet series which center about the mean-value theorem of Schnee do not, in general, hold if (44) is required instead of the more inclusive definition, (45), of $M_t\{\cdots\}$. Needless to say, the proofs of Theorems 1, 2, 3 hold, without any change, if (44) is replaced by (45) in (2), (3), (4) and (42)-(43). The resulting wordings of Theorems 1, 2, 3 are then, of course, independent of their original wordings. On the other hand, Theorem 5 then expresses a weaker statement than its original wording.

The situation with regard to Theorem 5 becomes quite different if (44) is replaced not by (45) but by

$$M_t\{g(t)\} = \lim_{B-A\to\infty} \frac{1}{B-A} \int_{A}^{B} g(t)\,dt, \tag{46}$$

where the limit process consists of an indefinite increase of the *length* of the integration domain $A \leqq t \leqq B$; a process which is compatible with either $A \to \infty$ or $B \to -\infty$. Let $f(t)$ be called almost periodic (W^2) if there exists a sequence (41) of trigonometric polynomials satisfying (42) and (43) in the sense of (46). It is easy to see that $f(t)$ is almost periodic (W^2) if and only if it is almost periodic in the sense of Weyl.[30] The proofs of Theorems 1, 2, 3 hold without change if (44) is replaced by (46) in (2), (3), (4) and (42)-(43) and, correspondingly, (B^2) in Theorem 3 by (W^2).

On the other hand, *Theorems 4, 5, 6 become false if* (44) *is replaced by* (46) *and, correspondingly,* (B^q) *by* (W^q). This is implied by the fact that a measure-preserving, (P, t)-measurable flow T^t on S may or may not be of the unrestricted type, that is to say such that $M_t\{f(T^tP)\}$ exists, for every function $f(P)$ of class (L) on S and for almost all P, if $M_t\{\cdots\}$ is meant in the sense (46), instead of the restricted sense (44) of Birkhoff's ergodic theorem.

---

[29] Cf. A. Wintner, "The almost periodic behavior of $1/\zeta(1+it)$," *Duke Mathematical Journal*, vol. 2 (1936), pp. 443-446; "Riemann's hypothesis and almost periodic behavior," *Universidad Mayor de San Marcos*, Lima, vol. 61 (1939), pp. 575-585.

[30] H. Weyl, "Integralgleichungen und fastperiodische Funktionen," *Mathematische Annalen*, vol. 97 (1926), pp. 338-356.

ADDENDUM.*

Since this paper was written, it has been possible to decide the alternative formulated at the end of § 4. The answer to the question turns out to be affirmative. In other words, the assertion of Theorem 2 holds without the assumption

(47) $$|f(t)| < \text{const.}$$

also:

THEOREM 2 bis. *If $f(t)$ has a correlation function, the x-th Fourier constant of $f(t)$ exists and vanishes for all those values of x which are continuity points of the periodogram of $f(t)$.*

In order to simplify the formulae occurring in the proof, let (1), where $-\infty < t < \infty$, be replaced by

(48) $$M_t\{g(t)\} = \lim_{u\to\infty} \frac{1}{u}\int_0^u g(t)\,dt,$$

where $0 \leqq t < \infty$. It is clear from the remark which follows (1), that this modification is unessential.

For sake of a further typographical simplification, let it be assumed that

(49) $$f(t) \equiv 0 \quad \text{for} \quad 0 \leqq t \leqq 1.$$

This assumption involves no loss of generality, since none of the t-averages M occurring in Theorem 2 bis is influenced by a change of the values of $f(t)$ on a bounded t-interval.

In view of (49), the function

(50) $$F^x(t) = \frac{1}{t}\int_0^t e^{-ixv} f(v)\,dv$$

is bounded on every bounded t-interval. Since the existence of the correlation function, (3), implies that

(51) $$M_t\{|f(t)|^2\} < \infty,$$

it follows from the Schwarz inequality that

(52) $$|F^x(t)| < \text{const.}$$

for $0 \leqq t < \infty$ and for every x.

The proof of Theorem 2 was based[10] on a Tauberian fact, according to which $M_t\{g(t)\}$ exists and vanishes whenever

* Received May 2, 1941.

(53) $$|g(t)| < \text{const.}$$

and

(54) $$\int_0^\infty g(ut)\,\frac{\sin^2 t}{t^2}\,dt \to 0 \quad \text{as} \quad u \to \infty.$$

It was shown *loc. cit.*[10] that if $f(t)$ has a correlation function, and if x is any continuity point of the periodogram of $f(t)$, then

(55) $$\lim_{\epsilon \to \infty} \epsilon \int_0^\infty e^{-ixt} f(t) \left(\frac{\sin \epsilon t}{\epsilon t}\right)^2 dt = 0,$$

whether the additional restriction $|f(t)| < \text{const.}$ is or is not satisfied. In fact, (55) becomes identical with the last formula line of § 5, *loc. cit.*[10], if one replaces the sequence $\{a_n\}$, considered there, by the present case of a function, $f(t)$. The restriction $|f(t)| < \text{const.}$ was there used, not in the proof of (55), but only in the passage from (55) to

(56) $$M_t\{e^{-ixt} f(t)\} = 0;$$

this passage having been based on the Tauberian theorem quoted before.

In order to prove Theorem 2 bis, it will now be shown that (51) and (55) imply (56) without the restriction $|f(t)| < \text{const.}$ also (it being understood that the *existence*, and not the vanishing, of the mean-value (56) is the essential part of the statement).

First, on writing t for ϵt in (55),

$$\int_0^\infty e^{-ixt/\epsilon} f\!\left(\frac{t}{\epsilon}\right) \frac{\sin^2 t}{t^2}\,dt \to 0 \quad \text{as} \quad \epsilon \to 0.$$

Hence, on placing $v = 1/\epsilon$ and averaging with respect to v between $v = 0$ and a large $v (= u)$,

$$\frac{1}{u} \int_0^u \left(\int_0^\infty e^{-ixtv} f(vt)\,\frac{\sin^2 t}{t^2}\,dt \right) dv \to 0 \quad \text{as} \quad u \to \infty,$$

and so, by Fubini's theorem,

$$\int_0^\infty \left(\frac{1}{u} \int_0^u e^{-ixtv} f(vt)\,dv \right) \frac{\sin^2 t}{t^2}\,dt \to 0 \quad \text{as} \quad u \to \infty$$

(in fact, (49) and (51) imply that the last iterated integral, when written as a double integral, exists absolutely for every fixed $u > 0$). Since

$$\frac{1}{u}\int_0^u e^{-ixtv}f(vt)\,dv \equiv \frac{1}{ut}\int_0^{ut} e^{-ixt}f(v)\,dv$$

is, by (50), identical with $F^x(ut)$, it follows that

$$\int_0^\infty F^x(ut)\,\frac{\sin^2 t}{t^2}\,dt \to 0 \quad \text{as} \quad u \to \infty.$$

The last relation and (52) show that both conditions, (54) and (53), of the Tauberian theorem are satisfied by $g = F^x$. Hence, $M_t\{g(t)\}$ exists and vanishes for $g = F^x$. This means, by (48) and (50), that

$$\frac{1}{u}\int_0^u \frac{dt}{t}\int_0^t e^{-ixv}f(v)\,dv \to 0 \quad \text{as} \quad u \to \infty.$$

Since (49) and Fubini's theorem imply that

$$\int_0^u\left(\int_0^t \frac{e^{-ixv}f(v)}{t}\,dv\right)dt = \int_0^u\left(\int_t^u \frac{e^{-ixt}f(t)}{v}\,dv\right)dt,$$

and since

$$\int_t^u \frac{e^{-ixt}f(t)}{v}\,dv = e^{-ixt}f(t)\log\frac{u}{t},$$

it follows that

$$(57)\qquad \frac{1}{u}\int_0^u e^{-ixt}f(t)\log\frac{u}{t}\,dt \to 0, \qquad \text{[cf. (49)]},$$

where $u \to \infty$.

The assertion of Theorem 2 bis is (56), that is, by (48),

$$(58)\qquad \frac{1}{u}\int_0^u e^{-ixt}f(t)\,dt \to 0.$$

While it is an obvious Abelian fact that (58) implies (57), it is clear that (57) in itself cannot imply (58). However, it will now be shown that (57) implies (58) in virtue of (51). This will complete the proof of Theorem 2 bis.

On placing $g(t) = e^{-ixt}f(t)$, where x is fixed, one sees from (49) that (57) and (58) become

$$(59) \qquad \frac{1}{u}\int_1^u g(t)\log\frac{u}{t}\,dt \to 0$$

and

$$(60) \qquad \frac{1}{u}\int_1^u g(t)\,dt \to 0$$

respectively; while (51) becomes

$$(61) \qquad M_t\{|g(t)|^2\} < \infty.$$

Thus all that is to be proved is that (59) implies (60), if (61) is satisfied. Actually, (59) *implies* (60) *whenever*

$$(62) \qquad \overline{\lim_{u\to\infty}}\frac{1}{u}\int_1^u |g(t)|^2 dt < \infty$$

(and even if only

$$(62\text{ bis}) \qquad \overline{\lim_{u\to\infty}}\frac{1}{u}\int_1^u |g(t)|^p\, dt < \infty$$

holds for some $p > 1$).

In order to prove this, let ϑ be a fixed number in the interval $0 < \vartheta < 1$. Then $u \to \infty$ is equivalent to $\vartheta u \to \infty$; so that, on writing ϑu for u in (59), one sees that

$$\frac{\log\vartheta}{\vartheta u}\int_1^{\vartheta u} g(t)\,dt + \frac{1}{\vartheta u}\int_1^{\vartheta u} g(t)\log\frac{u}{t}\,dt \to 0$$

as $u \to \infty$. Since

$$\frac{1}{\vartheta u}\int_1^{\vartheta u} g(t)\log\frac{u}{t}\,dt + \frac{1}{\vartheta u}\int_{\vartheta u}^{u} g(t)\log\frac{u}{t}\,dt \equiv \frac{1}{\vartheta u}\int_1^{u} g(t)\log\frac{u}{t}\,dt \to 0,$$

by (59), it follows that

$$\frac{\log\vartheta}{\vartheta u}\int_1^{\vartheta u} g(t)\,dt - \frac{1}{\vartheta u}\int_{\vartheta u}^{u} g(t)\log\frac{u}{t}\,dt \to 0;$$

so that, since

$$\left|\int_{\vartheta u}^{u} g(t)\log\frac{u}{t}\,dt\right| \leqq \int_{\vartheta u}^{u} |g(t)|\log\frac{u}{\vartheta u}\,dt = |\log\vartheta|\int_{\vartheta u}^{u}|g(t)|\,dt,$$

division by $|\log\vartheta|$ gives

$$\overline{\lim_{u\to\infty}}\left|\frac{1}{\vartheta u}\int_{1}^{\vartheta u} g(t)\,dt\right| \leqq \frac{1}{\vartheta}\,\overline{\lim_{u\to\infty}}\,\frac{1}{u}\int_{\vartheta u}^{u}|g(t)|\,dt.$$

But the upper limit on the left is independent of ϑ; so that, on estimating the integral on the right by the Schwarz inequality

$$\int_{\vartheta u}^{u}|g(t)|\,dt \leqq \left(\int_{\vartheta u}^{u} 1^2 dt\right)^{\frac{1}{2}}\left(\int_{\vartheta u}^{u}|g(t)|^2\right)^{\frac{1}{2}} = u(1-\vartheta)^{\frac{1}{2}}\left(\frac{1}{u}\int_{\vartheta u}^{u}|g(t)|^2\,dt\right)^{\frac{1}{2}},$$

one obtains the inequality

$$\overline{\lim_{u\to\infty}}\left|\frac{1}{u}\int_{1}^{u} g(t)\,dt\right| \leqq \frac{(1-\vartheta)^{\frac{1}{2}}}{\vartheta}\,\overline{\lim_{u\to\infty}}\left(\frac{1}{u}\int_{1}^{u}|g(t)|^2\,dt\right)^{\frac{1}{2}}.$$

On letting here $\vartheta\to 1-0$, one sees from the assumption (62), that the proof of (60) is complete.

It is clear from this proof that (62) can be generalized to (62 bis), since the inequality of Schwarz can be replaced by that of Hölder.

MASSACHUSETTS INSTITUTE OF TECHNOLOGY,
THE JOHNS HOPKINS UNIVERSITY.

Commentary on [41a, b]

W. A. Veech*

These papers are concerned with the spectral analysis and synthesis, in a certain sense, of square integrable, stationary stochastic processes (X_t, $t \in \Gamma$), where Γ is either the group of integers or the group of real numbers.

Let there be given a representation, $(\Gamma, \mathcal{X})$, of Γ by invertible, measure-preserving transformations (automorphisms) of a finite measure space, $\mathcal{X} = (X, \mathcal{B}, \mu)$ where it is assumed that if $\Gamma = \mathbb{R}$ (= real numbers), the mapping $(t, x) \to tx$ is measurable from $\Gamma \times X$ to X. If $f \in L^2(\mathcal{X})$, then for almost all x, f_x is a locally square integrable function on Γ, where $f_x(t) = f(tx)$, $t \in \Gamma$. Under suitable hypotheses it is proved that, for fixed f, "almost every" function f_x possesses an *amplitude function*,

$$a_x(\lambda) = M_s(f_x(s)e^{-i\lambda s}) = \lim_{\substack{B\to+\infty \\ A\to+\infty}} \frac{1}{B-A} \int_A^B f_x(s)e^{-i\lambda s}\, ds,$$

the limit existing for every $\lambda \in \mathbb{R}$. (To unify notations we use the integral sign for the Lebesgue integral on $\mathbb{R}$ and summation on $\mathbb{Z}$ (= integers).) This is the sense of spectral ("anharmonic") analysis for the papers under discussion. The problem then is to determine when almost all the functions f_x can be synthesized from their "Fourier series", $f_x \sim \Sigma_\lambda a_x(\lambda)e^{i\lambda t}$ (where the sum is over $\lambda \in [-\pi, \pi)$ if $\Gamma = \mathbb{Z}$ and in all cases at most a countable number of terms are nonzero). Synthesis is in the sense of B^2 approximation: that is, there should exist a sequence of trigonometric polynomials P_n^x (which can actually be taken to be partial sums of the Fourier series) such that

$$M_s(|f_x(s) - P_n^x(s)|^2) \to 0 .$$

Functions with this property are B^2*-almost periodic.*† It is proved, for example, that if $(\Gamma, \mathcal{X})$ is ergodic, then for all f, f_x is B^2-almost periodic for almost all x, if and only if $(\Gamma, \mathcal{X})$ has pure point spectrum. For further discussion it is natural to separate the question of spectral analysis from that of synthesis.

Spectral analysis.

To prove amplitude functions exist the authors make use of the Birkhoff

*Research supported by NSF-GP-18961 and Alfred P. Sloan Foundation.

†The "B" refers to A. S. Besicovitch.

ergodic theorem and the following three facts, the second of which appears, in its present generality, as an addendum to the second paper [41b].

I. We say the locally square integrable function g on Γ possesses a *correlation function*, $\sigma(t)$, if $M_s(g(s+t)\overline{g}(s))$ exists for each $t \in \Gamma$ and is continuous in t. Every correlation function has a unique Herglotz-Bochner representation,

$$\sigma(t) = \int_{-T}^{T} e^{-i\lambda t}\, \nu(d\lambda),$$

ν a positive measure on $[-\pi, \pi)$ if $\Gamma = \mathbb{Z}$, or on $(-\infty, \infty)$ if $\Gamma = \mathbb{R}$.

II. If g, σ, and ν are as above, then for every $\lambda \in [-T, T]$ such that $\nu(\{\lambda\}) = 0$, $M_s(g(s)e^{-i\lambda s}) = 0$. This amounts to justifying the interchange of limits in the equation $0 = M_t(\sigma(t)e^{-i\lambda t}) = M_t M_s(g(t+s)\overline{g}(s)e^{-i\lambda t}) = M_s M_t(g(t+s)\overline{g}(s)e^{-i\lambda t}) = |a(\lambda)|^2$. For this purpose the authors prove a Tauberian theorem.

III. If $f \in L^2$, then for almost all x, f_x possesses a correlation function, σ_x. This is an unpublished result of Wiener's (nontrivial for $\Gamma = \mathbb{R}$), for which a reference to E. Hopf is given.

Amplitude functions are now proved to exist under the following hypothesis (on $f \in L^2$) which we denote by (E): *There exists a fixed countable set* $E \subseteq [-T, T]$ *such that for almost all* x, $\nu_x(\{\lambda\}) = 0$ for all $\lambda \in [-T, T] \cap E^c$, *where* ν_x *is the representing measure for* σ_x. From this and II it follows that for almost all x, $M_s(f_x(s)e^{-i\lambda s}) = 0$, $\lambda \in [-T, T] \cap E^c$. Finally, by a countable number of applications of the ergodic theorem, one for each $\lambda \in E$, it is established that for almost all x, f_x possesses an amplitude function. Since the proof of II is more complicated than is necessary, and since the proof of III is not readily accessible, we make further comments on II and III.

Comments on II. Associate to a function g having a correlation function the vector space $\mathcal{H}_0$ of all finite linear combinations of translates of g. Because g has a correlation function, $\mathcal{H}_0$ can be endowed with an inner product $\langle h_1, h_2\rangle_0 = M_s(h_1(s)\overline{h}_2(s))$. Let $\mathcal{H}_{00} = \{h \in \mathcal{H}_0 | \langle h, h\rangle_0 = 0\}$. Both $\mathcal{H}_0$ and $\mathcal{H}_{00}$ are translation invariant, and therefore there is a canonical unitary representation $(\Gamma, \mathcal{H})$ on the Hilbert space completion, $\mathcal{H}$, of $\mathcal{H}_0/\mathcal{H}_{00}$. We claim if λ is not in the point spectrum of $(\Gamma, \mathcal{H})$, then $M_s(g(s)e^{-i\lambda s}) = 0$. For if the latter fails, either because the limit does not exist, or because it exists and is not 0, there will exist sequences $B_n \to +\infty$, $A_n \to -\infty$, such that

$$\Phi(h) = \lim_{n \to \infty} \frac{1}{B_n - A_n} \int_{A_n}^{B_n} h(s) e^{-is\lambda} \, ds$$

exists for all $h \in \mathcal{H}_0$ and $\Phi(g) \neq 0$. Obviously, $|\Phi(h)|^2 \leqslant \langle h, h \rangle_0$, and therefore there exists $w \in \mathcal{H}$ such that $\Phi(h) = \langle \pi h, w \rangle$, where $\pi: \mathcal{H}_0 \to \mathcal{H}$ is the canonical mapping, and $\langle . , . \rangle$ is the inner product on $\mathcal{H}$. Since $\Phi(th) = e^{i\lambda t} \Phi(h)$, $h \in \mathcal{H}_0$, it must be that $tw = e^{i\lambda t} w$, meaning that either $w = 0$ or else λ is in the point spectrum of $(\Gamma, \mathcal{H})$. Both alternatives lead to contradictions, and therefore $M_s(g(s)e^{-i\lambda s}) = 0$, as claimed.

Application 1. Assume (Γ, X) is *ergodic*. In this case $\sigma_x(t) \equiv (tf,f)$ for almost all x, where $(. , .)$ is the $L^2(X)$ inner product, by the ergodic theorem. (Verify this for a countable dense subset of Γ, and use the continuity of both sides.) Therefore, if $\mathcal{H}_x$ is to f_x as $\mathcal{H}$ is to g above, the representation $(\Gamma, \mathcal{H}_x)_B$ is, for almost all x, isomorphic to the representation $(\Gamma, \mathcal{H}_f)$ of Γ on the cyclic subspace of L^2 generated by f. Thus, for the countable set E above we take the (at most countable) point spectrum of $(\Gamma, \mathcal{H}_f)$.

Application 2. The point of application 1 is to show that the Herglotiz-Bochner representation is unnecessary in the presence of the ergodic hypothesis. The second application is to the proof of II. Indeed, simply note that $\mathcal{H}$ and $L^2(\nu)$ are naturally isomorphic and that "translation" by t on $\mathcal{H}$ corresponds to muliplication by e^{its} on $L^2(\nu)$. Therefore λ is in the point spectrum of $(\Gamma, \mathcal{H})$ if and only if $\nu(\{\lambda\}) \neq 0$.

Comments on III. Let Γ be a locally compact Abelian group, and let Γ_d be Γ with the discrete topology. We state two facts from Fourier analysis. The first is elementary and is all that is necessary to verify Wiener's result in the ergodic case; for the second, see Cartan and Godement.[1] (1) If σ_1 and σ_2 are positive definite on Γ_d, if $\sigma_1 = \sigma_2$ a. e. on Γ with respect to Haar measure, if $\sigma_1(e) = \sigma_2(e)$, and if σ_2 is continuous, then $\sigma_1 \equiv \sigma_2$. (Assume $\sigma_1(e) = 1$. Fix $t \in \Gamma$. Choose $t_\nu \to t$ in such a way that $\sigma_1(t_\nu) = \sigma_2(t_\nu)$, $\sigma_1(t - t_\nu) = \sigma_2(t - t_\nu)$, all ν, and set $\alpha = \sigma_1(t)$, $\beta = \sigma_2(t)$. Use the positive definiteness of σ_1 and the continuity of σ_2 to conclude that

$$\begin{pmatrix} 1 & \alpha & \beta \\ \bar{\alpha} & 1 & 1 \\ \bar{\beta} & 1 & 1 \end{pmatrix}$$

is nonnegative definite. An elementary computation shows $\alpha = \beta$.) (2) If σ is positive definite on Γ_d, if σ is measurable on Γ, and if for every $\epsilon > 0$ $\{t \mid |\sigma(t) - \sigma(e)| < \epsilon\}$ has positive measure in every neighborhood of e, then σ is continuous. (Indeed, there exists a continuous positive definite function, σ_1, on Γ such that $\sigma = \sigma_1$ a.e. Apply (1).)

We now prove III using (1) and (2) above. Fix $f \in L^2$, and assume $\mu(X) = 1$. Applying the ergodic theorem and the Fubini theorem, as Wiener indicates, we find for almost all x that $\sigma_x(t)$ exists for almost all t. Applying (1) in the ergodic case (in the ergodic case $\sigma_x(t) = (tf,f)$ for almost all x,t) or (2) in the general case we see for almost all x that $\sigma_x(\cdot)$ has a unique positive definite extension to Γ. (Using the ergodic theorem and the Schwarz inequality one has for every $t, \|\sigma_x(t) - \sigma_x(e)\|_1 \leqslant \|tf - f\|_2 \, \|f\|_2$. Since $t \to tf$ is strongly continuous, $\sigma_x(t) \to \sigma_x(e)$ in measure. In particular, (2) applies for almost all x.) On the other hand, if x is such that $\sigma_x(t)$ exists for almost all t, including $t = 0$, but fails to exist for at least one t_0, then there exist at least two positive definite extensions of σ_x to Γ. (Extensions are obtained by choosing nets (A_ν, B_ν), $A_\nu \to -\infty$, $B_\nu \to +\infty$ such that

$$\lim_{\nu} \frac{1}{B_\nu - A_\nu} \int_{A_\nu}^{B_\nu} f_x(s+t)\bar{f}_x(s)\, ds$$

exists for all t.) We conclude that for almost all x, $\sigma_x(t)$ exists for all t, as claimed.

Lebesgue spaces

It is mentioned in the first paper [41a] that with suitable "local topological" hypotheses on X it can be proved that amplitude functions exist for general μ by virtue of von Neumann's decomposition of μ into its irreducible components. The significance of this for any physical interpretation of the authors' result is that generally the "ergodic hypothesis" has not been verified. (For one instance in which it has, see Sinai.[2] For an account of the measure theory that is involved, see Rohlin.)[3] We will sketch the construction based on these results.

Assume X is a Lebesgue space. Define $\mathcal{B}_I \subseteq \mathcal{B}$ to be $\mathcal{B}_I = \{B \in \mathcal{B} \mid \mu(tB \Delta B) = 0, t \in \Gamma\}$, where Δ is symmetric difference. $\mathcal{B}_I$ is a σ-field, and so by the theory of Lebesgue spaces there exists a Lebesgue space $\mathcal{Y} = (Y, \mathcal{C}, \lambda)$ and a $\mathcal{B}$-measurable, μ-almost everywhere defined mapping $\pi: X \to Y$ with the following properties: (a) $\mu(\pi^{-1} C) = \lambda(C)$; (b) $\pi^{-1}\mathcal{C} = \mathcal{B}_I$, modulo null sets; (c) for λ-almost all $c \in \mathcal{Y}$ there exists a σ-field, $\mathcal{B}_c$, of subsets of $X_c = \pi^{-1} c$ and a measure μ_c with properties (i) $(X_c, \mathcal{B}_c, \mu_c)$ is a Lebesgue space, (ii) for each $B \in \mathcal{B}$ $B \cap X_c \in \mathcal{B}_c$ for almost all c, and (iii) $\mu(B) = \int_Y \mu_c(B \cap X_c)\lambda(dc)$, the integrand being measurable $\mathcal{C}$. It can further be shown that for almost all c $\mu(tA\Delta A) = 0, t \in \Gamma, A \in \mathcal{B}_c$, and therefore (Γ, X_c) is a representation of Γ. Given $f \in L^2$ we have for almost all x that $\sigma_{sx}(t) = \sigma_x(t), s, t \in \Gamma$, meaning for each t that $x \to \sigma_x(t)$ is $\mathcal{B}_I$ measurable. Thus, for almost all c we have $\sigma_x(t) \equiv \sigma_{x'}(t)$ for μ_c almost all pairs $x, x' \in X_c$. Now for almost all c, $f^{(c)} = f|_{X_c} \in L^2(X_c)$, and by what has just been proved, $f^{(c)}$ obeys hypothesis (E). It follows that for μ-almost all $x \in X$, f_x has an amplitude function.

Spectral synthesis

The authors prove that a locally square integrable function, g, on Γ is B^2-a.p. if and only if (a) it has an amplitude function, $a(\lambda)$; (b) it has a correlation function, $\sigma(t)$; (c) ν is purely atomic, where ν is the representing measure for σ (or, equivalently, σ is Bohr almost periodic); and (d) the formal interchange of limits $|a(\lambda)|^2 = M_s(\sigma(s)e^{-i\lambda s})$ is valid for every real number λ. (II above only asserts that this equality holds when the right side is 0). They go on to prove, given (Γ, X) and $f \in L^2(X)$, that for fixed λ (d) holds for almost all f_x. Therefore, if f satisfies (E) for example, if (Γ, X) is ergodic), then (a), (b), and (d) are already true for almost all x. With (E) then, one has that f_x is B^2-a.p. for almost all x if and only if ν_x is purely atomic for almost all x (if and only if σ_x is Bohr almost periodic for almost all x). In particular, if (Γ, X) is ergodic, then f_x is B^2-a.p. for almost all x if and only if $(\Gamma, \mathcal{H}_f)$ has pure point spectrum. ($\mathcal{H}_f$ is the cyclic subspace generated by f.) In the ergodic case (Γ, X) has pure point spectrum if and only if for every $f \in L^2$ f_x is B^2-a.p. for almost all x.

Remark. The statement that f_x is B^2-a.p. for almost all x if and only if σ_x is almost periodic for almost all x is valid for general μ, if X is a Lebesgue space, as one sees easily by a decomposition argument similar to the above.

Without some such hypothesis as (E) some of the last mentioned equivalences are false. (It was not clear to the reviewer whether the authors were, at the time of writing these papers, aware of this. See the second statement in the corollary to Theorem 5 and the statements of Theorems 6 and 7 in the second paper.) For example the transformations $t(s,s') = (s, s' + st)$ on the unit square, $0 \leqslant s < 1$, $0 \leqslant s' < 1$, in which $s' + st$ is interpreted modulo one, give rise to a representation of $\Gamma = \mathbb{R}$ which does not have pure point spectrum, but for any $f \in L^2$, f_x is B^2-a.p. for almost all x. Moreover, if $f(s, s') = e^{-2\pi i s'}$, then f_x is *Bohr* almost periodic for all x, but $(\Gamma, \mathcal{H}_f)$ has no point spectrum.

It may not have previously been noted that a weaker form of synthesis holds quite generally. If $g \in L^\infty(\Gamma)$ define the spectrum, S_g, in the sense of Beurling and Wiener, to be the set of $\lambda \in [-T, T]$ such that $\hat{\varphi}(\lambda) = 0$ whenever $\varphi \in L^1(\Gamma)$ and $\varphi * g \equiv 0$ ($\hat{\varphi}$ denotes Fourier transform and $*$ denotes convolution). g obeys *spectral synthesis* is $\hat{\varphi} = 0$ on S_g implies $\varphi * g \equiv 0$. (Equivalently, g is approximable weak-$*$ by trigonometric polynomials whose exponents lie in S_g.)

Given (Γ, X), $L^1(\Gamma)$ operates on $L^\infty(X)$, as follows. If $\varphi \in L^1(\Gamma), f \in L^\infty(X)$, define

$$\varphi \# f(x) = \int_\Gamma \varphi(-t) f(tx)\, dt.$$

Let $A_f = \{\varphi \# f \mid \varphi \in L^1(\Gamma)\}$. A_f is a Γ-invariant (not closed) subspace of $L^\infty(X)$. Also, if $g \in L^\infty(\Gamma)$, define $B_g = \{\varphi * g \mid \varphi \in L^1(\Gamma)\}$. B_g is a translation invariant subspace of $L^\infty(\Gamma)$.

Now assume (Γ, X) is ergodic. Then because $L^1(\Gamma)$ is separable, there exists a set $X_0 \in \mathcal{B}$, $\mu(X_0) = 0$, such that if $x \notin X_0$, then for all $\varphi \in L^1(\Gamma)$, $M_s(|\varphi * f_x|^2) = \|\varphi \# f\|_2^2$ and $\sup_t |\varphi * f_x(t)| = \text{ess sup}_{x'} |\varphi \# f(x')|$. ($\varphi \# f(sx) = \varphi * f_x(s)$.) We claim: *if $x \notin X_0$, then f_x obeys spectral synthesis.* To see this suppose $\hat{\varphi} = 0$ on S_{f_x}. Then it is known that $M_s(|\varphi * f_x|^2) = 0$. (See either Herz[4] or Veech.[5]) Thus, $\|\varphi \# f\|_2^2 = 0$, meaning $\varphi \# f = 0$ a.e., meaning $\varphi * f_x \equiv 0$. That is, f_x obeys synthesis. Using a decomposition argument it is possible to generalize this result to Lebesgue spaces X (and separable locally compact Abelian groups Γ).

The results also generalize in another direction. Let Γ be a locally compact, σ-compact topological group which is *amenable*. That is, there is assumed to exist a nontrivial, nonnegative, translation invariant linear functional on $L^\infty(\Gamma)$. According to Emerson and Greenleaf,[6] there exists a sequence of measurable sets U_n in Γ such that if $|\cdot|$ denotes Haar measure and $U_n k = \{uk \mid u \in U_n\}$, $k \in \Gamma$, then (1) $0 < |U_n| < \infty$, (2) $U_n \subseteq U_{n+1}$, $\bigcup_n U_n = \Gamma$, and (3) $\lim_{n\to\infty} |U_n k \Delta U_n|\, |U_n|^{-1} = 0$ uniformly for $k \in K$, K any compact subset of Γ. If g is locally integrable on Γ, define

$$M_s(g) = \lim_{n\to\infty} \frac{1}{|U_n|} \int_{U_n} g(s)\, ds,$$

provided the limit exists. If in addition to the above, sets U_n can be chosen to satisfy

$$\limsup_{n\to\infty} |U_n U_n^{-1}||U_n|^{-1} < \infty,$$

then according to Chatard,[7] there is an individual ergodic theorem for Γ. That is if (Γ, X) is a representation of Γ by automorphisms of X, and if $f \in L^1(X)$, then $M_s(f(sx)) = f^*(x)$ exists, both a.e. and in $L^1(X)$.

Using the results just cited together with the Godement decomposition $\sigma = \sigma_1 + \sigma_2$ of a continuous positive definite function on a locally compact group Γ into a continuous, almost periodic part, σ_1, and a part σ_2, where $M_s(|\sigma_2|^2) = 0$, it is possible to extend the Wiener-Wintner analysis to Γ.[8] These are the essential modifications: (a) Let $\{a_{ij}^{\lambda}\}$, $1 \leqslant i, j \leqslant d_\lambda$, $\lambda \in \Lambda$, be a complete set of inequivalent, irreducible finite dimensional, continuous, unitary representations of Γ. The amplitude function now is $a(i, j, \lambda) = M_s(a_{ij}^{\lambda}(s^{-1})g(s))$, provided the limits exist; (b) $\sigma(t) = M_s(g(st)\overline{g}(s))$ is the correlation function, and $\sigma = \sigma_1 + \sigma_2$ as above. $\nu(\{\lambda\}) = 0$ is replaced by $M_s(a_{ij}^{\lambda}(s^{-1})g(s)) = 0$, all i, j. The condition that the representing measure ν be purely atomic is replaced by the condition $\sigma_2 = 0$, or what is the same, the condition that σ be almost periodic; (c) hypothesis (E) is similar, except it now refers to a countable subset of Λ, the set which indexes the finite dimensional representations; (d) to say λ is in the "point spectrum" of a representation π of Γ is not to say the representation λ occurs in π; (e) g is B^2-a.p. on Γ if there exists a sequence $\{P_n\}$ of continuous almost periodic functions on Γ such that $M_s(|g - P_n|^2) \to 0$.

For more recent developments on the individual ergodic theorem for amenable groups, see Bewley,[9] Greenleaf,[10] and Emerson.[11]

References

1. H. Cartan and R. Godement, *Théorie de la dualité et analyse harmonique dans les groupes abéliens localement compacts,* Ann. Sci. École Norm. Sup. 64(1947), 79-99.

2. Ya. G. Sinai, *On the foundations of the ergodic hypothesis for a dynamical system of statistical mechanics,* Soviet Math. Dokl. 4(1963), 1818-1821.

3. V. A. Rohlin, *On the fundamental ideas of measure theory,* Amer. Math. Soc. Transl. 10, no. 1 (1962), 1-54.

4. C. S. Herz, *The spectral theory of bounded functions,* Trans. Amer. Math. Soc. 94(1960), 181-232, Theorem 2.2.

5. W. A. Veech, *Properties of minimal functions on abelian groups,* Amer. J. Math. 91(1969), 415-440, Proposition 1.

6. W. R. Emerson and F. P. Greenleaf, *Covering properties and Følner conditions for locally compact groups,* Math. Z. 102(1967), 370-384.

7. J. Chatard, *Applications des propriétés de moyenne d'un groupe localement compact à la théorie ergodique,* Ann. Inst. H. Poincaré 6(1970), 307-326; erratum, *ibid.* 7(1971), 81-82.

8. R. Godement, *Les fonctions de type positif et la théorie des groupes,* Trans. Amer. Math. Soc. 63(1948), 1-84.

9. T. Bewley, *Extension of the Birkhoff and von Neumann ergodic theorems to semi-group actions,* Ann. Inst. H. Poincaré 7(1971), 283-291.

10. F. P. Greenleaf, *Ergodic theorems and the construction of summing sequences in amenable locally compact groups,* Comm. Pure Appl. Math. 26(1973), 29-46.

11. W. R. Emerson, *The pointwise ergodic theorem for amenable groups,* Amer. J. Math. 96(1974), 472-487.

THE DEFINITION AND ERGODIC PROPERTIES OF THE STOCHASTIC ADJOINT OF A UNITARY TRANSFORMATION

by

N. Wiener and **E. J. Akutowicz**

1. INTRODUCTION

The concepts and results of this paper rest upon a combination of elementary Hilbert space notions with the theory of the Brownian motion. While the interplay between these theories has been considerably exploited in recent years by mathematical probabilists (Karhunen [1947], Loève [1948], Doob [1953], among others), it has not been applied in quite the manner, nor to the same purpose, as is done here. The present set-up is closely related to earlier work by Wiener, dating from 1924, which may be consulted in the last two chapters of the book by Paley and Wiener [1934], but does not depend upon it.

In the study of stationary time series it is a standard device to introduce the unitary operator generated by translation in time; this unitary operator acts in a linear manifold of random variables defined over the given underlying probability space. In the present paper we proceed in the opposite direction and begin with an arbitrary group of unitary transformations U^{λ} acting in a Hilbert space (which we take in its realization as square-integrable functions on the real line, but any other realization would do as well), and this space is in a sense «adjoint» to the space of complex Brownian motions. In terms of U^{λ} a group of measure-preserving point transformations T^{λ} is constructed in the latter space, and the ergodic properties of T^{λ} are determined.

The results of this paper contribute essentially to the mathematical foundation for a new interpretation of the statistical basis of quantum mechanics. (Wiener and Siegel [1953] and forthcoming publications).

2. THE COMPLEX BROWNIAN MOTION

Basic for the calculations to follow is the particular measure introduced by Wiener into the space of all real functions defined on the interval $-\infty < t < \infty$ and vanishing at the origin. Such functions are labeled by a parameter α and the measure of the generalized interval,

$$\mathrm{I} : a_k \leqslant x(t_k, \alpha) < b_k; \qquad k = 1, \ldots, N; \qquad 0 < t_1 < t_2 < \cdots < t_N,$$

is defined as

$$(2.1) \qquad \begin{cases} m(\mathrm{I}) = \dfrac{1}{\sqrt{(2\pi)^N t_1 (t_2 - t_1) \ldots (t_N - t_{N-1})}} \displaystyle\int_{a_1}^{b_1} \cdots \\ \cdots \displaystyle\int_{a_N}^{b_N} \exp\left[-\frac{u_1^2}{2t_1} - \sum_1^{N-1} \frac{(u_{k+1} - u_k)^2}{2(t_{k+1} - t_k)} \right] du_1 \ldots du_N, \end{cases}$$

with the obvious modifications if some or all of the t_k are negative. The function m can be extended uniquely to the Borel field generated by the generalized intervals, and integration theory can be built up following a familiar pattern. It is also well known that the space of α's with the above measure is measure-theoretically isomorphic to the unit interval with Lebesgue measure. The family of random variables $x(t, \alpha)$, $-\infty < t < \infty$, is known as the real Brownian motion process.

The real Brownian motion process can be briefly characterized as a real stochastic process $x(t, \alpha)$ with independent Gaussian increments, defined for $-\infty < t < \infty$, and satisfying

$$x(0, \alpha) = 0, \quad \text{all } \alpha,$$

$$E\{x(t, \alpha)\} = 0, \qquad -\infty < t < \infty,$$

$$E\{[x(s, \alpha) - x(t, \alpha)]^2\} = \sigma^2 |s - t|, \qquad -\infty < s,\ t < \infty,$$

where σ is a positive constant.

We shall require the complex Brownian motion process, whose real and imaginary parts are independent Brownian motions, both with variance σ^2 equal to $\frac{1}{2}$. This complex Gaussian process $X(t, \alpha)$ has the characteristic properties

(2.2) $X(0, \alpha) = 0, \quad \text{all} \quad \alpha,$

(2.3) $E\{X(t, \alpha)\} = 0, \qquad -\infty < t < \infty,$

(2.4) $E\{|X(t, \alpha) - X(s, \alpha)|^2\} = |s - t|, \qquad -\infty < s,\ t < \infty,$

(2.5) $E\{(X(t_2, \alpha) - X(t_1, \alpha))(\overline{X(s_2, \alpha) - X(s_1, \alpha)})\} = 0, \qquad t_1 \leqslant t_2 \leqslant s_1 \leqslant s_2$ (1).

3. INTEGRATION WITH RESPECT TO $X(t, \alpha)$

We shall make systematic use of random variables that are integrals of the form

(3.1)
$$\int_{-\infty}^{\infty} f(t)\, d_t X(t, \alpha),$$

where $f(t)$ belongs to $L^2(-\infty, \infty)$. We shall briefly recapitulate the definition and main properties of these integrals. For $f(t)$ equal to a constant c_j on non-overlapping intervals (t_j, t_{j+1}), $j = 1 \ldots N$, and vanishing otherwise, the integral (3.1) is defined as

$$\sum_{j=1}^{N} c_j (X(t_{j+1}, \alpha) - X(t_j, \alpha)).$$

Then

$$\int_0^1 \left| \int_{-\infty}^{\infty} f(t)\, dX(t, \alpha) \right|^2 d\alpha$$

$$= \sum_{j,k} c_j \bar{c}_k \int_0^1 (X(t_{j+1}, \alpha) - X(t_j, \alpha))(\overline{X(t_{k+1}, \alpha) - X(t_k, \alpha)})\, d\alpha$$

$$= \sum_{j=1}^{N} |c_j|^2 (t_{j+1} - t_j) = \int_{-\infty}^{\infty} |f(t)|^2\, dt.$$

Since the step-functions are everywhere dense in $L^2(-\infty, \infty)$ it follows that the integral exists as a limit in the mean with respect to α-integration for every f belonging to $L^2(-\infty, \infty)$, and furthermore that the relation

(3.2)
$$\int_0^1 d\alpha \int_{-\infty}^{\infty} f_1(t)\, dX(t, \alpha) . \overline{\int_{-\infty}^{\infty} f_2(t)\, dX(t, \alpha)} = \int_{-\infty}^{\infty} f_1(t) \overline{f_2(t)}\, dt$$

(1) Wiener [1934] referred to $X(t, \alpha)$ as «the fundamental random function». In the case when t varies over $0 \leq t \leq 1$ $X(t, \alpha)$ has a development into an almost surely convergent Fourier series; in the present case, the t-interval being $-\infty < t < \infty$, the corresponding harmonic analysis is

$$X(t, \alpha) = \frac{1}{\sqrt{2\pi}} \int_{-\infty}^{\infty} \frac{e^{i\lambda t} - 1}{i\lambda}\, dX(\lambda, \alpha),$$

where $X(\lambda, \alpha)$ satisfies (2.2)-(2.5), and where the integral is defined in Sect. 3 as a limit in the mean of suitable approximations.

is valid. It also follows that the Riemann-Stieltjes sums for (3.1) converge in the mean whenever $f(t)$ is Riemann integrable, and that the usual rules for manipulating integrals are valid. In particular, integration by parts is valid if f is of bounded variation.

4. EXISTENCE OF THE STOCHASTIC ADJOINT OF A UNITARY TRANSFORMATION

Given any unitary transformation U in the space $L^2(-\infty, \infty)$ we aim to associate with U a measure-preserving point transformation $T(=T_U)$ on the α's, which is one to one, if a null set of α's is excluded, and is such that for any φ belonging to $L^2(-\infty, \infty)$,

$$\int_{-\infty}^{\infty} U\varphi(t)\,dX(t, \alpha) = \int_{-\infty}^{\infty} \varphi(t)\,dX(t, T\alpha), \tag{4.1}$$

with probability one.

In the first place, T, if it exists, must be essentially unique. That is, there cannot exist T_1 and T_2 with the properties stated and different over a set of α's of a positive measure. Let ús put for $-\infty < \tau < \infty$,

$$\chi_\tau(t) = \begin{cases} \operatorname{sgn}\tau \text{ on the interval between } \tau \text{ and } 0, \\ 0 \quad \text{otherwise.} \end{cases} \tag{4.2}$$

Then, by (4.1)

$$\begin{aligned} \int_{-\infty}^{\infty} U\chi_{\tau_k}(t)\,dX(t, \alpha) &= \int_{-\infty}^{\infty} \chi_{\tau_k}(t)\,dX(t, T_1\alpha). \\ &= X(\tau_k, T_1\alpha) = X(\tau_k, T_2\alpha), \end{aligned} \tag{4.3}$$

simultaneously for a countable dense set of τ_k on $(-\infty, \infty)$ except for a null set of α's. Since the paths of the Brownian motion process are continuous except for a null set, we have from (4.3),

$$X(\tau, T_1\alpha) = X(\tau, T_2\alpha), \qquad -\infty < \tau < \infty, \tag{4.4}$$

except for a null set. But (4.4) means just that $T_1 = T_2$, except for a null set, which is our assertion of uniqueness.

To establish the existence of T we shall use the following

LEMMA. The stochastic process $Y(t, \alpha)$ defined by

$$Y(t, \alpha) = \int_{-\infty}^{\infty} U\chi_t(s)\,dX(s, \alpha), \qquad -\infty < t < \infty,$$

where χ_t *is given by* (4.2), *is a complex Brownian motion* ([1]).

Clearly, for each t, $Y(t, \alpha)$ is a complex Gaussian random variable with mean 0, and equal to 0 for $t = 0$, with probability one. The variance of $Y(t_2, \alpha) - Y(t_1, \alpha)$ is $|t_2 - t_1|$:

$$\int_0^1 \left| \int_{-\infty}^{\infty} (U\chi_{t_2} - U\chi_{t_1}) dX(s, \alpha) \right|^2 d\alpha = \int_{-\infty}^{\infty} (\chi_{t_2} - \chi_{t_1})^2 ds = |t_2 - t_1|.$$

The increments over non-overlapping t-intervals are independent:

$$\int_0^1 \left\{ \int_{-\infty}^{\infty} (U\chi_{t_2} - U\chi_{t_1}) dX(s, \alpha) . \overline{\int_{-\infty}^{\infty} (U\chi_{t_4} - U\chi_{t_3}) dX(s, \alpha)} \right\} d\alpha$$

$$= \int_{-\infty}^{\infty} (\chi_{t_2} - \chi_{t_1})(\chi_{t_4} - \chi_{t_3}) ds = 0.$$

The foregoing properties are sufficient to identify the process $Y(t, \alpha)$ as a complex Brownian motion.

From this lemma it follows that, except for a null set of α's, $Y(t, \alpha)$ is continuous in t, and therefore coincides with some path $X(t, T\alpha)$. $T\alpha$ is thus well-defined outside the null set of discontinuous paths. Since $Y(t, \alpha)$ and $X(t, T\alpha)$ are one and the same function, T is one to one where it is defined.

It remains to show that T is measure-preserving, and that (4.1) holds. That T is measure-preserving follows at once from the proof of the preceding lemma if we write $X(t, T\alpha)$ instead of $Y(t, \alpha)$. As for (4.1), we have

$$\text{(4.5)} \qquad \int_{-\infty}^{\infty} \chi_t(s) dX(s, T\alpha) = X(t, T\alpha) = \int_{-\infty}^{\infty} U\chi_t(s) dX(s, \alpha).$$

Since any φ in $L^2(-\infty, \infty)$ is a limit in the mean of linear combinations of functions of the form $\chi_{t''} - \chi_{t'}$, (4.5) easily implies (4.1).

We call T the *stochastic adjoint* of U.

5. ERGODICITY OF THE STOCHASTIC ADJOINT

Let U^λ, $-\infty < \lambda < \infty$, be a group of unitary transformations in $L^2(-\infty, \infty)$,

$$U^{\lambda_1+\lambda_2} = U^{\lambda_1} U^{\lambda_2}, \ U^0 = \text{Identity}.$$

From the fact that the stochastic adjoint of U is defined outside a null set which is independent of U, it follows that corresponding to the group U^λ we

([1]) Compare the Fourier integral appearing in the preceding footnote.

have a group T^λ of measure-preserving point transformations.

THEOREM 1. The point spectrum of U^λ is absent if and only the associated group T^λ is weakly mixing ([1]).

By Stone's theorem, for f, g in $L^2(-\infty, \infty)$,

$$(U^\lambda f, g) = \int_{-\infty}^{\infty} e^{i\lambda y} d(E(y)f, g), \tag{5.1}$$

where $E(y)$ is the spectral family of projections belonging to U^λ. Our assumption that the point spectrum is absent implies that $(E(y)f, g)$ is a continuous function of y.

Let $t_1 < t_2 < \cdots < t_N$ be given. By (5.1) it follows that

$$\begin{aligned}
\int_{-\infty}^{\infty} e^{i\lambda y} d(E(y)\chi_{t_i}, \chi_{t_j}) &= (U^\lambda \chi_{t_i}, \chi_{t_j}) \\
&= \int_{-\infty}^{\infty} U^\lambda \chi_{t_i}(s)\overline{\chi_{t_j}(s)}\, ds \\
&= \int_0^1 d\alpha \int_{-\infty}^{\infty} U^\lambda \chi_{t_i}(s)\, dX(s, \alpha) . \overline{\int_{-\infty}^{\infty} \chi_{t_j}(s)\, dX(s, \alpha)} \\
&= \int_0^1 \overline{X(t_j, \alpha)}\, d\alpha \int_{-\infty}^{\infty} \chi_{t_i}(s)\, dX(s, T^\lambda \alpha) \\
&= \int_0^1 X(t_i, T^\lambda \alpha)\overline{X(t_j, \alpha)}\, d\alpha \\
&= r_{ij}(\lambda),
\end{aligned}$$

where $(r_{ij}(\lambda))$ is the upper right hand quarter of the covariance matrix of the random vector

$$(X(t_1, T^\lambda \alpha), \ldots, X(t_N, T^\lambda \alpha), X(t_1, \alpha), \ldots, X(t_N, \alpha)).$$

([1]) Weak mixing is the condition that

$$\lim_{C\to\infty} \frac{1}{2c} \int_{-C}^{C} (m(A \cap T^\lambda B) - m(A)m(B))^2\, d\lambda = 0,$$

for every pair of measurable sets A, B. Weak mixing implies ergodicity, but the converse is false. Strong mixing is the condition that

$$\lim_{|\lambda|\to\infty} m(A \cap T^\lambda B) = m(A).m(B),$$

for every pair of measurable sets A, B.

The fact that $r_{ij}(\lambda)$ is a Fourier-Stieltjes transform of a continuous function of bounded variation implies that

$$\lim_{C\to\infty} \frac{1}{2c}\int_{-C}^{C} |r_{ij}(\lambda)|^2 d\lambda = 0. \tag{5.2}$$

Conversely, if

$$r(\lambda) = \int_{-\infty}^{\infty} e^{i\lambda y} dS(y),$$

where $S(y)$ is a complex function of bounded variation over (∞, ∞), and normalized by $S(y-0) = S(y)$, then

$$\lim_{C\to\infty} \frac{1}{2c}\int_{-C}^{C} |r(\lambda)|^2 d\lambda = \sum_{-\infty<y<\infty} |S(y+0) - S(y)|^2,$$

both sides being finite. Hence, if (5.2) holds, we have that $(E(y)\chi_{t_i}, \chi_{t_j})$ is continuous in y for all t_i, t_j. But this implies $E(y\pm 0) = E(y)$. For, suppose that for some y, $E(y+0) \neq E(y)$. Put $\Delta(\delta) = E(y+\delta) - E(y)$ and let φ be such that $\Delta(0+)\varphi = \varphi$, $\|\varphi\| = 1$. Let ψ_1 denote a linear combination of functions $\chi_{t''} - \chi_{t'}$, so that ψ_1 belongs to $L^2(-\infty, \infty)$. Then we have

$$(\Delta(\delta)\varphi, \varphi) = (\Delta(\delta)\psi_1, \psi_1) - (\Delta(\delta)\varphi, \psi_1 - \varphi) - (\Delta(\delta)(\psi_1 - \varphi), \psi_1)$$

identically in $\delta > 0$ and in ψ_1, and this is contradictory for $\delta \to 0$, if $\|\varphi - \psi_1\|$ is chosen sufficiently small, independent of δ. Therefore, we must have $E(y+0) = E(y) = E(y-0)$; i.e. the point spectrum is absent if and only if (5.2) holds.

Theorem 1 is then a consequence of

THEOREM 2. (1) *A stationary (wide sense) Gaussian process $x(t, \omega)$, $\omega \in \Omega$, $-\infty < t < \infty$, is weakly mixing if and only if for arbitrary N, $t_1, \ldots, t_N$, the elements,*

$$\rho_{ij}(\lambda) = \int_\Omega x(t_i + \lambda, \omega)\overline{x(t_j, \omega)}\, d\omega,$$

of the 2N-dimensional covariance matrix $P(\lambda)$ of

$$x(t_1, \omega), \ldots, x(t_N, \omega), x(t_1+\lambda, \omega), \ldots, x(t_N+\lambda, \omega) \tag{5.3}$$

(1) It is likely that Theorems 2 and 3 are known, though we were unable to locate them in the literature. See however U. Grenander [1950]. The corresponding results for discrete time are also valid.

satisfy the condition

(5.4) $$\lim_{C\to\infty}\frac{1}{2c}\int_{-C}^{C}|\rho_{ij}(\lambda)|^2\,d\lambda = 0.$$

We can assume $E(x(t,\omega)) = 0$, $-\infty < t < \infty$, so that the finite-dimensional distribution of the real and imaginary parts of the variables (5.3) is uniquely determined ([1]) by the covariance matrix $P(\lambda)$. This distribution has Gaussian density, which is clearly an analytic function of the real and imaginary parts of $\rho_{ij}(\lambda)$. Hence if I_1 and I_2 are generalized intervals based on a finite number of t_i's, and defined by the condition that the real and imaginary parts of $x(t_i, \omega)$ lie in finite intervals, then (5.4) implies that

$$\lim_{C\to\infty}\frac{1}{2c}\int_{-C}^{C}\{m(I_1\cap T^\lambda I_2) - m(I_1)m(I_2)\}^2\,d\lambda = 0$$

where $x(t+\lambda,\omega)$ belongs to $T^\lambda I_2$ if and only if $x(t,\omega)$ belong to I_2. This proves that $x(t,\omega)$ is weakly mixing.

Conversely, if $x(t,\omega)$ is weakly mixing, then, by a standard approximation argument, the quantity

$$\Theta(\lambda) \equiv E\{\Phi(x(t_1,\omega))\Psi(x(t_2+\lambda,\omega))\} - E\{\Phi(x(t_1,\omega))\}E\{\Psi(x(t_2,\omega))\}$$

has the property

$$\lim_{C\to\infty}\frac{1}{2c}\int_{-C}^{C}|\Theta(\lambda)|^2\,d\lambda = 0$$

for any Borel functions $\Phi(z)$, $\Psi(z)$. Taking $\Phi(z) = z$, $\Psi(z) = \bar{z}$ we obtain (5.4). This proves Theorem 2, and also Theorem 1.

THEOREM 3. A stationary (wide sense) Gaussian process $x(t,\omega)$ is strongly mixing if and only if

(5.5) $$\lim_{|\lambda|\to\infty}\rho_{ij}(\lambda) = 0,$$

where the ρ_{ij}'s have the same meaning as in Theorem 2.

This result follows by the proof of Theorem 2, if we everywhere substitute pointwise convergence for mean convergence.

THEOREM 4. If the spectrum of the unitary group U^λ is absolutely continuous, then the associated group T^λ is strongly mixing.

([1]) Doob [1953], pp. 72-73.

We need only remark that the Riemann-Lebesgue theorem implies

$$\lim_{|\lambda|\to\infty} r_{ij}(\lambda) = 0,$$

where $r_{ij}(\lambda)$ is as in the proof of Theorem 1.

We wish to point out that there exist unitary groups with non-absolutely continuous spectra for which the $r_{ij}(\lambda)$ tend to zero, so that the converse of Theorem 4 is false. Examples of such unitary groups can be obtained as follows. Wiener and Wintner [1938] have constructed singular functions $S(y)$ which are strictly increasing from $-\frac{\pi}{2}$ to $\frac{\pi}{2}$ as y varies from $-\pi$ to π and such that

$$r(\lambda) \equiv \int_{-\pi}^{\pi} e^{i\lambda y}\, dS(y) = o(|\lambda|^{-\delta}), \tag{5.6}$$

$|\lambda| \to \infty$, where δ may be taken anywhere on the interval $0 < \delta < \frac{1}{2}$, $S(y)$ depending upon the δ chosen. Let us write $\eta = S(y)$, so that $y = s(\eta)$ where s is the function inverse to S, $-\pi \leqslant y \leqslant \pi$ and $-\frac{\pi}{2} \leqslant \eta \leqslant \frac{\pi}{2}$. Define

$$V^{\lambda} g(x) = e^{i\lambda s(\arctan x)} g(x), \quad -\infty < \lambda < \infty,$$

for g belonging to $L^2(-\infty, \infty)$. We aim to show that V^{λ} is a unitary group of the required sort.

An isometric correspondence between $L^2(-\infty, \infty)$ and $L^2\left(-\frac{\pi}{2}, \frac{\pi}{2}\right)$ is set up by the equation

$$g(\tan \eta) \sec \eta = \varphi(\eta), \tag{5.7}$$

where g is in $L^2(-\infty, \infty)$ and φ in $L^2\left(-\frac{\pi}{2}, \frac{\pi}{2}\right)$. Then for g_1, g_2 arbitrary in $L^2(-\infty, \infty)$,

$$(V^{\lambda} g_1, g_2) = \int_{-\infty}^{\infty} e^{i\lambda s(\arctan x)} g_1(x) \overline{g_2(x)}\, dx$$

$$= \int_{-\frac{\pi}{2}}^{\frac{\pi}{2}} e^{i\lambda s(\xi)} g_1(\tan \xi) \overline{g_2(\tan \xi)} \sec^2 \xi\, d\xi.$$

If g_1 and g_2 correspond by (5.7) with χ_{t_1} and χ_{t_2},

$0 < t_1 \leqslant t_2 \leqslant \frac{\pi}{2}$ (1), we obtain

$$(V^\lambda g_1, g_2) = \int_{-\frac{\pi}{2}}^{\frac{\pi}{2}} e^{i\lambda s(\xi)} \chi_{t_1}(\xi) \chi_{t_2}(\xi) d\xi$$

$$= \int_{-\pi}^{\pi} e^{i\lambda y} \chi_{t_1}(S(y)) \chi_{t_2}(S(y)) dS(y)$$

$$= \int_{-\pi}^{\pi} e^{i\lambda y} (\chi_{s(t_1)}(y) - \chi_{s(0)}(y)) dS(y)$$

$$= \int_{s(0)}^{s(t_1)} e^{i\lambda y} dS(y).$$

The required properties of V^λ now follow from two remarks.

I. $(V^\lambda g, h) \to 0$ *for* g, h *varying over a basis of* $L^2(-\infty, \infty)$ *implies* $(V^\lambda \chi_{t_i}, \chi_{t_j}) = r_{ij}(\lambda) \to 0$.

II. $\int_{-\pi}^{\pi} e^{i\lambda y} dS(y) = o(|\lambda|^{-\delta})$, $\delta > 0$, *implies*

$$\int_a^b e^{i\lambda y} dS(y) = o(1), \quad -\pi \leqslant a \leqslant b \leqslant \pi.$$

Since I. is clearly true, it only remains to establish II. We have that

$$\frac{1}{\pi} \int_{A'}^{A''} e^{-i\left(y - \frac{a+b}{2}\right)\lambda} \frac{\sin (b-a)\frac{\lambda}{2}}{\lambda} d\lambda$$

converges boundedly as $A' \to -\infty$, $A'' \to \infty$ to the function which is unity for $a < y < b$ and 0 for $y < a$ and $y > b$. Hence

$$\int_a^b e^{i\lambda y} dS(y) = \int_{-\pi}^{\pi} e^{i\lambda y} \left\{ \lim_{\substack{A'' \to \infty \\ A' \to -\infty}} \frac{1}{\pi} \int_{A'}^{A''} e^{-i\left(y - \frac{a+b}{2}\right)\omega} \frac{\sin (b-a)\frac{\omega}{2}}{\omega} d\omega \right\} dS(y)$$

(1) This condition is inessential.

$$= \lim_{\substack{A' \to -\infty \\ A'' \to \infty}} \frac{1}{\pi} \int_{-\pi}^{\pi} e^{i\lambda y} \, dS(y) \int_{A'}^{A''} e^{-i\left(y - \frac{a+b}{2}\right)\omega} \frac{\sin (b-a)\frac{\omega}{2}}{\omega} \, d\omega$$

$$= \frac{1}{\pi} \int_{-\infty}^{\infty} e^{i\left(\frac{a+b}{2}\right)\omega} \frac{\sin (b-a)\frac{\omega}{2}}{\omega} r(\lambda - \omega) \, d\omega$$

$$= \frac{1}{\pi} \int_{-\infty}^{\infty} e^{i\left(\frac{a+b}{2}\right)(\lambda-\omega)} \frac{\sin \left(\frac{b-a}{2}\right)(\lambda - \omega)}{\lambda - \omega} r(\omega) \, d\omega$$

$$= \int_{-\infty}^{-A} + \int_{-A}^{A} + \int_{A}^{\infty}$$

where A is chosen so large that for $|\omega| \geqslant A$,

$$|\omega|^{\delta} . |r(\omega)| < 1.$$

With such A fixed,

$$\int_{-A}^{A} \to 0 \quad \text{as} \quad |\lambda| \to \infty,$$

this being implied by $r(\lambda) = o(1)$, merely.

Now $\int_{A}^{\infty} = o(1)$, $\lambda \to -\infty$, without difficulty and it remains to consider $\lambda \to +\infty$.

$$\left| \int_{A}^{\infty} \right| \leqslant \frac{1}{\pi} \int_{A}^{\infty} \left| \frac{\sin \left(\frac{b-a}{2}\right)(\lambda - \omega)}{\lambda - \omega} \right| \frac{d\omega}{\omega^{\delta}},$$

and

$$\int_{A}^{\lambda-1} \left| \frac{\sin \left(\frac{b-a}{2}\right)(\lambda - \omega)}{\lambda - \omega} \right| \frac{d\omega}{\omega^{\delta}} < \int_{A}^{\lambda-1} \frac{d\omega}{(\lambda - \omega)\omega^{\delta}} = O\left(\frac{\log \lambda}{\lambda^{1+\delta}}\right),$$

$$\int_{\lambda-1}^{\lambda+1} \left| \frac{\sin \left(\frac{b-a}{2}\right)(\lambda - \omega)}{\lambda - \omega} \right| \frac{d\omega}{\omega^{\delta}} = O\left(\frac{1}{\lambda^{\delta}}\right),$$

$$\int_{\lambda+1}^{\infty} \left| \frac{\sin \left(\frac{b-a}{2}\right)(\lambda - \omega)}{\lambda - \omega} \right| \frac{d\omega}{\omega^{\delta}} < \int_{\lambda+1}^{\infty} \frac{d\omega}{(\omega - \lambda)\omega^{\delta}} < \frac{1}{\delta(\lambda+1)^{\delta}} + \frac{\log(\lambda+1)}{(\lambda+1)^{\delta}}$$

The remaining integral, $\int_{-\infty}^{-A}$, behaves similarly and thus II, is established. We have used the order condition (5.6) in an essential, though not refined, way and it remains a moot point whether (5.6) can be replaced by $r(\lambda) = o\ (1)$.

THEOREM 5. The time shift in the Brownian motion is strongly mixing.

This follows because the spectrum of the particular unitary group

$$U^{\lambda} f(t) = f(t+\lambda), \quad -\infty < \lambda < \infty, \quad (f \in L^2(-\infty, \infty)),$$

is absolutely continuous and $X(t+\lambda, \alpha) = X(t, T^{-\lambda}\alpha)$.

6. *THE POINT SPECTRUM.* The ergodic property [1] of T^{λ} necessarily fails if the point spectrum of U^{λ} is not empty. For suppose that $e^{i\Lambda}$, Λ real, is an eigenvalue of U^{λ} belonging to the eigenfunction φ:

$$U^{\lambda}\varphi = e^{i\lambda\Lambda}\varphi, \quad -\infty < \lambda < \infty, \quad \varphi \in L^2(-\infty, \infty).$$

Then

$$\xi(\alpha) = \int_{-\infty}^{\infty} \varphi(t)\, dX(t, \alpha)$$

is a random variable on the space of the α's such that

$$\xi(T^{\lambda}\alpha) = e^{i\lambda\Lambda}\xi(\alpha),$$

with probability 1. It follows that α-sets of the form

$$|\xi(\alpha)| < r$$

are non-trivial invariant sets.

Massachusetts Institute of Technology.

December, 1956

[1] Also known as metric transitivity.

REFERENCES

Karhunen, K. 1947, Ann. Acad. Sci. Fennicae, Series A, No. 37.

Loève, M. 1948, Appendix to P. Levy, *Processus Stochastiques et Mouvement Brownien,* Paris.

Doob, J. L. 1953, *Stochastic Processes,* New York.

Wiener, N. 1930, *Generalized Harmonic Analysis,* Acta Math. 55.

Paley, R.E.A.C. and Wiener, N. 1934, *Fourier Transforms in the Complex Domain,* Chapters IX and X.

Wiener, N. and Siegel, A. 1953, *A new form of the statistical postulate of quantum mechanics,* The Physical Review, 91, No. 6.

Grenander, U. 1950, *Stochastic processes and statistical inference,* Arkiv för Mat., Band I.

Kallianpur, G. and Robbins, H. 1953, *Ergodic property of the Brownian motion process,* Proc. Nat. Acad. Sci. 39, No. 6.

Wiener, N. and Wintner, A. 1938, *Fourier-Stieltjes transforms and singular infinite convolutions,* Amer. J. Math. 60.

ERRATA-CORRIGE

Addendum to « The Definition and Ergodic Properties of the Stochastic Adjoint of a Unitary Transformation »

by **N. Wiener** and **E. J. Akutowicz,** (pag. 205)

The Brownian motion processes occurring throughout the above-named paper should be understood to be *separable* Brownian motion processes.

The separability condition is tacitly used in § 4 where it is asserted that the paths $X(t, \alpha)$ are continuous except for a null α-set. Because conditions (2.2)-(2.5) on p. 207 are used to identify a process as a Brownian motion process, this continuity would not follow without the additional hypothesis that $X(t, \alpha)$ is separable. The process $Y(t, \alpha)$ occurring in the LEMMA in § 4 will then necessarily be separable.

Comments on [57a]

K. Ito

Let $X(t, \omega)$ be a complex Brownian motion, that is, a complex valued stochastic process with continuous sample functions whose real and imaginary parts are real Brownian motions independent of each other. With no loss of generality one can assume that $X(0, \omega) \equiv 0$ and that $X(t, \omega) = X(t, \omega')$ for every t implies $\omega = \omega'$. Let U be a unitary operator acting on the Hilbert space $L^2(-\infty, \infty)$. Then they proved that there exists a unique measure-preserving transformation T on the ω-space such that

$$\int_{-\infty}^{\infty} U\varphi(t)\, dx(t, \omega) = \int_{-\infty}^{\infty} \varphi(t)\, dx(t, T\omega)$$

for every $\varphi \in L^2(-\infty, \infty)$. Suppose U^λ, $\lambda \in (-\infty, \infty)$ is a one-parameter group of unitary operators acting on $L^2(-\infty, \infty)$. The stochastic adjoints T^λ of U^λ, $-\infty < \lambda < \infty$, form a one-parameter group of measure-preserving transformations on the ω-space. The authors proved the following interesting relation between the ergodic property of $\{T^\lambda\}$ and the continuity of the spectrum of $\{U^\lambda\}$:

(1) (Theorem 1) $\{U^\lambda\}$ has no point spectrum if and only if $\{T^\lambda\}$ is weakly mixing

and

(2) (Theorem 4) If the spectrum of $\{U^\lambda\}$ is absolutely continuous, then $\{T^\lambda\}$ is strongly mixing.

The authors derived these results from similar ones (Theorem 2 and 3) on Gaussian stationary processes. Theorem 2 and 3 were known as Maruyama's theorems (*The harmonic analysis of stationary stochastic processes,* Mem. Fac. Sci. Kyusyu Univ. Ser. A, vol. 4 (1949), 45-106), although the author proved them independently in this paper.

Comments on [57a]

E. J. Akutowicz

Although this paper gives a new and seemingly natural turn to the notion of duality, it has remained isolated. There is recent work by L. Schwartz in course of publication (cf. L. Schwartz, Séminaire à l'Ecole Polytechnique, 1969-1970, mimeographed) on Radonifying mappings, which is clearly very powerful, and which develops and applies the duality theory involved in a most interesting way.